P9-DFN-048

ANNUAL REVIEW OF ASTRONOMY AND ASTROPHYSICS

EDITORIAL COMMITTEE (2002)

GEOFFREY BURBIDGE
HUGH HUDSON
KENNETH IRWIN KELLERMANN
JOSEPH S. MILLER
MARCIA R. RIEKE
ALLAN SANDAGE
FRANK H. SHU
GEORGE WALLERSTEIN

RESPONSIBLE FOR THE ORGANIZATION OF VOLUME 40
(EDITORIAL COMMITTEE, 2000)

KENNETH BRECHER
GEOFFREY BURBIDGE
DAVID J. HELFAND
KENNETH IRWIN KELLERMANN
JOSEPH S. MILLER
ALLAN SANDAGE
FRANK H. SHU
GEORGE WALLERSTEIN
GEORGE BLUMENTHAL (GUEST)
GEORGE FISHER (GUEST)

Production Editor: ANNE E. SHELDON
Bibliographic Quality Control: MARY A. GLASS
Color Graphics Coordinator: EMÉ O. AKPABIO
Electronic Content Coordinator: SUZANNE K. MOSES
Subject Indexer: BRUCE TRACY

ANNUAL REVIEW OF ASTRONOMY AND ASTROPHYSICS

VOLUME 40, 2002

GEOFFREY BURBIDGE, *Editor*
University of California at San Diego

ALLAN SANDAGE, *Associate Editor*
Observatories of the Carnegie Institution of Washington

FRANK H. SHU, *Associate Editor*
University of California at Berkeley

www.annualreviews.org   science@annualreviews.org   650-493-4400

ANNUAL REVIEWS
4139 El Camino Way • P.O. BOX 10139 • Palo Alto, California 94303-0139

ANNUAL REVIEWS
Palo Alto, California, USA

COPYRIGHT © 2002 BY ANNUAL REVIEWS, PALO ALTO, CALIFORNIA, USA. ALL RIGHTS RESERVED. The appearance of the code at the bottom of the first page of an article in this serial indicates the copyright owner's consent that copies of the article may be made for personal or internal use, or for the personal or internal use of specific clients. This consent is given on the condition that the copier pay the stated per-copy fee of $14.00 per article through the Copyright Clearance Center, Inc. (222 Rosewood Drive, Danvers, MA 01923) for copying beyond that permitted by Section 107 or 108 of the US Copyright Law. The per-copy fee of $14.00 per article also applies to the copying, under the stated conditions, of articles published in any *Annual Review* serial before January 1, 1978. Individual readers, and nonprofit libraries acting for them, are permitted to make a single copy of an article without charge for use in research or teaching. This consent does not extend to other kinds of copying, such as copying for general distribution, for advertising or promotional purposes, for creating new collective works, or for resale. For such uses, written permission is required. Write to Permissions Dept., Annual Reviews, 4139 El Camino Way, P.O. Box 10139, Palo Alto, CA 94303-0139 USA.

International Standard Serial Number: 0066-4146
International Standard Book Number: 0-8243-0940-5
Library of Congress Catalog Card Number: 63-8846

All Annual Reviews and publication titles are registered trademarks of Annual Reviews. ⊗ The paper used in this publication meets the minimum requirements of American National Standards for Information Sciences—Permanence of Paper for Printed Library Materials, ANSI Z39.48-1992.

Annual Reviews and the Editors of its publications assume no responsibility for the statements expressed by the contributors to this *Annual Review*.

TYPESET BY TECHBOOKS, FAIRFAX, VA
PRINTED AND BOUND IN THE UNITED STATES OF AMERICA

PREFACE

This volume was planned at a meeting held in San Francisco, California on May 13, 2000. Those who attended the meeting included Geoffrey Burbidge (Editor), Allan Sandage and Frank Shu (Associate Editors), K. Brecher, D. Helfand, K. Kellermann, J. Miller and G. Wallerstein (Editorial Committee members) and guests G. Blumenthal and G. Fisher. Others who attended the meeting were S. Gubins, Editor-in-Chief, Annual Reviews, Anne Sheldon, Production Editor, Annual Reviews, and K. Ericson (UCSD).

Once again I wish to thank the authors and the Associate Editors, The Production Editor, and K. Ericson, all of whom have worked hard to bring out another successful volume.

Last year I reported that 24 articles were scheduled for this volume. In fact, 16 are published here. For the next Volume 41 (2003), we have scheduled 25 articles.

Geoffrey Burbidge
Editor
April, 2002

Annual Review of Astronomy and Astrophysics
Volume 40, 2002

CONTENTS

FRONTISPIECE, *Edwin E. Salpeter*   xii

A GENERALIST LOOKS BACK, *Edwin E. Salpeter*   1

ULTRA-COMPACT HII REGIONS AND MASSIVE STAR FORMATION,
Ed Churchwell   27

KUIPER BELT OBJECTS: RELICS FROM THE ACCRETION DISK OF
THE SUN, *Jane X. Luu and David C. Jewitt*   63

THEORY OF GIANT PLANETS, *W. B. Hubbard, A. Burrows,
and J. I. Lunine*   103

THEORIES OF GAMMA-RAY BURSTS, *P. Mészáros*   137

COSMIC MICROWAVE BACKGROUND ANISOTROPIES, *Wayne Hu
and Scott Dodelson*   171

STELLAR RADIO ASTRONOMY: PROBING STELLAR ATMOSPHERES
FROM PROTOSTARS TO GIANTS, *Manuel Güdel*   217

MODIFIED NEWTONIAN DYNAMICS AS AN ALTERNATIVE TO
DARK MATTER, *Robert H. Sanders and Stacy S. McGaugh*   263

CLUSTER MAGNETIC FIELDS, *C. L. Carilli and G. B. Taylor*   319

THE ORIGIN OF BINARY STARS, *Joel E. Tohline*   349

RADIO EMISSION FROM SUPERNOVAE AND GAMMA-RAY
BURSTERS, *Kurt W. Weiler, Nino Panagia, Marcos J. Montes,
and Richard A. Sramek*   387

SHAPES AND SHAPING OF PLANETARY NEBULAE, *Bruce Balick
and Adam Frank*   439

THE NEW GALAXY: SIGNATURES OF ITS FORMATION, *Ken Freeman
and Joss Bland-Hawthorn*   487

THE EVOLUTION OF X-RAY CLUSTERS OF GALAXIES, *Piero Rosati,
Stefano Borgani, and Colin Norman*   539

LYMAN-BREAK GALAXIES, *Mauro Giavalisco*   579

COSMOLOGY WITH THE SUNYAEV-ZEL'DOVICH EFFECT,
John E. Carlstrom, Gilbert P. Holder, and Erik D. Reese   643

INDEXES
Subject Index   681
Cumulative Index of Contributing Authors, Volumes 29–40   709
Cumulative Index of Chapter Titles, Volumes 29–40   712

ERRATA
An online log of corrections to *Annual Review of Astronomy
and Astrophysics* chapters (if any, 1997 to the present)
may be found at http://astro.annualreviews.org/errata.shtml

RELATED ARTICLES

From the *Annual Review of Earth and Planetary Sciences*, Volume 30 (2002)

Modern Integrations of Solar System Dynamics, A. Morbidelli

Implications of Extrasolar Planets for Understanding Planet Formation, Peter Bodenheimer and D. N. C. Lin

Pluto and Charon: Formation, Seasons, Composition, Michael E. Brown

From the *Annual Review of Nuclear and Particle Science*, Volume 51 (2001)

Nuclear Astrophysics Measurements with Radioactive Beams, Michael S. Smith and K. Ernst Rehm

Neutrino Propagation in Dense Astrophysical Systems, Madappa Prakash, James M. Lattimer, Raymond F. Sawyer, and Raymond R. Volkas

ANNUAL REVIEWS is a nonprofit scientific publisher established to promote the advancement of the sciences. Beginning in 1932 with the *Annual Review of Biochemistry*, the Company has pursued as its principal function the publication of high-quality, reasonably priced *Annual Review* volumes. The volumes are organized by Editors and Editorial Committees who invite qualified authors to contribute critical articles reviewing significant developments within each major discipline. The Editor-in-Chief invites those interested in serving as future Editorial Committee members to communicate directly with him. Annual Reviews is administered by a Board of Directors, whose members serve without compensation.

2002 Board of Directors, Annual Reviews

Richard N. Zare, *Chairman of Annual Reviews*
  Marguerite Blake Wilbur, Professor of Chemistry, Stanford University
John I. Brauman, *J. G. Jackson–C. J. Wood Professor of Chemistry, Stanford University*
Peter F. Carpenter, *Founder, Mission and Values Institute*
W. Maxwell Cowan, *Bethesda, Maryland*
Sandra M. Faber, *Professor of Astronomy and Astronomer at Lick Observatory,*
  University of California at Santa Cruz
Susan T. Fiske, *Professor of Psychology, Princeton University*
Eugene Garfield, *Publisher*, The Scientist
Samuel Gubins, *President and Editor-in-Chief, Annual Reviews*
Daniel E. Koshland, Jr., *Professor of Biochemistry, University of California at Berkeley*
Joshua Lederberg, *University Professor, The Rockefeller University*
Sharon R. Long, *Professor of Biological Sciences, Stanford University*
J. Boyce Nute, *Palo Alto, California*
Michael E. Peskin, *Professor of Theoretical Physics, Stanford Linear Accelerator Ctr.*
Harriet A. Zuckerman, *Vice President, The Andrew W. Mellon Foundation*

Management of Annual Reviews

Samuel Gubins, President and Editor-in-Chief
Richard L. Burke, Director for Production
Paul J. Calvi, Jr., Director of Information Technology
Steven J. Castro, Chief Financial Officer
John W. Harpster, Director of Sales and Marketing

Annual Reviews of

Anthropology	Fluid Mechanics	Physiology
Astronomy and Astrophysics	Genetics	Phytopathology
Biochemistry	Genomics and Human Genetics	Plant Biology
Biomedical Engineering	Immunology	Political Science
Biophysics and Biomolecular	Materials Research	Psychology
Structure	Medicine	Public Health
Cell and Developmental	Microbiology	Sociology
Biology	Neuroscience	
Earth and Planetary Sciences	Nuclear and Particle Science	
Ecology and Systematics	Nutrition	SPECIAL PUBLICATIONS
Energy and the Environment	Pharmacology and Toxicology	Excitement and Fascination of
Entomology	Physical Chemistry	Science, Vols. 1, 2, 3, and 4

Edwin E. Salpeter

Annu. Rev. Astron. Astrophys. 2002. 40:1–25
doi: 10.1146/annurev.astro.40.060401.093901
Copyright © 2002 by Annual Reviews. All rights reserved

A GENERALIST LOOKS BACK

Edwin E. Salpeter

*J. G. White Distinguished Professor of Physical Sciences, Emeritus, Cornell University,
Ithaca, New York 14853; email: ees12@cornell.edu*

Key Words   autobiography, quantum electrodynamics, nuclear astrophysics,
interstellar medium, neurobiology, epidemiology

■ **Abstract**   I fled with my parents from Hitler's Austria to Australia and studied
physics at Sydney University. I obtained my Ph.D. in quantum electrodynamics with
Rudolf Peierls at Birmingham University and came to Cornell to work with Hans Bethe.
I have stayed at Cornell ever since, and I have essentially had only a single job in my
whole life, but have switched fields quite often. I worked in nuclear astrophysics and
in late-stellar evolution, estimated the Initial Mass Function for star formation and the
metal enrichment of the interstellar medium. I suggested black hole accretion as the
energy source for quasars, worked on molecule formation on dust grain surfaces, and
was involved in 21-cm studies of gas clouds and disk galaxies. I collaborated with my
wife on the neurobiology of the neuromuscular junction and with one of my daughters
on the epidemiology of tuberculosis.

INTRODUCTION

A generalist is a person who learns less and less about more and more subjects, until
he knows nothing about everything. I am getting close to that endstate, and this
essay will trace my evolution toward it. The definition above really refers only to
successive generalists—a professional basketball player giving up his profession
and turning to baseball, for instance. I will also have to deal in some cases with
simultaneous generalists applying expertise from one field of science to a problem
in another field.

Most essays in this series have stressed some unifying theme, some primary
goal, or some preferred scientific method. Presumably, most successful scientists
are golf-players, i.e., they map out their own strategy beforehand and can then write
a unified story. I belong to the minority of scientists who are ping-pong players, i.e.,
they react to influences from others on a short timescale without much systematic
planning on their own. I therefore cannot write a coherent scientific narrative (and
if I could, only a fraction of it would be astrophysics anyway) but will deal mainly
with sociology: A ping-pong player reacts to outside players, but has to decide
which of several possible influences to react to, so decision making is important.
It may be natural for an autobiographer to brag about his past, but I brag more

0066-4146/02/0922-0001$14.00   1 about the few right decisions I have made than about scientific achievements. I will also try to mention some of my failures in both areas (they may be more instructive to young players than the successes), but I am probably glossing over the most important ones (just as the characters did in the old Japanese film, "Rashomon"). I should also apologize for citing so many of my own papers and so few others.

SCHOOLING AND PARENTS

I was born in Vienna in December 1924, spent some time in Hungary, but returned to Austria for elementary school in Vienna. My first academic event came at the age of 10, when I failed my entrance exam into Gymnasium, the eight-year academic high school system in Austria. My mother claimed it was just due to the authorities being anti-Semitic. They were indeed anti-Semitic, but I felt I had flunked mainly because I read Karl May (Wild West adventure books especially for German boys) instead of studying for the exams. Gymnasium was essential for anything else you might want to do later in life in Austria, and I learned my lesson, at least temporarily: I had to learn Latin and other academic subjects on my own for one year, took another exam and got into second year of Gymnasium. Having to work very hard that year was actually an exhilarating experience, but my concentration did not last long without an external stimulus. We lived in a rather nonintellectual neighborhood in Vienna; there was no tracking in school of any kind, and I was soon the top of my class in all subjects in my rather mediocre gymnasium class. I soon returned to reading Karl May, had no preference for one academic subject over another, nor any ambitions for my future. This serenity (or indolence) was interrupted three years later by a traumatic event—the annexation of Austria to Nazi Germany in the spring of 1938.

In September 1938, all Jewish children were thrown out of the regular schools and only a very small fraction were able to go to a single Jewish high school in all of Vienna for the next academic year. The selection of the students was academic, so each of my new classmates had also been at the top of their class in everything the year before, and it was an exciting school atmosphere. Suddenly I was no longer the top of my class in anything except math and science. I also noticed that one half of the kids who were beating me in Latin, history, and other subjects, were girls—this school was one of the rare places where purely academic selection replaced male chauvinism. This exciting experience only lasted three months for me and was replaced by excitement of a different kind: On November 10, 1938, the infamous Kristalnacht, someone tried to arrest me even though I was only 13 years old (but I was proud of looking older). My parents took me out of school that day, and I had to go into hiding, even though we did not know why they had tried to arrest me. I did not attend any school again until we fled from Hitler and arrived in Australia in June 1939.

My stay in the Jewish school demonstrated that I was more gifted in some subjects than in others, before I had thought about which subjects I actually liked best.

It is curious that during·my hiding period, when I assessed that my probability of surviving to adulthood was small, I started thinking seriously what I might do if I did become an adult after all. Although I had not formulated any preferences, I decided it was safer to choose a profession according to one's abilities, rather than one's preferences. Much later I found out that I loved opera but could not hold a tune, so singing would not be my profession; of academic subjects, I found anthropology more interesting than the physical sciences, but I wasn't very good with people.

Apart from ability, my eventually going into the physical sciences was almost inevitable: Sons of professional sword swallowers often become sword swallowers, and my father was a physicist. Even my mother was a physicist—at any rate she had a Ph.D. in physics, although she stopped working after I was born. Although I seemingly did not have a close relation with my father, he had had an interesting career, and he had a strong influence on me. He was born in Galicia, as were many Austrian Jews, but moved to Vienna to study at the university. As a young student he was a close friend of Erwin Schrödinger. There even was a rumor that Schrödinger lost his virginity while borrowing my father's apartment, somewhat like the Jack Lemmon movie "The Apartment." My father was poor but had his own little apartment; Schrödinger was rich, but etiquette required that he live with his family (Moore 1989). They maintained some contact even after Schrödinger got his Nobel Prize. My father had an early interest in mathematics but also in applications, and he wrote a very early textbook on *Mathematical Methods for Scientists and Physicians*. Had he been a gentile, he would probably have ended up as a university professor, but Austrian anti-Semitism made that difficult. Industry was more liberal at the time, and he joined Elin, an Austrian electrical concern outside of the European cartel, and became the director of a factory making incandescent lamps. The main challenge was not making the lamps, but avoiding patent fights with the cartel by inventing alternative patents. This cartel/outsider rivalry indirectly saved our lives in 1938–1939 when an Australian outsider hired my father to start a lamp factory in Sydney, in competition with the Philips-led cartel. This enabled us to flee from Hitler and start life again in Sydney. Ironically, a few years later my father switched and joined Philips.

My new life in Sydney began at fourteen, just as World War II started and I was classified as a "Friendly Enemy Alien" (German passport, but Jewish). Even with the war going on, Australia was an ideal place for a teenager to grow up, and I had a happy life. I was accepted into Sydney Boys' High School, a selective high school with some tracking, and found it an excellent and liberal school. Academically, I had a problem at first—the numbering of school years was different from Austria, with the end of high school after fifth grade instead of eighth. Because I seemed bright (although I knew very little English), the school took a chance and put me in fourth grade—if it all worked out, I would graduate at sixteen instead of eighteen in spite of the half year of school I had missed! I had to make up one or two years' worth of schooling in all subjects, mostly on my own but with just the right amount of encouragement and advice from my teachers. I started in June 1939, and I certainly had to work harder the next 18 months than any time before or after.

My hard work in my mid-teen years paid off; in fact, I overcompensated a bit (as often happens), and I graduated from high school with honors. I had a great time and only regretted that I did not learn how to date girls, but I figured I could catch up in college. Having had to work hard in all academic subjects, instead of concentrating on one favorite, already predisposed me toward being a generalist. It also taught me how to put in the minimum effort for a successful conclusion, which is useful if you're working on several problems simultaneously. On the other hand, all this also predisposed me toward being a little bit sloppy, which is not so good. At any rate, I was able to start at Sydney University at the age of sixteen, majoring in mathematics and physics as my parents had expected me to.

My undergraduate years provided enjoyable distractions, especially bushwalking in the outback during major vacations and some minor involvement in left-wing scientific politics. However, these years also gave me a good basic education and were uneventful except for two excursions. I got interested in Chemistry, had summer jobs in chemistry labs, and almost switched my major. In the end I stayed in physics, but retained a love for chemistry. The other distraction had to do with Australia's war effort in radar, which university students majoring in the sciences were supposed to help with, instead of joining the army. As a consequence, I was involved in research and development in radiophysics, and what we would now call plasma physics, but I did not return to this field for about 15 years. Radiophysics pointing toward radio astronomy was in the air in Sydney in 1945, with some very able people at the Radiophysics Lab of CSIRO, just across the street from my physics department. I did not make use of the exciting opportunity there, but instead went into quantum electrodynamics, another field that was in the air in Europe and in the USA, even though not in Australia. I started with a pedagogical masters thesis on Field Theory—a modest effort but enough to get me a prestigious scholarship to go to graduate school in Great Britain.

Here I faced my first challenge to make the right decision in spite of overwhelming advice for the opposite. With my scholarship I could have gone to Oxford or Cambridge, which of course had great prestige overall. Furthermore, Cambridge had Dirac, one of the greatest figures in quantum electrodynamics, and most of my advisers were urging me to go there. Fortunately, a few people I trusted most, including Bert Corben in Melbourne, assured me that Rudolf Peierls had the best school for theoretical physics in Europe—even though he was at the less prestigious red-brick university in Birmingham. Fortunately, I followed this correct advice and went there.

QUANTUM ELECTRODYNAMICS: BIRMINGHAM TO CORNELL

I started in Birmingham University in the Fall of 1946 in Peierls' school, housed in temporary and uncomfortable quarters. Altogether, I was surprised to find that living in England in the postwar years was much more strenuous than the war years themselves had been in Australia. In spite of the food rationing and other discomforts, being a graduate student with Peierls was a glorious experience. He had a stable of bright young people and guided them in many topics—not just quantum electrodynamics, but nuclear theory, statistical mechanics, and solid state physics as well. Rudi Peierls was a great role model for becoming a generalist and even the vacations were a broadening experience. My overseas scholarship was financed and administered by "The Commission for the Exhibition of 1851" and this organization also helped to farm us students (possibly thought of as "the deserving poor from the colonies") out to the stately homes of England for vacations. These stays were all enjoyable, but one in Wales was also memorable—with Charles and Dorothea Singer—early and brilliant historians of medicine. Medicine as an academic venture seemed fascinating to me from then on.

The early postwar years in England were also memorable for observing the bigshots in theoretical physics at conferences that Peierls organized. It was particularly interesting to watch the first time Werner Heisenberg met Peierls, Bethe, Frisch, Oppenheimer, Pauli, Dirac, and others after the war. It was a strained atmosphere, but (unlike remarks in a recent play) none overtly refused to shake Heisenberg's hand (although some people who happened to be in another part of the room when he entered did not have to shake hands). I was a bit puzzled that Pauli, who normally made a point of being rude to people, was singularly deferential to Dirac. In any case, Pauli's rudeness was only the cover for a very kind man: When I asked him a physics question once to which I really needed the answer, he first shook his head sadly and said, "I don't know how people can be SO stupid," but then gave me the real answer patiently and in detail. I also made some pilgrimages to Cambridge—Dirac was impressive at answering questions, even though he would not have been a good thesis adviser. Incidentally, I had already met Tommy Gold, Fred Hoyle, and Herman Bondi in the late 1940s. My scientific interests did not yet overlap theirs, but their fertile and bubbling minds were most stimulating, and the trio became both role models and friends later on.

During the three years in Birmingham, I had already showed a tendency toward generalism by doing some work outside of quantum electrodynamics. One example was a paper on nuclear induction signals (Salpeter 1949), which was not only pedagogical, but modestly useful and has been quoted, or used in lab courses, off and on for some years later. By contrast, the paper based on my whole Ph.D. thesis is probably my least quoted paper (even I am not citing it here)—and for a good reason. My thesis was designed to take the singularity in the electrodynamic self-energy of the electron seriously and to calculate it rigorously by starting with a finite size and then proceeding to the limit of a point electron. In the meantime, it had become important to do a practical calculation of the Lamb shift in the hydrogen atom, in spite of the singularity, and Hans Bethe had done this in a simple, though approximate manner (Bethe 1947). Essentially, he just compared a free electron with one inside the atom and was able to circumvent the singularity. This beginning of renormalization theory was followed by two more rigorous and elegant formulations, one by Schwinger at Harvard and the other by Feynman at Cornell (plus later synthesis by Dyson). These elegant formulations were beyond my capacity as a graduate student, but my thesis work had put me in an excellent position to have done Bethe's simple calculation; I had an excellent opportunity to do it first but did not take it—I simply just did not think of it! This was my first, but not last, instance of missing the boat, and my thesis was out of date before it was out in print.

In spite of missing out on the big time in Birmingham, I had done enough so that I had the opportunity to go as a postdoc or research associate, either to Oppenheimer at the Princeton Institute for Advanced Study, or to Bethe at Cornell University. This gave me my second chance to make the right decision in spite of opposite advice. This was the heyday of Oppie and his high-powered group at the Institute, and I was assured that two years in this intense atmosphere would condition me to return to a good job in England (a two-year stay seemed to be the norm, as for Ladies Finishing Schools, although Peierls' wife Genia had already told me "within a year you will marry an American girl and you will never come back permanently"). There was then advice to choose Princeton, from many people but fortunately not from Peierls, and I myself had no doubt that I wanted to go to Cornell. Bethe was a close friend of Peierls, had a very similar scientific temperament, and had visited the Birmingham department a few times. The graduate student offices were lined up in a row like a railroad car, and Hans would go from one to the next, like a chess Master playing simultaneous games, and give advice. The advice was instant, but low-key, and I was sure I would find Bethe's group of theorists at Cornell a relaxing but stimulating place. I arrived at Newman Lab of Nuclear Studies in October 1949, one month late, because I had had trouble getting a U.S. immigration visa. I consequently got the worst of the offices for first-year postdocs (it even had a crack on the outside wall). To give away the punchline early in this essay—I have essentially had only one single job for my whole life, and I stayed in that office for exactly 50 years (at the end of 50 years the crack had not got smaller nor larger). Maybe the invariance of my geography led me to change fields often.

In early 1950, I met Mika Mark, an entering graduate student in psychobiology, and as Genia Peierls had predicted, we got married in June. Some years after her Ph.D., she switched to neurobiology and biophysics, which gave us an opportunity to collaborate, but in the 1950s she mainly helped me in decision making, because she had more common sense than me. With Richard Feynman and Freeman Dyson passing through Cornell, and with a certain rivalry between Feynman and Harvard's Julian Schwinger, Cornell was a hotbed of quantum electrodynamics, and I was involved off and on for a few years. (For one horrendous calculation of fourth-order vacuum polarization, three of us did the same calculation independently to check each other—one, a graduate student, Michel Baranger, the other two, Freeman Dyson and myself. Dyson and I got the same answer, the graduate student got a different answer—and his was the correct one.) However, I have to describe next one particular paper and one decision very soon after that.

RELATIVISTIC BOUND STATES TO
NUCLEAR ASTROPHYSICS

The early 1950s saw the beginning of elementary particle physics and high energy theory, including speculations about very strong, hypothetical, attractive forces, including a theory proposed by Fermi & Yang (1949). One extreme example of strong attraction would be the bound state of two Fermi-Dirac particles where the binding energy is so large as to cancel (fully or almost) the sum of the two rest mass energies. For such an artificially low-mass composite particle, a combination of special relativity and quantum mechanics would be needed. Such a combination for bound states did not yet exist, but Hans Bethe saw in 1951 how to develop a formalism, starting from techniques Feynnman had invented recently for quantum electrodynamics of scattering problems. Hans got me to work out the details and we soon had a plausibility argument for a new equation that incorporates relativity into bound state problems. It was a rather elegant equation, involving a mysterious relative time coordinate, and potentially useful in the future but not easy to manipulate. It soon became known as the "Bethe-Salpeter Equation," even before it was published (Salpeter & Bethe 1951), and I got a lot of attention from the theoretical physics profession. You might think that I am particularly proud of that paper and that I was all set to stay in the field of high energy theoretical physics and quantum electrodynamics for life. In reality I am proud, not of that paper, but of my decision to slowly get out of that field altogether.

I am proud of my realization in 1951–1952, in spite of my seeming success, that I just did not have the right abilities or temperament for quantum electrodynamics/high energy/elementary particle theory. This field required abstract mathematical thinking and a mind devoted to rigor, whereas I had a quick but sloppy mind. Even the Bethe-Salpeter equation itself was really invented by Hans—I had only filled in some details—and I could see that I would not get very far with applying it in the future to challenging, fully relativistic problems. My unsuitable temperament was illustrated at a colloquium I gave at the Princeton Institute for Advanced Study, on possible applications of the equation but with only a plausibility argument for it instead of a rigorous derivation. The whole audience felt that such a derivation was required and somebody, who looked young enough so that I thought he was a high school boy there by mistake, got up and (together with Francis Low) outlined the rigorous derivation of the equation on the spot. He was not quite a high school student, he was Murray Gell-Mann, and he got the Nobel Prize for other things a short while later, but he produced the derivation with rather little effort (Gell-Mann & Low 1951). However, the main warning for me was not so much that I had not given a derivation myself, but that with my temperament I had felt a plausibility argument, rather than a derivation, was good enough.

Over the next few years I still wrote some papers on quantum electrodynamics, especially on some precision atomic physics calculations that used low-brow applications of the Bethe-Salpeter equation (e.g., Salpeter 1952a). Hans and I also wrote a rather detailed monograph on Quantum Mechanics, which still sells one or two copies a year worldwide (Bethe & Salpeter 1957). Having written meticulous papers with highly accurate calculations (including the fourth order correction paper with Baranger & Dyson) would give me self-confidence later on, but I needed a change. For my scientific temperament I needed a field that was more controversial, more open-ended and new, where quick was useful and sloppy did not matter too much because it would all change soon anyway. Hans Bethe had invented such a field a few years earlier—nuclear astrophysics—by showing how the conversion of hydrogen into helium provided the energy source for the sun and other main-sequence stars (Bethe 1939, Bethe & Critchfield 1938). There were star types other than the main-sequence and plenty of elements other than helium, so nuclear astrophysics was sufficiently open-ended with plenty of uncertainties. Willy Fowler, a nuclear experimentalist at Caltech's Kellogg Radiation Lab, was doing the appropriate measurements, and he had the foresight to invite some young theoretical physicists to explore the theoretical aspects of nuclear astrophysics. Although I continued to stay at Cornell for the main academic year, I started to spend summers at Kellogg Lab, beginning in June 1951. To demonstrate my maturity in 1951: My first impression was of Willy's immense old age—he was having his 40th birthday when we arrived that summer—but his leadership qualities came through pretty quickly after that.

Some areas in nuclear astrophysics were fairly straightforward, and I worked on some of these in that summer and over the next few years—for instance, on the detailed completion of the proton-proton chain (Salpeter 1952b). However, the real challenge was to explore nuclear reactions that start from seemingly inert helium in red giant stars, after main-sequence stars have exhausted the hydrogen. I started to work on that also in the summer of 1951. This was a more controversial subject, which brings me to a particular paper, published the following year, plus its aftermath a few years later.

HELIUM-BURNING: A TALE OF TWO RESONANCES

Hans Bethe was a different role model for different situations, but for an open-ended and controversial situation the advice was three-pronged: (*a*) Be prepared to switch fields; (*b*) use only the minimum mathematical technique necessary; (*c*) in the face of uncertainty, be prepared to use conjectures and shortcuts and take risks—in other words—have CHUTZPAH! For my next paper, I had learned lessons (*a*) and (*b*), but not yet (*c*).

Nuclear astrophysics was needed for two related but different problems: (*a*) The source for energy production in red giant stars and other types of evolved stars and, (*b*) building up all the heavy elements in evolved stars and getting the isotope ratios right. The second of these was most exciting in the long run, partly because of the controversy as to whether the Big Bang Cosmology could produce all the heavy elements in the first few minutes (e.g., Gamow 1946) or whether evolved stars could, as is required by the Steady State Theory (reviewed by Bondi et al. 1995). In spite of its importance, this problem was too daunting for me (there are many more isotopes than there are types of stars), so I merely looked for energy-producing reactions in red giant stars. Stellar evolution theory was in its infancy, but calculations by Schwarzschild & Sandage, by Hoyle and by some others, (e.g., Sandage & Schwarzschild 1952), were already showing that red giant stars, in spite of their cool surface, have a hot and dense interior consisting mainly of helium. Be^8 was known not to be stable, so something more complicated than a single two-nuclei reaction was needed. Fortunately, my host Willy Fowler and his experimental colleagues had shown recently that Be^8 is almost stable, i.e., that its meta-stable groundstate provides a resonance level at a positive but quite low (and known) excitation energy for a pair of alpha-particles. Hence, one can form C^{12} from He^4 without needing an explicit three-body reaction, i.e., the meta-stable Be^8 state is in thermal equilibrium with helium and can then absorb another alpha-particle from the tail of the thermal distribution.

I calculated the rate for this indirect conversion of helium into carbon (much more rapid than the direct triple-alpha reaction without the beryllium resonance would have been) in the summer of 1951 and published it in the following year (Salpeter 1952c). I noted in that paper that my calculated rate could easily be too low by a factor of 1000, say, if there should be an appropriate resonance level in C^{12}, but I did not have the chutzpah (or guts) to do anything about it: My energy production rate for red giant stars required a central temperature that was within the rather uncertain range given by stellar evolution theory at the time; my calculation would lead to most of the helium being converted to oxygen and neon instead of carbon, but I just did not have the guts to think of resonance levels that had not been found yet! A short while later Fred Hoyle demonstrated both chutzpah and insight by using the known abundance ratios of C^{12}, O^{16}, and Ne^{20} to show that there JUST HAD to be an appropriate resonance level in C^{12}, and he was able to predict its energy (Hoyle 1954).

Willy Fowler and his colleagues soon looked for Hoyle's predicted resonance level and found it just where it should be (Cook et al. 1957). This made a believer in theoretical nuclear astrophysics out of Willy and has been a great object lesson to many ever since. However, history tends to oversimplify, and the spectacular prediction of the carbon resonance level has obscured the fact that the beryllium resonance level also was needed to increase the rate enormously. This level, and many others, was the experimental achievement of the Kellogg Radiation Lab (including Charlie and Tommy Lauritsen, Ward Whaling, and many others in addition to Willy Fowler) and a few other experimentalists elsewhere. One reason why experimentalists do not get enough credit may be that most of the review articles are written by theorists. The two resonances and the work of Ernst Öpik also illustrate the danger of being in the wrong place too early: Öpik was an Estonian but worked much of his life in a prestigious, but somewhat isolated, Irish Observatory. He actually published a paper on the triple-alpha reaction before mine or Hoyle's (Öpik 1951), but (*a*) got too low a reaction rate because he did not know about the beryllium resonance, and (*b*) his work was completely unknown in Britain and the USA until 1953. Even after that, Öpik did not get as much publicity as this remarkable, although gruff, man (and generalist) deserved. I also suffered the consequences of being 20 years too early in another topic: I wrote a paper on the burning of primordial deuterium during the contraction of a proto-star onto the main sequence for an interesting meeting in Belgium that I unfortunately was not able to attend (Salpeter 1953). This paper had two defects: (*a*) I missed out completely on the Hayashi Phase of contraction (Hayashi 1964), which drastically alters the contraction phase, and (*b*) I did not know the value of the primordial deuterium abundance. Deservedly, this became one of my least quoted papers even though deuterium burning became of interest later on.

THE INITIAL MASS FUNCTION AND THE ENRICHMENT OF THE INTERSTELLAR GAS

In the early 1950s I thought of my nuclear calculations as straight physics, to be applied to astrophysics mainly by others. Nevertheless, I was learning some real astronomy from experts, largely on visits to Princeton and Caltech, especially from observers like Baade, Greenstein, and Sandage in Pasadena and from theorists like Schwarzschild and Spitzer in Princeton. One special educational experience for me was the Ann Arbor summer school, or Michigan symposium on Astrophysics, in the summer of 1953. This summer school was the most formative meeting ever for me (and possibly for others as well; see Gingerich 1994), probably because it both had bigshots as role models and also had us youngsters lecturing to each other (reminiscent of Oppie's motto "what we don't understand, we explain to each other"). Walter Baade was the star of the show, explaining the two Stellar Populations and also introducing us to galaxies. George Gamow was equally impressive talking on everything, an important role model for me for being a generalist (fortunately his example of drinking Vodka from a water pitcher already at lunchtime was not followed by us youngsters). I also remember fondly Leo Goldberg and Allan Sandage; also George Batchelor, who lectured on turbulence. I myself talked on hydrogen and helium burning, but also on reactions between C^{12} and O^{16} in more evolved stars at about 10^9K, as well as subsequent photodisintegration. I had already mentioned this qualitatively in print (Salpeter 1952c), but I never followed it up with detailed calculations.

Just before the 1953 summer school our first daughter, Judy, was born and soon after it we went for one year to the newly founded Australian National University (ANU) in Canberra. Although nominally a one-year trip to the ANU, I was offered the Chairmanship of the new Theoretical Physics Department there. The main emphasis of this department would have to be High Energy Theory, and I would have a dilemma. In fact, I started spending more and more time in 1954 at Mt. Stromlo Observatory, which was quite close geographically, learning more about observational stellar astronomy and also hearing about galaxies from Gerard and Antoinette de Vaucouleur.

The Burbidges, Fowler, and Hoyle (Burbidge et al. 1953), as well as Al Cameron (e.g., Cameron 1958), were showing how heavy elements are built up in the deep interior of massive, and very massive, stars in various late stages of evolution, including supernova explosions. However, a practical question was how much enrichment of the interstellar gas there could have been from the breakup of these stars, given the fact that massive stars are very rare today. Walter Baade had made clear (mostly orally, rather than in print) the distinction between Stellar Population II (formed when the Galaxy was young) and Stellar Population I, with stars having been born continuously over the lifetime of the Galaxy. Observations had already been made for globular cluster stars in Stellar Population II (Arp et al. 1952), but we and the interstellar gas are situated in the Galactic Disk where Stellar Population I resides. Heavy elements today in the interstellar gas came mostly from stars that are already dead and the question was how many more stars were born and have died than the few that are still alive today, separately for different stellar masses.

To calculate how many massive stars were born and have died over the lifetime of our galaxy, the "initial mass function" for star formation (or the "birthrate function") was needed. This in turn required three ingredients: the lifetime of a star as a function of its mass, the mass-luminosity relation, and the luminosity function for Population I main-sequence stars. Each of these three was known only extremely approximately in 1954, but I had learned my lesson about the importance of chutzpah: I used what little I could find in print on each of these three functions. A clean separation of main-sequence stars from other stellar types was a bit difficult at the time, but the situation was made easier by the fact that stars spend much more time on the main-sequence than on the (interesting) later evolutionary phases. I fudged over the distinction between stars in the galactic plane versus those in the whole disk and over the time variation of the star formation rate from the young and gas-rich galaxy to the present-day gas-poor galaxy. With these two approximations in hand, I finally got a crude estimate of the Initial Mass Function, of the amount of heavy elements in the interstellar gas, and also of what fraction of stars should be White Dwarfs today. Of my 300-odd papers overall, the one I am most proud of was published in 1955 (Salpeter 1955).

I am proud, rather than ashamed, of the sloppy approximations I made in that paper, but I was also lucky that the several errors (corrected later by M. Schmidt and others) mostly canceled each other, rather than adding up. The result was also a particularly simple powerlaw for the Initial Mass Function, which is close to predicting logarithmic scale-invariance for the distribution of mass among different stars (a similar, but not identical, relation holds for the distribution of people into villages, towns, and cities). Developments (or lack thereof) in the 45 years since have been good for me personally, but bad for the science: I had hoped that my Initial Mass Function was only an average for today and that the actual function would vary strongly from quiescent to active regions and would also vary from the young galaxy to the galaxy today. Some such variations probably exist but are too weak for universal agreement, so "the Salpeter function" is still of some use today and gets quoted, but the theory of star formation has suffered greatly by not having clear-cut observational variations, which would have to be predicted by a correct theory.

My dilemma between accepting the Theoretical Physics Chair at the ANU and returning to Cornell was resolved fairly quickly: At the ANU I would have to concentrate on high-energy physics, but by now I was hooked on astrophysics and even on astronomy. Furthermore, male chauvinism was slightly less severe at Cornell than at Australian universities, especially later on after my wife had switched from psychology (the area of her Ph.D.) to neurobiology. In September 1954 we returned to Ithaca—myself to a tenured associate professorship in physics and my wife to two rush-jobs before returning to biology research in 1956: (*a*) she learned enough architectural skills to design a house and acted as contractor for its construction; (*b*) she had a second baby.

A GENERALIST'S DIVERSIONS: PLASMA PHYSICS AND HIGH PRESSURE STATISTICAL MECHANICS

I did not overlap in time with Sir Arthur Eddington, but my thesis advisor Rudi Peierls did, and he corrected a wrong paper of Eddington's (Eddington 1935; Peierls 1936). As I discuss in more detail in a Chandrasekhar memorial (Salpeter 1996), Eddington's error was in part due to his realization that Coulomb forces in the ionized stellar interior (i.e., plasma physics) have to be included in principle, in addition to gravity. It was mostly a red herring in Eddington's time, but this admonition led me to an early and approximate treatment of electron-screening corrections to stellar thermonuclear reaction rates (Salpeter 1954). In this field also, later developments have been good for me and bad for the science. Modern treatments of screening have been controversial, and results have tended to oscillate around mine, so that my old paper still gets quoted! However, by now it is pretty clear—at least for the sun where several inequalities are not very strong—very extensive numerical calculations will be needed for very precise results, and computers are just about up to that task now. The importance of plasma physics and statistical mechanics for stellar interior calculations was also clear to Evry Schatzman, who wrote an even earlier paper than mine on electron screening (Schatzman 1948), and to Al Cameron. Cameron wrote the first paper on pycnonuclear reactions, the real cold fusion (Cameron 1959), a topic that also inspired one of my later graduate students, Hugh Van Horn (Van Horn & Salpeter 1969).

My second encounter with plasma physics, a few years later, also got me involved in ionospheric research and came about by a comedy of errors. In 1958, Bill Gordon had proposed the Air Force fund the construction of the Arecibo Ionospheric Observatory in Puerto Rico for incoherent backscatter of radio waves from the free electrons in the ionized upper atmosphere. He pointed out, correctly, that the electrons would mostly follow in time the density fluctuations of the massive, slow positive ions, but he did not give a rigorous derivation (Gordon 1958). Somebody at the Advanced Research Projects Agency (ARPA) wanted to check up by using one of the early rigorous plasma theory papers by Bohm & Pines. That paper had replaced the positive ions by a fixed uniform background density, for simplicity in treating the free electrons alone. In this hypothetical world, there would be no ion density fluctuations at all, so the electrons could not follow the nonexistent ions, and the ARPA official concluded (incorrectly) that the backscattered intensity would be much smaller than Bill Gordon's estimate. I was associated with ARPA as a member of the JASON division and agreed to look into the controversy. I soon found out that most of the backscatter would indeed be qualitatively as Bill Gordon had predicted, and I was able to allay ARPA's misgivings. The controversy was laid to rest further when two other groups also published independent papers to this effect. However, the diversion about free electrons without positive ions got me curious about the smaller, but interesting, electron backscatter effects that do not involve the ions at all. I showed that this backscatter would give a frequency shift related to the plasma frequency, a kind of Raman scattering (Salpeter 1960a). I believe this paper turned out to be my most-quoted paper ever, not because of the ionosphere but because this kind of backscatter is a convenient way to measure the electron density in laboratory plasmas, and the number of plasma experimentalists is very large. A little while later, an expert graduate student, "Rip" Perkins, succeeded in measuring this kind of Raman scattering from the ionosphere's plasma oscillations (Perkins et al. 1962).

Plasma physics, and related subjects such as interplanetary scintillations, also fascinated many of my graduate students and some became much more professional as plasma physicists than I was, e.g., Rip Perkins, Nick Krall, Allen Boozer, Dick Lovelace, and Jonathan Katz. Besides plasma physics, the statistical mechanics for calculating high pressure equations of state was of interest, not only for the really high pressures in white dwarfs and neutron stars, but also for the intermediate high pressures in the Jovian planets. Some of the papers in these areas may seem a bit abstract (Salpeter 1960b, 1961; Zapolsky & Salpeter 1969), but had applications to real objects.

Statistical mechanics is also involved in various neutrino emission processes, both in the sun and in highly evolved stars. Shaviv and I wrote a few papers relevant to the solar neutrino controversy (e.g., Shaviv & Salpeter 1971), but this topic was covered more extensively by John Bahcall and his colleagues. Neutrinos from highly evolved stars, however, were a favorite Cornell topic with many competing emission processes to be considered. There were papers from Phil Morrison and a number of my graduate students, ex-students, and visitors, including Beaudet, Chiu, Petrosian, Silvestro, Stabler, and Zaidi (e.g., Beaudet et al. 1967).

MAPLE SYRUP, ADMINISTRATION, AND NATIONAL SERVICE

Starting in the 1960s I occasionally got job offers for chairmanships and other administrative positions elsewhere. Cornell salaries were rather low, and I was only a rank and file Professor, so these offers involved a substantial raise in salary. My wife and I once made a list of what we would do with the extra money, but most items were a mixed blessing, e.g., with two cars instead of one we would have to get registrations and snow tires twice as often. The only improvement with no drawbacks at all would be buying pure maple syrup instead of the cheaper imitation syrup we had used in the past. We decided that we could afford maple syrup even on a Cornell salary, have used it ever since, and never again thought of income as a determining factor. Of course, when considering an administrative job the real questions are in any case not the money, but how stimulating the challenge is and whether you are up to the job. Unfortunately I was not and am not gifted in such matters, but fortunately I realized that fact before the Peter Principle took over. Consequently I turned down all these offers, but I often suggested people who had the right abilities and sometimes they even were appointed.

My only administrative exception was my directorship of Cornell's Center for Radiophysics and Space Research for eight years, rather late in my career. Fortunately, by that time Tommy Gold and Yervant Terzian had established a smoothly working modus vivendi for astronomy and related fields and other people (e.g., E. Bilson, P. Gierasch, and Y. Terzian) did most of my job for me. My major achievement came when a professor wanted a salary for his secretary above the top of the range—I managed to invent a new job category, so there was no salary range to worry about!

When I returned from Australia in 1955, there was some worry about a missile gap, and the defense establishment was looking for technical consultants. By that time the Joe McCarthy hysteria was mostly over and, in spite of my left-wing "pinko" youth in Australia, I was able to get my security clearance. After some time consulting for private industry, I became a member of JASON, a group of young professors with their fulltime job at a university during the academic year but doing classified defense work during the summers. In contrast to my administrative ineptitude, I was fairly good at technical National Service, especially at analyzing work already done by others for the defense department and at evaluating claims. My JASON activity that I was and am most proud of had to do with anti-ballistic missile defense (ABMD) in the 1960s. Various schemes for ABMD had been proposed and my evaluations debunked a number of them, and I wrote a thick report that saved the defense department a lot of money. However, it was much more difficult to combat the atmosphere of "dishonesty without outright lies" that pervaded the ABMD community then (and now): I even spent a week on Kwajalene Atoll in the Marshall Islands where the ABM radars were, and are, located. It took me almost the whole week to discover all the information about the incoming missiles that the radars had been given, contrary to the propaganda I had been fed initially. Almost forty years have passed, but false or deceptive claims in favor of ABMD were remarkably similar then as now. I believe CNN televison had a problem recently, similar to mine on Kwajulene, to discover just what information the ABM radar and missile was given beforehand at a claimed ABM success. I resigned from JASON during the Vietnam war, but kept my security clearance. I became a member of a panel, appointed by the American Physical Society to evaluate directed energy weapons for ABMD during the Reagan Administration.

The unanimity of our panel, in spite of a wide range of political views amongst the members, was very gratifying. The fact that all panel members had security clearance gave more political credence to our report debunking ABMD (APS Study Panel 1987). A recent evaluation of our panel report makes it clear that a similar panel will be needed to look critically at current ABM plans, which seem at least as misguided as the two previous plans (Kubbig 2001).

A less satisfying piece of National Service was my membership on the National Science Board (the Board of Trustees for the NSF) from 1978 to 1984. That board was and is quite important (my only U.S. Presidential appointment), but I personally was not an effective member. During my tenure on the Board, the NSF short-changed astronomy in my opinion, especially by abandoning a NRAO project for a large millimeter wave radio astronomy dish. This was partly due to some unjustified badmouthing of observational cosmology in the literature, and I could not even neutralize these false accusations in the eyes of the Board and of the NSF. These six years illustrated for me a general feature of the academic life: Most of the time one has enough flexibility to work only on those things one is gifted for, but occasionally one has to tackle a job that highlights one's own inadequacies.

THE NEUROMUSCULAR JUNCTION

Fairly soon after our second daughter Shelley was born in 1955, my wife Miriam "Mika" Mark Salpeter went back to experimental research in neurobiology. Electron microscopes were becoming practical tools by the end of the 1950s and would be useful in studying synapses and other neurobiology structures that were rather too small for light microscopy. In the 1960s, Mika became an expert electron microscopist, and nominally a research associate in Cornell's engineering college, where Dale Corson was particularly sympathetic to female liberation, whereas biology was almost as male chauvinist as psychology (or Australian universities) before Cornell's Biology division was formed. My wife became particularly interested in the neuromuscular junction (NMJ), where the nerve releases quantal packets of about 10,000 acetylcholine molecules (ACh) each into a narrow space (the "primary cleft") facing the muscle surface. This surface has a high density of receptors for ACh on it. The ACh molecules diffuse in the cleft, get bound once or twice to the receptors, or get destroyed by esterases or unbind again, and so on. This process is rather reminiscent of the chain of nuclear reactions in a stellar interior, but here the main question is just how many doubly bound receptors cause ion channels to open up after some time delay. Such miniature endplate current pulses (MEPC) can be studied individually by electrophysiology, and their summation after a nerve impulse also triggers the muscle contraction.

Mika worked on various aspects of the NMJ throughout her life, because it is simpler and easier to manipulate than synapses in the central nervous system (CNS). In particular, a very stable snake toxin can be attached to the ACh receptors and can also be made radioactive. So, at least in principle, the receptors can be localized and their site density can be measured and also decreased in a controlled manner, and the reduced MEPCS can then be studied. My wife turned this principle into practice, especially by perfecting the technique of electron microscope autoradiography, where a photographic emulsion is first put on top of a thin biological section with radioactive receptor/toxin complexes (after some exposure time, the developed photographic grains are then visualized by electron microscopy). She elucidated many aspects of the NMJ (how a MEPC develops, the degradation and turnover of the receptors, the effect of myasthenia gravis, etc.), received several NIH career development awards, the Jacob Javits Distinguished Career Award, and was a professor in the Section of Neurobiology and Behavior and its Chair for two terms. I was involved in only three aspects of her multi-faceted career.

1. My first involvement may not sound very academic: Life was not made easy for a working mother with two children in those days. We usually had household help but there were emergencies, and I sometimes had to stay home and babysit. This had the advantage of my getting closer to my two daughters and of helping my wife's career indirectly, but what about its effect on my own career? Dealing with two children and working simultaneously tended to make my work even more qualitative than it would have been otherwise. This may have accentuated my sloppiness, but it also made it easier for me to switch fields rapidly and to work on several problems simultaneously.

2. The electron microscope has such a good intrinsic resolution that the scattering in the emulsion of the electrons released by radioactivity is the main bottleneck to localizing the source in autoradiography. My mentor and role model, Hans Bethe, had developed the theory of such scattering a long time before, and it was easy for me to calculate the scattering distribution function and the resolution for the autoradiographic technique. This was only one input into my wife's quantification of the technique, but it resulted in a long series of joint papers between 1969 (e.g., M.M. Salpeter et al. 1969) and 1978 (different distribution functions had to be calculated for different radioactive isotopes).

3. By the early 1980s, the electrophysiology of MEPCS under various conditions had got pretty quantitative (risetime, amplitude, falltime, shape, etc. could be measured), and it was interesting to carry out theoretical simulations of the time development of the current pulse. As mentioned, all this diffusing, binding, and unbinding is reminiscent of the nuclei in the interior of evolved stars. For a simplified geometry (nerve and muscle surface replaced by plane sheets), the development can be modeled by differential equations, and such calculations were carried out by research associate Bruce Land (Land et al. 1984). The actual space between nerve and muscle has a complicated three-dimensional geometry and a Monte Carlo code, "M Cell" for various simulations was developed by Tom Bartol and Joel Stiles to investigate the effect of geometric peculiarities (e.g., Stiles et al. 2001).

My wife Mika died tragically and unexpectedly in October 2000, and I had to take over running her lab. I also became the Principal Investigator for the remaining two years of her National Institute of Health research grant—not an easy task but at least made easier by my being a generalist already. The Monte Carlo code and the computers have become fast enough that one can now do an intensive parameter study. Such a study may sound esoteric, but two of the parameters (cleft width and receptor site density) are altered by a disease like myasthenia gravis, and a third parameter (esterase abundance) can be manipulated by a potential treatment for this disease. A parameter study can thus give some guidance for treatment variation. Tom Bartol and Bruce Land are just tooling up to do a similar (but lengthier) parameter study for CNS synapses.

BLACK HOLE ACCRETION

I attended the first Texas symposium in Dallas in December 1963 and was immersed in a lot of discussion and speculation on the recently discovered quasars. There were no detailed, realistic, rival models for the energy source of a quasar, but accretion of some matter onto some kind of black hole seemed to me like one plausible possibility. Soon after the Dallas symposium I explored one particular scenario for some hypothetical dense clusters, about ten times more massive than ordinary globular clusters but with similar velocity and location in disk galaxies like our own. While crossing the Galactic Disk, these objects would accrete interstellar matter and form massive black holes that would then accrete further interstellar gas (Salpeter 1964). What I did and did not do in that paper, and subsequently, may be a useful lesson for young players in the future, and I will describe it in some detail.

I am proud of three features in my 1964 paper: (*a*) I had to modify the Bondi-Hoyle-Lyttleton theory of accretion for motion through a uniform gas where standing shock fronts have to be considered (this exercise also kindled my interest in interstellar gas and dust on the one hand and in hydrodynamics on the other). (*b*) I needed to know the smallest radius of a circular orbit around a black hole where the orbit is still stable in General Relativity. It was long before Saul Teukolsky started a General Relativity group at Cornell, but I asked professional general relativists elsewhere for the answers. I got no answers, so I calculated it myself for that paper. However, to detract just a little from my glory, I have to add a qualifier: I mentioned my problem to Richard Feynman (who was not a relativist either) at a cocktail party and, without a moment's hesitation and without putting down his glass, he told me exactly how to start the calculation. (*c*) I showed that the increasing mass of the black hole, after reaching a critical value, would increase exponentially with a constant growth factor.

I am less than proud of having guessed wrong on two counts: (*a*) I had thought of condensed objects in the outer regions of disk galaxies, but densities are higher in the core of a galaxy, so a central location for black holes would have been worth exploring and is now known to be correct. (*b*) I guessed in 1964 that there would be many theoretical papers on black hole accretion very soon, with little observational information for a long time, and I just stopped working on the subject. In reality, there was only some independent and simultaneous Russian work (Zeldovich 1964) and then rather little interest until Lynden-Bell's paper on centrally located black holes in 1969 (Lynden-Bell 1969) and subsequent work by Martin Rees. My underestimating the potential of this subject also led to the worst advice I ever gave to any graduate student.

In 1965, Bruce Tarter had started on his own to work on the spiraling inwards of an accretion disk around a black hole, a subject surrounded by many uncertainties. I had told Bruce to stop working on such an unprofitable topic and to do a thesis on the interaction of X-rays with the interstellar medium, where one could hope for observational data soon. His work in this field was indeed well-received, (e.g., Tarter & Salpeter 1969), he became the director of the Livermore National Laboratory at a rather young age and generally did not suffer too much from my bad advice. Nevertheless, it took quite a while for others to get well beyond Bruce's unpublished work on accretion disks, and he might have had a more exciting youth without my timid, bad advice. Graduate students are too young to know what is impossible, so they will sometimes achieve the impossible. Such miracles may be rare, but a professor should think twice before discouraging his student.

THE INTERSTELLAR MEDIUM AND HYDRODYNAMICS

By the middle of the 1960s stellar structure and evolution had become a mature subject, but the interstellar medium was becoming a challenging field. By that time Tommy Gold, with his fertile imagination, had joined Cornell University and interstellar molecular hydrogen was one of his interests. We had early theoretical estimates of molecule formation on the surfaces of dust grains and of the abundance of H_2, before this molecule could be observed directly (Gould et al. 1963). Improvements to these calculations were carried out a few years later by Dave Hollenbach, Bill Watson, and Mike Werner. We were proud of the innovations in our 1963 paper, but even this paper was not innovative enough in one aspect: We did not stress that enormous variations could be expected in gas density and ultraviolet flux from one region to another, so we mainly discussed average conditions that lead to rather small predicted molecular abundances. We thus missed out on the extreme conditions that can lead to almost completely molecular clouds. John Bahcall and I speculated not just about interstellar matter, but also about intergalactic gas, and we predicted fairly sharp Lyα absorption lines in quasar spectra (Bahcall & Salpeter 1965, 1966). We guessed slightly wrong that HI would be clumped inside galaxy clusters, but at least it was a start. We were also able to show from observations that the fine structure constant, $\alpha = e^2/hc$, of quantum electrodynamics has changed by less than 5% per 10^{10} years, but that is a far cry from today's accuracy. Another, rare excursion far outside our own Galaxy in the 1960s had to do with cosmologies that used a cosmological constant to give a coasting period in the expansion (e.g., Petrosian et al. 1967b).

The upgrade in the 1970s of the Arecibo Radio Telescope enabled observations of interstellar HI through its 21-cm line and saw the beginning of my long collaboration with two radio astronomers, Yervant Terzian and John Dickey. In much of this work I only provided the theoretical underpinning (e.g., Dickey et al. 1979), but occasionally John Dickey managed to train me to do some actual observational work. The various phases of hydrogen in the Galactic disk and halo interact with interstellar dust grains, and the late 1970s also saw a lot of work on the physics of these grains, especially by Bruce Draine (e.g., Draine & Salpeter 1979).

Hydrodynamics impinges on astrophysics in many topics and the handover problem from radiative to convective heat transport in a star is one of them. An example of my long collaboration with Giora Shaviv is a paper on this topic (Shaviv & Salpeter 1973). Questions of solubility, nucleation of droplets, and gravitational separation of helium and hydrogen in the major planets are another topic (Salpeter 1973, Stevenson & Salpeter 1977). Energy production from gravitational layering is still a controversial question for Jovian planets and for white dwarf stars, as illustrated in some later work by Dave Stevenson. One speculative paper on hypothetical balloon animals floating and convecting in Jupiter's atmosphere was even published in an archival journal, in spite of references to some unusual forms of mating (Sagan & Salpeter 1976).

I have collaborated with Franco Pacini, on pulsars and exotic stars, since about 1968 (e.g., Pacini & Salpeter 1968). I also have benefitted greatly from my frequent trips to Arcetri Observatory (in Florence) and from stays at Cornell by many of Franco's younger Arcetri colleagues, including A. Natta, M. Salvati, G. Giovanardi, F. Palla, E. Corbelli, and P. Lenzuni.

The late 1970s also saw a lot of work on planetary nebulae where controversy and hydrodynamics overlap, and some work by Vic Mansfield on supernova remnants (Mansfield & Salpeter 1974). A similar overlap occurs in a series of papers on accretion onto neutron stars, starting with Shapiro & Salpeter (1975). In the 1980s and 1990s, my interests overlapped with those of Ira Wasserman, who was by then on the Cornell faculty. One result was a number of papers with two chiefs and one Indian, i.e., collaboration by both of us with one graduate student at a time. However, in some cases the real boss seemed to be the Indian, rather than either of the two chiefs (e.g., Bildsten et al. 1992).

GALAXIES AND DARK MATTER; NSF ASTRONOMY GRANTS, AND EPIDEMIOLOGY OF TUBERCULOSIS

As mentioned, by the late 1970s Arecibo was an ideal place for studying emission from HI gas, and the Virgo cluster of galaxies is conveniently located in the sky for the Arecibo fixed dish. I could not resist getting involved in Virgo cluster HI emission observations, at first with just one graduate student at a time, starting with Nathan Krumm and George Helou. We had some fairly early data on the nonplentiful gas in Sa galaxies (e.g., Krumm & Salpeter 1977) and then concentrated on the outermost HI disks of spiral galaxies and on the question of flat rotation curves in galaxies. Constant HI rotation velocities outside the star-containing galactic disk are very exciting because they indicate the presence of dark matter in this region of no stars and little gas—or else indicate some version of MOND, an alternative to Newtonian dynamics (e.g., Milgrom 1983). However, one can also get fooled by some fairly substantial sidelobes that the Arecibo beam had before the most recent upgrade. Instead of advising my graduate students to be doubly careful, as a professor should, I was sometimes gung-ho and urged rapid publication, which a professor should not do. As a consequence, at least one of our papers claimed a larger HI radius for some galaxies than is actually the case. Later observations by many groups on other disk galaxies still showed that the HI gas extends well beyond the stellar disk in many disk galaxies, so they are still exciting for studying dark matter halos and for the MOND controversy. Nevertheless, my indiscretion has meant that my armor has clanked somewhat ever since, just as Sir Gawain's armor did after his indiscretion in *Sir Gawain and the Green Knight*. I also employed the two chiefs and one Indian mode in the 1980s and 1990s with radio astronomy graduate students, making sure that the other chief was a genuine radio astronomy professor.

Because I was already an aging professor in the 1990s, it was in any case beneficial for each graduate student to also have a younger chief who would be around much later on: It is not always appreciated that, even quite a few years after a Ph.D., help from an ex–thesis-advisor is very useful. I have had enjoyable collaborations with Professors Terzian and Dickey on external galaxy observations, and a collaboration with Terzian is still continuing on observations of galaxy pairs that also give information on dark matter, although not uniquely (e.g., Nordgren et al. 1998). Another fruitful and continuing collaboration (with emphasis on dwarf galaxies) is with Professor Lyle Hoffman who combines a successful research career with teaching in a fairly small College.

The controversies on the distribution in the universe of dark matter (apart from its physical nature) has of course led to much theoretical work. I have had some fruitful collaborations with Jane Charlton and Edvige Corbelli on the related questions of Lyman-alpha clouds and the ionization-recombination equilibrium of hydrogen (e.g., Charlton et al. 1994, Corbelli et al. 2001). One dark matter controversy relates to the cosmological formation of large-scale structure and here the main stream and I diverged during the 1990s: Inspired in part by the success of three-dimensional stellar interior calculations, and encouraged by the ever-increasing speed of computers, the main stream theorists were carrying out elegant numerical three-dimensional structure formation calculations. These calculations became more and more precise, but still had to be based on a number of assumptions. I, on the other hand, was confused by what the assumptions should really be and was content with writing up toy models, conjectures, and speculations—some of them slightly outrageous. In a paper in honor of Antoinette de Vaucouleurs (Salpeter 1993), I even suggested that the outskirts of disk galaxies (dark matter and partially ionized hydrogen) might extend so far out that the rotation period is about one Hubbletime, so that the outer disk or Halo of Andromeda and of the Milky Way almost touch and provide some Lyman-alpha absorption. In the same paper I also speculated on "vanishing Cheshire Cat galaxies," where supernova-driven galactic fountains mostly evacuate the interior of a low surface density galaxy of its gas (Salpeter 1993). Only "the grin of the Cheshire Cat" is just barely detectable in the distant outskirts as gas.

My theoretical astrophysics grant proposals to the NSF were turned down in two successive years, partly because of the qualitative nature of my calculations. Fortunately, Fred Hoyle and I won the Crafoord Prize for nuclear astrophysics jointly in 1997; I am able to support my own theoretical work financially from a Crafoord Fund, and I have made no further NSF grant proposals since. It is probably reasonable not to give federal grants to old codgers and to leave more for main stream scientists, but I hope the main stream will at least listen to all the old codgers from time to time: The emperor's new clothes are usually described by the very young, but occasionally the very old might remind the main stream of some omissions. Possibly because of emphasis on elegance and beauty, most of the discussion in the 1990s of structure formation concentrated on regular galaxies and on giant galaxy clusters, both of which have beautiful, well-defined centers. This drew attention away from irregular structures like the Local Group and slightly larger galaxy groups—even though they far outnumber galaxies with a well-defined center in that luminosity range (not to mention dwarf irregular galaxies with ill-defined structure). When I talked to two junior (and young) NSF program officers, what hurt was not that I did not get the grant, but that these two sounded as though they themselves read papers only from the main stream and not from old codgers. I hope I misunderstood their attitude.

I had come full circle since I switched away from quantum electrodynamics because it was too mature a subject for my scientific temperament. Theoretical astrophysics had now become similarly mature, and I needed another subject that is more open-ended and controversial. Medicine and infectious diseases are of course not new, but the mathematical epidemiology of such diseases is still an open field, largely because there are so many uncertainties. My daughter Shelley is a physician and public health officer in California, where tuberculosis is a big problem. She is interested in the analysis of decisions between prophylactic chemotherapy, applied when an individual only has a latent infection and does not yet have the active disease, and other measures, applied only when patients have an active case of tuberculosis and can then infect others. The reproductive number of the disease is the average number of people who are first infected by a single person with the active disease and then themselves proceed from latent infection to active tuberculosis. This number is important for various cost-effectiveness studies, and we were able to estimate it for the USA from CDC data on active tuberculosis for the last 50 years and from skintest results for latent infection (Salpeter & Salpeter 1998).

My switch to epidemiology was not as radical a change as you might think—humans coughing tuberculosis mycobacteria into the air at different ages requires similar mathematical treatment as stars of different lifetimes disbursing heavy elements into the ISM do. It was easy to introduce a continuous distribution function for the delay time from latent infection to active tuberculosis that led to an improved calculation for the reproductive number. It is an eerie feeling to apply standard astrophysics techniques to medical problems and come up even with politically relevant results. For instance, we showed that the total probability for activation from latency is the same for all ethnicities prevalent in the USA, even though the infection rate and the active case rate is much larger for Asians than Whites. The next big epidemiological challenge for tuberculosis is its disastrous interaction with HIV/AIDS, especially in southern Africa where both diseases are increasing rapidly. Calculations here will be useful in the future but are hampered by sociological, as well as medical, uncertainties—e.g., do tuberculosis patients with HIV go to fewer parties so they do not cough up mycobacteria as much?

ODDS AND ENDS

Given the unstructured nature of my scientific career, I can't come up with a summary or resume either, but can add a few tidbits and gripes. Besides the summers with the nuclear physics group at Caltech, I have enjoyed a few sabbaticals and other visits, including a stay at Mount Wilson and Palomar offices in Pasadena in 1959—my one period surrounded by real optical observers. At an even earlier stage, I enjoyed overlapping with Vera Rubin (who was an undergraduate at Cornell), and Walter Baade and Jesse Greenstein were always an inspiration. In 1960 I spent time at my Alma Mater, Sydney University, with Stuart Butler, Harry Messel, and Bernie Mills. In 1961 (and again in 1969), I was an Overseas Fellow at Churchill College in Cambridge at the College's very beginning. The Fellows' meetings were most instructive, especially the superb administrative technique of Sir John Cockcroft, the first Master. The Fellows were young, the Master was very old and was often in the minority on issues but, without knowing how, we Fellows usually ended up voting with him. The trio of Herman Bondi, Fred Hoyle, and Tommy Gold provided both friendship and role models. The lively debates on Steady State versus Big Bang were always stimulating, with me favoring the Steady State. Many of us tried to salvage the Steady State from the onslaught of new data from the microwave background (e.g., Hazard & Salpeter 1969), but the data finally won. The champion for the Big Bang, Sir Martin Ryle, was also quite impressive but I resented the excessive secrecy that he imposed on all his "young men" regarding their observational results—conversations would have been more fun without that inhibition.

I have usually been pretty easygoing myself (I have been called a "half-assed equivocator") and made no waves. Consequently I have often been rewarded with getting more credit than I deserve, but I have a gripe because more controversial characters get too little credit. Ernst Öpik, who did many things first, and the Barnothys, who preached early about gravitational lensing, were mostly disregarded. Carl Sagan, who asked many of the important scientific questions before other people, was not disregarded but had detractors about mistakes in his scientific work. These detractors claim they are fair in quoting the mistakes, but everybody has made mistakes, even the Archangel Gabriel (and certainly I), and the detractors do not mention the Archangel's mistakes (not even mine)—so they are really not fair.

In looking back at my career, I don't even know whether to be proud or ashamed overall, but I am proud and gratified about the many graduate students, postdocs, and colleagues I have worked with. Although I never met him, I am grateful to my academic grandfather Arnold Sommerfeld, Hans Bethe's teacher, and I hope that some of my distant academic offsprings will remember me also. (I already have some academic great-grandchildren, mainly due to Hubert Reeves who started training students early.) What colleagues and students remember you for is more important than the published writing you leave behind. On the personal side, I particularly value my close relation to my children and grandchildren.

My friends tell me that I should not apologize for this essay being self-centered—that it is only natural. I apologize a little for being more morose than I usually am, but I am still in a state of shock from the loss of my wife. Apart from that, I have enjoyed my life and my wife's legacy sustains me still.

I also have enjoyed switching fields and even being sloppy. I've already shown how my astrophysics background has helped me to work in neurobiology and epidemiology, but the opposite is also true—or at least would be if I were a little younger. One example where epidemiology knowhow might have helped is my vanishing Cheshire Cat paper (Salpeter 1993) plus a forerunner speculation (Gerola et al. 1983). There we speculated that supernova-driven galactic winds in some galaxies might not only evacuate gas but, indirectly through dynamic effects, also lower the density of stars. Modern observations of dwarf spheroidals (Blitz & Robishaw 2000) are somewhat reminiscent of these two papers, but more complicated symbiotic relations must be involved. Such symbiotic relations are the bread and butter of the epidemiology of infectious diseases where humans and germs co-evolve—humans in crowded ghettos develop some immunity to tuberculosis, chemotherapy is beneficial but mycobacteria develop drug-resistance (with even more rapid evolution for HIV/AIDS). Experience in such biological co-evolution may yet help understand how gas clouds and low surface brightness galaxies interact.

I sometimes get asked how real neurobiologists or epidemiologists look at me, and I can answer that by showing how military officers looked at me long ago when I worked for the defense department as a part-time amateur: To go to Kwajalene Atoll I had to fly military air transport. I was first given a highly inflated simulated civil service rank, which was then translated into a highly inflated military rank—on that plane I was a simulated Major General! I bragged about this after my return until a friend remarked "Ed, to me you're a Major General, to your mother you're a Major General, but—BUT—to a Major General you're no Major General!"

The *Annual Review of Astronomy and Astrophysics* is online at
http://astro.annualreviews.org

LITERATURE CITED

APS Study Panel 1987. Science and technology of directed energy weapons. *Rev. Mod. Phys.* 59:S1–202

Arp H, Baum W, Sandage AR. 1952. *Astron. J.* 57:4

Bahcall JN, Salpeter EE. 1965. On the interaction of radiation from distant sources with the intervening medium. *Ap. J.* 142:1677

Bahcall JN, Salpeter EE. 1966. Absorption lines in the spectra of distant sources. *Ap. J.* 144:847

Beaudet G, Petrosian V, Salpeter EE. 1967. Energy losses due to neutrino processes. *Ap. J.* 150:979

Bethe HA. 1939. *Phys. Rev.* 55:434

Bethe HA. 1947. *Phys. Rev.* 72:339

Bethe HA, Critchfield CL. 1938. *Phys. Rev.* 54:248

Bethe HA, Salpeter EE. 1957. *Quantum Mechanics of One- and Two-Electron Atoms.* Berlin: Springer-Verlag

Bildsten L, Salpeter EE, Wasserman I. 1992. The fate of accreted CNO elements in neutron star atmospheres: X-ray bursts and gamma-ray lines. *Ap. J.* 384:143

Blitz L, Robishaw T. 2000. Gas-rich dwarf spheroidals. *Ap. J.* 541:675

Bondi H, Gold T, Hoyle F. 1995. Origins of steady-state theory. *Nature* 373:10

Burbidge EM, Burbidge GR, Fowler WA, Hoyle F. 1957. Synthesis of the elements in stars. *Rev. Mod. Phys.* 29:547

Cameron AGW. 1958. Nuclear astrophysics. *Annu. Rev. Nucl. Part. Sci.* 8:299

Cameron AGW. 1959. Pycnonuclear reactions and nova explosions. *Ap. J.* 130:916

Charlton JC, Salpeter EE, Linder SM. 1994. Competition between pressure and gravity confinement in Ly-alpha forest observations. *Ap. J. Lett.* 430:L29

Cook CW, Fowler WA, Lauritsen T. 1957. Boron, carbon and the red giants. *Phys. Rev.* 105:508

Corbelli E, Salpeter EE, Bandiera R. 2001. Sharp HI edges at high z: the gas distribution from damped Lyα to Lyman limit absorption systems. *Ap. J.* 550:26

Dickey JM, Salpeter EE, Terzian Y. 1979. Interpretation of neutral hydrogen absorption. *Ap. J.* 228:465

Draine BT, Salpeter EE. 1979. Destruction mechanisms for interstellar dust. *Ap. J.* 231:438

Eddington A. 1935. The pressure of a degenerate electron gas and related problems. *Proc. R. Soc. London Ser. A* 152:258

Fermi E, Yang CN. 1949. *Phys. Rev.* 76:1739

Gamow G. 1946. Expanding universe and the origin of elements. *Phys. Rev.* 70:572

Gell-Mann M, Low F. 1951. *Phys. Rev.* 84:350

Gerola H, Carnevali P, Salpeter EE. 1983. Dwarf elliptical galaxies. *Ap. J.* 268:L75

Gingerich O. 1994. The summer of 1953: a watershed in astrophysics. *Phys. Today* 47:34

Gordon WE. 1958. Incoherent scattering of radio waves by free electrons. *Proc. IRE* 46:1824

Gould RJ, Gold T, Salpeter EE. 1963. The interstellar abundance of the hydrogen molecule. II. Galactic abundance and distribution. *Ap. J.* 138:395

Hayashi C. 1966. Evolution of protostars. *Annu. Rev. Astron. Astrophys.* 4:171–92

Hazard C, Salpeter EE. 1969. Discrete sources and the microwave background in steady-state cosmologies. *Ap. J. Lett.* 157:L87

Hoyle F. 1946. *MNRAS* 106:343

Hoyle F. 1954. On nuclear reactions occurring in very hot stars. *Ap. J. Suppl.* 1:121

Krumm N, Salpeter EE. 1977. Rotation curves, mass distribution and total masses of some spiral galaxies. *Astron. Astrophys.* 56:456

Kubbig BW. 2001. *The American Physical Society's Directed Energy Weapons Study.* Frankfurt: Peace Res. Inst.

Land BR, Harris WV, Salpeter EE, Salpeter MM. 1984. Diffusion and binding constants for acetylcholine derived from the falling phase of miniature endplate currents. *Proc. Natl. Acad. Sci. USA* 81:1594

Lynden-Bell D. 1969. Galactic nuclei as collapsed old quasars. *Nature* 223:690

Mansfield VN, Salpeter EE. 1974. Numerical models for supernova remnants. *Ap. J.* 190:305

Milgrom M. 1983. A modification of the Newtonian dynamics. *Ap. J.* 270:371

Moore W. 1989. *Schrödinger.* Cambridge: Cambridge Univ. Press

Nordgren TE, Terzian Y, Salpeter EE. 1998. Very wide galaxy pairs of the northern and southern sky. *Ap. J. Suppl.* 115:43

Öpik EJ. 1951. Stellar models with variable composition. *Proc. R. Irish Acad. A* 54:49

Pacini F, Salpeter EE. 1968. Some models for pulsed radio sources. *Nature* 218:733

Peierls R. 1936. Note on the derivation of the equation of state for a degenerate relativistic gas. *MNRAS* 96:780

Perkins FW, Salpeter EE, Yngvesson KO. 1965. Incoherent scatter from plasma oscillations in the ionosphere. *Phys. Rev. Lett.* 14:579

Petrosian V, Beaudet G, Salpeter EE. 1967a. Photoneutrino energy loss rates. *Phys. Rev.* 154:1445

Petrosian V, Salpeter EE, Szekeres P. 1967b. Quasi-stellar objects in universes with non-zero cosmological constant. *Ap. J.* 147:1222

Sagan C, Salpeter EE. 1976. Particles, environments and possible ecologies in the Jovian atmosphere. *Ap. J. Suppl.* 32:737

Salpeter EE. 1949. Nuclear induction signals for long relaxation times. *Proc. Phys. Soc. A* 63:337

Salpeter EE. 1952a. Mass corrections to the fine structure of hydrogenlike atoms. *Phys. Rev.* 87:328

Salpeter EE. 1952b. Nuclear reactions in the stars. I. Proton-proton chain. *Phys. Rev.* 88:547

Salpeter EE. 1952c. Nuclear reactions in stars without hydrogen. *Ap. J.* 115:326

Salpeter EE. 1953. Reactions of light nuclei and young contracting stars. *Mem. Soc. R. Sci.* 13:113

Salpeter EE. 1954. Electron screening and thermonuclear reactions. *Aust. J. Phys.* 7:373

Salpeter EE. 1955. The luminosity function and stellar evolution. *Ap. J.* 121:161

Salpeter EE. 1960a. Electron density fluctuations in a plasma. *Phys. Rev.* 120:1528

Salpeter EE. 1960b. Matter at high densities. *Ann. Phys.* 11:393

Salpeter EE. 1964. Accretion of interstellar matter by massive objects. *Astrophys. J.* 140:796

Salpeter EE. 1973. On convection and gravitational layering in Jupiter and in stars of low mass. *Ap. J.* 181:L83

Salpeter EE. 1993. Large-scale structure and galaxy disks as Lyman-alpha clouds. *Astron. J.* 106:1265

Salpeter EE. 1996. Neutron stars before 1967 and my debt to Chandra. *J. Ap. Astron.* 17:77

Salpeter EE, Bethe HA. 1951. A relativistic equation for bound-state problems. *Phys. Rev.* 84:1232

Salpeter EE, Hamada T. 1961. Models for zero-temperature stars. *Ap. J.* 134:683

Salpeter EE, Reeves H. 1959. Nuclear reactions in stars. IV. Buildup from carbon. *Phys. Rev.* 116:1505

Salpeter EE, Salpeter SR. 1998. Mathematical model for the epidemiology of tuberculosis, with estimates of the reproductive number and infection-delay function. *Am. J. Epidemiol.* 142:398

Salpeter MM, Bachmann L, Salpeter EE. 1969. Resolution in electron microscope radioautography. *J. Cell Biol.* 41:1

Sandage AR, Schwarzschild M. 1952. *Ap. J.* 116:463

Schatzman E. 1948. *J. Phys. Radium* 9:46

Shapiro SL, Salpeter EE. 1975. Accretion onto neutron stars under adiabatic shock conditions. *Ap. J.* 198:671

Shaviv G, Salpeter EE. 1971. Solar-neutrino flux and stellar evolution with mixing. *Ap. J.* 165:171

Shaviv G, Salpeter EE. 1973. Convective overshooting in stellar interior models. *Ap. J.* 184:191

Stevenson DJ, Salpeter EE. 1977. The dynamics and helium distribution in hydrogen-helium fluid planets. *Ap. J. Suppl.* 35:239

Stiles JR, Bartol TM, Salpeter MM, Salpeter EE, Sejnowski TJ. 2000. Synaptic variability. New insights from reconstructions and Monte Carlo simulations with M cell. In *Synapses*, ed. WM Cowan, TC Südhof, CF Stevens. Baltimore: Johns Hopkins Univ. Press 792 pp.

Tarter CB, Salpeter EE. 1969. The interaction of x-ray sources with optically thick environments. *Ap. J.* 156:953

VanHorn HM, Salpeter EE. 1969. Nuclear reaction rates at high densities. *Astrophys. J.* 155:183

Zapolsky HS, Salpeter EE. 1969. The mass-radius relation for cold spheres of low mass. *Ap. J.* 158:809

Zel'dovich YB. 1964. *Sov. Phys. Dokl.* 9:195

Annu. Rev. Astron. Astrophys. 2002. 40:27–62
doi: 10.1146/annurev.astro.40.060401.093845
Copyright © 2002 by Annual Reviews. All rights reserved

ULTRA-COMPACT HII REGIONS AND MASSIVE STAR FORMATION

Ed Churchwell

*Department of Astronomy, University of Wisconsin, 475 N. Charter St., Madison,
Wisconsin 53706-1582; email: churchwell@astro.wisc.edu*

Key Words   prestellar cores, hot cores, accretion disks, bipolar outflows

■ **Abstract**   This review discusses three main topics: evolution to the ultra-compact
(UC) state; the properties of UC HII regions and their environments; and UC HII
regions as probes of Galactic structure. The evolution to UC HII regions begins in
giant molecular clouds that provide the natal material for prestellar cores that evolve
into hot cores, the precursors of UC HII regions. The properties of each evolutionary
phase are reviewed, with particular emphasis on those of hot cores. The observed
properties of UC HII regions and their environments are summarized with emphasis
on the physical processes that may produce the observed properties. The final section
summarizes the use of UC HII regions as probes of Galactic structure: in particular,
the Galactic population and distribution of newly formed massive stars, the location of
spiral arms, and the average galactocentric temperature and abundance gradients.

INTRODUCTION

Ultra-compact (UC) HII regions are manifestations of newly formed massive stars
that are still embedded in their natal molecular clouds. Dust in the molecular
cloud core renders UC HII regions observable only at radio, submillimeter, and
infrared wavelengths. Hofner et al. (2002) showed that deeply embedded stars in
the W3 massive star formation complex can be detected via hard X-rays (≥ 2.5 kev),
including the detection of the apparent ionizing stars of the compact HII regions
in W3. At this stage, the central ionizing star of a UC HII region is believed
to have ceased accreting matter and to have settled down for a short lifetime
on the main sequence. During the main sequence lifetime of the central star,
its HII region will evolve from a deeply embedded UC state to a much larger,
uncloaked, classical nebula. As the name suggests, UC HII regions are small
(diameters $\leq 10^{17}$ cm), very dense (typically $\geq 10^4$ cm^{-3}), and bright (emission
measures $>10^7$ pc cm^{-6}). They are among the brightest and most luminous objects
in the Galaxy at 100 μm because of their warm dust envelopes that convert the
entire stellar luminosity to far infrared radiation. Typically the infrared emission
peaks at $\sim$100 μm and is $\sim$3 to 4 orders of magnitude above the free-free emission
at this wavelength. To be detectable at radio wavelengths, the luminosity of UC

0066-4146/02/0922-0027$14.00

27

HII regions must be equivalent to a B3 or hotter main sequence star. At mid-infrared wavelengths where warm circumstellar dust, stellar photospheres, and nebular fine structure lines are bright, it will be possible to detect even cooler, less luminous, embedded, newly formed stars using sensitive modern facilities such as the Space Infrared Telescope Facility (SIRTF) and the Stratospheric Observatory for Infrared Astronomy (SOFIA). At wavelengths shorter than ~2 μm, detection of the ionizing star in most UC HII regions is difficult to impossible because the dust cocoon becomes optically thick.

The study of O stars and their associated HII regions inherently concerns the nature of massive star formation and the impact of newly formed massive stars on their environments. UC HII regions represent a key stage in the development of massive stars. This is the period between the rapid accretion phase of star formation, when the central protostar is being formed, and the period when the UC phase gives way to a larger, more diffuse, less obscured HII region either by destruction of the natal molecular core or by moving out of the core. UC HII regions provide a way to study the process of formation and early evolution of massive stars and the environments in which this occurs. They also have high luminosities and thus are important probes of global properties of the Galaxy because with modern facilities they can be detected throughout the entire Galaxy. They have been used to confirm the electron temperature and metallicity gradients in the Galaxy, establish the scale height in Galactic latitude and angular distribution in longitude of O stars in the Galaxy, infer the fraction of O stars that are obscured at visible wavelengths in their natal cocoons, and estimate the contribution of hot O and B stars to the radiant and mechanical energy budget of the Galaxy. Several reviews that address various aspects of UC HII regions are contained in the recently published compendium *Protostars and Planets IV*; the reviews in Parts I, II, and IV are of particular relevance to massive star formation and UC HII regions. The extensive reviews by Garay & Lizano (1999) and Churchwell (1999) summarize the state of our understanding of UC HII regions up to 1999.

This review concentrates on the evolution leading to UC HII regions (below), the physical properties and nature of UC HII regions as presently understood "UC HII Regions", and the Galactic population and distribution of UC HII regions "Galactic Population and Distribution".

EVOLUTION TO THE ULTRACOMPACT STATE

The Natal Material

UC HII regions do not represent the earliest stage of massive star formation, as is often claimed. The story of their origin begins much earlier in the dense cores of giant molecular clouds (GMCs). The clumpy structure of GMCs has been referred to as clumps and cores by Blitz and coworkers (Blitz 1991, Blitz & Williams 1999, and others), as fractal by Bazell & Désert (1988), Scalo (1990), and Falgarone et al. (1991), and as turbulent structures by Vázquez-Semadeni et al. (2000 and references therein). The structure of GMCs appear to be self-similar over a very wide range of sizes and masses (Williams et al. 2000 and references therein). The clumps in which massive stars and their associated lower mass cluster stars form can be quite massive (up to several thousand solar masses). Typically only those with mass ≥ 300–$500\ M_\odot$ are gravitationally bound and produce stars; those below $\sim 300\ M_\odot$ are not bound (Blitz 1991). The most massive clumps ($>10^3$–$10^4\ M_\odot$) contain most of the mass in GMCs, and they provide the natal material for the formation of stellar clusters (Blitz 1991, Blitz & Williams 1999). Cores are the substructure of clumps; they are smaller, denser, and have lower mass (a few tens of solar masses) than clumps. Cores are the sites of individual star formation within clumps. Crutcher (1999) found that the supersonic motions in molecular clouds are about equal to the Alfvén speeds and suggested that the supersonic motions are magneto hydrodynamic (MHD) waves. The magnetic energy in cores appears to be close to but slightly less than the gravitational energy (Crutcher 2001), so magnetic fields alone probably cannot prevent gravitational collapse.

The origin of the stellar initial mass function (IMF) is a fundamental unsolved problem in star formation. To first approximation, the IMF appears to be universal, independent of abundance differences in the Milky Way and in neighboring galaxies (Massey 1999, Massey & Hunter 1998, Meyer et al. 2000, Kroupa 2001). Several hypotheses have been proposed to explain the observed stellar IMF, some of which are, the mass spectrum produced by cloud fragmentation; the impact of the initial stars on the structure of the local interstellar medium (ISM); the fractal structure of molecular clouds; the MHD wave patterns in molecular clouds; shock patterns produced by cloud-cloud collisions, etc. Blitz & Williams (1999) and Williams et al. (2000) have postulated that the IMF is determined by the mass spectrum of the natal molecular cloud well before star formation begins. This would imply that the mass spectrum of the molecular cores out of which stars form have an approximate Salpeter slope ($dN/d\ln M \propto M^{-\alpha}$ where $\alpha \sim 1.35$). It is interesting that the mass spectra of embedded molecular cores in ρ Oph (Motte et al. 1998) and in Serpens (Testi & Sargent 1998) have slopes $\alpha > 1.1$, quite close to the Salpeter value. This is consistent with the hypothesis that the stellar IMF could be determined by the structure of the natal cloud prior to star formation activity but not sufficient for a proof (Meyer et al. 2000).

Why do cores become gravitationally unstable and begin to collapse? Myers & Fuller (1992) and others have called attention to the importance of nonthermal motions (mostly turbulence) in molecular cores. Turbulent dissipation seems to occur on shorter timescales than the gravitational free-fall time (Vázquez-Semadeni et al. 2000 and references therein). However, recent simulations by Cho et al. (J. Cho, A. Lazarian, E.T. Vishniac, submitted) find weak coupling between wave packets traveling in the direction of magnetic field lines, resulting in a slower decay of turbulent motions than reported by Vázquez-Semadeni et al. (2000). In the absence of either continuous mechanical energy injection and/or strong magnetic fields, short dissipation times would lead to wholesale cloud collapse.

Myers & Lazarian (1998), Nakano (1998), Williams & Myers (2000), and Williams (2001) have proposed the interesting hypothesis that turbulent cooling flows in molecular clouds may initiate collapse by producing low pressure regions that can be compressed by nearby high pressure regions. From an analysis of line dispersions toward several molecular cores, Williams & Myers (2000) found that inward motions predominate in quiescent cores (i.e., those with the narrowest line widths), while μ cores with the largest dispersions the motions are primarily outward. Williams (2001) suggests that in regions of high pressure (where turbulent motions are large), matter flows outward to regions of lower pressure, and in regions of low pressure (where turbulent motions have dissipated), matter flows inward. In regions of low turbulence (low pressure), gravity plus external pressure may initiate collapse. In this picture, the turbulent structure of molecular cores plays a critical role in the initiation of star formation. It also requires that turbulent structure in molecular clumps produce an approximate Salpeter stellar mass spectrum—an issue that remains to be determined.

Prestellar Cores

Prestellar cores (PSCs) are the earliest identifiable stage of a star in the process of forming. They are dense, gravitationally bound, molecular cloudlets that are undergoing quasistatic gravitational contraction. Formation of PSCs may be the most poorly understood stage of massive star formation (Garay & Lizano 1999). PSCs have not yet formed a central protostar and therefore will not appear as near-infrared (near-IR) ($\lambda \leq 2\ \mu$m) point sources. They are also generally optically thick at these wavelengths. Because PSCs are only heated by the ambient interstellar UV radiation field, they have temperatures of only 10–20 K, and their spectral energy distributions (SEDs) peak in the far infrared at $\sim$200 μm, as illustrated in Figure 1 by the low-mass PSC in L1544. Due to their low temperatures, PSCs can be detected at 4–8 μm in absorption against the bright Galactic plane background emission (Bacmann et al. 1998). At far-IR and sub-mm wavelengths ($\lambda > 150\ \mu$m) (Ward-Thompson et al. 1994, Launhardt & Henning 1997, André et al. 1996, Ward-Thompson & André 1998) they can be detected in emission. PSCs can also be detected in the rotational lines of CO and other molecules.

Proof that a cloudlet is a PSC requires molecular line measurements that show the characteristics of infall. Zhou et al. (1993) and Ward-Thompson et al. (1996) have demonstrated this toward B335 and NGC 1333-IRAS2, respectively. However, both of these clouds have already formed a protostar at their core. Lee et al. (1999), Gregersen et al. (2000), and Gregersen & Evans (2000) have identified over a dozen PSC candidates with infall signatures. Claims for other PSCs have been based on indirect arguments such as high densities, L_{bol} too low to be consistent with an accreting protostar, low temperatures (indicating no internal heating), masses approximately virial, no IR point source detected in the cloudlet, etc. This is rapidly changing with the efforts to detect infall by Myers, Evans, and their coworkers.

The distribution of H_2 column density with radius is an indicator of the support mechanism(s) of PSCs, which also has consequences for the process of star formation. For example, the "inside-out" collapse model of Shu (1977), Shu et al. (1987),

SPECTRAL ENERGY DISTRIBUTION OF L1544

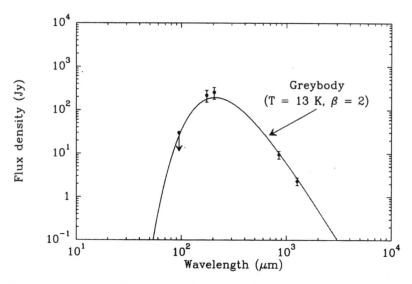

Figure 1 The spectral energy distribution of the prestellar core in L1544 from Bacmann et al. (1998). A greybody model of temperature 13 K (solid curve) is superimposed on the observed data points.

and Shu et al. (2000) predicts a power law distribution, $N_{H_2} \propto r^{-2}$, consistent with an isothermal sphere for the protostellar precursor cloud. Ward-Thompson et al. (1994), André et al. (1996), and Bacmann et al. (1998) have found PSC column density profiles to be approximately flat near the center and to approach $N_{H_2} \sim r^{-2}$ toward the outer boundaries. They argue that the observed profiles are consistent with magnetically supported clouds undergoing ambipolar diffusion. It is worth noting that the derived density profiles require several uncertain assumptions or extrapolations that could significantly alter the results. Also, the timescale for star formation via ambipolar diffusion seems to be too long (Lee & Myers 1998, Jijina et al. 1999).

The study of PSCs is still in its infancy, and much more work will be required to obtain a better understanding of them. PSCs probably do not remain in this state for more than about 10^6 years (Ward-Thompson et al. 1994, Fuller & Myers 1987) and therefore are rather rare and difficult to find. All PSCs detected so far are low-mass cloudlets that will give rise to low-mass stars. PSCs that produce massive stars have not been unambiguously identified, although one suspects that they may be among the more compact objects detected by Egan et al. (1998). The densities, sizes, and masses of the objects reported by Egan et al. (1998) are in the right range to give rise to massive stars and their associated lower-mass cluster members (Carey et al. 1998).

Hot Cores: Precursors to UC HII Regions

Hot cores (HCs) have been defined by Kurtz et al. (2000) as compact (diameters ≤ 0.1 pc), dense ($n_{H_2} \geq 10^7$ cm^{-3}), and warm (T ≥ 100 K) molecular cloud cores that have large molecular line optical depths and high line-brightness temperatures when resolved. This observationally based definition of HCs is broad enough to apply to any hot, dust/molecular gas–enshrouded object such as UC HII regions, evolved hot stars that have moved into a dense molecular cloud, or massive protostars still undergoing rapid accretion. In this section, I consider only the class of hot cores containing rapidly accreting, massive protostars. These are the precursors of UC HII regions, which I refer to as PUCHs (precursors of UC HIIs). PUCHs are internally heated by the central massive protostar plus any associated lower-mass cluster members. They are likely to be surrounded by an equatorial accretion disk and a massive bipolar outflow along their spin axes. Their outflow masses, momenta, and kinetic energies are much larger than those of low-mass protostars, and they are generally poorly collimated relative to those associated with low-mass stars. Because of rapid accretion, the protostar does not produce a detectable HII region, even though it has a large UV photon flux. PUCH lifetimes are quite short, typically $\leq 10^5$ yr.

The above statements implicitly assume that massive stars are formed via accretion of ambient interstellar matter through an equatorial disk. This assumption has been questioned by Bonnell et al. (1998) and Stahler et al. (2000) who have suggested that massive stars (M $> 10 \ M_\odot$) may be formed via coalescence of intermediate mass protostars. This hypothesis is supported by the central location of the most massive stars in young open clusters and the high stellar density of many open clusters, such as the Orion cluster whose mean stellar separation is less than the Jeans length for any reasonable initial natal cloud temperature. The coalescence hypothesis for massive star formation, in fact, is likely to be even more effective than argued by Bonnell et al. (1998) if three-body interactions had been included in which a bound binary pair of protostars interact with a third member of a forming stellar cluster. Models including such interactions have not been published but clearly need to be done. It is entirely possible that massive stars could be formed both by accretion and coalescence in the same cluster. The presence of massive bipolar molecular outflows (and reported detection of equatorial accretion disks toward a few objects) provide strong arguments in favor of formation via accretion, but there are equally strong arguments for the coalescence hypothesis. In much of the discussion of hot cores below, it is assumed that accretion produces massive stars. However, one should be aware that coalescence may be an important formation mechanism of massive stars.

TEMPERATURE STRUCTURE   Wolfire & Churchwell (1994) investigated the temperature structure of dust cocoons enshrouding UC HII regions for a range of different dust properties, cocoon density profiles, and stellar radiation fields. The temperature structure of these models is likely to apply to any type of hot star or protostar whose UV radiation is absorbed by a large circumstellar dust cocoon such as PUCHs. Wolfire & Churchwell (1994) found that the temperature profiles can generally be described by four zones around the central heat source. A central dust evacuated cavity around the ionizing star or protostar exists where the temperatures are higher than the dust sublimation temperature. Their models and others (e.g., Chini et al. 1987) indicate that the inner dust radius is generally substantially larger than the dust sublimation radius, probably owing to stellar winds and radiation pressure. A thin inner dust shell exposed to direct stellar UV radiation declines in average dust temperature with distance from the central protostar as $\bar{T}_d \propto e^{-\tau_{UV}}$, where $\tau_{UV}(r)$ is the optical depth at UV wavelengths to the stellar radiation field at distance r from the protostar. Most of the decrease in dust temperature can occur in this thin shell. At radii beyond where direct and scattered stellar UV radiation has been absorbed in the inner shell, the dust is heated by NIR radiation from hot dust at the inner edge of the shell. Due to dilution of this emission, the mean dust temperature, $\bar{T}_d$, declines with radius as $\bar{T}_d \propto r^{-\alpha}$, where $\alpha \approx 2/(\beta + 4)$ and β is the index of the FIR dust opacity ($K_\lambda \propto \lambda^{-\beta}$). At distances where extinction of the NIR emission of the inner shell becomes important, the temperature profiles become slightly steeper than $r^{-\alpha}$. Figure 2 shows the calculated temperature profile

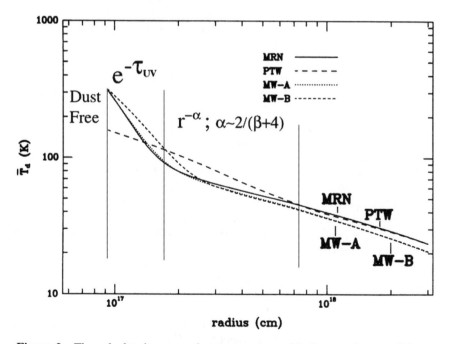

Figure 2   The calculated average dust temperature with distance from an O6 star for four different dust models from Wolfire & Churchwell (1994). Three different temperature zones are indicted by vertical lines.

of a UC HII region ionized by an O6 star from Wolfire & Churchwell (1994). This model and others (e.g., Chini et al. 1987) imply that dust hotter than ~100 K occupies a very small fraction (a few percent) of the volume of the dust envelope; most of the dust is well below 100 K, with average values of ~30 K. Thus PUCHs and UC HII regions are expected to be brightest at about 100 μm, at which most of the emission escapes.

SPECTRAL ENERGY DISTRIBUTIONS   The IR/sub-mm SEDs of PUCHs are likely to be quite similar to those of UC HII regions. The primary differences are the locations of the stellar wind terminal shock and the radius of the dust sublimation temperature. The large infall rates will push both boundaries closer to the protostar than would otherwise be the case. A central dust-free cavity must exist around the rapidly accreting protostar of a PUCH because its large luminosity will heat the dust to temperatures well above the sublimation temperature. For example, for $L \geq 10^4$ $L_\odot$, even graphite dust would be destroyed at radii less than ~850 solar radii; for $L = 10^5$ $L_\odot$, the sublimation radius becomes ~2700 $R_\odot$ or about 12.5 AU. Dust may exist inside the dust-free cavity in a dense, self-protective accretion disk and thereby provide some NIR emission from this region. Because of the temperature gradient with radius, the SEDs are broader than single temperture Planck functions. Osorio et al. (1999) have modeled four hot molecular cores with rapidly accreting massive protostars at their centers and found that the observed SEDs are better fit by envelopes with singular logatropic spheres ($\rho(r) \propto r^{-1}$) than singular isothermal spheres ($\rho(r) \propto r^{-2}$). Their model fits to the SEDs, and temperature distributions of four PUCHs are shown in Figure 3. These are, as expected, essentially identical to the SEDs of UC HII regions. One implication is that the radius of the wind terminal shock, the dust sublimation temperature, and the accretion disk are so small relative to the extent of the dust envelope that they make essentially no impact on the emergent spectrum. Osorio et al. (1999) also find that accretion provides the dominant source of luminosity at this stage of evolution.

RADIO FREE-FREE EMISSION   PUCHs are not expected to produce a detectable HII region, even though the central protostar produces a large UV photon flux, because these photons cannot travel far from the protostar before being absorbed by infalling matter. The result is a small Strömgren sphere around the central protostar. The critical mass infall rate at which all the stellar UV photons are absorbed by the infalling material is

$$\dot{M}_{crit} = \left[\frac{4\pi m_H^2 G L_{UV}^* M^*}{\alpha(2)} \right]^{1/2} \qquad (1)$$

(Walmsley 1995), where m_H is the mass of hydrogen, $\alpha(2)$ is the hydrogen recombination coefficient to all levels except $n = 1$, G is the gravitational constant, L_{UV}^* is the protostellar ionizing photon flux, and M^* is the mass of the central protostar. In Figure 4, log ($\dot{M}_{crit}$) versus T_{eff} (or spectral type) for main sequence O stars is plotted using the stellar data (M^* and L_{UV}^*) from Vacca et al. (1996). One sees

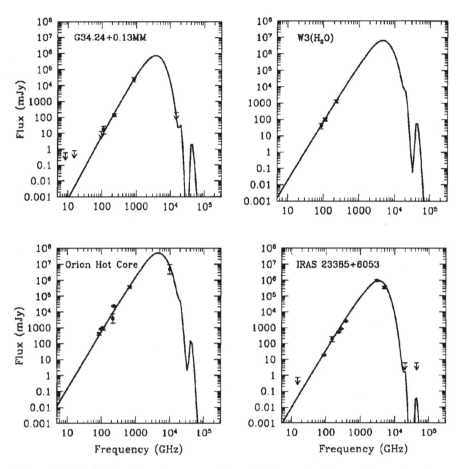

Figure 3   The SEDs of four hot cores with model fits to the observed data points from Osorio et al. (1999). These are very similar to those of UC HII regions.

from Figure 4 that for O4 to B0 main sequence stars, a critical mass infall rate of only 10^{-4} to 10^{-5} solar masses per year, respectively, is adequate to quench the circumstellar HII. The mass infall rate around massive protostars in their rapid accretion phase is likely to be $\sim 10^{-3}$ solar masses per year, or a factor of 10–100 times the critical infall rate. Thus, it is reasonable to expect that a HC harboring a massive protostar at its center, even if it is very luminous, will not have detectable radio free–free emission. It is important to note, however, that there will be a small dust-evacuated cavity around luminous protostars undergoing rapid accretion with radius R_c in which hydrogen will also be ionized. Temperatures will be larger than the dust sublimation temperature for $R_c \leq (L_*/4\pi\sigma T^4)^{1/2}$, where T is the dust sublimation temperature. For $L_* = 10^5\ L_\odot$, $R_c \leq 12$ AU. This is still too small and probably optically thick to be detected by current instrumentation.

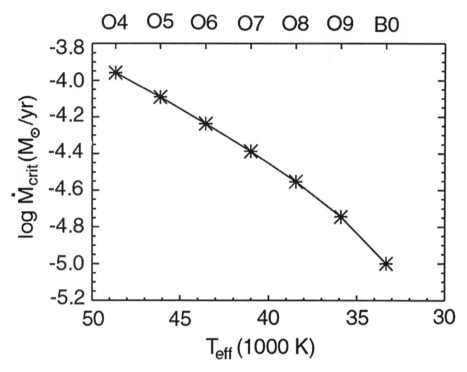

Figure 4 The critical mass infall rate that will absorb all stellar UV photons as a function of effective temperature (or spectral type). The stellar properties of Vacca et al. (1996) are assumed and the relation of Walmsley (1995) is plotted.

What is the evidence that massive protostars have such high accretion rates and short lifetimes? Several arguments lead us to this conclusion. First, the Kelvin-Helmholtz timescale, which is a strict lower limit on the time to produce a star of a given mass, is only a few thousand years for a 30 solar mass star. These short timescales require accretion rates of $\dot{M}_{acc} \approx M_*/\Delta t \leq 10^{-2}$ solar masses per year. A more accurate estimate of the timescale may be obtained using the properties of molecular outflows associated with massive star formation. Molecular outflows driven by massive protostars have dynamical ages of a few times 10^4 years, masses of a few tens to a couple of hundred solar masses, and mass outflow rates ranging from $10^{-0.6}$ to a few times $10^{-2.0}$ solar masses per year with a typical value of $\sim 10^{-3}$ solar masses per year (see Churchwell 1997). Presumably the mass accretion rate has to be somewhat larger than the outflow rate because some of the mass must go into building the star. Taking the conservative approximation that $\dot{M}_{acc} > \dot{M}_{out} \approx 10^{-3}$ solar masses per year requires that the timescale for massive star formation is in the range of a few times 10^4 to a few times 10^5 yr. These arguments require both a high mass accretion rate and a short timescale for massive

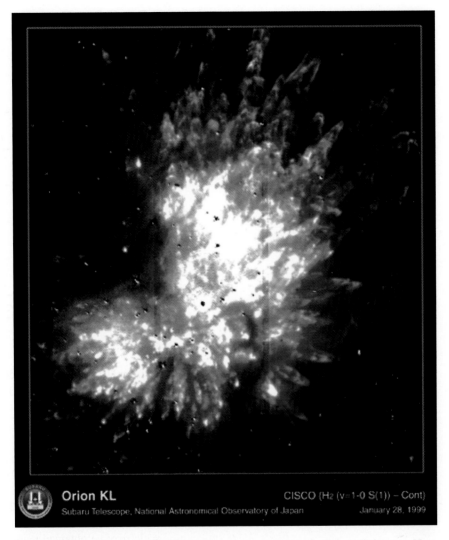

Orion KL   CISCO (H₂ (v=1-0 S(1)) − Cont)

Subaru Telescope, National Astronomical Observatory of Japan   January 28, 1999

Figure 7 The outflow in Orion IRc2 in H_2 emission courtesy the Subaru Telescope. Note the poor collimation of the outflow, the fact that it does not appear to originate at a point, and the jaggedness (bow shocks) at the ends of the outflow.

star formation, neither of which is particularly surprising because gravitational processes occur more rapidly in deeper gravitational potentials. Recent evolution models of star formation by Behrend & Maeder (2001) using mass-dependent accretion rates (increasing with increasing stellar mass) find that the average mass accretion timescale for formation of stars from 8 to 80 solar masses is $\sim 3 \times 10^5$ yr independent of the mass. They also find that low-mass stars form first and massive stars later using current best IMF models. A further argument for short PUCH lifetimes has been proposed by Wilner et al. (2001). They find about half as many PUCHs as UC HII regions in the giant massive star formation region W49 and suggest that the average PUCH lifetime may be about half that of UC HII regions.

OUTFLOWS AND ACCRETION DISKS   If massive stars form by accretion of ambient gas, then PUCHs must experience a period of rapid accretion accompanied by an equatorial accretion disk and a massive bipolar outflow normal to the disk according to the standard paradigm. Observational evidence for accretion disks around massive protostars exists for only a very few objects; Garay & Lizano (1999) list eight objects in their Table 3. A more recent reference for G192.16 − 3.82 is Shepherd et al. (2001). Accretion disks in massive star formation regions are very difficult to detect because of their large distances, the brightness of the central protostar, and the difficulty of distinguishing a possible disk from the dense, hot, circumstellar environment of the natal hot core. The natal core is bright and may have large velocity gradients other than Keplerian motions. The protostars with detected accretion disks have masses ranging from 10 to 370 $M_\odot$, radii from <500 to 10^4 AU, and all of them have observed outflows (Garay & Lizano 1999). High resolution observations of methanol masers by Norris et al. (1993, 1998) and Phillips et al. (1998) have been interpreted to originate in accretion disks, but this hypothesis is not universally accepted.

Molecular outflows have been observed toward numerous massive star formation regions (see Shepherd & Churchwell 1996, Ridge 2000). The molecular outflows from massive protostars have very large masses, mass fluxes, momenta, and mechanical luminosities (Churchwell 1997, Ridge 2000). They are also not as well collimated as those from low-mass protostars (see the closest one in Figure 7). There is a tight correlation between the outflow mass flux, $\dot{M}_{out}$, and the bolometric luminosity of protostars that holds over at least six orders of magnitude in L_{bol} from 1 to $\sim 10^6$ $L_\odot$ (see Cabrit & Bertout 1992, Shepherd & Churchwell 1996, Churchwell 1999 and references therein). Figure 5 shows a recent version of this relation with additional data from Henning et al. (2000). The continuity of the $\dot{M}_{out}$-L_{bol} relation has been used to argue that the formation of massive stars is simply an extension of the process of low-mass star formation to more massive stars. This seems unlikely for several reasons. Since the luminosity is a reflection of the mass of the central protostar and the mass determines the gravitational potential that governs the rate of accretion and other processes associated with star formation, it is not surprising that a continuous $\dot{M}_{out}$-L_{bol} relationship holds for all stellar luminosities. This correlation may have nothing to do with the actual

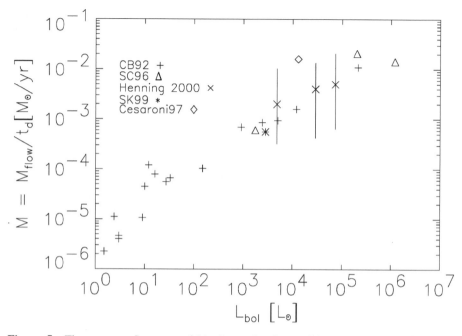

Figure 5   The mass outflow rate of bipolar molecular outflows driven by protostars of bolometric luminosity ranging from 1 to $10^6\ M_\odot\ yr^{-1}$. Data for this plot were assembled from Cabrit & Bertout (1992), Shepherd & Churchwell (1996), Henning et al. (2000), Cesaroni et al. (1997), and Shepherd & Kurtz (1999).

process of star formation but may simply reflect the central role gravity plays in star formation. Ridge (2000) finds that the mass and luminosity of outflows suffer from Malmquist bias. She argues that since more luminous protostars are embedded in more massive HCs, the mass in outflows only reflects the amount of material available in the HC. She concludes that "the outflow dynamics may tell us nothing about the mechanism that generated the flow, only that they have propagated through more material" (Ridge 2000). The sources with the highest luminosities in Figure 5 are unlikely to suffer Malmquist bias, as they could be detected at much greater distances, although the lower luminosity sources may well suffer such an effect. The $\dot{M}_{out}$-L_{bol} relationship has been used to infer L_{bol} for the central protostar from measured $\dot{M}_{out}$ values. However, most recently Beuther et al. (2001) found that the relation shown in Figure 5 only represents an upper envelope for $\log(L_{bol} > 3.5)$. This calls into question the determination of L_{bol} for massive protostars using the $\dot{M}_{out}$-L_{bol} relationship.

BIPOLAR MOLECULAR OUTFLOWS   Bipolar outflows provide a mechanism for accretion disks to shed angular momentum, thereby permitting matter in the disk to migrate to the central protostar. Since bipolar outflows appear to be intrinsic to the process of star formation, it is essential to understand them as a step toward understanding star formation. Unfortunately, there is no generally accepted theory for how tens to hundreds of solar masses of cold molecular gas can be accelerated and collimated on short timescales while matter is simultaneously rapidly accreting onto the protostar. The X-wind theory of Shu and coworkers (Shu et al. 1988, 1994, 2000) and other theories that invoke magnetic fields to redirect and collimate outflows were developed for low-mass protostars that involve small masses and probably are not applicable to massive outflows.

Massive outflows have some general properties that have to be explained by any successful theory of their origin. One, they are not well collimated relative to those associated with low-mass protostars. Two, most of the mass in the outflows have velocities of only a few tens of km s^{-1}, not hundreds. The outflow observed by Henning et al. (2000) toward IRAS12091-6129 is typical of the low collimation, low outflow velocities, and large outflow masses found in massive star formation regions. This outflow is shown in Figure 6. Three, the Orion IRc2 outflow shown in Figure 7, see color insert, (Subaru Telescope Facility) appears to originate from a surface much larger than a star, the opening angle is quite wide, and the outflow working surface is jagged (i.e., does not act like a smooth piston on the ambient medium). The large surface of origin, wide opening angle, and low outflow velocities suggest that much of the outflow material must have been accelerated outward at fairly large distances from the protostar where the escape velocity is commensurate with the flow velocities.

We know that radiation alone cannot drive the outflows because the radiation momentum flux is small relative to that in the outflow $L^*/c \ll \dot{M}_f v_f$, where f refers to the outflow. Additional driving mechanisms could be protostellar winds and/or rotation perhaps combined with a magnetic field and gravity. However, the physics of how bipolar molecular outflows are driven and collimated are not understood. There must be either an additional component of force normal to the disk, such as a disk wind, and/or a radiation field produced by a hot, shocked disk surface (as postulated by Neufeld & Hollenbach 1994, 1996). Perhaps magnetic fields normal to the disk play a role in aligning and redirecting infalling material outward along the spin axis. The poor collimation of the outflows, however, is not a good sign for magnetic alignment. Larson (2002) has postulated that tidal torques could play an important role in redistributing angular momentum during the process of massive star formation.

Although there is considerable uncertainty in the determination of outflow masses (see Ridge 2000), all investigators find very large outflow masses associated with massive protostars. This presents an interesting and unsolved problem. Churchwell (1997) noted that the outflow masses are generally greater than the inferred mass of the protostar that drives them, sometimes by a factor of ~ 8. He examined four mechanisms that may explain the large outflow masses and concluded that neither accumulated stellar winds nor broad piston-like outflow working surfaces are likely to be able to account for the observations. Turbulent entrainment of ambient HC gas may be able to account for a few solar masses of material,

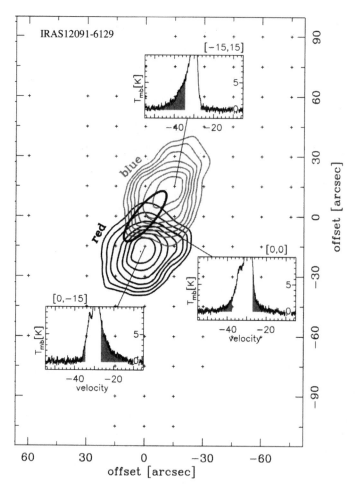

Figure 6   The CO outflow presumably driven by the FIR source IRAS12091-6129 from Henning et al. (2000). This figure illustrates three general properties of outflows driven by massive protostars: low outflow speeds, poor collimation, and large masses. The arbitrary cutoff velocities (shaded regions in the spectra) for the outflow also illustrate why outflow masses are uncertain and different authors disagree on the estimated masses.

but the entrainment efficiency appears to be too small to account for tens of solar masses. Churchwell (1997) concluded that the most likely mechanism is infalling material that is somehow diverted out again along the protostellar spin axis. An interesting implication of this scenario is that only a small fraction of the infalling matter actually reaches the central protostar; most is blown back out into the hot core. This would also imply that for a hot core to produce a star of a given mass, it would have to have an initial mass many times the mass of the most massive star in an OB cluster. Thus to form a 50 solar mass star, it may require a molecular clump of several thousand solar masses. Such clumps are rare and may be a primary reason why O stars are rare. A possible way to save the entrainment idea is by magneto-turbulent entrainment. The advantage of magneto-hydrodynamic entrainment is that it operates at Alfvén speeds, about a factor of 10 larger than the sound speed in hot cores. This is about the factor needed to account for the measured outflow masses. Observational support for this mechanism does not exist at this time.

Hyper-Compact HII Regions

A growing number of super compact and super dense objects have been discovered mostly at millimeter wavelengths (see the compilation of Johnson et al. 1998). They are typically about 10 times smaller and 100 times denser than UC HII regions. They are generally very weak or not detected at cm wavelengths, but their flux densities increase roughly proportional to frequency at mm wavelengths. Several have been observed in hydrogen radio recombination lines (H42α to H66α) and found to have astonishingly broad linewidths (FWHM $\sim$ 50 to 180 km s^{-1}); typical HII linewidths are $\sim$30 km s^{-1}. The nature of these objects is not understood: Neither the spectral indices nor the broad lines, nor their origin have adequate explanations. Hofner et al. (1996) reproduced the spectral index of G9.62E with a partially ionized stellar wind; unfortunately, no radio recombination lines exist for this source. With the exception of G25.5 + 0.2, all the known hyper-compact HII regions are located in massive star formation complexes. It is tempting to speculate that these objects may possibly be massive protostars near the end of their rapid accretion stage and are in the process of producing a detectable HII region. They are so dense that they are optically thick at cm wavelengths and only become detectable at mm wavelengths. The broad linewidths could be produced by ionized bipolar outflows that drive the observed molecular outflows associated with accretion. Other scenarios have been postulated, but at this point too little information is available to decide among them. G25.5 + 0.2 appears to be an evolved object (Subrahmanyan et al. 1993). Much more intensive study will be required before a better understanding of these objects emerges.

UC HII REGIONS

Massive stars begin burning hydrogen well before they reach the main sequence (i.e., while they are still accreting matter; see Bernasconi & Maeder 1996). However, they probably do not form detectable UC HII regions until they have ceased most accretion for the reasons given above (see Radio Free-Free Emission). Thus, we should expect that the ionizing star of a UC HII region is on the main sequence, no longer accreting significant (if any) mass, and basically obeys the statistical relations found for optically visible O stars such as the mass-luminosity relation ($L \propto M^3$) and mass loss rate ($\dot{M} \propto L^{1.7}$).

The Ionized Gas

MORPHOLOGIES AND LIFETIMES   UC HII regions have only a few recurring morphological types, which Wood & Churchwell (1989a) classified as cometary, core-halo, shell, and irregular and multiple peaked structures. An additional type, not included in the Wood & Churchwell (1989a) survey, is bipolar HII regions that comprise a very small number of objects: NGC7538 IRS1 (Campbell 1984, Turner & Matthews 1984), NGC6334A (Rodríguez et al. 1988), G45.48 + 0.13 (Garay et al. 1993), W49A-A (De Pree et al. 1997), and K3-50A (De Pree et al. 1994). S106 also has a bipolar morphology, but it is too large to qualify as a UC HII region. The bipolar nebulae have broad radio recombination lines, strong velocity gradients along the bipolar axes, and an hourglass shape when projected on the sky. Radio images shown in Figure 8 illustrate the four morphological types that have some degree of symmetry.

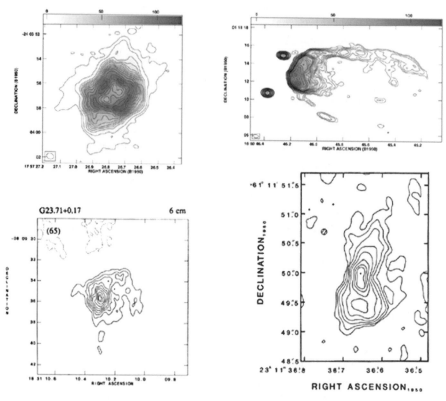

Figure 8   Examples of UC HII morphological types. Top left: G5.89 − 0.39, a shell-like structure. Top right: G34.26 + 0.15, a cometary nebula. Lower left: G23.71 + 0.17 from Wood & Churchwell (1989a), a core/halo morphology. Lower right: NGC7538 IRS1 from Campbell (1984), a bipolar nebula.

How do we explain the small number of morphological types, and how they are formed? The morphology of a UC HII region is a complicated function of its age, the dynamics of both the ionized and molecular gas, the density structure of the local ambient ISM, and the motion of the HII region relative to the ambient medium. Because the morphology depends on so many factors, it has not been easy to isolate the primary mechanism(s) that determine the morphology and maintain UC HII regions in a compact state for periods up to $\sim 10^5$ years (see "Galactic Population and Distribution" below). The morphology and lifetime in the UC state are related, and any successful theory that explains the morphology must also account for the long dwell time in the compact state. Churchwell (1999) discussed six proposed hypotheses to explain either the long lifetimes of UC HII regions (required to account for their large number) or their morphologies. These are champagne flow or blister models; infall models; photo-evaporating disks; pressure-confined nebulae; stellar wind–supported bow shocks; and mass-loaded stellar winds. Except for the bow shock hypothesis, each of the other hypotheses either addresses a particular morphology or the long lifetime issue but not both. The bow shock hypothesis can explain the long lifetimes and cometary, core/halo, and shell morphologies, but it does involve fine-tuning the ambient density, stellar velocity through the molecular core, and the aspect angle of observation to account for a given morphology and lifetime. Churchwell (1999) concluded that all the above hypotheses may be valid at different evolutionary stages in the lifetime of a UC HII region. For example, a champagne flow is likely to occur as a star exits its natal molecular cloud, photo-evaporating disks are likely to occur as long as a disk exists around a massive protostar, and if ambient pressure is greater than that of the UC HII region, it will provide confinement. Each of the hypotheses predicts different kinematics of the ionized gas, so they can be tested by observing high resolution velocity distributions of the gas via radio recombination lines and/or IR fine structure lines. Using high spatial resolution images of radio recombination lines toward the cometary UC HII region G29.96 − 0.02, Afflerbach et al. (1994) concluded that the distribution of line velocities and widths was consistent with a bow shock model. Using long slit echelle spectra of the Brγ line toward G29.96 − 0.02, Lumsden & Hoare (1996) concluded that the distribution of line velocities was inconsistent with the bow shock model and generally consistent with a champagne flow. This illustrates the difficulty of distinguishing between various models and the importance of obtaining more high-quality data.

LOW-DENSITY EXTENDED HALOS   Interferometric observations with lower resolution (i.e., sensitive to larger spatial structures) than the surveys of Wood & Churchwell (1989a) and Kurtz et al. (1994) show that many UC HII regions are surrounded by extended lower density ionized halos (Garay et al. 1993, Koo et al. 1996, Kurtz et al. 1999, Kim & Koo 2001). This is illustrated in Figure 9 from Kurtz et al. (1999). The panels from bottom to top are in order of increasing resolution and decreasing sensitivity to extended emission. The ratio of halo to core luminosities ranges from 10 to 20 with a mean of $\sim$15 (S. Kurtz, private

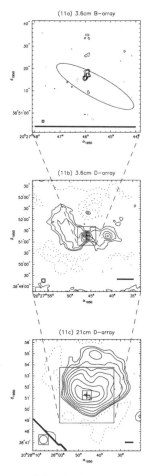

Figure 9   A montage of images of G77.9 − 0.0 obtained with different resolutions by Kurtz et al. (1999). This illustrates the extended halo that becomes quite apparent at lower spatial resolutions. Note the change in scale and field of view from bottom to top.

communication). Thus, the energy flux from the halos is quite large relative to that of the small UC HII cores. This clearly requires ionization by a cluster of hot stars because the energies are generally larger than that expected from single stars. Extended low-density gas associated with UC HII regions is also required to explain the detection of FIR fine structure emission lines of O^{++}, N^{++}, N^+, and S^{++} (Martín-Hernández et al. 2002, and references therein). Extended, lower-density, ionized hydrogen surrounding UC HII regions supposedly deeply embedded in dense molecular hydrogen cores does not easily fit the standard picture. Kim & Koo (2001) have suggested that extended halos can be understood in terms of a champagne flow combined with the clumpy structure of the

molecular core. A clumpy structure of the molecular core would permit ionizing radiation to penetrate further from the protostar than in a uniform medium.
Kim & Koo's idea also requires that the UC HII regions be located near the
boundary of their natal molecular core. Is this reasonable for the majority of
massive protostars? If UC HII regions are substantially older than their diameters suggest based on classical Strömgren theory, this could well be true. It
would also be consistent with the bow shock model of UC HII regions, which
predicts that they convert to champagne-type flows as they exit their natal molecular core. Formation of a halo during the bow shock phase is unlikely because
existing bow shock models predict that the ionization fronts are trapped behind
the bow shock and would therefore prevent formation of a halo in all directions
except the tail. The proposed Kim & Koo (2001) model is shown in Figure 10.
Kim & Koo (2001) have suggested that the existence of extended halos could
also resolve the lifetime problem of UC HII regions. However, it is also possible
that the halos are formed on or near light travel timescales. If champagne flows

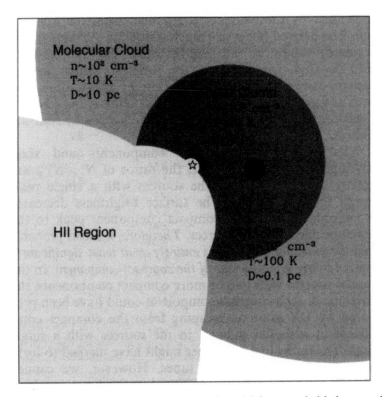

Figure 10 The Kim & Koo (2001) proposed model for extended halos associated
with UC HII regions. This model requires the ionizing star to be located near the
boundary of the hot natal molecular core and molecular clump.

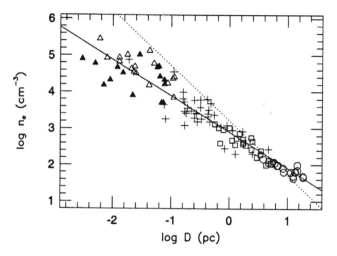

Figure 11   The electron density-size relation from Kim & Koo (2001). The dashed curve is the $D^{-3/2}$ relation expected for a spherical, uniform density nebula, and the solid curve is a least squares fit to the data points.

combined with clumpy molecular cores is the correct model, then this should be reflected in the morphology of the halos. The data so far available are not able to resolve this issue.

THE ELECTRON DENSITY-SIZE RELATION   Garay & Lizano (1999) and Kim & Koo (2001) have compiled the rms electron densities of HII regions ranging in diameter from 0.004 to 5 pc and found that n_e is approximately inversely proportional to the diameter. The Kim & Koo (2001) relation is shown in Figure 11. For a uniform-density nebula, the density should scale with diameter as $D^{-3/2}$. Garay & Lizano (1999) suggested that the observed shallower slope can be explained if UC HII regions are ionized by stars with lower luminosities than those ionizing more extended HII regions. This is not a compelling explanation because many UC HII regions are known to require $\geq$ the maximum UV photon flux that a single star can produce for its IR luminosity (Kurtz et al. 1994). Kim & Koo (2001) suggest that the more compact HII regions are located in denser parts of molecular cores, which results in proportionally smaller HII regions. The expectation of a $D^{-3/2}$ dependence is for a uniform, homogeneous medium. Natal molecular cloud cores are likely to be anything but uniform and homogeneous. It is interesting that Figure 11 implies that the column density of H^+ is $\sim 3 \times 10^{21}$ cm^{-2} for all D, implying $A_V(dust) \sim 3 \times 10^{21}(R/5.8 \times 10^{21}) \sim 1.7$ $(R = A_V/E_{B-V})$; $\tau_V(dust) \sim 1.5$; $\tau_{912A}(dust) \sim 8$; and $\tau_{912A}(dust\ absorption) \sim 5$. This implies that ionization of extended halos requires both a clumpy medium and the destruction of dust in the interclump medium. Models need to be calculated that include stellar winds and a clumpy ambient medium into which the HII region is expanding.

The Warm Dust Cocoons

UC HII regions heat the dust in their natal molecular cores out to large radii, causing them to be among the most luminous objects in our galaxy at far infrared wavelengths. The association of UC HII regions with luminous far infrared emission combined with obscuration of UC HII regions at visible and even NIR wavelengths leads to the conclusion that UC HII regions are embedded in their natal molecular cloud cores. The infrared SEDs of UC HII regions have been modeled by several groups (Chini et al. 1986, Hoare et al. 1988, Churchwell et al. 1990, Wolfire & Churchwell 1994). The predictions of these models have been reviewed by Churchwell (1991, 1999), and further discussion of them is not pursued here. The above models all assume a spherical, homogeneous medium. More sophisticated models should include a clumpy medium, nonspherical geometry, and the possible effects of remnant accretion disks.

The Molecular Environment

The molecular cores in which UC HII regions reside have rich molecular spectra. The spectra indicate a large variety of molecular species and isotopic substitutions as well as a wide range of excitation conditions. Molecular spectra provide a rich energy level structure and a wide range of excitation energies from which physical properties such as the distributions of density, temperature, and gas kinematics can be derived. They can also be used to probe the cosmic ray ionization rate and magnetic field strengths in dense, hot cores surrounding UC HII regions and in massive star formation clouds with hundreds of magnitudes of extinction.

PHYSICAL PROPERTIES   Wood & Churchwell (1989a) postulated that UC HII regions are deeply embedded in their natal molecular clouds, based on their coincidence with strong IR emission and small sizes, both indirect inferences. This was put on firmer ground by Churchwell et al. (1990), who observed 84 UC HII regions and found that $\sim$70% of the sample had $NH_3(1,1)$ and/or H_2O maser emission. Following this, a series of papers were published to establish the physical properties of the molecular clouds in which UC HII regions are formed. Churchwell et al. (1991) found from ^{13}CO, CS, and CH_3CN that the molecular clouds have large column densities ($\geq 5 \times 10^{23}$ cm^{-2}), are hot (≥ 100 K), dense ($\geq 10^5$ cm^{-3}), and lie within a few tenths of a pc of the ionizing star. Cesaroni et al. (1991a) confirmed these properties from $C^{34}S$ and showed that cloud masses determined assuming virial equilibrium, statistical equilibrium analyses, and CS column densities (all of which depend on distance to different powers) give about the same mass. They found cloud masses of $\sim$2000 $M_\odot$ and diameters of $\sim$0.4 pc. Cesaroni et al. (1991b) used observed inversion transitions of NH_3 from (1,1) to (5,5) inclusive to model the excitation of NH_3 associated with 16 UC HII regions. They found column densities in the range $N(H_2) \sim 10^{23}$–10^{24} cm^{-2}, densities $n(H_2) \sim 10^5$–10^6 cm^{-3}, kinetic temperatures 100–200 K, diameters of $\sim$0.4 pc, and virial masses $\sim$100 to 2000 $M_\odot$. High resolution VLA observations by Cesaroni et al. (1994) of the

$NH_3(4,4)$ emission toward G9.62, G10.47, G29.96, and G31.41 showed that the natal molecular clouds contain hot, compact NH_3 clumps located close to the UC HII regions. The NH_3 clumps are small (≤ 0.1 pc), dense ($\sim 10^7$ cm^{-3}), massive (~ 100 $M_\odot$), and luminous (10^4–10^6 $L_\odot$). They suggested that these luminous clumps may be the sites of newly emerging young massive stars; their spatial coincidence with water masers gives further support to this hypothesis. Cesaroni et al. (1994) also pointed out what they referred to as a "luminosity paradox" in which the luminosity (for spherical clouds) inferred from the NH_3 excitation analysis was about an order of magnitude larger than the luminosity obtained from the observed FIR flux densities. In a follow-up study with even higher spatial resolution (0.4"), Cesaroni et al. (1998) were able to resolve the NH_3 clumps for the first time and derive the velocity structure and temperature gradients across them. They found that $T(r) \propto r^{-3/4}$ and that the only way the temperature gradients could be satisfactorily modeled was with disk-like (oblate) clumps. Using oblate clumps rather than spheres in combination with the temperature gradients also resolved the luminosity paradox so that the luminosities inferred from the NH_3 temperature structure and from FIR fluxes are in agreement.

In summary, the natal clouds that harbor UC HII regions are small (≤ 0.5 pc), dense (typically $> 10^5$ cm^{-3}), hot (100–200 K), luminous (10^4–10^6 $L_\odot$), and massive (up to several $\times 10^3$ $M_\odot$). They are clumpy, with clumps of diameter ~ 0.1 pc, mass ~ 100 $M_\odot$, and densities of $\sim 10^7$ cm^{-3} located near the embedded UC HII region. The clumps are enhanced in NH_3/H_2 abundance by ≥ 100 over that in quiescent molecular clouds. This enhancement is believed to be due to evaporation of ice mantles on grains in the neighborhood of the UC HII regions.

CHEMISTRY   The chemistry of massive star formation regions (hot cores or HCs) has been recently reviewed by Ohishi (1997), van Dishoeck & van der Tak (2000), and Millar (2000). Detailed, high spatial, and spectral resolution studies show that there are few generalizations one can make about the chemistry in HCs. There are large relative abundance variations among HCs and large variations in distribution of different species within a given HC (van Dishoeck 2001 and references therein). Perhaps the main generalizations are that grain surface chemistry appears to play a critical role in establishing relative abundances; photochemistry and shock chemistry are likely to be important at different evolutionary phases and locations within a HC; and the chemistry is almost certainly time-dependent. The current chemical evolution paradigm involves formation of icy grain mantles by accretion of gas-phase atoms and molecules during the prestellar phase of cloud evolution, followed by grain-surface chemistry and evaporation of ices as the protostar begins to heat its environment. The evaporated molecules provide reactants for a high-temperature gas-phase chemistry, which can produce complex saturated organic molecules (Charnley et al. 1992, 1995; Millar 1997; van Dishoeck 2001). Understanding the evolution of HCs depends on establishing their ages. The relative abundances of certain species such as CH_3OCH_3/CH_3OH and others that show strong variations with time in chemical models may provide chemical clocks that will determine ages.

An example of the relative abundances of molecular species in HCs containing UC HII regions is provided by the work of Hatchell et al. (1998) who used the James Clark Maxwell Telescope (JCMT) to obtain spectra in the 230 GHz and 345 GHz bands of 14 UC HII regions previously shown to have high excitation NH_3 and CH_3CN emission. They found that the number of transitions varied by a factor of 20 between sources. Half of the sources showed only a few lines, the carriers of which were $C^{17}O$, $C^{18}O$, SO, $C^{34}S$, and CH_3OH. In the other half of their sample (the line rich sources), they detected over 150 lines, many of which were from high excitation lines of some of the more complex interstellar molecules such as $HCOOCH_3$, C_2H_5CN, and CH_3CCH. They found that the line rich sources require hot, dense, compact cores surrounded by an ambient cloud consisting of less dense, cooler gas. The molecular cores are <0.1 pc in size, with densities of $\sim 10^8$ cm^{-3} and temperatures >80 K. The line poor sources could be modeled with a temperature of 20–30 K, density of 10^5 cm^{-3}, and no hot core. The hot cores in the line rich sources have sizes about equivalent to their UC HII regions. In the line poor sources, it is unclear how NH_3 and CH_3CN can achieve high excitation if temperatures and densities are as low as claimed. Is it possible that the difference between the line poor and line rich sources is mainly an evolutionary one in which the line poor sources have mostly destroyed their molecular cocoons and are now surrounded by a small amount of molecular gas that can produce only weak lines of a few impervious molecules?

MASERS   Molecular masers are very bright and ubiquitous in the vicinity of UC HII regions and MSF regions. The most widespread masers associated with MSF regions are H_2O, OH, and CH_3OH; H_2CO and NH_3 have been detected toward fewer sources and only in a few transitions. The presence of protostars or newly formed massive stars provides photons and heat, both of which are probably necessary to main population inversion depending on the molecule. Masers are distinguishable from nonmasering clouds by brightness, temperature, size, temporal variability, polarization, number of spectral features, and line width. Masers are generally confined to very dense (10^6–10^9 cm^{-3}), small (a few tens of AUs) clumps, which makes them excellent probes of the kinematics and physical properties of MSF regions on scale sizes that are otherwise very difficult or impossible to observe directly.

H_2O masers are signposts of massive star formation; Churchwell et al. (1990) detected H_2O masers toward 67% of about 100 observed UC HII regions. H_2O masers occur in small clusters with extents of $\sim$10–100 AU. Their spectra usually show many velocity components spread over velocities of $\sim$50 to >100 km s^{-1}, and their luminosities are proportional to the integrated far infrared luminosity of the SFR (Moran 1990, Palagi et al. 1993). Precisely what dynamical features of massive star formation regions are probed by H_2O masers has been a topic of considerable debate. VLBI measurements of Genzel et al. (1981), Reid et al. (1988), Bloemhof et al. (1992) and others indicate that the H_2O masers are expanding away from the center in several MSF regions. Torrelles et al. (1996, 1997, 1998a,b) find that H_2O masers trace both molecular outflows and accretion disks (perpendicular to the molecular outflow). Because of the high excitation of the 1.3 cm H_2O maser

(>600 K above ground), it is generally believed that shocks are probably required to achieve the temperatures and densities necessary to produce the observed H_2O masers. In the case of outflows or expanding shells, these would occur at the interface of the outflow and the ambient medium. In accretion disks, they could be produced by infalling matter onto the disk.

The 1665 and 1667 MHz OH masers are closely associated with star formation and UC HII regions. Individual velocity components have sizes of $\sim 10^{14}$ cm and are distributed over regions of $\sim 10^{16}$ cm (about the radius of typical UC HII regions). The origin of OH masers in MSF regions is uncertain. Garay et al. (1985) have suggested that OH masers lie in an accreting envelope outside the advancing ionization front of the HII region. Bloemhof et al. (1992) have interpreted their proper motion measurements of OH masers toward W3(OH) as an expanding shell between the shock front and the ionization front of the HII region.

Methanol masers have been observed in many transitions toward massive star formation regions (see Garay & Lizano 1999 for an enumeration of the observed maser transitions). As with H_2O and OH masers, it is not clear if methanol masers probe exclusively outflows, accretion disks, or advancing shock fronts. Claims have been made for each in different sources. The NH_3 and H_2CO masers have not been observed in as many sources as the other masers, and general conclusions about their properties cannot be made.

Based on the typical relative positions of the masers with each other and with radio continuum contours, several investigators have postulated that H_2O, CH_3OH, and OH masers appear at different evolutionary stages of star formation (see the summary of Garay & Lizano 1999). They and others suggest that H_2O masers appear in the earliest stages of massive star formation during the rapid accretion phase. CH_3OH masers appear later during the period when the emerging protostar forms a detectable UC HII region, and OH masers appear last in a compressed, possibly infalling, circumnebular shell. These masers fade away as the UC HII region evolves into a diffuse HII region (Codella et al. 1994). Although compelling because of its simplicity and potential for establishing relative ages, this scenerio is based on shaky age determinations. Also, numerous examples are known where all three masers are observed in what appears to be the same star formation region.

GALACTIC POPULATION AND DISTRIBUTION

If the number, spatial distribution, and ages of newly formed O stars in the Galaxy were known, they would permit a direct determination of the current rate of massive star formation in the Galaxy (independent of uncertainties in the IMF of O stars), provide estimates of the mechanical and radiative energy and momentum input to molecular clouds, allow a determination of the contribution of massive stars to the global energetics of the Galaxy, show the locations of massive star formation relative to spiral arms, and delineate the spiral structure of the Galaxy. To estimate the contribution of O stars to the global energetics, we must know what fraction of the total O star population resides in molecular clouds hidden from optical and UV observations. Distances are required to use O stars as tracers of galactic structure.

The Distribution and Number of UC HII Regions

Estimates of the number and distribution of O stars in the Milky Way have been made by Wood & Churchwell (1989b), Hughes & McLeod (1989), Zoonematkermani et al. (1990), White et al. (1991), Helfand et al. (1992), and Becker et al. (1994). All the above groups used far infrared (FIR) colors from the Infrared Astronomical Satellite (IRAS) Point Source Catalog to select HII regions. All the above groups found close to the same far infrared colors for galactic HII regions, including UC HIIs. The small differences in the precise color boundaries chosen, in the lower limit to the flux density for a given wavelength band, and whether the IRAS source coincides with a radio source are the basic reasons why each group gets somewhat different numbers of HII regions and different scale heights in Galactic latitude. Hughes & McLeod (1989) used optical HII regions to establish IRAS colors [$\log(S_{25\,\mu m}/S_{12\,\mu m}) \geq 0.4$ and $\log(S_{60\,\mu m}/S_{25\,\mu m}) \geq 0.25$] and found $\sim$2300 HII region candidates lying between $\pm 3^0$ in Galactic latitude. Wood & Churchwell (1989b) used selection criteria $S_{12\,\mu m}$ and $S_{25\,\mu m} \geq 10$ Jy, $\log(S_{60\,\mu m}/S_{12\,\mu m}) \geq 1.30$, and $\log(S_{25\,\mu m}/S_{12\,\mu m}) \geq 0.57$, based on the colors of radio-identified UC HII regions. They identified $\sim$1650 UC HII region candidates in the Galaxy with a scale height of $0.6^0 \pm 0.05^0$ in latitude, corresponding to about 90pc at a distance of 8.5 kpc. They also showed that most UC HII region candidates lie in quadrants I and IV of the Galactic plane. White et al. (1991), Helfand et al. (1992), and Becker et al. (1994) derived IRAS colors based on the coincidence of IRAS point sources with identified thermal radio sources. They identified all together <1000 candidate UC HII regions in the fraction of the Galactic plane observable with the VLA. They find the distribution in latitude to have a full-width at half-maximum (FWHM) $\sim$15′ [a factor of 0.4 times that found by Wood & Churchwell (1989b)], which implies a latitude scale height for O stars of only 36 pc at a distance of 8.5 kpc. They argue that the sample of Wood & Churchwell (1989b) is contaminated by UC HII regions ionized by B stars, which have a wider latitude distribution than O stars. Both estimates imply that O stars are very tightly confined to the Galactic plane with a rapidly declining chance of finding young massive stars more than a few tens of parsecs from the plane. Also, O stars are far more likely to be found in the inner galaxy than outside the solar circle. The concentration of UC HII regions to the Galactic plane and to the inner Galaxy is illustrated by the FIR color–selected plot of candidate UC HII regions shown in Figure 12 [taken from Wood & Churchwell (1989b)]. This has implications both for massive star formation and the location of luminous UV energy sources in the Galaxy. In particular, the conditions required to form massive stars are far more likely to occur at midplane in the Galaxy than at latitudes only a few parsecs above or below it. The distribution of UC HII candidates and molecular gas, as traced by CO, are very similar as illustrated in Figure 13 taken from Wood & Churchwell (1989b).

All attempts to obtain the number of newly formed O stars (represented by UC HII regions) using IRAS colors find about 2000 ± 40% integrated over the entire Galactic plane. Wood & Churchwell (1989b) found that about 10% of all O stars within about 2.5 kpc of the Sun are in the ultracompact state. Assuming that

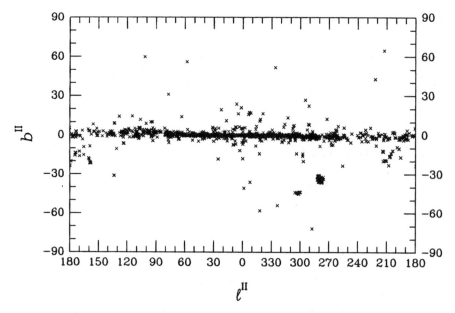

Figure 12   The distribution in Galactic coordinates of all candidate UC HII regions selected from the IRAS Point Source Catalog according to the FIR color-color criteria of Wood & Churchwell (1989b). Note the tight confinement to the Galactic plane, the crowding in the inner galaxy, and the conspicuousness of the LMC and SMC.

this percentage is about constant over the Galactic disk, this implies that the total number of O stars in the Galaxy is about 20,000 ± 80% (including generous errors introduced by assumptions). Wood & Churchwell (1989b) estimated that O stars have an average integrated luminosity of $\sim 10^9 L_\odot$ or $\sim 30\%$ of the FIR luminosity of molecular clouds and $\sim 8\%$ of the FIR luminosity of the entire Galaxy (Sodroski 1988), hardly a dominant contribution to the total luminosity of the Galaxy. However, the above estimates are likely to be substantially increased by 2MASS and SIRTF surveys, which will go much deeper and have better spatial resolution than IRAS. The fraction of O stars with nebulae in the UC state is also an indication of the fraction of the time that they remain in the UC state. Wood & Churchwell (1989b) estimated that UC HII regions have typical lifetimes of about 10^5 years. This is substantially longer than the sound crossing time and the timescale predicted by classical Strömgren theory in the absence of significant confining pressure. This dilemma was discussed above ("Morphologies and Lifetimes") and is often referred to as the lifetime problem.

All the above attempts to determine the population and angular distribution of O stars in the Galaxy suffer from the poor spatial resolution and limited sensitivity of IRAS. Higher spatial resolution and better sensitivity will more precisely define the IR colors and better determine the total number of O stars in the Galaxy. With

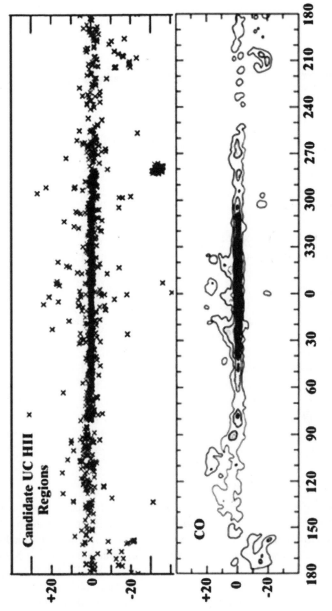

Figure 13   A comparison of the distributions of UC HII regions from Wood & Churchwell (1989b) and CO emission from Dame et al. (1987). This visually demonstrates the close association of molecular gas and young massive stars.

18″ spatial resolution, the Mid-Course Space Experiment (MSX) can take a large step in this direction. However, both the 2 Micron All Sky Survey (2MASS) and the Space Infrared Telescope Facility Infrared Array Camera (SIRTF/IRAC) will achieve more than two orders of magnitude greater sensitivity and spatial resolution than IRAS. The combination of 2MASS and SIRTF surveys are expected to fine-tune the infrared (NIR-MIR) colors of O and B stars (including embedded ones) and detect all those not confused or hidden by background emission. Together these two large databases should resolve many of the remaining questions related to the number, distribution, formation rate, and energetics of massive stars in the Galaxy.

Spiral Structure Using UC HII Regions

Spiral arms in other galaxies are defined by the distribution of O stars; thus we expect one of the best tracers of spiral arms in our Galaxy to be newly formed O stars or their HII regions. As early as 1952, Morgan et al. (1952) reported evidence for two and possibly three spiral arms in the solar neighborhood using optically visible HII regions. More recent attempts to determine the spiral arms of the Galaxy by using HII regions as tracers were Georgelin & Georgelin (1976), who used visible, evolved HII regions, and Kurtz et al. (1994), who used radio-detected UC HII regions. Both of these were of limited success in delineating the Galaxy's spiral structure; without lines drawn to guide the eye, different observers could have easily drawn quite different spiral arms. Kurtz et al. (1994) postulated that the combination of near/far distance ambiguities in the inner Galaxy plus peculiar motions masked the spiral arm structure. Recently, Araya et al. (2001) measured both radio recombination line emission and H_2CO in absorption toward 21 selected UC HII regions using the Arecibo 305 m telescope. The velocity of the recombination line established the velocity of the UC HII region, and the pattern of H_2CO absorption velocities was used to resolve the distance ambiguity for each source. With this limited number of UC HII regions, they found surprisingly good agreement with previous spiral arm determinations. This is illustrated in Figure 14. The accuracy of this picture will be tested when more sources are measured using this promising technique.

Galactic Temperature and Abundance Gradients

The mean electron temperatures of evolved Galactic HII regions depend on their galactocentric distance, D_G, in the sense that temperatures increase with D_G (Churchwell & Walmsley 1975, Shaver et al. 1983, and others). This has been in-terpreted to imply a corresponding gradient in metallicity because metals provide the primary coolants of HII regions and determine their equilibrium temperatures. In a classical paper by Shaver et al. (1983), the relationship between metal abun-dances and mean temperatures of HII regions was quantified by determining both metal abundances and temperatures independently for a large number of evolved HII regions whose distances were well known.

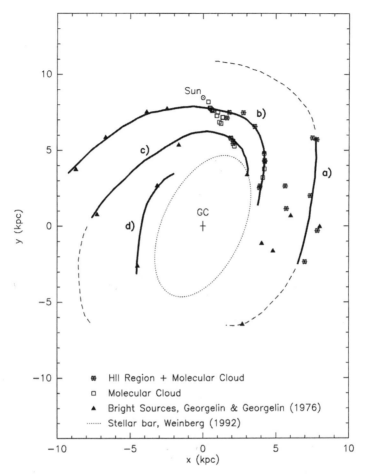

Figure 14   The spiral arms suggested by Georgelin & Georgelin (1976) (solid curves) with the distribution of UC HII regions from Araya et al. (2001). With the exception of two points that fall between arms *a* and *b*, the UC HII regions trace quite closely the end portions of arms *a*, *b*, and *c*.

UC HII regions have very recently ionized their natal molecular cores and provide an opportunity to determine if the abundances and electron temperatures in these rather special regions of the interstellar medium are similar to those of evolved HII regions (where the gas and dust have been exposed to UV radiation for long periods and probably mixed with ambient diffuse interstellar matter). Afflerbach et al. (1996) made such a test by using high spatial resolution observations in the radio recombination lines H42α, H66α, H76α, and H93α toward 17 UC HII regions distributed in galactocentric radius from 4 kpc to 11 kpc. Nonlocal thermodynamic equilibrium (NonLTE) models were calculated for all nebulae from which electron temperatures and densities were derived. They found that the electron temperature increases with D_G as $T_e(K) = (5540 \pm 390) + (320 \pm 64)D_G(kpc)$ with a correlation coefficient of 0.82. The data with the linear least squares fit superimposed is shown in Figure 15. Also shown for comparison is the Shaver et al. (1983) fit to their diffuse HII data. The Afflerbach et al. (1996) T_e values are on average about

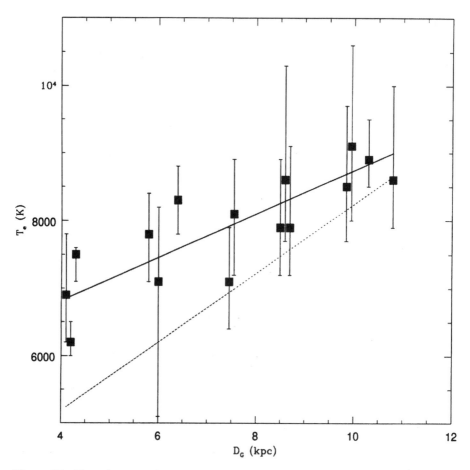

Figure 15   The galactocentric electron temperature gradient derived for UC HII regions by Afflerbach et al. (1996) using non-LTE analyses of radio recombination line observations. The dashed curve is the gradient found by Shaver et al. (1983) from diffuse optically visible HII regions. The solid curve is the least squares fit to the UC HII data. The systematically higher temperatures of UC HII regions than of diffuse HII regions is consistent with collisional quenching of coolants in UC HII regions.

1500 K higher than those of Shaver et al. (1983); this is entirely consistent with increased collisional quenching of the coolants in UC HII regions. From the T_e-D_G relation, Afflerbach et al. (1996) inferred a Galactic O/H abundance gradient of the form

$$\frac{d(O/H)}{dD_G} = -0.047 \pm 0.009 \; dex \; \text{kpc}^{-1},$$

which is about the same as found by Shaver et al. (1983) using optical line observations of diffuse HII regions. This is also in reasonable agreement with galactocentric abundance gradients derived from far infrared fine structure line observations of Lester et al. (1987), Simpson et al. (1995), Afflerbach et al. (1997), and Rudolph et al. (1997). The far infrared fine structure and radio recombination line data together indicate that (1) the mean temperatures of UC HII regions systematically increase with D_G by about 300 K per kpc; (2) the average abundances of O, N, and S decrease with D_G by $\sim$0.065 dex per kpc in UC HII regions; and (3) the abundances in UC HII regions are about the same as those in classical diffuse HII regions at the same galactocentric distance, which indicates that the abundances (gas plus dust) in natal molecular cloud cores must also have the same abundance gradient as other constituents of the interstellar medium.

SUMMARY AND FUTURE PROSPECTS

A brief summary is given of our present understanding of the evolution to the UC HII region state. This is based on observationally identified properties of giant molecular clouds that provide the natal material for prestellar cores, which are the precursors of hot cores that probably produce hypercompact HII regions postulated to quickly evolve into UC HII regions. The properties of hot cores were emphasized because most is known about this stage of evolution. The stages about which least is known are prestellar cores and hypercompact HII regions. Both these stages of evolution are expected to be very short and the number of objects in each state is therefore expected to be very small. In fact, no prestellar core that will produce an O star is known. The discussion of prestellar cores is based on the properties of low-mass star forming prestellar cores.

The properties of UC HII regions and their associated molecular and dust environments are discussed in some detail. It is argued that the limited number of observed morphological types and the lifetime of UC HII regions are related and that any theory that can account for their morphologies must also account for their extended lifetimes. The existence of extended halos around UC HII regions requires either a very clumpy molecular core or a modification of our present picture of UC HII regions with sharply truncated boundaries. An electron density-size relation is one in which $n_e \propto D^{-1}$ exists for HII regions with diameters from 0.004 pc to 5 pc. Uniform density, spherical HII regions are expected to scale as $D^{-3/2}$. The shallower dependence on size is not understood, but it is suggested that departures from uniform density, nonspherical morphologies, and ionizing stars that have strong winds probably play a role. The $n_e \propto D^{-1}$ relation implies H^+ column density of about 3×10^{21} cm^{-2}, independent of diameter. We also infer that dust must be destroyed in the interclump gas in UC HII regions to be consistent with observed ionized halos. The properties of the warm dust cocoons and molecular cores that enshroud UC HII regions were briefly discussed, including the chemistry and maser emission in the hot environments around UC HII regions.

Finally, the use of UC HII regions as probes of global properties of the Galaxy was summarized. In particular, the distribution of UC HII regions in the Galaxy and implications for the UV and kinetic energy contributions to the interstellar medium in the Galaxy were reviewed. The attempts to determine spiral structure as well as temperature and abundance gradients in the Galaxy were also discussed.

All evolutionary stages preceding the UC state are not well understood, and future efforts are likely to concentrate on these, especially prestellar cores and hypercompact HII regions. The physics of how massive bipolar outflows are driven and collimated is basically still a mystery and needs much more observational and theoretical attention. The same can be said for the interdependence of accretion disks and outflows. No viable theoretical models of massive star formation are available that can account for all the observed properties of the early evolutionary stages of massive star formation. Higher resolution and more sensitive observations of the thermal dust emission, ionized gas, and associated molecular gas are also needed to better constrain models. New facilities that can provide incisive observational constraints are the Space Infrared Telescope Facility (SIRTF), the Stratospheric Observatory for Infrared Astronomy (SOFIA), the Submillimeter Array (SMA), the Atacama Large Millimeter Array (ALMA), the Square Kilometer Array (SKA), the Combined Array for Research in Millimeter-Wave Astronomy (CARMA), the Large Millimeter Telescope (LMT), and the Chandra X-Ray Observatory (CHANDRA). Although there is still much to be done before massive star formation will be understood, the future looks quite promising with the array of facilities currently online or under construction.

ACKNOWLEDGMENTS

I am indebted to all my coworkers who have contributed to much of the work reported here and from whom I have learned so much; in particular, Malcolm Walmsley, Riccardo Cesaroni, Doug Wood, Peter Hofner, Stan Kurtz, Deborah Shepherd, Andrew Afflerbach, Jerry Acord, Christer Watson, and Marta Sewilo. I also thank John Mathis, who gave critical comments on the manuscript; I have benefited greatly from his deep understanding of this subject. I am indebted to Christer Watson for help with some of the figures. Finally, figures were taken from several sources authored by numerous people to whom I am indebted for giving permission to reproduce images. This review was supported in part by NSF grant AST-9986548.

The *Annual Review of Astronomy and Astrophysics* is online at
http://astro.annualreviews.org

LITERATURE CITED

Afflerbach A, Churchwell E, Acord JM, Hofner P, Kurtz S, DePree CG. 1996. *Ap. J. Suppl.* 106:423–46

Afflerbach A, Churchwell E, Hofner P, Kurtz S. 1994. *Ap. J.* 437:697–704

Afflerbach A, Churchwell E, Werner MW. 1997. *Ap. J.* 478:190–205

André P, Ward-Thompson D, Motte F. 1996. *Astron. Astrophys.* 314:625–35

Araya E, Hofner P, Churchwell E, Kurtz S. 2002. *Ap. J. Suppl.* 138:63–74

Bacmann A, André P, Abergel A, Puget J-L, Bontemps S, et al. 1998. In *The Universe as Seen by ISO*, ed. P Cox, MF Kessler, pp. 467–70. Noordwijk, The Netherlands: ESA

Bazell D, Désert FX. 1988. *Ap. J.* 333:353–58

Becker RH, White RL, Helfand DJ, Zoonematkermani S. 1994. *Ap. J. Suppl.* 91:347–87

Behrend R, Maeder A. 2001. *Astron. Astrophys.* 373:190–98

Bernasconi PA, Maeder A. 1996. *Astron. Astrophys.* 307:829–39

Beuther H, Schilke P, Walmsley CM, Sridharan TK, Wyrowski F, Menten KM. 2001. In *Massive Star Birth*, ed. P Crowther, ASP Conf. Ser. In press

Blitz L. 1991. In *The Physics of Star Formation and Early Evolution*, NATO Sci. Ser., ed. CJ Lada, ND Kylafis, 342:3–33

Blitz L, Williams JP. 1999. In *The Origin of Stars and Planetary Systems*, NATO Sci. Ser., ed. CJ Lada, ND Kylafis, 540:3–28

Bloemhof EE, Reid MJ, Moran JM. 1992. *Ap. J.* 397:500–19

Bonnell IA, Bate MR, Zinnecker H. 1998. *MNRAS* 295:93–102

Cabrit S, Bertout C. 1992. *Astron. Astrophys.* 26:274–84

Campbell B. 1984. *Ap J.* 282:L27–30

Carey SJ, Clark FO, Egan M, Price SD, Shipman RF, Kuchar TA. 1998. *Ap. J.* 508:721–28

Cesaroni R, Churchwell E, Hofner P, Walmsley CM, Kurtz S. 1994. *Astron. Astrophys.* 288:903–20

Cesaroni R, Felli M, Testi L, Walmsley CM, Olmi L. 1997. *Astron. Astrophys.* 325:725–44

Cesaroni R, Hofner P, Walmsley CM, Churchwell E. 1998. *Astron. Astrophys.* 331:709–725

Cesaroni R, Walmsley CM, Churchwell E. 1991b. *Astron. Astrophys.* 256:618–30

Cesaroni R, Walmsley CM, Kömpe C, Churchwell E. 1991a. *Astron. Astrophys.* 252:278–90

Charnley SB, Kress ME, Tielens AGGM, Millar TJ. 1995. *Ap. J.* 448:232–39

Charnley SB, Tielens AGGM, Millar TJ. 1992. *Ap. J.* 399:L71–74

Chini R, Krügel E, Kreysa E. 1986. *Astron. Astrophys.* 167:315–24

Chini R, Krügel E, Wargau W. 1987. *Astron. Astrophys.* 181:378–82

Churchwell E. 1991. In *The Physics of Star Formation and Early Stellar Evolution*, ed. CJ Lada, ND Kylafis, pp. 221–68. Dordrecht: The Netherlands: Kluwer

Churchwell E. 1997. *Ap. J. Lett.* 479:L59–61

Churchwell E. 1999. In *The Origin of Stars and Planetary Systems*, ed. CJ Lada, ND Kylafis, pp. 515–52. Dordrecht: The Netherlands: Kluwer

Churchwell E, Walmsley CM. 1975. *Astron. Astrophys.* 38:451–54

Churchwell E, Walmsley CM, Cesaroni R. 1990. *Acta Astron. Sinica.* 83:119–44

Churchwell E, Walmsley CM, Wood DOS. 1991. *Astron. Astrophys.* 253:541–56

Churchwell E, Wolfire M, Wood DOS. 1990. *Ap. J.* 354:247–61

Codella C, Felli M, Natale V, Palagi F, Palla F. 1994. *Astron. Astrophys.* 291:261–70

Crutcher RM. 1999. *Ap. J.* 520:706–13

Crutcher RM. 2001. In *Massive Star Birth*, ed. P Crowther. *ASP Conf. Ser.* In press

Dame TM, Ungerechts H, Cohen RS, de Geus EJ, Grenier IA, et al. 1987. *Ap. J.* 322:706–20

De Pree CG, Mehringer DM, Goss WM. 1997. *Ap. J.* 482:307–33

De Pree CG, Goss WM, Palmer P, Rubin RH. 1994. *Ap. J.* 428:670–79

Egan MP, Shipman RF, Price SD, Carey SJ, Clark FO, Cohen M. 1998. *Ap. J.* 494:L199–202

Falgarone E, Phillips TG, Walker CK. 1991. *Ap. J.* 378:186–201

Fuller GA, Myers PC. 1987. In *Physical Processes in Interstellar Clouds*, ed. M Scholer, pp. 137–60. Dordrecht: Reidel

Garay G, Lizano S. 1999. *Publ. Astron. Soc. Pac.* 111:1049–87

Garay G, Reid MJ, Moran JM. 1985. *Ap. J.* 289:681–97

Garay G, Rodríguez LF, Moran JM, Churchwell E. 1993. *Ap. J.* 418:368–85

Genzel R, Reid MJ, Moran JM, Downes D. 1981. *Ap. J.* 244:884–902

Georgelin YM, Georgelin YP. 1976. *Astron. Astrophys.* 49:57–79

Gregersen EM, Evans NJ II. 2000. *Ap. J.* 538:260–67

Gregersen EM, Evans NJ II, Mardones D, Myers PC. 2000. *Ap. J.* 533: 440–53

Hatchell J, Thompson MA, Millar TJ, MacDonald GH. 1998. *Astron. Astrophys. Suppl.* 133:29–49

Helfand DJ, Zoonematkermani S, Becker RH, White RL. 1992. *Ap. J. Suppl.* 80:211–55

Henning TH, Schreyer K, Launhardt R, Burkert A. 2000. *Astron. Astrophys.* 353:211–26

Hoare MG, Glencross WM, Roche PF, Clegg RES. 1988. In *Dust in the Universe*, ed. ME Bailey, DA Williams, pp. 107–11. Cambridge, UK: Cambridge Univ. Press

Hofner P, Delgado H, Whitney B, Churchwell E. 2002. *Ap. J.* In press

Hofner P, Kurtz S, Churchwell E, Walmsley CM, Cesaroni R. 1996. *Ap. J.* 460:359–71

Hughes VA, McLeod GC. 1989. *Astron. J.* 97:786–800

Jijina J, Myers PC, Adams FC. 1999. *Ap. J. Suppl.* 125:161–236

Johnson C, De Pree C, Goss WM. 1998. *Ap. J.* 500:302–10

Kim K-T, Koo B-C. 2001. *Ap. J.* 549:979–96

Koo B-C, Kim K-T, Lee H-G, Yun M-S, Ho PT. 1996. *Ap. J.* 456:662–76

Kroupa P. 2001. *MNRAS* 322:231–46

Kurtz S, Cesaroni R, Churchwell E, Hofner P, Walmsley CM. 2000. See Mannings et al. 2000, pp. 299–326

Kurtz S, Churchwell E, Wood DOS. 1994. *Ap. J. Suppl.* 91:659–712

Kurtz SE, Watson AM, Hofner P, Otte B. 1999. *Ap. J.* 514:232–48

Larson RB. 2002. *MNRAS.* In press

Launhardt R, Henning Th. 1997. *Astron. Astrophys.* 326:329–46

Lumsden SL, Hoare MG. 1996. *Ap. J.* 464:272–85

Lee CW, Myers PC. 1998. *Ap. J. Suppl.* 123: 233–50

Lee CW, Myers PC, Taffalla M. 1999. *Ap. J.* 526:788–805

Lester DF, Dinerstein HL, Werner MW, Watson DM, Genzel R, Storey JWV. 1987. *Ap. J.* 320:573–85

Mannings V, Boss AP, Russell SS, eds. *Protostars and Planets IV.* Tucson: Univ. Ariz. Press

Martín-Hernández NL, Peeters E, Morisset C, Tielens AGGM, Cox P, et al. 2002. *Astron. Astrophys.* 381:606–27

Massey P. 1999. *IAU Cir. Symp. #190*, ed. Y-H Chu, N Suntzeff, J Hesser et al., pp. 173–80

Massey P, Hunter DA. 1998. *Ap. J.* 493:180–94

Meyer MR, Adams FC, Hillenbrand LA, Carpenter JM, Larson RB. 2000. See Mannings et al. 2000, pp. 121–49

Millar TJ. 1997. In *Molecules in Astrophysics: Probes and Processes, IAU Cir. Symp.*, ed. EF van Dishoeck, 178:75–88

Millar TJ. 2000. In *Science with the Atacama Large Millimeter Array*, ASP Conf. Ser., ed. A Wootten. In press

Moran JM. 1990. In *Molecular Astrophysics*, ed. T Hartquist, pp. 397–423. Cambridge, MA: Cambridge Univ. Press

Morgan WW, Sharpless S, Osterbrock D. 1952. *Astron. J.* 57:3

Motte F, André P, Neri R. 1998. *Astron. Astrophys.* 336:150–72

Myers PC, Fuller GA. 1992. *Ap. J.* 396:631–42

Myers PC, Lazarian A. 1998. *Ap. J.* 507:L157–60

Nakano T. 1998. *Ap. J.* 494:587–604

Neufeld DA, Hollenbach DJ. 1994. *Ap. J.* 428:170–85

Neufeld DA, Hollenbach DJ. 1996. *Ap. J.* 471:L45–48

Norris RP, Byleveld SE, Diamond PJ, Ellingsen SP, Ferris RH, et al. 1998. *Ap. J.* 508:275–85

Norris RP, Whiteoak JB, Caswell JL, Wieringa MH, Gough RG. 1993. *Ap. J.* 412:222–32

Ohishi M. 1997. In *Molecules in Astrophysics: Probes and Processes. IAU Cir. Symp. 178*, ed. EF van Dishoeck, pp. 61–74. Dordrecht: Kluwer

Osorio M, Lizano S, D'Alessio P. 1999. *Ap. J.* 525:808–20

Palagi F, Cesaroni R, Comoretto G, Felli M, Natale V. 1993. *Astron. Astrophys. Suppl.* 101:153–93

Phillips CJ, Byleveld SE, Ellingsen SP, McCulloch PM. 1998. *MNRAS* 300:1131–57

Reid MJ, Schneps MH, Moran JM, Gwinn CR, Genzel R, et al. 1988. *Ap. J.* 330:809–16

Ridge NA. 2000. *The dynamics of outflows from young stellar objects*. PhD thesis. John Moores Univ. Liverpool, UK

Rodríguez LF, Cantó J, Moran JM. 1988. *Ap. J.* 333:801–5

Rudolph AL, Simpson JP, Haas MR, Erickson EF, Fich M. 1997. *Ap. J.* 489:94–101

Scalo J. 1990. In *Physical Processes in Fragmentation and Star Formation*, ed. R Capuzzi-Dolcetta C Chiosi, A Di Fazio, et al., pp. 151–76. Dordrecht: Kluwer

Shaver PA, McGee RX, Newton LN, Danks AC, Pottash SR. 1983. *MNRAS* 204:53–112

Shepherd D, Churchwell E. 1996. *Ap. J.* 472:225–39

Shepherd D, Claussen MJ, Kurtz SE. 2001. *Science* 292:1513–18

Shepherd DS, Kurtz SE. 1999. *Ap. J.* 523:690–700

Shu FH, Lizano S, Ruden SP, Nagita J. 1988. *Ap. J.* 328:L19–23

Shu FH, Nagita J, Ostriker E, Wilkin F, Ruden S, Lizano S. 1994. *Ap. J.* 429:781–96

Shu FH, Najita JR, Shang H, Li Z-Y. 2000. See Mannings et al. 2000, pp. 789–813

Simpson JP, Colgan SWJ, Rubin RH, Erickson EF, Haas MR. 1995. *Ap. J.* 444:721–38

Sodroski TJ. 1988. *The galactic large-scale far-infrared emission observed by IRAS: implications for the morphology, physical conditions, and energetics of dust in the interstellar medium*. PhD thesis, Univ. Maryland

Stahler SW, Palla F, Ho PTP. 2000. See Mannings et al. 2000, pp. 327–51

Subrahmanyan R, Ekers R, Wilson W, Goss WM, Allen D. 1993. *MNRAS* 263:868–74

Testi L, Sargent AI. 1998. *Ap. J. Lett.* 508:L91–94

Torrelles JM, Gómez JF, Garay G, Rodriguez LF, Curiel S, Cohen RJ, Ho PTP. 1998b. *Ap. J.* 509:262–69

Torrelles JM, Gómez JF, Rodriguez LF, Curiel S, Anglada G, Ho PTP. 1998a. *Ap. J.* 505:756–65

Torrelles JM, Gómez JF, Rodriguez LF, Curiel S, Ho PTP, Garay G. 1996. *Ap. J.* 457:L107–11

Torrelles JM, Gómez JF, Rodriguez LF, Ho PTP, Curiel S, Vázquez R. 1997. *Ap. J.* 489:744–52

Turner BE, Matthews HE. 1984. *Ap. J.* 277:164–80

Vacca WD, Garmany CD, Shull JM. 1996. *Ap. J.* 460:914–31

van Dishoeck EF. 2001. In *Galactic Structure, Stars and the Interstellar Medium*, ASP Conf. Ser., ed. CE Woodward, MD Bicay, JM Shull, 231:244–64

van Dishoeck EF, van der Tak FFS. 2000. In *Astrochemistry: from Molecular Clouds to Planetary Systems*, IAU Symp., ed. YC Minh, EF van Dishoeck. 197:97–112

Vázquez-Semadeni E, Ostriker EC, Passot T,

Gammie CF, Stone JM. 2000. See Mannings et al. 2000, pp. 3–28

Walmsley CM. 1995. *Rev. Mex. Astron. Astrofis. Ser. de Conf.* 1:137–48

Ward-Thompson D, André P. 1998. In *The Universe as Seen by ISO*, ed. P Cox, MF Kessler, pp. 463–66. Nooᵣdwijk, The Neth.: ESA

Ward-Thompson D, Buckley HD, Greaves JS, Holland WS, André P. 1996. *MNRAS* 281: L53–56

Ward-Thompson D, Scott PF, Hills RE, André P. 1994. *MNRAS* 268:276–90

White RL, Becker RH, Helfand DJ. 1991. *Astron. J.* 371:148–62

Williams JP. 2001. In *Massive Star Birth*, ed. P Crowther. *ASP Conf. Ser.*: In press

Williams JP, Blitz L, Mc Kee CF. 2000. See Mannings et al. 2000, pp. 97–120

Williams JP, Myers PC. 2000. *Ap. J.* 537:891–903

Wilner DJ, DePree CG, Welch WJ, Goss WM. 2001. *Ap. J. Lett.* 550:L81–85

Wolfire MG, Churchwell E. 1994. *Ap. J.* 427: 889–97

Wood DOS, Churchwell E. 1989a. *Ap. J. Suppl.* 69:831–95

Wood DOS, Churchwell E. 1989b. *Ap. J.* 340: 265–72

Zhou S, Evans NJ II, Kömpe C, Walmsley CM. 1993. *Ap. J.* 404:232–46

Zoonematkermani S, Helfand DJ, Becker RH, White RL, Perley RA. 1990. *Ap. J. Suppl.* 74:181–224

Annu. Rev. Astron. Astrophys. 2002. 40:63–101
doi: 10.1146/annurev.astro.40.060401.093818
Copyright © 2002 by Annual Reviews. All rights reserved

KUIPER BELT OBJECTS: Relics from the Accretion Disk of the Sun

Jane X. Luu

*Lincoln Laboratory, Massachusetts Institute of Technology, Lexington, Massachusetts
02420; email: luu@ll.mit.edu*

David C. Jewitt

*Institute for Astronomy, University of Hawaii, Honolulu, Hawaii 96822;
email: jewitt@ifa.hawaii.edu*

Key Words   Kuiper Belt, comets, solar system formation

■ **Abstract**   The Kuiper Belt consists of a large number of small, solid bodies in
heliocentric orbit beyond Neptune. Discovered as recently as 1992, the Kuiper Belt
objects (KBOs) are thought to hold the keys to understanding the early solar system,
as well as the origin of outer solar system objects, such as the short-period comets
and the Pluto-Charon binary. The KBOs are probably best viewed as aged relics of
the Sun's accretion disk. Dynamical structures in the Kuiper Belt provide evidence
for processes operative in the earliest days of the solar system, including a phase of
planetary migration and a clearing phase, in which substantial mass was lost from the
disk. Dust is produced to this day by collisions between KBOs. In its youth, the Kuiper
Belt may have compared to the dust rings observed now around such stars as GG Tau
and HR 4796A. This review presents the basic physical parameters of the KBOs and
makes connections with the disks observed around nearby stars.

INTRODUCTION

The perceived extent of the planetary region has changed greatly in the last few
centuries. The planets out to Saturn (at 10 AU) are visible to the naked eye and
were known to the ancients. Uranus (20 AU) and Neptune (30 AU) are fainter
and their discoveries, in 1781 and 1846, respectively, required the aid of telescopes. Pluto, whose semimajor axis is 39 AU but whose eccentric orbit brings
it to perihelion at 28 AU, was discovered photographically in 1930 following
dynamical calculations purportedly accounting for deviations in the motion of
Neptune (these deviations are now known to have been spurious). Very recently,
observations with highly sensitive charge-coupled device (CCD) detectors have
revealed a vast swarm of bodies beyond Neptune (Figure 1, see color insert),
now often referred to as the Kuiper Belt. Continued work has shown that Pluto
is simply the largest known of these trans-Neptunian objects and that its original

classification as a planet obscures its true significance as a member of the Kuiper Belt.

At the same time, the new discoveries have shown that the Solar System still retains a great many bodies that may be relatively unaltered products of accretion in the circumsolar disk. The Kuiper Belt itself is a remnant of the solar nebula, with its larger members primordial fossils that have escaped collisional disruption or ejection from the solar system. In addition, the Belt is most likely the common link between our solar system and other younger planetary systems. The Belt is sufficiently populated that its members collide with each other, and these collisions have generated dust since the Belt's earliest days. This collisionally produced dust should form a large, albeit distended, disk surrounding the planetary region, thus providing our solar system with its own circumstellar disk—much like a local analog of the dusty disks around some young and main-sequence stars. Indeed, future technology might one day detect the extrasolar analogs of Kuiper Belt bodies in these external systems.

A number of individuals speculated in print in the early twentieth century about the possible existence of objects beyond the planetary region. These included the Irish aristocrat Kenneth Edgeworth (Edgeworth 1943, 1949; see also MacFarland 1996). Musing about the origin of short-period comets, Edgeworth suggested that they might originate beyond Pluto—a hypothesis that is now widely accepted. Later, the Dutch-American astronomer Gerard Kuiper (1951) discussed a ring of small bodies beyond Pluto, the latter thought by him to be a massive planet (its mass is, in fact, 0.002 $M_\oplus$ where $M_\oplus = 6 \times 10^{24}$ kg is the mass of the Earth). He postulated that small, comet-sized bodies might have formed in the trans-Neptunian region, subsequently to be scattered to the Oort cloud by Pluto. Comets in the Oort cloud can then return to the planetary region through perturbations of passing stars (Oort 1950). In this way, Kuiper used the trans-Neptunian space to supply comets to the much more distant Oort Cloud. For reasons that remain the subject of widespread speculation, Kuiper did not refer to the prior investigations of Edgeworth. Finally, neither Edgeworth nor Kuiper cited the even earlier writing of Leonard (1930), who correctly speculated that Pluto would prove to be the first of many objects to be discovered beyond Neptune. Was the Kuiper Belt predicted? In the modern sense in which a physically based theory is required to make observationally testable predictions, the answer is clearly "no." However, several people correctly guessed that the region beyond Pluto might not be empty.

The Kuiper Belt is sometimes also referred to as the Edgeworth-Kuiper Belt in an effort to recognize the contribution of Edgeworth (perhaps the Leonard-Edgeworth-Kuiper Belt is not far behind). However, in this review, for the sake of brevity, we will use the name Kuiper Belt and refer to members of the Belt as Kuiper Belt objects or KBOs for short. Other general reviews of the KBOs may be found in Jewitt (1999), Jewitt & Luu (2000), and Schulz (2001). Reviews of specific aspects include (dynamics) Malhotra et al. (2000) and (collisions) Farinella et al. (2000). A technical Kuiper Belt book has been published (Fitzsimmons et al. 2000), as has a popular one (Davies 2001).

STRUCTURE OF THE KUIPER BELT

Figure 2 (see color insert) depicts the KBO orbits as seen from the North ecliptic pole. At the time of writing (August 2001), about 400 KBOs have been reported, the results of several telescopic surveys carried out all over the world. The interested reader is referred to the Web sites below for KBO orbits and additional information:

http://cfa-www.harvard.edu/cfa/ps/lists/TNOs.html

ftp://ftp.lowell.edu/pub/elgb/astorb.html

http://www.ifa.hawaii.edu/faculty/jewitt/kb.html

Neptune (semimajor axis $a = 30$ AU) defines the inner edge of the Kuiper Belt. Almost all the known KBOs were found beyond Neptune, most of them with semimajor axes between 30 and 50 AU, although a few, the Scattered KBOs, have semimajor axes of hundreds of AU (see below). It is also clear that the Kuiper Belt possesses a complex structure: The KBOs are not spread out uniformly in the Belt, but are clustered into subgroups, attesting to the fact that the Belt has been extensively sculpted by the gravitational influence of the giant planets (particularly Neptune). From the orbits of known KBOs, three distinct dynamical subgroups have emerged: (*a*) the Classical KBOs, (*b*) the Resonant KBOs, and (*c*) the Scattered KBOs. The details that distinguish the KBO dynamical classes can be seen in Figures 3 and 4, and are discussed below.

Classical KBOs and the Inclination Distribution

The Classical KBOs make up the majority (~two-thirds) of known KBOs, and are characterized by semimajor axes $42 < a < 48$ AU, relatively small eccentricity e ($e \sim 0.1$, Figure 3), perihelia $q > 35$ AU, and the fact that they lie outside resonances. Numerical simulations show that objects with such large semimajor axes and perihelia are relatively immune to Neptune's gravitational influence on Gyr timescales (Holman & Wisdom 1993, Duncan et al. 1995, Morbidelli et al. 1995). It was initially suspected that the Classical KBOs formed in situ and have largely kept their primordial orbits, but the unexpectedly broad inclination (*i*) and eccentricity (*e*) distributions (see Figures 3 and 4) show that the orbits have been excited since formation.

The inclination distribution of the KBOs is a measure of the velocity dispersion among these objects and gives lie to the excited state of the Kuiper Belt. Measurement of the inclination distribution is made difficult by observational bias: Most surveys to date have been conducted near the ecliptic, but the fraction of each orbit a KBO spends near the ecliptic scales roughly as the inverse of its inclination, so that high-inclination KBOs are discriminated against by these surveys. Furthermore, efforts to map the thickness of the Kuiper Belt are hindered by the Belt's relatively low off-ecliptic surface density. Detection of KBOs with $i \sim 30°$ in early ecliptic surveys already showed that the inclination distribution must be broad

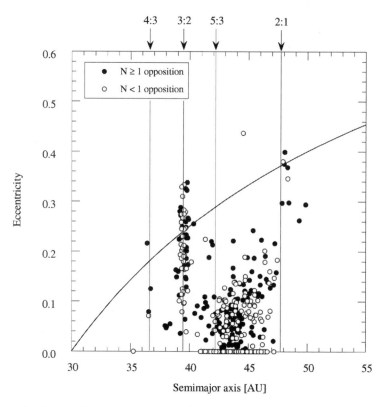

Figure 3 Semimajor axis *a* vs. eccentricity *e* for Classical and Resonant KBOs. Filled circles denote multi-opposition orbits, open circles denote orbits computed from astrometry taken within a single opposition. The curve marks perihelion $q = 30$ AU; objects above the line are Neptune-crossers. Vertical lines mark the locations of mean-motion resonances with Neptune. Orbits with $e = 0$ are assumed, not accurately measured.

(Jewitt et al. 1996), with a Half Width at Half Max (HWHM) $\Delta i \geq 20°$. Subsequently, a very large-format 12k-CCD camera (12 edge-abutted 4096×2048 pixel CCDs, Cuillandre et al. 2000) on the Canada-France-Hawaii Telescope (CFHT) made possible targeted observations at ecliptic latitudes $0°$, $10°$, and $20°$, yielding $\Delta i \sim 20°^{+6°}_{-4°}(1\sigma)$, (Trujillo et al. 2001a, Brown 2001). This thick disk is very different from the one in which the KBOs are believed to have accreted (Kenyon & Luu 1999a). The mean velocity dispersion is thus $\Delta V \approx 1.2$ km s^{-1}, which is large compared to the escape velocities of all but the largest KBOs.

A number of hypotheses have been offered to explain the relatively high e's and i's of the Classical KBOs, such as scattering by Earth-sized bodies in the early Kuiper Belt (Morbidelli & Valsecchi 1997), and partial trapping by sweeping of the 2:1 resonance (Hahn & Malhotra 1999). Alternatively, orbits of KBOs beyond

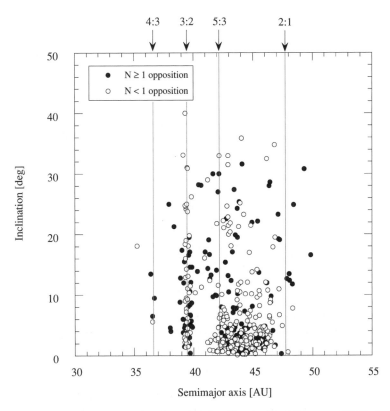

Figure 4 Inclination i vs. semimajor axis a for Classical and Resonant KBOs. Filled circles denote multi-opposition orbits, open circles denote orbits computed from astrometry taken within a single opposition. Vertical lines mark the locations of mean-motion resonances with Neptune.

42 AU may have been excited by stellar encounters with periapse on the order of 150–200 AU (Ida et al. 2000b), or by sweeping secular resonances (Nagasawa & Ida 2000). The former hypothesis also predicts that e and i rapidly increase beyond 50 AU (Figure 5), inhibiting planetesimal growth and thus producing a steep drop in the KBO number density. This seems consistent with the observed boundary of Classical KBOs at 50 AU (see "Radial Extent of the Classical Kuiper Belt" below), but this hypothesis, like the others, needs further testing. Luckily, all these models predict different distributions of e and i, so observations targeted at the region beyond 50 AU should allow meaningful discrimination between them.

Resonant KBOs

The Resonant KBOs occupy mean motion resonances with Neptune, i.e., their orbital periods and that of Neptune form a ratio of integers. Due to the resonance dynamics, the Resonant KBOs generally have larger e's and i's than the average

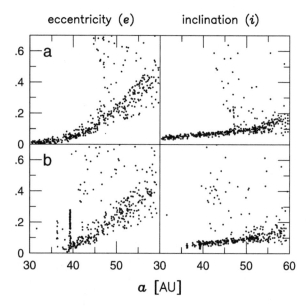

eccentricity (e)   inclination (i)

a [AU]

Figure 5 Predicted orbital distribution of KBOs: (*a*) distribution after stellar encounter (before sweeping resonances), (*b*) distribution after including sweeping resonances. Figure from Ida et al. (2000b).

Classical KBO (see Table 1). The most densely populated resonance is the 3:2, located at semimajor axis 39.4 AU (Figures 3 and 4). About 100 KBOs are known to occupy this resonance, as does Pluto. The dynamical similarities between Pluto and the other (smaller) 3:2 objects have led to the classification of these bodies as Plutinos. Some Plutinos, like Pluto itself, have perihelia $q \leq 30$ AU, i.e., their orbits cross that of Neptune. However, like Pluto, they are protected from destabilizing close encounters with this planet by their location in the resonance (Malhotra 1995), even if their immediate environs are highly unstable (Holman & Wisdom 1993, Levison & Duncan 1993, Duncan et al. 1995). But not all Plutino orbits are stable over the age of the solar system. Depending on their eccentricities, some can be driven out of the 3:2 resonance by Pluto and into close encounters with

TABLE 1   Mean orbital elements of KBOs

Dynamical group	a [AU]	$\bar{e}$	$\bar{i}$ [deg]
Classical	>42	0.09	7.11
4:3 resonance	36.4	0.22	7.81
3:2 resonance	39.4	0.36	12.87
2:1 resonance	47.8	0.14	9.83
Scattered	>30	0.49	14.08

Neptune (Yu & Tremaine 1999). The latter may represent one mechanism by which the Kuiper Belt supplies the short-period comets to the inner solar system.

The apparent fraction of Plutinos in the KBO sample is about 25%. However, this fraction is over-estimated, the result of a discovery bias in favor of Plutinos: With smaller orbits than most other KBOs, Plutinos are, on average, brighter and over-represented in flux-limited surveys. The true abundance of Plutinos is perhaps closer to 10% (Jewitt et al. 1998, Trujillo et al. 2001a). Extrapolating to the whole sky, we estimate that there are roughly 1500 Plutinos larger than 100 km (see Trujillo et al. 2001a). Other mean motion resonances also appear to be populated, notably the 2:1 at 47.8 AU and the 4:3 at 36.4 AU, although small-number statistics as yet prevent their absolute populations from being determined.

How might the mean motion resonances have become populated? According to Malhotra (1993, 1995), the answer lies in the radial migration of the giant planets. It has long been suspected that, during the final stages of planet formation, the giant planets migrated radially due to their exchange of angular momentum with the planetesimals that they scattered (Fernandez & Ip 1984). In this scenario, exchange of angular momentum among the gas giant planets during the disk-clearing phase led to a net outward migration of Saturn, Uranus, and Neptune, at the expense of a slight inward migration of massive Jupiter (Fernandez & Ip 1984). As Neptune moved outwards, its mean motion resonances swept ahead of it through the planetesimal disk and could have captured Pluto, along with the Plutinos, into the 3:2 resonance. The other mean motion resonances would also have captured objects from the planetesimal disk, although usually with a lower efficiency (Malhotra 1995).

The magnitude of the radial migration of Neptune can be estimated from the eccentricity of the trapped objects, via

$$ e_f^2 = e_i^2 + \left(\frac{1}{j+1} \right) \ln \left(\frac{a_{N,f}}{a_{N,i}} \right), \tag{1} $$

where e_i, e_f are the initial and final eccentricities before and after trapping in the resonance, j is an integer ($j = 2$ for the 3:2 resonance), and $a_{N,i}$ and $a_{N,f}$ are the initial and final semimajor axes of Neptune (Malhotra 1995). Assuming $e_i = 0$, $a_{N,f} = 30$ AU and noting that the most eccentric Plutinos have $e = 0.33$ (Figure 3), we solve Equation 1 to find $a_{N,i} = 21.6$ AU, corresponding a total outward migration $\Delta a \approx 8$ AU.

In principle, the relative populations of different resonances can be predicted directly by the capture theory. Malhotra (1995) predicted that the populations of the 3:2 and 2:1 resonances should be of the same order. (We regard this prediction as noteworthy because this one was made before the key observations were in hand to test it).

However, the capture model in its initial form is highly simplistic—dynamical effects of collisions are ignored and the migration of Neptune is taken to be smooth and continuous—and depends on several unknown parameters such as the time-dependent mass of growing Neptune and the form of its migration rate (Malhotra 1995, Hahn & Malhotra 1999). Pointing out that the resonant capture theory overestimated the 2:1 resonance population, Ida et al. (2000a) suggested that capture into the 2:1 resonance required much longer Neptune migration timescales, and explored mechanisms that might result in such long timescales. They proposed two mechanisms, involving (*a*) the tidal effects of the solar nebula on Neptune's orbit, and (*b*) Neptune's close scatterings of planetesimals. Friedland (2001) proposed yet another mechanism that would inhibit capture into the 2:1 resonance: the effect of the Sun's rotation around the center of mass. The relative resonance populations determined by carefully debiased surveys might provide a method to measure the migration timescale of Neptune, and perhaps also estimate the total mass of planetesimals scattered, but the results may not be as clear-cut as was initially hoped.

Scattered KBOs

The third dynamical class in the Kuiper Belt, the Scattered KBOs (SKBOs), was established in October 1996 with the discovery of 1996 TL_{66} (Luu et al. 1997). SKBOs are distinguished by their large, highly eccentric and highly inclined orbits (Figure 6). Evidently, SKBOs represent a separate type of trans-Neptunian

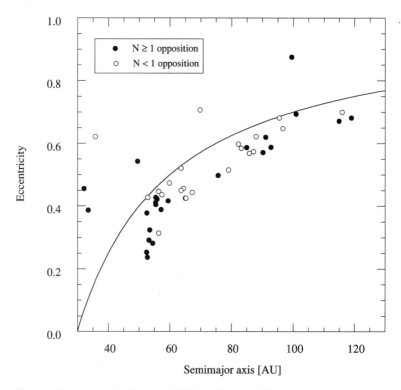

Figure 6 *a* vs. *e* for Scattered KBOs. Same as Figure 2, but on a larger scale to show the Scattered KBOS.

dynamics, and provide the first direct evidence that the trans-Neptunian population extends much further than the $\sim$50 AU region sampled by previous surveys. A crude estimate of the SKBO population scattered by Neptune (with diameter $D \geq 100$ km) is $3.1^{+1.9}_{-1.3} \times 10^4$ (Trujillo et al. 2000), which is comparable to the number of the Classical KBOs. The population of SKBOs with perihelion too large to permit Neptune scattering could potentially be much larger.

The SKBOs are hypothesized to be the population of scattered planetesimals discussed earlier by Ip (1989), Torbett (1989), and Ip & Fernandez (1991). In this scenario, during the late stages of planet formation, Uranus and Neptune scattered many of the nearby planetesimals to large distances, subsequently to form the Oort cloud. Planetesimals scattered to more modest distances formed the Scattered Kuiper Belt and remained weakly interacting with Neptune (Duncan & Levison 1997). The recent discovery of 2000 CR$_{105}$, a SKBO with perihelion ($q \sim 44$) outside the domain controlled by Neptune, raises the possibility that some SKBO orbits are not the result of direct gravitational scattering by this giant planet (B. Gladman et al., unpublished manuscript). A much larger SKBO sample is needed before firm conclusions can be drawn about their origin.

All the dynamical groups in the Kuiper Belt suffer from observational biases, but to different degrees. For example, the Plutinos are favored over the 2:1 resonant objects due to their smaller heliocentric distances, and the SKBOs are discriminated against most of all because of their very large orbits. The current best estimates of the ratios of the (bias-corrected) populations of various dynamical groups in the Kuiper Belt are as follows (Trujillo et al. 2001a):

Classical : Scattered : Plutino: 2:1 Resonant = 1.0 : 0.8 : 0.4 : 0.07

POPULATION CHARACTERISTICS

Surface Density

Published ground-based surveys of KBOs include Tombaugh (1961), Luu & Jewitt (1988, 1998), Kowal (1989), Levison & Duncan (1990), Irwin et al. (1995), Jewitt et al. (1996, 1998), Gladman et al. (1998), Trujillo & Jewitt (1998), Chiang & Brown (1999), Sheppard et al. (2000), Larsen et al. (2001), Gladman et al. (2001), and Trujillo et al. (2001a). The only survey carried out from space, using the Hubble Space Telescope at red magnitude 28.1 (Cochran et al. 1995) yielded controversial results (Brown et al. 1997, Cochran et al. 1998) and will be left out of this review.

The number most directly measured by all surveys is the cumulative surface density of KBOs, which is the number of KBOs per square degree brighter than a given apparent magnitude (Figure 7). Poisson statistics are usually assumed, and are roughly equal to Gaussian statistics for all data points representing more than a few detections. From these measurements one can calculate the cumulative

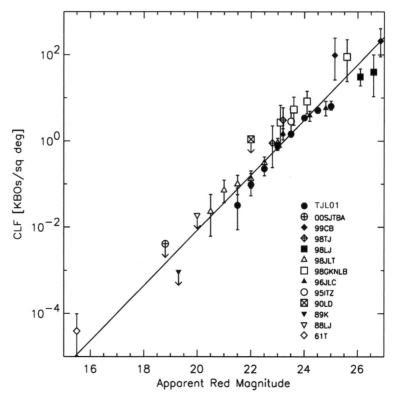

Figure 7   The cumulative surface density of KBOs. The data show the number of KBOs per square degree brighter than a given apparent red magnitude, as measured by various surveys. The legends are as follows: TJL01 = Trujillo et al. (2001a), 00SJTBA = Sheppard et al. (2000), 99CB = Chiang & Brown (1999), 98TJ = Trujillo & Jewitt (1998), 98LJ = Luu & Jewitt (1998), 98JLT = Jewitt et al. (1998), 98GKNLB = Gladman et al. (1998), 96JLC = Jewitt et al. (1996), 95ITZ = Irwin et al. (1995), 90LD = Levison & Duncan (1990), 89K = Kowal (1989), 88LJ = Luu & Jewitt (1988), T61 = Tombaugh (1961). Figure adapted from Trujillo et al. (2001a).

luminosity function (CLF) of the Kuiper Belt by fitting the surface density Σ with the power law

$$\log[\Sigma(m_R)] = \alpha(m_R - m_0), \qquad (2)$$

where m_R is the red apparent magnitude, m_0 is the (red) magnitude at which the surface density $\Sigma = 1\,\mathrm{deg}^{-2}$, and 10^α gives the slope of the CLF. However, since the CLF at the faint end is highly correlated with that at the bright end, fitting the CLF would count the bright bodies more than once and thus underestimate the number of faint bodies. Arguably a more meaningful fit can be obtained by fitting not the CLF, but the differential luminosity function, which represents the number of bodies in each magnitude bin. Fitting data from our latest surveys, which span the magnitude range $m_R = 21$–26 (Trujillo et al. 2001a), yields $\alpha = 0.64^{+0.11}_{-0.10}$ and $m_0 = 23.23^{+0.15}_{-0.20}$. In other words, there is 1 KBO per square degree at red magnitude $m_R \approx 23.2$, increasing by $\sim$4 per magnitude. This result is consistent with other works (see Table 2). The luminosity function at $m_R < 20$ and $m_R > 26$ may be different, and remains to be assessed.

The luminosity function at $m_R < 20$ is ill constrained due to the low surface density of bright KBOs. In spite of their scarcity, recent years have seen surveys targeted at large KBOs, such as Sheppard et al. (2000; detection limit $m_R = 18.8$), Larsen et al. (2001; detection limit $m_R \sim 21.5$), Trujillo et al. (2001b; detection limit $m_R = 21.1$), and the QUEST survey (Ferrin et al. 2001; detection limit $m_R = 19.6$). These surveys are adding valuable data to the bright end of the KBO size distribution (Figure 7), and rule out the previously reported limit of 1 KBO brighter than $m_R \sim 19.5$ per thousand square degrees of Kowal (1989).

Size Distribution

The KBOs are usually assumed to obey a differential power law size distribution of the form $n(r)dr = \Gamma r^{-q}dr$, where $n(r)\,dr$ is the number of objects in the radius range r to $r + dr$, and Γ and q are constants. The best-fit value of q (Table 2) can be determined by means of a maximum likelihood simulation of the survey, which corrects for the effect of heliocentric distance R, geocentric distance Δ, and the detection efficiency. Doing so yields (Trujillo et al. 2001a)

$$q = 4.0^{+0.6}_{-0.5}(1\sigma). \tag{3}$$

Table 2 shows that various survey results are in general agreement with each other. The Classical Belt thus contains $N(r > 1\ \mathrm{km}) \sim 10^{10}$, $N(r > 50\ \mathrm{km}) \sim 3 \times 10^4$, and $N(r > 1000\ \mathrm{km}) \sim 10$ KBOs. There is more than enough to supply the Jupiter-family comets (see "Relationship to Other Solar System Bodies" below) (Levison

TABLE 2   Selected size distribution measurements of KBOs

Discovery m_R	Number of KBOs found	q	Reference
21.1–24.6	86	$4.0^{+0.6}_{-0.5}$	Trujillo et al. (2001a)
23.4–26.0	17	4.4 ± 0.3	Gladman et al. (2001)
25.5–27.2[a]	2	3.6 ± 0.1	Chiang & Brown (1999)[b]
23.8–26.7	6	3.7 ± 0.2	Luu & Jewitt (1998)[b]
23.0–25.8	5	$4.8^{+0.5}_{-0.6}$	Gladman et al. (1998)[b]
20.6–23.0	13	4.0 ± 0.5	Jewitt et al. (1998)[b]

[a]V magnitude.
[b]Extrapolated from CLF slope α, according to $q = 5\alpha + 1$.

& Duncan 1994, Duncan & Levison 1997), and up to 10 Pluto-size bodies—all, presumably, much more distant than Pluto.

Total Mass

The total mass of the Kuiper Belt can be estimated by integrating over the size distribution, between a minimum radius r_{min} and a maximum radius r_{max}. Current observations are far from probing the small end of the Kuiper Belt size distribution, so r_{min} remains unconstrained, but the best-fitting r_{max} can be determined by fitting Monte Carlo simulations of the KBO population to the data. Existing data rule out $r_{max} \leq 250$ km at the 3σ level, and can be fitted with $r_{max} \geq 500$ km (Figure 8, see Trujillo et al. 2001b). Formally, however, there is no evidence for a maximum-size

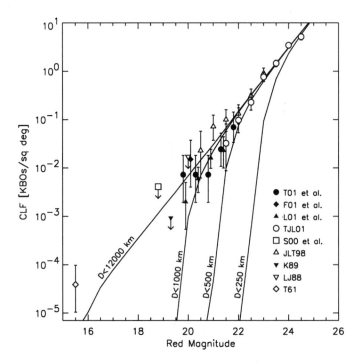

Figure 8   The observed cumulative luminosity function (CLF) for KBOs, adopting the $q = 4.2$ size distribution. The legend is as follows: T01 et al. = Trujillo et al. (2001b), F01 et al. = Ferrin et al. (2001), L01 et al. = Larsen et al. (2001), TJL01 = Trujillo et al. (2001a), S00 et al. = Sheppard et al. (2000) (upper limit), JLT98 = Jewitt et al. (1998), K89 = Kowal (1989) (upper limit), LJ88 = Luu & Jewitt (1988) (upper limit), T61 = Tombaugh (1961). The lines correspond to model KBO populations with (left to right) $r_{max} = 6000$ km, 500 km, 250 km, and 125 km. The data rule out $r_{max} \leq 250$ km, but cannot distinguish between $r_{max} = 500$ km and larger. Figure adapted from Trujillo et al. (2001b).

cutoff; if such a cutoff exists, it must be ≥ 1000 km. We note that $r_{max} = 1000$ km would be consistent with Pluto ($r = 1100$ km) being the largest KBO known, but reiterate that a larger r_{max} cannot be ruled out.

The total mass of a $q = 4$ distribution is

$$M = \frac{4 \times 10^9 \pi \rho \Gamma}{3} \left(\frac{0.04}{p_R} \right)^{3/2} \ln \left(\frac{r_{max}}{r_{min}} \right), \tag{4}$$

where ρ [kg m^{-3}] is the bulk density and p_R is the red geometric albedo. The bulk density of KBOs is unknown, but can be guessed from related solar system objects. KBOs are believed to be precursors of short-period comets, whose nuclei have $\rho = 500$–1000 kg m^{-3} (Rickman et al. 1987). At the other end, the (gravitationally self-compressed) density of Pluto and Charon—the two largest known KBOs—is $\rho \sim 2000$ kg m^{-3}. Adopting the plausible values $r_{min} = 1$ km, $r_{max} = 1000$ km, $\rho = 1000$ kg m^{-3} (Jewitt & Sheppard 2002) albedo $p_R = 0.07$ (as measured for the KBO 20,000 Varuna, Jewitt et al. 2001; see "Albedos" below) yields a total mass $M = 0.08 \ M_{\oplus}$.

The 7% albedo of 20,000 Varuna may or may not be representative of the KBO population. Size deduced from optical data scales as $p_R^{-1/2}$ and mass as $P_R^{-3/2}$, so higher albedos would lower the mass estimate above (for example, a factor-of-10 increase in the albedo results in a factor-of-30 decrease in mass). While we think it very likely that the KBO albedos are generally close to the $p_R = 0.07$ assumed, it will be important to secure measurements of p_R for a large sample of KBOs when possible [e.g., with the Space Infrared Telescope Facility (SIRTF)].

The optical mass is consistent with earlier dynamical limits of about 1 $M_{\oplus}$ (Hamid et al. 1968, Hogg et al. 1991), but is much smaller than would be expected from extrapolating the surface mass density of the solar system. In particular, this mass is far too small to form the larger KBOs within a $\sim 10^8$ year timescale imposed by the nearby accretion of Neptune (Lissauer et al. 1996) and implies a much more massive original Kuiper Belt (see "Formation of the Kuiper Belt" above). Accretion models predict that the original Kuiper Belt was roughly 100 times more massive than the current Belt (Kenyon & Luu 1999a). Scattering by Neptune may have helped to deplete the original Belt, but could not be the major culprit (Holman & Wisdom 1993). Collisional grinding might have been an effective mass loss mechanism (Stern 1996, Kenyon & Luu 1999a). However, KBOs larger than ~ 100 km are largely immune to collisional disruption, and should be original remnants from the earliest days of the solar system.

RADIAL EXTENT OF THE CLASSICAL KUIPER BELT

The apparent brightness of a given KBO varies as R^{-4}, where R is the heliocentric distance, imposing a daunting bias against the detection of very distant KBOs. Nevertheless, several independent survey results are consistent with the KBO surface density of Classical KBOs dropping steeply near 50 AU (Jewitt et al. 1996,

1998; Dones 1997; Allen et al. 2001; Trujillo et al. 2001a). Trujillo & Brown (2001) narrowed the location of the edge to 47 ± 1 AU, assuming a differential size distribution with power law index $q = 4$ and geometric albedo 0.04. This edge persists even for other size distributions ($3 < q < 5$) and other albedo assumptions. Gladman et al. (1998) argued that a steep power law size distribution could produce an apparent outer edge to the belt, but the data already showed that sufficiently steep distributions ($q \sim 5$) were not viable (Table 2). Gladman et al. (2001) also cited recent discoveries of KBOs beyond 48 AU to argue against the density drop-off distances, although they themselves admitted that it was not clear whether these distant KBOs were members of the Classical Kuiper Belt. In short, the reliable data convincingly show that $\sim$50 AU marks the edge of the Classical Kuiper Belt. (The SKBOs are excluded, even though they have large e's and a's that carry them as far as several hundred AUs at aphelion.)

What could be the cause of this unexpected outer boundary to the Classical Kuiper Belt? Suggested explanations include (Jewitt et al. 1998)

■ systematically lower albedos of KBOs beyond 50 AU, rendering more distant objects harder to detect. This explanation is ad hoc, and has no clear physical foundation. Indeed, one might naively expect that the albedos of more distant KBOs of a given size should be higher, owing to the increased stability of ice at larger distances and lower temperatures.

■ a decrease in the maximum size of KBOs (r_{max}) with increasing semimajor axis. Since all published surveys are magnitude-limited, only the largest KBOs can be detected at larger distances, and a decrease in r_{max} with increasing semi-major axis would give rise to an apparent edge to the KBO distribution. The rate of growth of the largest body of mass M in a swarm of planetesimals is (Wetherill 1980):

$$\frac{dM}{dt} = \pi r^2 \rho_0 v \left(1 + 2\theta\right), \tag{5}$$

where ρ_0 is the mean density of matter in the swarm and v is the mean velocity of the swarm. The term $(1 + 2\theta)$ represents the gravitational enhancement of the collisional cross section:

$$(1 + 2\theta) = 1 + \left(v_e^2 / v^2\right), \tag{6}$$

where v_e is the escape velocity; θ is often called the Safronov number and has values 2–5 for most situations in planetary growth (Safronov 1969). In terms of the body's radius r and density ρ, Equation 5 becomes

$$\frac{dr}{dt} = \frac{\rho_0 v}{4\rho}(1 + 2\theta). \tag{7}$$

The swarm density ρ_0 is related to the surface density σ_0 by (Safronov 1969)

$$\rho_0 = \frac{4\sigma_0}{vP}, \tag{8}$$

where P is the body's orbital period. In a nebula whose surface density σ_0 scales with distance as R^{-2}, the growth timescale τ is thus

$$\tau = r/\dot{r} \propto \frac{r\rho R^{3.5}}{(1 + 2\theta)}.$$ (9)

In a given time, the size of the largest objects grown by accretion thus should scale as $R^{-3.5}$. All other things being equal, the largest objects at 50 AU should be smaller than those at 40 AU by a factor $(40/50)^{3.5} = 0.46$. This is too small a factor to account for the deficiency of objects discovered at $R \geq 50$ AU. So, while the sense of this effect is correct, the magnitude appears too small to account for the observations. In support of this conclusion we note that there is no significant trend of maximum size with heliocentric distance across the inner Kuiper Belt (from 30 AU to 50 AU).

■ the steep density drop-off in the Kuiper Belt near 50 AU corresponding to a real edge. One possibility that has been numerically explored is tidal truncation by a passing star (Ida et al. 2000b). According to these simulations, a passing star with periapse at 150–200 AU would truncate the disk at 50 AU, as is observed. Ida et al. further claimed that the resonant KBOs at 39 AU could survive such a stellar close encounter. The likelihood of a close stellar passage in the present environment of the Sun is extremely remote, but if the Sun were born in a cluster, the high stellar densities and short encounter times might render this scenario more plausible (Eggers et al. 1997, Fernandez & Brunini 2000). We note that the smallest measured Orion proplyds have sizes comparable to the inferred diameter of the KBO belt, perhaps because they were tidally truncated (McCaughrean & O'Dell 1996). The contamination of the protoplanetary disk by ejecta from nearby exploding supernovae provides independent evidence that the Sun was born in a cluster containing perhaps 2000 stars (Adams & Laughlin 2001).

PHYSICAL PROPERTIES

Albedos

The only reliable KBO albedos available are those of Pluto (0.5–0.7, Tholen & Buie 1997), Charon (~0.4, Tholen & Buie 1997), and 20,000 Varuna (0.07, Jewitt et al. 2001). The large albedo of Pluto is produced by the interaction between its surface and its atmosphere, whereby gases in the atmosphere condense onto the surface as frost and give rise to a large albedo. By that argument, the albedos of those KBOs too small to sustain atmospheres are not expected to be high. This is supported by indirect evidence provided by the Centaurs and the known comet nuclei (Figure 9). Centaurs are objects on unstable, planet-crossing orbits between Jupiter and Neptune, and their transient nature suggests that they are ex-KBOs that have been scattered out of the Kuiper Belt (see "Relationship to Other Solar

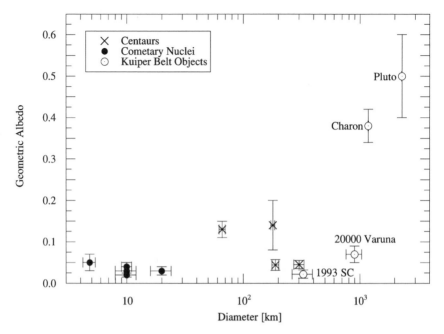

Figure 9   Plot of albedos vs. diameter for comet nuclei, Centaurs, Pluto-Charon, and the KBOs 20000 Varuna and 1993 SC. Figure adapted from Jewitt et al. (2001).

System Bodies" below). Figure 9 suggests a trend towards higher albedos at larger sizes: The small cometary nuclei are the darkest, whereas Pluto and Charon are the most reflective. Varuna, with a radius 40% that of Pluto, lies between these two extremes because its gravity is insufficient to retain an atmosphere. However, given the fact that water ice has been detected on several KBOs (see "Spectral Properties" below), a patchy frost distribution on the surface of Varuna cannot be excluded. An initial search for CO outgassing from KBOs has proved negative (Bockelee-Morvan et al. 2001).

Colors

The colors of KBOs provide a first-order measure of surface composition, albeit compounded with effects caused by size-dependent scattering from regolith particles (Moroz et al. 1998). The optical colors are the most easily measured: colors of about 3 dozen KBOs are available (Luu & Jewitt 1996; Tegler & Romanishin 1998, 2000; Green et al. 1997; Barucci et al. 2000; Jewitt & Luu 2001). Figure 10 shows that KBO colors are widely distributed, with $0.7 \leq B\text{-}V \leq 1.2$ and $0.4 \leq V\text{-}R \leq 0.8$ (for comparison, solar colors are $B\text{-}V = 0.67$, and $V\text{-}R = 0.36$). The B-V, V-R, and R-I colors are strongly correlated in a statistical sense: Objects that are red at 4500 Å are also red at 8000 Å. J-band photometry

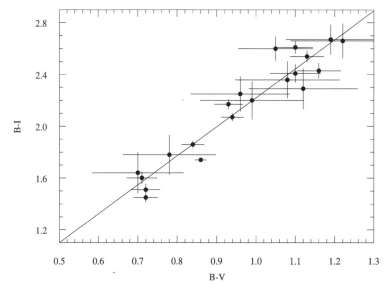

Figure 10   B-V vs. B-I for KBOs (with $\sigma_{B-V} < 0.15$ mag) observed by Jewitt & Luu (2001). The straight line is a least-squares fit to guide the eye. Figure from Jewitt & Luu (2001).

available is also available for a dozen KBOs, showing that this correlation extends into the infrared, as can be seen in Figure 11. These correlations suggest that a single agent is responsible for the reddened colors from optical to infrared wavelengths; this property may help identify the agent among the low-albedo minerals with similar colors.

The color diversity of the KBOs (Luu & Jewitt 1996) came as a surprise since it was expected that all KBOs should suffer from long-term irradiation, and should all sport irradiation mantles (e.g., Strazzulla & Johnson 1991). When carbon-containing ices are irradiated by high energy particles prevalent in interplanetary space such as cosmic rays and solar wind particles, a solid residue is created that is rich in complex carbon compounds and reddish in color. The penetration depths of GeV protons in solid ice are near 1 m (Strazzulla & Johnson 1991). An irradiation mantle of this thickness should grow on timescales $\sim 10^{8\pm1}$ yr (Shul'man 1972). After 4.5 billion years, all KBOs are thus expected to sport a red, nonvolatile, organic-rich crust, but this is clearly not the case. Could different dynamics (see "Structure of the Kuiper Belt" above) or physical properties be responsible for the color diversity? Jewitt & Luu (2001) searched for and found no correlation between the optical colors and absolute magnitude or any orbital parameter (heliocentric distance, semimajor axis, etc.).

The origin of color diversity could reflect intrinsic compositional differences among the KBOs, or it could be a result of collisional resurfacing (Luu & Jewitt

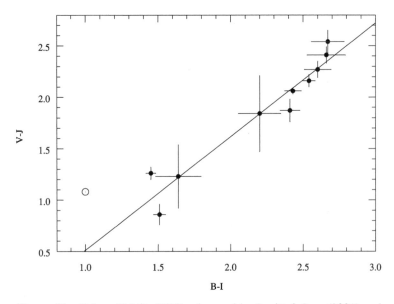

Figure 11   B-I vs. V-J for KBOs observed by Jewitt & Luu (2001) and Davies et al. (2000). The straight line is a least-squares fit to guide the eye. The color of the Sun is marked. Figure from Jewitt & Luu (2001).

1996). In the latter case, the color diversity would result from competition between two counteracting processes: irradiation reddening and collisional resurfacing. Occasional strong collisions may shower the surface with impact debris which would be unirradiated and possibly ice-rich. If this freshly excavated material were of a different composition and color from the irradiation mantle, the net result would be a resetting of the color by each large impact. Depending on how much of the surface is covered with impact debris, resurfacing may change the overall surface color. However, this resurfacing model works well only when (*a*) the irradiated and buried materials have quite different colors, and (*b*) the timescale for collisional resurfacing is comparable to the timescale for irradiation damage of the exposed surface material.

The resurfacing scenario makes several predictions. First, since the collision timescale depends on the cross-section of a KBO, the globally averaged color should also be related to KBO size. This is not observed in the currently available data. Second, many KBOs should display rotational color/compositional variations, due to incomplete blanketing of their surfaces by impact excavated matter. However, the general agreement between colors measured by different investigators at random rotational phases suggests that color anisotropy is rare (see Table 5 of Jewitt & Luu 2001). Detailed measurements of Centaur 8405 Asbolus (Kern et al. 2000), on the other hand, show evidence for azimuthal compositional variations that are consistent with resurfacing. Recently, Trujillo & Brown (2002a) reported a significant correlation between color and inclination among the Classical KBOs but not among the Plutinos. No correlation was found between color and any other orbital parameter. A fair statement would be that, although the resurfacing hypothesis is not yet ruled out by the available data, there is presently no compelling support for it. The correct explanation for the KBO color distribution may involve a complex combination of both evolutionary and cosmogonic effects.

Tegler & Romanishin (1998, 2000) have reported that the distribution of KBO colors was bimodal. If correct, this finding would have important implications for the origin of color diversity in the Kuiper Belt. However, bimodality has not been independently confirmed (Green et al. 1997, Barucci et al. 2000, Davies et al. 2000), and was not borne out by statistical tests applied to Tegler & Romanishin's 1998 and 2000 data sets (either separately or combined, Jewitt & Luu 2001). One can thus conclude that evidence for a bimodal color distribution is lacking, and that initial indications to the contrary result from a pathologically small sample.

The V-R color histograms of the KBOs are compared with those of other small solar system bodies in Figure 12. The KBOs have a large color range that is matched only by the Centaurs, which are believed to have originated in the Kuiper Belt (see "Relationship to Other Solar System Bodies" below). In particular, the most extreme red colors (V-$R > 0.7$) belong to the KBOs and the Centaurs. The facts that KBOs and Centaurs have similar color distributions, and that their color distributions are distinct from those of known asteroids and comet nuclei, suggest 1. a common origin for KBOs and Centaurs, and 2. the presence of indigenous primitive material on the surface of KBOs and Centaurs that may correspond to the predicted irradiation mantle (Luu & Jewitt 1996). As shown in laboratory experiments, the mantle thickness depends on the actual irradiation flux and may be sufficiently thick (~ 1 m) to survive a Centaur's first entry into the inner solar system, but this primitive surface is gradually lost/modified once KBOs leave the Kuiper Belt and join the comet population (Jewitt 2002). It is of interest that the extreme red color as well as the near-IR spectrum of the Centaur 5145 Pholus have been reproduced by complex hydrocarbons (Wilson et al. 1994, Cruikshank et al. 1998).

Spectral Properties

Few high quality spectra of KBOs exist and even fewer spectral features have been reported. KBOs with reported features include 1996 TO_{66} (Brown et al. 1999), 20,000 Varuna (Licandro et al. 2001), and 1999 DE_9 (Jewitt & Luu 2001). The first two show the familiar 1.5 and 2 μm features of water ice, but the DE_9 spectrum appears more complex. Figure 13 shows the 1999 DE_9 spectrum, with apparent absorption features near 1.4 μm, 1.6 μm, and 2 μm, and a weak absorption feature at ~ 2.25 μm. The reflectivity also drops shortward of 1.4 μm down to the limit of the spectral coverage at 1.0 μm. Water ice has well-known diagnostic features at 1.55 μm, 1.65 μm, and 2.02 μm (Clark 1981a), and in the case of fine grain frost, the 1.55 μm and 1.65 μm features become blended and weaker

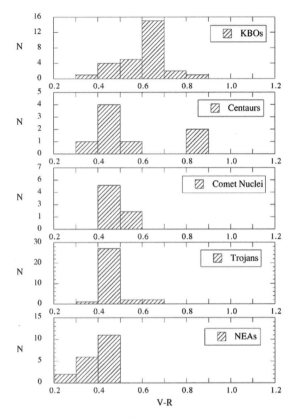

Figure 12 Comparison of Kuiper Belt V-R colors with those of other small solar system bodies: near-Earth asteroids (NEAs), Trojan asteroids, comet nuclei. Figure adapted from Luu & Jewitt (1996).

than the 2 μm feature (Clark 1981a), much like the DE_9 spectrum. The water ice high-overtone band at 1.25 μm is not seen in the DE_9 spectrum, and this is possibly a consequence of the presence of other minerals in addition to water ice (Clark 1981a). The weakness of the water ice features suggests a low abundance of water; the DE_9 spectrum is very similar to that of a mixture of Mauna Kea red cinder and 1% water ice (Clark 1981b), also shown in Figure 13.

The features at 1.4 μm and 2.25 μm are usually diagnostic of the presence of the OH stretch (Hunt 1977), and the exact location of the 2.25 μm feature can be diagnostic of the compound (either Al or Mg) associated with the OH stretch. However, the quality of the DE_9 spectrum is not good enough for this identification. The spectral drop toward 1 μm may be indicative of an absorption feature, more specifically, the 1 μm feature due to the ferrous ion (Fe^{2+}) in sixfold coordination, a feature well known from spectra of silicates such as olivine and

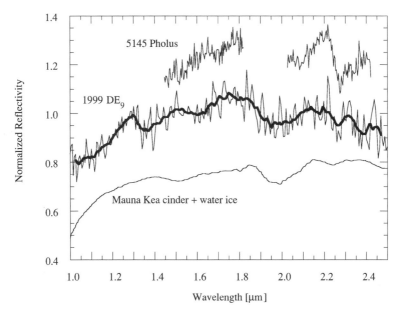

Figure 13   Reflection spectrum of KBO 1999 DE$_9$, compared with Mauka Kea cinder +H$_2$O ice, and the Centaur 5145 Pholus. The thick line through DE$_9$ is a running mean added to guide the eye. DE$_9$ shows absorption features due to water ice (2 μm), silicates (1 μm), and the OH stretch (1.4 and 2.25 μm). Figure adapted from Jewitt & Luu (2001).

pyroxene (Hunt 1977). We rule out pyroxene as a possible match for DE$_9$ because pyroxene usually shows a strong feature near 1.8 μm which is not seen in DE$_9$. The olivine forsterite (Mg$_2$SiO$_4$) has been found in comets (Crovisier et al. 2000), and would fit the 1 μm absorption in DE$_9$'s spectrum (Figure 13), if the feature is confirmed.

Other remarkably featureless KBO spectra exist, e.g., 2000 EB$_{173}$ (Brown et al. 2000), 1996 TL66 (Jewitt & Luu 1998), and 1993 SC (Luu & Jewitt 1996). Based on this handful of results, it appears that KBO spectra (thus far) may be roughly divided into at least two groups: one characterized by nearly featureless neutral/bluish spectra, and the other by spectral features indicating the presence of water ice and/or hydroxyl. Brown et al. (2000) attempt to link the presence of the water ice features to the albedo, but this is probably overly naive since laboratory experiments have indicated that water ice features can be very apparent even when the albedo is low (Clark 1981b). Luu et al. (2000) speculate that water ice is ubiquitous on KBOs but may not always be detectable due to its low abundance or high degree of contamination. The spectral diversity exhibited by the few well observed KBOs thus seems to mimic the optical color diversity.

Rotation

The rotation states of KBOs, like those of asteroids, can be determined from time-resolved measurements of the scattered light. The larger KBOs are thought to be primary products of accretion and, as such, may preserve the angular momentum they acquired at formation. Measurements of the distribution of angular momenta are thus of great interest as a potential constraint on the physics of accretion. By contrast, the smaller KBOs and almost all main-belt asteroids have been heavily processed by collisions, and their current spin states and angular momenta are unlikely to be primordial.

Unfortunately, the rotational states of most KBOs are, as yet, poorly defined. Romanishin & Tegler (1999) reported lightcurve variations for 1995 QY_9, 1994 VK_8, and 1994 TB, and reported a correlation between lightcurve amplitude and absolute magnitude (a proxy for size). However, Collander-Brown et al. (1999) independently observed 1994 VK_8 but detected no clear rotational signature. Hainaut et al. (2000) reported a 6.25 hr periodicity in 1996 TO_{66} and an amplitude varying from 0.12 to 0.33 mag, but the latter result is contradicted by Romanishin & Tegler (1999) who set a 0.1 mag upper limit to the amplitude. One of the few convincing and remarkable lightcurves is that of (20,000) Varuna (Figure 14), which has a double-peak period of 6.34 hr and amplitude 0.42 mag (Jewitt & Sheppard 2002). The rapid rotation and large photometric range of Varuna suggests a body

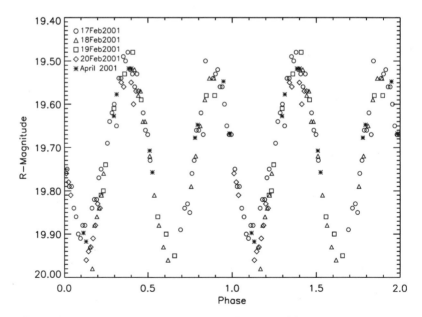

Figure 14   Phased lightcurve (in R band) of KBO (20000) Varuna, assuming a double-peak rotation period $P_{rot} = 6.3436$ hours. Error bars are 0.02 mag (not plotted for the sake of clarity). Figure from Jewitt & Sheppard (2002).

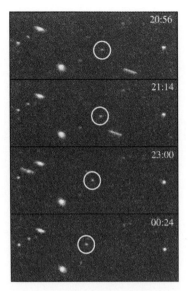

Figure 1 Discovery images of the first Kuiper Belt object 1992 QB$_1$, taken on 1992 August 30, at the University of Hawaii 2.2 m telescope. The four CCD images (UT time indicated in the upper right hand corner) show QB$_1$ (circled) slowly moving left against the background stars and galaxies. The elongated trail in the top 3 images is an asteroid (note the slow motion of QB$_1$ compared to the motion of the asteroid). Figure from Jewitt & Luu (1993).

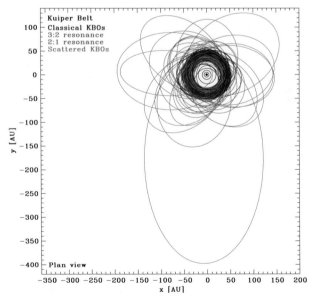

Figure 2 Plan view of the Kuiper Belt (as seen from the North ecliptic pole) showing the orbits of KBOs as of June 2001. The orbits are color-coded according to the legend: *blue* = Classical KBOs, *red* = 3:2 resonance, *green* = 2:1 resonance, *pink* = Scattered KBOs. Innermost circle is Jupiter's orbit. The region shown is 560 × 560 AU.

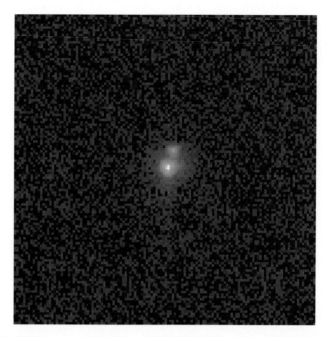

Figure 15 CCD image of the binary KBO 1999 TC$_{36}$ taken with the Hubble Space Telescope, showing the binary and secondary clearly separated by ~ 0.37 arcsec. Image courtesy of C. Trujillo (Caltech).

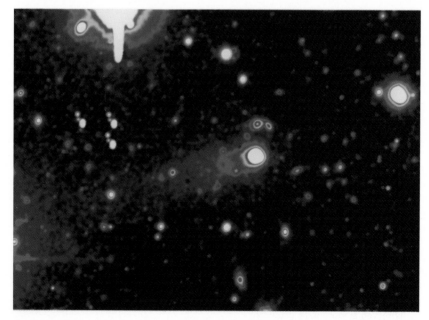

Figure 16 Centaur 2060 Chiron, with cometary tail (pointing toward lower left hand corner). Image is ~ 2 arcmin wide.

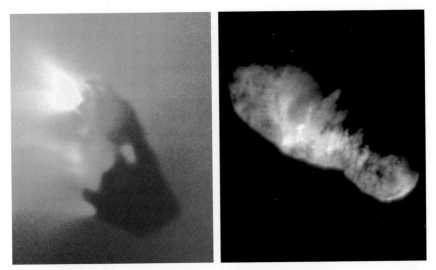

Figure 17 (*a*) The 16 × 8 × 8 km nucleus of comet Halley. Photo taken by the Halley Multicolour Camera onboard ESA's Giotto spacecraft (copyright Max-Planck-Institut fuer Aeronomie, Lindau/Harz, Germany). Figure courtesy of Dr. H. U. Keller. (*b*) The 8 km-long nucleus of comet Borrelly. Photo taken by the Deep Space 1 spacecraft in September 2001. Figure courtesy of NASA/JPL.

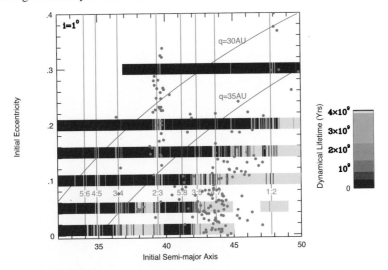

Figure 20 Dynamical lifetimes in the Kuiper Belt. Each narrow vertical strip at a given semimajor axis *a* and eccentricity *e* represents the lifetime of a test particle with the given initial *a* and *e* (initial inclination *i* = 1°). The strips are color-coded according to the lifetimes. *Yellow* denotes objects that survive for 4 Gyr, and progressively darker regions denote progressively shorter dynamical lifetimes. The *green circles* represent known KBOs observed at more than 1 opposition (as of 2001 Aug 1). The long *blue vertical lines* mark the locations of the important Neptune mean motion resonances. The *red curves* denote constant perihelion distance *q*. Figure adapted from Duncan et al. (1995).

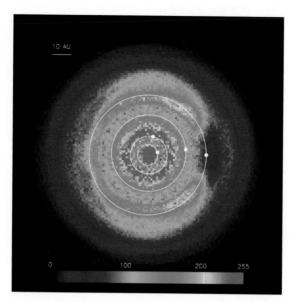

Figure 22 The calculated brightness of 23 μm dust in the solar system (face-on view) in false colors. The orbits and positions of the four giant planets are shown in *yellow*. Colors correspond to the number of dust particles per AU^2 (*red* is brightest, *purple* darkest). Figure courtesy of Liou & Zook (1999).

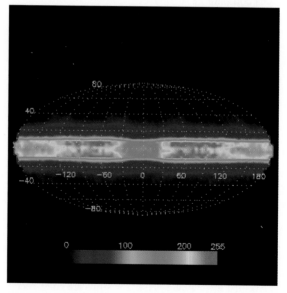

Figure 23 Brightness variations of 23 μm Kuiper Belt dust grains outside 20 AU, as viewed from the Sun. The vertical axis is the ecliptic latitude, and the horizontal axis the longitude with respect to Neptune (longitude 0°). Notice the two enhancements at 50°–100° ahead and behind Neptune. Figure courtesy of Liou & Zook (1999).

that is rotationally distorted/deformed from spherical shape by its own specific angular momentum. This object, like most large main-belt asteroids, is probably structurally weak as a result of repeated energetic impacts. The resulting rubble pile structure enables Varuna to adopt a near-equilibrium form. Modelled as an equilibrium rotational figure (a Jacobi ellipsoid), Varuna is found to have a bulk density $\rho \geq 1000$ kg m^{-3} (Jewitt & Sheppard 2002). This low density suggests that Varuna must be porous (at the 5–30% level) in order to allow for a cosmochemically normal admixture of refractory compounds (rocks). Many of the icy satellites of Saturn and Uranus also possess densities near that of water ice: They may also be porous (Kossacki & Leliwa-Kopystynski 1993).

INTERNAL STRUCTURE

It is likely that, even though they are ice-rich, the KBOs contain a significant or even dominant component of refractory material incorporated from the dusty protosolar disk. Measurements of the outgassed products of near-sun comets yield dust/gas production rate ratios, $F \gtrsim 1$ (Lisse et al. 1998, Jewitt & Matthews 1999), which may be indicative of the volatile fraction in KBOs. In any case, the refractories will include radioactive nuclei (K, Th, U) whose decay constitutes a source of internal heat. In small KBOs, this heat can be conducted to the surface and lost to space by radiation. The larger KBOs, however, cannot conduct radiogenic heat on timescales comparable to their age. Therefore, we expect that the larger KBOs should be internally heated, perhaps leading to partial differentiation, mobilization of internal gases, outgassing, and even cryogenic volcanism.

The dividing line between objects that are able to cool conductively and those that are too large to lose radiogenic heat occurs when the conduction cooling time, $\tau = r^2/\kappa$, equals the age of the solar system, t_{SS}. This occurs for critical radius

$$r = \sqrt{\kappa t_{SS}}. \tag{10}$$

For the thermal diffusivity of solid ice, $\kappa = 10^{-6}$ m^2 s^{-1}, and $t_{SS} = 4.6$ Gyr = 3×10^{17} s, Equation 10 gives $r \sim 500$ km. For diffusivity appropriate to slightly porous ice, $\kappa \sim 10^{-7}$ m^2 s^{-1}, the critical radius is near 100 km. Approximately 5000 KBOs are larger than 100 km in radius. All these objects will be internally heated by radioactive decay, and the magnitude of the heating can be estimated by comparing the radioactive energy liberated over t_{SS} with the heat capacity of the KBO. The specific energy generation due to radioactivity in chondritic material is $H \sim 10^{-12}$ W kg^{-1} (Stacey 1969). In KBOs, the heating rate will be diluted by mixing with (nonradioactive) ice, so we adopt $H = 5 \times 10^{-13}$ W kg^{-1} to be conservative. For a body at the critical radius (Equation 10) the globally averaged temperature rise is of order

$$\Delta T = \frac{H t_{SS}}{c_P}, \tag{11}$$

where c_P [J kg^{-1} K^{-1}] is the specific heat capacity. With $c_P \sim 10^3$ J kg^{-1} K^{-1}, we obtain $\Delta T \approx 150$ K. This is too small to melt water ice but sufficient to mobilize supervolatiles such as CO and N_2, and to initiate crystalline phase transitions in water ice. We would thus expect KBO interiors to have a layered structure, with the more volatile ices concentrated outside the warm core.

Real KBO interiors are likely to be much more complex than implied by the simple calculation above. The complete collisional history of the Kuiper Belt, particularly towards the end of the 100 Myr accretion period during which the density of objects was $\sim$100 times larger than now (Kenyon & Luu 1999a), may have produced a rubble-pile structure in many bodies. Internal void space may be common (Jewitt & Sheppard 2002) and could have significant impact on the thermal and mechanical properties of KBOs. Heat transfer within a porous, volatile body can produce startling structural complexity (Prialnick 2000). De Sanctis et al. (2001) investigated the internal evolution of KBOs under the combined effect of radiogenic heating and solar irradiation, and found several possible outcomes. For example, KBOs could be very differentiated: The top several hundred meters are depleted in the most volatile ices (e.g., CO), followed by interlaced layers of depleted and enriched volatiles produced by the sublimation of supervolatiles from the inner regions and the recondensation of these volatiles in cooler layers. Alternatively, KBOs could also preserve undifferentiated cores, depending on the kind and amount of radiogenic elements present. The particular outcome depends strongly on the particular radiogenic material and the assumed physical parameters of the KBO (e.g., thermal conductivity, porosity, radius).

BINARY KBOs

Toth (1999) briefly speculated on the observability of KBO satellites, but the existence of binary KBOs (other than Pluto-Charon) was first established by Veillet et al. (2000). At the time of writing, 5 binary KBOs are known: Pluto-Charon (Tholen & Buie 1997), 1998 WW$_{31}$ (Veillet et al. 2001), 2001 QT$_{297}$ (Elliot et al. 2001), 2001 QW$_{322}$ (Kavelaars et al. 2001), and 1999 TC$_{36}$ (Trujillo & Brown 2002b). TC$_{36}$ is the first binary trans-Neptunian that has been discovered in 3:2 mean-motion resonance with Neptune since Pluto-Charon. Images from the Hubble Space Telescope clearly show 1999 TC$_{36}$ to be accompanied by a secondary, separated by $\sim$0.37 arcsec (Figure 15, see color insert), implying an orbital distance of $\geq$8,000 km. The detection of binary KBOs is important because such a configuration potentially allows the direct determination of the masses of the components.

The Kuiper Belt binaries are remarkable for their wide separations. For example, the KBO binary 2001 QW$_{322}$ has a separation near 100,000 km, while the components are probably not much larger than 200 km in diameter (based on their absolute magnitude). The stability of such wide binaries is partly a result of the large scale of the respective Hill spheres of the KBOs. Long-term stability also places constraints on the mutual perturbation rate amongst KBOs.

The formation of the binaries remains mysterious. The best characterized pair, Pluto-Charon, possesses a combined angular momentum per unit mass that is larger than the maximum possible for a stable, self-gravitating, strengthless body (McKinnon 1989). As a result, it is widely supposed that Charon was formed when Pluto was struck by a massive interloper. Charon would have accreted from the fraction of the debris that remained trapped in bound orbits after the collision, in much the same way that the Earth's Moon is thought to have formed following an early collision. The angular momenta of the other Kuiper Belt binaries are, pending orbit determinations, unknown, and the formation mechanisms are essentially unconstrained.

RELATIONSHIP TO OTHER SOLAR SYSTEM BODIES

Pluto-Charon and Triton

Much of the interest in the Kuiper Belt stems from its link to a remarkably diverse group of objects with origin in the outer solar systerm. The most scientifically appropriate identification of Pluto-Charon as the two largest known Kuiper Belt objects has already been discussed (see "Binary KBOs" above). An origin in the Kuiper Belt can probably be extended also to the captured Neptunian satellite Triton, given that this object is a near-twin of Pluto in terms of bulk physical properties. The identification of these large outer solar system bodies as KBOs provides a simple and consistent framework for their formation in the outer solar system, ruling out more contrived theories for their origin (Lyttleton 1936, Harrington & van Flandern 1979, Farinella et al. 1979, Woolfson 1999). In this picture, Pluto and Triton formed in the trans-Neptunian region and became two of its largest members. However, this framework does not solve the puzzle of how Triton was captured into a satellite orbit around Neptune. Suggested capture mechanisms include a collision with an original satellite (Goldreich et al. 1989), or gas drag in a proto-Neptunian nebula (McKinnon & Leith 1995). Further understanding of the population and dynamics of the Kuiper Belt may shed light on Triton's dynamical history.

Comets and Centaurs

At the small end of the size distribution in the outer solar system are the Jupiter-family comets (comets whose orbits cross or closely approach Jupiter's orbit and which are therefore strongly influenced by the planet), believed to have originated in the Kuiper Belt. The nuclei of the Jupiter-family comets have sublimation-limited lifetimes of $\sim 10^4$ yr and dynamically limited lifetimes of $\sim 10^5$ yr (due to ejection by the planets or impact with planets or the sun) (Levison & Duncan 1994). To maintain a steady state population over longer timescales the Jupiter-family comets must be replenished from a more stable reservoir. Whereas it was previously thought that the Jupiter-family comets could be derived from the long-period

(Oort Cloud) comets by gravitational capture by the gas giant planets, detailed calculations show this path to the Jupiter-family comets to be inefficient. Instead, it is more likely that the Jupiter-family comets are escaped KBOs that have been scattered inward by the planets to the vicinity of Earth (Fernandez 1980, Duncan et al. 1988). The instability that triggers the ejection of KBOs from long-lived orbits beyond Neptune may be either dynamical chaos (Holman & Wisdom 1993, Levison & Duncan 1994) or collisional scattering (Ip & Fernandez 1997). Once Neptune-crossing, the timescale for transfer to the inner planetary region is 10^7 to 10^8 years (Hahn & Bailey 1990, Dones et al. 1999).

If Jupiter-family comets originate in the Kuiper Belt and eventually end up with orbits not much larger than Jupiter's, one would expect to find comets-in-transition, i.e., those objects in the process of making the journey from the Kuiper Belt into the inner solar system. These transition objects have probably been found in the Centaurs. Roughly 25 Centaurs are known as of August 2001. Some Centaurs already show cometary activity: The well-studied Centaur 2060 Chiron possesses ice on its surface (Luu et al. 2000), and has displayed a persistent coma for roughly a decade (e.g., Hartmann et al. 1990, Luu & Jewitt 1990, Meech & Belton 1990, Marcialis & Buratti 1993) (Figure 16, see color insert). Others (like 5145 Pholus) lack coma, making the nucleus directly accessible for study and reducing the likelihood that outgassing has altered the surface properties from the pristine state. As noted in "Colors" above, broadband photometry and spectral observations of the brightest Centaurs show a wide diversity, from neutral/blue 2060 Chiron to ultra-red 5145 Pholus, much like the color diversity of the KBOs. This color diversity, plus evidence for spectral similarity between the Centaurs and KBOs (Figure 13), are consistent with an origin in the Kuiper Belt for the Centaurs.

Measurements of the physical properties of cometary nuclei are complicated by the presence of near-nucleus coma, and reliable data exist for only a few. The most consistent and distinguishing property is the geometric albedo, for which very low values $0.02 \leq p_V \leq 0.04$ have been measured in 15 well-measured comets (Figure 1 of Fernandez et al. 2001), including Halley [Figure 17a, (see color insert), and comet Borrelly (Figure 17b)]. Spatially resolved images of the nucleus of P/Halley (Figure 17a) show that outgassing is largely confined to collimated jets that occupy perhaps 10% of the nucleus surface. The remaining 90% is blanketed by dark, inert material that has been widely interpreted as a refractory mantle composed of fallback debris. The low albedo seen in both Halley and Borrelly suggests a carbon-rich composition, perhaps consisting of complex hydrocarbon polymers (Moroz et al. 1998). The measured short-period comet nuclei resemble elongated spheroids, with effective radii in the range $1 \leq r \leq 10$ km (Jewitt 1990, Lamy et al. 2000) and axis ratios up to 2:1. Larger nuclei exist but are rare. Most comet nuclei are probably collision fragments, while KBOs larger than 100 km are likely to have survived collisional disruption (Farinella & Davis 1996, Kenyon & Luu 1999a). If so, the short-period comets may not be as primitive as originally believed. The form of the comet nucleus size distribution has not been accurately measured because of practical difficulties by coma contamination. Measurements of the size distribution of small impact craters on the surfaces of the satellites of the outer planets provide the best, although admittedly indirect, measure of the nucleus size distribution. These give a projectile differential size distribution index $q = 3$ (McKinnon et al. 1991).

FORMATION OF THE KUIPER BELT

In all likelihood, the Kuiper Belt is a processed remnant of the protoplanetary disk, and therein lies its enormous scientific interest. Efforts to reconstruct the growth and dynamical evolution of the KBOs will probably be the most fruitful steps taken in the next decade toward understanding accretion in the solar system. The problem of the formation of the KBOs was tackled early on by Stern (1996) and Stern & Colwell (1997), using the particle-in-a-box paradigm where the collisions of particles are treated in a statistical manner. These models allowed for the possibility of fragmentation during the accretion process, but did not fully model the velocity evolution of the KBOs. This omission is important because the distribution of planetesimal velocities plays a crucial role in determining the growth rate. Nevertheless, they recognized that the present mass of the Belt ($\sim$0.1 $M_{\oplus}$, see "Total Mass" above) was too small to build up the currently observed distribution via binary accretion.

Kenyon and Luu Model

Kenyon & Luu (1998, 1999a) performed the most detailed simulations yet of the growth of KBOs, with the 1999 work following simultaneously the evolution of both planetesimal masses and velocities, and allowing for fragmentation during the accretion process. Their model is based closely on Earth-accretion models of Wetherill & Stewart (1993), also adopting the particle-in-a-box approach (Safronov 1969). The calculations typically begin with an initial distribution of $\sim$100 m planetesimals located in a single annulus around the Sun. As planetesimals collide with each other, they may grow larger due to accretion, smaller due to fragmentation, or become completely pulverized in a catastrophic collision. The mass distribution is recorded in n mass bins, and the mass and velocity in each bin is recorded at each time step, after calculating the collision rates, the outcome of each collision, and the velocity changes due to collisions and long-range gravitational interactions. The calculations are usually carried out for $\sim$10^8 yr, the expected formation timescale for Neptune (Lissauer et al. 1996), since a full grown Neptune is expected to excite the orbits of KBOs and thus inhibit further growth.

Adopting an annulus width $\Delta a = 6$ AU, centered at heliocentric distance $a = 35$ AU, containing 80 m radius planetesimals with mass density $\rho_0 = 1500$ kg m^{-3} and initial eccentricities $e_0 = 10^{-4}$–10^{-2}, the results can be summarized as follows:

1. It is possible to form the observed trans-Neptunian distribution within a 10^8-yr timescale with an initial annulus mass $M_0 = 3$–$30\ M_\oplus$. This initial mass is 30 to 300 times larger than the current mass of the Kuiper Belt, but is roughly consistent with to the mass of the Minimum Mass Solar Nebula (M_{MMSN}, Hayashi et al. 1985), which predicts $M_{MMSN} = 7$–$15\ M_\oplus$ in the 32–38 AU annulus.

2. The final size distribution is determined by two competing physical processes: growth by mergers and erosion by fragmentation. It can be described by a broken (differential) power law with two power indices: $q = 3.5$, for $r \le 0.1$–1 km (representing the fragment regime, Dohnanyi 1969), and $q = 3.75$–4.25 for $r > 1$–10 km (merger regime). Several Pluto-sized bodies are naturally produced on the 10^8-yr timescale, depending on the initial mass.

3. Fragmentation limits the maximum size reached by KBOs: r_{max} ranges from 450 km to 3000 km, depending the tensile strength (a low tensile strength facilitates fragmentation, inhibiting growth).

4. Due to fragmentation, most of the initial mass ends up in smaller (0.1–10 km) objects that can be collisionally depleted over the age of the solar system. The present Kuiper Belt contains only 1–2% of the initial mass.

The final results, especially the final size distribution, are surprisingly insensitive to most input parameters (initial size distribution, initial planetesimal size, etc.). A typical outcome of the calculations is shown in Figure 18.

Figure 19 compares the model luminosity functions of KBOs with observations: The left panel shows models with $e_0 = 10^{-3}$ and different M_0, whereas the right panel shows models with a MMSN M_0 with different e_0. The models were originally computed to fit the Kowal (1989) data point, which has recently been invalidated (Trujillo et al. 2001a), but the models fit the remaining observations very well. It should be noted that the KL99a model is over-simplified: For example, it does not allow for the migration of KBOs, and it does not fully account for the dynamical effects of the largest bodies upon the planetesimal swarm. Nevertheless, it provides a self-consistent model for accretion and fragmentation in the early Kuiper Belt, and makes several testable predictions.

EVOLUTION OF THE KUIPER BELT

The picture of the Kuiper Belt that has emerged from dynamical and accretion models is that of a heavily attenuated population, eroded dynamically and physically by a variety of mass loss mechanisms. Long-term gravitational effects of the planets and overlapping secular resonances induce instabilities in large regions interior to 42 AU (Figure 20, see color insert). The Figure shows the good agreement between the dynamical calculations and the observed orbital distribution of KBOs: Most known KBOs are indeed located in stable regions. The current Kuiper Belt

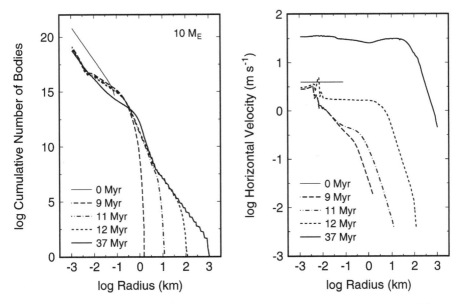

Figure 18 Size and velocity distributions for a model with initial mass $M_0 = 10\ M_\oplus$, initial eccentricity $e_0 = 10^{-3}$, tensile strength $S_0 = 2 \times 10^6$ ergs g^{-1}. *Left*, cumulative size distribution; *right*, horizontal velocity as a function of time. Figure adapted from Kenyon & Luu (1999a).

contains far too little mass to form the observed population in the $\sim 10^8$ yr available prior to the disruptive emergence of Neptune; instead, the original mass must have been ~ 100 times the current mass. Much of the original mass was ground to dust by collisions, then lost from the system (see "Kenyon and Luu Model" above). However, KBOs larger than ~ 100 km are largely immune to collisional disruption (Farinella & Davis 1996); they are thus likely to be true survivors from the earliest days of the solar system.

Other more violent mass loss mechanisms may have also occurred: Morbidelli & Valsecchi (1997) and Petit et al. (1999) showed that Mars-sized bodies in eccentric orbits in the Kuiper Belt could excite the e's and i's of primordial KBOs to the current observed values. Yet another line of thought attributes the excited orbits of the Kuiper Belt to the birth environment of the Sun. Pointing out that stars are generally born in clusters and stay there for $\sim 10^8$ yr (Kroupa 1995, 1998), Ida et al. (2000b) proposed close stellar encounters as the mechanism to excite the KBOs and erode the Kuiper Belt beyond 50 AU. Kobayashi & Ida (2001) elaborated on this idea and showed that perturbations from close stellar encounters (pericenter distance ~ 200 AU) could have increased the e's and i's in the outer region of the planetesimal disk to the point where planetary accretion beyond a maximum semimajor axis a_{planet} was inhibited. For a dense stellar cluster, a_{planet}

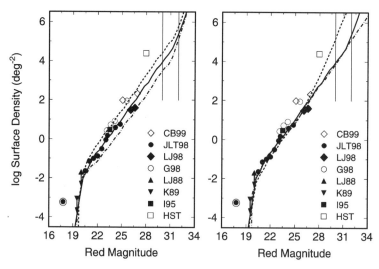

Figure 19 Comparison of model cumulative luminosity functions of KBOs with observations. The open circle with central dot indicates the position of Pluto for an adopted albedo of 4%. The source of other observations are as follows: HST = Cochran et al. (1998), I95 = Irwin et al. (1995), K89 = Kowal (1989), LJ88 = Luu & Jewitt (1988), G98 = Gladman et al. (1998), LJ98 = Luu & Jewitt (1998), JLT98 = Jewitt et al. (1998), CB99 = Chiang & Brown (1999). For clarity, error bars for each datum (typically a factor of 2–3) are not shown. $M_{MMSN} = 7$–$17 M_\oplus$ = mass of the Minimum Mass Solar Nebula at 35 AU. (*Left*) Results from model with $e_0 = 10^{-3}$ and $M_0 = 0.3 M_{MMSN}$ (*dot-dashed curve*), $M_0 = 1 M_{MMSN}$ (*solid curve*), $M_0 = 3 M_{MMSN}$ (*dashed curve*). (*Right*) Results from a model with $M_0 = 1 M_{MMSN}$ and $e_0 = 10^{-2}$ (*dashed curve*), $e_0 = 10^{-3}$ (*solid curve*), and $e_0 = 10^{-4}$ (*dot-dashed curve*). Figure adapted from Kenyon & Luu (1999b).

was roughly the size of the solar system's planetary region (∼30–40 AU). Adams & Laughlin (2001) investigated the effects on the Kuiper Belt if the solar system was formed in a stellar group large enough to contain a massive star, and found that all KBOs beyond ∼50 AU would have attained large eccentricities from external perturbations ($e > 0.2$ for semimajor axis $a = 50$ AU, $e > 0.4$ for $a = 70$ AU). As a result, 40% of the KBOs would have been removed by close encounters with Neptune. All these hypotheses, whether invoking planetary or stellar perturbers, share the outcome that the excited e's and i's of KBOs would have resulted in violent collisions and scattering events, adding to the collisional mass loss.

A rough history of the Kuiper Belt is thus one where the planetesimals originally had low relative velocities and nearly circular orbits (in order for growth to proceed). The velocity dispersion was of order $\Delta V = (i^2 + e^2)^{1/2} V_K$, where V_K is the local Keplerian orbital velocity. With $e = i$ [rad] = 0.001, we find $\Delta V \approx$

5 m s^{-1}, similar to the escape velocity from a 5-km-radius planetesimal. These planetesimals grew by binary collisions, and as they grew, their velocity dispersion also increased due to mutual scattering. The increased velocity dispersion also meant stronger collisions and therefore a larger debris production rate. A significant fraction of this collision debris was in dust form, implying that the trans-Neptunian belt was originally very dusty, perhaps like the dust rings seen around some nearby main-sequence stars. Shortly after 10^7 yrs, there were now bodies large enough (radius ~300 km) to excite the velocities of smaller planetesimals. At this point the dust content should rapidly increase, since collisions became disruptive instead of accumulative. The peak dust content should occur at ~10^8 years, right before nearby Neptune reached its final mass and began to erode the Kuiper Belt.

In short, it is clear that the Kuiper Belt has also undergone substantial dynamical and collisional erosion, and hypotheses abound that try to recreate its eventful past. Unexpectedly, the Kuiper Belt has turned out to be a test bed for theories of the Sun's birth environment.

RELATIONSHIP WITH CIRCUMSTELLAR DISKS

Since the discovery of the Kuiper Belt, it has often been postulated that dust disks around main sequence stars (age ~100 Myr, e.g., β Pic, Vega) are extrasolar analogs of the Kuiper Belt (e.g., Backman & Paresce 1993). Dust rings of dimensions comparable to the Kuiper Belt have been resolved around several stars, such as HR4796A (Figure 21, see also Koerner et al. 1998) and ϵ Eridani (Greaves et al. 1998). Finally, dust has already been detected in our own Kuiper Belt by the spacecraft Voyager 1 and 2 (Gurnett et al. 1997). Both spacecraft detected dust impacts in the 30–50 AU region, and although Gurnett et al. did not identify the source of the impacts, Kuiper Belt dust seems to be the most plausible source. The dust production rate implied by the Voyager detections is ~3 × 10^3 kg s^{-1} and the normal optical depth ~10^{-7} (Jewitt & Luu 1997).

There is thus strong evidence to link the Kuiper Belt to circumstellar disks, but since the evolution of circumstellar disks is still not well understood, it is not clear how the Kuiper Belt fits into this evolutionary scheme. The connection between the Kuiper Belt and circumstellar disks would be substantially elucidated if, for example, extrasolar KBOs were detected in circumstellar disks, or short of that, if the Kuiper Belt dust distribution could be compared with observations of circumstellar disks. According to studies of the orbital dynamics of Kuiper Belt dust (Liou et al. 1996, Liou & Zook 1999), the latter might be feasible in the near future. Very small dust grains (≤1 μm) originating in the Kuiper Belt are quickly blown out of the Solar System by radiation pressure, while the rest slowly spiral toward the Sun due to Poynting-Robertson and solar wind drag. During their journey they gravitationally interact with planets and can be scattered out of the

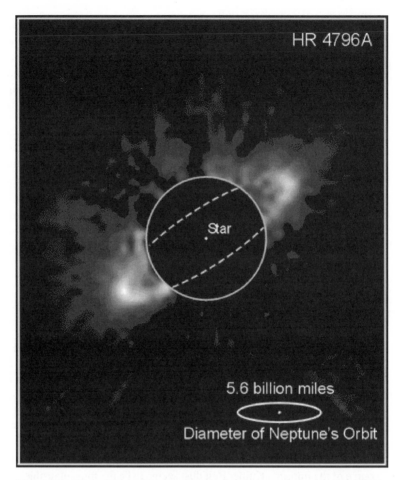

Figure 21   Near-infrared image of a dust ring around the star HR4796A, taken by NASA's Hubble Space Telescope. Figure courtesy of Brad Smith (University of Hawaii), Glenn Schneider (University of Arizona), and NASA.

solar system, trapped into mean motion resonances (Liou et al. 1996), or shattered by collisions with other Kuiper Belt grains or with interstellar dust (Yamamoto et al. 1998). Small (<9 μm) dust grains generally reach the Sun and contribute to the zodiacal light. Intermediate-size (9–50 μm radius) grains are susceptible to collisional shattering by interstellar dust along the way, but much larger grains (>50 μm radius) may survive these collisions and eventually also reach the Sun. Statistically, 80% of the grains are ejected from the Solar System, while 20% eventually reach the Sun and contribute to the zodiacal light.

Liou & Zook (1999) pointed out that the distribution of circumstellar dust may provide another method by which to detect extrasolar planets (Figure 22, see color insert). Dust tends to be trapped in mean motion resonances exterior to the outermost planet, and thus forms a giant circumsolar dust ring. The dust column density varies along the ring because different resonances have different resonance angles. The immediate vicinity of the planets should also be cleared of dust, due to ejection by the planets. It should be noted that the azimuthal dust variations may be washed out by interparticle collisions if the dust density is high. However, the optical depth of our Kuiper Belt dust is very low (Backman et al. 1995, Jewitt & Luu 1997), so the brightness variations should still be preserved. In our solar system, the presence of the Kuiper Belt dust ring thus may be deduced by comparing the dust brightness at different longitudes (Figure 23, see color insert). As seen from a space telescope, the minimum brightness would coincide with the direction where Neptune is located, and the maxima would appear close to the ecliptic, at 50–100° ahead of or behind Neptune. A search for the Kuiper Belt dust signature (temperature $\sim$42 K) in the COBE infrared sky map proved negative (Backman et al. 1995), but such longitudinal brightness variations may have been detected in the dust disk around the star ϵ Eri (Greaves et al. 1998, Liou et al. 2000).

Finally, Melnick et al. (2001) and Ford & Neufeld (2001) pointed out that the vaporization of Kuiper Belt comets might explain the unexpected presence of detectable water vapor around late-type AGB stars. The ice mass ($\sim$10 $M_\oplus$) required to explain the water vapor abundance is much larger than the current mass of the Kuiper Belt, but is comparable to the inferred original mass (Kenyon & Luu 1999a). Confirmation of this vaporization hypothesis would suggest that massive Kuiper Belts could survive for the entire main sequence lifetime of the central stars, even escaping the dramatic clearing phase that apparently removed 99% of the mass from our Kuiper Belt.

FUTURE WORK

Many external disks have large radial extents compared to the 50 AU scale of the classical Kuiper Belt. In this regard, it is interesting to consider the possible structure of the outer Kuiper Belt, beyond the region that has so far been optically probed. One possibility, as already noted, is that the disk-like structure of the Kuiper Belt is simply absent beyond 50 AU, perhaps as a consequence of an ancient stellar encounter. Alternatively, it is also possible (e.g., Stern 1996) that the surface density of the Kuiper Belt drops at 50 AU only to rise towards its primordial level at some larger radius. This speculation is not specific about the physical factors controlling the surface density, and offers no prediction of the distance at which a return to a larger density is expected. Naturally, it is difficult to rule it out. Observations show only that any wall in the surface density cannot be closer than $\sim$100 AU. Because of the R^{-4} dependence of the scattered sunlight, it will be difficult to study KBOs at 100's of AU and beyond. However, the planned Taiwan-America Occultation Survey (TAOS) experiment should provide an interesting constraint on the outer Belt and on the possible existence of a wall (Chen 1999). This experiment will photometrically monitor thousands of stars to search for occultations by intervening KBOs. The distance dependence of detectable occultations is modest (the size of the projected Fresnel diffraction spot, and hence of the smallest detectable KBO, varies only as $R^{1/2}$). Halley-sized (15 km in diameter) KBOs should be detectable many hundreds of AU from the sun, if they are present in sufficient abundance.

At the time of writing (August 2001), the Kuiper Belt has been amply probed in the magnitude range $20 \leq m_R \leq 25$, with sky coverage of a few hundred square degrees. Survey observations with the 8-m diameter SUBARU telescope on Mauna Kea have started, and should reach $\sim$28 magnitude, probing Halley-nucleus sized objects in their natal region. At the other extreme, KBOs brighter than $m_R \sim 19$ are so rare (the expected surface density for $m_R \sim 19$ is $\sim 10^{-3}$ KBO per square degree; see Figure 7) that large amounts of time on wide field telescopes are needed to find them.

A simple extrapolation from the current luminosity function to the entire area within $60°$ of the ecliptic predicts that $\sim$a few $\times$ 100 KBOs brighter than $m_R = 20$ will be found. Establishing the luminosity function at bright magnitudes provides clues to the accretion environment in the Kuiper Belt, especially in the late stages (Bailey 1994, Kenyon & Luu 1999a), but, perhaps more importantly, the discovered bright KBOs provide the only opportunities to study the physical nature of KBOs as a group (see "Physical Properties" above). The fitted size distribution admits the possibility of finding more Pluto-sized (2000 km in diameter or larger) KBOs. If these objects (perhaps $\sim$5 in number) still reside in the Solar System, they presumably inhabit the outer parts of the Kuiper Belt, because they would have been found otherwise.

The Large Synoptic Survey Telescope now being planned in the United States has the exploration of the Kuiper Belt as a major scientific priority (McKee & Taylor 2001). This facility will survey the optical sky to magnitude 24 at least once every week. Through the use of sophisticated software, it will provide a continuing source of astrometric and photometric data on $\sim$10,000 KBOs. These data will provide exquisite knowledge of the dynamical structure and contents of the trans-Neptunian solar system, rivalling that now available for the nearby main asteroid belt.

Ground-based observations have done much to reveal many of the gross properties of the Kuiper Belt, but such observations nevertheless suffer from limitations that result from being Earth-bound. Fortunately, these limitations can be rectified in the near future with space missions such as SIRTF and GAIA. SIRTF will allow the detection of the thermal radiation from $\sim$100 KBOs. With simultaneous measurements at optical wavelengths, it will be possible to determine albedos and so to search for correlations between albedo, color and orbital properties (see Jewitt & Luu 2001). Such correlations (or their absence) will tell us about the competing effects of collisions, cosmic ray bombardment, outgassing and compositional variations in the Kuiper Belt. GAIA is an all-sky astrometric and photometric survey that will create a precise three-dimensional map of the Milky Way and the Solar System. By dint of repeated astrometric and photometric observations of all objects brighter than $m_V \sim 20$, GAIA will yield the complete sample of all bright KBOs, free of all directional observational bias, and thus reveal the true three-dimensional distribution of the Kuiper Belt. GAIA's all-sky coverage will confirm the presence or absence of other Plutos and Charons in the Kuiper Belt, while its extremely precise astrometry of detected KBOs could reveal the presence of Kuiper Belt satellites from wobble about the barycenter. The sample of bright KBOs will be invaluable for detailed physical studies.

ACKNOWLEDGMENT

This work was supported by NASA's Origins of Solar Systems and Planetary Astronomy programs. We thank R. Hoogerwerf for help with the figures, and J. Kleyna for working during his vacation.

The *Annual Review of Astronomy and Astrophysics* is online at
http://astro.annualreviews.org

LITERATURE CITED

Adams FC, Laughlin G. 2001. *Icarus* 150:151–62

Allen RL, Bernstein GM, Malhotra R. 2001. *Astrophys. J.* 549:L241–44

Backman DE, Dasgupta A, Stencel RE. 1995. *Astrophys. J.* 450:L35–38

Backman DE, Paresce F. 1993. Main-sequence stars with circumstellar solid material—The VEGA phenomenon. In *Protostars and Planets III*, ed. E Levy, J Lunine, pp. 1253–304. Tucson: Univ. Ariz. Press

Bailey ME. 1994. Formation of the outer solar system bodies: Comets and planetesimals. In *Asteroids, Comets, Meteors 1993. Proc. IAU Symp. 160*, ed. A Milani, M Di Martino, A Cellino, pp. 443–59. Dordrecht: Kluwer

Barucci MA, Romon J, Doressoundiram A, Tholen DJ. 2000. *Astron. J.* 120:496–500

Bockelee-Morvan D, Lellouch E, Biver N, Paubert G, Bauer J, et al. 2001. *Astron. Astrophys.* 377:343–53

Brown ME. 2001. *Astron. J.* 121:2804–14

Brown ME, Blake GA, Kessler JE. 2000. *Astrophys. J.* 543:L163–65

Brown RH, Cruikshank DP, Pendleton Y. 1999. *Astrophys. J.* 519:L101–4

Brown ME, Kulkarni SR, Liggett TJ. 1997. *Astrophys. J.* 490:L119–22

Chen WP. 1999. Census of Kuiper Comets-the TAOS Project. In *Observational Astrophysics in Asia and its Future. Proc. East Asian Meetg. Astron., 4th, Kunming*, ed. PS Chen, pp. 181. Yunnan Observatory. Beijing: Chinese Acad. Sci.

Chiang EI, Brown ME. 1999. *Astron. J.* 118:1411–22

Clark RN. 1981a. *J. Geo. Res.* 86:3087–96

Clark RN. 1981b. *J. Geo. Res.* 86:3074–86

Cochran AL, Levison HF, Stern SA, Duncan MJ. 1995. *Astrophys. J.* 455:342–46

Cochran AL, Levison HF, Tamblyn P, Stern SA, Duncan MJ. 1998. *Astrophys. J.* 503:L89–93

Collander-Brown SJ, Fitzsimmons A, Fletcher E, Irwin MJ, Williams IP. 1999. *MNRAS* 308:588–92

Crovisier J, Brooke TY, Leech K, Bockelee-Morvan D, Lellouch E, et al. 2000. The thermal infrared spectra of comets Hale-Bopp and 103P/Hartley 2 observed with the Infrared Space Observatory. In *Thermal Emission Spectroscopy and Analysis of Dust, Disks, and Regoliths, Astron. Soc. Pac. Conf. Ser.*, ed. M Sitko, A Sprague, D Lynch, 196:109–17

Cruikshank DP, Roush TL, Bartholomew MJ, Moroz LV, Geballe TR, et al. 1998. *Icarus* 135:389–407

Cuillandre J-C, Luppino G, Starr B, Isani S. 2000. Performance of the CFH12K: a 12K by 8K CCD mosaic camera for the CFHT prime focus. In *Optical and IR Telescope Instrumentation and Detectors, Proc. SPIE*, ed. M Iye, AF Moorwood, 4008:1010–21. Bellingham, WA: SPIE

Davies JK. 2001. *Beyond Pluto: Exploring the Outer Limits of the Solar System*. Cambridge: Cambridge Univ. Press

Davies JK, Green S, McBride N, Muzzerall E, Tholen DJ, et al. 2000. *Icarus* 146:253–62

De Sanctis MC, Capria MT, Coradini A. 2001. *Astron. J.* 121:2792–99

Dones L. 1997. Origin and evolution of the Kuiper Belt. In *From Stardust to Planetesimals*, Astron. Soc. Pac. Conf. Ser., Vol. 122, ed. YJ Pendleton, AGGM Tielens, ML Savage, pp. 347–65. San Francisco: Astron. Soc. Pac.

Dones L, Gladman B, Melosh HJ, Tonks WB, Levison HF, Duncan M. 1999. *Icarus* 142:509–24

Dohnanyi JS. 1969. *J. Geophys. Res.* 74:2531–54

Duncan M, Quinn T, Tremaine S. 1988. *Astrophys. J.* 328:69–73

Duncan MJ, Levison HF. 1997. *Science* 276:1670–72

Duncan MJ, Levison HF, Budd SM. 1995. *Astron. J.* 110:3073–81

Edgeworth KE. 1943. *J. Br. Astron. Soc.* 53:181–88

Edgeworth KE. 1949. *MNRAS* 109:600–9

Eggers S, Keller HU, Kroupa P, Markiewicz WJ. 1997. *Planet. Space Sci.* 45:1099–104

Elliot JL, Kern SD, Osip DJ, Burles S. 2001. IAU circular 1733

Farinella P, Davis D. 1996. *Science* 273:938–41

Farinella P, Davis DR, Stern SA. 2000. Formation and Collisional Evolution of the Edgeworth-Kuiper Belt. See Mannings et al. 2000, pp. 1255–82

Farinella P, Milani A, Nobili AM, Valsacchi GB. 1979. *Moon Planets* 20:415–21

Fernandez JA. 1980. *MNRAS* 192:481–91

Fernandez JA, Brunini A. 2000. *Icarus* 145:580–90

Fernandez JA, Ip W-H. 1984. *Icarus* 58:109–20

Fernandez YR, Jewitt DC, Sheppard SS. 2001. *Astrophys. J.* 553:197–200

Ferrin I, Rabinowitz D, Schaefer B, Snyder J, Ellman N, et al. 2001. *Astrophys. J.* 548:L243–47

Fitzsimmons A, Jewitt DC, West R. 2000. *Minor Bodies in the Outer Solar System*. Berlin: Springer-Verlag

Ford K, Neufeld D. 2001. *Astrophys. J.* 557:113–16

Friedland L. 2001. *Astrophys. J.* 547:L75–79

Gladman B, Kavelaars JJ, Nicholson PD, Loredo TJ, Burns JA. 1998. *Astron. J.* 116:2042–54

Gladman B, Kavelaars JJ, Petit J-M, Morbidelli A, Holman MJ, Loredo T. 2001. *Astron. J.* 122:1051–66

Goldreich P, Murray N, Longaretti PY, Banfield D. 1989. *Science* 245:500–4

Greaves JS, Holland WS, Moriarty-Schieven G, Jenness T, Dent WRF, et al. 1998. *Astrophys. J.* 506:L133–37

Green SF, McBride N, O'Ceallaigh DP, Fitzsimmons A, Williams IP, Irwin MJ. 1997. *MNRAS* 290:186–92

Gurnett DA, Ansher JA, Kurth WS, Granroth LJ. 1997. *Geophys. Res. Lett.* 24:3125–28

Hahn G, Bailey ME. 1990. *Nature* 348:132–36

Hahn JM, Malhotra R. 1999. *Astron. J.* 117:3041–53

Hainaut OR, Delahodde CE, Boehnhardt H, Dotto E, Barucci MA, et al. 2000. *Astron. Astrophys.* 356:1076–88

Hamid SE, Marsden BG, Whipple FL. 1968. *Astron. J.* 73:727–28

Harrington RS, van Flandern TC. 1979. *Icarus* 39:131–36

Hartmann W, Tholen DJ, Meech K, Cruikshank DP. 1990. *Icarus* 83:1–15

Hayashi C, Nakazawa K, Nakagawa Y. 1985. Formation of the solar system. In *Protostars and Planets II*, ed. DC Black, MS Matthews, pp. 1100–53. Tucson: Univ. Ariz. Press

Hogg DW, Quinlan GD, Tremaine S. 1991. *Astron. J.* 101:2274–86

Holman M, Wisdom J. 1993. *Astron. J.* 105:1987–99

Hunt GR. 1977. *Geophysics* 42:501–13

Ida S, Bryden G, Lin DNC, Tanaka H. 2000a. *Astrophys. J.* 534:428–45

Ida S, Larwood J, Burkert A. 2000b. *Astrophys. J.* 528:351–56

Ip W-H. 1989. *Icarus* 80:167–78

Ip W-H, Fernandez JA. 1991. *Icarus* 92:185–93

Ip W-H, Fernandez JA. 1997. *Astron. Astrophys.* 324:778–84

Irwin M, Tremaine S, Zytkow AN. 1995. *Astron. J.* 110:3082–92

Jewitt DC. 1990. In *Comets in the Post-Halley Era*, ed. R Newburn, M Neugebauer, J Rahe, pp. 19–65. Dordrecht: Kluwer

Jewitt DC. 1999. *Annu. Rev. Earth Planet. Sci.* 27:287–312

Jewitt DC. 2002. *Astron J.* In press

Jewitt DC, Aussel H, Evans A. 2001. *Nature* 411:446–47

Jewitt DC, Luu JX. 1993. *Nature* 362:730–32

Jewitt DC, Luu JX. 1997. The Kuiper Belt. In *From Stardust to Planetesimals*, ed. Y Pendleton, AGG Tielens, pp. 335–46. San Francisco: Astron. Soc. Pac.

Jewitt DC, Luu JX. 2000. Physical Nature of the Kuiper Belt. See Mannings et al. 2000, pp. 1201–29

Jewitt DC, Luu JX. 2001. *Astron. J.* 122:2099–114

Jewitt DC, Luu JX, Chen J. 1996. *Astron. J.* 112:1225–38

Jewitt DC, Luu JX, Trujillo C. 1998. *Astron. J.* 115:2125–35

Jewitt DC, Matthews H. 1999. *Astron. J.* 117:1056–62

Jewitt DC, Sheppard SS. 2002. *Astron. J.* In press

Kavelaars J, Petit J, Gladman B, Holman M. 2001. IAU Circ. 7749

Kenyon SJ, Luu JX. 1998. *Astron. J.* 115:2136–60

Kenyon SJ, Luu JX. 1999a. *Astron. J.* 118:1101–19

Kenyon SJ, Luu JX. 1999b. *Astrophys. J.* 526:465–70

Kern SD, McCarthy DW, Buie MW, Brown RH, Campins H, Rieke M. 2000. *Astrophys. J.* 542:L155–59

Kobayashi H, Ida S. 2001. The effects of a stellar encounter on a planetesimal disk. In *Am. Astron. Soc., 33rd Div. of Planet Sci. Meet.* 15.06 (Abstr.)

Koerner DW, Ressler ME, Werner MW, Backman DE. 1998. *Astrophys. J.* 503:L83–87

Kossacki KJ, Leliwa-Kopystynski J. 1993. *Planet. Space Sci.* 41:729–41

Kowal C. 1989. *Icarus* 77:118–23

Kroupa P. 1995. *MNRAS* 277:1507–21

Kroupa P. 1998. *MNRAS* 298:231–42

Kuiper GP. 1951. On the origin of the solar system. In *Astrophysics*, ed. JA Hynek, pp. 357–424. New York: McGraw-Hill

Lamy PL, Toth I, Weaver HA, Delahodde C, Jorda L, A'Hearn MF. 2000. The nucleus of 13 short-period comets. In *Am. Astron. Soc., 32nd Div. Planet. Sci. Meet.* 36.04 (Abstr.)

Larsen JA, Gleason AE, Danzl NM, Descour AS, McMillan RS, et al. 2001. *Astron. J.* 121:562–79

Leonard FC. 1930. The New Planet Pluto. *Leaflet Astron. Soc. Pacific* No. 30, Aug., pp. 121–24

Levison HF, Duncan MJ. 1990. *Astron. J.* 100:1669–75

Levison HF, Duncan MJ. 1993. *Astrophys. J.* 406:L35–38

Levison HF, Duncan MJ. 1994. *Icarus* 108:18–36

Licandro J, Oliva E, Di Martino M. 2001. *Astron. Astrophys.* 373:L29–32

Liou J-C, Zook HA. 1999. *Astron. J.* 118:580–90

Liou J-C, Zook HA, Dermott SF. 1996. *Icarus* 124:429–40

Liou J-C, Zook HA, Greaves JS, Holland WS. 2000. Does a planet exist in ε Eridani? A comparison between observations numerical simulations. In *31st Ann. Lunar Planet. Sci. Conf.* 1416 (Abstr.)

Lissauer JJ, Pollack JB, Wetherill GW, Stevenson DJ. 1996. In *Neptune and Triton*, ed. DP Cruikshank, MS Matthews, AM Schumann, pp. 37–108. Tucson: Univ. Ariz. Press

Lisse C, A'Hearn M, Hauser M, Kelsall T, Lien D, et al. 1998. *Astrophys. J.* 496:971–91

Luu JX, Jewitt DC. 1988. *Astron. J.* 95:1256–62

Luu JX, Jewitt DC. 1990. *Astron. J.* 100:913–32

Luu JX, Jewitt DC. 1996. *Astron. J.* 112:2310–18

Luu JX, Jewitt DC. 1998. *Astrophys. J.* 502:91–94

Luu JX, Jewitt DC, Trujillo CA. 2000. *Astrophys. J.* 531:L151–54

Luu JX, Marsden B, Jewitt DC, Trujillo CA, Hergenrother C, et al. 1997. *Nature* 387:573–75

Lyttleton RA. 1936. *MNRAS* 97:108–15

MacFarland J. 1996. *Vistas Astron.* 40:343–54

Malhotra R. 1993. *Nature* 365:819

Malhotra R. 1995. *Astron. J.* 110:420–29

Malhotra R, Duncan MJ, Levison HJ. 2000. Dynamics of the Kuiper Belt. See Mannings et al. 2000, pp. 1231–54

Mannings V, Boss AP, Russell SS, eds. 2000. *Protostars and Planets IV*. Tucson: Univ. Ariz. Press

Marcialis RL, Buratti BJ. 1993. *Icarus* 104:234–43

McCaughrean M, O'Dell R. 1996. *Astron. J.* 111:1977–86

McKee C, Taylor J. 2001. *Astronomy & Astrophysics in the New Millennium*. Washington, DC: Natl. Acad. Press

McKinnon WB. 1989. *Astrophys. J.* 344:L41–43

McKinnon W, Chapman C, Housen K. 1991. In *Uranus*, ed. J Bergstrahl, E Miner, M Matthews, pp. 620–92. Tucson: Univ. Arizona Press

McKinnon WB, Leith AC. 1995. *Icarus* 118:392–413

Meech K, Belton MJS. 1990. *Astron. J.* 100:1323–38

Melnick GJ, Neufeld DA, Ford KES, Hollenbach DJ, Ashby MLN. 2001. *Nature* 412:160–63

Merline WJ, Close LM, Dumas C, Shelton JC, Menard F, et al. 2000. Discovery of Companions to Asteroids 762 Pulcova and 90 Antiope by Direct Imaging. In *Am. Astron. Soc. 32nd Div. Planet. Sci. Mtg.* 13.06 (Abstr.)

Morbidelli A, Thomas F, Moons M. 1995. *Icarus* 118:322–40

Morbidelli A, Valsecchi GB. 1997. *Icarus* 128:464–68

Moroz LV, Arnold G, Korochantsev AV, Wésch R. 1998. *Icarus* 134:253–68

Nagasawa M, Ida S. 2000. *Astron. J.* 120:3311–22

Oort JH. 1950. *Bull. Astron. Inst. Netherlands* 11:91–110

Petit J-M, Morbidelli A, Valsecchi GB. 1999. *Icarus* 141:367–87

Prialnik D. 2000. See Fitzsimmons et al. 2000. pp. 33–50

Prialnik D, Podolak M. 1995. *Icarus* 117:420–30

Rickman H, Kamel L, Festou MC, Froeschle C. 1987. Estimates of masses, volumes and densities of short-period comet nuclei. In *Proc. Int. Symp. Diversity Similarity Comets*, pp. 471–81. Noordwijk: Eur. Space Agency

Romanishin W, Tegler SC. 1999. *Nature* 398:129–32

Safronov VS. 1969. *Evolution of the Protoplanetary Cloud and Formation of the Earth and Planets*. Moscow: Nauka. Transl. Israel Prog. Sci. Transl., 1972. NASA TT F–677

Shul'man LM. 1972. The chemical composition of cometary nuclei. In *The Motion, Evolution of Orbits, and Origin of Comets. Proc. IAU Symp. 145*, ed. G Chebotarev, E Kazimirchak-Polonskaia, B Marsden, pp. 265–70. Dordrecht: Reidel

Schulz R. 2001. *Astron. Astrophys. Rev.* In press

Sheppard SS, Jewitt DC, Trujillo CA, Brown

MJI, Ashley MCBA. 2000. *Astron. J.* 120: 2687–94

Stacey FD. 1969. *Physics of the Earth*, pp. 239–62. New York: Wiley

Stern SA. 1996. *Astron. J.* 112:1203–11

Stern SA, Colwell JE. 1997. *Astron. J.* 114:841–49

Strazzulla G, Johnson RE. 1991. In *Comets in the Post-Halley Era*, ed. R Newburn Jr, M Neugebauer, J Rahe, pp. 243–76. Astrophys. Space Sci. Lib. Vol. 167. Dordrecht: Kluwer Acad.

Tegler SC, Romanishin W. 1998. *Nature* 392: 49–50

Tegler SC, Romanishin W. 2000. *Nature* 407: 979–81

Tholen D, Buie M. 1997. Bulk properties of Pluto and Charon. In *Pluto and Charon*, ed. SA Stern, DJ Tholen, pp. 193–220. Tucson: Univ. Ariz. Press

Tombaugh C. 1961. The trans-Neptunian planet search. In *Planets and Satellites*, ed. G Kuiper, B Middlehurst, pp. 12–30. Chicago: Univ. Chicago Press

Torbett M. 1989. *Astron. J.* 98:1477–81

Toth I. 1999. *Icarus* 141:420–25

Trujillo CA, Brown ME. 2001. *Astrophys. J.* 554:L95–98

Trujillo CA, Brown ME. 2002a. *Astrophys. J. Lett.* In press

Trujillo CA, Brown ME. 2002b. *IAC Circ. 7787*

Trujillo CA, Jewitt DC. 1998. *Astron. J.* 115: 1680–87

Trujillo CA, Jewitt DC, Luu JX. 2000. *Astrophys. J.* 529:L103–6

Trujillo CA, Jewitt DC, Luu JX. 2001a. *Astron. J.* 122:457–73

Trujillo CA, Luu JX, Bosh AS, Elliot JL. 2001b. *Astron. J. Astron. J.* 122:2740–48

Veillet C, Doressoundiram A, Shapiro J, Morbidelli A. 2001. *IAU Circ. 7610*

Wetherill GW. 1980. *Annu. Rev. Astron. Astrophys.* 18:77–113

Wetherill GW, Stewart GR. 1993. *Icarus* 106: 190–209

Wilson PD, Sagan C, Thompson WR. 1994. *Icarus* 107:288–303

Woolfson MM. 1999. *MNRAS* 304:195–98

Yamamoto S, Mukai T. 1998. *Astron. Astrophys.* 329:785–91

Yu Q, Tremaine S. 1999. *Astron. J.* 118:1873–81

Annu. Rev. Astron. Astrophys. 2002. 40:103–36
doi: 10.1146/annurev.astro.40.060401.093917
Copyright © 2002 by Annual Reviews. All rights reserved

THEORY OF GIANT PLANETS

W. B. Hubbard,[1] A. Burrows,[2] and J. I. Lunine[1]

[1]*Lunar and Planetary Laboratory, The University of Arizona, Tucson, Arizona
85721-0092; email: hubbard@lpl.arizona.edu; jlunine@lpl.arizona.edu*
[2]*Department of Astronomy and Steward Observatory, The University of Arizona,
Tucson, Arizona 85721; email: burrows@as.arizona.edu*

Key Words   Jupiter, Saturn, metallic hydrogen, brown dwarfs, extrasolar giant
planets

■ **Abstract**   Giant planet research has moved from the study of a handful of solar
system objects to that of a class of bodies with dozens of known members. Since the
original 1995 discovery of the first extrasolar giant planets (EGPs), the total number
of known examples has increased to ~80 (circa November 2001). Current theoretical
studies of giant planets emphasize predicted observable properties, such as luminosity, effective temperature, radius, external gravity field, atmospheric composition, and
emergent spectra as a function of mass and age. This review focuses on the general
theory of hydrogen-rich giant planets; smaller giant planets with the mass and composition of Uranus and Neptune are not covered. We discuss the status of the theory
of the nonideal thermodynamics of hydrogen and hydrogen-helium mixtures under
the conditions found in giant-planet interiors, and the experimental constraints on it.
We provide an overview of observations of extrasolar giant planets and our own giant
planets by which the theory can be validated.

INTRODUCTION AND HISTORICAL BACKGROUND

In the mid-twentieth century, the pioneering work of Demarcus (1958) established
that the solar system's two largest planets Jupiter and Saturn are of solar or near-
solar composition. At the time, Demarcus' important result elicited little curiosity
as to why the Sun and its two largest planets, separated by three orders of magnitude
in mass, should have similar composition. Conventional models of star formation
held, and still hold, that hydrogen-rich objects can form only via direct gravitational collapse from gas clouds with mass considerably larger than $\sim 10^{-3}$ $M_\odot$
($M_\odot = 1$ solar mass). The discovery of Jupiter's intrinsic luminosity (Low 1966)
added another intriguing similarity to stellar objects. The theoretical discovery of
objects later known as brown dwarfs (Hayashi & Nakano 1963, Kumar 1963) led
to questions about how brown dwarfs, at that time hypothetical, might be related
to giant planets (Hubbard 1969).

It is as yet unclear how nature forms gravitationally bound masses of hydrogen as small as Jupiter and Saturn. Over the past decade, we have learned that the

0066-4146/02/0922-0103$14.00   **103**

Galaxy possesses a continuum of objects with masses intermediate between Jupiter and Saturn ($\sim 10^{-3}$ $M_\odot$), and the lowest-mass stars capable of sustained thermonuclear fusion of hydrogen (~ 0.075 $M_\odot$). Objects in this continuum obey similar physics: convective, liquid-metallic-hydrogen (electron-degenerate) interiors, and cool atmospheres rich in molecular compounds and condensates, regulating the escape of primordial heat from the deep interior. We have also learned that a number of the known extrasolar giant planets (EGPs) orbit their primary stars at distances as small as 0.04 AU, a fact not predicted by any standard theory of giant-planet formation.

On the theoretical front, researchers have developed a unified theory describing the evolution of sub–main-sequence hydrogen-rich objects, allowing a classification of EGPs (and our own giant planets), brown dwarfs, and similar objects according to mass, age, luminosity, and atmospheric properties. Jupiter and Saturn, whose age, luminosity, radius, atmospheric composition, and other fundamental properties are knowable in vastly greater detail than the properties of any EGP, now play the role of fundamental local calibrators for studies of EGPs and brown dwarfs. In an interesting reprise of the history of astronomy from a century ago, provisional spectral sequences are being developed for hydrogen-rich objects, but this time for luminosities ranging from $\sim 10^{-5} L_\odot$ to $\sim 10^{-9} L_\odot$ (where $L_\odot =$ solar luminosity). Such empirical classifications (the new spectral sequences, starting below M-dwarfs, are being called L- and T-dwarfs) will ultimately be based upon an underlying theory for the evolution of these faint objects, a theory that is under active development today. At this frontier area of both astrophysics and planetary science, studies of Jupiter and Saturn assume the importance of studies of the Sun during the early twentieth century.

Our review focuses on the theory of hydrogen-rich masses in the range from roughly one third of a Jupiter mass (Saturn) to some tens of Jupiter masses. We do not review the theory of the smaller giant planets Uranus and Neptune, which are primarily nonsolar in composition and which do not yet have clearly detected extrasolar counterparts.

Thermal Properties Based on Heat Flow Measurements

Hubbard (1968) showed that the observed intrinsic luminosity of Jupiter implies a fully convective structure in Jupiter's metallic-hydrogen layers (the best current value for Jupiter's intrinsic luminosity is $L = 10^{-9.062} L_\odot$; up-to-date values for luminosities of all four of the solar system's giant planets are given in Conrath et al. 1989). Subsequent investigations (Guillot et al. 1994a,b) have confirmed that the opacity in Jupiter and Saturn at thermal wavelengths is generally high enough to cause convective instability in the outer molecular layers as well. However, in a zone ranging from $T = 1300$ to 2800 K, the opacity from collision-induced absorption (CIA) in H_2 is at a minimum at thermal wavelengths, and there is a chance for the temperature gradient to go subadiabatic in that zone, unless additional opacity contributions from alkali metals and grains maintain a superadiabatic temperature gradient. And, as we discuss in "Hydrogen–Theory Compared with Experiment" below, new revisions to the H_2 equation of state at pressures on the order of $\sim$100 Kbar admit the possibility of a radiative zone at somewhat deeper layers than the possible radiative zone proposed by Guillot et al. (1994b).

The above caveats notwithstanding, generally speaking the temperature profile in both Jupiter and Saturn is thought to be essentially adiabatic, as implied by efficient convection, and thus a common theory can be applied to the calculation of evolutionary histories for fully convective EGPs, brown dwarfs, and indeed lower–main-sequence stars: the radiative zone of the atmosphere acts as a lid, regulating the escape of internal heat. A grid of model atmospheres extending down to the deepest convective zone, coupled with a table of interior models at specified specific entropies, suffices to map the evolution.

As we discuss in "Application of Cooling Theory to Jupiter and Saturn" below, the general cooling theory works well when applied to Jupiter, suggesting that complications such as subadiabatic zones, chemical separation, and late accretion are not highly significant in the evolution of this giant planet. The smaller giant planet Saturn does not fit the cooling theory; discussion of this problem is also presented in "Application of Cooling Theory to Jupiter and Saturn" below.

Discovery of EGPs

The radial velocity technique, by which the induced wobble of a central star is measured, has revealed more than 80 EGPs in orbit around nearby stars (Butler et al. 1997, 1998, 1999; Cochran et al. 1997, Delfosse et al. 1998, Fischer et al. 1999, Henry et al. 2000, Korzennik et al. 2000, Latham et al. 1989, Marcy & Butler 1996, Marcy et al. 1998, 1999, 2000a,b; Mayor & Queloz 1995, Mazeh et al. 2000, Noyes et al. 1997a,b; Queloz et al. 2000, Santos et al. 2000, Udry et al. 2000, Vogt et al. 2000). Up-to-date listings of EGP discoveries can be found at http://exoplanets.org/news.shtml and http://www.obspm.fr/planets.

The minimum mass of these companions ranges from a bit less than 0.25 M_J (M_J = Jupiter's mass) to greater than 10 M_J, they are found around stars with spectral types from M to F, their orbital eccentricity distribution is flat from zero to $\sim$0.95, and the average metallicity of their central stars is super-solar. Such characteristics and variety explode the notion that our solar system is typical, although radial-velocity observations to date have been biased against planetary systems with the large semimajor axes and long periods of Jupiter and Saturn. Nevertheless, it is clear that the properties of the extrasolar giant planets we are now encountering in abundance considerably expands the range of effective temperatures T_{eff}, masses M, radii a, compositions, and (perhaps) rotation rates that are relevant to the comprehensive study of giant planets in the Galaxy.

HIGH-PRESSURE PHASE DIAGRAMS

A principal feature that distinguishes giant planets and brown dwarfs from stars is the nonideality of the thermodynamics of the hydrogen-helium mixture in the interiors of the former. As has long been known, the pressure-temperature profiles of typical giant planets and brown dwarfs pass through a region of the hydrogen phase diagram where thermal ionization and dissociation of hydrogen molecules and atoms compete with pressure ionization and dissociation. The unmodified Saha equations for thermal ionization and dissociation give nonsensical results in this region, because the particles are typically separated by distances comparable to their molecular dimensions. Much of the progress in understanding this difficult region of the hydrogen phase diagram of giant planets has resulted from a combination of theoretical studies and experiments conducted over the last two decades, as we discuss in "Hydrogen—Theory Compared with Experiment" and "Hydrogen and Helium—Theory" below.

Hydrogen—Theory Compared with Experiment

Figure 1, to which we refer extensively in this section, shows the region of the hydrogen pressure (P) vs. temperature (T) phase diagram that is of particular interest for giant planets.

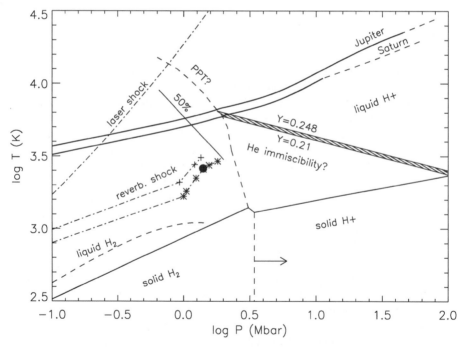

Figure 1   The phase diagram of hydrogen, showing regions of stability of liquid and solid molecular hydrogen (H_2), and of liquid and solid metallic (pressure-ionized) hydrogen (H+). Also shown are trajectories of experimental shock-compression experiments (*dashed-dot lines*) and trajectories of the interiors of Jupiter and Saturn at the present epoch (*heavy solid lines*). See text for discussion of further details of this figure.

The most widely used thermodynamic description of hydrogen at giant-planet pressures and densities is that of Saumon et al. (1995; SCVH in the following). Their code is built upon earlier work by DeWitt & Hubbard (1976), Slattery & Hubbard (1976), Hubbard & DeWitt (1985), Chabrier et al. (1992), and Ross et al. (1981) among others, and incorporates, as do its predecessors, an accurate description of liquid metallic hydrogen at high pressure ($P \gg 1$ Mbar), results fitted to experimentally determined intermolecular potentials in the nonideal region where H_2 predominates, and correct limiting behavior at low P and high T.

The treatment of the transition from molecular to metallic hydrogen (pressure ionization of hydrogen), is handled differently by various models and is still not well understood either experimentally or theoretically. The SCVH code is built upon a chemical description of dense hydrogen, wherein the abundance of various proton-bearing molecules is computed by minimizing the free energy, the latter being determined by internal molecular degrees of freedom, translational (ideal) degrees of freedom, and intermolecular interaction potentials (nonideal). The abundances of the various molecular species thus arise as a byproduct of the calculation. A peculiarity of the SCVH model is that it gives rise to a locus in $P-T$ space where the abundances of species (in particular, molecular hydrogen H_2 and atomic hydrogen H) change discontinuously, the so-called plasma phase transition, or PPT, as shown by a dashed line in Figure 1. The PPT is simply the latest version of a series of estimates as to the location of the point of metallization of hydrogen, long thought to be realized by a first-order phase transition in both solid and liquid phases of hydrogen (Wigner & Huntington 1935, Ebeling & Richert 1985, Chabrier 1994).

Whereas there is no question that hydrogen eventually pressure-ionizes at sufficiently large P, an accurate determination of the locus of pressure-ionization on the phase diagram remains elusive despite many decades of theoretical and experimental research. Diamond-anvil cell (DAC) measurements on the solid phase of H_2, mostly at room temperature, along the lower axis of Figure 1, have to date reached a maximum pressure of 3.42 ± 0.10 Mbar without any evidence of metallization (Narayana et al. 1998), as shown by an arrow in Figure 1. Measurements of polymorphic phase transitions in the solid-H_2 stability field (not shown in Figure 1 as they are irrelevant to giant planets), when fitted to theoretical models, have given rise to a detailed theoretical model for the melting temperature of H_2 as a function of P, shown as a solid line in Figure 1 (Cui et al. 2001). Somewhat different results for the H_2 melting line are obtained by fitting DAC measurements directly, as shown by the dashed line in Figure 1 (Datchi et al. 2000). The phase boundary between liquid and solid metallic hydrogen (we denote metallic hydrogen as H+ in the following) is considered well understood on the basis of extensive numerical experiments on the freezing of a proton plasma in a fluid of electrons (Stringfellow et al. 1990). The SCVH equation of state is, generally speaking, applicable to the regions of the $T-P$ plane that lie above the stability fields of the solid phases of hydrogen.

Dynamical compression, rather than DAC experiments, provides the most relevant tests of the SCVH equation of state. The most recent and relevant of the former are shown as dashed-dotted lines in Figure 1. Typically, dynamical compression experiments begin with a sample of hydrogen or deuterium at $P = 1$ bar and $T = 20$ K. The sample is then shock-compressed, either along a single Hugoniot, or by multiple reverberations yielding a much lower final entropy. The upper dashed-dotted curve in Figure 1 shows the most ambitious laser implosion experiment carried out to date (Collins et al. 1998). Coincidentally, this experiment reached a maximum pressure of 3.4 Mbar, the same as the maximum DAC pressure to date. The lower dashed-dotted curves show the shock-reverberation measurements of Nellis et al. (1999). The upper curve of these (pluses) shows the measurement points for D_2, while the lower curve (asterisks) shows H_2 points.

The lower heavy line in Figure 1, the quasi-isentrope of Saturn (Hubbard et al. 1999) represents, roughly, the lowest range of T (as a function of P) relevant to the structure of giant planets, although some extrasolar giant planets (EGPs) may be of sufficient age to have slightly lower interior entropies. As we see, the recent shock measurements roughly bracket the relevant T range up to $P \sim 1$ Mbar. The result of these measurements has been to call into question the SCVH equation of state in this pressure range.

The shock data do not provide a complete set of equation-of-state measurements: P, T, and mass density ρ. Rather, the Hugoniot equations (Mostovych et al. 2001) give, from the known shock and particle flow velocities u_s and u_p, the compressed P, ρ, and internal energy E. A theoretical model must be used to deduce T. The theoretical model that has been used to give the shock $P-T$ points in Figure 1 is not SCVH, but rather a newer model proposed by Ross (1998). The latter is better able to reproduce the general trend of both sets of shock data. The shock data indicate a pronounced lowering of shock pressures as compared with expectations for a model of compressed H_2 without dissociation. Consequently, the Ross model hypothesizes that H_2 dissociation occurs as a continuous process, at pressures and temperatures substantially below and to the left of the PPT line shown in Figure 1. The Ross model is a simple extension of previous work. Rather than using a model for pure interacting H_2 to the left of the PPT and a model for pure interacting H+ to the right, Ross creates a model for the Helmholtz free energy F of a mixture of molecular and metallic hydrogen as a simple average:

$$F = (1 - x)F_{\text{mol}} + xF_{\text{met}} - TS_{\text{mix}}, \tag{1}$$

where F_{mol} and F_{met} are the Helmholtz free energies of pure H_2 and pure H+, respectively, x is the mole fraction of dissociated molecules, and S_{mix} is the ideal mixing entropy of H_2 and H. Ross determines x for a given T and P by minimizing F. The line labeled "50%" in Figure 1 shows, approximately, the locus of points where Ross' theory predicts $x = 0.5$.

Ross' resulting equation of state for hydrogen differs strikingly from SCVH in the region approaching the putative PPT, as shown in Figure 2, which plots three isochores.

Although Ross' model appears to be successful in explaining the shock data, it is clearly ad hoc. As yet, no general, first-principles model exists for the equation

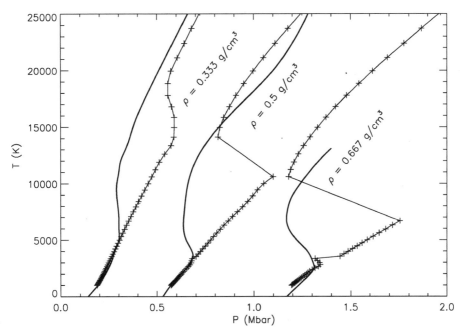

Figure 2   Isochores for hydrogen, comparing the model of Ross (1998; heavy solid lines) with predictions from the SCVH model (*plusses*). The big discontinuities in the SCVH isochores for $\rho = 0.5$ and 0.667 g/cm^3 result from crossing the PPT; the smaller discontinuity in the SCVH $\rho = 0.667$ g/cm^3 isochore is a numerical artifact.

of state of hydrogen in the difficult region where thermal and pressure dissociation and ionization occur essentially simultaneously, the region in the vicinity of the PPT and the 50% loci in Figure 1. The best hope for such a model comes from the approach of Militzer & Ceperley (2001) using quantum Monte Carlo simulations. Militzer & Ceperley (2001) find good agreement with SCVH in the region they investigate, which is, however, confined to the upper left-hand corner of Figure 1, generally above the laser shock line, and for $\rho \leq 0.153$ g/cm^3. Larger differences with SCVH, and possibly closer agreement with the Ross model, may be found when the quantum Monte Carlo calculations are extended to higher pressures.

As is the case for the DAC measurements, no shock data have yet provided direct evidence for full metallization of hydrogen. In the reverberation-shock measurements, Nellis (2000) finds that the electrical conductivity σ of the shocked sample increases steadily with pressure, reaching a plateau of 2000 $(\Omega$ cm$)^{-1}$ at $P = 1.4$ Mbar (point marked with a solid dot in Figure 1). This value of σ is still about 1.5 orders of magnitude smaller than the value predicted for H+ at megabar pressures (Hubbard & Lampe 1969, Stevenson & Salpeter 1977a).

The new picture of hydrogen dissociation and metallization as a continuous process has two important implications for the theory of giant planets. First, the

PPT has been thought to be present in Jupiter and Saturn, and indeed in many EGPs ranging in mass and age up to Gl 229B (Hubbard et al. 1997). Such a first-order phase transition would impose, in thermodynamic equilibrium, a condition on the abundances of helium and other trace species dissolved in the hydrogen on either side of the PPT. Atmospheric and interior abundances, on either side of a PPT, must be different in general, in accordance with the Gibbs phase rule. On the other hand, absence of a first-order phase transition in hydrogen at megabar pressures, as implied by Ross' theory and the new shock data, suggests that chemical abundances can be continuous in giant planets in this pressure range.

A second important implication of the Ross model has to do with convection in jovian planets (Nellis 2000). We start with the standard Schwarzschild criterion: a fluid of constant composition is unstable to convection if

$$\left(\frac{\partial P}{\partial \rho}\right)_S < \frac{dP}{d\rho}, \tag{2}$$

where $dP/d\rho$ is the actual variation of P with ρ in the planet. Employing the equation of state $P = P(\rho, T)$, we expand both sides of the inequality:

$$\frac{dP}{d\rho} = \left(\frac{\partial P}{\partial \rho}\right)_T + \left(\frac{\partial P}{\partial T}\right)_\rho \frac{dT}{d\rho}, \tag{3}$$

and

$$\left(\frac{\partial P}{\partial \rho}\right)_S = \left(\frac{\partial P}{\partial \rho}\right)_T + \frac{T}{\rho^2 C_V}\left(\frac{\partial P}{\partial T}\right)_\rho^2, \tag{4}$$

where C_V is the heat capacity per unit mass. Thus, the Schwarzschild criterion becomes

$$\frac{T}{\rho^2 C_V}\left(\frac{\partial P}{\partial T}\right)_\rho^2 < \left(\frac{\partial P}{\partial T}\right)_\rho \frac{dT}{d\rho}. \tag{5}$$

In the deeper parts of the atmosphere, at pressures ~ 0.1 to 1 Mbar, the intrinsic heat flow insures that $dT/d\rho$ is always positive. For a normal fluid $(\partial P/\partial T)_\rho$ is likewise positive, so in this case, the criterion simply says that if the actual temperature gradient $dT/d\rho$ is big enough, the fluid will convect. However, as we see from Figure 2, in pure-hydrogen planets at pressures ~ 1 Mbar, Ross' theory predicts that $(\partial P/\partial T)_\rho$ is negative in some layers. In these layers, the fluid is unconditionally stable to convection, regardless of the steepness of the temperature gradient (unless $dT/d\rho$ changes sign). The SCVH theory has no such regions, except across the PPT.

Admixture of sufficient He is likely to cause the value of $(\partial P/\partial T)_\rho$ to become everywhere positive again. Perhaps a solar fraction of He would suffice. However, detailed modeling studies are not yet available.

The laser implosion data of Collins et al. (1998) have subsequently been called into question by results from very recent shock compression experiments of Knudson et al. (2001). The latter employ intense magnetic pressure to launch a flyer plate to strike a liquid deuterium sample. This technique shocks the deuterium sample to a pressure range of 0.3 to 0.7 Mbar. Knudson et al. find disagreement with the Hugoniot of Collins et al. at pressures above 0.4 Mbar, and specifically find no evidence for anomalously high hydrogen compressibility in the pressure range of 0.4 to 0.7 Mbar. Thus, the theoretical model of Ross (1998) and its consequences for the interior structure of giant planets must be regarded as highly controversial as of early 2002.

Hydrogen and Helium—Theory

Salpeter (1973) originally proposed that helium might have limited miscibility in hydrogen fluid under conditions in the interiors of giant planets. This suggestion was further quantified in a series of papers (Stevenson 1975, Stevenson & Salpeter 1977a,b). The original theory is based on the assumption that neutral atoms of helium will be insoluble in metallic hydrogen (Smoluchowski 1967). Stevenson (1975) carried out an analysis for metallic-hydrogen and helium mixtures based on free-electron perturbation theory. An independent analysis (Hubbard & DeWitt 1985) obtained similar results. The shaded line in Figure 1 shows results of these perturbation theories: Above the upper boundary of the shaded region, it is thermodynamically allowed to have a homogeneous mixture of hydrogen and helium with a helium mass fraction $Y = 0.248$ or less. Similarly, above the lower boundary of the shaded region, mixtures with $Y = 0.21$ or less can exist. Below the critical line of an indicated composition, He-rich droplets will form and sink to deeper layers in the planet. It has been frequently noted that the critical line for solar, or near-solar composition, nearly intersects the present Jupiter and Saturn $P-T$ profiles. Discussion of the implications of this for Jupiter and Saturn evolution is given in "Application of Cooling Theory to Jupiter and Saturn" below.

It is important to note that the predictions of Stevenson and of Hubbard & DeWitt are based upon perturbation theory applied to fully ionized nuclei of hydrogen and helium. Whereas perturbation corrections to hydrogen are not large even in the megabar pressure range, it is extremely doubtful that helium can be accurately treated in this manner. Subsequent research has been directed toward improving the description of dense helium in particular. Results presented by Klepeis et al. (1991) suggested a large upward revision in the temperature of the critical line for a solar mixture of H and He. Figure 3 shows the approximate location (upper shaded region) of the phase-excluded region predicted by Klepeis et al. Were this calculation correct, major redistribution of He would occur in the interiors of both Jupiter and Saturn.

Klepeis et al's calculations were subsequently criticized by Pfaffenzeller et al. (1995), who noted that Klepeis et al. failed to properly account for the structural adjustment that results when helium atoms are introduced into metallic hydrogen. Pfaffenzeller et al. find that helium is much more soluble when such an adjustment is properly allowed for. The Pfaffenzeller phase-excluded region for solar

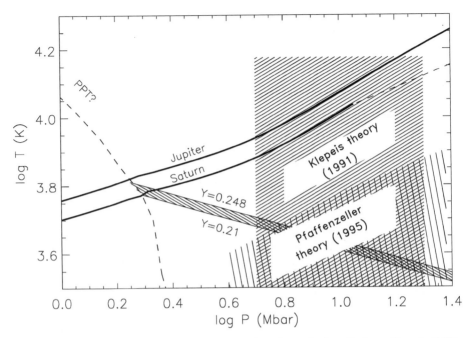

Figure 3   Closeup view of Figure 1, showing various theoretical predictions of immiscibility of a solar mixture of hydrogen and helium.

composition is shown as the lower shaded region in Figure 3. Interestingly, the critical line deduced by Pfaffenzeller et al. has increasing temperature with increasing pressure, opposite to the results for the older perturbation theory. The Pfaffenzeller critical line lies roughly parallel to the present Saturn adiabat, but almost a factor of two lower in temperature.

GRAVITY FIELD AS A WINDOW TO THE INTERIOR

The equation of state of a hydrogen-helium mixture, when coupled with a thermal model for the interior giving $T = T(P)$, yields a prediction of the relation $P = P(\rho)$ in the planet's interior (assuming that the planet is barotropic). The equation of hydrostatic equilibrium can then be integrated for a planet of given mass M to predict the distribution of mass density as a function of radius. The even zonal gravity harmonics of a rapidly-rotating giant planet provides a constraint on this interior mass distribution, and, when coupled to an accurate equation of state, has been used to restrict the class of admissible interior compositions. As a result of spacecraft encounters with the solar system's four giant planets, supplemented by information from ring precession rates for Saturn and Uranus, the first few zonal harmonics J_2, J_4, and for Jupiter and Saturn, J_6, are known. Of these, J_2 and J_4

are of the most utility in constraining interior models (Zharkov & Gudkova 1993). Coupled with atmospheric abundance measurements and with assumptions about the interior dynamical state of the giant planet, the zonal harmonics can be used to set limits on the size of a dense (nonhydrogen-helium) core in both Jupiter and Saturn (Guillot et al. 1997). Guillot et al. assume that the modest enrichment of carbon detected by the Galileo entry probe in Jupiter's atmosphere (about three times solar) translates into a similar enhancement of the abundant elements C and O throughout the jovian envelope. When models with such a mass density enhancement are fitted to the jovian J_2, J_4, J_6, with allowance for uncertainties in the effects of jovian differential rotation, Guillot et al. find that even jovian models with no dense core at all are admissable. Such interior structure would require a reassessment of the standard model of formation of Jupiter via gravitational capture of nebular hydrogen-helium gas onto a dense protoplanetary core of mass $\sim$10 Earth masses (Mizuno 1980).

Measurements of the external gravity harmonics of giant planets are still extremely crude in comparison with the data available for terrestrial planets, where many tens of zonal and tesseral harmonic coefficients are measured. What more could we learn about the jovian interior with a low-periapse Jupiter orbiter, capable of yielding gravity data comparable to those available for Venus or Mars?

Connection to Classical Theories of Rotating Bodies

Classical theories of a rotating gas-liquid body (Tassoul 1978) predict the structure, external shape, and hence, zonal harmonics of the body, as a function of the rotation state. The astrophysical literature is rich in methods for performing the calculation for a state of hydrostatic equilibrium. However, the traditional assumption of hydrostatic equilibrium plus barotropic structure must break down at some level, even for Jupiter. The relevant parameter of the classical perturbation theory, developed in great detail by Zharkov & Trubitsyn (1978), is the dimensionless measure of the importance of rotation, $q = \omega^2 a^3/GM$, where ω is the planet's solid-body rotation rate, a is its radius, M is its mass, and G is the gravitational constant. For the solar system's giant planets, $q \sim 10^{-1}$ (Jupiter and Saturn) to 10^{-2} (Uranus and Neptune), and the planets' J_2s are of similar magnitude, as they essentially represent the linear response of the planets to rotation. As shown by the analytic theory, the higher even zonal harmonics correspond to progressively higher powers of q: e.g., $J_4 \propto q^2$, $J_6 \propto q^3$, etc. They represent the planet's nonlinear response to rotation. Figure 4 shows an example of Jupiter's predicted gravity harmonics to high order, assuming exact hydrostatic equilibrium and solid-body rotation (Hubbard 1999).

Tests of Expected Interior Rotation and Convection State

The tidy zonal harmonic spectrum of Figure 4, with no odd harmonics and a roughly geometric decrease of even harmonics with increasing degree n, may not apply to Jupiter. As was first noticed by Vasilev et al. (1978), higher-order zonal

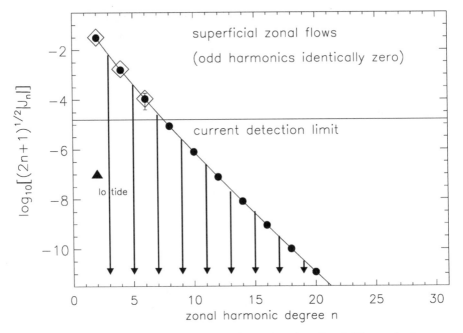

Figure 4 Jupiter zonal gravity harmonics, observed and predicted. Observed values are plotted as diamonds. Predicted values, for solid-body rotation, are shown as dots. Horizontal line shows the present limit of detection.

harmonics can also arise as a linear response to nonuniform rotation. Depending on the degree of nonuniform rotation, in principle the higher-degree linear response can be much larger than in the case of uniform rotation. Thus, measurement of giant-planet gravity harmonics out to high degree and order may give information about interior dynamics.

Schubert & Zhang (2000) have investigated the style of interior convection in Jupiter and EGPs as a function of luminosity L and rotation rate ω. They find that an important parameter is the Rayleigh number, which they define as

$$Ra = \frac{\alpha g L d^4}{k \kappa \nu},\qquad(6)$$

where α is the thermal expansivity, g the surface gravity, $d \sim a$, k is the thermal conductivity, κ is the thermal diffusivity, and ν is the kinematic viscosity. Schubert & Zhang estimate that, in Jupiter, $Ra \sim 10^{12}$ if turbulent diffusivities are used to estimate Ra, and $Ra \sim 10^{24}$ for the case of molecular diffusivities. They argue that under these circumstances, convection in rapidly rotating Jupiter is likely to occur in the form of convective rolls with axes parallel to the planet's rotation axis, as originally proposed by Busse (1976). Figure 5 shows observed zonal winds in Jupiter's atmosphere; this pattern of winds is, for the most part, very stable in

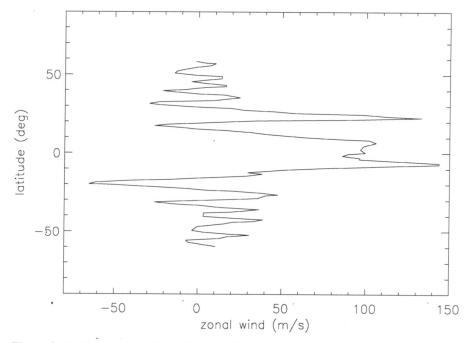

Figure 5   Jupiter zonal wind speeds (referred to a frame in which the magnetic field is stationary), as a function of Jupiter latitude.

time over years and possibly many decades (García-Melendo & Sánchez-Lavega 2001). If Busse and Schubert & Zhang are correct, these zonal winds could represent the surface expression of deep cylindrical zonal flows. Under these circumstances, the Poincaré-Wavre theorem may apply (Tassoul 1978). According to this theorem, if Jupiter (or Saturn) is barotropic throughout as a consequence of efficient convection, then any differential rotation that is present must be constant on cylinders. Figure 6 shows Hubbard's (1999) estimate of Jupiter's zonal harmonic spectrum, assuming that the planet rotates on cylinders at rates corresponding to the observed zonal winds in the northern and southern hemispheres, respectively.

A suitably instrumented low-periapse Jupiter orbiter could measure Jupiter's external gravity potential with an accuracy $\sim 10^{-9}$ for the vertical scale of Figures 4 and 6, and could thus shed light on the style of convection in Jupiter (Reitsema et al. 2001).

A determination of the style of convection in Jupiter may lead to a better understanding of transport mechanisms in more luminous EGPs and brown dwarfs. Schubert & Zhang argue that the much larger values of Ra in the latter, even for ω similar to Jupiter's, may lead to less well-defined belts and zones than in Jupiter and Saturn.

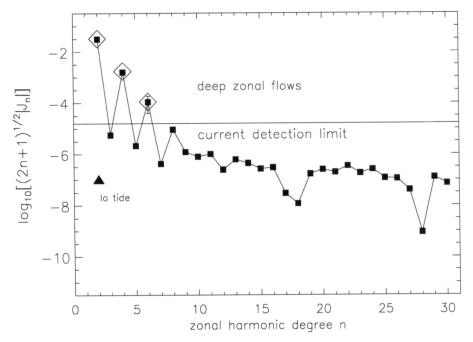

Figure 6   Jupiter zonal gravity harmonics, observed and predicted. Observed values are plotted as diamonds. Predicted values, for differential rotation on cylinders, are shown as dots.

As Figures 4 and 6 also make clear, an alternate determination of Jupiter's response to a quadrupole gravitational perturbation could be made by measuring the planet's tidal response to Io (and, probably, to the other Galilean satellites). This would be an alternative method for measuring Jupiter's degree of central condensation and the possible presence of a core, and should be feasible with a low-periapse orbiter with a sensitivity to gravitational perturbations comparable to that of the Galileo spacecraft. Measurements of Jupiter's response to satellite tides, in contrast to its response to rotation, should be relatively uncontaminated by interior dynamics.

FREE OSCILLATIONS?

Detection of free oscillations would provide a far more direct and detailed test of interior models of Jupiter and Saturn, as is done for the Sun via helioseismology. Numerous detailed theoretical predictions of expected jovian and saturnian modes and eigenfrequencies are available (Marley 1991, Mosser 1994, Gudkova & Zharkov 1999), but to date there is no unambiguous observational detection of free oscillations of any giant planet (Mosser et al. 2000).

One defines an acoustic characteristic frequency ν_0 in terms of the reciprocal travel time for a sound wave to propagate entirely through the planet:

$$\nu_0 = \left[2 \int_0^a \frac{dr}{c} \right]^{-1}, \tag{7}$$

where the sound speed c as a function of radius in the planet r is given in terms of the equation of state:

$$c^2 = \left(\frac{\partial P}{\partial \rho} \right)_S. \tag{8}$$

Then, (Mosser 1994), the eigenfrequencies of radial order n and spherical-harmonic degree ℓ are, crudely,

$$\nu_{n,\ell} \propto \left[n + \frac{\ell}{2} \right] \nu_0, \tag{9}$$

and for typical Jupiter interior models, the theoretical prediction is $\nu_0 = 152$ to 160 μHz. For Saturn, Mosser predicts $\nu_0 = 111$ μHz (coincidentally, the solar value is 136 μHz).

Mosser et al. (2000) find no direct identifications of jovian modes in their data, but they claim evidence for an oscillation characteristic frequency $\nu_0 = 142 \pm 3$ μHz. In order to be detectable with the spectroscopic technique used by Mosser et al., Jovian oscillations would need to have amplitudes of ~hundreds of meters. A possible model for excitation of Jovian oscillations is discussed by Bercovici & Schubert (1987), but predicted amplitudes are somewhat below the detection limit of Mosser et al.

Marley (1991) investigated the possibility that nonradial oscillations of Saturn might open gaps in the rings at resonant orbital positions. The modes that have potential ring resonances are at very low frequencies, a few tens of μHz. The absence of significant ring gaps at these resonances places a stringent upper limit on the surface amplitudes of these modes at ~1 meter.

ATMOSPHERES AND THEIR RELATION TO THE INTERIOR

The total heat flux rate from a giant planet's interior is characterized by an effective temperature T_{eff}, considered to be a composite value determined by the intrinsic heat flux from the deep interior, plus thermalized energy input from the primary star. Although both contributions are generally treated as arising from layers much deeper than the planet's effective photosphere, this approximation may not be valid in the case of "roasters" (see "Roasters"). The heat flux corresponding to the planet's intrinsic luminosity produces an extensive troposphere penetrating deep into the planet's core. The upper boundary of the troposphere is usually located close to the planet's photosphere, and is thus accessible to observation. The radiative properties of the atmosphere determine, for a given chemical composition, T_{eff}, and gravity g, the specific entropy of the adiabatic deeper layers of the planet.

Expected Compositions

Even for overall solar composition, and even without consideration of partitioning processes such as He immiscibility discussed above, the chemical composition of giant-planet atmospheres is appreciably altered by processes of condensation and equilibration between molecular species. Figure 7 (see color insert) depicts some of the relevant considerations (see Marley et al. 1996, Hubbard et al. 1997, Burrows et al. 1997). The black curves show, all for a surface gravity $g = 2200$ cm/s^2 (approximately the value for Jupiter), the $P-T$ profiles for five solar-composition atmospheres with T_{eff} decreasing from 1000 K to 128 K (close to the jovian value). Solid black curves show radiative portions of the atmosphere, whereas short-dashed black curves show convective regions. The black dots show the approximate location of the photosphere at thermal wavelengths. This progression of five curves shows, roughly, a sequence of profiles traversed by a proto-Jupiter as it evolves from high interior entropy to lower values. The deep interior entropy is represented parametrically, in the theory of Burrows et al. (1997), by T_{10}, the temperature at a pressure of 10 bars that would lie on the deep-interior adiabat if the latter were extended to a pressure of 10 bars. The actual atmosphere may or may not lie on the deep adiabat at 10 bars, and may or may not have an actual temperature of T_{10} at 10 bars (present-day Jupiter does). Thus, in order to obtain the relevant value of T_{10}, it may be necessary to integrate a model atmosphere to much higher pressures than 10 bars, depending on where it becomes fully convective and stays convective.

As shown in Figure 7 (see color insert), the molecules appearing in the giant planet's atmospheric spectrum will be affected by chemistry. Dashed green curves show the locus of points where a solar abundance of a given molecule will produce a precipitate of a condensed phase of that molecule. To the left of these dashed curves, the molecule will be depleted. Some of the most important species for giant planets are shown: ammonia, water, enstatite, and iron. Also shown as solid green lines are the loci of approximately equal abundances of nitrogen-bearing molecules (competing between NH_3 on the left and N_2 on the right), and approximately equal abundances of carbon-bearing molecules (competing between CH_4 on the left and CO on the right). Shown in red in Figure 7 (see color insert) are theoretical $P-T$ profiles for the brown dwarf Gl 229B and the roaster HD 209458b (discussed in "Roasters").

$T_{\text{eff}}-T_{10}$ Relation

Nongray model atmospheres (Marley et al. 1996, Burrows et al. 1997) have been calculated for a broad range of surface gravities g and effective temperatures T_{eff} to define a surface, as shown in Figure 8. This surface can be coupled to a thermodynamic code for the interior of a giant planet or brown dwarf to yield a complete description of the expected values of T_{eff} and radius a as a function of

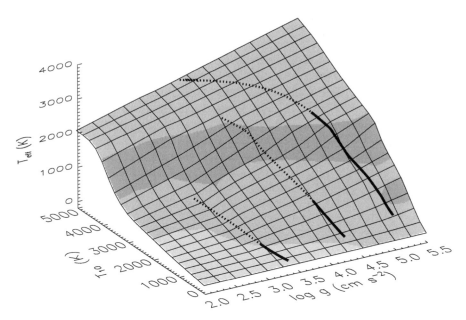

Figure 8 The T_{eff}–T_{10}–g surface used for calculating evolution of isolated giant planets. Evolutionary trajectories for 1, 10, and 40 M_J planets are shown. The early parts of their evolution, for ages less than 0.1 Gyr, are shown dashed.

mass M and age t. As depicted in Figure 8, the evolution of our own giant planets Jupiter and Saturn is now predicted as a subset of a general theory for all objects in the rough mass range $10^{-3}\ M_\odot \geq M \geq 10^{-1} M_\odot$.

Figures 7 and 9 illustrate how the deeper P–T profiles of more massive objects at later ages (or, in the case of HD 209458b, at arrested development) reprise the P–T profiles of Jupiter or Saturn at very young ages. In Figure 9, oriented like Figure 8 with depth in the atmosphere increasing downward, we see the evolution of the center of the brown dwarf Gl 229B (dot-dashed line) and the P–T profiles of Gl 229B at $t = 0.007$ and 2.0 Gyr (heavy short-dashed lines). The corresponding evolution of the center of Jupiter (other dot-dashed line) and profiles for Jupiter at $t = 0.1$ Gyr and 4.5 Gyr (present) are shown as lighter short-dashed lines. The present P–T profile of Saturn is also shown.

Condensates

Condensation is possible in an atmosphere when the Gibbs free energy of a substance (atomic, or molecular) in the condensed phase becomes less than the Gibbs free energy in the gas phase. When the condensate is composed of multiple species that exist separately or as different compounds in the gas phase, then the Gibbs free energy of each of the components in the condensed phase must be less than that in the gas phase (so-called heterogeneous condensation). This strict

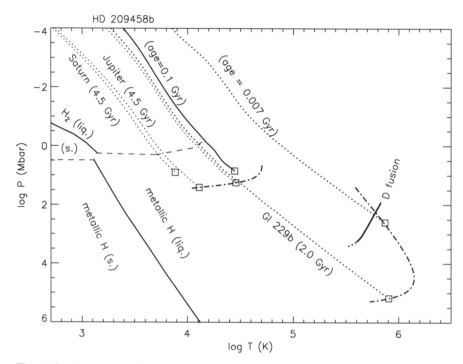

Figure 9   Trajectories of giant planets and brown dwarfs on the high-pressure hydrogen phase diagram (see Figure 1).

thermodynamic criterion collapses to the meteorologically more familiar criterion for a single condensing species, that the partial pressure of the species in the gas phase must exceed the saturation vapor pressure at the given temperature (homogeneous condensation). It governs the formation of water clouds in the troposphere of the Earth, whereas the stricter multi-component Gibbs condition applies to a variety of observed and predicted clouds formed heterogeneously in planetary atmospheres (Rossow 1978).

The criteria imposed above generally set an ordering to clouds of given compositional classes in all atmospheres of similar composition. Those substances whose bond strength is weakest tend to form clouds at the lowest temperature. Thus, ionically and covalently bonded species ranging from calcium-aluminum silicates through iron tend to form clouds in the warmest atmospheres, or for a given atmospheric temperature profile, they form clouds at the deepest levels. The hydrogen-bonded compounds of sulfur, oxygen, and nitrogen (hydrosulfides, water, and ammonia) form clouds at lower temperature, hence at a higher altitude in a given atmosphere. Finally, van der Waals bonded species like methane require very low temperatures for cloud formation, and are seen only in cold giant planets and at relatively high altitude. Figure 7 (see color insert) gives the condensation curves of a few of the major cloud-forming species compared with the temperature-pressure profiles of objects of various effective temperature. Cloud formation deep in a giant planet atmosphere can play a fundamental role in affecting abundances in the observable atmosphere, an effect we discuss for water in Jupiter in "Galileo Probe Results" below. But clouds forming at atmospheric levels corresponding to the observable atmosphere can have a big influence on the radiative properties and appearance of giant planets. For water and ammonia, this occurs for giant planets with effective temperatures in the 100–200 K range, whereas for iron and magnesium silicates, effective temperatures around 1000 to 2000 K are required.

Cloud formation can be inhibited by lack of appropriate nucleation sites to which atoms or molecules can adhere, so that vapor pressures well above the saturation value might occur under relatively dust-free or very cold conditions. However, at all atmospheric levels where the temperature drops with altitude approximately like an adiabat, cloud formation is not long delayed by nucleation issues, because the supersaturation (ratio of condensable partial pressure to saturation pressure) rises steeply with increasing altitude. Thus the vapor pressure curves of Figure 7 (see color insert), are a good guide to where the cloud base occurs (for enstatite, the curve is not strictly a vapor pressure curve, but should simply be intepreted as the locus of $P-T$ points representing the boundary of cloud formation).

The positions and compositions of clouds in Jupiter and Saturn were predicted from their thermodynamical properties over a quarter of a century ago (Weidenschilling & Lewis 1973). In Jupiter, and to a lesser extent in Saturn, the ammonia and water clouds play key roles in, respectively, the amount of sunlight absorbed by the atmosphere as well as the spectral distribution of the infrared emission. The well-known variability in the emission of thermal radiation from Jupiter at 5 μm is due to the presence of relatively cloud-free regions, extending down below the ammonia cloud through the water cloud region. At times large holes form, apparently under the influence of mesoscale subsidence or downdrafts. These increase the observed 5-μm emission, and one such hole appeared at the target point of the Galileo entry probe (Ortiz et al. 1998). In consequence, the nephelometer on Galileo detected little or no cloud material in regions that normally should have thick cloud decks. There were compositional implications, also, discussed in "Galileo Probe Results" below.

We can directly observe clouds and cloud morphology on the giant planets of our own solar system, but not yet on EGPs. However, EGP clouds must be modeled if accurate models of the spectrum and atmospheric energy balance are to be obtained. In particular, the cloud particle size distribution as a function of altitude, and the extent of horizontal heterogeneity, must be computed. Complicated models exist for Jupiter (Yair et al. 1995), but these are too cumbersome to be used with radiative-convective codes, and insufficient information exists to choose parameters for regimes that are well away from those of Jupiter. A simple model of cloud particle growth was applied to silicate clouds in brown dwarfs over a decade ago (Lunine et al. 1989), and has been progressively refined since then. The model compares timescales for particle growth, sedimentation, and (where appropriate) convective upwelling to determine the maximum grain size. Such models predict dramatic albedo changes for giant planets with, versus without, clouds (Sudarsky et al. 2000, Marley et al. 1999, Burrows et al. 1998) as a function of the cloud properties themselves. They also show that the assumption made by some modelers, of an interstellar particle size distribution for silicate clouds in hot Jupiters, is not correct. The most elaborate recent models include provision for partial rainout of cloud droplets (Ackerman & Marley 2001). The next frontier in development of general cloud models for giant planets is to account for horizontal heterogeneities on the base of plume development in convectively overturning atmospheres, but using models simple enough (Stoker 1986) to be incorporated in full-up radiative-convective codes.

Galileo Probe Results

In December 1995, the Galileo probe descended through the atmosphere of Jupiter and returned data to pressure levels slightly exceeding 20 bars. The Probe Mass Spectrometer provided the abundances of noble gases and selected heavy elements; for technical reasons the ammonia, hence nitrogen, abundance could not be determined from the mass spectrometer data (Mahaffy et al. 2000). The data and error bars are tabulated in those references; here we provide a narrative summary, expressed in terms of the ratio relative to hydrogen. The Galileo Mass Spectrometer (Niemann et al. 1996) and (separately) the Helium Abundance Detector (Von Zahn & Hunten 1996) both measured a helium mass fraction Y that is depleted by about 5% relative to the accepted protosolar value of Y. Such depletion may be consistent with models of Jupiter's thermal evolution constrained by its present luminosity and presumed age, if it is assumed to be due to the formation and rainout to the deep interior of a separate helium-rich fluid phase. However, a slightly better simultaneous fit to both Jupiter and Saturn is obtained with no helium depletion in the former (Hubbard et al. 1999; see also "Application of Cooling Theory to Jupiter and Saturn" below), so we take the 5% depletion number as the upper limit to the helium depletion in Jupiter. Since the mass of helium in Jupiter is approximately 5×10^{29} g, and at most 5% of the mass fraction of helium has differentiated, then the total mass of immiscible helium available as a solvent for other noble gases is 2.5×10^{28} g.

The Galileo mass spectrometer measured an abundance of neon in the atmosphere of only 0.1 times solar, a strong depletion that has been interpreted as direct evidence of the solvation of the noble gases in the helium-rich phase (Owen et al. 1999). For a solar neon abundance of 2×10^{-2} relative to hydrogen (mass fraction), most of the neon—roughly 4×10^{28} g—must be sequestered in the helium phase. This corresponds to a mole fraction of 0.3. The required solubility is large, but is not ruled out by the statistical mechanical calculations of noble gas solubility done to date (Roulston & Stevenson 1995). Alternatively, the neon might have been removed from the upper layers as a separate immiscible phase. Unfortunately, the high-pressure properties of the noble gases, and their solubility in metallic hydrogen and high-pressure helium, are very poorly known. The remaining noble gases (argon, krypton, xenon) have abundances relative to hydrogen that are enriched, approximately uniformly, by a factor of three relative to solar. Thus, formally these elements have identical enrichments in Jupiter, though we point out that the error bars are sufficiently large that the ratios of these species span a fair range: Ar/Xe, for example, could range from roughly 0.5 to unity. The enrichment over the solar value relative to hydrogen implies that at least half of the Jovian inventory of these gases was acquired through a solid carrier phase, the solar nebula itself being capable of supplying just a solar abundance worth of these gases. More of these gases may be sequestered deep in the Jovian interior in a helium-rich phase. Indeed, if the solubilities of these species in helium are comparable to that of neon, then these species are many times super-solar in Jupiter, and small differences in their relative solubilities could greatly alter the relationship between the measured atmospheric and bulk values. Carbon (as methane) and sulfur (as H_2S) are also enriched by roughly a factor of three over solar. The abundance of elemental nitrogen in the Jovian atmosphere is determined from ground-based microwave observations of ammonia to be roughly half solar at pressure levels of a couple of bars (de Pater et al. 2001), but three to four times solar beyond 8 bars based on attenuation by NH_3 of the probe radio signal. While the probe nephelometer detected little or no cloud material, the pattern suggests that global condensation of ammonia or ammonium hydrosulfide clouds, coupled with the regional downdraft within which the probe fell, alter the ammonia abundance profile. Water, which condenses in deeper clouds, is strongly subsolar in the Probe Mass Spectrometer data, but increasing slightly in abundance at deeper levels. The probe's descent through a dry region of subsidence makes oxygen determination impossible from the water abundance, and much deeper measurements (down to 100 bars?) will be required.

The interpretation of the enriched noble gases and major element abundances rests on the plausibility that the giant planets accreted large amounts of icy planetesimals in addition to the surrounding solar nebula gas, a plausibility established by constraints on the interior already discussed. It also requires us to suspend, for the moment, concerns that differing solubilities of noble gases in the Jovian interior have fortuitously created a uniform enrichment pattern in the upper envelope. Under these assumptions, two endmember points of origin for the planetesimals can be invoked. The uniform enrichment pattern of noble gases is reproduced in the lab by codepositing amorphous water ice and trapped gases on a cold finger at temperatures of 25–30 K (Owen et al. 1999). This is equivalent to forming Jupiter from icy planetesimals that were never heated much above that temperature, and plausibly formed in the solar system's nascent giant molecular cloud with little further alteration. Such planetesimals would dominate the noble gas and major element (C, N, S) pattern in Jupiter if the giant planet formed much farther out than 5 AU, perhaps 30 AU initially, or if preferential accretion of material from much greater orbital distances occurred in the late stages of Jupiter's formation. Dynamically, the latter is more plausible than the former.

Alternatively, the noble gas and major element pattern can be reproduced by assuming that Jupiter accreted icy planetesimals from the 5 AU region over a 2-million-year period of cooling of the nebula, during which trapping of the volatiles in successive stages of clathrate formation occurs (Gautier et al. 2001). In this model, temperatures as low as 35 K are required, but all of the accretion is of local planetesimals. This picture required that the nebula at 5 AU became very cold during its later stages (Drouart et al. 1999), which is plausible if disk accretion slowed or ceased before the nebular gas was dissipated (Chiang & Goldreich 1997). This possibility can be tested when SIRTF observes the spectral energy distributions of young stars possessed of gaseous circumstellar disks (Chiang et al. 2001).

The two hypotheses for the enrichment of Jupiter's envelope can be distinguished by the predicted oxygen abundance deep in Jupiter, since water was the carrier phase for the other volatile species. The amorphous ice model predicts an enrichment of oxygen close to that of the other elements, about four times solar, while the clathrate model requires a factor of eight times solar. Thus, measurement of the deep-water abundance in Jupiter, the primary oxygen-bearing molecule, is fundamental to deciding how Jupiter obtained its inventory of heavy elements.

EGP Atmospheres

This is a huge subject, and is covered extensively in another very recent review (Burrows et al. 2001). For EGPs close to a Jupiter mass formed in a disk of material orbiting the parent star, regardless of whether they form by nucleated or direct collapse, it is likely their atmospheres will exhibit a metallicity enrichment relative to the parent star. Much more massive EGPs may not exhibit such metallicity enrichment because they formed by direct collapse in the absence of a gas and particulate disk, or they are so massive that the available solid material does not enrich the envelope very much, or both. But even in a system of metallicity depressed from solar by one or two orders of magnitude, the atoms and molecules of the heavy elements will determine the atmospheric opacity and dominate the spectral appearance of these objects (Burrows et al. 1997). Over restricted regimes of effective temperature and gravity, grains will also affect or dominate the transfer of radiation and appearance of the atmospheres. Therefore, modeling of EGP atmospheres is extremely complex.

Figure 10 shows model EGP spectra at low resolution, for a 1 M_J object at different ages, which corresponds principally to a decline in the effective temperature and much less to an increase in the gravity at the photospheric surface. Jupiter's signature 5-μm flux enhancement is preserved at higher effective temperatures, though to less dramatic effect. A similar model of the near-infrared, based on the observations of GL229 B, is shown in Marley et al. (1996). The latter object is close to the maximum effective temperature at which methane dominates the infrared spectrum, and is thus at the top of a sequence of EGPs whose lower limit

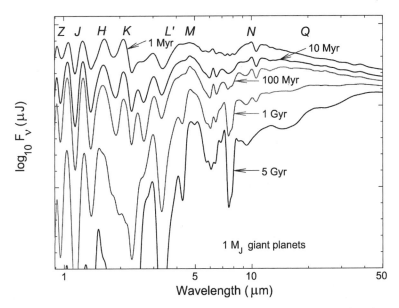

Figure 10   Synthetic spectra of isolated Jupiter-mass planets as a function of age.

corresponds to Jupiter and Saturn. The presence of water condensation clouds in objects as warm as 400 K effective temperature, down through Saturn with an effective temperature of about 90 K, strongly suppresses the effect of water vapor bands on the infrared spectrum. However, methane remains a signature feature of the giant planet until it too becomes a condensation cloud in the effective temperature range of Uranus and Neptune (70 K).

The complexity of these atmospheres requires the inclusion of hundreds of millions of lines to treat the gas opacity, and the inclusion of reasonably intricate models of grain growth and vertical distribution of cloud particles. These models therefore demand high computational power and memory, and must be constrained by both nearby low-temperature giant planets (Jupiter and Saturn) and warmer EGPs.

THERMAL EVOLUTION OF GIANT PLANETS

The availability of numerical techniques and sufficient computing power for accurate calculation of nongray model atmospheres in the region of $T_{eff}-g$ space relevant to brown dwarfs and giant planets has allowed the development of a unified theory for the evolution of brown dwarfs, EGPs, and Jupiter and Saturn. Furthermore, the theory can be calibrated by comparing the predicted evolution of Jupiter and Saturn with the evolution of observed EGPs and brown dwarfs.

Brown Dwarf-like Evolution

Since EGPs are predominantly metallic hydrogen/helium mixtures and have fully convective cores, they are the low-mass end of a continuum of objects, the substellar-mass objects (SMOs), that includes brown dwarfs with masses up to $\sim$0.075 $M_\odot$ ($\sim$80 M_J; 1.0 $M_\odot \equiv$ 1047 M_J). Hence, the same general theory that addresses the structure and evolution of brown dwarfs also applies to the study of giant planets (Burrows et al. 2001). Planets and brown dwarfs would ideally be distinguished on the basis of mode of formation, something not easily determined, and different origins might entail different heavy-element compositions, rotation rates, and mass functions for a class. Nevertheless, a suite of evolutionary and spectral calculations that encompasses a mass range from 0.2 M_J to 80 M_J, an age range from 10^6 to 10^{10} years, and T_{eff}s ranging from 80 K to 3000 K will encompass EGP behavior.

Due to the competition of Coulomb and electron-degeneracy effects, for a broad range of SMO masses from 0.3 M_J to 70 M_J, to within about 30% the radii of old objects are independent of mass. Coulomb effects set a fixed density and interparticle scale ($\sim$1 Å), which would make the radius a weakly increasing function of mass ($a \propto M^{1/3}$), and electron degeneracy effects would make the radius decrease with mass ($a \propto M^{-1/3}$), as for the family of white dwarfs. The resulting competition renders the radius nearly constant over roughly two orders of magnitude in mass near the radius of Jupiter ($\sim$0.1 solar radius), as long as the age is large enough for the thermal contribution to the total pressure P to be small compared with P (in Jupiter, the thermal contribution to the pressure is typically $\sim$10% of P, but in the EGP HD209458b, the contribution is much larger).

Figures 9 through 13, 14 (see color insert), and 15 show various aspects of the evolution of giant planets and low-mass brown dwarfs, based on the assumption of complete convective structure and homogeneous composition. In a provisional definition of planethood by an IAU working group struggling with the matter, giant planets are distinguished from brown dwarfs by the absence of deuterium fusion in the former. As is evident in Figure 15, the boundary of deuterium burning lies at a well-defined mass of 13 M_J.

Application of Cooling Theory to Jupiter and Saturn

Figures 12 and 13 show the predicted variation of T_{eff} with age for isolated giant planets of various masses. Figure 16 shows a close-up of the predictions of the same cooling theory for Jupiter and Saturn (Hubbard et al. 1999). The present T_{eff} s of Jupiter and Saturn are shown with horizontal lines, and the uncertainty in these values (Conrath et al. 1989) is shown with error bars at the planets' present age. The cooling curves marked with ×s are for isolated planets, the same as in Figures 12 and 13. The curves marked with dots show the effects of an approximate correction for irradiation from the Sun. Curves marked with ○s show the correction with solar luminosity fixed at the present value for all ages of Jupiter and Saturn, while curves marked with ●s allow for the predicted variation of solar luminosity at earlier epochs (lower than the present value). The dashed segments for Jupiter

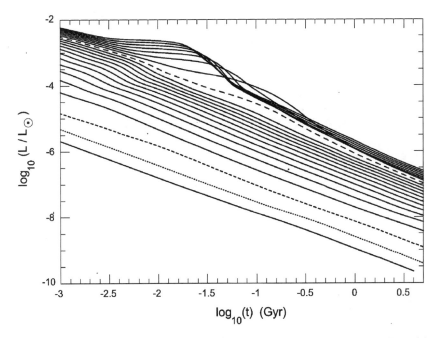

Figure 11 The luminosity (in solar units) versus age (in Gigayears) for EGPs with masses of (starting with bottom curve) 0.3 M_J, 0.5 M_J, 1.0 M_J and from 2.0 to 20.0 M_J in steps of 1.0 M_J. Epochs from 10^6 to 5×10^9 years are portrayed. Note that the models are all assumed to be homogeneous with solar composition and that no core of heavy elements has been incorporated. The evolution of the 13 M_J EGP/brown dwarf is depicted with a dashed line. The bumps near a few $\times 10^7$ years are consequences of deuterium burning, first manifest at $\sim$13.0 M_J. In these calculations, we have assumed a fractional baryon abundance for deuterium (Y_d) of 2×10^{-5}. Varying this abundance changes the minimum mass for deuterium burning only slightly (from $\sim$14.0M_J at very low Y_d to $\sim$12.5M_J for twice our reference value). However, the duration of the deuterium-burning phase scales almost directly with Y_d. Note that at 10^6 years the bolometric luminosity of a "Jupiter" is only a few factors below that of an old star with a mass of $\sim$0.75 solar masses at the stellar main sequence edge.

show the last time step, with no correction for chemical separation in Jupiter. The general cooling theory used for EGPs and brown dwarfs works almost perfectly for Jupiter. There is no need to invoke any additional energy source in Jupiter. And, the theory for Jupiter is consistent with the H-He phase separation curves shown in Figure 3, either the curves based on DeWitt & Hubbard (1976) or on Pfaffenzeller et al. (1995).

As is evident in Figure 16, the homogeneous cooling theory predicts a T_{eff} for Saturn much lower than the observed value at present. Hubbard et al. (1999) made an approximate evaluation of the amount of He separation required to sufficiently

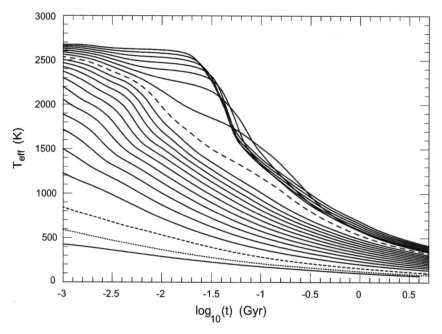

Figure 12 The effective temperature (T_{eff}) in Kelvin versus age in Gigayears for the models depicted in Figure 10. A young "Jupiter" has a T_{eff} above 800 K at an age of 10^6 years. Without stellar insolation or element differentiation, Jupiter and Saturn would have T_{eff}s of $\sim$106 K and $\sim$72 K, respectively.

prolong the cooling of Saturn, and estimated that a reduction of the atmospheric helium mass fraction from $Y \approx 0.27$ to $Y \approx 0.20$ might suffice, as indicated by the abrupt change in slope in the Saturn cooling curve at the last time step.

After Hubbard et al. (1999) was published, Conrath & Gautier (2000) re-evaluated Voyager measurements and concluded that the Saturn atmospheric He mass fraction might lie in the range $Y = 0.18$ to 0.25. The new observational upper limit is actually compatible with solar helium abundance, so another source of mass separation in Saturn might be needed to explain the present T_{eff}. This would be consistent with the Pfaffenzeller et al. (1995) phase diagram.

"Roasters"

As with brown dwarfs, the atmospheres of EGPs are dominated by molecules such as H_2, H_2O, CH_4, CO, and NH_3. Furthermore, at the higher effective temperatures encountered for massive and/or young EGPs, neutral alkali metals, clouds of water droplets, clouds of silicate grains, and clouds of iron droplets might play observable roles in spectrum formation and early evolution. Silicate and iron grains are expected at the high-pressure depths of even cool giant planets, such as Jupiter

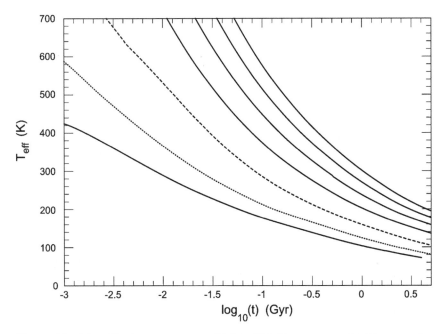

Figure 13 The effective temperature (T_{eff}) in Kelvin versus age in Gigayears for EGPs with masses of (starting from bottom) 0.3 M_J, 0.5 M_J, 1.0 M_J, 2.0 M_J, 3.0 M_J, 4.0 M_J, and 5.0 M_J. This figure is a close-up of Figure 11 for the lowest EGP masses.

and Saturn, but may also be manifest in the upper atmospheres of hot "roasters" in proximity to their central stars and in the atmosphere of a young proto-Jupiter. Hence, chemistry and molecular opacities play central roles in the theory of giant planet structure, evolution, spectra, and albedos.

In proximity to its central star, an EGP can achieve and maintain high surface temperatures that might exceed 1000 K. Furthermore, the day-night effect and phase dependence in both the near infrared and the optical can be pronounced. Close-in EGPs ("roasters") such as HD187123b, HD209458b, τ Boo b, HD75289b, 51 Peg b, υ And b, and HD217107b (Mayor & Queloz 1995; Butler et al. 1997, 1998, 1999; Fischer et al. 1999; Charbonneau et al. 2000; Udry et al. 2000) fall into this category and will radiate far in excess of Jupiter or Saturn (with T_{eff} s of only ~125 K and ~95 K, respectively). These hot EGPs should look more like brown dwarfs than Jovian planets.

Proximity not only leads to elevated surface temperatures, but can alter the structures and radii of giant planets. This is most clearly demonstrated with HD209458b, which has been observed to transit its primary (Charbonneau et al. 2000, Henry et al. 2000, Mazeh et al. 2000, Jha et al. 2000, Brown 2001). The transit by a giant planet of a main sequence star will lead to a photometric dip of 1–2%, easily measured from the ground. However, from the precise magnitude of this dip, one can

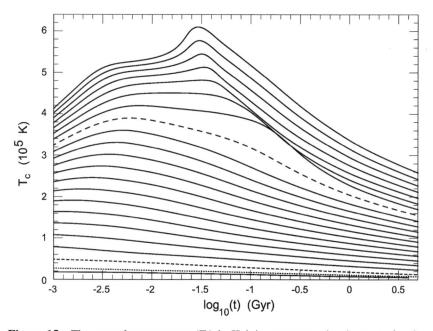

Figure 15 The central temperature (T_c) in Kelvin versus age in gigayears for the homogeneous models shown in Figure 10. The lowest three curves show, starting from the bottom, EGPs with masses of 0.3 M_J, 0.5 M_J, and 1.0 M_J, respectively. As in Figure 10, the higher dashed curve is for a model with a mass of 13.0 M_J. The peaks are associated with deuterium burning, which requires a core temperature above $\sim 4 \times 10^5$ Kelvin.

estimate the planet's radius, which for HD209458b is derived to be ~ 1.35 Jupiter's radius. Detailed modeling of the transit lightcurve yields a radius for HD209458b at a pressure of 1 bar equal to 94430 km, with an uncertainty of ~ 500 km (Hubbard et al. 2001). Radial-velocity measurements yield its mass ($\sim 0.69\,M_J$) and with both a mass and a radius the basic physics of the object can be studied.

On first glance, a radius for HD209458b that is 35% larger than Jupiter's should be problematic, since at the inferred T_{eff} of its skin (~ 1000–1500 K) the atmospheric scale height is still only $\sim 1\%$ of the planet's total radius a. However, irradiation from the primary, HD209458, over millions of years, moves the radiative/convective boundary deeper to higher pressures and flattens the inner atmospheric temperature gradient. Since radiative diffusion of heat is slower at higher pressures and is driven by temperature gradients, the upshot is drastically reduced heat loss from the planet's core. Therefore, the planet's core entropy is kept higher longer; it is the entropy and the mass that determine the planet's radius. The large stellar flux does not inflate the planet, but retards its otherwise inexorable contraction from a more extended configuration at birth. In fact, stellar irradiation can be

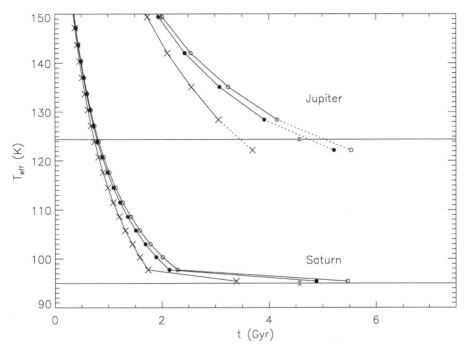

Figure 16   Final stages of evolution of Jupiter and Saturn, with effects of solar irradiation and He phase separation schematically included.

responsible for maintaining a giant planet's radius at a value 20% to 80% larger than that of Jupiter itself (Guillot et al. 1996, Burrows et al. 2000).

Since the scale-height effect is small, the $a-t$ trajectory of the isolated EGP immediately suggests that if HD209458b were allowed to reside at large orbital distances (≥ 0.5 A.U.) for more than a few $\times 10^7$ years, its observed radius could not be explained (Burrows et al. 2000). It is at such epochs that the radius of an isolated $0.69\text{-}M_J$ EGP falls below HD209458b's current radius. This implies either that HD209458b formed near its current orbital distance or that it migrated in from larger distances (≥ 0.5 A.U.), no later than a few times 10^7 years after its birth.

ORBITAL EVOLUTION OF GIANT PLANETS

The detection of dozens of giant planets spanning orbits from 0.05 AU out to several AU (Marcy et al. 2000) speaks of a formation process that either allows growth of the gaseous envelope at small orbital distances, or requires inward migration of newly formed giant planets. Formation at small orbital distances requires severe assumptions (Wuchterl et al. 2000), and is likely not a plausible mechanism for explaining the large number of close-in giant planets. A number of mechanisms for inward migration have been proposed, though interaction with the primordial gaseous disk (Lin et al. 1996) or with a later massive particulate disk (Murray et al. 1998) probably are more common fates than gravitational scattering among multiple giant planets (Weidenschilling & Marzari 1996).

Whatever the cause, the structure of HD209458b, based on the transit-observed radius and informed by modeling of the sort described above, requires that migration be complete within a few tens of millions of years, if not earlier (Burrows et al. 2000). Hence, migration is a process that in general is likely undergone early and often by giant planets, and we do not know if the stately distribution of the giant planets of our own solar system, from Jupiter at 5 AU to Neptune at 30 AU, is anomalous. Since the orbits of terrestrial planets in the 1 AU habitable zone around G-type stars are not dynamically stable if a giant planet exists at less than 3 AU in the same system (Jones et al. 2001), the answer to this question is a fundamental one in the search for habitable planets (Beichman et al. 1999). The answer is unlikely to come from theory, because migration–particularly in a gaseous disk–depends sensitively on the disk properties, giant planet masses, and timing (Ward & Hahn 2000). Instead, techniques to detect giant planets in orbits of 5 AU and greater must be brought to bear to remove the observational bias toward small semi-major axes that characterizes Doppler spectroscopic detection (Lunine 1995). Simple migration models (D.E. Trilling, unpublished; see Lunine 2001) do suggest however, that the cohort of Doppler spectroscopic detections represents perhaps one third of the total population of giant planets around nearby F, G, and K stars. If so, a wealth of giant planets may await detection by astrometry or direct imaging.

In our own solar system, the extent to which the giant planets migrated is limited weakly by dynamical studies as well as by the implausibility of forming Uranus and Neptune at 100 AU instead of 20–30 AU. The presence of remnant bodies in resonances in the Kuiper Belt could be the result of resonance sweeping as the orbits of the giant planets relaxed, or spread out, after formation (Malhotra 1995). Such migration represents only 10% or so of the current orbital semi-major axes, and there is little other evidence for more substantial migration of the giant planets in our own planetary system (Malhotra et al. 2001).

As discussed above, the environment and properties of close-in giant planets ("roasters") is so different from that of giant planets in looser orbits that their study is in some respects a subfield of EGP research unto itself. Thus, despite the failure of theory to anticipate them, the abundance of roasters supplied by migration is a happy circumstance for modelers whose tastes run to the extreme.

CONCLUSIONS

Prior to about ten years ago, our solar system's giant planets Jupiter and Saturn were astronomical curiosities, dismissed as "gas bags" by many planetary scientists, their masses too low to be of great interest to many astrophysicists. The observational discovery of bona fide brown dwarfs and EGPs has changed our perspective. Even the definition of planet has become controversial, as we face the possibility that ~ 1 M_J free-floaters may exist, and that some stars may be simultaneously

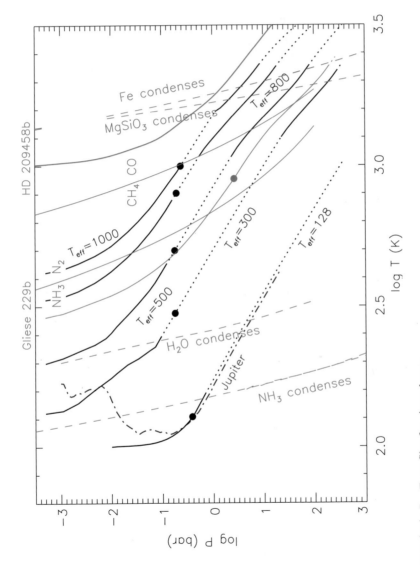

Figure 7 Atmospheric $P - T$ profiles for giant planets.

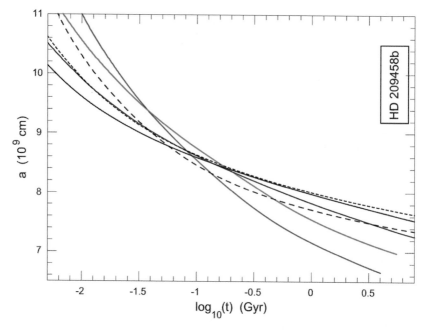

Figure 14 The radius a (in 10^9 centimeters) versus age t (in gigayears) for EGP masses of 0.3 M_J (*red*), 0.5 M_J (*green*), 1.0 M_J (*blue*), 2.0 M_J (*solid black*), 5.0 M_J (*dashed black*), and 10. M_J (*long-dashed black*). Note that the radius of an EGP is not a monotonic function of mass at any age and that the radius evolution curves cross (for Jupiter, $a \sim 7 \times 10^9$ centimeters). The box on the right shows the approximate transit-derived radius of the "roaster" HD 209458b, whose radius at $P = 1$ bar is inferred to be 94430 ± 500 km (Hubbard et al. 2001).

orbited by brown dwarfs and giant planets. To be sure, we do not know that all EGPs are hydrogen-rich, but the "roaster" HD209458b is incontrovertibly hydrogen-rich, and it seems likely that most EGPs more massive than $\sim 1\ M_J$ are as well.

There are good prospects that, within the next decade, we will have data on the spectra and atmospheric compositions of EGPs. As Figure 7 (see color insert) makes clear, the younger, higher-entropy EGPs may display in their atmospheres species such as H_2O that are buried, or mostly so, under the clouds of Jupiter and Saturn. The EGP HD209458b is not quite hot enough to offer the possibility of detecting atmospheric silicate clouds (Hubbard et al. 2001), but it is not far removed from that point. In Jupiter, the putative silicate clouds are buried at kilobar pressures. Determination of the location and mass of such refractories, and whether they are primarily concentrated in massive cores or distributed throughout the interior, will be fundamental for testing theories of the origin of giant planets. Indeed, it may be fundamental for determining whether they are formed from stellar-like collapse from hydrogen gas, or from planet-like accretion onto refractory cores.

It is encouraging that the general properties of giant planets and brown dwarfs can be understood within the framework of a unified theory. However, there are many unanswered questions concerning their interior heat transport processes and interior chemistry. Answering these questions will require further theoretical studies as well as high-pressure experiments on light elements. Intensive investigations of Jupiter and Saturn, making in situ and close-in remote sensing measurements of their deep interior compositions, as well as high-resolution measurements of their external gravitational and magnetic fields, will be essential for understanding this newly-important class of objects.

ACKNOWLEDGMENT

Some of the research mentioned in this review was supported by Grants NAG5-7211, NAG5-7499, and NAG5-10629 (NASA Origins Program), NAG5-4214 (NASA Planetary Astronomy), and NAG5-7073 and NAG5-10760 (NASA Astrophysics Theory).

The *Annual Review of Astronomy and Astrophysics* is online at
http://astro.annualreviews.org

LITERATURE CITED

Ackerman A, Marley M. 2001. *Astrophys. J.* 556:872–84

Beichman C, Woolf NJ, Lindensmith CA. 1999. *The Terrestrial Planet Finder.* Washington, DC: NASA

Bercovici D, Schubert G. 1987. *Icarus* 69:557–65

Brown TM. 2001. *Astrophys. J.* 553:1006–26

Burrows A, Guillot T, Hubbard WB, Marley M, Saumon D, et al. 2000. *Astrophys. J.* 534: L97–100

Burrows A, Hubbard WB, Lunine JI, Liebert J. 2001. *Rev. Mod. Phys.* 73:719–65

Burrows A, Marley MS, Hubbard WB, Lunine JI, Guillot T, et al. 1997. *Astrophys. J.* 491:856–75

Burrows A, Marley M, Hubbard WB, Sudarsky D, Sharp C, et al. 1998. The spectral character of giant planets and brown dwarfs. In *Cambridge Workshop Cool Stars, Stellar Systems and the Sun, 10th, Astron. Soc. Pac. Conf. Ser.*, ed. RA Donahue, J Bookbinder, pp. 27–46. San Francisco: Astron. Soc. Pac. 549 pp.

Busse FH. 1976. *Icarus* 29:255–60

Butler RP, Marcy GW, Fischer DA, Brown TM, Contos AR, et al. 1999. *Astrophys. J.* 526:916–27

Butler RP, Marcy GW, Vogt SS, Apps K. 1998. *Astron. Soc. Pac.* 110:1389–93

Butler RP, Marcy GW, Williams E, Hauser H, Shirts P. 1997. *Astrophys. J.* 474:L115–18

Chabrier G. 1994. The equation of state of fluid hydrogen at high density. In *The Equation of State in Astrophysics*, See Chabrier & Schatzman 1994, pp. 287–305. Cambridge, UK: Cambridge Univ. Press. 621 pp.

Chabrier G, Saumon D, Hubbard WB, Lunine JI. 1992. *Astrophys. J.* 391:817–26

Chabrier G, Schatzman E. eds. 1994. *The Equation of State in Astrophysics*, Cambridge, UK: Cambridge Univ. Press. 621 pp.

Charbonneau D, Brown TM, Latham DW, Mayor M. 2000. *Astrophys. J.* 529:L45–48

Chiang EI, Goldreich P. 1997. *Astrophys. J.* 490:368–76

Chiang EI, Joung MK, Creech-Eakman MJ, Qi C, Kessler JE, et al. 2001. *Astrophys. J.* 547:1077–89

Cochran WD, Hatzes AP, Butler RP, Marcy GW. 1997. *Astrophys. J.* 483:457–63

Collins GW, Da Silva LB, Celliers P, Gold DM, Foord ME, et al. 1998. *Science* 281:1178–81

Conrath BJ, Gautier D. 2000. *Icarus* 144:124–34

Conrath BJ, Hanel RA, Samuelson RE. 1989. Thermal structure and heat balance of the outer planets. In *Origin and Evolution of Planetary and Satellite Atmospheres*, ed. SK Atreya, JB Pollack, MS Matthews, pp. 513–38. Tucson: Univ. Ariz. Press. 881 pp.

Cui T, Takada Y, Cui Q, Ma Y, Zon G. 2001. *Phys. Rev. B* 6402:4108

Datchi F, Loubeyre P, LeToullec R. 2000. *Phys. Rev. B* 61:6535–46

Delfosse X, Forveille T, Mayor M, Perrier C, Naef D, Queloz D. 1998. *Astron. Astrophys.* 338:L67–70

Demarcus WC. 1958. *Astron. J.* 63:2–28

de Pater I, Dunn D, Romani P, Zahnle K. 2001. *Icarus* 149:66–78

DeWitt HE, Hubbard WB. 1976. *Astrophys. J.* 205:295–301

Drouart A, Dubrulle B, Gautier D, Robert F. 1999. *Icarus* 140(129):155

Ebeling W, Richert W. 1985. *Phys. Lett. A* 108:80–82

Fischer DA, Marcy GW, Butler RP, Vogt SS, Apps K. 1999. *Publ. Astron. Soc. Pac.* 111:50–56

García-Melendo E, Sánchez-Lavega A. 2001. *Icarus* 152:316–30

Gautier D, Hersant F, Mousis O, Lunine JI. 2001. *Astrophys. J.* 550:L227–30. Erratum. 2001. *Astrophys. J.* 559:L183

Gudkova TV, Zharkov VN. 1999. *Planet. Space Sci.* 47:1211–24

Guillot T, Burrows A, Hubbard WB, Lunine JI, Saumon D. 1996. *Astrophys. J.* 459:L35–38

Guillot T, Chabrier G, Morel P, Gautier D. 1994a. *Icarus* 112:354–67

Guillot T, Gautier D, Chabrier G, Mosser B. 1994b. *Icarus* 112:337–53

Guillot T, Gautier D, Hubbard WB. 1997. *Icarus* 130:534–39

Hayashi C, Nakano T. 1963. *Prog. Theor. Phys.* 30:460–74

Henry GW, Marcy GW, Butler RP, Vogt SS. 2000. *Astrophys. J.* 529:L41–44

Hubbard WB. 1968. *Astrophys. J.* 152:745–54

Hubbard WB. 1969. Convection in degenerate stars. In *Low-Luminosity Stars*, ed. SS Kumar, pp. 259–65. New York: Gordon & Breach. 542 pp.

Hubbard WB. 1999. *Icarus* 137:357–59

Hubbard WB, DeWitt HE. 1985. *Astrophys. J.* 290:388–93

Hubbard WB, Fortney JJ, Lunine JI, Burrows A, Sudarsky D, Pinto P. 2001. *Astrophys. J.* 560:413–19

Hubbard WB, Guillot T, Lunine JI, Burrows A, Saumon D, et al. 1997. *Phys. of Plasmas* 4:2011–15

Hubbard WB, Lampe M. 1969. *Astrophys. J. Suppl.* 18:297–346

Hubbard WB, Guillot T, Marley MS, Burrows A, Lunine JI, Saumon DS. 1999. *Planet. Space Sci.* 47:1175–82

Jha S, Charbonneau D, Garnavich PM, Sullivan DJ, Sullivan T, et al. 2000. *Astrophys. J.* 540:L45–48

Jones BW, Sleep PN, Chambers JE. 2001. *Astron. Astrophys.* 366:254–62

Klepeis JE, Schafer KJ, Barbee TW III, Ross M. 1991. *Science* 254:986–89

Knudson MD, Hanson DL, Bailey JI, Hall CA, Asay JR, et al. 2001. *Phys. Rev. Lett.* 87: 225501

Korzennik SG, Brown TM, Fischer DA, Nisenson P, Noyes RW. 2000. *Astrophys. J.* 533: L147–50

Kumar SS. 1963. *Astrophys. J.* 137:1121–25

Latham DW, Stefanik RP, Mazeh T, Mayor M, Burki G. 1989. *Nature* 339:38–40

Lin DNC, Bodenheimer P, Richardson DC. 1996. *Nature* 380:606–7

Low FJ. 1966. *Astron. J.* 71:391

Lunine JI. 1995. The frequency of planetary systems in the Galaxy. In *Extraterrestrials: Where Are They?*, ed. B Zuckerman, MH Hart, pp. 192–214. New York: Cambridge Univ. Press. 251 pp.

Lunine JI. 2001. *Proc. Natl. Acad. Sci. USA* 98:809–14

Lunine JI, Hubbard WB, Burrows A, Wang Y-P, Garlow K. 1989. *Astrophys. J.* 338:314–37

Mahaffy PR, Niemann HB, Alpert A, Atreya SK, Demick J, et al. 2000. *J. Geophys. Res.* 105:15061–72

Malhotra R. 1995. *Astron. J.* 110:420–29

Malhotra R, Holman M, Ito T. 2001. *Proc. Natl. Acad. Sci. USA.* In press

Mannings V, Boss AP, Russell SS, eds. 2000. *Protostars and Planets IV.* Tucson: Univ. Ariz. Press. 1422 pp.

Marcy GW, Butler RP. 1996. *Astrophys. J.* 464:L147–51

Marcy GW, Butler RP, Vogt SS. 2000a. *Astrophys. J.* 536:L43–46

Marcy GW, Butler RP, Vogt SS, Fischer D, Lissauer JJ. 1998. *Astrophys. J.* 505:L147–49

Marcy GW, Butler RP, Vogt SS, Fischer D, Liu MC. 1999. *Astrophys. J.* 520:239–47

Marcy GW, Cochran WD, Mayor M. 2000b. Extrasolar planets around main-sequence stars. See Mannings et al. 2000, pp. 1285–311

Marley MS. 1991. *Icarus* 94:420–35

Marley MS, Saumon D, Guillot T, Freedman RS, Hubbard WB, et al. 1996. *Science* 272:1919–21

Marley MS, Gelino C, Stephens D, Lunine JI, Freedman R. *Astrophys. J.* 513:879–93

Mayor M, Queloz D. 1995. *Nature* 378:355–59

Mazeh T, Naef D, Torres G, Latham DW, Mayor M, et al. 2000. *Astrophys. J.* 532:L55–58

Militzer B, Ceperley DM. 2001. *Phys. Rev. E* 6306:6404

Mizuno H. 1980. *Prog. Theor. Phys.* 64:544–57

Mosser B. 1994. Jovian seismology. In *The Equation of State in Astrophysics*, See Chabrier & Schatzman 1994, pp. 481–511. Cambridge, UK: Cambridge Univ. Press. 621 pp.

Mosser B, Maillard JP, Mékarnia D. 2000. *Icarus* 144:104–13

Mostovych AN, Chan Y, Lehecha T, Phillips L, Schmitt A, Sethian JD. 2001. *Phys. Plasmas* 8:2281–86

Murray N, Hansen B, Holman M, Tremaine S. 1998. *Science* 279:69–72

Narayana C, Luo H, Orloff J, Ruoff al. 1998. *Nature* 393:46–49

Nellis WJ. 2000. *Planet. Space Sci.* 48:671–77

Nellis WJ, Weir ST, Mitchell AC. 1999. *Phys. Rev. B* 59:3434–49

Niemann HB, Atreya SK, Carignan GR, Donahue TM, Haberman JA, et al. 1996. *Science* 272:846–49

Noyes RW, Jha S, Korzennik SG, Krockenberger M, Nisenson P, et al. 1997a. *Astrophys. J.* 483:L111–14

Noyes RW, Jha S, Korzennik SG, Krockenberger M, Nisenson P, et al. 1997b. *Astrophys. J.* 487:L195

Ortiz JL, Orton GS, Friedson AJ, Stewart ST, Fisher BM, Spencer JR. 1998. *J. Geophys. Res.* 103:23051–70

Owen T, Mahaffy P, Niemann HB, Atreya SK,

Donahue TM, et al. 1999. *Nature* 402:269–70

Pfaffenzeller O, Hohl D, Ballone P. 1995. *Phys. Rev. Lett.* 74:2599–602

Queloz D, Mayor M, Weber L, Blécha A, Burnet M, et al. 2000. *Astron. Astrophys.* 354:99–102

Reitsema HJ, Smith EJ, Spilker T, Reinert R. 2001. Significant science at Jupiter using solar power. *Forum Innov. Approaches Outer Planet. Explor. 2001–2020*, Houston, Tex., (Abstr. 4115)

Ross M. 1998. *Phys. Rev. B* 58:669–77

Ross M, Dewitt HE, Hubbard WB. 1981. *Phys. Rev. A* 24:1016–20

Rossow WV. 1978. *Icarus* 36:1–50

Roulston MS, Stevenson DJ. 1995. *EOS* 76:343 (Abstr.)

Salpeter EE. 1973. *Astrophys. J.* 181:L83–86

Santos NC, Mayor M, Naef D, Pepe F, Queloz D, et al. 2000. *Astron. Astrophys.* 356:599–602

Saumon D, Chabrier G, van Horn HM. 1995. *Astrophys. J. Suppl.* 99:713–41

Schubert G, Zhang K. 2000. Dynamics of giant planet interiors. In *From Giant Planets to Cool Stars*, ed. CA Griffith, MS Marley, pp. 210–22. San Francisco: Astron. Soc. Pac. 353 pp.

Slattery WL, Hubbard WB. 1976. *Icarus* 29:187–92

Smoluchowski R. 1967. *Nature* 215:691–95

Stevenson DJ. 1975. *Phys. Rev. B* 12:3999–4007

Stevenson DJ, Salpeter EE. 1977a. *Astrophys. J. Suppl.* 35:221–37

Stevenson DJ, Salpeter EE. 1977b. *Astrophys. J. Suppl.* 35:239–61

Stoker CR. 1986. *Icarus* 67:106–25

Stringfellow GS, DeWitt HE, Slattery WL. 1990. *Phys. Rev. A* 41:1105–11

Sudarsky D, Burrows A, Pinto P. 2000. *Astrophys. J.* 538:885–903

Tassoul JL. 1978. *Theory of Rotating Stars*. Princeton: Princeton Univ. Press. 506 pp.

Udry S, Mayor M, Naef D, Pepe F, Queloz D, et al. 2000. *Astron. Astrophys.* 356:590–98

Vasilev PP, Efimov AB, Trubitsyn VP. 1978. *Soviet Astron.* 22:83–87

Vogt SS, Marcy GW, Butler RP, Apps K. 2000. *Astrophys. J.* 536:902–14

Von Zahn U, Hunten DM. 1996. *Science* 272:849–50

Ward WR, Hahn JM. 2000. Disk-planet interactions and the formation of planetary systems. See Mannings et al. 2000, pp. 1135–55

Weidenschilling SJ, Lewis JS. 1973. *Icarus* 20:465–76

Weidenschilling SJ, Marzari F. 1996. *Nature* 384:619–21

Wigner E, Huntington HB. 1935. *J. Chem. Phys.* 3:764–70

Wuchterl G, Guillot T, Lissauer JJ. 2000. Giant planet formation. See Mannings et al. 2000, pp. 1081–109

Yair Y, Levin Z, Tzivion S. 1995. *Icarus* 114:278–99

Zharkov VN, Gudkova TV. 1993. *Astron. Lett.* 19:61–63

Zharkov VN, Trubitsyn VP. 1978. *Physics of Planetary Interiors*. Tucson, AZ: Pachart. 388 pp.

Annu. Rev. Astron. Astrophys. 2002. 40:137–69
doi: 10.1146/annurev.astro.40.060401.093821
Copyright © 2002 by Annual Reviews. All rights reserved

THEORIES OF GAMMA-RAY BURSTS

P. Mészáros

*Department of Astronomy and Astrophysics and Department of Physics, 525 Davey
Laboratory, Pennsylvania State University, University Park, Pennsylvania 16803;
email: nnp@lonestar.astro.psu.edu*

Key Words high energy, cosmology, neutrinos

■ **Abstract** The gamma ray burst phenomenon is reviewed from a theoretical point
of view, with emphasis on the fireball shock scenario of the prompt emission and
the longer wavelength afterglow. Recent progress and issues are discussed, including spectral-temporal evolution, localizations, jets, spectral lines, environmental and
cosmological aspects, as well as some prospects for future experiments in both electromagnetic and nonelectromagnetic channels.

INTRODUCTION

Gamma-ray bursts (GRB) were first detected in the late 1960s by military satellites
monitoring for compliance with the nuclear test ban treaty. This became public
information only several years later, with the publication of the results from the
Vela satellites (Klebesadel et al. 1973), which were quickly confirmed by data
from the Soviet Konus satellites Mazets et al. 1974). Their nature and origin
remained thereafter a mystery for more than two decades, largely due to the fact
that during this period they remained detectable only for tens of seconds, almost
exclusively at gamma-ray energies (e.g., Hurley 1992), with occasional reports at
X-ray energies (e.g., Murakami et al. 1988, Yoshida et al. 1989, Connors & Hueter
1998). Various satellites continued to accumulate data on hundreds of GRB over
the years, attracting an increasing amount of attention and leading to a large variety
of theoretical models (e.g., Ruderman 1975, Liang 1989).

A new era in GRB research opened in 1991 with the launch of the Compton
Gamma-Ray Observatory (CGRO), whose ground-breaking results have been summarized in Fishman & Meegan 1995. The most significant results came from the
all-sky survey by the Burst and Transient Experiment (BATSE) on CGRO, which
recorded over 2700 bursts, complemented by data from the Oriented Scintillation
Spectrometer Experiment (OSSE), Compton Telescope (Comptel), and Energetic
Gamma Ray Experiment Telescope (EGRET) on the CGRO. BATSE's earliest
and most dramatic result was that it showed that GRB were essentially isotropically distributed in the sky, with no significant dipole or quadrupole moments,
suggesting a cosmological distribution (Meegan et al. 1992). The spectra were

0066-4146/02/0922-0137$14.00

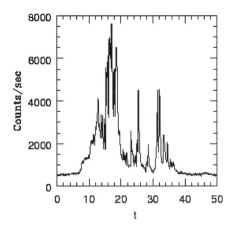

Figure 1   Typical GRB lightcurve observed with BATSE, showing photon count rate (0.05–0.5 MeV) versus time (s). No γ-rays are detected either before or after the burst trigger (Fishman & Meegan 1995).

nonthermal, the number of photons per unit photon energy varying typically as a power-law $N(\epsilon) \propto \epsilon^{-\alpha}$, where $\alpha \sim 1$ at low energies changes to $\alpha \sim 2$–3 above a photon energy $\epsilon_0 \sim 0.1$–1 MeV (Band et al. 1993). This spectral power law dependence was found to extend in several bursts up to at least GeV energies (Schneid et al. 1995, Hurley 1994). The gamma-ray light curves show a time dependence ranging from a smooth, fast-rise and quasi-exponential decay, through curves with several peaks, to variable curves with many peaks, and substructure sometimes down to milliseconds (Figure 1). The durations at MeV energies range from 10^{-3} s to about 10^3 s, with a well-defined bimodal distribution for bursts longer or shorter than $t_b \sim 2$ s (Kouveliotou et al. 1993). There is also an anticorrelation between spectral hardness and duration, the short one being harder, (e.g., Fishman & Meegan 1995). The pulse distribution is complex, and the time histories of the emission as a function of energy can provide clues for the geometry or physics of the emitting regions (e.g., Fenimore et al. 1990, Beloborodov et al. 1998). The results from BATSE sharpened the debate on whether the GRB were of a galactic or extragalactic origin, e.g., (Lamb 1995, Paczyński 1995), but the accumulating evidence increasingly swung the balance in favor of the cosmological interpretation.

A decisive watershed was reached in 1997 when the Italian-Dutch satellite Beppo-SAX succeeded in obtaining the first high resolution X-ray images (Costa et al. 1997) of the fading afterglow of a burst, GRB 970228, which had been expected on theoretical grounds. This discovery was promptly followed by an increasing list of other burst detections by Beppo-SAX, at the approximate rate of 10 per year. These X-ray detections, after a 4–6 hour delay needed for data processing, led to arc-minute accuracy positions, which finally made possible the optical detection and the follow-up of the GRB afterglows at longer wavelengths (e.g., van Pardijsa et al. 1997, Frail et al. 1997). This paved the way for the measurement of redshift distances, the identification of candidate host galaxies, and the confirmation that they were at cosmological distances (Metzger et al. 1997, Kulkarni et al. 1999b, etc.). Over 40 GRB afterglows have been located as of late 2001 in X-rays and optical, and more than a dozen in radio (Frail et al. 1999, Weiler 2002). Some afterglows have been followed over timescales of many months to over a year, and in the majority of cases (over 30) they have also resulted in the identification of the likely host galaxy (Bloom et al. 2001, Djorgovski 2001). A recent review of the observations and phenomenology of GRB afterglows is in Van Paradijs et al. 2000.

GAMMA-RAY BURST PHENOMENOLOGY: THE FIREBALL SHOCK SCENARIO

At cosmological distances the observed GRB fluxes imply energies of $\lesssim 10^{54}$ erg, if the emission is isotropic (see however "Jets and Limb-Brightening Effects" below), and from causality this must be liberated inside regions whose size is $\lesssim 100$ kilometers on timescales $\lesssim$ seconds. Independently of the nature and details of the progenitor and the trigger, such an intense, localized, and brief explosion implies the formation of an $e^{\pm}$, γ fireball (Cavallo & Rees 1978). In the context of a cosmological model, the fireball would be expected to expand relativistically (Paczyński 1986, Goodman 1986, Paczyński 1990). This hypothesis is natural, since most of the spectral energy is observed at $\gtrsim 0.5$ MeV, so the optical depth against $\gamma\gamma \to e^{\pm}$ is huge, and the expansion follows from the highly super-Eddington value of the luminosity. Because many bursts emit a large fraction of their luminosity at photon energies $\epsilon_{\gamma} \gg 1$ MeV, the flow must somehow be able to avoid the process $\gamma\gamma \to e^{\pm}$ degrading the observed photons to just below 0.511 MeV. A highly relativistic expansion is, in fact, strongly supported by the fact that it provides a natural explanation for the observed photons with $\epsilon_{\gamma} \gg 0.5$ MeV (Fenimore et al. 1993, Harding & Baring 1994). This is because in this case the relative angle at which the photons collide must be less than the inverse of the bulk Lorentz factor γ^{-1} and the effective threshold energy for pair production is correspondingly reduced. Roughly, the Lorentz factor must satisfy

$$\gamma \gtrsim 10^2 (\epsilon_{\gamma}/10\,\text{GeV})^{1/2}(\epsilon_t/\text{MeV})^{1/2}, \tag{1}$$

in order for photons with $\epsilon_{\gamma} \gtrsim 10$ GeV to escape annihilation against target photons with $\epsilon_t \sim 1$ MeV. (A more detailed calculation is in Lithwick & Sari 2001.)

From general considerations (Shemi & Piran 1990), a relativistic outflow arising from an initial energy E_o imparted to a mass $M_o \ll E_0/c^2$ starting out from a radius r_l leads to an expansion, as the gas converts its internal energy into bulk kinetic energy. Initially, the bulk Lorentz factor $\gamma \simeq r/r_l \propto r$, while the comoving temperature drops $\propto r^{-1}$. Clearly, γ cannot increase beyond $\gamma_{max} \sim \eta \sim E_o/M_oc^2$, which occurs at a saturation radius $r_s \gtrsim r_l\eta$, beyond which the flow continues to coast with $\gamma \sim \eta \sim$ constant. The simplicity of the original fireball picture, however, led to some serious difficulties. Among these are that the expansion of the fireball should lead to a conversion of most of its internal energy into kinetic energy of the entrained baryons, rather than into photon luminosity, hence it would be energetically very inefficient. Furthermore, it would produce a quasi-thermal photon spectrum, instead of the observed power-law spectra; and the typical timescales over which these photons escape is comparable to that during which the flow makes a transition to optical thinness (milliseconds), which could not explain the many events lasting much longer than that.

These efficiency, timescale, and spectrum problems can be solved with the fireball shock model, in its external (Rees & Mészáros 1992) and internal (Rees & Mészáros 1994) versions. This is based on the fact that shocks are likely to occur in such an outflow, and if these occur after the fireball has become optically thin, these shocks would reconvert the kinetic energy of the baryons into nonthermal particle and photon energy.

External shocks (Mészáros & Rees 1993a) will occur, unavoidably, in any outflow of total energy E_o in an external medium of average particle density n_o at a radius and on a timescale

$$r_{dec} \sim 10^{17} E_{53}^{1/3} n_o^{-1/3} \eta_2^{-2/3} \text{ cm,}$$

$$t_{dec} \sim r_{dec}/(c\gamma^2) \sim 3 \times 10^2 E_{53}^{1/3} n_o^{-1/3} \eta_2^{-8/3} \text{ s,} \qquad (2)$$

where (in the impulsive, or thin shell approximation) the lab-frame energy of the swept-up external matter ($\gamma^2 m_p c^2$ per proton) equals the initial energy E_o of the fireball, and $\eta = \gamma = 10^2 \eta_2$ is the final bulk Lorentz factor of the ejecta.

The external shock synchrotron spectra (Mészáros & Rees 1993a, Kartz 1994b) and combined synchrotron-IC spectra (Mészáros et al. 1993, 1994) reproduce in a general manner the observed gamma-ray spectral properties, as do the predicted spectral-temporal correlations (Sari et al. 1996, Panaitescu & Mészáros 1998b, Dermer et al. 1999, Böttcher & Dermer 2000; see also Liang et al. 1999). (However, internal shocks present an alternative for the brief burst of gamma-ray emission, motivated by variability issues, see below.) External shocks also serve as the model of choice for the afterglow radiation (see "Blast Wave Model of GRB Afterglows" below). The typical observer-frame dynamic time of the shock is $t_{dec} \sim r_{dec}/c\gamma^2 \sim$ seconds, for typical parameters, and $t_b \sim t_{dec}$ would be the burst duration (the impulsive assumption requires that the initial energy input occurs in a time shorter than t_{dec}). Variability on timescales shorter than t_{dec} may occur on the cooling timescale or on the dynamic timescale for inhomogeneities in the external medium, but this is not widely favored for reproducing highly variable profiles. (Sari & Piran 1998; see also Dermer & Mitman 1999). They could, however, reproduce bursts with several peaks (Panaitescu & Mészáros 1998a) and may therefore be applicable to the class of long, smooth bursts.

Internal shocks (Rees & Mészáros 1994) address another problem, posed by some of the rapidly variable γ-ray light curves, which for total durations of tens to hundreds of seconds are, sometimes, endowed with variability down to milliseconds or less (Fishman & Meegan 1995). One ingredient in solving this problem is to postulate a central engine (Fenimore et al. 1993), which ejects energy at a variable rate. This could be, e.g., magnetic flares in a transient accretion disk around a central compact object resulting from the disruption of a merging compact binary (Narayan et al. 1992). By itself, such a variable central engine is however not enough to explain the variable light curves, since a relativistic outflow is inevitable, and even if intermittent, this outflow will be on average optically thick to Compton scattering out to very large radii, leading to a smoothing-out of the light curve. This difficulty, however, is solved with the introduction of the internal shock model (Rees & Mészáros 1994), in which the time-varying outflow from the central engine leads to successive shells ejected with different Lorentz factors. Multiple shocks form as faster shells overtake slower ones, and the crucial point is that for a range of plausible parameters, this occurs above the Compton photosphere. These shocks are called internal because they arise from the flow interacting with itself, rather than with the external environment.

One can model the central engine outflow as a wind of duration t_w, whose average dynamics is similar to that of the impulsive outflows described previously, with an average lab-frame luminosity $L_o = E_o/t_w$ and average mass outflow $\dot{M}_o$, and mean saturation Lorentz factor $\gamma \sim \eta = L_o/\dot{M}_o c^2$. Significant variations of order $\Delta\gamma \sim \gamma \sim \eta$ occurring over timescales $t_{var} \ll t_w$ will lead then to internal shocks (Rees & Mészáros 1994) at radii r_{dis} above the photosphere r_{phot},

$$r_{dis} \sim c t_{var}\eta^2 \sim 3 \times 10^{14} t_{var}\eta_2^2 \text{ cm,}$$

$$r_{phot} \sim \dot{M}\sigma_T/(4\pi m_p c\eta^2) \sim 10^{11} L_{50}\eta_2^{-3} \text{ cm.} \tag{3}$$

The above assumes the photosphere to be above the saturation radius $r_s \simeq r_o\eta$, so that most of the energy comes out in the shocks, rather than in the photospheric quasi-thermal component (such photospheric effects are discussed in Mészáros & Rees (2000b). For shocks above the photosphere, large observable γ-ray variations are possible on timescales $t_{var} \gtrsim t_{var,min} \sim 10^{-3}(M_c/M_\odot)^{3/2}$, for an outflow originating from a central object of mass M_c at radii $\gtrsim r_o \sim c t_{var,min}/2\pi$. The internal shock model was specifically designed to allow an arbitrarily complicated light curve (Rees & Mészáros 1994) on timescales down to ms, the optically thin shocks producing the required nonthermal spectrum. Numerical calculations (Kobayashi et al. 1999, Daigne & Mochkovitch 2000, Spada et al. 2000) confirm that the light curves can indeed be as complicated as observed by BATSE in extreme cases. (By contrast, in external shocks the variations are expected to be smoothed out by relativistic time delays, e.g., Sari & Piran 1998, at most a few peaks being possible, e.g., Panaitescu & Mészáros 1998a. An alternative view invoking large variability from blobs in external shocks is discussed by Dermer & Mitman 1999). The observed power density spectra of GRB light curves (Beloborodov et al. 2000) provide an additional constraint on the dynamics of the shell ejection by the central engine and the efficiency of internal shocks (Spada et al. 2000, Guetta et al. 2001).

When internal shocks occur, these are generally expected to be followed (Mészáros & Rees 1994, 1997) by an external shock, a sequential combination sometimes referred to as the internal-external shock scenario (Piran & Sari 1998). The GRB external shocks, similar to what is observed in supernova remnants, consist of a forward shock or blast wave moving into the external medium ahead of the ejecta, and a reverse shock moving back into the ejecta as the latter is decelerated by the inertia of the external medium. The internal shocks would consist of forward and reverse shocks of a more symmetrical nature. As in interplanetary shocks studied with spacecraft probes, the internal and external shocks in GRB are tenuous and expected to be collisionless, i.e., mediated by chaotic electric and magnetic fields. The minimum random Lorentz factor of protons going through the shocks should be comparable to the relative bulk Lorentz factor, whereas that of the electrons may exceed this by a factor of up to the ratio of the proton to the electron mass. The energy of the particles can be further boosted by diffusive shock acceleration (Blandford & Eichler 1987) as they scatter repeatedly across the shock interface, acquiring a power law distribution $N(\gamma_e) \propto \gamma_e^{-p}$, where $p \sim 2\text{–}3$. In the presence of turbulent magnetic fields built up behind the shocks, the electrons produce a synchrotron power-law radiation spectrum (Mészáros & Rees 1993a, Rees & Mészáros 1994) similar to that observed by Band et al. (1993), while the inverse Compton (IC) scattering of these synchrotron photons extends the spectrum into the GeV range (Mészáros et al. 1994). Comparisons of a synchrotron hypothesis for the MeV radiation with data have been made by, e.g., Tavani 1996, Preece et al. 2000, Eichler & Levinson 2000, Mészáros & Rees 2000b, Medvedev 2000, Panaitescu & Mészáros 2000, and Lloyd & Petrosian 2001a. The effects of pair production and IC on the prompt spectra are discussed in the section "Radiation Processes, Efficiencies, and Pairs."

It is worth stressing that the fireball shock scenario, whether internal or external, is fairly generic: It is largely independent of the details of the progenitor. Although it is somewhat geometry dependent, the central engine generally lies enshrouded and out of view inside the optically thick outflow. Even after the latter becomes optically thin, the progenitor's remnant emission should be practically undetectable compared to the emission of the fireball shock that is its main manifestation (see, however, "Cosmological Setting, Galactic Hosts and Environment" below).

BLAST WAVE MODEL OF GRB AFTERGLOWS

The external shock becomes important when the inertia of the swept up external matter leads to an appreciable slowing down of the ejecta. As the fireball continues to plow ahead, it sweeps up an increasing amount of external matter, made up of interstellar gas plus possibly gas that was previously ejected by the progenitor star. As the external shock builds up, for high radiative efficiency its bolometric luminosity rises approximately as $L \propto t^2$. This follows from equating in the contact discontinuity frame the kinetic flux $L/4\pi r^2$ to the external ram pressure $\rho_{ext}\gamma^2$ during the initial phase while $\gamma \sim$ constant, $r \propto t$ (Rees & Mészáros 1992; see also

Sari 1998). After peaking, or plateauing in the thick shell limit, as the Lorentz factor decreases one expects a gradual dimming $L \propto t^{-1+q}$ (from energy conservation $L \propto E/t$ under adiabatic conditions, q takes into account radiative effects or bolometric corrections). At the deceleration radius (2) the fireball energy and the bulk Lorentz factor decrease by a factor ~ 2 over a timescale $t_{dec} \sim r_{dec}/(c\gamma^2)$, and thereafter the bulk Lorentz factor decreases as a power law in radius,

$$\gamma \propto r^{-g} \propto t^{-g/(1+2g)}, r \propto t^{1/(1+2g)}, \tag{4}$$

with $g = (3, 3/2)$ for the radiative (adiabatic) regime, in which $\rho r^3 \gamma \sim$ constant ($\rho r^3 \gamma^2 \sim$ constant). At late times, a similarity solution (Blandford & McKee 1976a, 1976b) with $g = 7/2$ may be reached. The spectrum of radiation is likely to be due to synchrotron radiation, whose peak frequency in the observer frame is $\nu_m \propto \gamma B' \gamma_e^2$, and both the comoving field B' and the minimum electron Lorentz factor $\gamma_{e,min}$ are likely to be proportional to γ (Mészáros & Rees 1993a). This implies that as γ decreases, so will ν_m, and the radiation will move to longer wavelengths. Consequences of this are the expectation that the burst would leave a radio remnant (Paczyński & Rhoads 1993) after some weeks, and before that an optical (Katz 1994b) transient.

The first self-consistent afterglow calculations (Mészáros & Rees 1997a) took into account both the dynamical evolution and its interplay with the relativistic particle acceleration; a specific relativistically beamed radiation mechanism resulted in quantitative predictions for the entire spectral evolution, going through the X-ray, optical, and radio range. For a spherical fireball advancing into an approximately smooth external environment, the bulk Lorentz factor decreases as in inverse power of the time (asymptotically $t^{-3/8}$ in the adiabatic limit), and the accelerated electron minimum random Lorentz factor and the turbulent magnetic field also decrease as inverse power laws in time. The synchrotron peak energy corresponding to the time-dependent minimum Lorentz factor and magnetic field then move to softer energies as $t^{-3/2}$. These scalings can be generalized in a straightforward manner also for the radiative regime, or in presence of density gradients. The radio spectrum is initially expected to be self-absorbed and becomes optically thin after a few days. For times beyond about one hour, the dominant radiation is from the forward shock, for which the flux at a given frequency and the synchrotron peak frequency decay as (Mészáros & Rees 1997a)

$$F_\nu \propto t^{-(3/2)\beta}, \nu_m \propto t^{-3/2}, \tag{5}$$

as long as the expansion is relativistic. This is referred to as the standard (adiabatic) model, where $g = 3/2$ in Equation 4 and β is the photon spectral energy slope ($F_\nu \propto \nu^{-\beta}$). The transition to the nonrelativistic regime has been discussed, e.g., by Wijers et al. 1997, Dai & Lu 1999, and Livio & Waxman 2000. More generally, (Mészáros & Rees 1999), the relativistic forward shock flux and frequency peak are given by $F_\nu \propto t^{[3-2g(1+2\beta)]/(1+2g)}$ and $\nu_m \propto t^{-4g/(1+2g)}$. (A reverse shock component is also expected (Mészáros & rees 1993b, 1997a) with high initial optical brightness but much faster decay rate than the forward shock, see "Prompt Flashes

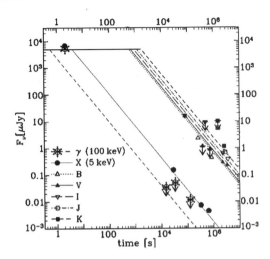

Figure 2 Comparison (Wijers et al. 1997) of the observed light curves of the afterglow of GRB 970228 at various wavelengths with the simple blast wave model predictions (Mészáros & Rees 1997a).

and Reverse Shocks" below). It is remarkable, however, that the simple standard model where reverse shock effects are ignored is a good approximation for modeling observations starting a few hours after the trigger, as was the case during 1997–1998.

The predictions of the fireball shock afterglow model (Mészáros & Rees 1997a) were made in advance of the first X-ray detections by Beppo-SAX (Costa et al. 1997) allowing subsequent follow-ups (van Paradijs et al. 1997, Metzger et al. 1997, Frail et al. 1999) over different wavelengths, which showed a good agreement with the standard model, (e.g., Vietri 1997a, Wijers et al. 1997, Tavani 1997, Waxman 1997a, Reichart 1997) (Figure 2). The comparison of increasingly sophisticated versions of this theoretical model (e.g., Sari et al. 1998, Wijers & Galama 1999, Piran 1999, Dermer et al. 2000, Granot et al. 2000b) against an increasingly detailed array of observations (e.g., as summarized in van Paradijs et al. 2000) has provided confirmation of this generic fireball shock model of GRB afterglows.

A snapshot spectrum of the standard model at any given time consists of a three-segment power law with two breaks. At low frequencies there is a steeply rising synchrotron self-absorbed spectrum up to a self-absorption break ν_a, followed by a $+1/3$ energy index spectrum up to the synchrotron break ν_m corresponding to the minimum energy γ_m of the power-law accelerated electrons, and then a $-(p-1)/2$ energy spectrum above this break, for electrons in the adiabatic regime (where γ_e^{-p} is the electron energy distribution above γ_m). A fourth segment is expected at energies above that where the electron cooling time becomes short compared to the expansion time, with a spectral slope $-p/2$ above that, with a corresponding third cooling break ν_b (Mészáros et al. 1998, Sari et al. 1998). The observations (e.g., van Paradijs et al. 2000) are compatible with an electron spectral index $p \sim 2.2$–2.5 (Gallant et al. 1999), which is typical of shock acceleration, e.g., Waxman 1997a, Sari et al. 1998, Wijers & Galama 1999. As the remnant expands, the photon spectrum moves to lower frequencies, and the flux in a given band decays as a power law in time, whose index can change as breaks move through it. Snapshot spectra have been deduced by extrapolating measurements at different wavelengths and times; assuming spherical symmetry and using the model time dependences (Waxman (1997b), Wijers & Galama 1999), fits were obtained for the different physical parameters of the burst and environment, e.g., the total energy E, the magnetic and electron-proton coupling parameters ϵ_B and ϵ_e and the external density n_o (see right panel of Figure 3, color insert). These lead to typical values $n_o \sim 10^{-2}$–10 cm^{-3}, $\epsilon_B \sim 10^{-2}$, $\epsilon_e \sim 0.1$–0.5 and $E \sim 10^{52}$–10^{54} ergs (if spherical; but see "Jets and Limb-Brightening Effects").

STANDARD MODEL DEVELOPMENTS AND ISSUES

The standard afterglow model is based on the following approximations: (a) spherical outflow; (b) a homogeneous external medium $n \sim n_o$; (c) highly relativistic expansion in the adiabatic approximation; (d) an impulsive energy input E_o and a single $\gamma_o = \eta = E_o/M_o c^2$; ($e$) line-of-sight scaling relations assumed valid for the entire visible hemisphere; (f) time-independent shock acceleration parameters p, ε_B, ε_e (electron energy index, magnetic to proton, and electron to proton energy ratios); and (g) only the forward shock radiation is included. The significant success of this model in explaining many of the observations in the first years after GRB 970228 indicates that these approximations are robust, at least in a broad sense and over a range of timescales. However, they are clearly simplifications and are expected to be appropriate only within certain limits.

Density, Angle, and Time-dependent Injection

Departures from the simplest standard model occur, e.g., if the external medium is inhomogeneous. For instance, for $n \propto r^{-d}$, the energy conservation condition is $\gamma^2 r^{3-d} \sim$ constant, which changes significantly the temporal decay rates (Mészáros et al. 1998). Such a power law dependence is expected if the external medium is a wind, say from an evolved progenitor star, and the light curve of some bursts fit better with such a hypothesis (Chevalier & Li 2000), whereas in many objects a homogeneous medium seems a better fit (Frail et al. 2001, Panaitescu & Kumar 2001b) (for a critical discussion see Li & Chevalier 2001, 2001b). Another obvious nonstandard effect is a departure from a simple impulsive injection approximation (i.e., an injection that is not a delta or a top-hat function with a single value for E_o and γ_o in time). An example is if the mass and energy injected during the burst duration t_w (say tens of seconds) obeys $M(>\gamma) \propto \gamma^{-s}$, $E(>\gamma) \propto \gamma^{1-s}$, i.e., more energy emitted with lower Lorentz factors at later times, but still shorter than the gamma-ray pulse duration (refreshed shocks). This would drastically change the temporal decay rate and extend the afterglow lifetime in the relativistic regime, providing a late "energy refreshment" to the blast wave on timescales comparable to the afterglow timescale (Rees & Mészáros 1998, Kumar & Piran 2000, Dai & Lu 2000, Sari & Mészáros 2000). These examples lead to nonstandard decay rates

$$\gamma \propto r^{-g} \propto \begin{cases} r^{-(3-d)/2}; n \propto r^{-d}; \\ r^{-(3-d)/(1+s)}; E(>\gamma) \propto \gamma^{1-s}, n \propto r^{-d}. \end{cases} \tag{6}$$

An additional complication occurs if the outflow has a transverse (θ-dependent) gradient in its properties such as energy per solid angle or Lorentz factor, e.g., as some power law θ^{-j}, θ^{-k} (Mészáros et al. 1998). Expressions for the temporal decay index $\alpha(\beta, s, d, j, k, \ldots)$ in $F_\nu \propto t^\alpha$ are given by (Mészáros et al. 1998, Sari & Mészáros 2000), which now depend also on s, d, j, k, and so on (and not just on β as in the standard relation of Equation 5). The result is that the decay can be flatter (or steeper, depending on s, d, etc) than the simple standard $\alpha = (3/2)\beta$,

$$F_\nu \propto t^\alpha \nu^\beta, \quad \text{with} \quad \alpha = \alpha(\beta, d, s, j, k, \ldots). \tag{7}$$

Thus, a diversity of behaviors is not unexpected. What is more remarkable is that, in many cases, the simple standard relation (5) is sufficient to describe the gross overall behavior at late times.

Strong evidence for departures from the simple standard model is provided by, e.g., sharp rises or humps in the light curves followed by a renewed decay, as in GRB 970508 (Pedersen et al. 1998, Piro et al. 1998a). Detailed time-dependent model fits (Panaitescu et al. 1998) to the X-ray, optical, and radio light curves of GRB 970228 and GRB 970508 indicate that, in order to explain the humps, a nonuniform injection is required. Other ways to get a lightcurve bump after $\sim$ days is through microlensing (Garnavich 2000), late injection (Zhang & Mészáros 2001a), or inverse Compton effects (Zhang & Mészáros 2001b, Harrison et al. 2001).

Jets and Limb-Brightening Effects

The spherical assumption is valid even when considering a relativistic outflow collimated within some jet of solid angle $\Omega_j < 4\pi$, provided the observer line of sight is inside this angle, and $\gamma \gtrsim \Omega_j^{-1/2}$ (Mészáros et al. 1993), so the light-cone is inside the jet boundary (causally disconnected) and the observer is unaware of what is outside the jet. As deceleration proceeds and the Lorentz factor drops below this value (in $\sim$ days), a change is expected in the dynamics and in the light curve (Rhoads 1997, 1999). The first effect after $\gamma < \Omega_j^{-1/2}$ is that, whereas before the effective transverse emitting area increased as $(r_\parallel/\gamma)^2 \propto t^2\gamma^2$, thereafter it grows more slowly as $r_\parallel \Omega_j^{-1/2} \propto t^2\gamma^4$, i.e., one expects a faster decay by $\gamma^2 \propto t^{-3/4}$ (Mészáros & Rees 1999), which in fact is the magnitude of the break seen, e.g., in GRB 990123 (Kulkarni et al. 1999, Fruchter A et al. 1999, Castro-Tirado et al. 1999). Soon after this, sideways expansion of the jet would lead to an even steeper decay, $\propto t^{-p}$ (Rhoads 1997, Panaitescu & Mészáros 1998d), possibly complicated by jet anisotropy (Dai & Gou 2001). Variable optical linear polarization can also be expected (Sari 1999, Ghisellini & Lazzati 1999). An example of the lightcurve break and a snapshot fit is shown in Figure 3 (see color insert). Numerical simulations of jet development (e.g., Granot et al. 2000a) are complicated due to the need for both high dimensionality and relativistic effects, and comparison between such models and phenomenological fits (Frail et al. 20001, Panaitescu & Kumar 2001a,b) still requires caution.

If the burst energy were emitted isotropically, the energy requirements spread over many orders of magnitude, $E_{\gamma,iso} \sim 10^{51}$–$10^{54}$ erg (Kulkarni et al. 1999b). However, taking into account the evidence for jets (Panaitescu & Kumar 2001a,b, Frail et al. 2001) the inferred spread in the total γ-ray energy is reduced to one order of magnitude, around a much less demanding mean value of $E_{\gamma,tot} \sim 8 \times 10^{50}$ erg (see also Rossi et al. 2002, Zhang & Mészáros 2002). This is not significantly larger than the kinetic energies in core-collapse supernovae, although GRB differ from the latter by being concentrated in the gamma-ray range, and by being initially substantially more collimated than supernovae (see, however, Höfflich et al. 1999). Radiative inefficiencies and the additional energy that must be associated with the proton and magnetic field components increase this value, but it would still be well within the theoretical energetics $\lesssim 10^{53.5}$–10^{54} erg achievable in either NS–NS, NS–BH mergers (Mészáros & Rees 1997b) or in hypernova/collapsar models (Paczyński 1998, Popham et al. 1999) using MHD extraction of the spin energy of a disrupted torus and/or a central fast spinning BH. It is worth stressing that the presence of jets does not invalidate the usefulness of snapshot spectral fits, since these constrain only the energy per solid angle (Mészáros et al. 1999).

An interesting property, which arises even in spherical outflows, is that the effective emitting region seen by the observer resembles a ring (Waxman (1997b), Goodman 1997, Panaitescu & Mészáros 1998b, Sari 1998, Granot et al. 1999a). This effect is thought to be implicated in giving rise to the radio diffractive scintillation pattern seen in several afterglows, since this requires the emitting source to be of small dimensions (the ring width), e.g., in GRB 970508 (Waxman et al. 1998). This provided an important observational check, giving a direct confirmation of the relativistic source expansion and a direct determination of the (expected) source size (Waxman et al. 1998, Katz et al. 1998).

Prompt Flashes and Reverse Shocks

A remarkable discovery was the observation (Akerlof et al. 1999) of a prompt and extremely bright ($m_v \sim 9$) optical flash in the burst GRB 990123, 15 s after the GRB started (and while it was still going on). A prompt multi-wavelength flash, contemporaneous with the γ-ray emission and reaching such optical magnitude levels is an expected consequence of the reverse component of external shocks (Mészáros & Rees 1993b). The prompt optical flash of GRB 990123 is generally interpreted (Sari & Piran 1999, Mészáros & Rees 1999) as the radiation from a reverse (external) shock, although a prompt optical flash could be expected either from an internal shock or from the reverse external shock (Mészáros & Rees 1997a). The decay rate of the optical flux from reverse shocks is much faster (and that of internal shocks is faster still) than that of forward shocks, so the emissions of the latter dominate after tens of minutes. Such bright prompt flashes, however, appear to be rare, since they have not so far been detected from other bursts, either using upgraded versions of the original Robotic Optical Transient Search Experiment, or ROTSE camera (Kehoe et al. 2001) or other similar systems (Park et al. 2001, Boer 2001). This is further discussed by Soderberg & Ramirez-Ruiz 2001.

Radiation Processes, Efficiencies, and Pairs

Pair-production due to $\gamma\gamma$ interactions among intra-shock photons satisfying $\epsilon_1 \epsilon_2 \gtrsim m_e c^2$ can be important when the compactness parameter $\ell \sim n'_\gamma \sigma_T \Delta R_{com} >$ 1, where ΔR_{com} is the comoving shock width and $n'_\gamma \propto L'_\gamma$ is the comoving photon density. This can affect the spectrum of external shocks (Mészáros et al. 1994, Baring 2000, Lithwick & Sari 2001) above GeV energies. An external shock occurring beyond a preceding internal shock (Mészáros & Rees 1994) is a possible model for the EGRET GeV observations (e.g., Hurley 1994) of 1–20 GeV photons in several GRBs. Internal shocks, occurring at smaller radii (Equation 3) than external shocks (Equation 2) will have larger compactness parameters, and pair formation can be more important (Rees & Mészáros 1994, Papathanassiou & Mészáros 1996, Pilla & Loeb 1998). For close-in shocks and high luminosities, pair-breakdown could lead to a self-regulating moderate pair optical thickness and subrelativistic pair temperature leading to a comptonized spectrum (Ghisellini & Cellotti 1999). Comptonization in a generic context has also been advocated by, e.g., Crider et al. 1997, Liang et al. 1999.

Low energy γ-ray spectral indices that appear steeper than predicted by a synchrotron mechanism has been reported by, e.g., Preece et al. 2000. Possible explanations include a fireball photospheric component, photospheric bulk, and pair-breakdown comptonization (Mészáros & Rees 2000b), and references therein). Other possibilities are synchrotron self-absorption in the X-ray (Granot et al. 1999b) or in the optical range upscattered to X-rays (Panaitescu & Mészáros 2000), low-pitch angle scattering (Medvedev 2000), or time-dependent acceleration and radiation (Lloyd-Ronning & Petrosian 2001b), where the latter also point out that low-pitch angle acceleration of electrons in a strong magnetic field may be preferred and can explain high energy indices steeper than predicted by an isotropic electron distribution.

A related problem is that of the radiative efficiency. For internal shocks, this is estimated to be moderate in the bolometric sense (10–30%), higher values being obtained if the shells have widely differing Lorentz factors (Spada et al. 2000, Beloborodov 2000, Kobayashi & Sari 2001). The total efficiency is substantially affected by inverse Compton losses (Papathanassiou & Mészáros 1996, Pilla & Loeb 1998, Ghisellini et al. 2000). The efficiency for emitting in the BATSE range is typically low $\sim$2–5%, both when the MeV break is due to synchrotron (Kumar 1999, Spada et al. 2000, Guetta et al. 2001) and when it is due to inverse Compton (Panaitescu & Mészáros 2000). This inefficiency is less of a concern when a jet is present (e.g., with typical values $\theta_{jet} \sim 3$ degrees and required total energies $E_0 \sim 10^{50}$–10^{51} erg; e.g., Fril et al. 2001, Panaitescu & Kumar 2001a).

Pair formation can also arise when γ-rays back-scattered by the external medium interact with the original γ-rays (Dermer & Böttcher 2000). This may lead to a cascade and acceleration of the pairs (Thompson & Madau 2000, Madau & Thompson 2000, Madau et al. 2000). For a model where γ-rays are produced in internal shocks, analytical estimates (Mészáros et al. 2000) indicate that even for modest external densities a pair cloud forms ahead of the fireball ejecta, which can accelerate to Lorentz factors $\gamma_{\pm} \lesssim 30$–50. These pairs produce a radio signal when they are swept-up by the ejecta; when the pair-enriched ejecta is in turn decelerated by the external medium, its radiative efficiency is increased. The afterglow reverse shock shares the same energy among a larger number of leptons so that its spectrum is softened towards the IR (Mészáros et al. 2001; see also Beloborodov 2002), compared with the optical/UV flash expected in the absence of pairs; this may contribute to the rarity of prompt optical detections.

Inverse Compton scattering can be an important energy loss mechanism in external shocks (Mészáros et al. 1993) and is the likeliest mechanism for producing GeV radiation (Mészáros et al. 1994, Mészáros & Rees 1994). Its effects on afterglows were considered by Waxman (1997b), and observational manifestations in afterglows were investigated more carefully by Panaitescu & Kumar 2000 and Sari & Esin 2001. This mechanism may be responsible for X-ray bumps after days in some afterglow light curves (Zhang & Mészáros 2001b, Harrison et al. 2001), alternative possibilities being microlensing (Garnavich 2000) or late injection (Zhang & Mészáros 2001a).

Shock Physics

The nonthermal spectrum in the fireball shock model is based on assuming that Fermi acceleration (e.g., Blandford & Eichler 1987) accelerates electrons to highly relativistic energies following a power law $N(\gamma_e) \propto \gamma_e^{-p}$, with $p \sim 2$–2.5 (Gallant et al. 1999). To get reasonable efficiencies, the accelerated electron to total energy ratio $\epsilon_e \lesssim 1$ must not be far below unity (Mészáros & Rees 1993a, Kumar 2000), whereas the implied magnetic to total energy ratio $\epsilon_b < 1$ depends on whether the synchrotron or the IC peak represents the observed MeV break (Papathanassiou & Mészáros 1996). The radiative efficiency and the electron power law minimum Lorentz factor also depend on the fraction $\zeta < 1$ of swept-up electrons injected into the acceleration process (Bykov & Mészáros 1996, Daigne & Mochkovitch 2000). Whereas many afterglow snapshot or multi-epoch fits can be done with time-independent values of the shock parameters ϵ_b, ϵ_e, p (e.g., Wijers & Galama 1999), in some cases the fits indicate that the shock physics may be a function of the shock strength. For instance, p, ϵ_b, ϵ_e, or the electron injection fraction ζ may change in time (Panaitescu et al. 1998, Panaitescu & Kumar 2001a). Whereas these are, in a sense, time-averaged shock properties, specifically time-dependent effects would be expected to affect the electron energy distribution and photon spectral slopes, leading to time-integrated observed spectra that could differ from those in the simple time-averaged picture (Medvedev 2000, Llyod & Petrosian 2001a). The back-reaction of protons accelerated in the same shocks (see "Cosmic Rays, Neutrinos, GeV–TeV Photons, and Gravity Waves") and magnetic fields may also be important, as in supernova remnants (e.g., Ellison et al. 2000). Turbulence may be important for the electron-proton energy exchange (Bykov & Mészáros 1996, Schlicheiser & Dermer 2000), whereas reactions leading to neutrons and vice versa (Rachen & Mészáros 1998) can influence the escaping proton spectrum.

Other Effects

Two potentially interesting developments are the possibility of a relationship between the differential time lags for the arrival of the GRB pulses at different energies and the luminosity (Norris et al. 2000), and between the degree of variability or spikyness of the gamma-ray light curve and the luminosity (Fenimore & Ramirez-Ruiz 2001, Reichart et al. 2000). Attempts at modeling the spectral lags have relied on observer-angle dependences of the Doppler boost (Nakamura 2000, Salmonson 2001b). In these correlations the isotropic equivalent luminosity was used, in the absence of jet signatures, and they must be considered tentative for now. However, if confirmed, they could be invaluable for independently estimating GRB redshifts.

SOME ALTERNATIVE MODELS

Whereas space limitations preclude a comprehensive review of many alternative models, a partial list includes precessing jets from pulsars (Blackman et al. 1996; see also Fargion 1999); jets (Cen 1997) or cannonballs from supernovae (Dado et al. 2001); magnetar bubble collapse (Gnedin & Kiikov 2000); neutron star collapse to a strange star (Cheng & Dai 1996), or collapse to a black hole caused by accretion (Vietri & Stella 1998) or by capture of a primordial black hole (Derishev et al. 1999c); supermassive black hole formation (Fuller & Shi 1998), and evaporating black holes (Halzen et al. 1991, Belyanin et al. 1996, Cline & Hong 1996).

PROGENITORS

Currently, the most widely held view is that GRBs arise in a very small fraction of stars ($\sim 10^{-6}$, or somewhat larger depending on beaming), which undergo a catastrophic energy release event toward the end of their evolution. One class of candidates involves massive stars whose core collapses (Woosley 1993, Paczyński 1998, MacFadyen & Woosley 1999), probably in the course of merging with a companion, often referred to as hypernovae or collapsars. Another class of candidates consists of neutron star (NS) binaries or neutron star-black hole (BH) binaries (Paczyński 1986, Goodman 1986, Eichler et al. 1989, Mészáros & Rees 1997b), which lose orbital angular momentum by gravitational wave radiation and undergo a merger. Both of these progenitor types are expected to have as an end result the formation of a few solar mass black holes, surrounded by a temporary debris torus whose accretion can provide a sudden release of gravitational energy, with similar total energies, sufficient to power a burst. An important point is that the overall energetics from these various progenitors do not differ by more than about one order of magnitude (Mészáros et al. 1999). The duration of the burst in this model is related to the fall-back time of matter to form an accretion torus around the BH (Fryer et al. 1999, Popham et al. 1999) or the accretion time of the torus (Narayan et al. 2001). Other possible alternatives include, e.g., the tidal disruption of compact stars by 10^5–$10^6 M_\odot$ black holes (R. Blandford, J. Ostriker, P. Mészáros, unpublished manuscript), and the formation from a stellar collapse of a fast-rotating ultra-high magnetic field neutron star (Usov 1994, Thompson 1994, Spruit 1999, Wheeler et al. 2000, Ruderman 2000).

Two large reservoirs of energy are available in such BH systems: the binding energy of the orbiting debris (Woosley 1993) and the spin energy of the black hole (Mészáros & Rees 1997b). The first can provide up to 42% of the rest mass energy of the disk, for a maximally rotating black hole, whereas the second can provide up to 29% of the rest mass of the black hole itself. The question is how to extract this energy.

One energy extraction mechanism is the $\nu\bar{\nu} \to e^+e^-$ process (Eichler et al. 1989), which can tap the thermal energy of the torus produced by viscous dissipation. To be efficient, the neutrinos must escape before being advected into the hole; on the other hand, the efficiency of conversion into pairs (which scales with the square of the neutrino density) is low if the neutrino production is too gradual. Estimates suggest a fireball of $\lesssim 10^{51}$ erg (Ruffert et al. 1997, Fryer & Woosley 1998, MacFadyen & Woosley 1999), or in the collapsar case (Popham et al. 1999) possibly $10^{52.3}$ ergs [see also higher estimates in the NS–NS case by (Salmonson et al. 2001a)]. If the fireball is collimated into a solid angle Ω_j, then of course the apparent isotropized energy would be larger by a factor $(4\pi/\Omega_j)$. Using the recent total energy estimates (corrected for jet collimation) $E_{\gamma, tot} \sim 10^{51}$ erg deduced from jet data by Frail et al. 2001 and Panaitescu & Kumar 2001b, neutrino annihilation would appear to be a likelier possibility than it did before these analyses. An alternative and more efficient mechanism for tapping the energy of the torus may be through dissipation of magnetic fields generated by the differential rotation in the torus (Paczyński 1991, Narayan et al. 1992, Mészáros & Rees 1997b, Katz 1997). Even before the BH forms, a NS–NS merging system might lead to winding up of the fields and dissipation in the last stages before the merger (Mészáros & Rees 1992, Vietri 1997a).

The black hole itself, being more massive than the disk, could represent an even larger source of energy, especially if formed from a coalescing compact binary, because then it is guaranteed to be rapidly spinning. The energy extractable in principle through MHD coupling to the rotation of the hole by the B-Z (Blandford & Znajek 1977) mechanism could then be even larger than that contained in the orbiting debris (Mészáros & Rees 1997b, Paczyński 1998). (Less conventional and more specific related BH energization of jets is discussed e.g., by van Putten 2000,

Li 2000, Ruffini et al. 2001). Collectively, such MHD outflows have been referred to as Poynting jets.

The various stellar progenitors differ slightly in the mass of the BH and somewhat more in that of the debris torus, but they can differ markedly in the amount of rotational energy contained in the BH. Strong magnetic fields, of order 10^{15} G, are needed to carry away the rotational or gravitational energy in a timescale of tens of seconds (Usov 1994, Thompson 1994), which may be generated on such timescales by a convective dynamo mechanism, the conditions for which are satisfied in freshly collapsed neutron stars or neutron star tori (Duncan & Thompson 1992, Kluzniak & Ruderman 1998). If the magnetic fields do not thread the BH, a Poynting outflow can at most carry the gravitational binding energy of the torus. This is

$$E_t \simeq \epsilon_m q 0.42 M_d c^2 \lesssim 8 \times 10^{53} \epsilon_m q (M_d/M_\odot) \, \text{ergs}, \qquad (8)$$

where $\epsilon_m \lesssim 0.3$ is the efficiency for converting gravitational into MHD jet energy, q is in the range $[1, 1/7]$ for [fast, slow] rotating BHs, and the mass M_d of the torus or disk in a NS–NS merger is (Ruffert & Janka 1998) $\sim 10^{-1}$–$10^{-2} M_\odot$, while in NS–BH, He–BH, WD–BH mergers or a binary WR collapse it may be (Paczyń Ski 1998, Fryer & Woosley 1998) $\sim 1 M_\odot$.

If the magnetic fields in the torus thread the BH, the spin energy of the BH that can be extracted e.g., through the B-Z or related mechanisms is (Mészáros & Rees 1997b, Mészáros et al. 1999)

$$E_{bh} \simeq \epsilon_m f(a) M_{bh} c^2 \lesssim 5 \times 10^{53} \epsilon_m (M_{bh}/M_\odot) \, \text{ergs}, \qquad (9)$$

where $f(a) = 1 - ([1 + \sqrt{1 - a^2}]/2)^{1/2} \leq 0.29$ is the rotational efficiency factor, $a = Jc/GM^2 =$ rotation parameter ($a = 1$ for a maximally rotating BH). The rotational factor is small unless a is close to 1, so the main requirement is a rapidly rotating black hole, $a \gtrsim 0.5$. Rapid rotation is guaranteed in a NS–NS merger, because (especially for a soft equation of state) the radius is close to that of a black hole, and the final orbital spin period is close to the required maximal spin rotation period. The central BH mass (Ruffert et al. 1997, Ruffert & Janka 1998) is ~ 2.5 $M_\odot$, so a NS–NS merger could power a jet of up to $E_{NS-NS} \lesssim 1.3 \times 10^{54} \epsilon_m$ ergs. A maximal rotation rate may also be possible in a He–BH merger, depending on what fraction of the He core gets accreted along the rotation axis as opposed to along the equator (Fryer & Woosley 1998). For a rotating He star, recent calculations (Lee et al. 2001) indicate that a BH rotation parameter $a = 0.7 - 0.9$ is achievable. A similar end result may apply to the binary fast-rotating WR scenario, which probably does not differ much in its final details from the He–BH merger. For a fast rotating BH of 2.5–3 $M_\odot$ threaded by the magnetic field, the maximal energy carried out by the jet is then similar to or somewhat larger than in the NS–NS case. The scenarios less likely to produce a fast rotating BH are the NS–BH merger (where the rotation parameter could be limited to $a \leq M_{ns}/M_{bh}$, unless the BH is already fast-rotating) and the failed SNe Ib (where the last material to fall in would have maximum angular momentum, but the material that was initially close to the hole has less angular momentum). Recent calculations of collapsar central BH mass, rotation rates, and disk masses have been discussed by Fryer & Kalogera 2001, MacFadyen et al. 2001, Fryer et al. 1999, Janka et al. 1999, and Zhang & Fryer 2001. The magnetic interaction between a rotating hole and disk is further discussed in van Puten & Ostriker (2001).

The total jet energetics differ between the various BH formation scenarios at most by a factor 20 for Poynting jets powered by the torus binding energy, and at most by factors of a few for Poynting jets powered by the BH spin energy, depending on the rotation parameter. For instance, allowing for a total efficiency of 50%, a NS–NS merger whose jet is powered by the torus binding energy would require a beaming of the γ-rays by a factor $(4\pi/\Omega_j) \sim 100$, or beaming by a factor ~ 10 if the jet is powered by the B-Z mechanism, to produce the equivalent of an isotropic energy of 4×10^{54} ergs. These beaming factors are compatible with the values derived from observations (Frail et al. 2001) (albeit so far available for long bursts only).

In all cases, including solar mass BHs and magnetar central objects, an $e^{\pm}$, γ fireball would be expected to arise from the heating and dissipation associated with the transient accretion event, in addition to MHD stresses. Even if the latter are not dominant, values in excess of 10^{15} Gauss can provide the driving stresses leading to highly relativistic $\gamma_j \gg 1$ expansion. The fireball would also be likely to involve some fraction of baryons, and uncertainties in this "baryon pollution" (Paczyński 1990) remain difficult to dispel until 3D MHD calculations capable of addressing baryon entrainment become available. In spherical symmetry, general considerations give insights into the development of the Lorentz factor in a shock wave as it propagates down the density gradient of a stellar envelope (Sari et al. 2000, Tan et al. 2001). The expectation that the fireball is likely to be substantially collimated is prevalent especially if the progenitor is a massive star, due to the constraint provided by an extended, fast-rotating envelope, which provides a natural fireball escape route along the rotation axis. The development of a jet and its Lorentz factor in a collapsar is discussed analytically in Mészáros & Rees 2001 and numerically in, e.g., Aloy et al. 2000 and Zhang et al. 2001 (see Figure 4), whereas the case of a magnetar jet is discussed by Wheeler et al. 2000. In the case of NS–NS or BH–NS mergers, a weaker degree of collimation would be expected, due to the lack of an extended envelope [unless magnetic or hydrodynamic self-collimation occurs, e.g., Levinson & Eichler (2000)].

An interesting question is whether the long bursts arise from a different parent population as the short bursts. A current hypothesis is that whereas massive stars (e.g., via the collapsar scenario) appear implicated in long bursts, NS–NS mergers might possibly lead to short bursts (Katz & Canel 1996, Popham et al. 1999), as also discussed in the next section (see van Putten & Ostriker 2001 for an alternative view in which both long and short bursts originate in collapsars).

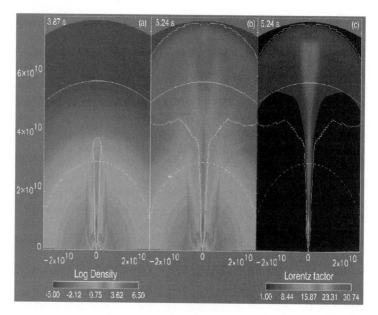

Figure 4   Jet development in a $14\,M_\odot$ collapsar (Aloy et al. 2000) after substantial envelope mass loss. Contours of the logarithm of density after 3.87 s and 5.24 s (*left two panels*), and of the Lorentz factor (*right panel*) after 5.24 s. X and Y axis measure distance in centimeters. Dashed and solid arcs mark the stellar surface and the outer edge of the exponential atmosphere.

COSMOLOGICAL SETTING, GALACTIC HOSTS AND ENVIRONMENT

For the long GRB afterglows localized so far, a host galaxy has been found in most cases ($\gtrsim 80\%$ out of over 30 optically identified), (Bloom et al. 2001). The GRB hosts are typically low mass, sub-L_* galaxies, with the blue colors and atomic lines indicative of active star formation (Fruchter 2000, Bloom et al. 2001, Frail et al. 2001; see also Schaefer 2000). Many of them are obscured, far-infrared luminous galaxies, some of which appear tidally disturbed (Chary et al. 2001). The redshifts of the hosts, with one exception, are in the range $0.43 \lesssim z \lesssim 4.5$, i.e., comparable to that of the most distant objects detected in the Universe. The observed number of bursts per unit photon flux can be fitted by cosmological distribution models, with a somewhat better fit if one assumes that the burst rate scales proportionally to the observed star-formation rate as a function of redshift (Wijers et al. 1998, Totani 1999, Blain & Natarajan 2000, Böttcher & Dermer 2000, Stern et al. 2001). The spread in the inferred isotropic-equivalent luminosities extends over three orders of magnitude, i.e., far from standard candles for the purposes of testing cosmological models (Mao & Mo 1998). However, this spread in luminosities is considerably reduced to less than one order of magnitude (Panaitescu & Kumar 2001a,b, Frail et al. 2001, Piran et al. 2001) if allowance is made for jet-like collimation. The sample of bursts for which this is possible is still too small ($\lesssim 10$) to do cosmology with them.

The bursts for which the intrinsic brightness is known from their measured redshifts would, in principle, be detectable out to much larger redshifts $z \lesssim 15$–20 with present detectors (Lamb & Reichart 2000). Within the first minutes to hours after the burst, the afterglow optical light is expected to be in the range $m_v \sim 10$–15, far brighter than quasars, albeit for a short time. Thus, promptly localized GRB could serve as beacons, which, shining through the pregalactic gas, provide information about much earlier epochs in the history of the Universe. The presence of iron or other x-ray lines provides an additional tool for measuring GRB distances, which may be valuable for investigating the small but puzzling fraction of bursts that have been detected only in X-rays but not optically, perhaps due to a high dust content in the host galaxy.

Accurate localizations and host galaxies have, so far, been restricted to the class of long bursts (γ-ray durations $t_b \sim 10$–10^3 s), because Beppo-SAX is mostly sensitive to bursts longer than about 5–10 s. One exception is a recent short burst localization, which led to optical upper limits $R > 22.3$ and $I > 21.2$ about 20 hours after the trigger (Gorosabel et al. 2001). For the long bursts, the fading x-ray and optical afterglow emission is predominantly localized within the optical image of the host galaxy. In most cases it is offset from the center, but in a few cases (out of a total of about twenty) it is near the center of the galaxy (Bloom et al. 2001). This is in disagreement with current simple calculations of NS–NS mergers that suggest (Bloom et al. 1999; also Narayan et al. 1992) that high spatial velocities would take these binaries, in more than half of the cases, outside of the confines of the host galaxy before they merge and produce a burst. These calculations, however, are uncertain, because they are sensitive to a number of poorly known parameters (e.g., distribution of initial separations, etc). On the other hand, theoretical estimates (Fryer et al. 1999) suggest that NS–NS and NS–BH mergers will lead to shorter bursts ($\lesssim 5$ s), beyond the capabilities of Beppo-SAX but expected to be detectable with the recently launched HETE-2 spacecraft (HETE homepage). More effectively, short as well as long bursts should be detected at the rate of 200–300 per year with the Swift multi-wavelength GRB afterglow mission (Swift hompage) currently under construction and scheduled for launch in 2003. Swift will be equipped with γ-ray, x-ray, and optical detectors for on-board follow-up and will be capable to slew within 30–70 seconds its arc-second resolution X-ray camera onto GRBs acquired with their large field-of-view gamma-ray monitor, relaying to the ground the burst coordinates within less than a minute from the burst trigger. This will permit much more detailed studies of the burst environment, the host galaxy, and the intergalactic medium.

Hydrogen Lyman α absorption from intervening newly formed galaxies would be detectable as the GRB optical/UV continuum light shines through them (Loeb & Barkana 2001, Lamb & Reichart 2000). Whereas the starlight currently detected is thought to come mostly from later, already metal-enriched generations of star formation, GRB arising from the earliest generation of stars may be detectable; and if this occurs before galaxies have gravitationally assembled, it would provide a glimpse into the pregalactic phase of the Universe. At a given observed wavelength and a given observed time delay, the observed brightness of a burst afterglow decreases more slowly at higher redshifts, since the afterglow is observed at an earlier source time and at a higher frequency where it is brighter (Ciardi & Loeb 2000, Lamb & Reichart 2000). The high redshift afterglows shining through their host or intervening galaxies would be expected to provide valuable information in the near IR, whereas in the far IR and sub-mm they would provide invaluable information about the dust content in high redshift environments (Venemans & Blain 2001). Dust affects the colors of the light curves and contains information about the metallicity as a function of redshift (Reichart 2001). Bursts that are highly dust-obscured in the optical would generally be detectable in γ-rays and X-rays, and quantitative information about the dust content may be obtained through the detection of a hump accompanied by a spectral softening in the keV X-ray light curve (Mészáros & Gruzinov 2000), caused by small-angle forward scattering on the dust grains, accompanied by a late brightening in the near-IR.

Most of the host galaxies of the long bursts detected so far show signs of active star formation, implying the presence of young, massive stars forming out of dense gaseous clouds. The diffuse gas around a GRB is expected to produce time-variable O/UV atomic absorption lines in the first minutes to hours after a burst (Perna & Loeb 1998). There is also independent evidence from the observation of 0.5–2 keV absorption in the x-ray afterglow spectra, attributed to metals in a high column density of gas in front of the burst (Galama & Wijers 2001). This appears to be higher than expected from optical extinction measures, which may be due to dust destruction by UV photons (Waxman & Draine 2000, Esin & Blandford 2000, Fruchter et al. 2001).

It is interesting that, at least in a few bursts so far, there appears to be evidence for an approximately coincident supernova explosion. There is good spatial-temporal coincidence for one burst, GRB 980425, associated with the unusually bright SN Ib/Ic 1998bw (Galama et al. 1998, Bloom et al. 1998, van Paradijs 1999). At a measured redshift of 0.0085, the association would imply an abnormally faint GRB luminosity ($\sim 10^{47}$ ergs), although it can be argued that the jet appears fainter due to being seen by chance almost close to its edge (e.g., Höfflich et al. 1999). For SN 1998bw, a mildly relativistic and quasi-spherical shock break-out is also a good model (Waxman & Loeb 1999, Tan et al. 2001). In at least three other localized long GRB, there is circumstantial evidence for a supernova remnant in the form of a bump and reddening in the GRB afterglow optical light curve after several weeks (Lazzati et al. 2001b, Galama et al. 2000, Reichart 1999, Bloom et al. 1999). Alternative explanations based on dust sublimation and scattering have been proposed by Esin & Blandford (2000) and Waxman & Draine (2000). The hypothesis of a generic association of GRB and supernovae ("hypernovae")

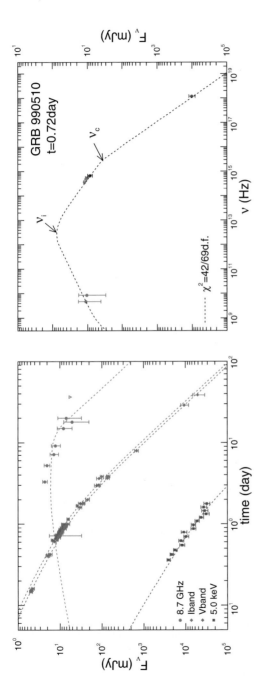

Figure 3 Model light curves at various energies (*left panel*) and snapshot spectral fit at 0.72 days (*right panel*) for GRB 990510, compared against the data (Panaitescu & Kumar 2001a). The model shown has $\chi^2 = 42$ for 69 df, and parameters: $E_0 = 3.0 \times 10^{50}$, erg, $\theta_{jet} = 2.7°$, $n_0 = 0.14$ cm^{-3}, $\varepsilon_e = 0.046$, $\varepsilon_B = 8.6 \times 10^{-4}$, $p = 2.01$. The steepening of the optical decay is due to the effect of a jet.

has been discussed by Paczyński 1998 (see also Woosley 1993) and by Wheeler et al. 2000, whereas multiple sub-jets are discussed by (Nakamura 2000).

X-ray atomic edges and resonance absorption lines are expected to be detectable from the gas in the immediate environment of the GRB, and in particular from the remnants of a massive progenitor stellar system (Mészáros & Rees 1998b, Weth et al. 2000, Böttcher & Fryer 2001). Observations with the Chandra ACIS X-ray spectrographic camera have provided evidence, at a moderate $\gtrsim 4\sigma$ confidence level, for iron K-α line and edge features in at least one burst (GRB 991216, Piro et al. 2000), and there are at least four other detections at the $\sim 3\sigma$ level with Beppo-SAX and ASCA (e.g., Amati et al. 2000, Yoshida et al. 1999, van Paradijs et al. 2000). The observed frequency of the iron lines appears displaced from the laboratory frequency by the right amount expected from the measured optical redshift, when available, indicating that the material producing the lines is expanding at $v/c \lesssim 0.1$ (Piro et al. 2000). The presence of iron line features would again strongly suggest a massive stellar progenitor, but the details remain model dependent.

One possible interpretation of the iron emission lines ascribes the approximate one day delay between the burst and the Fe line peak to light-travel time effects, a specific example postulating an Fe-enriched supernova remnant situated outside the burst region, which is illuminated by X-rays from the afterglow leading to Fe recombination line emission (Figure 5). This would require about $10^{-1}-1\,M_\odot$ of Fe in the shell from, e.g., a supernova (SN) explosion by the progenitor occurring weeks before the burst, which might be due from the accretion-induced

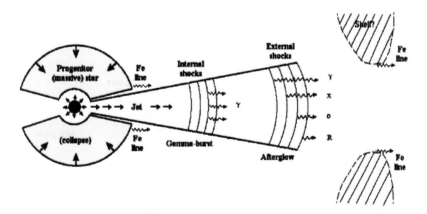

Figure 5 Schematic GRB from internal shocks and afterglow from external shock, arising from a relativistic jet emerging from a massive progenitor collapse (similar jets could arise from other progenitors). Internal shocks produce γ-rays and PeV neutrinos, external shocks produce γ-rays, X-rays, optical, radio and EeV neutrinos. Fe X-ray lines may arise from X-ray illumination of a pre-ejected supernova remnant (Piro et al. 2000) or from continued X-ray irradiation of the outer stellar envelope (Rees & Mészáros 2000; see also Figure 6 of this paper).

collapse of the NS remnant left behind by the SN (Piro et al. 2000, Vietri et al. 2001). A similar interpretation is made (Lazzati et al. 2001c) in the one reported Beppo-SAX case that appears as prompt ($\lesssim$40 s) Fe absorption feature (Amati et al. 2000). A delay of weeks is required to allow a SNR shell to travel out to a light-day distance and for the Ni in the explosion to decay to Fe. If the lines are ascribed to Ni or Co (Lazzati et al. 2001a), the shell velocity must match the difference to the Fe line energies. In either case, some fine-tuning appears necessary.

A less demanding Fe line model is possible if the GRB, after its usual initial outburst, continues to eject a progressively weaker jet for a few days, at a rate that does not violate the observed light-curve (Rees & Mészáros 2000). This jet may be fed, e.g., through continued fall-back at low rates on the BH, or through spin-down if the central object is a magnetar. A decaying jet with a luminosity $L \sim 10^{47}$ erg/s at one day impinging on the outer layers near $\sim 10^{13}$ cm of the progenitor envelope leads to Fe recombination line emission at the observed rate, requiring only solar abundances or a total of $\sim 10^{-5} M_{\odot}$ of Fe (Figure 5). However, the most plausible model may be one based upon the aftereffects of the cocoon of waste heat pumped into the lower envelope as the relativistic jet makes its way through the progenitor envelope (Mészáros & Rees 2001). This bubble (Figure 6) of waste heat, after the jet has emerged and produced the burst, rises slowly by buoyancy and emerges through the outer envelope on timescales of a day after the burst. Its structure is likely to be highly inhomogeneous, resulting in nonthermal X-rays produced by synchrotron in the low density medium between much denser photoionized filaments. This can produce the observed Fe line luminosity through recombination, requiring a modest $\sim 10^{-5} M_{\odot}$ of Fe, which can be easily supplied from the core of the star as the jet develops. In this type of nearby ($\lesssim 10^{13}$ cm)

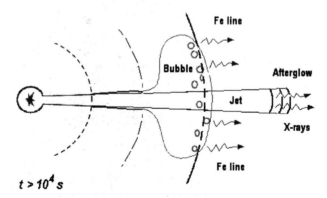

Figure 6 Schematic GRB afterglow from a jet emerging from a massive progenitor, followed hours later by emergence of a bubble of waste heat producing additional nonthermal X-ray and reprocessed Fe Kα recombination from dense filaments in the bubble and envelope (Mészáros & Rees 2001).

line production models, the Fe line energies could be more naturally mimicked by down-scattering of Ni or Co lines (McLaughlin et al. 2001).

The simple picture of an origin in star-forming regions, at least for the long ($t_b \gtrsim 5$ s) bursts, is complicated by the fact that the observed optical absorption is less than expected for the corresponding x-ray absorption. Also, standard afterglow model fits indicate an ambient gas density generally lower than that expected in star-forming clouds (Galama & Wijers 2001, Panaitescu & Kumar 2001b). These contradictions may possibly be reconcilable, e.g., through dust sublimation by x-ray/UV radiation (Waxman & Drainen 2000, Esin & Blandford 2000, Fruchter et al. 2001) or the blowing out of a cavity by a progenitor wind (Wijers 2000).

Whereas it is unclear whether there is one or more classes of GRB progenitors, e.g., corresponding to short and long bursts, there is a general consensus that they would all lead to the generic fireball shock scenario. Much of the current effort is dedicated to understanding the different progenitor scenarios and trying to determine how the progenitor and the burst environment can affect the observable burst and afterglow characteristics.

COSMIC RAYS, NEUTRINOS, GeV–TeV PHOTONS, AND GRAVITY WAVES

There are several other, as yet unconfirmed, but potentially interesting observing channels for GRBs, relating to the baryonic component of the outflow, the shock physics, and the progenitor collapse dynamics.

Among these, cosmic rays are perhaps most directly implicated in the fireball shock mechanism, thought to accelerate the electrons responsible for the nonthermal γ-rays in GRB. The same shocks should also accelerate protons, based on experience from interplanetary shocks. Both the internal and the external reverse shocks are mildly relativistic and are expected to lead to relativistic proton energy spectra of the form $dN_p/d\epsilon_p \propto \gamma_p^{-2}$ (Blandford & Eichler 1987; see also Kirk et al. 2000, and Lloyd-Ronning & Petrosian 2001b). The maximum proton energies achievable in GRB shocks are $E_p \sim 10^{20}$ eV (Waxman 1995, Vietri 1995, Dermer & Humi 2001), comparable to the highest energies measured with large cosmic ray ground arrays (e.g., Takeda et al. 1999). The condition for this is that the acceleration time up to that energy is shorter than the radiation or adiabatic loss time as well as the escape time from the acceleration region. The resulting constraints on the magnetic field and the bulk Lorentz factor (Waxman 1995) are close to those required to obtain efficient gamma-ray emission at ~ 1 MeV. If the accelerated electrons that produce the γ-rays and the protons carry a similar fraction of the total energy (a conservative assumption, based on interplanetary collisionless shock acceleration measurements), the GRB cosmic ray energy production rate at 10^{20} eV throughout the universe is of order 10^{44} erg/Mpc^3/yr, comparable to the observationally required rate from γ-ray observations and from the observed diffuse cosmic ray flux (Waxman 1995; see also Stecker 2000a). These numbers depend on uncertainties in the burst total energy and beaming fraction, as well as on the poorly constrained burst rate evolution with redshift. The highest energy protons would need to have arrived from within less than about 50–100 Mpc, to avoid interaction with the microwave background, and reasonable intergalactic magnetic field strengths can ensure time dispersions in excess of a few hundred years, needed to achieve compatibility with the estimated burst rate of $\sim 10^{-6}$/galaxy/year, as well as with arrival from clustered sources (Bahcall & Waxman 2000). The unknown strength and correlation length of the field could lead to anisotropies constraining both GRB models and competing AGN or other discrete source origin models, an issue that will be addressed by future large-area ground cosmic-ray arrays such as, e.g., Auger and HiRes.

Any stellar origin mechanism (whether collapsar, neutron star merger, or other) would lead to a very large ($\sim M_\odot c^2$) luminosity in thermal neutrinos and antineutrinos with energies $\sim$ few MeV, as in core-collapse supernova. However, at MeV energies the neutrino detection cross section is $\sim 10^{-44}$ cm^2, and as shown by the low count rates in the supernova SN 1987a detection from $\sim$50 Kpc, even larger detectors at these energies (super-Kamiokande, Sudbury, etc) would be insensitive to sources such as GRB with typical distances $\gtrsim$100 Mpc.

A mechanism leading to higher (GeV) energy neutrinos in GRB is inelastic nuclear collisions. Proton-proton collisions at internal shock radii $\sim 10^{14}$ cm could lead to $\sim$GeV neutrinos in the observer frame through charged pion decay (Paczyński & Xu 1994), with low efficiency due to the low densities at these large radii and small relative velocities between protons. However, proton-neutron inelastic collisions are expected, even in the absence of shocks, at much lower radii, due to the decoupling of neutrons and protons in the fireball or jet (Derishev et al. 1999a). Provided the fireball has a substantial neutron/proton ratio, as expected in most GRB progenitors, the collisions become inelastic and their rate peaks when the nuclear scattering time becomes comparable to the expansion time. This occurs when the n and p fluids decouple, their relative drift velocity becoming comparable to c, which is easier due to the lack of charge of the neutrons. Inelastic n, p collisions then lead to charged pions, and GeV muon and electron neutrinos (Bahcall & Mészáros 2000). The early decoupling and saturation of the n also leads to a somewhat higher final p Lorentz factor (Derishev et al. 1999a, Bahcall & Mészáros 2000, Fuller et al. 2000, Pruet et al. 2001), implying a possible relation between the n/p ratio and the observable fireball dynamics, relevant for the progenitor question and burst timescales. Inelastic p, n collisions leading to neutrinos can also occur in fireball outflows with transverse inhomogeneities in the bulk Lorentz factor, where the n can drift sideways into regions of different bulk velocity flow, or in situations where internal shocks involving n and p occur close to the saturation radius or below the photon photosphere (Mészáros & Rees 2000a). The typical n, p neutrino energies are in the 5–10 GeV range, which could be detectable in coincidence with observed GRBs for a sufficiently close photo-tube spacing in future km^3 detectors such as ICECUBE (Halzen 2000).

In addition, neutrinos with energies $\gtrsim$PeV can be produced in p, γ photo-meson interactions involving highly relativistic protons accelerated in the fireball internal or external shocks. A high collision rate is ensured here by the large density of photons in the fireball shocks. The most obvious case is the interaction between MeV photons produced by radiation from electrons accelerated in internal shocks (see Figure 5), and relativistic protons accelerated by the same shocks (Waxman & Bahcall 1997), leading to charged pions, muons, and neutrinos. This p, γ reaction peaks at the energy threshold for the photo-meson Δ resonance in the fluid frame moving with γ, or

$$\epsilon_p \epsilon_\gamma \gtrsim 0.2 \, \text{GeV}^2 \gamma^2. \tag{10}$$

For observed 1 MeV photons this implies $\gtrsim 10^{16}$ eV protons, and neutrinos with $\sim 5\%$ of that energy, $\epsilon_\nu \gtrsim 10^{14}$ eV in the observer frame. Above this threshold, the fraction of the proton energy lost to pions is $\sim 20\%$ for typical fireball parameters, and the typical spectrum of neutrino energy per decade is flat, $\epsilon_\nu^2 \Phi_\nu \sim$ constant. Synchrotron and adiabatic losses limit the muon lifetimes (Rachen & Mészáros 1998), leading to a suppression of the neutrino flux above $\epsilon_\nu \sim 10^{16}$ eV. In external shocks (Figure 5), another copious source of targets are the O/UV photons in the afterglow reverse shock (e.g., as deduced from the GRB 990123 prompt flash of Akerlof et al. 1999). In this case, the resonance condition implies higher energy protons, leading to neutrinos of 10^{17}–10^{19} eV (Waxman & Bahcall 1999a, Vietri 1998). These neutrino fluxes are expected to be detectable above the atmospheric neutrino background with the planned cubic kilometer ICECUBE detector (Halzen 2000, Alvarez-Muñiz et al. 2000). Useful limits to their total contribution to the diffuse ultrahigh energy neutrino flux can be derived from observed cosmic ray and diffuse gamma-ray fluxes (Waxman & Bahcall 1999b, Bahcall & Waxman 2001, Mannheim 2001). Whereas the p, γ interactions leading to $\gtrsim 100$ TeV energy neutrinos provide a direct probe of the internal shock acceleration process, as well as of the MeV photon density associated with them, the $\gtrsim 10$ PeV neutrinos would probe the reverse external shocks, as well as the photon energies and energy density there.

The most intense neutrino signals, however, may be due to p, γ interactions occurring inside collapsars while the jet is still burrowing its way out of the star (Mészáros & Waxman 2001), before it has had a chance to break through (or not) the stellar envelope to produce a GRB outside of it. While still inside the star, the buried jet produces thermal X-rays of ~ 1 keV that interact with $\gtrsim 10^5$ GeV protons that could be accelerated in internal shocks occurring well inside the jet/stellar envelope terminal shock, producing $\sim$ few TeV neutrinos for tens of seconds, which penetrate the envelope (Figure 7). This energy is close to the maximum sensitivity for detection, and the number of neutrinos is also larger for the same total energy output. The rare bright, nearby, or high γ collapsars could occur at the rate of ~ 10/year, including both γ-ray bright GRBs (where the jet broke through the envelope) and γ-ray dark events where the jet is choked (failed to break through), and both such γ-bright and dark events could have a TeV neutrino fluence of ~ 10/neutrinos/burst, detectable by ICECUBE in individual bursts.

GeV to TeV photon production is another consequence of the photo-pion and inelastic collisions responsible for the ultrahigh energy neutrinos (Waxman &

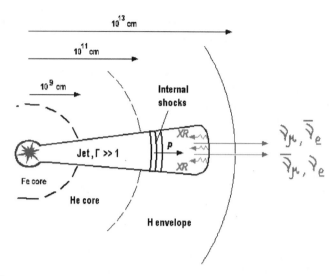

Figure 7 Sketch of TeV neutrino production by photomeson interactions in internal shocks before a relativistic jet has broken through the progenitor envelope (Mészáros & Waxman 2001). Neutrinos would be expected whether the jet is choked off (γ-dark) or emerges to make a GRB.

Bahcall 1997, Böttcher & Dermer 1998, Derishev et al. 1999a, Bahcall & Mészáros 2000). This is in addition to the GeV emission from electron inverse Compton in internal (Papathanassiou & Mészáros 1996) and external shocks (Mészáros et al. 1994, Derishev et al. 2001) and afterglows (Zhang & Mészáros 2001b). In these models, due to the high photon densities implied by GRB models, $\gamma\gamma$ absorption within the GRB emission region must be taken into account (see also Baring 2000, Lithwick & Sari 2001). A tentative $\gtrsim 0.1$ TeV detection of an individual GRB has been reported with the water Cherenkov detector Milagrito (Atkins et al. 2000), and better sensitivity is expected from its larger version MILAGRO as well as from atmospheric Cherenkov telescopes under construction such as VERITAS, HESS, MAGIC, and CANGAROO-III (Weekes 2000). GRB detections in the TeV range are expected only for rare nearby events, since at this energy the mean free path against $\gamma\gamma$ absorption on the diffuse IR photon background is $\sim$ few hundred Mpc (Coppi & Aharonian 1997, Stecker 2000b). The mean free path is much larger at GeV energies, and based on the handful of GRB reported in this range with EGRET (Schneid et al. 1995), several hundred should be detectable with large area space-based detectors such as GLAST (Gehrels & Michelson 2000, Zhang & Mészáros 2001b), in coincidence with the neutrino pulses and the usual MeV γ-ray event. Their detection would provide important constraints on the emission mechanism and the progenitors of GRBs.

GRB are also expected to be sources of gravitational waves. A time-integrated luminosity of the order of a solar rest mass ($\sim 10^{54}$ erg) is predicted from merging

NS–NS and NS–BH models (Narayan et al. 1992, Kochanek & Piran 1993, Ruffert & Janka 1998, Oohara & Nakamura 1999). The luminosity from collapsar models is more model-dependent, but expected to be lower (Fryer et al. 2001, Dimmelmeier et al. 2001; see also van Putten 2001). The rates of gravitational wave events (Finn et al. 2000) detectable by the Laser Interferometric Gravitational Wave Observatory (LIGO, currently under construction) from compact binary mergers, in coincidence with GRBs, have been estimated at a few/year for the initial LIGO, and up to 10–15 per year after the upgrades planned 2–4 years after first operations. The observation of such gravitational waves would be greatly facilitated by coincident detections in other channels, either electromagnetic or neutrinos. Detection of gravity wave pulses fitting the templates for compact binary mergers (or collapsars), in coincidence with positive GRB localizations, would have a great impact on our understanding of GRB progenitor systems.

PROSPECTS

In conclusion, major advances have been made in the understanding of GRB since their discovery almost 30 years ago. However, many questions remain, while new ones arise in the wake of the increasingly sophisticated and extensive observations. These questions will be addressed with new space missions and ground experiments dedicated to GRB studies that will come on-line in the near future. Based on past experience, the chances are high that these will bring not only answers but also new surprises and challenges.

ACKNOWLEDGMENTS

It is a pleasure to thank M.J. Rees, as well as C. Dermer, V. Kocharovsky, R. Narayan, R. Sari, E. Waxman, R. Wijers, B. Zhang for collaborations and/or comments and NASA grant NAG5-9192, NSF AST 0098416 for support.

The *Annual Review of Astronomy and Astrophysics* is online at
http://astro.annualreviews.org

LITERATURE CITED

Akerlof K, Balsano R, Barthelemey S, Bloch J, Butterworth P, et al. 1999. *Nature* 398:400

Aloy M, Mueller E, Ibanez J, Marti J, Mac-Fadyen A. 2000. *Ap. J. Lett.* 531:L119

Alvarez-Mūniz J, Halzen F, Hooper DW. 2000. *Phys. Rev. D* 620:3015

Amati L, Frontera F, Vietri M, in't Zand Jean, Soffitta P, et al. 2000. *Science* 290:958

Atkins R, Benbow W, Berley D, Chen M, Coyne D, et al. 2000. *Ap. J. Lett.* 533:L119

Bahcall JN, Mészáros P. 2000. *Phys. Rev. Lett.* 85:1362

Bahcall JN, Waxman E. 2000. *Ap. J.* 542:542–47

Bahcall JN, Waxman E. 2001. *Phys. Rev. D* 6402:3002

Band D, Matteson J, Ford L, Schaefer B, Palmer D, et al. 1993. *Ap. J.* 413:281

Baring MG. 2000. *GeV-TeV Gamma Ray Astrophysics Workshop*. ed. BL Dingus, MH

Salamon, DB Dieda, pp. 238–42. *AIP Conf. Proc.* 515, New York: AIP

Beloborodov A. 2000. *Ap. J. Lett.* 539:L29

Beloborodov A. 2002. *Ap. J.* In press (astro-ph/0103321)

Beloborodov A, Stern B, Svensson R. 1998. *Ap. J. Lett.* 508:L25

Beloborodov A, Stern B, Svensson R. 2000. *Ap. J.* 535:158

Belyanin AA, Kocharovsky VV, Kocharovsky VIV. 1996. *MNRAS* 283:626

Blackman E, Yi I, Field GB. 1996. *Ap. J. Lett.* 473:L79

Blain A, Natarajan P. 2000. *MNRAS* 312:L35

Blandford R, Eichler D. 1987. *Phys. Rep.* 154:1

Blandford R, McKee C. 1976a. *Phys. Fluids* 19:1130

Blandford R, McKee C. 1976b. *MNRAS* 180:343

Blandford R, Znajek R. 1977. *MNRAS* 179:433

Bloom J, Kulkarni S, Harrison F, Prince T, Phinney ES, Frail D. 1998. *Ap. J. Lett.* 506:L105

Bloom J, Kulkarni S, Djorgovski S, Eichelberger A, Cote P, et al. 1999. *Nature* 401:453

Bloom J, Kulkarni R, Djorgovski G. 2001. *Ap. J.* In press (astro-ph/0010176)

Bloom J, Sigurdsson S, Pols O. 1999. *MNRAS* 305:763

Boer M, Atteia JL, Bringer M, Gendre B, Klotz A, et al. 2001. *Astron. Astrophys.* In press (astro-ph/0109065)

Böttcher M, Dermer CD. 1998. *Ap. J.* 499:L131–34

Böttcher M, Dermer CD. 2000. *Ap. J.* 529:635–43

Böttcher M, Fryer CL. 2001. *Ap. J.* 547:338–44

Bykov A, Mészáros P. 1996. *Ap. J. Lett.* 461:L37

Castro-Tirado AJ, Zapatero-Osorio MR, Caon N, Marina Cairos L, Hjorth J, et al. 1999. *Science* 283:2069

Cavallo G, Rees MJ. 1978. *MNRAS* 183:359

Cen R. 1997. *Ap. J. Lett.* 507:L131

Chary R, Becklin E, Armus L. 2001. *Ap. J.* In press (astro-ph/0110010)

Cheng KS, Dai ZG. 1996. *Phys. Rev. Lett.* 77:1210

Chevalier R, Li Z. 2000. *Ap. J.* 536:195

Ciardi B, Loeb A. 2000. *Ap. J.* 540:687

Cline D, Hong W. 1996. *Astropart. Phys.* 5:175

Connors A, Hueter GJ. 1998. *Ap. J.* 501:307

Coppi P, Aharonian F. 1997. *Ap. J. Lett.* 487:L9

Costa E, Frontera F, Heise J, Feroci M, in't Zand J, et al. 1997. *Nature* 387:783

Crider A, Liang E, Smith I, Preece R, Briggs M, et al. 1997. *Ap. J. Lett.* 479:L39

Dado S, Dar A, de Rújula A. 2001. (astro-ph/0107367)

Dai Z, Gou L. 2001. *Ap. J.* 552:72

Dai Z, Lu T. 1999. *Ap. J. Lett.* 519:L155

Dai Z, Lu T. 2000. *Ap. J.* 537:803

Daigne F, Mochkovitch R. 2000. *Astron. Astrophys.* 358:1157

Derishev EV, Kocharovsky VV, Kocharovsky VIV. 1999a. *Ap. J.* 521:640

Derishev EV, Kocharovsky VV, Kocharovsky VIV. 1999b. *Astron. Astrophys.* 345:L51

Derishev EV, Kocharovsky VV, Kocharovsky VIV. 1999c. *JETP Lett.* 70:652

Derishev EV, Kocharovsky VV, Kocharovsky VIV. 2001. *Astron. Astrophys.* 372:1071

Dermer CD, Böttcher M. 2000. *Ap. J.* 534:L155–58

Dermer CD, Böttcher M, Chiang J. 1999. *Ap. J.* 515:L49–52

Dermer CD, Böttcher M, Chiang J. 2000. *Ap. J.* 537:255–60

Dermer CD, Humi M. 2001. *Ap. J.* 556:479–93

Dermer CD, Mitman KE. 1999. *Ap. J.* 513:L5–8

Dimmelmeier H, Font J, Mueller E. 2001. *Ap. J. Lett.* 560:L163

Djorgovski G, Kulkarni S, Bloom J, Frail D, Harrison F, et al. 2001. In *Gamma-Ray Bursts in the Afterglow Era: 2nd Workshop*, ed. N Masetti, *ESO Astrophys. Symp.* Berlin: Springer-Verlag. In press

Duncan R, Thompson C. 1992. *Ap. J. Lett.* 392:L9

Eichler D, Levinson A. 2000. *Ap. J.* 529:146

Eichler D, Livio M, Piran T, Schramm D. 1989. *Nature* 340:126

Ellison D, Berezhko E, Baring M. 2000. *Ap. J.* 540:292

Esin A, Blandford R. 2000. *Ap. J. Lett.* 534:L151

Fargion D. 1999. *Astron. Astrophys. Suppl.* 138: 507

Fenimore EE, Epstein RI, Ho C. 1993. *Astron. Astrophys. Suppl.* 97:59–62

Fenimore EE, Epstein RI, Ho C, Klebaesadel RW, Lacey C, et al. 1993. *Nature* 366:40–42

Fenimore EE, Ramirez-Ruiz E. 2001. *Ap. J.* Submitted (astro-ph/0004176)

Fenimore EE, Ramirez-Ruiz E, Wu BB. 1999. *Ap. J. Lett.* 518:L73–76

Finn LS, Mohanty S, Romano J. 1999. *Phys. Rev. D* 60:121101

Fishman GJ, Meegan CA. 1995. *Annu. Rev. Astron. Astrophys.* 33:415–58

Frail D, Kulkarni SR, Nicastro L, Feroci M, Taylor GB. 1997. *Nature* 389:261–63

Frail DA, Kulkarni SR, Wieringa M, Taylor G, Moriarty-Schieven G, et al. 1999. In *Gamma-Ray Bursts. AIP Conf. Proc. 526*, ed. M Kippen, pp. 298–302. New York: AIP

Frail DA, Kulkarni SR, Sari R, Djorgovski S, Bloom J, et al. 2001. (astro-ph/0102282)

Fruchter A. 2000. In *Gamma-Ray Bursts in the Afterglow Era: 2nd Workshop, Rome, Italy. ESO Astrophys. Symp.* In press

Fruchter A, Thorsett S, Metzger M, Sahu K, Petro L, et al. 1999. *Ap. J. Lett.* 519:L13

Fruchter A, Krolik J, Rhoads J. 2001. (astro-ph/0106343)

Fryer C, Kalogera V. 2001. *Ap. J.* 554:548

Fryer C, Woosley S. 1998. *Ap. J. Lett.* 502:L9

Fryer C, Woosley S, Hartmann D. 1999. *Ap. J.* 526:152

Fryer C, Woosley S, Heger A. 2001. *Ap. J.* 550:327

Fuller G, Pruet J, Kevork A. 2000. *Phys. Rev. Lett.* 85:2673

Fuller G, Shi X. 1998. *Ap. J. Lett.* 502:L5

Galama T, Vreeswijk P, van Paradijs J, Kouveliotou C, Augusteijn T, et al. 1998. *Nature* 395:670

Galama T, Tanvir N, Vreeswijk P, Wijers R, Groot P, et al. 2000. *Ap. J.* 536:185

Galama T, Wijers R. 2001. *Ap. J. Lett.* 549:L209

Gallant Y, Achterberg A, Kirk JG. 1999. *Astron. Astrophys. Suppl.* 138:549

Garnavich P, Loeb A, Stanek K. 2000. *Ap. J. Lett.* 544:L11

Gehrels N, Michelson P. 1999. *Astropart. Phys.* 11:277

Ghisellini G, Celotti A. 1999. *Ap. J. Lett.* 511: L93

Ghisellini G, Lazzati D. 1999. *MNRAS* 309:L13

Ghisellini G, Lazzati D, Celotti A. 2000. *MNRAS* 313:L1

Gnedin Yu, Kiikov S. 2000. *MNRAS* 318:1277

Goodman J. 1986. *Ap. J. Lett.* 308:L47

Goodman J. 1997. *New. Astron.* 2:449

Gorosabel J, Andersen M, Hjorth J, Pedersen H, Jensen B, et al. 2001. *Astron. Astrophys.* Submitted (astro-ph/0110009)

Granot J, Miller M, Piran T, Suen W-M, Hughes P. 2000a. In *Gamma-Ray Bursts in the Afterglow Era: 2nd Workshop, Rome, Italy. ESO Astrophys. Symp.* In press

Granot J, Piran T, Sari R. 1999a. *Ap. J.* 513:679

Granot J, Piran T, Sari R. 1999b. *Ap. J.* 527:236

Granot J, Piran T, Sari R. 2000b. *Ap. J. Lett.* 534:L163

Guetta D, Spada M, Waxman E. 2001. *Ap. J.* 555:101

Halzen F. 2000. *Weak Interactions and Neutrinos, Proc. Int. Workshop, 17th, Singapore*, pp. 123. Singapore: World Sci.

Halzen F, Zas E, McGibbon J, Weekes T. 1991. *Nature* 353:807

Harding A, Baring M. 1994. In *Gamma-Ray Bursts. AIP Conf. Proc. 307*, ed. G Fishman, pp. 520. New York: AIP

Harrison F, Yost S, Sari R, Berger E, Galama T, et al. 2001. *Ap. J.* Submitted (astro-ph/0103377)

HETE http://space.mit.edu/HETE/

Höfflich P, Wheeler JC, Wang L. 1999. *Ap. J.* 521:179

Hurley K. 1992. In *Gamma-Ray Bursts, AIP Conf. Proc. 265*, ed. W Paciesas, G Fishman, pp. 3. New York: AIP

Hurley K. 1994. *Nature* 372:652

Janka H-T, Eberl T, Ruffert M, Fryer C. 1999. *Ap. J. Lett.* 572:L39

Katz J. 1994. *Ap. J. Lett.* 432:L107

Katz J. 1997. *Ap. J.* 490:633

Katz J, Canel L. 1996. *Ap. J.* 471:915

Katz J, Piran T, Sari R. 1998. *Phys. Rev. Lett.* 80:1580

Kehoe R, et al. 2001. *Ap. J. Lett.* Submitted (astro-ph/0104208)

Kirk J, et al. 2000. *Ap. J.* 542:235

Klebesadel R, Strong I, Olson R. 1973. *Ap. J. Lett.* 182:L85

Kluzniak W, Ruderman M. 1998. *Ap. J. Lett.* 508:L113

Kobayashi S, Piran T, Sari R. 1999. *Ap. J.* 513: 669

Kobayashi S, Sari R. 2001. *Ap. J.* 551:934

Kochanek C, Piran T. 1993. *Ap. J. Lett.* 417:L17

Kouveliotou C, et al. 1993. *Ap. J. Lett.* 413:L101

Kulkarni S, Djorgovski SG, Odewahn S, Bloom J, Gal R, et al. 1999a. *Nature* 398:389

Kulkarni S, Berger E, Bloom J, Chaffee F, Diercks A, et al. 1999b. In *Gamma-Ray Bursts. AIP Conf. Proc.* 526, ed M Kippen, R Mallozzi, G Fishman, p. 277. New York: AIP

Kumar P. 1999. *Ap. J. Lett.* 523:L113

Kumar P. 2000. *Ap. J. Lett.* 538:L125

Kumar P, Piran T. 2000. *Ap. J.* 532:286

Lamb DQ. 1995. *Publ. Astron. Soc. Pac.* 107: 1152

Lamb DQ, Reichart D. 2000. *Ap. J.* 536:1

Lazzati D, Perna R, Ghisellini G. 2001a. *MNRAS* 325:L19

Lazzati D, Covino S, Ghisellini G, Fugazza D, Campana S, et al. 2001b. *Astron. Astrophys.* In press (astro-ph/0109287)

Lazzati D, Ghisellini G, Amati L, Frontera F, Vietri M, Stella L. 2001c. *Ap. J.* 556:L471 (astro-ph/0104062)

Lee C-H, Brown G, Wijers R. 2001. (astro-ph/0109538)

Levinson A, Eichler D. 2000. *Phys. Rev. Lett.* 85:236

Li L-X. 2000. *Ap. J.* 544:375

Li L-Zh, Chevalier R. 2001. *Ap. J.* 551:940

Li L-Zh, Chevalier R. 2001b. In *Supernovae and Gamma-Ray Bursts*, ed. K Weiler, Cambridge, UK: Cambridge Univ. Press. In press

Liang E, 1989. In *Proc. Gamma Ray Obs. Sci. Workshop*, ed. WN Johnson, pp. 4–397. Greenbelt, MD: NASA

Liang E, Crider A, Böttcher M, Smith I. 1999. *Ap. J. Lett.* 519:L21

Lithwick Y, Sari R. 2001. *Ap. J.* In press (astro-ph/001508)

Livio M, Waxman E. 2000. *Ap. J.* 538:187

Lloyd N, Petrosian V. 2001a. *Ap. J.* 543:722

Lloyd-Ronning N, Petrosian V. 2001b. (astro-ph/0109340)

Loeb A, Barkana R. 2001. *Annu. Rev. Astron. Astrophys.* 39:19–66

MacFadyen A, Woosley S. 1999. *Ap. J.* 524:262

MacFadyen A, Woosley S, Heger A, 2001. *Ap. J.* 550, 372

Madau P, Blandford R, Rees MJ. 2000. *Ap. J.* 541:712

Madau P, Thompson C. 2000 *Ap. J.* 534:239

Mannheim K. 2001. In *High Energy Gamma-Ray Astronomy. AIP Conf. Proc.* 558, Heidelberg, 2000, ed. FA Aharonian, HJ Völk, pp. 417. New York: AIP

Mao S, Mo HJ. 1998. *Astron. Astrophys.* 339:L1

Mazets EP, Golenetskii SV, Ilinskii VN. 1974. *JETP Lett.* 19:77

McLaughlin C, Wijers R, Brown G, Bethe H. 2001. (astro-ph/0110614)

Medvedev A. 2000. *Ap. J.* 540:704

Meegan C, Fishman G, Wilson R, Paciesas W, Pendleton G, et al. 1992. *Nature* 355:143

Mészáros P, Gruzinov A. 2000. *Ap. J. Lett.* 543:L35

Mészáros P, Laguna P, Rees MJ. 1993. *Ap. J.* 415:181

Mészáros P, Ramirez-Ruiz E, Rees MJ. 2001. *Ap. J.* 554:660

Mészáros P, Rees MJ. 1992. *Ap. J.* 397:570

Mészáros P, Rees MJ. 1993a. *Ap. J.* 405:278

Mészáros P, Rees MJ. 1993b. *Ap. J. Lett.* 418:L59

Mészáros P, Rees MJ. 1994. *MNRAS* 269:L41

Mészáros P, Rees MJ. 1997a. *Ap. J.* 476:232

Mészáros P, Rees MJ. 1997b. *Ap. J. Lett.* 482:L29

Mészáros P, Rees MJ 1998b. *MNRAS* 299:L10

Mészáros P, Rees MJ. 1999. *MNRAS* 306: L39–L43

Mészáros P, Rees MJ. 2000a. *Ap. J. Lett.* 541:L5

Mészáros P, Rees MJ. 2000b. *Ap. J.* 530:292

Mészáros P, Rees MJ. 2001. *Ap. J. Lett.* 556:L37

Mészáros P, Rees MJ, Papathanassiou H. 1994. *Ap. J.* 432:181

Mészáros P, Rees MJ, Wijers R. 1998. *Ap. J.* 499:301

Mészáros P, Rees MJ, Wijers R. 1999. *New Astron.* 4:313–323

Mészáros P, Waxman E. 2001. *Phys. Rev. Lett.* 87:171102

Metzger M, Djorgovski SG, Kulkarni S, Steidel C, Adelberger K, et al. 1997. *Nature* 387:878

Murakami T, et al. 1988. *Nature* 335:234

Nakamura T. 2000. *Ap. J. Lett.* 534:L159

Narayan R, Paczyński B, Piran T. 1992. *Ap. J. Lett.* 395:L8

Narayan R, Piran T, Kumar P. 2001. *Ap. J.* 557:949

Norris J, Marani G, Bonnell J. 2000. *Ap. J.* 534:248

Oohara K, Nakamura T. 1999. *Prog. Theor. Phys. Suppl.* 136:270

Paczyński B. 1986. *Ap. J. Lett.* 308:L43

Paczyński B. 1990. *Ap. J.* 363:218

Paczyński B. 1991. *Acta Astron.* 41:257

Paczyński B. 1995. *Publ. Astron. Soc. Pac.* 107:1167

Paczyński B. 1998. *Ap. J. Lett.* 494:L45

Paczyński B, Rhoads J. 1993. *Ap. J. Lett.* 418:L5

Paczyński B, Xu G. 1994. *Ap. J.* 427:708

Panaitescu A, Kumar P. 2000. *Ap. J.* 543:66

Panaitescu A, Kumar P. 2001. *Ap. J.* 554:667

Panaitescu A, Kumar P. 2001. *Ap. J. Lett.* 560:L49

Panaitescu A, Mészáros P. 1998a. *Ap. J.* 492:683

Panaitescu A, Mészáros P. 1998b. *Ap. J.* 501:772

Panaitescu A, Mészáros P. 1998c. *Ap. J. Lett.* 493:L31

Panaitescu A, Mészáros P. 1998d. *Ap. J.* 526:707

Panaitescu A, Mészáros P. 2000. *Ap. J. Lett.* 544:L17

Panaitescu A, Mészáros P, Rees MJ. 1998. *Ap. J.* 503:314

Papathanassiou H, Mészáros P. 1996. *Ap. J. Lett.* 471:L91

Park HS, Ables E, Porrata R, Ziock K, Williams G, Bradshaw M, et al. 2001. *Bull. Am. Astron. Soc.* 198:380

Pedersen H, Jaunsen A, Grav T, Ostensen R, Andersen M, et al. 1998. *Ap. J.* 496:311

Perna R, Loeb A. 1998. *Ap. J. Lett.* 508:L135

Pilla R, Loeb A. 1998. *Ap. J. Lett.* 494:L167

Piran T. 1999. *Phys. Rep.* 314:575

Piran T, Kumar P, Panaitescu P, Piro L. 2001. *Ap. J. Lett.* 560:L167

Piran T, Sari R. 1998. In *Gamma-Ray Bursts. AIP Conf. Proc. 428*, ed. C Meegan, R Preece, T Koshut, p. 662. New York: AIP

Piro L, Amati L, Antonelli L, Butler R, Costa E, et al. 1998. *Astron. Astrophys.* 331:L41

Piro L, Garmire G, Garcia M, Stratta G, Costa E, et al. 2000. *Science* 290:955

Popham R, Woosley S, Fryer C. 1999. *Ap. J.* 518:356

Preece R, Briggs M, Mallozzi R, Pendleton G, Paciesas W, Band D. 2000. *Ap. J. Suppl.* 126:19

Pruet J, Kevork A, Fuller G. 2001. *Phys. Rev. D* 64:063002

Rachen J, Mészáros P. 1998. *Phys. Rev. D* 58:123005

Rees MJ, Mészáros P. 1992. *MNRAS* 258:P41

Rees MJ, Mészáros P. 1994. *Ap. J. Lett.* 430:L93

Rees MJ, Mészáros P. 1998. *Ap. J. Lett.* 496:L1

Rees MJ, Mészáros P. 2000. *Ap. J. Lett.* 545:L73

Reichart D. 1997. *Ap. J. Lett.* 485:L57

Reichart D. 1999. *Ap. J. Lett.* 521:L111

Reichart D. 2001. *Ap. J.* 553:235

Reichart D, et al. 2000. *Ap. J.* Submitted (astro-ph/0004302)

Rhoads J. 1997. *Ap. J. Lett.* 487:L1

Rhoads J. 1999. *Ap. J.* 525:737

Rossi E, Lazzati D, Rees MJ. 2001. *MNRAS.* In press (astro-ph/0112083)

Ruderman M. 1975. *Ann. NY Acad. Sci.* 262:164

Ruderman M, Tao L, Kluzniak W. 2000. *Ap. J.* 542:243

Ruffert M, Janka HT. 1998. *Astron. Astrophys.* 338:53

Ruffert M, Janka HT, Takahashi K, Schaefer G. 1997. *Astron. Astrophys.* 319:122

Ruffini R, Bianco C, Fraschetti F, Xue S-Sh, Chardonnet P. 2001. *Ap. J. Lett.* 551:L107

Salmonson J, Wilson J. 2001. *Ap. J. Lett.* 544:L11

Salmonson J, Wilson J, Mathews G. 2001. *Ap. J.* 553:471

Sari R. 1998. *Ap. J. Lett.* 494:L49

Sari R. 1999. *Ap. J. Lett.* 524:L43

Sari R, Esin A. 2001. *Ap. J.* 548:787

Sari R, Mészáros P. 2000. *Ap. J. Lett.* 535:L33

Sari R, Narayan R, Piran T. 1996. *Ap. J.* 473:204

Sari R, Piran T. 1998. *Ap. J.* 485:270

Sari R, Piran T. 1999. *Ap. J. Lett.* 517:L109

Sari R, Piran T, Halpern J. 1999. *Ap. J. Lett.* 519:L17

Sari R, Piran T, Narayan R. 1998. *Ap. J. Lett.* 497:L17

Sari R, Waxman E, Shvarts D. 2000. *Ap. J. Suppl.* 127:475

Schaefer B. 2000. *Ap. J. Lett.* 533:L21

Schlickeiser R, Dermer CD. 2000. *Astron. Astrophys.* 360:789

Schneid E, Bertsch D, Fichtel C, Hartman R, Hunter S, et al. 1995. In *Proc. 17th Texas Symp. Relativistic Astrophys.*, ed. H Böhringer, G Morfill, J Trümper, *Ann. NY Acad. Sci.* 759:421

Shemi A, Piran T. 1990. *Ap. J. Lett.* 365:L55

Soderberg A, Ramirez-Ruiz E. 2001. (astro-ph/0110519)

Spada M, Panaitescu A, Mészáros P. 2000. *Ap. J.* 537:824

Spruit H. 1999. *Astron. Astrophys.* 341:L1

Stecker FW. 2000a. *Astropart. Phys.* 14:207

Stecker FW. 2000b. In *IAU Symp. 204.* In press (astro-ph/0010015)

Stern B, Tikhomirova Ya, Svensson R. 2001. (astro-ph/0108303)

Swift http://swift.gsfc.nasa.gov/

Takeda M, Hayashida N, Honda K, Inoue N, Kadota K, et al. 1999. *Ap. J.* 522:225

Tan J, Matzner CD, McKee CF. 2001. *Ap. J.* 551:946

Tavani M. 1996. *Ap. J.* 466:768

Tavani M. 1997. *Ap. J. Lett.* 483:L87

Thompson C. 1994. *MNRAS* 270:480

Thompson C, Madau P. 2000. *Ap. J.* 538:105

Totani T. 1999. *Ap. J.* 511:41

Usov V. 1994. *MNRAS* 267:1035

van Paradijs J. 1999. *Science* 286:693

van Paradijs J, Groot P, Galama T, Kouveliotou C, Strom R, et al. 1997. *Nature* 386:686

van Paradijs J, Kouveliotou C, Wijers RAMJ. 2000. *Annu. Rev. Astron. Astrophys.* 38:379–425

van Putten M. 2000. *Phys. Rev. Lett.* 84:3752

van Putten M. 2001. *Phys. Rev. Lett.* 87:1101

van Putten M, Ostriker E. 2001. *Ap. J. Lett.* 552:L31

Venemans B, Blaine A. 2001. *MNRAS* 325:1477

Vietri M. 1995. *Ap. J.* 453:883

Vietri M. 1997a. *Ap. J. Lett.* 478:L9

Vietri M. 1998. *Phys. Rev. Lett.* 80:3690

Vietri M, Ghisellini G, Lazzati D, Fiore F, Stella L. 2001. *Ap. J. Lett.* 550:L43, *Ap. J. Lett.* 507:L45

Waxman E. 1995. *Phys. Rev. Lett.* 75:386

Waxman E. 1997a. *Ap. J. Lett.* 485:L5

Waxman E. 1997b. *Ap. J. Lett.* 489:L33

Waxman E, Bahcall JN. 1997. *Phys. Rev. Lett.* 78:2292

Waxman E, Bahcall JN. 1999a. *Ap. J.* 541:707

Waxman E, Bahcall JN. 1999b. *Phys. Rev. D* 590:3002

Waxman E, Draine B. 2000. *Ap. J.* 537:796

Waxman E, Frail D, Kulkarni S. 1998. *Ap. J.* 497:288

Waxman E, Kulkarni S, Frail D. 1998. *Ap. J.* 497:288

Waxman E, Loeb A. 1999. *Ap. J.* 515:721

Weekes T. 2000. In *High Energy Gamma-Ray Astronomy. AIP Conf. Proc.* 558, Heidelberg, 2000, ed. FA Aharonian, HJ Völk, pp. 15. New York: AIP

Weiler K. 2002. *Annu. Rev. Astron. Astrophys.* 40:XX

Weth C, Mészáros P, Kallman T, Rees MJ. 2000. *Ap. J.* 534:581

Wheeler JC, Yi I, Höfflich P, Wang L. 2000. *Ap. J.* 537:810

Wijers RAMJ. 2000. In *Gamma-Ray Bursts in the Afterglow Era: 2nd Workshop, Rome, Italy. ESO Astrophys. Symp.* In press

Wijers RAMJ, Bloom JS, Bagla JS, Natarajan P. 1998. *MNRAS* 294:L13–17

Wijers RAMJ, Galama TJ. 1999. *Ap. J.* 523:177–86

Wijers RAMJ, Rees MJ, Mészáros P. 1997. *MNRAS* 288:L51–56

Woosley SE. 1993. *Ap. J.* 405:273–77

Yoshida A, Murakami T, Itoh M, Nishimura J,

Tsuchiya T, et al. 1989. *Publ. Astron. Soc. Jpn.* 41:509

Yoshida A, Namiki M, Otani C, Kawai N, Murakami T, et al. 1999. *Astron. Astrophys. Suppl.* 138:433

Zhang B, Mészáros P. 2001a. *Ap. J. Lett.* 552:L35

Zhang B, Mészáros P. 2001b. *Ap. J.* 559: 110

Zhang B, Mészáros P. 2002. *Ap. J.* In press (astro-ph/0112118)

Zhang W, Fryer C. 2001. *Ap. J.* 550:357

Zhang W, Woosley S, McFadyen A. 2001. *Bull. Am. Astron. Soc.* 198:3803

Annu. Rev. Astron. Astrophys. 2002. 40:171–216
doi: 10.1146/annurev.astro.40.060401.093926
Copyright © 2002 by Annual Reviews. All rights reserved

COSMIC MICROWAVE BACKGROUND ANISOTROPIES

Wayne Hu[1,2,3] and Scott Dodelson[2,3]

[1]Center for Cosmological Physics, University of Chicago, Chicago, Illinois 60637;
email: whu@background.uchicago.edu
[2]NASA/Fermilab Astrophysics Center, P.O. Box 500, Batavia, Illinois 60510;
email: dodelson@fnal.gov
[3]Department of Astronomy and Astrophysics, University of Chicago, Chicago,
Illinois 60637

Key Words background radiation, cosmology, theory, dark matter, early universe

■ **Abstract** Cosmic microwave background (CMB) temperature anisotropies have and will continue to revolutionize our understanding of cosmology. The recent discovery of the previously predicted acoustic peaks in the power spectrum has established a working cosmological model: a critical density universe consisting of mainly dark matter and dark energy, which formed its structure through gravitational instability from quantum fluctuations during an inflationary epoch. Future observations should test this model and measure its key cosmological parameters with unprecedented precision. The phenomenology and cosmological implications of the acoustic peaks are developed in detail. Beyond the peaks, the yet to be detected secondary anisotropies and polarization present opportunities to study the physics of inflation and the dark energy. The analysis techniques devised to extract cosmological information from voluminous CMB data sets are outlined, given their increasing importance in experimental cosmology as a whole.

INTRODUCTION

The field of cosmic microwave background (CMB) anisotropies has dramatically advanced over the past decade (see White et al. 1994), especially on its observational front. The observations have turned some of our boldest speculations about our Universe into a working cosmological model: namely, that the Universe is spatially flat, consists mainly of dark matter and dark energy, with the small amount of ordinary matter necessary to explain the light element abundances, and all the rich structure in it formed through gravitational instability from quantum mechanical fluctuations when the Universe was a fraction of a second old. Observations over the coming decade should pin down certain key cosmological parameters with unprecedented accuracy (Knox 1995, Jungman et al. 1996, Bond et al. 1997, Zaldarriaga et al. 1997, Eisenstein et al. 1999). These determinations will have profound implications for astrophysics, as well as other disciplines. Particle physicists, for example, will be able to study neutrino masses, theories of inflation impossible to test at accelerators, and the mysterious dark energy or cosmological constant.

For the 28 years between the discovery of the CMB (Penzias & Wilson 1965) and the COBE DMR detection of 10^{-5} fluctuations in its temperature field across the sky (Smoot et al. 1992), observers searched for these anisotropies but found none except the dipole induced by our own motion (Smoot et al., 1977). They learned the hard way that the CMB is remarkably uniform. This uniformity is in stark contrast to the matter in the Universe, organized in very nonlinear structures like galaxies and clusters. The disparity between the smooth photon distribution and the clumpy matter distribution is due to radiation pressure. Matter inhomogeneities grow owing to gravitational instability, but pressure prevents the same process from occuring in the photons. Thus, even though both inhomogeneities in the matter in the Universe and anisotropies in the CMB apparently originated from the same source, these appear very different today.

Because the photon distribution is very uniform, perturbations are small, and linear response theory applies. This is perhaps the most important fact about CMB anisotropies. Because they are linear, predictions can be made as precisely as their sources are specified. If the sources of the anisotropies are also linear fluctuations, anisotropy formation falls in the domain of linear perturbation theory. There are then essentially no phenomenological parameters that need to be introduced to account for nonlinearities or gas dynamics or any other of a host of astrophysical processes that typically afflict cosmological observations.

CMB anisotropies in the working cosmological model, which we briefly review in "Observables," fall almost entirely under linear perturbation theory. The most important observables of the CMB are the power spectra of the temperature and polarization maps. Theory predicted, and now observations confirm, that the temperature power spectrum has a series of peaks and troughs. In "Acoustic Peaks," we discuss the origin of these acoustic peaks and their cosmological uses. Although they are the most prominent features in the spectrum and are the focus of the current generation of experiments, future observations will turn to even finer details, potentially revealing the physics at the two opposite ends of time. Some of these are discussed in "Beyond the Peaks." Finally, the past few years have witnessed important new advances, introduced in "Data Analysis," from a growing body of CMB data analysts on how best to extract the information contained in CMB data. Some of the fruits of this labor have already spread to other fields of astronomy.

OBSERVABLES

Standard Cosmological Paradigm

Whereas a review of the standard cosmological paradigm is not our intention (see Narkilar & Padmanabhan 2001 for a critical appraisal), we briefly introduce the observables necessary to parameterize it.

The expansion of the Universe is described by the scale factor $a(t)$, set to unity today, and by the current expansion rate, the Hubble constant $H_0 = 100\,h$ km sec^{-1} Mpc^{-1}, with $h \simeq 0.7$ (Freedman et al. 2001; compare Theureau et al. 1997, Sandage et al. 2000). The Universe is flat (no spatial curvature) if the total density is equal to the critical density, $\rho_c = 1.88\,h^2 \times 10^{-29}$ g cm^{-3}; it is open (negative curvature) if the density is less than this and closed (positive curvature) if greater. The mean densities of different components of the Universe control $a(t)$ and are typically expressed today in units of the critical density Ω_i, with an evolution with a specified by equations of state $w_i = p_i/\rho_i$, where p_i is the pressure of the ith component. Density fluctuations are determined by these parameters through the gravitational instability of an initial spectrum of fluctuations.

The working cosmological model contains photons, neutrinos, baryons, cold dark matter, and dark energy with densities proscribed within a relatively tight range. For the radiation, $\Omega_r = 4.17 \times 10^{-5}\,h^{-2}\,(w_r = 1/3)$. The photon contribution to the radiation is determined to high precision by the measured CMB temperature, $T = 2.728 \pm 0.004$ K (Fixsen et al. 1996). The neutrino contribution follows from the assumption of three neutrino species, a standard thermal history, and a negligible mass $m_\nu \ll 1$ eV. Massive neutrinos have an equation of state $w_\nu = 1/3 \to 0$ as the particles become nonrelativistic. For $m_\nu \sim 1$ eV this occurs at $a \sim 10^{-3}$ and can leave a small but potentially measurable effect on the CMB anisotropies (Ma & Bertschinger 1995, Dodelson et al. 1996).

For the ordinary matter or baryons, $\Omega_b \approx 0.02\,h^{-2}\,(w_b \approx 0)$, with statistical uncertainties at about the 10% level determined through studies of the light element abundances (for reviews, see Boesgaard & Steigman 1985, Schramm & Turner 1998, Tytler et al. 2000). This value is in strikingly good agreement with that implied by the CMB anisotropies themselves, as we shall see. There is very strong evidence that there is also substantial nonbaryonic dark matter. This dark matter must be close to cold ($w_m = 0$) for the gravitational instability paradigm to work (Peebles 1982), and when added to the baryons gives a total in nonrelativistic matter of $\Omega_m \simeq 1/3$. Because the Universe appears to be flat, the total Ω_{tot} must equal one. Thus, there is a missing component to the inventory, dubbed *dark energy*, with $\Omega_\Lambda \simeq 2/3$. The cosmological constant ($w_\Lambda = -1$) is only one of several possible candidates, but we generally assume this form unless otherwise specified. Measurements of an accelerated expansion from distant supernovae (Riess et al. 1998, Perlmutter et al. 1999) provide entirely independent evidence for dark energy in this amount.

The initial spectrum of density perturbations is assumed to be a power law with a power law index or tilt of $n \approx 1$ corresponding to a scale-invariant spectrum. Likewise the initial spectrum of gravitational waves is assumed to be scale-invariant, with an amplitude parameterized by the energy scale of inflation E_i and compared with the initial density spectrum (see "Gravitational Waves"). Finally the formation of structure will eventually reionize the Universe at some redshift $6 \lesssim z_{\text{ri}} \lesssim 20$.

Many of the features of the anisotropies will be produced even if these parameters fall outside the expected range or even if the standard paradigm is incorrect. Where appropriate, we try to point these out.

Cosmic Microwave Background Temperature Field

The basic observable of the CMB is its intensity as a function of frequency and direction on the sky $\hat{n}$. Because the CMB spectrum is an extremely good blackbody (Fixsen et al. 1996) with a nearly constant temperature across the sky T, we generally describe this observable in terms of a temperature fluctuation $\Theta(\hat{n}) = \Delta T/T$.

If these fluctuations are Gaussian, then the multipole moments of the temperature field

$$\Theta_{\ell m} = \int d\hat{n} Y^*_{\ell m}(\hat{n})\Theta(\hat{n}) \tag{1}$$

are fully characterized by their power spectrum

$$\langle \Theta^*_{\ell m}\Theta_{\ell'm'}\rangle = \delta_{\ell\ell'}\delta_{mm'}C_\ell, \tag{2}$$

whose values as a function of ℓ are independent in a given realization. For this reason predictions and analyses are typically performed in harmonic space. On small sections of the sky where its curvature can be neglected, the spherical harmonic analysis becomes ordinary Fourier analysis in two dimensions. In this limit ℓ becomes the Fourier wavenumber. Because the angular wavelength $\theta = 2\pi/\ell$, large multipole moments corresponds to small angular scales with $\ell \sim 10^2$ representing degree scale separations. Likewise, because in this limit the variance of the field is $\int d^2\ell C_\ell/(2\pi)^2$, the power spectrum is usually displayed as

$$\Delta_T^2 \equiv \frac{\ell(\ell+1)}{2\pi}C_\ell T^2, \tag{3}$$

the power per logarithmic interval in wavenumber for $\ell \gg 1$.

Figure 1 (*top*) (see color insert) shows observations of Δ_T along with the prediction of the working cosmological model, complete with the acoustic peaks mentioned in the "Introduction" and discussed extensively in "Acoustic Peaks," below. Whereas COBE first detected anisotropy on the largest scales (*inset*), observations in the past decade have pushed the frontier to smaller and smaller scales (*left to right* in the figure). The Microwave Anisotropy Probe (MAP) satellite, launched in June 2001, will go out to $\ell \sim 1000$, whereas the European satellite, Planck, scheduled for launch in 2007, will go a factor or two higher (see Figure 1, *bottom*).

The power spectra shown in Figure 1 all begin at $\ell = 2$ and exhibit large errors at low multipoles. This is because the predicted power spectrum is the average power in the multipole moment ℓ an observer would see in an ensemble of universes. However, a real observer is limited to one universe and one sky with its one set of $\Theta_{\ell m}$'s, $2\ell + 1$ numbers for each ℓ. This is particularly problematic for the monopole and dipole ($\ell = 0, 1$). If the monopole were larger in our vicinity than its average value, we would have no way of knowing it. Likewise for the dipole, we have no way of distinguishing a cosmological dipole from our own peculiar motion with respect to the CMB rest frame. Nonetheless, the monopole and dipole—which we will often call simply Θ and v_γ—are of the utmost significance in the early Universe. It is precisely the spatial and temporal variation of these quantities, especially the monopole, which determine the pattern of anisotropies we observe today. A distant observer sees spatial variations in the local temperature or monopole, at a distance given by the lookback time, as a fine-scale angular anisotropy. Similarly, local dipoles appear as a Doppler-shifted temperature that is viewed analogously. In the jargon of the field this simple projection is referred to as the *freestreaming* of power from the monopole and dipole to higher multipole moments.

How accurately can the spectra ultimately be measured? As alluded to above, the fundamental limitation is set by "cosmic variance," the fact that there are only $2\ell + 1$ m-samples of the power in each multipole moment. This leads to an inevitable error of

$$\Delta C_\ell = \sqrt{\frac{2}{2\ell + 1}} C_\ell. \tag{4}$$

Allowing for further averaging over ℓ in bands of $\Delta\ell \approx \ell$, we see that the precision in the power spectrum determination scales as ℓ^{-1}, i.e., $\sim 1\%$ at $\ell = 100$ and $\sim 0.1\%$ at $\ell = 1000$. It is the combination of precision predictions and prospects for precision measurements that gives CMB anisotropies their unique stature.

There are two general caveats to these scalings. The first is that any source of noise, instrumental or astrophysical, increases the errors. If the noise is also Gaussian and has a known power spectrum, one simply replaces the power spectrum on the right-hand side of Equation 4 with the sum of the signal and noise power spectra (Knox 1995). This is why the errors for the Planck satellite increase near its resolution scale in Figure 1 (*bottom*). Because astrophysical foregrounds are typically non-Gaussian, it is usually also necessary to remove heavily contaminated regions, e.g., the galaxy. If the fraction of sky covered is f_{sky}, then the errors increase by a factor of $f_{\text{sky}}^{-1/2}$ and the resulting variance is usually dubbed "sample variance" (Scott et al. 1994). An $f_{\text{sky}} = 0.65$ was chosen for the Planck satellite in Figure 1.

Cosmic Microwave Background Polarization Field

Whereas no polarization has yet been detected, general considerations of Thomson scattering suggest that up to 10% of the anisotropies at a given scale are polarized. Experimenters are currently hot on the trail, with upper limits approaching the expected level (Hedman et al. 2001, Keating et al. 2001). Thus, we expect polarization to be an extremely exciting field of study in the coming decade.

The polarization field can be analyzed in a way very similar to the temperature field, save for one complication. In addition to its strength, polarization also has an orientation, depending on relative strength of two linear polarization states. The classical literature has tended to describe polarization locally in terms of the Stokes parameters Q and U^1, but recently cosmologists (Seljak 1997, Kamionkowski et al. 1997, Zaldarriaga & Seljak 1997) have found that the scalar E and pseudoscalar B, linear but nonlocal combinations of Q and U, provide a more useful description. Postponing the precise definition of E and B until "Polarization," below, we can, in complete analogy with Equation 1, decompose each of them in terms of multipole moments and then, following Equation 2, consider the power spectra,

$$\left\langle E^*_{\ell m} E_{\ell' m'} \right\rangle = \delta_{\ell \ell'} \delta_{mm'} C_\ell^{EE},$$

$$\left\langle B^*_{\ell m} B_{\ell' m'} \right\rangle = \delta_{\ell \ell'} \delta_{mm'} C_\ell^{BB},$$

$$\left\langle \Theta^*_{\ell m} E_{\ell' m'} \right\rangle = \delta_{\ell \ell'} \delta_{mm'} C_\ell^{\Theta E}. \tag{5}$$

Parity invariance demands that the cross correlation between the pseudoscalar B and the scalars Θ or E vanishes.

The polarization spectra shown in Figure 1 (*bottom*, plotted in μK following Equation 3) have several notable features. First, the amplitude of the EE spectrum of Equation 5 is indeed down from the temperature spectrum by a factor of 10. Second, the oscillatory structure of the EE spectrum is very similar to the temperature oscillations, although they are apparently out of phase but correlated with each other. Both of these features are a direct result of the simple physics of acoustic oscillations, as shown in "Acoustic Peaks" below. The final feature of the polarization spectra is the comparative smallness of the BB signal. Indeed, density perturbations do not produce B modes to first order. A detection of substantial B polarization, therefore, would be momentous. Whereas E polarization effectively doubles our cosmological information, supplementing that contained in C_ℓ, B detection would push us qualitatively forward into new areas of physics.

ACOUSTIC PEAKS

When the temperature of the Universe was ~3000 K at a redshift $z_* \approx 10^3$, electrons and protons combined to form neutral hydrogen, an event usually known as recombination (Peebles 1968, Zel'dovich et al. 1969; see Seager et al. 2000 for recent refinements). Before this epoch, free electrons acted as glue between the photons and the baryons through Thomson and Coulomb scattering, so the cosmological plasma was a tightly coupled photon-baryon fluid (Peebles & Yu 1970). The spectrum depicted in Figure 1 can be explained almost completely by analyzing the behavior of this prerecombination fluid.

In "Basics," below, we start from the two basic equations of fluid mechanics and derive the salient characteristics of the anisotropy spectrum: the existence of

---

[1]There is also the possibility in general of circular polarization, described by the Stokes parameter V, but this is absent in cosmological settings.

peaks and troughs, the spacing between adjacent peaks, and the location of the peaks. These properties depend in decreasing order of importance on the initial conditions, the energy contents of the Universe before recombination, and those after recombination. Ironically, the observational milestones have been reached in almost the opposite order. Throughout the 1990s constraints on the location of the first peak steadily improved, culminating with precise determinations from the Toco (Miller et al. 1999), Boomerang (de Bernardis et al. 2000), and Maxima-1 (Hanany et al. 2000) experiments (see Figure 1, *top*). In the working cosmological model it shows up right where it should be if the present energy density of the Universe is equal to the critical density, i.e., if the Universe is flat. The skeptic should note that the working cosmological model assumes a particular form for the initial conditions and energy contents of the Universe before recombination, which we shall see have only recently been tested directly (with an as yet much lower level of statistical confidence) with the higher peaks.

In "Initial Conditions," below, we introduce the initial conditions that apparently are the source of all clumpiness in the Universe. In the context of ab initio models the term "initial conditions" refers to the physical mechanism that generates the primordial small perturbations. In the working cosmological model this mechanism is inflation and it sets the initial phase of the oscillations to be the same across all Fourier modes. Remarkably, from this fact alone comes the prediction that there will be peaks and troughs in the amplitude of the oscillations as a function of wavenumber. Additionally, the inflationary prediction of an approximately scale-invariant amplitude of the initial perturbations implies roughly scale-invariant oscillations in the power spectrum, and inflation generically predicts a flat Universe. These are all falsifiable predictions of the simplest inflationary models, and they have withstood the test against observations to date.

The energy contents of the Universe before recombination all leave their distinct signatures on the oscillations, as discussed in "Gravitational Forcing" and the two following sections. In particular, the cold dark matter and baryon signatures have now been seen in the data (Halverson et al. 2001, Netterfield et al. 2001, Lee et al. 2001). The coupling between electrons and photons is not perfect, especially as one approaches the epoch of recombination. As discussed in "Damping," below, this imperfect coupling leads to damping in the anisotropy spectrum: Very small scale inhomogeneities are smoothed out. The damping phenomenon has now been observed by the Cosmic Background Imager (CBI) experiment (Padin et al. 2001). Importantly, fluid imperfections also generate linear polarization as covered in "Polarization," below. Because the imperfection is minimal and appears only at small scales, the polarization generated is small and has not been detected to date.

After recombination the photons essentially travel freely to us today, so the problem of translating the acoustic inhomogeneities in the photon distribution at recombination to the anisotropy spectrum today is simply one of projection. This projection depends almost completely on one number, the angular diameter distance between us and the surface of last scattering. This number depends on the energy contents of the Universe after recombination through the expansion rate.

The hand-waving projection argument of "Basics" is formalized in "Integral Approach" in the process introducing the popular code used to compute anisotropies, CMBFAST. Finally, we discuss the sensitivity of the acoustic peaks to cosmological parameters in "Parameter Sensitivity."

Basics

For pedagogical purposes, let us begin with an idealization of a perfect photon-baryon fluid and neglect the dynamical effects of gravity and the baryons. Perturbations in this perfect fluid can be described by a simple continuity and a Euler equation that encapsulate the basic properties of acoustic oscillations.

The discussion of acoustic oscillations takes place exclusively in Fourier space. For example, we decompose the monopole of the temperature field into

$$\Theta_{\ell=0,m=0}(\mathbf{x}) = \int \frac{d^3k}{(2\pi)^3} e^{i\mathbf{k}\cdot\mathbf{x}} \Theta(\mathbf{k}) \tag{6}$$

and omit the subscript $_{00}$ on the Fourier amplitude. Because perturbations are very small, the evolution equations are linear, and different Fourier modes evolve independently. Therefore, instead of partial differential equations for a field $\Theta(\mathbf{x})$, we have ordinary differential equations for $\Theta(\mathbf{k})$. In fact, owing to rotational symmetry, all $\Theta(\mathbf{k})$ for a given k obey the same equations. Here and in the following sections we omit the wavenumber argument k when no confusion with physical space quantities will arise.

Temperature perturbations in Fourier space obey

$$\dot{\Theta} = -\frac{1}{3}kv_\gamma. \tag{7}$$

This equation for the photon temperature Θ, which does indeed look like the familiar continuity equation in Fourier space (derivatives ∇ become wavenumbers $i\mathbf{k}$), has a number of subtleties hidden in it, owing to the cosmological setting. First, the over-dot derivative here is with respect to conformal time $\eta \equiv \int dt/a(t)$. Because we are working in units in which the speed of light $c = 1$, η is also the maximum comoving distance a particle could have traveled since $t = 0$. It is often called the comoving horizon or more specifically the comoving particle horizon. The physical horizon is a times the comoving horizon.

Second, the photon fluid velocity, v_γ, has been written as a scalar instead of a vector. In the early universe only the velocity component parallel to the wavevector $\mathbf{k}$ is expected to be important because it alone has a source in gravity. Specifically, $\mathbf{v}_\gamma = -iv_\gamma\hat{\mathbf{k}}$. In terms of the moments introduced in "Observables," above, v_γ represents a dipole moment directed along $\mathbf{k}$. The factor of 1/3 comes about because continuity conserves photon number, not temperature, and the number density $n_\gamma \propto T^3$. Finally, we reiterate that, for the time being, we are neglecting the effects of gravity.

The Euler equation for a fluid is an expression of momentum conservation. The momentum density of the photons is $(\rho_\gamma + p_\gamma)v_\gamma$, where the photon pressure

$p_\gamma = \rho_\gamma/3$. In the absence of gravity and viscous fluid imperfections, pressure gradients $\nabla p_\gamma = \nabla \rho_\gamma/3$ supply the only force. Because $\rho_\gamma \propto T^4$, this becomes $4k\Theta\bar{\rho}_\gamma/3$ in Fourier space. The Euler equation then becomes

$$\dot{v}_\gamma = k\Theta. \tag{8}$$

Differentiating the continuity equation and inserting the Euler equation yields the most basic form of the oscillator equation,

$$\ddot{\Theta} + c_s^2 k^2 \Theta = 0, \tag{9}$$

where $c_s \equiv \sqrt{\dot{p}/\dot{\rho}} = 1/\sqrt{3}$ is the sound speed in the (dynamically baryon-free) fluid. What this equation says is that pressure gradients act as a restoring force to any initial perturbation in the system, which thereafter oscillates at the speed of sound. Physically these temperature oscillations represent the heating and cooling of a fluid that is compressed and rarefied by a standing sound or acoustic wave. This behavior continues until recombination. Assuming negligible initial velocity perturbations, we have a temperature distribution at recombination of

$$\Theta(\eta_*) = \Theta(0)\cos(ks_*), \tag{10}$$

where $s = \int c_s \, d\eta \approx \eta/\sqrt{3}$ is the distance sound can travel by η, usually called the sound horizon. Asterisks denote evaluation at recombination z_*.

In the limit of scales that are large compared with the sound horizon $ks \ll 1$, the perturbation is frozen into its initial conditions. This is the gist of the statement that the large-scale anisotropies measured by COBE directly measure the initial conditions. On small scales the amplitude of the Fourier modes exhibit temporal oscillations, as shown in Figure 2 [with $\Psi = 0$, $\Psi_i = 3\Theta(0)$ for this idealization]. Modes that are caught at maxima or minima of their oscillation at recombination correspond to peaks in the power, i.e., the variance of $\Theta(k, \eta_*)$. Because sound takes half as long to travel half as far, modes corresponding to peaks follow a harmonic relationship $k_n = n\pi/s_*$, where n is an integer (see Figure 2a).

How does this spectrum of inhomogeneities at recombination appear to us today? Roughly speaking, a spatial inhomogeneity in the CMB temperature of wavelength λ appears as an angular anisotropy of scale $\theta \approx \lambda/D$, where $D(z)$ is the comoving angular diameter distance from the observer to redshift z. We address this issue more formally in "Integral Approach," below. In a flat universe $D_* = \eta_0 - \eta_* \approx \eta_0$, where $\eta_0 \equiv \eta(z=0)$. In harmonic space the relationship implies a coherent series of acoustic peaks in the anisotropy spectrum, located at

$$\ell_n \approx n\ell_a, \quad \ell_a \equiv \pi D_*/s_*. \tag{11}$$

To get a feel for where these features should appear, note that in a flat matter-dominated universe $\eta \propto (1+z)^{-1/2}$ so that $\eta_*/\eta_0 \approx 1/30 \approx 2°$. Equivalently, $\ell_1 \approx 200$. Notice that because we are measuring ratios of distances, the absolute distance scale drops out; we see in "Radiation Driving," below, that the Hubble

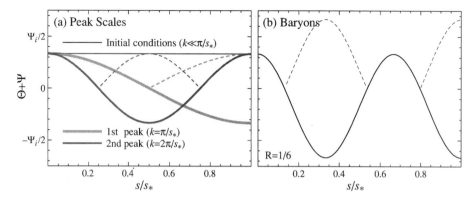

Figure 2 Idealized acoustic oscillations with time measured as the extent of the sound horizon relative to its value at recombination (s/s$_*$). (*a*) Peak scales: The wavemode that completes half an oscillation by recombination sets the physical scale of the first peak. Both minima and maxima correspond to peaks in power (*dashed lines*, absolute value), so higher peaks are integral multiples of this scale with equal height. Plotted here is the idealization of Equation 15 (constant potentials, no baryon loading). (*b*) Baryon loading: Baryon loading boosts the amplitudes of every other oscillation. Plotted here is the idealization of Equation 16 (constant potentials and baryon loading $R = 1/6$ for the third peak).

constant sneaks back into the problem because the Universe is not fully matter-dominated at recombination.

In a spatially curved universe the angular diameter distance no longer equals the coordinate distance, making the peak locations sensitive to the spatial curvature of the Universe (Doroshkevich et al. 1978, Kamionkowski et al. 1994). Consider first a closed universe with radius of curvature $R = H_0^{-1} |\Omega_{\rm tot} - 1|^{-1/2}$. Suppressing one spatial coordinate yields a two-sphere geometry with the observer situated at the pole (see Figure 3). Light travels on lines of longitude. A physical scale λ at fixed latitude given by the polar angle θ subtends an angle $\alpha = \lambda/R \sin \theta$. For $\alpha \ll 1$, a Euclidean analysis would infer a distance $D = R \sin \theta$, even though the coordinate distance along the arc is $d = \theta R$; thus

$$D = R \sin(d/R). \tag{12}$$

For open universes, simply replace sin with sinh. The result is that objects in an open (closed) universe are closer (further) than they appear, as if seen through a lens. In fact one way of viewing this effect is as the gravitational lensing caused by the background density (see "Gravitational Lensing," below). A given comoving scale at a fixed distance subtends a larger (smaller) angle in a closed (open) universe than a flat universe. This strong scaling with spatial curvature indicates that the observed first peak at $\ell_1 \approx 200$ constrains the geometry to be nearly spatially flat. We implicitly assume spatial flatness in the following sections unless otherwise stated.

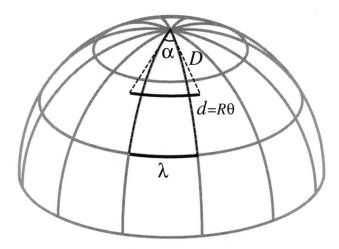

Figure 3   Angular diameter distance. In a closed universe objects are further than they appear to be from Euclidean (flat) expectations, corresponding to the difference between coordinate distance d and angular diameter distance D. Consequently, at a fixed coordinate distance, a given angle corresponds to a smaller spatial scale in a closed universe. Acoustic peaks therefore appear at larger angles or lower ℓ in a closed universe. The converse is true for an open universe.

Finally, in a flat dark energy-dominated universe the conformal age of the universe decreases approximately as $\eta_0 \to \eta_0 (1 + \ln \Omega_m^{0.085})$. For reasonable Ω_m, this causes only a small shift of ℓ_1 to lower multipoles (see Figure 4, color insert) relative to the effect of curvature. Combined with the effect of the radiation near recombination, the peak locations provide a means to measure the physical age t_0 of a flat universe (Hu et al. 2001).

Initial Conditions

As suggested above, observations of the location of the first peak strongly point to a flat universe. This is encouraging news for adherents of inflation, a theory that initially predicted $\Omega_{tot} = 1$ at a time when few astronomers would sign on to such a high value (see Liddle & Lyth 1993 for a review). However, the argument for inflation goes beyond the confirmation of flatness. In particular, the discussion of the previous subsection begs the question: Whence $\Theta(0)$, the initial conditions of the temperature fluctuations? The answer requires the inclusion of gravity and considerations of causality that point to inflation as the origin of structure in the Universe.

The calculations of the typical angular scale of the acoustic oscillations in the previous section are familiar in another context: the horizon problem. Because the sound speed is near the speed of light, the degree scale also marks the extent of a causally connected region or particle horizon at recombination. For the picture in the previous section to hold, the perturbations must have been laid down while the scales in question were still far outside the particle horizon[2]. The recent observational verification of this basic peak structure presents a problem potentially more serious than the original horizon problem of approximate isotropy: The mechanism that smooths fluctuations in the Universe must also regenerate them with superhorizon-sized correlations at the 10^{-5} level. Inflation is an idea that solves both problems simultaneously.

The inflationary paradigm postulates that an early phase of near exponential expansion of the Universe was driven by a form of energy with negative pressure. In most models this energy is usually provided by the potential energy of a scalar field. The inflationary era brings the observable universe to a nearly smooth and spatially flat state. Nonetheless, quantum fluctuations in the scalar field are unavoidable and are carried to large physical scales by the expansion. Because an exponential expansion is self-similar in time, the fluctuations are scale-invariant, i.e., in each logarithmic interval in scale the contribution to the variance of the fluctuations is equal. Because the scalar field carries the energy density of the Universe during inflation, its fluctuations induce variations in the spatial curvature (Guth & Pi 1985, Hawking 1982, Bardeen et al. 1983). Instead of perfect flatness, inflation predicts that each scale will resemble a very slightly open or closed universe. This fluctuation in the geometry of the Universe is essentially frozen while the perturbation is outside the horizon (Bardeen 1980).

Formally, curvature fluctuations are perturbations to the space-space piece of the metric. In a Newtonian coordinate system, or gauge, where the metric is diagonal the spatial curvature fluctuation is called $\delta g_{ij} = 2a^2 \Phi \delta_{ij}$ (see, e.g., Ma & Bertschinger 1995). The more familiar Newtonian potential is the time-time fluctuation $\delta g_{tt} = 2\Psi$ and is approximately $\Psi \approx -\Phi$. Approximate scale invariance then says that $\Delta_\Phi^2 \equiv k^3 P_\Phi(k)/2\pi^2 \propto k^{n-1}$, where $P_\Phi(k)$ is the power spectrum of Φ and the tilt $n \approx 1$.

Now let us relate the inflationary prediction of scale-invariant curvature fluctuations to the initial temperature fluctuations. Newtonian intuition based on the Poisson equation $k^2\Phi = 4\pi Ga^2\delta\rho$ tells us that on large scales (small k) density and hence temperature fluctuations should be negligible compared with Newtonian potential. General relativity says otherwise because the Newtonian potential is also a time-time fluctuation in the metric. It corresponds to a temporal shift of $\delta t/t = \Psi$. The CMB temperature varies as the inverse of the scale factor, which in turn depends on time as $a \propto t^{2/[3(1+p/\rho)]}$. Therefore, the fractional change in the CMB temperature

$$\Theta = -\frac{\delta a}{a} = -\frac{2}{3}\left(1 + \frac{p}{\rho}\right)^{-1}\frac{\delta t}{t}. \tag{13}$$

---

[2]Recall that the comoving scale k does not vary with time. At very early times, then, the wavelength k^{-1} is much larger than the horizon η.

Thus, a temporal shift produces a temperature perturbation of $-\Psi/2$ in the radiation-dominated era (when $p = \rho/3$) and $-2\Psi/3$ in the matter-dominated epoch ($p = 0$) (Sachs & Wolfe 1967, Peacock 1991, White & Hu 1997). The initial temperature perturbation is therefore inextricably linked with the initial gravitational potential perturbation. Inflation predicts scale-invariant initial fluctuations in both the CMB temperature and the spatial curvature in the Newtonian gauge.

Alternate models that seek to obey causality can generate curvature fluctuations only inside the particle horizon. Because the perturbations are then not generated at the same epoch independent of scale, there is no longer a unique relationship between the phase of the oscillators. That is, the argument of the cosine in Equation 10 becomes $ks_* + \phi(\mathbf{k})$, where ϕ is a phase that can in principle be different for different wavevectors, even those with the same magnitude k. This leads to temporal incoherence in the oscillations and hence a washing out of the acoustic peaks (Albrecht et al. 1996), most notably in cosmological defect models (Allen et al. 1997, Seljak et al. 1997). Complete incoherence is not a strict requirement of causality because there are other ways to synch up the oscillations. For example, many isocurvature models, in which the initial spatial curvature is unperturbed, are coherent because their oscillations begin with the generation of curvature fluctuations at horizon crossing (Hu & White 1996). Still they typically have $\phi \neq 0$ (compare Turok 1996). Independent of the angular diameter distance D_*, the ratio of the peak locations gives the phase: $\ell_1 : \ell_2 : \ell_3 \sim 1 : 2 : 3$ for $\phi = 0$. Likewise independent of a constant phase, the spacing of the peaks $\ell_n - \ell_{n-1} = \ell_A$ gives a measure of the angular diameter distance (Hu & White 1996). The observations, which indicate coherent oscillations with $\phi = 0$, therefore, have provided a nontrivial test of the inflationary paradigm and supplied a substantially more stringent version of the horizon problem for contenders to solve.

Gravitational Forcing

We saw above that fluctuations in a scalar field during inflation get turned into temperature fluctuations via the intermediary of gravity. Gravity affects Θ in more ways than this. The Newtonian potential and spatial curvature alter the acoustic oscillations by providing a gravitational force on the oscillator. The Euler equation (8) gains a term on the right-hand side owing to the gradient of the potential $k\Psi$. The main effect of gravity then is to make the oscillations a competition between pressure gradients $k\Theta$ and potential gradients $k\Psi$ with an equilibrium when $\Theta + \Psi = 0$.

Gravity also changes the continuity equation. Because the Newtonian curvature is essentially a perturbation to the scale factor, changes in its value also generate temperature perturbations by analogy to the cosmological redshift $\delta\Theta = -\delta\Phi$, and so the continuity equation (7) gains a contribution of $-\dot{\Phi}$ on the right-hand side. These two effects bring the oscillator equation (9) to

$$\ddot{\Theta} + c_s^2 k^2 \Theta = -\frac{k^2}{3}\Psi - \ddot{\Phi}. \tag{14}$$

In a flat universe and in the absence of pressure Φ and Ψ are constant. Also, in the absence of baryons, $c_s^2 = 1/3$, so the new oscillator equation is identical to Equation 9 with Θ replaced by $\Theta + \Psi$. The matter-dominated epoch is then

$$[\Theta + \Psi](\eta) = [\Theta + \Psi](\eta_{md}) \cos(ks)$$
$$= \frac{1}{3}\Psi(\eta_{md}) \cos(ks), \tag{15}$$

where η_{md} represents the start of the matter-dominated epoch (see Figure 2a). We have used the matter-dominated "initial conditions" for Θ given in the previous section, assuming large scales, $ks_{md} \ll 1$.

The results from the idealization of "Basics," above, carry through with a few exceptions. Even without an initial temperature fluctuation to displace the oscillator, acoustic oscillations would arise by the infall and compression of the fluid into gravitational potential wells. Because it is the effective temperature $\Theta + \Psi$ that oscillates, they occur even if $\Theta(0) = 0$. The quantity $\Theta + \Psi$ can be thought of as an effective temperature in another way: After recombination, photons must climb out of the potential well to the observer and thus suffer a gravitational redshift of $\Delta T/T = \Psi$. The effective temperature fluctuation is therefore also the observed temperature fluctuation. We now see that the large scale limit of Equation 15 recovers the famous Sachs-Wolfe result that the observed temperature perturbation is $\Psi/3$ and overdense regions correspond to cold spots on the sky (Sachs & Wolfe 1967). When $\Psi < 0$, although Θ is positive, the effective temperature $\Theta + \Psi$ is negative. The plasma begins effectively rarefied in gravitational potential wells. As gravity compresses the fluid and pressure resists, rarefaction becomes compression and rarefaction again. The first peak corresponds to the mode that is caught in its first compression by recombination. The second peak at roughly half the wavelength corresponds to the mode that went through a full cycle of compression and rarefaction by recombination. We will use this language of the compression and rarefaction phase inside initially overdense regions, but one should bear in mind that there are an equal number of initially underdense regions with the opposite phase.

Baryon Loading

So far we have been neglecting the baryons in the dynamics of the acoustic oscillations. To see whether this is a reasonable approximation consider the photon-baryon momentum density ratio $R = (p_b + \rho_b)/(p_\gamma + \rho_\gamma) \approx 30\Omega_b h^2(z/10^3)^{-1}$. For typical values of the baryon density this number is of order unity at recombination, and so we expect baryonic effects to begin appearing in the oscillations just as they are frozen in.

Baryons are conceptually easy to include in the evolution equations because their momentum density provides extra inertia in the joint Euler equation for pressure and potential gradients to overcome. Because inertial and gravitational mass are equal, all terms in the Euler equation except the pressure gradient are multiplied by $1 + R$, leading to the revised oscillator equation (Hu & Sugiyama, 1995)

$$c_s^2 \frac{d}{d\eta}\left(c_s^{-2}\dot{\Theta}\right) + c_s^2 k^2 \Theta = -\frac{k^2}{3}\Psi - c_s^2 \frac{d}{d\eta}\left(c_s^{-2}\dot{\Phi}\right), \qquad (16)$$

where we have used the fact that the sound speed is reduced by the baryons to $c_s = 1/\sqrt{3(1 + R)}$.

To get a feel for the implications of the baryons take the limit of constant R, Φ, and Ψ. Then $d^2(R\Psi)/d\eta^2(= 0)$ may be added to the left-hand side to again put the oscillator equation in the form of Equation 9 with $\Theta \to \Theta + (1 + R)\Psi$. The solution then becomes

$$[\Theta + (1 + R)\Psi](\eta) = [\Theta + (1 + R)\Psi](\eta_{md})\cos(ks). \qquad (17)$$

Aside from the lowering of the sound speed, which decreases the sound horizon, baryons have two distinguishing effects: They enhance the amplitude of the oscillations and shift the equilibrium point to $\Theta = -(1 + R)\Psi$ (see Figure 2b). These two effects are intimately related and are easy to understand because the equations are exactly those of a mass $m = 1 + R$ on a spring in a constant gravitational field. For the same initial conditions, increasing the mass causes the oscillator to fall further in the gravitational field, leading to larger oscillations and a shifted zero point.

The shifting of the zero point of the oscillator has significant phenomenological consequences. Because it is still the effective temperature $\Theta + \Psi$ that is the observed temperature, the zero-point shift breaks the symmetry of the oscillations. The baryons enhance only the compressional phase, i.e., every other peak. For the working cosmological model these are the first, third, fifth . . . . Physically, the extra gravity provided by the baryons enhances compression into potential wells.

These qualitative results remain true in the presence of a time-variable R. An additional effect arises owing to the adiabatic damping of an oscillator with a time-variable mass. Because the energy/frequency of an oscillator is an adiabatic invariant, the amplitude must decay as $(1 + R)^{-1/4}$. This can also be understood by expanding the time derivatives in Equation 16 and identifying the $\dot{R}\dot{\Theta}$ term as the remnant of the familiar expansion drag on baryon velocities.

Radiation Driving

We have hitherto also been neglecting the energy density of the radiation in comparison to the matter. The matter-to-radiation energy-density ratio scales as $\rho_m/\rho_r \approx 24\Omega_m h^2(z/10^3)^{-1}$ and so is also of order unity at recombination for reasonable parameters. Moreover, fluctuations corresponding to the higher peaks entered the sound horizon at an earlier time, during radiation domination.

Including the radiation changes the expansion rate of the Universe and hence the physical scale of the sound horizon at recombination. It introduces yet another potential ambiguity in the interpretation of the location of the peaks. Fortunately,

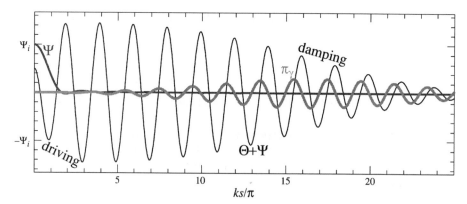

Figure 5   Radiation driving and diffusion damping. The decay of the potential Ψ drives the oscillator in the radiation-dominated epoch. Diffusion generates viscosity π_γ, i.e., a quadrupole moment in the temperature, which damps oscillations and generates polarization. Plotted here is the numerical solution to Equation 18 and Equation 19 for a mode with wavelength much smaller than the sound horizon at decoupling, $ks_* \gg 1$.

the matter-radiation ratio has another effect in the power spectrum by which it can be distinguished. Radiation drives the acoustic oscillations by making the gravitational force evolve with time (Hu & Sugiyama 1995). Matter does not.

The exact evolution of the potentials is determined by the relativistic Poisson equation. However, qualitatively, we know that the background density is decreasing with time, so unless the density fluctuations in the dominant component grow unimpeded by pressure, potentials will decay. In particular, in the radiation-dominated era once pressure begins to fight gravity at the first compressional maxima of the wave, the Newtonian gravitational potential and spatial curvature must decay (see Figure 5).

This decay actually drives the oscillations: It is timed to leave the fluid maximally compressed with no gravitational potential to fight as it turns around. The net effect is doubled because the redshifting from the spatial metric fluctuation Φ also goes away at the same time. When the Universe becomes matter dominated, the gravitational potential is no longer determined by photon-baryon density perturbations but by the pressureless cold dark matter. Therefore, the amplitudes of the acoustic peaks increase as the cold dark matter-to-radiation ratio decreases (Seljak 1994, Hu & Sugiyama 1995). Density perturbations in any form of radiation will stop growing around horizon crossing and lead to this effect. The net result is that across the horizon scale at matter-radiation equality ($k_{eq} \equiv (4 - 2\sqrt{2})/\eta_{eq}$) the acoustic amplitude increases by a factor of 4–5 (Hu & Sugiyama 1996). By eliminating gravitational potentials, photon-baryon acoustic oscillations eliminate the alternating peak heights from baryon loading. The observed high third peak (Halverson et al. 2001) is a good indication that cold dark matter both exists and dominates the energy density at recombination.

Damping

The photon-baryon fluid has slight imperfections corresponding to shear viscosity and heat conduction in the fluid (Silk 1968, Weinberg 1971). These imperfections damp acoustic oscillations. To consider these effects, we now present the equations of motion of the system in their full form, including separate continuity and Euler equations for the baryons. Formally the continuity and Euler equations follow from the covariant conservation of the joint stress-energy tensor of the photon-baryon fluid. Because photon and baryon numbers are separately conserved, the continuity equations are unchanged,

$$\dot{\Theta} = -\frac{k}{3}v_\gamma - \dot{\Phi}, \quad \dot{\delta}_b = -kv_b - 3\dot{\Phi}, \tag{18}$$

where δ_b and v_b are the density perturbation and fluid velocity of the baryons. The Euler equations contain qualitatively new terms:

$$\dot{v}_\gamma = k(\Theta + \Psi) - \frac{k}{6}\pi_\gamma - \dot{\tau}(v_\gamma - v_b),$$

$$\dot{v}_b = -\frac{\dot{a}}{a}v_b + k\Psi + \dot{\tau}(v_\gamma - v_b)/R. \tag{19}$$

For the baryons the first term on the right accounts for cosmological expansion, which makes momenta decay as a^{-1}. The third term on the right accounts for momentum exchange in the Thomson scattering between photons and electrons (protons are very tightly coupled to electrons via Coulomb scattering), with $\dot{\tau} \equiv n_e \sigma_T a$, the differential Thomson optical depth, and is compensated by its opposite in the photon Euler equation. These terms are the origin of heat conduction imperfections. If the medium is optically thick across a wavelength, $\dot{\tau}/k \gg 1$ and the photons and baryons cannot slip past each other. As it becomes optically thin, slippage dissipates the fluctuations.

In the photon Euler equation there is an extra force on the right-hand side owing to anisotropic stress gradients or radiation viscosity in the fluid, π_γ. The anisotropic stress is directly proportional to the quadrupole moment of the photon temperature distribution. A quadrupole moment is established by gradients in v_γ as photons from, for example, neighboring temperature crests meet at a trough (see Figure 6, "*damping and polarization*"; color insert). However, it is destroyed by scattering. Thus, $\pi_\gamma = 2(kv_\gamma/\dot{\tau})A_v$, where the order unity constant can be derived from the Boltzmann equation as $A_v = 16/15$ (Kaiser 1983). Its evolution is shown in Figure 5. With the continuity Equation 7, $kv_\gamma \approx -3\dot{\Theta}$, so viscosity takes the form of a damping term. The heat conduction term can be shown to have a similar effect by expanding the Euler equations in $k/\dot{\tau}$. The final oscillator equation including both terms becomes

$$c_s^2 \frac{d}{d\eta}\left(c_s^{-2}\dot{\Theta}\right) + \frac{k^2 c_s^2}{\dot{\tau}}[A_v + A_h]\dot{\Theta} + c_s^2 k^2 \Theta = -\frac{k^2}{3}\Psi - c_s^2 \frac{d}{d\eta}\left(c_s^{-2}\dot{\Phi}\right), \tag{20}$$

where the heat conduction coefficient $A_h = R^2/(1+R)$). Thus, we expect the inhomogeneities to be damped by an exponential factor of order $e^{-k^2\eta/t}$ (see Figure 5). The damping scale k_d is thus of order $\sqrt{t/\eta}$, corresponding to the geometric mean of the horizon and the mean free path. Damping can be thought of as the result of the random walk in the baryons that takes photons from hot regions into cold and vice versa (Silk 1968). Detailed numerical integration of the equations of motion are required to track the rapid growth of the mean free path and damping length through recombination itself. These calculations show that the damping scale is of order $k_d s_* \approx 10$, leading to a substantial suppression of the oscillations beyond the third peak.

How does this suppression depend on the cosmological parameters? As the matter density $\Omega_m h^2$ increases, the horizon η_* decreases because the expansion rate goes up. Because the diffusion length is proportional to $\sqrt{\eta_*}$, it too decreases as the matter density goes up but not as much as the angular diameter distance D_*, which is also inversely proportional to the expansion rate. Thus, more matter translates into more damping at a fixed multipole moment; conversely, it corresponds to slightly less damping at a fixed peak number. The dependence on baryons is controlled by the mean free path, which is in turn controlled by the free electron density: The increase in electron density caused by an increase in the baryons is partially offset by a decrease in the ionization fraction caused by recombination. The net result under the Saha approximation is that the damping length scales approximately as $(\Omega_b h^2)^{-1/4}$. Accurate fitting formulae for this scale in terms of cosmological parameters can be found in (Hu & White 1997c).

Polarization

The dissipation of the acoustic oscillations leaves a signature in the polarization of CMB in its wake (see e.g., Hu & White 1997a and references therein for a more complete treatment). Much like reflection off of a surface, Thomson scattering induces a linear polarization in the scattered radiation. Consider incoming radiation in the $-\mathbf{x}$ direction scattered at right angles into the $\mathbf{z}$ direction (see Figure 7, *top left*; see color insert). Heuristically, incoming radiation shakes an electron in the direction of its electric field vector or polarization $\hat{e}'$, causing it to radiate with an outgoing polarization parallel to that direction. However, because the outgoing polarization $\hat{e}$ must be orthogonal to the outgoing direction, incoming radiation that is polarized parallel to the outgoing direction cannot scatter, leaving only one polarization state, more generally, the Thomson differential cross section $d\sigma_T/d\Omega \propto |\hat{e}' \cdot \hat{e}|^2$.

Unlike the reflection of sunlight off a surface, the incoming radiation comes from all angles. If it were completely isotropic in intensity, radiation coming along the $\hat{\mathbf{y}}$ would provide the polarization state that is missing from that coming along $\hat{\mathbf{x}}$, leaving the net outgoing radiation unpolarized. Only a quadrupole temperature anisotropy in the radiation generates a net linear polarization from Thomson scattering. As we have seen, a quadrupole can only be generated causally by the motion of photons and only if the Universe is optically thin to Thomson scattering across this scale (i.e., it is inversely proportional to $\dot{\tau}$). Polarization generation suffers from a Catch-22: The scattering that generates polarization also suppresses its quadrupole source.

The fact that the polarization strength is of order the quadrupole in amplitude explains the shape and height of the polarization spectra in Figure 1b. The monopole and dipole Θ and v_γ are of the same order of magnitude at recombination, but their oscillations are $\pi/2$ out of phase as follows from Equation 9 and Equation 10. Because the quadrupole is of order $kv_\gamma/\dot{\tau}$ (see Figure 5), the polarization spectrum should be smaller than the temperature spectrum by a factor of order $k/\dot{\tau}$ at recombination. As in the case of the damping, the precise value requires numerical work (Bond & Efstathiou 1987) because $\dot{\tau}$ changes so rapidly near recombination. Calculations show a steady rise in the polarized fraction with increasing l or k to a maximum of about 10% before damping destroys the oscillations and hence the dipole source. Because v_γ is out of phase with the monopole, the polarization peaks should also be out of phase with the temperature peaks. Indeed, Figure 1b shows that this is the case. Furthermore, the phase relation also tells us that the polarization is correlated with the temperature perturbations. Because the correlation power $C_\ell^{\Theta E}$ is the product of the two, it exhibits oscillations at twice the acoustic frequency.

Until now, we have focused on the polarization strength without regard to its orientation. The orientation, like a two-dimensional vector, is described by two components, E and B. The E and B decomposition is simplest to visualize in the small scale limit, where spherical harmonic analysis coincides with Fourier analysis (Seljak 1997). The wavevector $\mathbf{k}$ picks out a preferred direction against which the polarization direction is measured (see Figure 7, *top right*). Because the linear polarization is a "headless vector" that remains unchanged upon a 180° rotation, the two numbers E and B that define it represent polarization aligned or orthogonal with the wavevector (positive and negative E) and crossed at $\pm45°$ (positive and negative B).

In linear theory scalar perturbations such as the gravitational potential and temperature perturbations have only one intrinsic direction associated with them, that provided by $\mathbf{k}$, and the orientation of the polarization inevitably takes it cue from that one direction, thereby producing an E-mode. The generalization to an all-sky characterization of the polarization changes none of these qualitative features. The E-mode and the B-mode are formally distinguished by the orientation of the Hessian of the Stokes parameters that define the direction of the polarization itself. This geometric distinction is preserved under summation of all Fourier modes as well as the generalization of Fourier analysis to spherical harmonic analysis.

The acoustic peaks in the polarization appear exclusively in the EE power spectrum of Equation 5. This distinction is very useful, as it allows a clean separation of this effect from those occuring beyond the scope of the linear perturbation theory of scalar fluctuations: in particular, gravitational waves). Moreover, in the working cosmological model, the polarization peaks and correlation are precise predictions of the temperature peaks, as they depend on the same physics. As such, their detection would represent a sharp test of the implicit assumptions of the working model, especially its initial conditions and ionization history.

Integral Approach

The discussion in the previous sections suffices for a qualitative understanding of the acoustic peaks in the power spectra of the temperature and polarization anisotropies. To refine this treatment we must consider more carefully the sources of anisotropies and their projection into multipole moments.

Because the description of the acoustic oscillations takes place in Fourier space, the projection of inhomogeneities at recombination onto anisotropies today has an added level of complexity. An observer today sees the acoustic oscillations in effective temperature as they appeared on a spherical shell at $x = D_* \hat{n}$ at recombination, where $\hat{n}$ is the direction vector and $D_* = \eta_0 - \eta_*$ is the distance light can travel between recombination and the present (see Figure 6). Having solved for the Fourier amplitude $[\Theta + \Psi](k, \eta_*)$, we can expand the exponential in Equation 6 in terms of spherical harmonics, so the observed anisotropy today is

$$\Theta(\hat{n}, \eta_0) = 4\pi \sum_{\ell m} Y_{\ell m}(\hat{n}) \left[(-i)^\ell \int \frac{d^3 k}{(2\pi)^3} a_\ell(k) Y_{\ell m}^*(\hat{k}) \right], \tag{21}$$

where the projected source $a_\ell(k) = [\Theta + \Psi](\mathbf{k}, \eta_*) j_\ell(k D_*)$. Because the spherical harmonics are orthogonal, Equation 1 implies that $\Theta_{\ell m}$ today is given by the integral in square brackets today. A given plane wave actually produces a range of anisotropies in angular scale, as is obvious from Figure 6. The one-to-one mapping between wavenumber and multipole moment described in "Basics," above, is only approximately true and comes from the fact that the spherical Bessel function $j_\ell(k D_*)$ is strongly peaked at $k D_* \approx \ell$. Notice that this peak corresponds to contributions in the direction orthogonal to the wavevector where the correspondence between ℓ and k is one-to-one (see Figure 6).

Projection is less straightforward for other sources of anisotropy. We have hitherto neglected the fact that the acoustic motion of the photon-baryon fluid also produces a Doppler shift in the radiation that appears to the observer as a temperature anisotropy as well. In fact, we argued above that $v_b \approx v_\gamma$ is of comparable magnitude but out of phase with the effective temperature. If the Doppler effect projected in the same way as the effective temperature, it would wash out the acoustic peaks. However, the Doppler effect has a directional dependence as well because it is only the line-of-sight velocity that produces the effect. Formally, it is a dipole source of temperature anisotropies and hence has an $\ell = 1$ structure. The coupling of the dipole and plane-wave angular momenta implies that the projection of the Doppler effect involves a combination of $j_{\ell \pm 1}$ that may be rewritten as $j_\ell'(x) \equiv d j_\ell(x)/dx$. The structure of j_ℓ' lacks a strong peak at $x = \ell$. Physically this corresponds to the fact that the velocity is irrotational and hence has no component in the direction orthogonal to the wavevector (see Figure 6). Correspondingly, the Doppler effect cannot produce strong peak structures (Hu & Sugiyama 1995). The observed peaks must be acoustic peaks in the effective temperature, not Doppler peaks.

There is one more subtlety involved when passing from acoustic oscillations to anisotropies. Recall from "Radiation Driving," above, that radiation leads to decay of the gravitational potentials. Residual radiation after decoupling therefore implies that the effective temperature is not precisely $[\Theta + \Psi](\eta_*)$. The photons actually have slightly shallower potentials to climb out of and lose the perturbative analogue of the cosmological redshift, so the $[\Theta + \Psi](\eta_*)$ overestimates the difference between the true photon temperature and the observed temperature. This effect of course is already in the continuity equation for the monopole Equation 18, so the source in Equation 21 gets generalized to

$$a_\ell(k) = [\Theta + \Psi](\eta_*) j_l(kD_*) + v_b(k, \eta_*) j_\ell'(kD_*) + \int_{\eta_*}^{\eta_0} d\eta (\dot{\Psi} - \dot{\Phi}) j_l(kD). \quad (22)$$

The last term vanishes for constant gravitational potentials but is nonzero if residual radiation driving exists, as it will in low $\Omega_m h^2$ models. Note that residual radiation driving is particularly important because it adds in phase with the monopole: The potentials vary in time only near recombination, so the Bessel function can be set to $j_l(kD_*)$ and removed from the η integral. This complication has the effect of decreasing the multipole value of the first peak ℓ_1 as the matter-radiation ratio at recombination decreases (Hu & Sugiyama 1995). Finally, we mention that time varying potentials can also play a role at very late times owing to nonlinearities or, a cosmological constant. Those contributions, to be discussed more in "Integrated Sachs-Wolfe Effect," are sometimes referred to as late integrated Sachs-Wolfe effects and do not add coherently with $[\Theta + \Psi](\eta_*)$.

Putting these expressions together and squaring, we obtain the power spectrum of the acoustic oscillations

$$C_\ell = \frac{2}{\pi} \int \frac{dk}{k} k^3 a_\ell^2(k). \quad (23)$$

This formulation of the anisotropies in terms of projections of sources with specific local angular structure can be completed to include all types of sources of temperature and polarization anisotropies at any given epoch in time, linear or nonlinear: the monopole, dipole, and quadrupole sources arising from density perturbations, vorticity, and gravitational waves (Hu & White 1997b). In a curved geometry one replaces the spherical Bessel functions with ultraspherical Bessel functions (Abbott & Schaefer 1986, Hu et al. 1998). Precision in the predictions of the observables is then limited only by the precision in the prediction of the sources. This formulation is ideal for cases in which the sources are governed by nonlinear physics, even though the CMB responds linearly, as we see in "Beyond the Peaks," below.

Perhaps more importantly, the widely used CMBFAST code (Seljak & Zaldarriaga 1996) exploits these properties to calculate the anisotropies in linear perturbation efficiently. It numerically solves for the smoothly varying sources on a sparse grid in wavenumber, interpolating in the integrals for a handful of ℓ's in the smoothly varying C_ℓ. It has largely replaced the original ground breaking codes (Wilson & Silk 1981, Bond & Efstathiou 1984, Vittorio & Silk 1984) based on tracking the rapid temporal oscillations of the multipole moments that simply reflect structure in the spherical Bessel functions themselves.

Parameter Sensitivity

The phenomenology of the acoustic peaks in the temperature and polarization is essentially described by four observables and the initial conditions (Hu et al. 1997). These are the angular extents of the sound horizon $\ell_a \equiv \pi D_*/s_*$, the particle horizon at matter radiation equality $\ell_{eq} \equiv k_{eq} D_*$, and the damping scale $\ell_d \equiv k_d D_*$, as well as the value of the baryon-photon momentum density ratio R_*. ℓ_a sets the spacing between of the peaks; ℓ_{eq} and ℓ_d compete to determine their amplitude through radiation driving and diffusion damping. R_* sets the baryon loading and, along with the potential well depths set by ℓ_{eq}, fixes the modulation of the even and odd peak heights. The initial conditions set the phase, or equivalently the location of the first peak in units of ℓ_a, and an overall tilt n in the power spectrum.

In the model of Figure 1, these numbers are $\ell_a = 301$ ($\ell_1 = 0.73\ell_a$), $\ell_{eq} = 149$, $\ell_d = 1332$, $R_* = 0.57$, and $n = 1$, and in this family of models the parameter sensitivity is approximately (Hu et al. 2001)

$$\frac{\Delta \ell_a}{\ell_a} \approx -0.24 \frac{\Delta \Omega_m h^2}{\Omega_m h^2} + 0.07 \frac{\Delta \Omega_b h^2}{\Omega_b h^2} - 0.17 \frac{\Delta \Omega_\Lambda}{\Omega_\Lambda} - 1.1 \frac{\Delta \Omega_{tot}}{\Omega_{tot}},$$

$$\frac{\Delta \ell_{eq}}{\ell_{eq}} \approx 0.5 \frac{\Delta \Omega_m h^2}{\Omega_m h^2} - 0.17 \frac{\Delta \Omega_\Lambda}{\Omega_\Lambda} - 1.1 \frac{\Delta \Omega_{tot}}{\Omega_{tot}},$$

$$\frac{\Delta \ell_d}{\ell_d} \approx -0.21 \frac{\Delta \Omega_m h^2}{\Omega_m h^2} + 0.20 \frac{\Delta \Omega_b h^2}{\Omega_b h^2} - 0.17 \frac{\Delta \Omega_\Lambda}{\Omega_\Lambda} - 1.1 \frac{\Delta \Omega_{tot}}{\Omega_{tot}}, \quad (24)$$

and $\Delta R_*/R_* \approx 1.0 \Delta \Omega_b h^2 / \Omega_b h^2$. Current observations indicate that $\ell_a = 304 \pm 4$, $\ell_{eq} = 168 \pm 15$, $\ell_d = 1392 \pm 18$, $R_* = 0.60 \pm 0.06$, and $n = 0.96 \pm 0.04$ (Knox et al. 2001; see also Wang et al. 2001, Pryke et al. 2001, de Bernardis et al. 2002) if gravitational wave contributions are subdominant and the reionization redshift is low, as assumed in the working cosmological model (see "Standard Cosmological Paradigm," above).

The acoustic peaks therefore contain three rulers for the angular diameter distance test for curvature, i.e., deviations from $\Omega_{tot} = 1$. However, contrary to popular belief, any one of these alone is not a standard ruler whose absolute scale is known even in the working cosmological model. This is reflected in the sensitivity of these scales to other cosmological parameters. For example, the dependence of ℓ_a on $\Omega_m h^2$, and hence on the Hubble constant, is quite strong. However, in combination with a measurement of the matter-radiation ratio from ℓ_{eq}, this degeneracy is broken.

The weaker degeneracy of ℓ_a on the baryons can likewise be broken from a measurement of the baryon-photon ratio R_*. The damping scale ℓ_d provides an additional consistency check on the implicit assumptions of the working model, e.g., recombination and the energy contents of the Universe during this epoch. What makes the peaks so valuable for this test is that the rulers are standardizable and contain a built-in consistency check.

There remains a weak but perfect degeneracy between Ω_{tot} and Ω_Λ because they both appear only in D_*. This is called the angular diameter distance degeneracy in the literature and can readily be generalized to dark energy components beyond the cosmological constant assumed here. Because the effect of Ω_Λ is intrinsically so small, it creates a correspondingly small ambiguity in Ω_{tot} for reasonable values of Ω_Λ. The down side is that dark energy can never be isolated through the peaks alone because it only takes a small amount of curvature to mimic its effects. The evidence for dark energy through the CMB comes about by allowing for external information. The most important is the nearly overwhelming direct evidence for $\Omega_m < 1$ from local structures in the Universe. The second is the measurements of a relatively high Hubble constant $h \approx 0.7$; combined with a relatively low $\Omega_m h^2$ that is preferred in the CMB data, it implies $\Omega_m < 1$, but at low significance currently.

The upshot is that precise measurements of the acoustic peaks yield precise determinations of four fundamental parameters of the working cosmological model: $\Omega_m h^2$, $\Omega_m h^2$, D_*, and n. More generally, the first three can be replaced by ℓ_a, ℓ_{eq}, ℓ_d, and R_* to extend these results to models in which the underlying assumptions of the working model are violated.

BEYOND THE PEAKS

Once the acoustic peaks in the temperature and polarization power spectra have been scaled, the days of splendid isolation of CMB theory, analysis, and experiment will have ended. Beyond and beneath the peaks lies a wealth of information about the evolution of structure in the Universe and its origin in the early universe. As CMB photons traverse the large-scale structure of the Universe on their journey from the recombination epoch, they pick up secondary temperature and polarization anisotropies. These depend on the intervening dark matter, dark energy, baryonic gas density, and temperature distributions, and even the existence of primordial gravity waves, so the potential payoff of their detection is enormous. The price for this extended reach is the loss of the ability both to make precise predictions, owing to uncertain and/or nonlinear physics, and to make precise measurements, owing to the cosmic variance of the primary anisotropies and the relatively greater importance of galactic and extragalactic foregrounds.

We begin in the following section with a discussion of the matter power spectrum to set the framework for the discussion of secondary anisotropies. Secondaries can be divided into two classes: those caused by gravitational effects and those induced by scattering off of electrons. The former are treated in "Gravitational

Secondaries" and the latter in "Scattering Secondaries." Secondary anisotropies are often non-Gaussian, so they show up not only in the power spectra of "Observables," above, but in higher point functions as well. We briefly discuss non-Gaussian statistics in "Non-Gaussianity," below. All of these topics are subjects of current research to which this review can only serve as an introduction.

Matter Power Spectrum

The same balance between pressure and gravity that is responsible for acoustic oscillations determines the power spectrum of fluctuations in nonrelativistic matter. This relationship is often obscured by focusing on the density fluctuations in the pressureless cold dark matter itself, so we instead consider the matter power spectrum from the perspective of the Newtonian potential.

PHYSICAL DESCRIPTION   After recombination, without the pressure of the photons, the baryons simply fall into the Newtonian potential wells with the cold dark matter, an event usually referred to as the end of the Compton drag epoch. We claimed in "Radiation Driving," above, that above the horizon at matter-radiation equality the potentials are nearly constant. This follows from the following dynamics: Where pressure gradients are negligible, infall into some initial potential causes a potential flow of $v_{tot} \sim (k\eta)\Psi_i$ (see Equation 19) and causes density enhancements by continuity of $\delta_{tot} \sim -(k\eta)v_{tot} \sim -(k\eta)^2\Psi_i$. The Poisson equation says that the potential at this later time $\Psi \sim -(k\eta)^{-2}\delta_{tot} \sim \Psi_i$, so this rate of growth is exactly right to keep the potential constant. Formally, this Newtonian argument only applies in general relativity for a particular choice of coordinates (Bardeen 1980), but the rule of thumb is that if what is driving the expansion (including spatial curvature) can also cluster unimpeded by pressure, the gravitational potential will remain constant.

Because the potential is constant in the matter-dominated epoch, the large-scale observations of COBE set the overall amplitude of the potential power spectrum today. Translated into density, this is the well-known COBE normalization. It is usually expressed in terms of δ_H, the matter density perturbation at the Hubble scale today. Because the observed temperature fluctuation is approximately $\Psi/3$ (Sachs & Wolfe 1967),

$$\frac{\Delta_T^2}{T^2} \approx \frac{1}{9}\Delta_\Psi^2 \approx \frac{1}{4}\delta_H^2,$$

(25)

where the second equality follows from the Poisson equation in a fully matter-dominated universe with $\Omega_m = 1$. The observed COBE fluctuation of $\Delta_T \approx 28\ \mu K$ (Smoot et al. 1992) implies $\delta_H \approx 2 \times 10^{-5}$. For corrections for $\Omega_m < 1$, where the potential decays because the dominant driver of the expansion cannot cluster, see Bunn & White (1997).

On scales below the horizon at matter-radiation equality, we have seen in "Radiation Driving," above, that pressure gradients from the acoustic oscillations themselves impede the clustering of the dominant component, i.e., the photons, and lead to decay in the potential. Dark matter density perturbations remain but grow only logarithmically from their value at horizon crossing, which (just as for large scales) is approximately the initial potential, $\delta_m \approx -\Psi_i$. The potential for modes that have entered the horizon already will therefore be suppressed by $\Psi \propto -\delta_m/k^2 \sim \Psi_i/k^2$ at matter domination (neglecting the logarithmic growth), again according to the Poisson equation. The ratio of Ψ at late times to its initial value is called the *transfer function*. On large scales, then, the transfer function is close to one, and it falls off as k^{-2} on small scales. If the baryons fraction ρ_b/ρ_m is substantial, baryons alter the transfer function in two ways. First, their inability to cluster below the sound horizon causes further decay in the potential between matter-radiation equality and the end of the Compton drag epoch. Second, the acoustic oscillations in the baryonic velocity field kinematically cause acoustic wiggles in the transfer function (Hu & Sugiyama 1996). These wiggles in the matter power spectrum are related to the acoustic peaks in the CMB spectrum, like twins separated at birth, and are actively being pursued by the largest galaxy surveys (Percival et al. 2001). For fitting formulas for the transfer function that includes these effects see Eisenstein & Hu (1998).

COSMOLOGICAL IMPLICATIONS   The combination of the COBE normalization, the matter transfer function, and the near scale-invariant initial spectrum of fluctuations tells us that by the present, fluctuations in the cold dark matter or baryon density fields will have gone nonlinear for all scales $k \gtrsim 10^{-1}h\,\mathrm{Mpc}^{-1}$. It is a great triumph of the standard cosmological paradigm that there is just enough growth between $z_* \approx 10^3$ and $z = 0$ to explain structures in the Universe across a wide range of scales.

In particular, because this nonlinear scale also corresponds to galaxy clusters, measurements of their abundance yield a robust measure of the power near this scale for a given matter density Ω_m. The agreement between the COBE normalization and the cluster abundance at low $\Omega_m \sim 0.3$–0.4 was pointed out immediately following the COBE result (e.g., White et al. 1993, Bartlett & Silk 1993) and is one of the strongest pieces of evidence for the parameters in the working cosmological model (Ostriker & Steinhardt 1995, Krauss & Turner 1995).

More generally, the comparison between large-scale structure and the CMB is important in that it breaks degeneracies between effects owing to deviations from power law initial conditions and the dynamics of the matter and energy contents of the Universe. Any dynamical effect that reduces the amplitude of the matter power spectrum corresponds to a decay in the Newtonian potential that boosts the level of anisotropy (see "Radiation Driving," above, and "Integrated Sachs-Wolfe Effect," below). Massive neutrinos are a good example of physics that drives the matter power spectrum down and the CMB spectrum up.

The combination is even more fruitful in the relationship between the acoustic peaks and the baryon wiggles in the matter power spectrum. Our knowledge of the physical distance between adjacent wiggles provides our best calibrated standard ruler for cosmology (Eisenstein et al. 1998). For example, at very low z the radial distance out to a galaxy is cz/H_0. The unit of distance is therefore h^{-1} Mpc, and a knowledge of the true physical distance corresponds to a determination of h. At higher redshifts the radial distance depends sensitively on the background cosmology (especially the dark energy), so a future measurement of baryonic wiggles at, for instance, $z \sim 1$ would be a powerful test of dark energy models. To a lesser extent, the shape of the transfer function, which mainly depends on the matter-radiation scale in h Mpc^{-1}, i.e., $\Omega_m h$, is another standard ruler (see, e.g., Tegmark et al. 2001 for a recent assessment), more heralded than the wiggles but less robust owing to degeneracy with other cosmological parameters.

For scales corresponding to $k \gtrsim 10^{-1} h$ Mpc^{-1}, density fluctuations are nonlinear by the present. Numerical N-body simulations show that the dark matter is bound up in a hierarchy of virialized structures or halos (see Bertschinger 1998 for a review). The statistical properties of the dark matter and the dark matter halos have been extensively studied in the working cosmological model. Less certain are the properties of the baryonic gas. We see that both enter into the consideration of secondary CMB anisotropies.

Gravitational Secondaries

Gravitational secondaries arise from two sources: the differential redshift from time-variable metric perturbations (Sachs & Wolfe 1967) and gravitational lensing. There are many examples of the former, one of which we have already encountered in "Integral Approach" in the context of potential decay in the radiation-dominated era. Such gravitational potential effects are usually called the integrated Sachs-Wolfe (ISW) effect in linear perturbation theory (see next section), the Rees-Sciama ("Rees-Sciama and Moving Halo Effects," below) effect in the nonlinear regime, and the gravitational wave effect for tensor perturbations ("Gravitation Waves," below). Gravitational waves and lensing also produce B-modes in the polarization (see "Polarization," below), by which they may be distinguished from acoustic polarization.

INTEGRATED SACHS-WOLFE EFFECT   As we have seen in the previous section, the potential on a given scale decays whenever the expansion is dominated by a component whose effective density is smooth on that scale. This occurs at late times in an $\Omega_m < 1$ model at the end of matter domination and the onset dark energy (or spatial curvature) domination. If the potential decays between the time a photon falls into a potential well and when it climbs out, it gets a boost in temperature of $\delta\Psi$ owing to the differential gravitational redshift and $-\delta\Phi \approx \delta\Psi$ owing to an accompanying contraction of the wavelength (see "Gravitational Forcing," above).

Potential decay owing to residual radiation was introduced in "Integral Approach," above, but that owing to dark energy or curvature at late times induces much different changes in the anisotropy spectrum. What makes the dark energy or curvature contributions different from those owing to radiation is the longer length of time over which the potentials decay, on order of the Hubble time today. Residual radiation produces its effect quickly, so the distance over which photons feel the effect is much smaller than the wavelength of the potential fluctuation. Recall that this meant that $j_l(kD)$ in the integral in Equation 23 could be set to $j_l(kD_*)$ and removed from the integral. The final effect then is proportional to $j_l(kD_*)$ and adds in phase with the monopole.

The ISW projection, indeed the projection of all secondaries, is much different (see Figure 6). Because the duration of the potential change is much longer, photons typically travel through many peaks and troughs of the perturbation. This cancellation implies that many modes have virtually no impact on the photon temperature. The only modes that do have an impact are those with wavevectors perpendicular to the line of sight, so that along the line of sight the photon does not pass through crests and troughs. What fraction of the modes contributes to the effect then? For a given wavenumber k and line of sight instead of the full spherical shell at radius $4\pi k^2 dk$, only the ring $2\pi k dk$ with $\mathbf{k} \perp \mathbf{n}$ participates. Thus, the anisotropy induced is suppressed by a factor of k (or $\ell = kD$ in angular space). Mathematically, this arises in the line-of-sight integral of Equation 23 from the integral over the oscillatory Bessel function $\int dx\, j_\ell(x) \approx (\pi/2\ell)^{1/2}$ (see also Figure 6).

The ISW effect thus generically shows up only at the lowest ℓ's in the power spectrum (Kofman & Starobinskii 1985). This spectrum is shown in Figure 7 (*bottom left*). Secondary anisotropy predictions in this figure are for a model with $\Omega_{\text{tot}} = 1, \Omega_\Lambda = 2/3, \Omega_b h^2 = 0.02, \Omega_m h^2 = 0.16, n = 1$, and $z_{\text{ri}} = 7$ and an inflationary energy scale $E_i \ll 10^{16}$ GeV. The ISW effect is especially important in that it is extremely sensitive to the dark energy: its amount, equation of state, and clustering properties (Coble et al. 1997, Caldwell et al. 1998, Hu 1998). Unfortunately, being confined to the low multipoles, the ISW effect suffers severely from the cosmic variance in Equation 4 in its detectability. Perhaps more promising is its correlation with other tracers of the gravitational potential [e.g., X-ray background (Boughn et al. 1998) and gravitational lensing (see below)].

This type of cancellation behavior and corresponding suppression of small-scale fluctuations is a common feature of secondary temperature and polarization anisotropies from large-scale structure and is quantified by the Limber equation (Limber 1954) and its CMB generalization (Hu & White, 1996; Hu, 2000a). It is the central reason why secondary anisotropies tend to be smaller than the primary ones from $z_* \approx 10^3$ despite the intervening growth of structure.

REES-SCIAMA AND MOVING HALO EFFECTS   The ISW effect is linear in the perturbations. Cancellation of the ISW effect on small scales leaves second-order and nonlinear analogues in its wake (Rees & Sciama 1968). From a single isolated structure, the potential along the line of sight can change not only from evolution in the density profile but more importantly from its bulk motion across the line of sight. In the context of clusters of galaxies, this is called the moving cluster effect (Birkinshaw & Gull 1983). More generally, the bulk motion of dark matter halos of all masses contributes to this effect (Tuluie & Laguna 1995, Seljak 1996b), and their clustering gives rise to a low level of anisotropies on a range of scales but is never the leading source of secondary anisotropies on any scale (see Figure 7, *bottom left*).

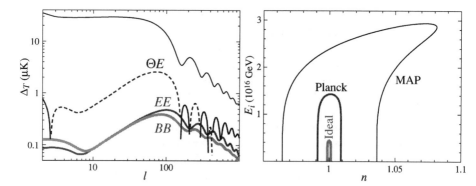

Figure 8 Gravitational waves and the energy scale of inflation E_i. (*Left*) Temperature and polarization spectra from an initial scale-invariant gravitational wave spectrum with power $\propto E_i^4 = (4 \times 10^{16}\ \text{GeV})^4$. (*Right*) 95% confidence upper limits statistically achievable on E_i and the scalar tilt n by the MAP and Planck satellites as well as an ideal experiment out to $\ell = 3000$ in the presence of gravitational lensing B-modes.

GRAVITATIONAL WAVES   A time-variable tensor metric perturbation similarly leaves an imprint in the temperature anisotropy (Sachs & Wolfe 1967). A tensor metric perturbation can be viewed as a standing gravitational wave and produces a quadrupolar distortion in the spatial metric. If its amplitude changes, it leaves a quadrupolar distortion in the CMB temperature distribution (Polnarev 1985). Inflation predicts a nearly scale-invariant spectrum of gravitational waves. Their amplitude depends strongly on the energy scale of inflation,[3] (power $\propto E_i^4$) (Rubakov et al. 1982, Fabbri & Pollock 1983), and its relationship to the curvature fluctuations discriminates between particular models for inflation. Detection of gravitational waves in the CMB therefore provides our best hope to study the particle physics of inflation.

Gravitational waves, like scalar fields, obey the Klein-Gordon equation in a flat universe, and their amplitudes begin oscillating and decaying once the perturbation crosses the horizon. Although this process occurs even before recombination, rapid Thomson scattering destroys any quadrupole anisotropy that develops (see "Damping," above). This fact dicates the general structure of the contributions to the power spectrum (see Figure 8, *left panel*): They are enhanced at $\ell = 2$ or the present quadrupole and sharply suppressed at a multipole larger than that of the first peak (Abbott & Wise 1984, Starobinskii 1985, Crittenden et al. 1993). As is the case for the ISW effect, confinement to the low multipoles means the isolation of gravitational waves is severely limited by cosmic variance.

The signature of gravitational waves in the polarization is more distinct. Because gravitational waves cause a quadrupole temperature anisotropy at the end of recombination, they also generate a polarization. The quadrupole generated by

---

[3] $E_i^4 \equiv V(\phi)$, the potential energy density associated with the scalar field(s) driving inflation.

a gravitational wave has its main angular variation transverse to the wavevector itself (Hu & White 1997a). The resulting polarization has components directed both along or orthogonal to the wavevector and at 45° angles to it. Gravitational waves therefore generate a nearly equal amount of E and B mode polarization when viewed at a distance that is much greater than a wavelength of the fluctuation (Kamionkowski et al. 1997, Zaldarriaga & Seljak 1997). The B-component presents a promising means of measuring the gravitational waves from inflation and hence the energy scale of inflation (see Figure 8, *right panel*). Models of inflation correspond to points in the n, E_i plane (Dodelson et al. 1997). Therefore, the anticipated constraints discriminate among different models of inflation, probing fundamental physics at scales well beyond those accessible in accelerators.

GRAVITATIONAL LENSING   The gravitational potentials of large-scale structure also lens the CMB photons. Because lensing conserves surface brightness, it only affects anisotropies and hence is second order in perturbation theory (Blanchard & Schneider 1987). The photons are deflected according to the angular gradient of the potential projected along the line of sight with a weighting of $2(D_* - D)/(D_* D)$. Again the cancellation of parallel modes implies that it is mainly the large-scale potentials that are responsible for deflections. Specifically, the angular gradient of the projected potential peaks at a multipole $\ell \sim 60$, corresponding to scales of a $k \sim$ few 10^{-2} Mpc^{-1} (Hu 2000b). The deflections are therefore coherent below the degree scale. The coherence of the deflection should not be confused with its rms value, which in the model of Figure 1 is a few arcminutes.

This large coherence and small amplitude ensures that linear theory in the potential is sufficient to describe the main effects of lensing. Because lensing is a one-to-one mapping of the source and image planes, it simply distorts the images formed from the acoustic oscillations in accord with the deflection angle. This warping naturally also distorts the mapping of physical scales in the acoustic peaks to angular scales (see "Integral Approach," above) and hence smooths features in the temperature and polarization (Seljak 1996a). The smoothing scale is the coherence scale of the deflection angle $\Delta\ell \approx 60$ and is sufficiently wide to alter the acoustic peaks with $\Delta\ell \sim 300$. The contributions, shown in Figure 7 (*bottom left*), are therefore negative (*dashed*) on scales corresponding to the peaks.

For the polarization, the remapping not only smooths the acoustic power spectrum but actually generates B-mode polarization (see Figure 1 and Zaldarriaga & Seljak 1998). Remapping by the lenses preserves the orientation of the polarization but warps its spatial distribution in a Gaussian random fashion and hence does not preserve the symmetry of the original E-mode. The B-modes from lensing set a detection threshold for gravitational waves for a finite patch of sky (Hu 2002).

Gravitational lensing also generates a small amount of power in the anisotropies on its own, but this is only noticeable beyond the damping tail, where diffusion has destroyed the primary anisotropies (see Figure 7, *bottom left*). On these small scales, the anisotropy of the CMB is approximately a pure gradient on the sky, and the inhomogeneous distribution of lenses introduces ripples in the gradient on the scale of the lenses (Seljak & Zaldarriaga 2000). In fact, the moving halo effect of "Rees-Sciama and Moving Halo Effects," above, can be described as the gravitational lensing of the dipole anisotropy owing to the peculiar motion of the halo (Birkinshaw & Gull, 1983).

Because the lensed CMB distribution is not linear in the fluctuations, it is not completely described by changes in the power spectrum. Much of the recent work in the literature has been devoted to utilizing the non-Gaussianity to isolate lensing effects (Bernardeau 1997, 1998; Zaldarriaga & Seljak 1999; Zaldarriaga 2000) and their cross-correlation with the ISW effect (Goldberg & Spergel 1999, Seljak & Zaldarriaga 1999). In particular, there is a quadratic combination of the anisotropy data that optimally reconstructs the projected dark matter potentials for use in this cross-correlation (Hu 2001b). The cross-correlation is especially important in that in a flat universe it is a direct indication of dark energy and can be used to study the properties of the dark energy beyond a simple equation of state (Hu 2002).

Scattering Secondaries

From the observations both of the lack of a Gunn-Peterson trough (Gunn & Peterson 1965) in quasar spectra and its preliminary detection (Becker et al. 2001), we know that hydrogen was reionized at $z_{ri} \gtrsim 6$. This is thought to occur through the ionizing radiation of the first generation of massive stars (see, e.g., Loeb & Barkana 2001 for a review). The consequent recoupling of CMB photons to the baryons causes a few percent of them to be rescattered. Linearly, rescattering induces three changes to the photon distribution: suppression of primordial anisotropy, generation of large angle polarization, and a large angle Doppler effect. The latter two are suppressed on small scales by the cancellation highlighted in "Integrated Sachs-Wolfe Effect", above. Nonlinear effects can counter this suppression; these are the subject of active research and are outlined in "Modulated Doppler Effect," below.

PEAK SUPPRESSION   Like scattering before recombination, scattering at late times suppresses anisotropies in the distributions that have already formed. Reionization therefore suppresses the amplitude of the acoustic peaks by the fraction of photons rescattered, approximately the optical depth $\sim \tau_{ri}$ (see Figure 7, *bottom right*, *dotted line* and negative, *dashed line*; contributions corresponding to $|\delta \Delta_T^2|^{1/2}$ between the $z_{ri} = 7$ and $z_{ri} = 0$ models). Unlike the plasma before recombination, the medium is optically thin, so the mean free path and diffusion length of the photons is of order the horizon itself. New acoustic oscillations cannot form. On scales approaching the horizon at reionization, inhomogeneities have yet to be converted into anisotropies (see "Integral Approach," above) so large angle fluctuations are not suppressed. Whereas these effects are relatively large compared with the expected precision of future experiments, they mimic a change in the overall normalization of fluctuations except at the lowest, cosmic variance limited, multipoles.

LARGE-ANGLE POLARIZATION   The rescattered radiation becomes polarized because, as discussed in "Integral Approach," temperature inhomogeneities become anisotropies by projection, passing through quadrupole anisotropies when the perturbations are on the horizon scale at any given time. The result is a bump in the power spectrum of the E-polarization on angular scales corresponding to the horizon at reionization (see Figure 1). Because of the low optical depth of reionization and the finite range of scales that contribute to the quadrupole, the polarization contributions are on the order of tenths of μK on scales of $\ell \sim$ few. In a perfect, foreground free world, this is not beyond the reach of experiments and can be used to isolate the reionization epoch (Hogan et al. 1982, Zaldarriaga et al. 1997). As in the ISW effect, cancellation of contributions along the line of sight guarantees a sharp suppression of contributions at higher multipoles in linear theory. Spatial modulation of the optical depth owing to density and ionization (see "Modulated Doppler Effects," below) does produce higher order polarization but at an entirely negligible level in most models (Hu 2000a).

DOPPLER EFFECT   Naively, velocity fields of order $v \sim 10^{-3}$ ($c = 1$; see, e.g., Strauss & Willick 1995 for a review) and optical depths of a few percent would imply a Doppler effect that rivals the acoustic peaks themselves. That this is not the case is the joint consequence of the cancellation described in "Integrated Sachs-Wolfe Effect," above, and the fact that the acoustic peaks are not "Doppler peaks" (see "Integral Approach," above). Because the Doppler effect comes from the peculiar velocity along the line of sight, it retains no contributions from linear modes with wavevectors perpendicular to the line of sight. However, as we have seen, these are the only modes that survive cancellation (see Figure 6 and Kaiser 1984). Consequently, the Doppler effect from reionization is strongly suppressed and is entirely negligible below $\ell \sim 10^2$ unless the optical depth in the reionization epoch approaches unity (see Figure 7, *bottom right*).

MODULATED DOPPLER EFFECTS   The Doppler effect can survive cancellation if the optical depth has modulations in a direction orthogonal to the bulk velocity. This modulation can be the result of either density or ionization fluctuations in the gas. Examples of the former include the effect in clusters and linear as well as nonlinear large-scale structures.

Cluster modulation   The strongly nonlinear modulation provided by the presence of a galaxy cluster and its associated gas leads to the kinetic Sunyaev-Zel'dovich effect. Cluster optical depths on order of 10^{-2} and peculiar velocities of 10^{-3} imply signals in the 10^{-5} regime in individual arcminute-scale clusters, which are of course rare objects. Although this signal is reasonably large, it is generally dwarfed by the thermal Sunyaev-Zel'dovich effect (see below) and has yet to be detected with high significance (see Carlstrom et al. 2001 and references therein). The kinetic Sunyaev-Zel'dovich effect has negligible impact on the power spectrum of anisotropies, owing to the rarity of clusters, and can be included as part of the general density modulation effect below.

Linear modulation   At the opposite extreme, linear density fluctuations modulate the optical depth and give rise to a Doppler effect, as pointed out by Ostriker & Vishniac (1986) and calculated by Vishniac (1987) (see also Efstathiou & Bond 1987). The result is a signal at the μK level peaking at $\ell \sim$ few $\times\ 10^3$ that increases roughly logarithmically with the reionization redshift (see Figure 7, *bottom right*).

General density modulation   Both the cluster and linear modulations are limiting cases of the more general effect of density modulation by the large-scale structure of the Universe. For the low reionization redshifts currently expected ($z_{ri} \approx 6$–7), most of the effect comes neither from clusters nor the linear regime but from intermediate-scale dark matter halos. An upper limit to the total effect can be obtained by assuming the gas density traces the dark matter density (Hu 2000a) and implies signals on the order of $\Delta_T \sim$ few μK at $\ell > 10^3$ (see Figure 7, *bottom right*). Based on simulations, this assumption should hold in the outer profiles of halos (Pearce et al. 2001, Lewis et al. 2000), but gas pressure will tend to smooth out the distribution in the cores of halos and reduce small scale contributions. In the absence of substantial cooling and star formation, these net effects can be modeled under the assumption of hydrostatic equilibrium (Komatsu & Seljak 2001) in the halos and included in a halo approach to the gas distribution (Cooray 2001).

Ionization modulation   Finally, optical depth modulation can also come from variations in the ionization fraction (Aghanim et al. 1996, Gruzinov & Hu 1998, Knox et al. 1998). Predictions for this effect are the most uncertain, as it involves both the formation of the first ionizing objects and the subsequent radiative transfer of the ionizing radiation (Bruscoli et al. 2000, Benson et al. 2001). It is, however, unlikely to dominate the density-modulated effect except perhaps at very high multipoles of $\ell \sim 10^4$ (crudely estimated, following Gruzinov & Hu 1998) (Figure 7, *bottom right*).

SUNYAEV-ZEL'DOVICH EFFECT   Internal motion of the gas in dark matter halos also gives rise to Doppler shifts in the CMB photons. As in the linear Doppler effect, shifts that are first order in the velocity are canceled as photons scatter off of electrons moving in different directions. At second order in the velocity, there is a residual effect. For clusters of galaxies where the temperature of the gas can reach T_e ~ 10 keV, the thermal motions are a substantial fraction of the speed of light $v_{rms} = (3\ T_e/m_e)^{1/2} \sim 0.2$. The second-order effect represents a net transfer of energy between the hot electron gas and the cooler CMB and leaves a spectral distortion in the CMB where photons on the Rayleigh-Jeans side are transferred to the Wien tail. This effect is called the thermal Sunyaev-Zel'dovich (SZ) effect (Sunyaev & Zel'dovich 1972). Because the net effect is of order $\tau_{cluster} T_e/m_e \propto n_e T_e$, it is a probe of the gas pressure. Like all CMB effects, once imprinted, distortions relative to the redshifting background temperature remain unaffected by cosmological dimming, so one might hope to find clusters at high redshift using the SZ effect. However, the main effect comes from the most massive clusters because of the strong temperature weighting, and these have formed only recently in the standard cosmological model.

Great strides have recently been made in observing the SZ effect in individual clusters, following pioneering attempts that spanned two decades (Birkinshaw 1999). The theoretical basis has remained largely unchanged save for small relativistic corrections as T_e/m_e approaches unity. Both developments are comprehensively reviewed by Carlstrom et al. (2001). Here we instead consider the implications of the SZ effect as a source of secondary anisotropies.

The SZ effect from clusters provides the most substantial contribution to temperature anisotropies beyond the damping tail. On scales much larger than an arcminute, in which clusters are unresolved, contributions to the power spectrum appear as uncorrelated shot noise ($C_\ell = $ const. or $\Delta_T \propto \ell$). The additional contribution owing to the spatial correlation of clusters turns out to be almost negligible in comparison because of the rarity of clusters (Komatsu & Kitayama 1999). Below this scale, contributions turn over as the clusters become resolved. Though there has been much recent progress in simulations (Refregier et al. 2000, Seljak et al. 2001, Springel et al. 2001), dynamic range still presents a serious limitation.

Much recent work has been devoted to semi-analytic modeling following the technique of Cole & Kaiser (1988), in which the SZ correlations are described in terms of the pressure profiles of clusters, their abundance, and their spatial correlations [now commonly referred to as an application of the "halo model" (see Komatsu & Kitayama 1999, Atrio-Barandela & Mücket 1999, Cooray 2001, Komatsu & Seljak 2001)]. We show the predictions of a simplified version in Figure 7, (*bottom right*, Rayleigh-Jeans limit), where the pressure profile is approximated by the dark matter halo profile and the virial temperature of the halo. Although this treatment is comparatively crude, the inaccuracies that result are dwarfed by "missing physics" in both the simulations and more sophisticated modeling, e.g., the nongravitational sources and sinks of energy that change the temperature and density profile of the cluster, often modeled as a uniform "preheating" of the intercluster medium (Holder & Carlstrom 2001).

Although the SZ effect is expected to dominate the power spectrum of secondary anisotropies, it does not necessarily make the other secondaries unmeasurable or contaminate the acoustic peaks. Its distinct frequency signature can be used to isolate it from other secondaries (see, e.g., Cooray et al. 2000). Additionally, it mainly comes from massive clusters that are intrinsically rare. Hence, contributions to the power spectrum are non-Gaussian and concentrated in rare, spatially localized regions. Removal of regions identified as clusters through X rays and optical surveys or ultimately through high-resolution CMB maps themselves can greatly reduce contributions at large angular scales at which they are unresolved (Persi et al. 1995, Komatsu & Kitayama 1999).

Non-Gaussianity

As we have seen, most of the secondary anisotropies are not linear in nature and hence produce non-Gaussian signatures. Non-Gaussianity in the lensing and SZ signals is important for their isolation. The same is true for contaminants such as galactic and extragalactic foregrounds. Finally, the lack of an initial non-Gaussianity in the fluctuations is a testable prediction of the simplest inflationary models (Guth & Pi 1985, Bardeen et al. 1983). Consequently, non-Gaussianity in the CMB is a very active field of research. The primary challenge in studies of non-Gaussianity is in choosing the statistic that quantifies it. Non-Gaussianity tells us what the distribution is not, not what it is. The secondary challenge is to optimize the statistic against the Gaussian "noise" of the primary anisotropies and instrumental or astrophysical systematics.

Early theoretical work on the bispectrum, the harmonic analogue of the three-point function, addressed its detectability in the presence of the cosmic variance of the Gaussian fluctuations (Luo 1994) and showed that the inflationary contribution is not expected to be detectable in most models (Allen et al. 1987, Falk et al. 1993, Gangui et al. 1994). The bispectrum is defined by a triplet of multipoles, or configuration, that defines a triangle in harmonic space. The large cosmic variance in an individual configuration is largely offset by the great number of possible triplets. Interest was spurred by reports of significant signals in specific bispectrum configurations in the COBE maps (Ferreira et al. 1998) that turned out to be caused by systematic errors (Banday et al. 2000). Recent investigations have focused on the signatures of secondary anisotropies (Goldberg & Spergel 1999, Cooray & Hu 2000). These turned out to be detectable with experiments that have both high resolution and angular dynamic range but require the measurement of a wide range of configurations of the bispectrum. Data analysis challenges for measuring the full bispectrum largely remain to be addressed (see Heavens 1998, Spergel & Goldberg 1999, Phillips & Kogut 2001).

The trispectrum, the harmonic analogue of the four-point function, also has advantages for the study of secondary anisotropies. Its great number of configurations are specified by a quintuplet of multipoles that correspond to the sides and diagonal of a quadrilateral in harmonic space (Hu 2001a). The trispectrum is important in that it quantifies the covariance of the power spectrum across multipoles that are often very strong in nonlinear effects, e.g., the SZ effect (Cooray 2001). It is also intimately related to the power spectra of quadratic combinations of the temperature field and has been applied to study gravitational lensing effects (Bernardeau 1997, Zaldarriaga 2000, Hu 2001a).

The bispectrum and trispectrum quantify non-Gaussianity in harmonic space and have clear applications for secondary anisotropies. Tests for non-Gaussianity localized in angular space include the Minkowski functionals (including the genus) (Winitzki & Kosowsky 1997), the statistics of temperature extrema (Kogut et al. 1996), and wavelet coefficients (Aghanim & Forni 1999). These may be more useful for examining foreground contamination and trace amounts of topological defects.

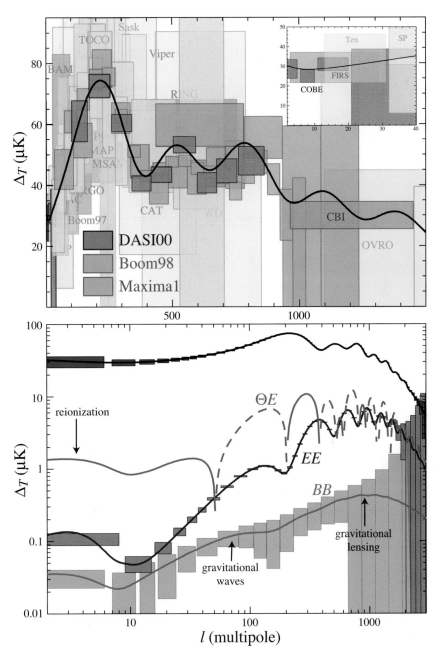

Figure 1 (*Top*) Temperature anisotropy data with boxes representing 1-σ errors and approximate *l*-bandwidth. (*Bottom*) Temperature and polarization spectra for $\Omega_{tot} = 1$, $\Omega_\Lambda = 2/3$, $\Omega_b h^2 = 0.02$, $\Omega_m h^2 = 0.16$, $n = 1$, $z_{ri} = 7$, $E_i = 2.2 \times 10^{16}$ GeV. *Dashed lines* represent negative cross correlation and *boxes* represent the statistical errors of the Planck satellite.

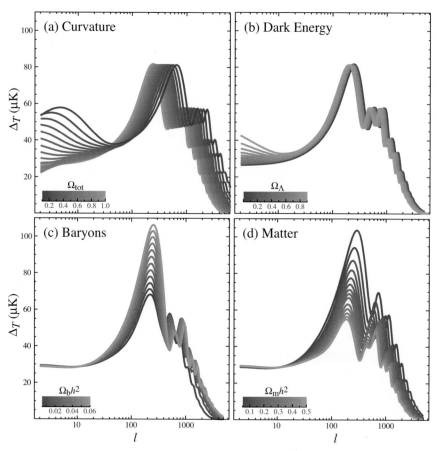

Figure 4 Sensitivity of the acoustic temperature spectrum to four fundamental cosmological parameters. (*a*) The curvature as quantified by Ω_{tot}. (*b*) The dark energy as quantified by the cosmological constant Ω_Λ ($w_\Lambda = -1$). (*c*) The physical baryon density $\Omega_b h^2$. (*d*) The physical matter density $\Omega_m h^2$. All are varied around a fiducial model of $\Omega_{tot} = 1$, $\Omega_\Lambda = 0.65$, $\Omega_b h^2 = 0.02$, $\Omega_m h^2 = 0.147$, $n = 1$, $z_{ri} = 0$, $E_i = 0$.

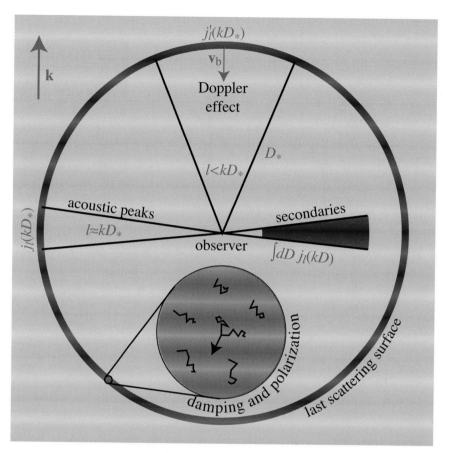

Figure 6 Integral approach. Cosmic Microwave Background anisotropies can be thought of as the line-of-sight projection of various sources of plane wave temperature and polarization fluctuations: the acoustic effective temperature and velocity or Doppler effect (see §3.8), the quadrupole sources of polarization (see §3.7) and secondary sources (see §4.2, §4.3). Secondary contributions differ in that the region over which they contribute is thick compared with the last scattering surface at recombination and the typical wavelength of a perturbation.

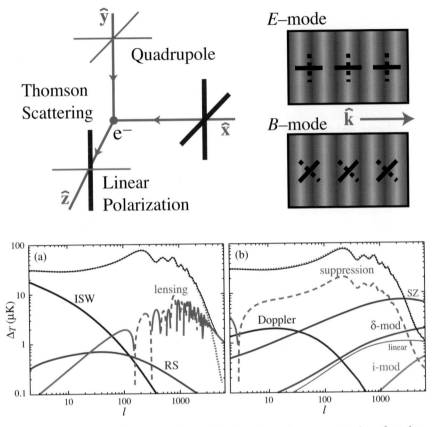

Figure 7 (*Top*) Polarization generation and classification. Thomson scattering of quadrupole temperature anisotropies generates linear polarization. The component of the polarization that is parallel or perpendicular to the wavevector **k** is called the *E*-mode, and the one at 45° angles is called the *B*-mode. (*Bottom*) Secondary anisotropies. (*a*) Gravitational secondaries: Integrated Sachs-Wolfe, lensing, and Rees-Sciama (moving halo) effects. (*b*) Scattering secondaries: Doppler, density (δ), and ionization (i) modulated Doppler, and the Sunyaev-Zel'dovich effects. Curves and model are described in the text.

DATA ANALYSIS

The very large CMB data sets that have begun arriving require new, innovative tools of analysis. The fundamental tool for analyzing CMB data—the likelihood function—has been used since the early days of anisotropy searches (Readhead et al. 1989, Bond et al. 1991, Dodelson & Jubas 1993). Brute-force likelihood analyses (Tegmark & Bunn 1995) were performed even on the relatively large COBE data set, with 6000 pixels in its map. Present data sets are a factor of 10 larger, and this factor will soon increase by yet another factor of 100. The brute-force approach, the time for which scales as the number of pixels cubed, no longer suffices.

In response, analysts have devised a host of techniques that move beyond the early brute-force approach. The simplicity of CMB physics—owing to linearity—is mirrored in analysis by the apparent Gaussianity of both the signal and many sources of noise. In the Gaussian limit, "optimal statistics" are easy to identify. These compress the data so that all of the information is retained, but the subsequent analysis—because of the compression—becomes tractable.

The Gaussianity of the CMB is not shared by other cosmological systems because gravitational nonlinearities turn an initially Gaussian distribution into a non-Gaussian one. Nontheless, many of the techniques devised to study the CMB have been proposed for studying the 3D galaxy distribution (Tegmark et al. 1998), the 2D galaxy distribution (Efstathiou & Moody 2001, Huterer et al. 2001), the Lyman alpha forest (Hui et al. 2001), and the shear field from weak lensing (Hu & White 2001), among others. Indeed, these techniques are now indispensible, powerful tools for all cosmologists, and we would be remiss not to at least outline them in a disussion of the CMB, the context in which many of them were developed.

Figure 9 summarizes the path from the data analyis starting point, a timestream of data points, to the end, the determination of cosmological parameters. Preceding this starting point comes the calibration and the removal of systematic errors from the raw data, but being experiment specific, we do not attempt to cover such

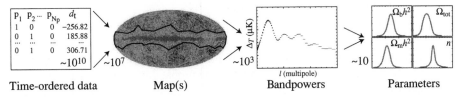

| Time-ordered data | Map(s) | Bandpowers | Parameters |

Figure 9   Data pipeline and radical compression. Maps are constructed for each frequency channel from the data timestreams, combined, and cleaned of foreground contamination by spatial (represented here by excising the galaxy) and frequency information. Bandpowers are extracted from the maps and cosmological parameters from the bandpowers. Each step involves a substantial reduction in the number of parameters needed to describe the data, from potentially $10^{10} \to 10$ for the Planck satellite.

issues here.[4] Each step radically compresses the data by reducing the number of parameters used to describe it. Although this data pipeline and our discussion below focus on temperature anisotropies, similar steps have been elucidated for polarization (Bunn 2001, Tegmark & de Oliveira-Costa 2001, Lewis et al. 2001).

Mapmaking

An experiment can be characterized by the data d_t taken at many different times; a "pointing matrix," P_{ti}, relating the data timestream to the underlying signal at pixelized positions indexed by i; and a noise matrix $C_{d,tt'}$ characterizing the covariance of the noise in the timestream. A model for the data then is $d_t = P_{ti}\Theta_i + n_t$ (with implicit sum over the repeating index i); it is the sum of signal plus noise. Here n_t is drawn from a distribution (often Gaussian) with mean zero and covariance $\langle n_t n_{t'} \rangle = C_{d,tt'}$. In its simplest form the pointing matrix $\mathbf{P}$ contains rows—which correspond to a particular time—with all zeroes in it except for one column with a one (see Figure 9). That column corresponds to the particular pixel observed at the time of interest. Typically, a pixel will be scanned many times during an experiment, so a given column will have many ones in it, corresponding to the many times the pixel has been observed.

Given this model, a well-posed question is, What is the optimal estimator for the signal Θ_i? In effect, what is the best way to construct a map? The answer stems from the likelihood function $\mathcal{L}$, defined as the probability of getting the data, given the theory $\mathcal{L} \equiv P[\text{data}|\text{theory}]$. In this case the theory is the set of parameters Θ_i,

$$\mathcal{L}_\Theta(d_t) = \frac{1}{(2\pi)^{N_t/2}\sqrt{\det \mathbf{C}_d}} \exp\left[-\frac{1}{2}(d_t - P_{ti}\Theta_i)C_{d,tt'}^{-1}(d_{t'} - P_{t'j}\Theta_j)\right]. \quad (26)$$

That is, the noise, the difference between the data and the modulated signal, is assumed to be Gaussian with covariance $\mathbf{C}_d$.

Two important theorems are useful in the construction of a map and more generally in each step of the data pipeline (Tegmark et al. 1997). The first is Bayes' theorem. In this context, it says that $P[\Theta_i|d_t]$, the probability that the temperatures are equal to Θ_i given the data, is proportional to the likelihood function times a prior $P(\Theta_i)$. Thus, with a uniform prior,

$$P[\Theta_i|d_t] \propto P[d_t|\Theta_i] \equiv \mathcal{L}_\Theta(d_t), \quad (27)$$

with the normalization constant determined by requiring the integral of the probability over all Θ_i to be equal to one. The probability on the left is the one of interest. The most likely values of Θ_i therefore are those that maximize the likelihood function. Because the log of the likelihood function in question, Equation 26, is quadratic in the parameters Θ_i, it is straightforward to find this maximum point. Differentiating the argument of the exponential with respect to Θ_i and setting to

---

[4]Aside from COBE, experiments to date have had a sizable calibration error ($\sim$5–10%) that must be factored into the interpretation of Figure 1.

zero leads immediately to the estimator

$$\hat{\Theta}_i = C_{N,ij} P_{jt} C_{d,tt'}^{-1} d_{t'},$$ (28)

where $\mathbf{C}_N \equiv (\mathbf{P}^{tr} \mathbf{C}_d^{-1} \mathbf{P})^{-1}$. As the notation suggests, the mean of the estimator is equal to the actual Θ_i, and the variance is equal to $\mathbf{C}_N$.

The second theorem states that this maximum likelihood estimator is also the minimum variance estimator. The Cramer-Rao inequality says no estimator can measure the Θ_i with errors smaller than the diagonal elements of $\mathbf{F}^{-1}$, where the Fisher matrix is defined as

$$F_{\Theta,ij} \equiv \left\langle -\frac{\partial^2 \ln \mathcal{L}_\Theta}{\partial \Theta_i \partial \Theta_j} \right\rangle.$$ (29)

Inspection of Equation 26 shows that, in this case, the Fisher matrix is precisely equal to $\mathbf{C}_N^{-1}$. Therefore, the Cramer-Rao theorem implies that the estimator of Equation 28 is optimal: It has the smallest possible variance (Tegmark 1997a). No information is lost if the map is used in subsequent analysis instead of the timestream data, but huge factors of compression have been gained. For example, in the recent Boomerang experiment (Netterfield et al. 2001), the timestream contained 2×10^8 numbers, and the map had only 57,000 pixels. The map resulted in compression by a factor of 3500.

There are numerous complications that must be dealt with in realistic applications of Equation 28. Perhaps the most difficult is estimation of $\mathbf{C}_d$, the timestream noise covariance. This typically must be done from the data itself (Ferreira & Jaffe 2000, Stompor et al. 2001). Even if $\mathbf{C}_d$ were known perfectly, evaluation of the map involves inverting $\mathbf{C}_d$, a process that scales as the number of raw data points cubed. For both of these problems, the assumed "stationarity" of $C_{d,tt'}$ (it depends only on $t - t'$) is of considerable utility. Iterative techniques to approximate matrix inversion can also assist in this process (Wright et al. 1996). Another issue that has received much attention is the choice of pixelization. The CMB community has converged on the Healpix pixelization scheme (http://www.eso.org/science/healpix/), now freely available.

Perhaps the most dangerous complication arises from astrophysical foregrounds, both within and from outside the Galaxy, the main ones being synchrotron, bremmsstrahlung, dust, and point source emission. All of the main foregrounds have spectral shapes different than the blackbody shape of the CMB. Modern experiments typically observe at several different frequencies, so a well-posed question is, How can we best extract the CMB signal from the different frequency channels (Bouchet & Gispert 1999)? The blackbody shape of the CMB relates the signal in all the channels, leaving one free parameter. Similarly, if the foreground shapes are known, each foreground comes with just one free parameter per pixel. A likelihood function for the data can again be written down and the best estimator for the CMB amplitude determined analytically. In the absence of foregrounds, one would extract the CMB signal by weighting the frequency channels according to inverse noise. When foregrounds are present, the optimal combination of different frequency maps is a more clever weighting that subtracts out the foreground contribution (Dodelson 1997). One can do better if the pixel-to-pixel correlations of the foregrounds can also be modeled from power spectra (Tegmark & Efstathiou 1996) or templates derived from external data.

This picture is complicated somewhat because the foreground shapes are not precisely known, varying across the sky, e.g., from a spatially varying dust temperature. This too can be modeled in the covariance and addressed in the likelihood analysis (Tegmark 1998, White 1998). The resulting cleaned CMB map is obviously noisier than if foregrounds were not around, but the multiple channels keep the degradation managable. For example, the errors on some cosmological parameters coming from Planck may degrade by almost a factor of 10 as compared with the no-foreground case. However, many errors will not degrade at all, and even the degraded parameters will still be determined with unprecedented precision (Knox 1999, Prunet et al. 2000, Tegmark et al. 2000).

Many foregrounds tend to be highly non-Gaussian, in particular well-localized regions of the map. These pixels can be removed from the map as was done for the region around the galactic disk for COBE. This technique can also be highly effective against point sources. Indeed, even if there is only one frequency channel, external foreground templates set the form of the additional contributions to C_N, which, when properly included, immunize the remaining operations in the data pipeline to such contaminants (Bond et al. 1998). The same technique can be used with templates of residual systematics or constraints imposed on the data, from, e.g., the removal of a dipole.

Bandpower Estimation

Figure 9 indicates that the next step in the compression process is extracting bandpowers from the map. What is a bandpower and how can it be extracted from the map? To answer these questions, we must construct a new likelihood function, one in which the estimated Θ_i are the data. No theory predicts an individual Θ_i, but all predict the distribution from which the individual temperatures are drawn. For example, if the theory predicts Gaussian fluctuations, then Θ_i is distributed as a Gaussian with mean zero and covariance equal to the sum of the noise covariance matrix C_N and the covariance due to the finite sample of the cosmic signal C_S. Inverting Equation 1 and using Equation 2 for the ensemble average leads to

$$C_{S,ij} \equiv \langle \Theta_i \Theta_j \rangle = \sum_\ell \Delta_{T,\ell}^2 W_{\ell,ij}, \qquad (30)$$

where $\Delta_{T,\ell}^2$ depends on the theoretical parameters through C_ℓ (see Equation 3). Here W_ℓ, the window function, is proportional to the Legendre polynomial $P_\ell(\hat{n}_i \cdot \hat{n}_j)$ and a beam and pixel smearing factor b_ℓ^2. For example, a Gaussian beam of width σ dictates that the observed map is actually a smoothed picture of true signal, insensitive to structure on scales smaller than σ. If the pixel scale is much smaller than the beam scale, $b_\ell^2 \propto e^{-\ell(\ell+1)\sigma^2}$. Techniques for handling asymmetric beams have also recently been developed (Wu et al. 2001, Wandelt & Gorski 2001, Souradeep & Ratra 2001, Fosalba et al. 2001). Using bandpowers corresponds to assuming that $\Delta_{T,\ell}^2$ is constant over a finite range, or band, of ℓ, equal to B_a for $\ell_a - \delta\ell_a/2 < \ell < \ell_a + \delta\ell_a/2$. Figure 1 gives a sense of the width and number of bands N_b probed by existing experiments.

For Gaussian theories, then, the likelihood function is

$$
\mathcal{L}_B(\Theta_i) = \frac{1}{(2\pi)^{N_p/2}\sqrt{\det \mathbf{C}_\Theta}} \exp\left(-\frac{1}{2}\Theta_i C_{\Theta,ij}^{-1}\Theta_j\right), \tag{31}
$$

where $\mathbf{C}_\Theta = \mathbf{C}_S + \mathbf{C}_N$ and N_p is the number of pixels in the map. As before, $\mathcal{L}_B$ is Gaussian in the anisotropies Θ_i, but in this case Θ_i are not the parameters to be determined; the theoretical parameters are the B_a, upon which the covariance matrix depends. Therefore, the likelihood function is not Gaussian in the parameters, and there is no simple, analytic way to find the point in parameter space (which is multidimensional depending on the number of bands being fit) at which $\mathcal{L}_B$ is a maximum. An alternative is to evaluate $\mathcal{L}_B$ numerically at many points in a grid in parameter space. The maximum of the $\mathcal{L}_B$ on this grid then determines the best fit values of the parameters. Confidence levels on, e.g., B_1 can be determined by finding the region within which $\int_a^b dB_1[\Pi_{i=2}^{N_b}\int dB_i]\mathcal{L}_B = 0.95$, e.g., for 95% limits.

This possibility is no longer viable owing to the sheer volume of data. Consider the Boomerang experiment with $N_p = 57{,}000$. A single evaluation of $\mathcal{L}_B$ involves computation of the inverse and determinant of the $N_p \times N_p$ matrix $\mathbf{C}_\Theta$, both of which scale as N_p^3. Whereas this single evaluation might be possible with a powerful computer, a single evaluation does not suffice. The parameter space consists of $N_b = 19$ bandpowers equally spaced from $l_a = 100$ up to $l_a = 1000$. A blindly placed grid on this space would require at least 10 evaluations in each dimension, so the time required to adequately evaluate the bandpowers would scale as $10^{19}N_p^3$. No computer can do this. The situation is rapidly getting worse (better) because Planck will have of order 10^7 pixels and be sensitive to of order 10^3 bands.

It is clear that a "smart" sampling of the likelihood in parameter space is necessary. The numerical problem, searching for the local maximum of a function, is well posed, and a number of search algorithms might be used. $\mathcal{L}_B$ tends to be sufficiently structureless that these techniques suffice. Bond et al. (1998) proposed the Newton-Raphson method, which has become widely used. One expands the derivative of the log of the likelihood function—which vanishes at the true maximum of $\mathcal{L}_B$—around a trial point in parameter space, $B_a^{(0)}$. Keeping terms second order in $B_a - B_a^{(0)}$ leads to

$$
\hat{B}_a = \hat{B}_a^{(0)} + \hat{F}_{B,ab}^{-1}\frac{\partial \ln \mathcal{L}_B}{\partial B_b}, \tag{32}
$$

where the curvature matrix $\hat{F}_{B,ab}$ is the second derivative of $-\ln \mathcal{L}_B$ with respect to B_a and B_b. Note the subtle distinction between the curvature matrix and the Fisher matrix in Equation 29, $\mathbf{F} = \langle\hat{\mathbf{F}}\rangle$. In general, the curvature matrix depends on the data, on the Θ_i. In practice, though, analysts typically use the inverse of the Fisher matrix in Equation 32. In that case, the estimator becomes

$$\hat{B}_a{}' = \hat{B}_a^{(0)} + \frac{1}{2}F_{B,ab}^{-1}\left(\Theta_i C_{\Theta,ij}^{-1}\frac{\partial C_{\Theta,jk}}{\partial B_b}C_{\Theta,ki}^{-1}\Theta_i - C_{\Theta,ij}^{-1}\frac{\partial C_{\Theta,ji}}{\partial B_b}\right),\qquad(33)$$

quadratic in the data Θ_i. The Fisher matrix is equal to

$$F_{B,ab} = \frac{1}{2}C_{\Theta,ij}^{-1}\frac{\partial C_{\Theta,jk}}{\partial B_a}C_{\Theta,kl}^{-1}\frac{\partial C_{\Theta,li}}{\partial B_b}.\qquad(34)$$

In the spirit of the Newton-Raphson method, Equation 33 is used iteratively but often converges after just a handful of iterations. The usual approximation is then to take the covariance between the bands as the inverse of the Fisher matrix evaluated at the convergent point $\mathbf{C}_B = \mathbf{F}_B^{-1}$. Indeed, Tegmark (1997b) derived the identical estimator by considering all unbiased quadratic estimators and identifying this one as the one with the smallest variance.

Although the estimator in Equation 33 represents a $\sim 10^{N_b}$ improvement over brute force coverage of the parameter space—converging in just several iterations—it still requires operations that scale as N_p^3. One means of speeding up the calculations is to transform the data from the pixel basis to the so-called signal-to-noise basis, based on an initial guess as to the signal and throwing out those modes that have low signal to noise (Bond 1995, Bunn & Sugiyama 1995). The drawback is that this procedure still requires at least one N_p^3 operation and potentially many, as the guess at the signal improves by iteration. Methods to truly avoid this prohibitive N_p^3 scaling (Oh et al. 1999, Wandelt & Hansen 2001) have been devised for experiments with particular scan strategies, but the general problem remains. A potentially promising approach involves extracting the real space correlation functions as an intermediate step between the map and the bandpowers (Szapudi et al. 2001). Another involves consistently analyzing coarsely pixelized maps with finely pixelized submaps (Dore et al. 2001).

Cosmological Parameter Estimation

The huge advantage of bandpowers is that they represent the natural meeting ground of theory and experiment. The above two sections outline some of the steps involved in extracting them from the observations. Once they are extracted, any theory can be compared with the observations without knowledge of experimental details. The simplest way to estimate the cosmological parameters $\{c_i\}$ is to approximate the likelihood as

$$\mathcal{L}_c(\hat{B}_a) \approx \frac{1}{(2\pi)^{N_c/2}\sqrt{\det \mathbf{C}_B}}\exp\left[-\frac{1}{2}(\hat{B}_a - B_a)C_{B,ab}^{-1}(\hat{B}_b - B_b)\right]\qquad(35)$$

and evaluate it at many points in parameter space (the bandpowers depend on the cosmological parameters). Because the number of cosmological parameters in the working model is $N_c \sim 10$, this represents a final radical compression of information in the original timestream, which recall has up to $N_t \sim 10^{10}$ data points.

In the approximation that the bandpower covariance C_B is independent of the parameters c, maximizing the likelihood is the same as minimizing χ^2 the argument of the exponential. This has been done by dozens of groups over the past few years, especially since the release of CMBFAST (Seljak & Zaldarriaga, 1996), which allows fast computation of theoretical spectra. Even after all the compression summarized in Figure 9, these analyses are still computationally cumbersome owing to the large numbers of parameters varied. Various methods of speeding up spectra computation have been proposed (Tegmark & Zaldarriaga 2000), based on the understanding of the physics of peaks outlined in "Acoustic Peaks," above, and Monte Carlo explorations of the likelihood function (Christensen et al. 2001).

Again the inverse Fisher matrix gives a quick and dirty estimate of the errors. Here the analogue of Equation 29 for the cosmological parameters becomes

$$F_{c,ij} = \frac{\partial B_a}{\partial c_i} C_{B,ab}^{-1} \frac{\partial B_b}{\partial c_j}. \tag{36}$$

In fact, this estimate has been widely used to forecast the optimal errors on cosmological parameters, given a proposed experiment and a band covariance matrix C_B that includes diagonal sample and instrumental noise variance. The reader should be aware that no experiment to date has even come close to achieving the precision implied by such a forecast!

As we enter the age of precision cosmology, a number of caveats will become increasingly important. No theoretical spectra are truly flat in a given band, so the question of how to weight a theoretical spectrum to obtain B_a can be important. In principle, one must convolve the theoretical spectra with window functions (Knox 1999) distinct from those in Equation 30 to produce B_a. Among recent experiments, Degree Angular Scale Interferometer (DASI) (Pryke et al. 2001), among others, have provided these functions. Another complication arises because the true likelihood function for B_a is not Gaussian, i.e., not of the form in Equation 35. The true distribution is skewed: The cosmic variance of Equation 4 leads to larger errors for an upward fluctuation than for a downward fluctuation. The true distribution is closer to log-normal (Bond et al. 2000), and several groups have already accounted for this in their parameter extractions.

DISCUSSION

Measurements of the acoustic peaks in the CMB temperature spectrum have already shown that the Universe is nearly spatially flat and began with a nearly scale-invariant spectrum of curvature fluctuations, consistent with the simplest of inflationary models. In a remarkable confirmation of a prediction of big bang nucleosynthesis, the CMB measurements have now verified that baryons account for about 4% of the critical density. Further, they suggest that the matter density is about 10 times higher than this, implying the existence of nonbaryonic dark matter and dark energy.

Future measurements of the morphology of the peaks in the temperature and polarization should determine the baryonic and dark matter content of the Universe with exquisite precision. Beyond the peaks, gravitational wave imprint on the polarization, the gravitational lensing of the CMB, and gravitational and scattering secondary anisotropies hold the promise of elucidating the physics of inflation and the impact of dark energy on structure formation.

The once and future success of the CMB anisotropy enterprise rests on three equally important pillars: advances in experimental technique, precision in theory, and development of data analysis techniques. The remarkable progress in the field over the past decade owes much to the efforts of researchers in all three disciplines. That much more effort will be required to fulfill the bright promise of CMB suggests that the field will remain active and productive for years to come.

ACKNOWLEDGMENTS

W.H. thanks Fermilab, where this review was written, for their hospitality. W.H. was supported by NASA NAG5-10840 and the DOE OJI program. S.D. was supported by the DOE, by NASA grant NAG 5-10842 at Fermilab, and by NSF Grant PHY-0079251 at Chicago.

The *Annual Review of Astronomy and Astrophysics* is online at
http://astro.annualreviews.org

LITERATURE CITED

Abbott LF, Schaefer RK. 1986. *Ap. J.* 308:546–62

Abbott LF, Wise MB. 1984. *Nucl. Phys.* B244:541–48

Aghanim N, Desert FX, Puget JL, Gispert R. 1996. *Astron. Astrophys.* 311:1–11

Aghanim N, Forni O. 1999. *Astron. Astrophys.* 347:409–18

Albrecht A, Coulson D, Ferreira P, Magueijo J. 1996. *Phys. Rev. Lett.* 76:1413–16

Allen B, Caldwell RR, Dodelson S, Knox L, Shellard EPS, et al. 1997. *Phys. Rev. Lett.* 79:2624–27

Allen TJ, Grinstein B, Wise MB. 1987. *Phys. Lett.* B197:66–70

Atrio-Barandela F, Mücket JP. 1999. *Ap. J.* 515:465–70

Banday AJ, Zaroubi S, Górski KM. 2000. *Ap. J.* 533:575–87

Bardeen JM. 1980. *Phys. Rev. D* 22:1882–905

Bardeen JM, Steinhardt PJ, Turner MS. 1983. *Phys. Rev. D* 28:679–93

Bartlett JG, Silk J. 1993. *Ap. J. Lett.* 407:L45–48

Becker R, Fan X, White R, Strauss M, Narayanan V, et al. 2001. *Astron. J.* 122:2850–57

Benson AJ, Nusser A, Sugiyama N, Lacey CG. 2001. *MNRAS* 320:153–76

Bernardeau F. 1997. *Astron. Astrophys.* 324:15–26

Bernardeau F. 1998. *Astron. Astrophys.* 338:767–76

Bertschinger E. 1998. *Annu. Rev. Astron. Astrophys.* 36:599–654

Birkinshaw M. 1999. *Phys. Rep.* 310:97–195

Birkinshaw M, Gull SF. 1983. *Nature* 302:315–17

Blanchard A, Schneider J. 1987. *Astron. Astrophys.* 184:1–2

Boesgaard AM, Steigman G. 1985. *Annu. Rev. Astron. Astrophys.* 23:319–78

Bond JR, Efstathiou G, Lubin P, Meinhold P. 1991. *Phys. Rev. Lett.* 66:2179–82

Bond JR. 1995. *Phys. Rev. Lett.* 74:4369–72

Bond JR, Efstathiou G. 1984. *Ap. J. Lett.* 285: L45–48

Bond JR, Efstathiou G. 1987. *MNRAS* 226:655–87

Bond JR, Efstathiou G, Tegmark M. 1997. *MNRAS* 291:L33–41

Bond JR, Jaffe AH, Knox L. 1998. *Phys. Rev. D* 57:2117–37

Bond JR, Jaffe AH, Knox L. 2000. *Ap. J.* 533: 19–37

Bouchet F, Gispert R. 1999. *New Astron.* 4:443–79

Boughn SP, Crittenden RG, Turok NG. 1998. *New Astron.* 3:275–91

Bruscoli M, Ferrara A, Fabbri R, Ciardi B. 2000. *MNRAS* 318:1068–72

Bunn EF. 2001. *Phys. Rev. D* 65:043003

Bunn EF, Sugiyama N. 1995. *Ap. J.* 446:49–53

Bunn EF, White M. 1997. *Ap. J.* 480:6–21

Caldwell RR, Dave R, Steinhardt PJ. 1998. *Phys. Rev. Lett.* 80:1582–85

Carlstrom J, Joy M, Grego L, Holder G, Holzapfel W, et al. 2001. *IAP Conf.* In press (astro-ph/0103480)

Christensen N, Meyer R, Knox L, Luey B. 2001. *Class. Quantum Gravity* 18:2677

Coble K, Dodelson S, Frieman JA. 1997. *Phys. Rev. D* 55:1851–59

Cole S, Kaiser N. 1988. *MNRAS* 233:637–48

Cooray A. 2001. *Phys. Rev. D* 64:063514

Cooray A, Hu W. 2000. *Ap. J.* 534:533–50

Cooray A, Hu W, Tegmark M. 2000. *Ap. J.* 540:1–13

Crittenden R, Bond JR, Davis RL, Efstathiou G, Steinhardt PJ. 1993. *Phys. Rev. Lett.* 71:324–27

de Bernardis P, Ade P, Bock J, Bond J, Borrill J, et al. 2002. *Ap. J.* 564:559–66

de Bernardis P, Ade PAR, Bock JJ, Bond JR, Borrill J, et al. 2000. *Nature* 404:955–59

Dodelson S. 1997. *Ap. J.* 482:577–87

Dodelson S, Gates E, Stebbins A. 1996. *Ap. J.* 467:10–18

Dodelson S, Jubas JM. 1993. *Phys. Rev. Lett.* 70:2224–27

Dodelson S, Kinney WH, Kolb EW. 1997. *Phys. Rev. D* 56:3207–15

Dore O, Knox L, Peel A. 2001. *Phys. Rev. D* 64:083001

Doroshkevich AG, Zel'Dovich YB, Sunyaev RA. 1978. *Sov. Astron.* 22:523–28

Efstathiou G, Bond JR. 1987. *MNRAS* 227: 33P–38

Efstathiou G, Moody SJ. 2001. *MNRAS* 325: 1603–15

Eisenstein DJ, Hu W, Tegmark M. 1998. *Ap. J. Lett.* 504:L57–60

Eisenstein DJ, Hu W. 1998. *Ap. J.* 496:605–14

Eisenstein DJ, Hu W, Tegmark M. 1999. *Ap. J.* 518:2–23

Fabbri R, Pollock MD. 1983. *Phys. Lett.* B125: 445–48

Falk T, Rangarajan R, Srednicki M. 1993. *Ap. J. Lett.* 403:L1–3

Ferreira PG, Jaffe AH. 2000. *MNRAS* 312:89–102

Ferreira PG, Magueijo J, Gorski KM. 1998. *Ap. J. Lett.* 503:L1–4

Fixsen DJ, Cheng ES, Gales JM, Mather JC, Shafer RA, et al. 1996. *Ap. J.* 473:576–87

Fosalba P, Dore O, Bouchet FR. 2001. *Phys. Rev. D* 65:063003

Freedman WL, Madore BF, Gibson BK, Ferrarese L, Kelson DD, et al. 2001. *Ap. J.* 553:47–72

Gangui A, Lucchin F, Matarrese S, Molkenach S. 1994. *Ap. J.* 430:447–57

Goldberg DM, Spergel DN. 1999. *Phys. Rev. D* 59:103002

Gruzinov A, Hu W. 1998. *Ap. J.* 508:435–39

Gunn JE, Peterson BA. 1965. *Ap. J.* 142:1633–36

Guth AH, Pi SY. 1985. *Phys. Rev. D* 32:1899–920

Halverson NW, Leitch EM, Pryke C, Kovac J, Carlstrom JE, et al. 2001. *Ap. J.* In press (astro-ph/0104489)

Hanany S, Ade P, Balbi A, Bock J, Borrill J, et al. 2000. *Ap. J. Lett.* 545:L5–9

Hawking SW. 1982. *Phys. Lett.* B115:295–97

Heavens AF. 1998. *MNRAS* 299:805–8

Hedman MM, Barkats D, Gundersen JO, Staggs ST, Winstein B. 2001. *Ap. J. Lett.* 548:L111–14

Hogan CJ, Kaiser N, Rees MJ. 1982. *Philos. Trans. R. Soc. London Ser.* 307:97–109

Holder G, Carlstrom J. 2001. *Ap. J.* 558:515–59

Hu W. 1998. *Ap. J.* 506:485–94

Hu W. 2000a. *Ap. J.* 529:12–25

Hu W. 2000b. *Phys. Rev. D* 62:043007

Hu W. 2001a. *Phys. Rev. D* 64:083005

Hu W. 2001b. *Ap. J. Lett.* 557:L79–83

Hu W. 2002. *Phys. Rev. D* 65:023003

Hu W, Fukugita M, Zaldarriaga M, Tegmark M. 2001. *Ap. J.* 549:669–80

Hu W, Seljak U, White M, Zaldarriaga M. 1998. *Phys. Rev. D* 57:3290–301

Hu W, Sugiyama N. 1995. *Ap. J.* 444:489–506

Hu W, Sugiyama N. 1996. *Ap. J.* 471:542–70

Hu W, Sugiyama N, Silk J. 1997. *Nature* 386: 37–43

Hu W, White M. 1996. *Ap. J.* 471:30–51

Hu W, White M. 1997a. *New Astron.* 2:323–44

Hu W, White M. 1997b. *Phys. Rev. D* 56:596–615

Hu W, White M. 1997c. *Ap. J.* 479:568–79

Hu W, White M. 2001. *Ap. J.* 554:67–73

Hui L, Burles S, Seljak U, Rutledge RE, Magnier E, et al. 2001. *Ap. J.* 552:15–35

Huterer D, Knox L, Nichol RC. 2001. *Ap. J.* 555:547–57

Jungman G, Kamionkowski M, Kosowsky A, Spergel DN. 1996. *Phys. Rev. D* 54:1332–44

Kaiser N. 1983. *MNRAS* 202:1169–80

Kaiser N. 1984. *Ap. J.* 282:374–81

Kamionkowski M, Kosowsky A, Stebbins A. 1997. *Phys. Rev. D* 55:7368–88

Kamionkowski M, Spergel DN, Sugiyama N. 1994. *Ap. J. Lett.* 426:L57–60

Keating B, O'Dell C, de Oliveira-Costa A, Klawikowski S, Stebor N, et al. 2001. *Ap. J. Lett.* 560:L1–4

Knox L. 1995. *Phys. Rev. D* 52:4307–18

Knox L. 1999. *MNRAS* 307:977–83

Knox L, Christensen N, Skordis C. 2001. *Ap. J. Lett.* 563:L95–98

Knox L, Scoccimarro R, Dodelson S. 1998. *Phys. Rev. Lett.* 81:2004–7

Kofman LA, Starobinskii AA. 1985. *Sov. Astron. Lett.* 11:271–74

Kogut A, Banday AJ, Bennett CL, Gorski KM, Hinshaw G, et al. 1996. *Ap. J. Lett.* 464:L29–33

Komatsu E, Kitayama T. 1999. *Ap. J. Lett.* 526: L1–4

Komatsu E, Seljak U. 2001. *MNRAS* 327:1353–66

Krauss LM, Turner MS. 1995. *Gen. Relativ. Gravity* 27:1137–44

Lee A, Ade P, Balbi A, Bock J, Borill J, et al. 2001. *Ap. J. Lett.* 561:L1–5

Lewis A, Challinor A, Turok N. 2001. *Phys. Rev. D* 65:023505

Lewis GF, Babul A, Katz N, Quinn T, Hernquist L, et al. 2000. *Ap. J.* 536:623–44

Liddle AR, Lyth DH. 1993. *Phys. Rep.* 231:1–105

Limber DN. 1954. *Ap. J.* 119:655–81

Loeb A, Barkana R. 2001. *Annu. Rev. Astron. Astrophys.* 39:19–66

Luo X. 1994. *Ap. J. Lett.* 427:L71–74

Ma C, Bertschinger E. 1995. *Ap. J.* 455:7–25

Miller AD, Caldwell R, Devlin MJ, Dorwart WB, Herbig T, et al. 1999. *Ap. J. Lett.* 524: L1–4

Narkilar J, Padmanabhan T. 2001. *Annu. Rev. Astron. Astrophys.* 39:211–48

Netterfield C, Ade P, Bock J, Bond J, Borrill J, et al. 2001. *Ap. J.* In press (astro-ph/0104460)

Oh SP, Spergel DN, Hinshaw G. 1999. *Ap. J.* 510:551–63

Ostriker JP, Steinhardt PJ. 1995. *Nature* 377: 600–2

Ostriker JP, Vishniac ET. 1986. *Ap. J. Lett.* 306: L51–54

Padin S, Cartwright JK, Mason BS, Pearson TJ, Readhead ACS, et al. 2001. *Ap. J. Lett.* 549:L1–5

Parodi BR, Saha A, Sandage A, Tamman GA. 2000. *Ap. J.* 540:634–51

Peacock JA. 1991. *MNRAS* 253:1P–5P

Pearce FR, Jenkins A, Frenk CS, White SDM, Thomas PA, et al. 2001. *MNRAS* 326:649–66

Peebles PJE. 1968. *Ap. J.* 153:1–11

Peebles PJE. 1982. *Ap. J. Lett.* 263:L1–5

Peebles PJE, Yu JT. 1970. *Ap. J.* 162:815–36

Penzias AA, Wilson RW. 1965. *Ap. J.* 142:419–21

Percival W, Baigj CM, Bland-Hawthorn J, Bridges T, Cannon R, et al. 2001. *MNRAS* 327:1297–1306

Perlmutter S, Aldering G, Goldnater G, Knop RA, Nugent P, et al. 1999. *Ap. J.* 517:565–86

Persi FM, Spergel DN, Cen R, Ostriker JP. 1995. *Ap. J.* 442:1–9

Phillips NG, Kogut A. 2001. *Ap. J.* 548:540–49

Polnarev AG. 1985. *Sov. Astron.* 29:607

Prunet S, Sethi SK, Bouchet FR. 2000. *MNRAS* 314:348–53

Pryke C, Halverson N, Leitch E, Kovac J, Carlstrom J, et al. 2001. *Ap. J.* In press (astro-ph/0104490)

Readhead A, Lawrence C, Myers S, Sargent W, Hardebeck H, et al. 1989. *Ap. J.* 346:566–87

Rees M, Sciama D. 1968. *Nature* 217:511

Refregier A, Komatsu E, Spergel DN, Pen UL. 2000. *Phys. Rev. D* 61:123001

Riess A, Fillipenko AV, Challis P, Clocchiatti A, Diercks A, et al. 1998. *Ap. J.* 116:1009

Rubakov VA, Sazhin MV, Veryaskin AV. 1982. *Phys. Lett. B* 115:189–92

Sachs RK, Wolfe AM. 1967. *Ap. J.* 147:73–90

Sandage A, Tamman GA, Saha A. 1998. *Phys. Rep.* 307:1–14

Schramm DN, Turner MS. 1998. *Rev. Mod. Phys.* 70:303–18

Scott D, Srednicki M, White M. 1994. *Ap. J. Lett.* 421:L5–7

Seager S, Sasselov DD, Scott D. 2000. *Ap. J. Suppl.* 128:407–30

Seljak U. 1994. *Ap. J. Lett.* 435:L87–90

Seljak U. 1996a. *Ap. J.* 463:1–7

Seljak U. 1996b. *Ap. J.* 460:549–55

Seljak U. 1997. *Ap. J.* 482:6–16

Seljak U, Burwell J, Pen UL. 2001. *Phys. Rev. D* 63:063001

Seljak U, Pen U, Turok N. 1997. *Phys. Rev. Lett.* 79:1615–18

Seljak U, Zaldarriaga M. 1996. *Ap. J.* 469:437–44

Seljak U, Zaldarriaga M. 1999. *Phys. Rev. D* 60:043504

Seljak U, Zaldarriaga M. 2000. *Ap. J.* 538:57–64

Silk J. 1968. *Ap. J.* 151:459–71

Smoot GF, Bennett CL, Kogut A, Wright EL, Aymon J, et al. 1992. *Ap. J. Lett.* 396:L1–5

Smoot GF, Gorenstein MV, Muller RA. 1977. *Phys. Rev. Lett.* 39:898–901

Souradeep T, Ratra BV. 2001. *Ap. J.* 560:28–40

Spergel D, Goldberg D. 1999. *Phys. Rev. D* 59:103001

Springel V, White M, Hernquist L. 2001. *Ap. J.* 549:681–87

Starobinskii A. 1985. *Sov. Astron. Lett.* 11:133–36

Stompor R, Balbi A, Borrill J, Ferreira P, Hanany S, et al. 2001. *Ap. J.* In press (astro-ph/0106451)

Strauss MA, Willick JA. 1995. *Phys. Rep.* 261:271–431

Sunyaev R, Zel'dovich Y. 1972. *Comments Astrophys. Space Phys.* 4:173

Szapudi I, Prunet S, Pogosyan D, Szalay AS, Bond JR. 2001. *Ap. J. Lett.* 548:L115–18

Tegmark M. 1997a. *Ap. J. Lett.* 480:L87–90

Tegmark M. 1997b. *Phys. Rev. D* 55:5895–907

Tegmark M. 1998. *Ap. J.* 502:1–6

Tegmark M, Bunn EF. 1995. *Ap. J.* 455:1–6

Tegmark M, de Oliveira-Costa A. 2001. *Phys. Rev. D* 64:063001

Tegmark M, Efstathiou G. 1996. *MNRAS* 281:1297–314

Tegmark M, Eisenstein DJ, Hu W, de Oliveira-Costa A. 2000. *Ap. J.* 530:133–65

Tegmark M, Hamilton AJS, Strauss MA, Vogeley MS, Szalay AS. 1998. *Ap. J.* 499:555–76

Tegmark M, Taylor AN, Heavens AF. 1997. *Ap. J.* 480:22–35

Tegmark M, Zaldarriaga M. 2000. *Ap. J.* 544:30–42

Tegmark M, Zaldarriaga M, Hamilton AJS. 2001. *Phys. Rev. D* 63:043007

Theureau G, Hanski M, Ekholm T, Bottinelli L, Goriguenheim U, et al. 1997. *Astron. Astrophys.* 322:730

Tuluie R, Laguna P. 1995. *Ap. J. Let.* 445:L73–76

Turok N. 1996. *Phys. Rev. Lett.* 77:4138–41

Tytler D, O'Meara JM, Suzuki N, Lubin D. 2000. *Physica Scripta Vol. T* 85:12–31

Vishniac ET. 1987. *Ap. J.* 322:597–604

Vittorio N, Silk J. 1984. *Ap. J. Lett.* 285:L39–43

Wandelt B, Hansen F. 2001. *Phys. Rev. D* In press (astro-ph/0106515)

Wandelt BD, Gorski KM. 2001. *Phys. Rev. D* 63:123002

Wang X, Tegmark M, Zaldarriaga M. 2001. *Phys. Rev. D* In press (astro-ph/0105091)

Weinberg S. 1971. *Ap. J.* 168:175–94

White M. 1998. *Phys. Rev. D* 57:5273–75

White M, Hu W. 1997. *Astron. Astrophys.* 321:8–9

White M, Scott D, Silk J. 1994. *Annu. Rev. Astron. Astrophys.* 32:319–70

White SDM, Efstathiou G, Frenk CS. 1993. *MNRAS* 262:1023–28

Wilson ML, Silk J. 1981. *Ap. J.* 243:14–25

Winitzki S, Kosowsky A. 1997. *New Astron.* 3:75–100

Wright EL, Hinshaw G, Bennett CL. 1996. *Ap. J. Lett.* 458:L53–55

Wu JHP, Balbi A, Borrill J, Ferreira PG, Hanany S, et al. 2001. *Ap. J. Suppl.* 132:1–17

Zaldarriaga M. 2000. *Phys. Rev. D* 62:063510

Zaldarriaga M, Seljak U. 1997. *Phys. Rev. D* 55:1830–40

Zaldarriaga M, Seljak U. 1998. *Phys. Rev. D* 58:023003

Zaldarriaga M, Seljak U. 1999. *Phys. Rev. D* 59:123507

Zaldarriaga M, Spergel DN, Seljak U. 1997. *Ap. J.* 488:1–13

Zel'dovich Y, Kurt V, Sunyaev R. 1969. *Sov. Phys.–JETP* 28:146

Annu. Rev. Astron. Astrophys. 2002. 40:217–61
doi: 10.1146/annurev.astro.40.060401.093806
Copyright © 2002 by Annual Reviews. All rights reserved

STELLAR RADIO ASTRONOMY: Probing Stellar Atmospheres from Protostars to Giants

Manuel Güdel

*Paul Scherrer Institut, Würenlingen & Villigen, CH-5232 Villigen PSI, Switzerland;
email: guedel@astro.phys.ethz.ch*

Key Words radio stars, coronae, stellar winds, high-energy particles, nonthermal radiation, magnetic fields

■ **Abstract** Radio astronomy has provided evidence for the presence of ionized atmospheres around almost all classes of nondegenerate stars. Magnetically confined coronae dominate in the cool half of the Hertzsprung-Russell diagram. Their radio emission is predominantly of nonthermal origin and has been identified as gyrosynchrotron radiation from mildly relativistic electrons, apart from some coherent emission mechanisms. Ionized winds are found in hot stars and in red giants. They are detected through their thermal, optically thick radiation, but synchrotron emission has been found in many systems as well. The latter is emitted presumably by shock-accelerated electrons in weak magnetic fields in the outer wind regions. Radio emission is also frequently detected in pre–main sequence stars and protostars and has recently been discovered in brown dwarfs. This review summarizes the radio view of the atmospheres of nondegenerate stars, focusing on energy release physics in cool coronal stars, wind phenomenology in hot stars and cool giants, and emission observed from young and forming stars.

Eines habe ich in einem langen Leben gelernt, nämlich, dass unsere ganze Wissenschaft, an den Dingen gemessen, von kindlicher Primitivität ist—und doch ist es das Köstlichste, was wir haben.

One thing I have learned in a long life: that all our science, measured against reality, is primitive and childlike—and yet it is the most precious thing we have.

A. Einstein 1951, in a letter to H. Mühsam, Einstein Archive 36-610

INTRODUCTION

Stellar radio astronomy has matured over the past two decades, driven in particular by discoveries made with the largest and most sensitive radio interferometers. Radio emission is of great diagnostic value, as it contains telltale signatures not available from any other wavelength regime. Some of the detected radio emission represents the highest-energy particle populations (MeV electrons) yet accessible on stars, the shortest (subsecond) detectable time scales of variability and energy release, and probably refers most closely to the primary energy release responsible for coronal heating. This review is to a large extent devoted to demonstrating the ubiquity of high-energy processes in stars as revealed by radio diagnostics.

Stellar radio sources include thermal and nonthermal magnetic coronae, transition regions and chromospheres, stars shedding winds, colliding-wind binaries, pre–main sequence stars with disks and radio jets, and embedded young objects visible almost exclusively by their radio and millimeter-wave emission. Most recent additions to the zoo of objects are brown dwarfs, and with the increasingly blurred transition from stars to brown dwarfs to giant planets like Jupiter and Saturn, even the magnetospheres of the latter may have to be considered a manifestation of magnetic activity in the widest sense. To keep the discussion somewhat focused, this paper concentrates on physical processes in magnetic coronae, but includes, in a more cursory way, atmospheres of young and forming stars and winds of hot stars. We do not address in detail the large phenomenology of extended and outflow-related sources such as radio jets, HII regions, masers, Herbig-Haro objects, and the diverse millimeter/submillimeter phenomenology, e.g., molecular outflows and dust disks. Compact stellar objects (white dwarfs, neutron stars, black holes) are not considered here.

Inevitably—and fortunately—much of the knowledge gained in stellar astronomy is anchored in solar experience. The privilege of having a fine specimen—and even an exemplary prototype—next door is unique among various fields of extrasolar astrophysics, being shared since recently only by the related field of extrasolar planetary astronomy. Detailed solar studies, even in situ measurements of the solar wind, are setting high standards for studies of stellar atmospheres, with a high potential reward. Solar radio astronomy has been reviewed extensively in the literature. For detailed presentations, we refer the reader to Dulk (1985) and Bastian et al. (1998).

The maturity of stellar radio astronomy is demonstrated by a number of review articles on various subjects; a nonexhaustive list for further reference includes André (1996), Bastian (1990, 1996), Bookbinder (1988, 1991), Dulk (1985, 1987), Güdel (1994), Hjellming (1988), Kuijpers (1985, 1989a), Lang (1990, 1994), Lestrade (1997), Linsky (1996), Melrose (1987), Mullan (1985, 1989), Mutel (1996), Phillips (1991), Seaquist (1993), van den Oord (1996), and White (1996, 2000). A natural starting point for this review, even if not strictly adhered to, is Dulk's comprehensive 1985 *Annual Reviews* article that summarizes the pre- and early-Very Large Array (VLA) view of stellar (and solar) radio emission. Meanwhile, two dedicated conferences, the first one in Boulder in 1984 [proceedings edited by R.M. Hjellming & D.M. Gibson (1985)] and the second held in Barcelona in 1995 [proceedings edited by A.R. Taylor & J.M. Paredes (1996)], provided a rich forum to discuss new developments; together, they beautifully illustrate the progress made over the past decades.

RADIO SURVEYS AND THE RADIO HERTZSPRUNG-RUSSELL DIAGRAM

Figure 1 (see color insert) presents a radio Hertzsprung-Russell diagram (HRD) based on stellar radio detections between 1 and 10 GHz, reported in the catalog of Wendker (1995); for other examples, see White (1996, 2000). The luminosity is the logarithmic average of all reported detections. The accuracy of the location of some stars on the HRD is compromised by the limited quality of distance and color measurements, or in multiple systems by the uncertain attribution of the radio emission to one of the components. Nevertheless, almost all of the usual features of an HRD are recovered, testifying to the importance and ubiquity of radio emission.

The many nearby M dwarfs in Figure 1 were among the first radio stars surveyed. The typically larger distances to earlier-type cool dwarfs made their discovery more of an adventure, but the samples now include late-type binaries, K stars, solar analog G stars, and a few F stars. Most radio-detected dwarf stars are strong X-ray sources and young, rapid rotators.

The cool half of the subgiant and giant area is dominated by the large and radio-strong sample of RS CVn and Algol-type close binaries. Other sources in this area include the vigorous coronal radio sources of the FK Com class, chromospheric radio sources, symbiotic stars, and thermal-wind emitters. Additionally, a very prominent population of thermal or nonthermal sources just above the main sequence is made up of various classes of pre–main sequence stars, such as classical and weak-lined T Tauri stars.

Common to most cool star radio emitters is their nonthermal nature, which especially in its quiescent form, constitutes one of the most significant—and unexpected—discoveries in stellar radio astronomy, suggesting the presence of magnetic fields and high-energy electrons. Moving toward A-type stars on the HRD, one expects, and finds, a dearth of radio detections owing to the absence of magnetic dynamo action. However, this is also the region of the chemically peculiar Ap/Bp stars that possess strong magnetic fields and many of which are now known to be nonthermal radio sources as well. Some pre–main sequence Herbig Ae/Be stars in this area are also prominent radio emitters.

Finally, the hot-star region is heavily populated by luminous wind-shedding O and B stars and Wolf-Rayet (WR) stars, both classes showing evidence for either thermal or nonthermal radio emission.

THEORY OF RADIO EMISSION FROM STELLAR ATMOSPHERES

Elementary Formulae for Radiation and Particles

We summarize below some handy formulae that have become standard for most radio-stellar interpretational work; (the interested reader is advised to consult the original literature as well). In thermodynamic equilibrium, the emissivity of a plasma of temperature T_{eff}, η_ν (in erg s^{-1} cm^{-3} Hz^{-1} sr^{-1}) and the absorption coefficient κ_ν (in cm^{-1}) are related by Kirchhoff's law; for combined modes of polarization, unity spectral index, and a frequency $\nu \ll 10^{10} T_{eff}$ (ν in Hz, T_{eff} in K; Rayleigh-Jeans approximation):

$$\frac{\eta_\nu}{\kappa_\nu} = \frac{2kT_{eff}\nu^2}{c^2} \approx 3.1 \times 10^{-37} T_{eff}\nu^2. \tag{1}$$

Here $k = 1.38 \times 10^{-16}$ erg K^{-1} is the Boltzmann constant. The brightness temperature T_b is

$$T_b = \begin{cases} \tau T_{eff}, & \tau \ll 1 \\ T_{eff}, & \tau \gg 1 \end{cases}, \tag{2}$$

with $\tau = \int \kappa d\ell$ being the optical depth along the line of sight ℓ. The spectral radio flux density S_ν then is

$$S_\nu = \frac{2kT_b\nu^2}{c^2}\frac{A}{d^2} \approx 0.1 \left(\frac{T_b}{10^6\,\text{K}}\right)\left(\frac{\nu}{1\,\text{GHz}}\right)^2\left(\frac{r}{10^{11}\,\text{cm}}\right)^2\left(\frac{1\,\text{pc}}{d}\right)^2 \text{mJy}, \tag{3}$$

where A is the cross-sectional source area perpendicular to the line of sight, and the approximation on the right-hand side is for a circular source with $A = \pi r^2$. In nonthermal astrophysical plasmas, electron volumetric number densities are often observed to follow a power-law form

$$n(\epsilon) = N(\delta - 1)\epsilon_0^{\delta-1}\epsilon^{-\delta}\left[\text{cm}^{-3}\,\text{erg}^{-1}\right], \tag{4}$$

where $\epsilon = (\gamma-1)m_e c^2$ is the kinetic particle energy, γ is the Lorentz factor, and $\delta > 1$ has been assumed so that N is the total nonthermal electron number density above ϵ_0. A fundamental frequency of a plasma is its plasma frequency, given by

$$\nu_p \equiv \frac{\omega_p}{2\pi} = \left(\frac{n_e e^2}{\pi m_e}\right)^{1/2} \approx 9000\, n_e^{1/2}\,[\text{Hz}]. \tag{5}$$

Here $m_e = 9.1 \times 10^{-28}$ g is the electron rest mass, $e = 4.8 \times 10^{-10}$ esu is the electron charge, and n_e is the total electron number density (cm^{-3}).

Bremsstrahlung

A thermal plasma emits free-free emission (bremsstrahlung) across the electromagnetic spectrum. For cosmic abundances, the absorption coefficient is then approximately given by (in the usual units; $T \equiv T_{eff}$, mix of 90% H and 10% He; upper equation for singly ionized species, lower equation for fully ionized plasma, and $\nu \gg \nu_p$) (Dulk 1985, Lang 1999):

$$\kappa_\nu \approx 0.01 n_e^2 T^{-3/2}\nu^{-2} \times \begin{cases} \ln(5 \times 10^7 T^{3/2}/\nu), & T \lesssim 3.2 \times 10^5\,\text{K} \\ \ln(4.7 \times 10^{10} T/\nu), & T \gtrsim 3.2 \times 10^5\,\text{K} \end{cases}, \tag{6}$$

and the emissivity η_ν follows from Equation 1. Optically thick bremsstrahlung shows a ν^2 dependence, whereas optically thin flux is nearly independent of ν. A homogeneous optically thin magnetized source is polarized in the sense of the magnetoionic x-mode, whereas an optically thick source is unpolarized (Dulk 1985).

Gyromagnetic Emission

Electrons in magnetic fields radiate gyromagnetic emission. The gyrofrequency in a magnetic field of strength B is

$$\nu_c \equiv \frac{\Omega_c}{2\pi} = \frac{eB}{2\pi m_e c} \approx 2.8 \times 10^6 B \text{ [Hz],} \tag{7}$$

where $c = 3 \times 10^{10} \text{ cm s}^{-1}$ is the speed of light, and the magnetic field strength B is given in Gauss. The relativistic gyrofrequency is $\nu_{c,\text{rel}} = \nu_c/\gamma$. For large pitch angles, spectral power is predominantly emitted around a harmonic s with

$$\nu_{\max} = s\nu_{c,\text{rel}} \approx \gamma^3 \nu_{c,\text{rel}} \approx 2.8 \times 10^6 B\gamma^2 \text{ [Hz].} \tag{8}$$

Depending on γ, the emission is termed gyroresonance or cyclotron emission ($s < 10$, $\gamma \approx 1$, nonrelativistic, typically thermal electrons), gyrosynchrotron ($s \approx 10\text{--}100$, $\gamma \lesssim 2\text{--}3$, mildly relativistic electrons), or synchrotron emission ($s > 100$, $\gamma \gg 1$, relativistic electrons). Because relativistic effects play an increasingly important role in increasing s, the fundamental properties (e.g., directivity, bandwidth, polarization) change between the three categories of gyromagnetic emission. The total power emitted by an electron is (Lang 1999)

$$P = 1.6 \times 10^{-15} \beta^2 \gamma^2 B^2 \sin^2\alpha \text{ [erg s}^{-1}\text{],} \tag{9}$$

where $\beta = v/c = (1 - \gamma^{-2})^{1/2}$ and α is the pitch angle of the electron.

Simplified approximate expressions for separate magnetoionic modes of gyromagnetic emission have been given by Dulk & Marsh (1982), Robinson & Melrose (1984), and Klein (1987). For stellar observations, the angle θ between the line of sight and the magnetic field is often unknown and should be averaged. We give handy, simplified expressions for η_ν, κ_ν, the turnover frequency ν_{peak}, and the degree of circular polarization r_c derived from Dulk (1985) for $\theta = \pi/3$, with exponents for the model parameters B, N, T, and the characteristic source scale along the line of sight, L. For more comprehensive expressions, we refer the reader to the original work by Dulk & Marsh (1982) and Dulk (1985).

GYRORESONANCE EMISSION If s is the harmonic number, $s^2 T \ll 6 \times 10^9$ K and $\nu \gtrsim \nu_c$, then for each magnetoionic mode (for unity refractive index),

$$\kappa_\nu(s) = 1020(1 \pm 0.5)^2 \frac{n_e}{\nu T^{1/2}} \frac{s^2}{s!} \left(\frac{s^2 T}{1.6 \times 10^{10}}\right)^{s-1} \exp\left[-\frac{(1 - s\nu_c/\nu)^2}{8.4 \times 10^{-11} T}\right], \tag{10}$$

$$\eta_\nu(s) = \frac{kT\nu^2}{c^2}\kappa_\nu(s), \tag{11}$$

(for the x-mode, and marginally for the o-mode) [see Dulk (1985)], where n_e is the ambient electron density. Properties: Emission is strongly concentrated in emission "lines" at harmonic frequencies $s\nu_c$, and κ_ν depends on the magnetoionic mode (lower sign for o-mode, upper sign for x-mode). The emission at a given harmonic comes from a layer of constant B with a thickness $L = 2\Lambda_B\beta_0\cos\theta \approx \Lambda_B\beta_0$ determined by the magnetic scale length $\Lambda_B = B/\nabla B$ and $\beta_0 \equiv [kT/(m_ec^2)]^{1/2} \approx 1.3 \times 10^{-5}\,T^{1/2}$.

GYROSYNCHROTRON EMISSION FROM A THERMAL PLASMA   If $10^8 \lesssim T \lesssim 10^9$ K and $10 \lesssim s \lesssim 100$, then for the x-mode (and r^c for $\tau \ll 1$)

$$\kappa_\nu \approx 21 n_e T^7 B^9 \nu^{-10} \tag{12}$$

$$\eta_\nu \approx 3.2 \times 10^{-36} n_e T^8 B^9 \nu^{-8} \tag{13}$$

$$\nu_{peak} \approx 1.3(n_e L/B)^{0.1} T^{0.7} B \tag{14}$$

$$r_c \approx 2.9 \times 10^4 T^{-0.138}(\nu/B)^{-0.51}. \tag{15}$$

Properties: The spectral power is $\propto\nu^2$ on the optically thick side but $\propto\nu^{-8}$ on the optically thin side of the spectrum for a homogeneous source.

GYROSYNCHROTRON EMISSION FROM A POWER-LAW ELECTRON DISTRIBUTION: For isotropic pitch angle electron distributions according to Equation 4, with $\epsilon_0 = 10$ keV $= 1.6 \cdot 10^{-8}$ erg, $2 \lesssim \delta \lesssim 7$, and $10 \lesssim s \lesssim 100$, and for the x-mode (and r_c for $\tau \ll 1$),

$$\kappa_\nu \approx 10^{-0.47+6.06\delta} NB^{0.3+0.98\delta} \nu^{-1.3-0.98\delta} \tag{16}$$

$$\eta_\nu \approx 10^{-31.32+5.24\delta} NB^{-0.22+0.9\delta} \nu^{1.22-0.9\delta} \tag{17}$$

$$\nu_{peak} \approx 10^{3.41+0.27\delta}(NL)^{0.32-0.03\delta} B^{0.68+0.03\delta} \tag{18}$$

$$r_c \approx 10^{3.35+0.035\delta}(\nu/B)^{-0.51}. \tag{19}$$

Properties: Broad spectra with intermediate circular polarization degree on the optically thin side. Optically thick spectral power is approximately $\propto\nu^{5/2}$, and the optically thin side is a power law with an index $\alpha = 1.22 - 0.9\delta$ for a homogeneous source.

SYNCHROTRON EMISSION FROM A POWER-LAW ELECTRON DISTRIBUTION (HOMO-GENEOUS AND ISOTROPIC)   For $\gamma \gg 1$, i.e., $s \gg 100$, in each of the magnetoionic modes

$$\kappa_\nu \approx 10^{5.89+1.72\delta}(\delta - 1)NB^{(\delta+2)/2}\nu^{-(\delta+4)/2} \tag{20}$$

$$\eta_\nu \approx 10^{-24.7+1.57\delta}(\delta - 1)NB^{(\delta+1)/2}\nu^{-(\delta-1)/2}, \tag{21}$$

$$\nu_{\text{peak}} \approx \left[10^{11.77+3.44\delta} (\delta - 1)^2 N^2 L^2 B^{\delta+2} \right]^{1/(\delta+4)}. \qquad (22)$$

Properties: Continuous and broad spectrum, important harmonics $s \approx \gamma^3$. For a homogeneous, optically thin source, the degree of linear polarization is $r = (\delta + 1)/(\delta + 7/3)$. The optically thick (self-absorbed) spectral power is $\propto \nu^{5/2}$, and the optically thin side of the spectrum has a power-law spectral index of $\alpha = -(\delta - 1)/2$.

Coherent Emission

The brightness temperature of synchrotron emission is limited to $T_b \lesssim 10^{12}$ K by inverse Compton scattering (Kellermann & Pauliny-Toth 1969). If higher T_b is observed, then a coherent radiation mechanism should be considered. Two mechanisms have received most attention both for solar and stellar coherent bursts (for details, see Benz 2002).

Plasma radiation is emitted at the fundamental or the second harmonic of ν_p (Equation 5) (for propagation parallel to the magnetic field), or of the upper hybrid frequency $\nu_{\text{UH}} = (\nu_p^2 + \nu_c^2)^{1/2}$ (for perpendicular propagation, with $\nu_p > \nu_c$). It is thus a useful means to approximately determine the electron density in the source. It can account for high brightness temperatures (up to 10^{18} K) (Melrose 1989) and small bandwidth, and it is frequently observed in the Sun (Bastian et al. 1998) and in low-frequency stellar flares (Bastian 1990). Owing to increasing free-free absorption with increasing ν (because $n_e \propto \nu_p^2$ in Equation 6), and possibly also owing to gyroresonance absorption in hot plasma, fundamental emission is best observed below 1 GHz, although in very-high-temperature environments such as coronae of RS CVn binaries, the absorption is milder and the limitations are more relaxed (White & Franciosini 1995). The escape of fundamental emission is also alleviated in a highly structured medium with sharp gradients in density (Aschwanden 1987, Benz et al. 1992).

Electron cyclotron maser emission (Melrose & Dulk 1982, 1984) is emitted mostly at the fundamental and the second harmonic of ν_c (Equation 7) and can therefore be used to determine the magnetic field strength in the source. The requirement for radiation propagation, $s\nu_c > \nu_p$, also sets an upper limit to the electron density in the source and along the line of sight to the observer. The cyclotron maser mechanism accounts for the observed high T_b ($\lesssim 10^{20}$ K for $s = 1$, $\lesssim 10^{16-17}$ K for $s = 2$), and polarization degrees up to 100%.

Other mechanisms have occasionally been proposed. For example, plasma maser emission from a collection of double layers was initially calculated by Kuijpers (1989b) and applied to highly polarized stellar flare emission by van den Oord & de Bruyn (1994).

Wind Emission

Olnon (1975), Panagia & Felli (1975), and Wright & Barlow (1975) calculated the radio spectrum of a spherically symmetric, ionized wind. If $\dot{M}$ is the mass-loss rate in units of $M_\odot$ yr^{-1}, v the terminal wind velocity in km s^{-1}, d_{pc} the distance to the star in pc, and T the wind temperature in K, the optically thick radius is (we assume a mean atomic weight per electron of 1.2 and an average ionic charge of 1)

$$R_{thick} = 8 \times 10^{28} (\dot{M}/v)^{2/3} T^{-0.45} v^{-0.7} \text{ [cm]}. \tag{23}$$

Then, the following formulae apply (R_* in cm):

$$S_\nu = \begin{cases} 9 \times 10^{10} (\dot{M}/v)^{4/3} T^{0.1} d_{pc}^{-2} v^{0.6} \text{ [mJy]}, & \text{if } R_{thick} \geq R_* \\ 5 \times 10^{39} (\dot{M}/v)^2 T^{-0.35} R_*^{-1} d_{pc}^{-2} v^{-0.1} \text{ [mJy]}, & \text{if } R_{thick} < R_*. \end{cases} \tag{24}$$

Loss Times

Relativistic electrons in an ambient gas of density n_e lose energy by Coulomb collisions with the ions. The energy loss rate and the corresponding life time are (under typical coronal conditions) (Petrosian 1985)

$$-\dot{\gamma}_{coll} = 5 \times 10^{-13} n_e \text{ [s}^{-1}], \quad \tau_{coll} = 2 \times 10^{12} \frac{\gamma}{n_e} \text{ [s]}. \tag{25}$$

For electrons in magnetic fields, the synchrotron loss rate and the life time are given by (Petrosian 1985) (pitch angle $= \pi/3$)

$$-\dot{\gamma}_B = 1.5 \times 10^{-9} B^2 \gamma^2 \text{ [s}^{-1}], \quad \tau_B = \frac{6.7 \times 10^8}{B^2 \gamma} \text{ [s]}. \tag{26}$$

RADIO FLARES FROM COOL STARS

Incoherent Flares

Many radio stars have been found to be flaring sources, and a considerable bibliography on stellar radio flares is now available. As in the Sun, two principal flavors are present: incoherent and coherent radio flares. Incoherent flares with time scales of minutes to hours, broad-band spectra, and moderate degrees of polarization are thought to be the stellar equivalents of solar microwave bursts. Like the latter, they show evidence for the presence of mildly relativistic electrons radiating gyrosynchrotron emission in magnetic fields. Many flares on single F/G/K stars are of this type (Vilhu et al. 1993; Güdel et al. 1995a, 1998), as are almost all radio flares on M dwarfs above 5 GHz (Bastian 1990, Güdel et al. 1996b), on RS CVn binaries (Feldman et al. 1978, Mutel et al. 1985), on contact binaries (Vilhu et al. 1988), and on other active subgiants and giants, although some of the latter perhaps stretch the solar analogy beyond the acceptable limit: The FK Com-type giant HD 32918 produced flare episodes lasting 2–3 weeks, with a radio luminosity of 6×10^{19} erg s^{-1} Hz^{-1} (Slee et al. 1987b, Bunton et al. 1989). Its integrated microwave luminosity is thus about 1000 times larger than the total X-ray output of the nonflaring Sun!

Coherent Radio Bursts

Bursts that exhibit the typical characteristics of coherent emission (see "Coherent Emission," above) probably represent stellar equivalents of metric and decimetric solar bursts that themselves come in a complex variety (Bastian et al. 1998). Some exceptionally long-lasting ($\sim$1 h) but highly polarized bursts require a coherent mechanism as well (examples in Lang & Willson 1986a, White et al. 1986, Kundu et al. 1987, Willson et al. 1988, van den Oord & de Bruyn 1994). Coherent flares are abundant on M dwarfs at longer wavelengths (20 cm) (Kundu et al. 1988, Jackson et al. 1989). There are also some interesting reports on highly polarized, possibly coherent bursts in RS CVn binaries (Mutel & Weisberg 1978, Brown & Crane 1978, Simon et al. 1985, Lestrade et al. 1988, White & Franciosini 1995, Jones et al. 1996, Dempsey et al. 2001). After many early reports on giant metric flares observed with single-dish telescopes (Bastian 1990), the metric wavelength range has subsequently been largely neglected, with a few notable exceptions (Kundu & Shevgaonkar 1988, van den Oord & de Bruyn 1994).

Coherent bursts carry profound information in high-time resolution light curves. Radio "spike" rise times as short as 5–20 ms have been reported, implying source sizes of $r < c\Delta t \approx$ 1500–6000 km and brightness temperatures up to $T_b \approx 10^{16}$ K, a clear proof of the presence of a coherent mechanism (Lang et al. 1983, Lang & Willson 1986b, Güdel et al. 1989a, Bastian et al. 1990). Quasi-periodic oscillations were found with time scales of 32 ms to 56 s (Gary et al. 1982, Lang & Willson 1986b, Bastian et al. 1990), and up to $\approx$5 min during a very strong flare (Brown & Crane 1978) (although this emission was proposed to be gyrosynchrotron radiation).

Radio Dynamic Spectra

If the elementary frequencies relevant for coherent processes, ν_p and ν_c, evolve in the source, or the radiating source itself travels across density or magnetic field gradients, the emission leaves characteristic traces on the $\nu - t$ plane, i.e., on dynamic spectra (Bastian et al. 1998). A rich phenomenology has been uncovered, including (a) short, highly polarized bursts with structures as narrow as $\Delta\nu/\nu = 0.2\%$ suggesting either plasma emission from a source of size $\sim$3 $\times$ 10^8 cm or a cyclotron maser in magnetic fields of $\sim$250 G (Bastian & Bookbinder 1987, Güdel et al. 1989a, Bastian et al. 1990); (b) evidence for spectral structure with positive drift rates of 2 MHz s^{-1} around 20 cm wavelength, taken as evidence for a disturbance propagating downward in the plasma emission interpretation (Jackson et al. 1987); and (c) in solar terminology, rapid broadband pulsations, sudden (flux) reductions, and positive and negative drift rates of 250–1000 MHz s^{-1} (Bastian et al. 1990; Abada-Simon et al. 1994, 1997) (Figure 2, see color insert).

The smallest spectral bandwidths were found to be in the 1% range for some bursts. Conservatively assuming a magnetic scale height of $\Lambda_B = 1R_*$, the source size can be estimated to be $r \approx (\Delta\nu/\nu)\Lambda_B \approx$ a few 1000 km, again implying very high T_b (Lang & Willson 1988a, Güdel et al. 1989a, Bastian et al. 1990).

QUIESCENT EMISSION FROM CORONAE OF COOL STARS

Phenomenology

The discovery of quiescent radio emission from magnetically active stars was one of the most fundamental and one of the least anticipated achievements in radio astronomy of cool stars because there simply is no solar counterpart. The Sun emits steady, full-disk, optically thick thermal radio emission at chromospheric and transition region levels of a few 10^4 K. However, one derives from Equation 3 that such emission cannot be detected with present-day facilities, except for radiation from the very nearest stars or giants subtending a large solid angle (see "Radio Emission from Chromospheres and Winds," below). The radio luminosity of the corona caused by optically thin bremsstrahlung is proportional to the observed X-ray emission measure, but again the calculated radio fluxes are orders of magnitude below observed fluxes (Gary & Linsky 1981, Topka & Marsh 1982, Kuijpers & van der Hulst 1985, Borghi & Chiuderi Drago 1985, Massi et al. 1988, Güdel & Benz 1989, van den Oord et al. 1989). Obviously, another mechanism is in charge, and its characterization requires a proper description of the phenomenology.

Quiescent emission is best defined by the absence of impulsive, rapidly variable flare-like events. Common characteristics of quiescent emission are (a) slow variations on time scales of hours and days (Pallavicini et al. 1985, Lang & Willson 1986a, 1988a, Willson et al. 1988, Jackson et al. 1989), (b) broad-band microwave spectra (Mutel et al. 1987, Güdel & Benz 1989), (c) brightness temperatures in excess of coronal temperatures measured in X-rays (Mutel et al. 1985, Lang & Willson 1988a, White et al. 1989a), and often (d) low polarization degrees (Kundu et al. 1988, Jackson et al. 1989, Drake et al. 1992). Occasionally, strong steady polarization up to 50% (Linsky & Gary 1983, Pallavicini et al. 1985, Willson et al. 1988, Jackson et al. 1989) or unexpectedly narrow-band steady emission (Lang & Willson 1986a, 1988a) is seen on M dwarfs. Quiescent emission has been reported at frequencies as low as 843 MHz (Vaughan & Large 1986, Large et al. 1989) and as high as 40 GHz (White et al. 1990).

Gyromagnetic Emission Mechanisms

Because active stars show high coronal temperatures and large magnetic filling factors (Saar 1990) that prevent magnetic fields from strongly diverging with increasing height, the radio optical depth can become significant at coronal levels owing to gyroresonance absorption (Equation 10). Such emission is also observed above solar sunspots. The emission occurs at low harmonics of ν_c, typically at $s = 3$–5 (Güdel & Benz 1989, van den Oord et al. 1989, White et al. 1994). Rising spectra between 5 and 22 GHz occasionally observed on stars, support such an interpretation (Cox & Gibson 1985) and imply the presence of $\sim 10^7$ K plasma in strong, kG magnetic fields (Güdel & Benz 1989) (Figure 3). However, in most cases, high-frequency fluxes are lower than predicted from a uniform, full-disk gyroresonance layer, constraining the filling factor of strong magnetic fields

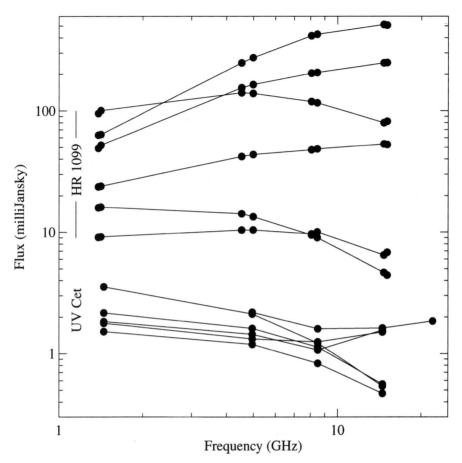

Figure 3   Radio spectra of the RS CVn binary HR 1099 (*upper set*) and the dMe dwarf UV Cet (*lower set*) at different flux levels. The gently bent spectra are indicative of gyrosynchrotron emission, and the high-frequency part of the U-shaped spectra for UV Cet has been interpreted as a gyroresonance component (HR 1099 spectra: courtesy of S.M. White).

immersed in hot plasma. It is possible that the hot plasma resides in low-B field regions where the gyroresonance absorption is negligible, for example between strong fields from underlying active regions, or at large heights (White et al. 1994, Lim et al. 1996c). Uniform magnetic structures of $\sim$600 G containing hot plasma can, on average, not reach out to beyond 1–2R_* on M dwarfs (Leto et al. 2000).

The gyroresonance mechanism cannot apply to lower-frequency radio emission because the radius of the optically thick layer, still at $s = 3$–5, would have to be $>3R_*$ for dMe stars. However, an extrapolation of the corresponding magnetic fields of more than 100 G down to photospheric levels would result in photospheric fields much stronger than observed (Gary & Linsky 1981, Topka & Marsh 1982,

–   Linsky & Gary 1983, Pallavicini et al. 1985, Lang & Willson 1986a, Willson et al. 1988, Güdel & Benz 1989).

The only remedy is to allow for much higher T_{eff}. Then the optically thick layer shifts to harmonics above 10, the range of gyrosynchrotron radiation, and the optically thick source sizes become more reasonable for M dwarfs, $R \approx R_*$ (Linsky & Gary 1983, Drake et al. 1989, 1992). For a thermal plasma, however, the spectral power drops like ν^{-8} at high frequencies (Equation 13), far from the rather shallow $\nu^{-(0.3...1)}$ spectra of active stars (Massi & Chiuderi-Drago 1992) (Figure 3). On the other hand, an optically thick thermal contribution requires strong magnetic fields ($\sim$200 G) over such large source areas that inferred photospheric magnetic fields again become unrealistically large (Kuijpers & van der Hulst 1985, Drake et al. 1992).

The situation is much more favorable for a nonthermal electron energy distribution such as a power law, analogous to distributions inferred for solar and stellar microwave flares (Kundu & Shevgaonkar 1985, Pallavicini et al. 1985). This model is supported by measured high brightness temperatures (Lestrade et al. 1984b, Mutel et al. 1985, Umana et al. 1991, Benz et al. 1995). Comprehensive spectral modeling suggests mildly relativistic electrons in $\sim$100 G fields with power-law indices of $\delta \approx$ 2–4, matching observed broad spectra with turnover frequencies in the 1–10 GHz range (White et al. 1989b; Slee et al. 1987a, 1988; Umana et al. 1998) (Figure 3). The nonthermal model is now quite well established for many classes of active stars and provokes the question of how these coronae are continuously replenished with high-energy electrons.

Occasionally, steady narrow-band or strongly polarized coherent emission has been observed at low frequencies. An interesting possibility is large numbers of unresolved, superimposed coherent bursts (Pallavicini et al. 1985; Lang & Willson 1986a, 1988a; White & Franciosini 1995; Large et al. 1989).

RADIO FLARES AND CORONAL HEATING

Is Quiescent Emission Composed of Flares?

The question of the nature of stellar quiescent radio emission has defined one of the most fascinating aspects of stellar radio astronomy. A number of observations seem to suggest that flares play an important role: Quiescent emission could simply be made up of unresolved flares. Inevitably, this question relates to the physics of coronal heating and the presence of X-ray coronae.

Very long baseline interferometry (VLBI) studies of RS CVn binaries suggest that flare cores progressively expand into a large-scale magnetosphere around the star, radiating for several days (flare remnant emission). The electron distributions then evolve from initial power-law distributions as they are subject to collisional losses affecting the low-energy electrons and to synchrotron losses affecting the high-energy electrons. Time-dependent calculations predict rather flat spectra similar to those observed (Chiuderi Drago & Franciosini 1993, Franciosini & Chiuderi Drago 1995, Torricelli-Ciamponi et al. 1998).

Microflaring at Radio Wavelengths

Very long quiescent episodes with little flux changes impose challenges for flare-decay models. Limitations on the collisional losses (Equation 25) require a very-low-ambient electron density (Massi & Chiuderi-Drago 1992). Frequent electron injection at many sites may be an alternative. Based on spectral observations, White & Franciosini (1995) suggest that the emission around 1.4 GHz is composed of a steady, weakly polarized broad-band gyrosynchrotron component plus superimposed, strongly and oppositely polarized, fluctuating plasma emission that is perceived as quasi-steady but that may occasionally evolve into strong, polarized flare emission (Simon et al. 1985, Lestrade et al. 1988, Jones et al. 1996, Dempsey et al. 2001; Figure 4).

RS CVn and Algol-type binaries also reveal significant, continuous gyrosynchrotron variability on time scales of ~10–90 min during ~30% of the time, with an increasing number of events with decreasing flux (Lefèvre et al. 1994). These flare-like events may constitute a large part of the quiescent gyrosynchrotron emission. However, microbursts or nanoflares with durations on the order of seconds to a few minutes are usually not detected at available sensitivities (Kundu et al. 1988, Rucinski et al. 1993), although Willson & Lang (1987) found some rapid

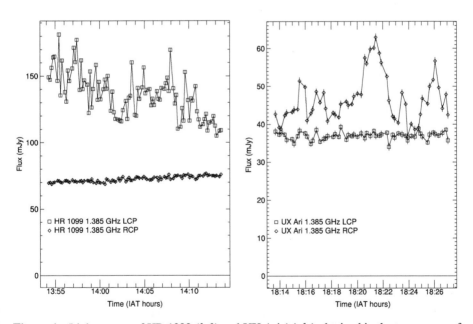

Figure 4   Light curves of HR 1099 (*left*) and UX Ari (*right*) obtained in the two senses of circular polarization. The brighter of the two polarized fluxes varies rapidly and has been interpreted as 100% polarized coherent emission superimposed on a gradually changing gyrosynchrotron component (White & Franciosini 1995) (figures from S.M. White).

variability on time scales between 30 s and 1 h that they interpreted as the result of variable absorption by thermal plasma.

Radio Flares and the Neupert Effect

The chromospheric evaporation scenario devised for many solar flares assumes that accelerated coronal electrons precipitate into the chromosphere where they lose their kinetic energy by collisions, thereby heating the cool plasma to coronal flare temperatures and evaporating it into the corona. The radio gyrosynchrotron emission from the accelerated electrons is roughly proportional to the injection rate of electrons, whereas the X-ray luminosity is roughly proportional to the accumulated energy in the hot plasma. To first order, one expects

$$L_R(t) \propto \frac{d}{dt} L_X(t), \tag{27}$$

a relation that is known as the Neupert effect (Neupert 1968) and that has been well observed on the Sun in most impulsive and many gradual flares (Dennis & Zarro 1993). The search for stellar equivalents has been a story of contradictions if not desperation (Figure 5). A first breakthrough came with simultaneous

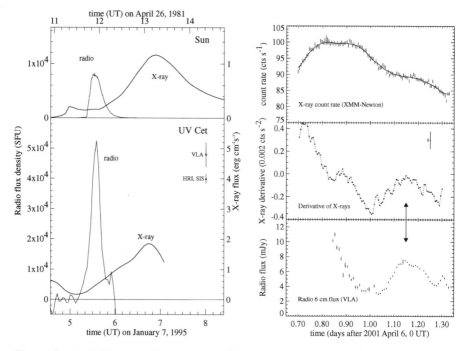

Figure 5 (*Left*) Neupert effect seen in an M dwarf star, compared with a solar example in the upper panel (Güdel et al. 1996b). (*Right*) Neupert effect seen in an RS CVn binary during a large flare (Güdel et al. 2002).

EUV (a proxy for X-rays) and optical observations (a proxy for the radio emission) of a flare on AD Leo (Hawley et al. 1995) and radio plus X-ray observations of flares on UV Cet (Güdel et al. 1996b; Figure 5a). The latter revealed a relative timing between the emissions that is very similar to solar flares. Also, the energy ratios seen in nonthermal and thermal emissions are similar to solar analogs, but perhaps more interesting, they are also similar to the corresponding ratio for the quiescent losses. These observations suggest that high-energy particles are the ultimate cause for heating through chromospheric evaporation not only in flares but also in the quiescent state. In retrospect, further suggestive examples of the Neupert effect can be found in the published literature, most notably in Vilhu et al. (1988), Stern et al. (1992), Brown et al. (1998), or Ayres et al. (2001).

It is important to note that the Neupert effect is observed neither in each solar flare (50% of solar gradual flares show a different behavior) (Dennis & Zarro 1993), nor in each stellar flare. Stellar counter-examples include an impulsive optical flare with following gradual radio emission (van den Oord et al. 1996), gyrosynchrotron emission that peaks after the soft X-rays (Osten et al. 2000), and an X-ray depression during strong radio flaring (Güdel et al. 1998). Note also that complete absence of correlated flaring has been observed at radio and UV wavelengths (e.g., Lang & Willson 1988b).

The Correlation Between Quiescent Radio and X-Ray Emissions

Radio detections are preferentially made among the X-ray brightest stars (White et al. 1989a), a result that is corroborated by new, unbiased all-sky surveys (Helfand et al. 1999). For RS CVn and Algol binaries, one finds a correlation between the radio and X-ray luminosities, $L_R \propto L_X^{1.0-1.3}$ (Drake et al. 1986, 1989, 1992; Fox et al. 1994; Umana et al. 1998), whereas for late-type dwarfs, a rather tight linear correlation appears to apply (Güdel et al. 1993). In fact, several classes of active stars and solar flares follow a similar relation. Overall, for 5–8 GHz emission,

$$L_X/L_R \approx 10^{15\pm1} \text{ [Hz]}, \tag{28}$$

(Güdel & Benz 1993, Benz & Güdel 1994) (Figure 6). Phenomenologically, the relation simply suggests that both radio and X-ray emissions are activity indicators that reflect the level of magnetic activity. The most straightforward physical model, proposed by Drake et al. (1989, 1992), assumes that hot plasma emits both thermal X-rays and nonthermal gyrosynchrotron radiation, but this model predicts steep ($\propto \nu^{-8}$) optically thin spectra that are not observed.

Chiuderi Drago & Franciosini (1993) argued that the build-up of magnetic energy in magnetic loops is $\propto B^2$ and proportional to the gradient of the turbulent velocity (assumed to be similar in all stars). Because most of the energy in the magnetic fields is eventually radiated away in X-rays, $L_X \propto B^2 V$, where V is the source volume. A power-law electron population in the same loop emits optically

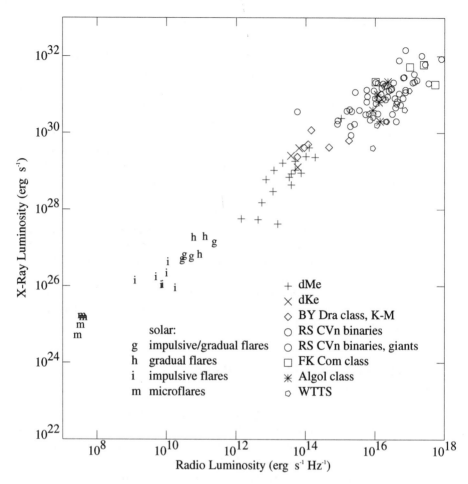

Figure 6 Correlation between quiescent radio and X-ray luminosities of magnetically active stars (*symbols*) and solar flares (*letters*) (after Benz & Güdel 1994).

thin radio radiation, as $L_R \propto B^{-0.22+0.9\delta}V$ for given ν (Equation 17). Therefore, $L_R \propto L_X^{0.45\delta-0.11}V^{1.11-0.45\delta}$. For $\delta = 3$, the dependence on V is small and $L_R \propto L_X^{1.24}$, close to the reported relations.

If the quiescent emission is made up of numerous solar-like coronal flares, then the high-energy particles carry their energy downward to heat chromospheric plasma. For a steady-state situation in which the electron injection rate is balanced by an unspecified electron loss mechanism, one derives a relation between the synchrotron losses, the particle lifetime, and the energy losses in X-rays. Using the empirical correlation, Güdel & Benz (1993) estimated an electron lifetime that indeed corresponds quite well to the observed radio variability time scales. Holman (1985, 1986), and Airapetian & Holman (1998) presented detailed model calculations for the alternative situation of simultaneous heating and particle acceleration in current sheets.

STELLAR MAGNETIC FIELDS

Stellar radio astronomy has provided invaluable information on stellar coronal magnetic fields not accessible by any other methods. We briefly describe a few principal results. Electron cyclotron maser emission can be used to estimate the field strength in the source region, namely in converging, lower-coronal fields. Because the radiation is emitted at the fundamental ($s = 1$) or the second harmonic ($s = 2$) of the gyrofrequency, $B = \nu_{Hz}/(s \times 2.8 \times 10^6) \approx 250$–500 G (Güdel et al. 1989a, Bastian et al. 1990).

Coronal gyroresonance emission sometimes detected at 8–23 GHz is optically thick at harmonics $s = 3$–5 of ν_B (see "Gyromagnetic Emission Mechanisms," above); hence, $B = \nu_{Hz}/(s \times 2.8 \times 10^6) \approx 600$–2070 G in low-lying loops in active regions (Güdel & Benz 1989).

Large coronal sizes as seen by VLBI restrict field strengths to typically 10–200 G if the approximate dipolar extrapolation to the photosphere should not produce excessive surface field strengths (Benz et al. 1998, Mutel et al. 1998, Beasley & Güdel 2000).

Very high synchrotron brightness temperatures T_b constrain, together with the power-law index δ, the characteristic harmonic numbers s and, therefore, the field strength (Dulk & Marsh 1982). Lestrade et al. (1984a, 1988), Mutel et al. (1984, 1985), and Slee et al. (1986) found B between ≈ 5 G and several tens of G for RS CVn halo sources and a few tens to several hundred G for the core. Further, the electron energy is $\epsilon = kT_b$ for optically thick emission and $\epsilon > kT_b$ for optically thin emission. Equation 8 for synchrotron emission ($s \gg 1$) then implies

$$B \leq \frac{2\pi m_e^3 c^5 \nu}{e(m_e c^2 + kT_b)^2} \approx 3.6 \times 10^{-7} \frac{\nu}{(1 + 1.7 \times 10^{-10} T_b)^2} \, [\text{G}], \qquad (29)$$

where the $<$ sign applies for optically thin emission. Alternatively, from Dulk & Marsh (1982) one finds for gyrosynchrotron emission $T_b \leq T_{eff} = 10^{6.15-0.85\delta}$ $(\nu/B)^{0.5+0.085\delta}$. Such arguments lead to field strengths of a few tens to a few hundred G in RS CVn binaries (Mutel et al. 1985, Lestrade et al. 1988).

The synchrotron turnover frequency ν_{peak} determines B through $B = 2.9 \times 10^{13}\nu_{peak}^5 \, \theta^4 S^{-2}$, where θ is the source diameter in arcsec and S is the radio flux density in mJy at ν_{peak}, which is given in units of GHz here (Lang 1999, after Slish 1963). For a typical RS CVn or dMe coronal source with a size of 1 milliarcsec (mas) as measured by VLBI, a turnover at $\nu_{peak} \approx 5$ GHz, and a flux of 10 mJy, magnetic fields of up to a few hundred G are inferred. For dipolar active region gyrosynchrotron models, White et al. (1989b) derived $B \approx 150\nu_{peak}^{1.3}$ (ν_{peak} in GHz), which again implies field strengths of several hundred G for turnover frequencies in the GHz range.

Full spectral modeling of the magnetic field strength B, the nonthermal particle density, the geometric size, and the electron power-law index δ can constrain some of these parameters. Umana et al. (1993, 1999) and Mutel et al. (1998) found $B \approx$ 10–200 G in their models for core plus halo structures in Algol-type stars.

RADIO CORONAL STRUCTURE

Introduction

Although the photospheric filling factor of kG magnetic fields on the Sun is small, it can exceed 50% on late-type active stars. On the Sun magnetic fields can rapidly expand above the transition region and thus drop below kG levels. On magnetically active stars such divergence is prevented by adjacent magnetic flux lines so that strong magnetic fields may also exist at coronal levels. The effective scale height of the coronal magnetic field, however, also largely depends on the structure of the magnetic flux lines, i.e., on whether they are compact loops with short baselines, long loops connecting distant active regions, or large dipolar magnetospheres anchored at the stellar poles. This issue is unresolved. Arguments for small-scale coronal active regions as well as for star-sized global magnetospheres have been put forward (see discussions in White et al. 1989b, Morris et al. 1990, Storey & Hewitt 1996).

Radio Eclipses and Rotational Modulation

Eclipses and rotational modulation offer reliable information on radio source geometries, but neither are frequently seen. The radio sources may be much larger than the eclipsing star or they may not be within the eclipse zone. Complete absence of radio eclipses has, for example, been reported for AR Lac (Doiron & Mutel 1984), Algol (van den Oord et al. 1989, Mutel et al. 1998) and YY Gem (Alef et al. 1997). Positive detections include the Algol system V505 Sgr (Gunn et al. 1999) and the precataclysmic system V471 Tau (Patterson et al. 1993, Lim et al. 1996b). The former, surprisingly, shows both a primary and a secondary eclipse, although one of the components is supposed to be inactive. The authors suggested that a radio coronal component is located between the two stars, a conjecture that gives rise to interesting theoretical models such as diffusion of magnetic fields from the active to the inactive star (on unreasonably long time scales, however), radio emission from the optically thin mass-accretion stream (which is found to be too weak), or field shearing by the inactive companion. The radio emission of V471 Tau may originate from magnetic loops that extend to the white dwarf, where they interact with its magnetic field (Lim et al. 1996b).

Radio rotational modulation is often masked by intrinsic variability (Rucinski 1992). Well-documented examples are the RS CVn binaries CF Tuc (Budding et al. 1999), for which Gunn et al. (1997) suggested the presence of material in the intrabinary region, and the RS CVn binary UX Ari, which appears to have radio-emitting material concentrated above magnetic spots on the hemisphere of the K subgiant that is invisible from the companion (Neidhöfer et al. 1993, Elias et al. 1995, Trigilio et al. 1998). The clearest main-sequence example is AB Dor, with two emission peaks per rotation seen repeatedly over long time intervals. Both peaks probably relate to preferred longitudes of active regions (Slee et al. 1986, Lim et al. 1992, Vilhu et al. 1993). The shape of the maxima suggests that the radiation is intrinsically directed in the source, which is plausible for synchrotron emission by ultrarelativistic electrons (Lim et al. 1994).

Very Long Baseline Interferometry

Interferometry at large baselines has long been a privilege of radio astronomy and has proven versatile for numerous applications (Mutel 1996). Various intercontinental very long baseline interferometry (VLBI) networks have been arranged, and dedicated networks such as the U.S. Very Long Baseline Array (VLBA) or the U.K. MERLIN network are routinely available. Apart from astrometric applications with sub-mas accuracy (Lestrade et al. 1990, 1993, 1995, 1999; Guirado et al. 1997), VLBI provides coronal mapping at $\lesssim 1$ mas angular resolution.

One of the principal, early VLBI results for RS CVn and Algol-like binaries relates to evidence for a compact core plus extended halo radio structure of a total size similar to the binary system size (Mutel et al. 1984, 1985, Massi et al. 1988; Trigilio et al. 2001). Compact cores appear to be flaring sites, whereas halo emission corresponds to quiescent, low-level radiation, perhaps from decaying electrons from previous flares. There is evidence for nonconcentric, moving, or expanding sources (Lestrade et al. 1988, 1999; Trigilio et al. 1993; Lebach et al. 1999; Franciosini et al. 1999).

Pre–main sequence stars are attractive VLBI targets. Although some sources are overresolved at VLBI baselines (Phillips et al. 1993), others have been recognized as components in close binaries such as HD 283447 (Phillips et al. 1996); but most of the nearby weak-lined T Tau stars (see "Evolution to the Main Sequence," below) reveal radio structures as large as $10R_*$, indicative of extended, global, probably dipolar-like magnetospheres somewhat similar to those seen in RS CVn binaries (Phillips et al. 1991, André et al. 1992). Perhaps also related to youth, magnetospheres around Bp/Ap stars have also been observed with VLBI and have strongly supported global magnetospheric models (Phillips & Lestrade 1988, André et al. 1991; see "Stars at the Interface Between Hot Winds and Coronae," below).

VLBI techniques have been more demanding for single late-type dwarf stars, owing both to lower flux levels and apparently smaller coronal sizes. Some observations with mas angular resolutions show unresolved quiescent or flaring sources, thus constraining the brightness temperature to $T_b > 10^{10}$ K (Benz & Alef 1991, Benz et al. 1995), whereas others show evidence for extended coronae with coronal sizes up to several times the stellar size (Alef et al. 1997, Pestalozzi et al. 2000).

The dMe star UV Cet is surrounded by a pair of giant synchrotron lobes, with sizes up to 2.4×10^{10} cm and a separation of 4–5 stellar radii along the putative rotation axis of the star, suggesting very extended magnetic structures above the

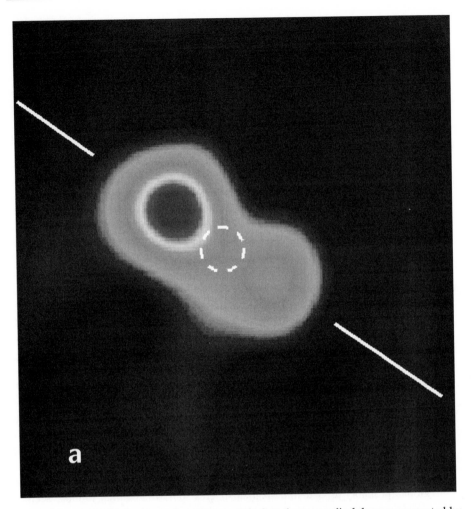

Figure 7 (*a*) VLBA image of the dMe star UV Cet; the two radio lobes are separated by about 1.4 mas, and the best angular resolution reaches 0.7 mas. The straight line shows the orientation of the putative rotation axis, assumed to be parallel to the axis of the orbit of UV Cet around the nearby Gl 65 A. The small circle gives the photospheric diameter to size, although the precise position is unknown (after Benz et al. 1998). (*b*) VLBA image of the Algol binary. The most likely configuration of the binary components is also drawn. The radio lobes show opposite polarity (Mutel et al. 1998) (reproduced with permission of the AAS).

magnetic poles (Benz et al. 1998) (Figure 7*a*). VLBA imaging and polarimetry of Algol reveals a similar picture with two oppositely polarized radio lobes separated along a line perpendicular to the orbital plane by more than the diameter of the K star (Mutel et al. 1998) (Figures 7*b*, 8*b*). Global polarization structure is also suggested in UX Ari, supporting the view that the magnetic fields are large scale and well ordered (Beasley & Güdel 2000).

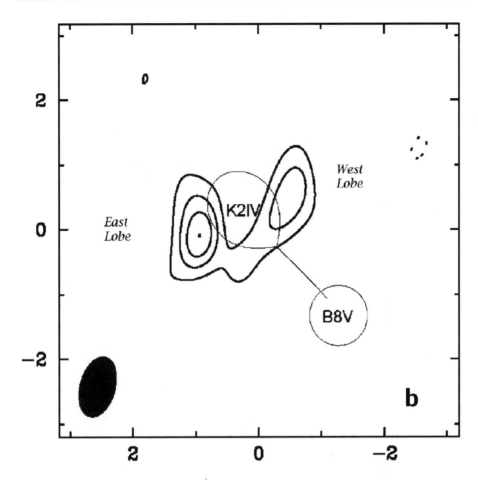

Figure 7   (*Continued*)

Magnetospheric Models

Quite detailed geometric models of large magnetospheres around RS CVn and
Algol binaries, T Tau stars, and magnetic Bp/Ap stars have been designed based
on VLBI results, radio spectra, and polarization. Common to all is a global, dipole-
like structure somewhat resembling the Earth's Van Allen belts (Figure 8a). Stellar
winds escaping along magnetic fields draw the field lines into a current sheet
configuration in the equatorial plane. Particles accelerated in that region travel
back and are trapped in the dipolar-like, equatorial magnetospheric cavity. Variants
of this radiation belt model, partly based on theoretical work of Havnes & Goertz
(1984), have been applied to RS CVn binaries (Slee et al. 1987b, Morris et al. 1990,
Jones et al. 1994, Storey & Hewitt 1996), in an optically thick version to Bp/Ap
stars (Drake et al. 1987a, Linsky et al. 1992), and in an optically thin version to a
young B star (André et al. 1988).

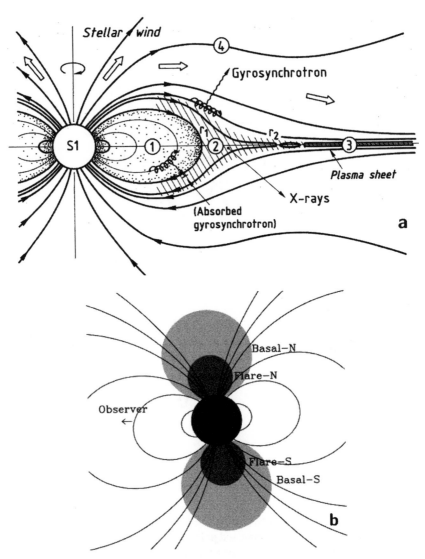

Figure 8   (*a*) Equatorial model for the magnetosphere of the young B star S1 in ρ Oph (André et al. 1988). (*b*) Sketch for radio emission from a global dipole consistent with the VLBI observation shown in Figure 7*b* (Mutel et al. 1998) (reproduced with permission of the AAS).

Such models are particularly well supported by polarization measurements in RS CVn binaries. For a given system, the polarization degree p at any frequency is anticorrelated with the luminosity, whereas the sense of polarization changes between lower and higher frequencies. For a stellar sample, p is inversely correlated with the stellar inclination angle such that low-inclination ("pole-on") systems show the strongest polarization degrees (Mutel et al. 1987, 1998; Morris et al. 1990).

When flares are in progress, the two-component core plus halo model appears to correctly describe radio spectral properties. Stronger magnetic fields of 80–200 G are inferred for the core, and weaker fields of 10–30 G for the halo (Umana et al. 1993, 1999; Mutel et al. 1998; Trigilio et al. 2001). The frequency-dependent optical depth makes the source small at high frequencies (size $\approx R_*$, above 10 GHz) and large at small frequencies (size comparable to the binary system size at 1.4 GHz; Klein & Chiuderi-Drago 1987, Jones et al. 1995). This effect correctly explains the relatively flat, optically thick radio spectra seen during flares.

STELLAR ROTATION, BINARITY, AND MAGNETIC CYCLES

Radio Emission and Stellar Rotation

There is little doubt that rotation is largely responsible for the level of magnetic activity in cool stars. It is clearly the young, rapidly rotating stars or spun-up evolved stars in tidally interacting binaries that define the vast majority of radio-emitting cool stars detected to date. The rather peculiar class of FK Com stars, single giants with unusually short rotation periods of only a few days, are among the most vigorous emitters of gyrosynchrotron emission (Hughes & McLean 1987, Drake et al. 1990, Rucinski 1991, Skinner 1991), including extremely luminous and long flares (Slee et al. 1987b, Bunton et al. 1989). However, large surveys of RS CVn binary systems find at best a weak correlation between radio luminosity and rotation parameters (Mutel & Lestrade 1985; Morris & Mutel 1988; Caillault et al. 1988; Drake et al. 1989, 1992), and it may even depend on the optical luminosity class considered. Although Stewart et al. (1988) reported a trend of the form $L_R \propto R_*^{2.5} v^{2.5}$ (where v is the equatorial velocity) for different luminosity classes, this result was later challenged (Drake et al. 1989) and may be related to using peak fluxes and flares. The absence of a correlation may be related to the coronal saturation regime known in X-rays (Vilhu & Rucinski 1983, Mutel & Lestrade 1985, White et al. 1989a): Almost all radio-detected stars emit at the maximum possible and rotation-independent X-ray level—hence perhaps also at the maximum possible radio level. There is obviously a large potential for stellar radio astronomy in the less exotic regime below saturation!

Moving to tighter binary systems in which the components are in (near-)contact, we would expect magnetic activity to increase or at least to stay constant. This is, however, not the case. The radio emission of such systems is significantly weaker than that of RS CVn binaries or active single stars, a trend that is also seen at other wavelengths (Hughes & McLean 1984, Rucinski & Seaquist 1988, Vilhu et al. 1988). Possible physical causes include (*a*) a reduced differential rotation, and hence a weaker dynamo action (Beasley et al. 1993), (*b*) a shallower convection zone in contact binary systems, and (*c*) influence by the energy transfer between contacting stars (Rucinski 1995).

Activity Cycles

Radio polarization may be a telltale indicator for magnetic activity cycles on stars, analogous to the solar 11-year cycle (or 22 years, if the magnetic polarity reversal is considered). Long-term measurements in the Ca H&K lines (the HK project) (Baliunas 1998) indicate the presence of activity cycles with durations of several years in low-activity stars, but irregular long-term variations on active stars. In any case, one would expect reversals of the dominant sense of radio polarization to accompany any of these quasi-cycles. The contrary is true, to an embarrassing level: After decades of monitoring, many active stars show a constant sign of radio polarization throughout, both in quiescence and during flares (Gibson 1983; White et al. 1986; Mutel et al. 1987, 1998; Kundu & Shevgaonkar 1988; Jackson et al. 1989; White & Franciosini 1995), with few exceptions (Bastian & Bookbinder 1987, Willson et al. 1988, Lang 1990). This suggests the presence of some stable magnetospheric structures or a predominance of strong magnetic fields in one polarity. A concerted and ongoing effort to look for polarity reversals is negative at the time of this writing (S. White, private communication).

Perhaps the most convincing case yet reported in favor of a magnetic cycle is HR 1099, in which the average radio flux density correlates with the spot coverage, revealing a possible periodicity of 15–20 years (Umana et al. 1995). Massi et al. (1998) reported a surprisingly rapid quasi-periodicity in microwave activity ($P \approx 56 \pm 4$ days) and the accompanying sense of polarization ($P \approx 25$ days) in the RS CVn binary UX Ari. Whether such rapid oscillations are related to an internal magnetic dynamo cycle remains to be shown.

RADIO EMISSION FROM CHROMOSPHERES AND WINDS

Lower Atmospheres of Cool Stars

Stellar transition regions and chromospheres are expected to be radio sources as well (see "Quiescent Emission from Coronae of Cool Stars," above). Among cool main-sequence stars the slightly evolved mid-F star Procyon is the only such source in the solar vicinity detected to date (Drake et al. 1993). In the red giant area, however, outer atmospheres fundamentally change their characteristics. Chromospheres become as large as several stellar radii. Such sources are now well detected at radio wavelengths (Knapp et al. 1995) and spatially resolved

(Skinner et al. 1997), albeit with some surprises. Because their outer atmospheres are optically thick at radio wavelengths, spatially resolved observations provide a direct temperature measurement. Reid & Menten (1997) inferred optically thick radio photospheres at about $2R_*$ but at subphotospheric/nonchromospheric temperatures. Indeed, contrary to chromospheric UV measurements, the radio-derived temperature in the nearby supergiant Betelgeuse is seen to drop systematically from optical-photospheric levels outward (Lim et al. 1998) (Figure 9). This cool material completely dominates the outer atmosphere. The authors suggested that cool, photospheric material is elevated in giant convection cells; dust formation in this environment could then drive Betelgeuse's outflow.

Winds from Cool Stars

Could dwarf stars lose mass by a substantial stellar wind (Doyle & Mathioudakis 1991, Mullan et al. 1992)? Such a wind, if (partially) ionized, would be a radio source. The very fact that coronal radio flares are seen at GHz frequencies implies that the extended winds must be optically thin down to coronal heights, and this places stringent limits on the mass-loss rate. Sensitive millimeter measurements and theoretical arguments have constrained such mass loss for M dwarfs to $\lesssim 10^{-13}\ M_\odot\ \mathrm{yr}^{-1}$ or $\lesssim 10^{-12}\ M_\odot\ \mathrm{yr}^{-1}$ for a 10^4 K or a 10^6 K wind, respectively (Lim & White 1996, Lim et al. 1996c, van den Oord & Doyle 1997), in agreement with upper limits derived from UV observations (Wood et al. 2001). Stringent upper limits to the radio emission of solar analogs also help confine the mass-loss history of the Sun or solar-like stars (Drake et al. 1993, Gaidos et al. 2000). An upper limit to the mass loss during the Sun's main-sequence life of 6% indicates that the Sun was never sufficiently luminous to explain the "Young Sun Paradox," i.e., the suggestion that the young Sun was more luminous in early times, given the apparently much warmer climate on Mars (Gaidos et al. 2000).

Winds become progressively more important toward the red giants, in particular beyond the corona–cool wind dividing line. Ionized-mass losses between 10^{-10}–$10^{-9}\ M_\odot\ \mathrm{yr}^{-1}$ are indicated, increasing toward cooler and more luminous stars (Drake & Linsky 1986). However, the surface mass-loss flux is similar in all stars, and is in fact also similar to the solar wind mass-loss flux. The coolest, late-M and C (super)giants still support massive winds, but the much weaker radio emission indicates that the ionization fraction drops by at least an order of magnitude compared with earlier M giants, i.e., their chromospheres must be cool, and their optical depth at radio wavelengths becomes small (Drake et al. 1987b, 1991b; Luttermoser & Brown 1992). The same holds for technetium-deficient S and M-S giants, although white dwarf companions of some of them may sufficiently ionize the outer atmosphere of the giant to become visible at radio wavelengths (Drake et al. 1991a). This latter mechanism is probably also relevant in the interacting "symbiotic binaries" that usually consist of a red giant and a white dwarf companion and that are radio sources with considerable luminosities. The white dwarf is sufficiently UV strong to ionize part of the cool-star wind (Seaquist et al. 1984, 1993, Taylor & Seaquist 1984, Seaquist & Taylor 1990). No appreciable radio

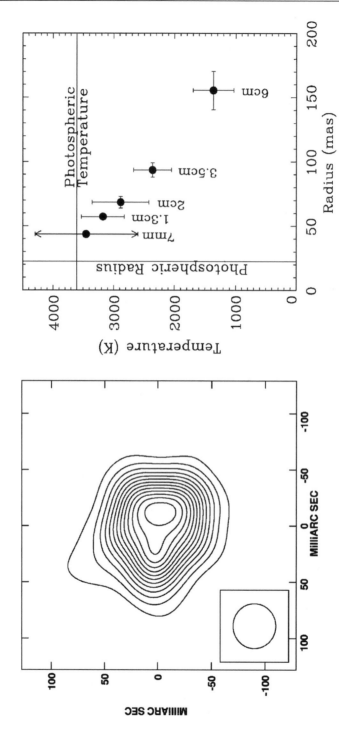

Figure 9 (a) Betelgeuse observed at 43 GHz. The radio photosphere of this star is resolved, with an angular resolution of 40 mas. (b) The atmospheric temperature of Betelgeuse as a function of radius, observed at different frequencies. (From Lim et al. 1998) (reproduced with the permission of *Nature*.)

emission is, however, detected from the similar class of G-K–type Barium stars (showing overabundances of Ba and other nuclear-processed elements), although they show evidence of white-dwarf companions (Drake et al. 1987c).

Ionized Winds and Synchrotron Emission from Hot Stars

Because OB and Wolf-Rayet stars shed strong ionized winds, they emit thermal-wind radio emission (Equations 24). Some of the hot-star radio sources are very large and resolved by the VLA because the optically thick surface (Equation 23) is located at hundreds of stellar radii. Under the assumption of a steady, spherically symmetric wind, wind mass-loss rates of $\dot{M} \approx 10^{-6}$–$10^{-5} M_\odot \, \mathrm{yr}^{-1}$ are inferred for O and B stars (Scuderi et al. 1998), and $\dot{M} \approx 2$–$4 \times 10^{-5} M_\odot \, \mathrm{yr}^{-1}$ for WR stars (Bieging et al. 1982; Leitherer et al. 1995, 1997). The inferred mass-loss rate is closely correlated with the bolometric luminosity, $\log \dot{M} = (1.15 \pm 0.2) \log L + C$ (Scuderi et al. 1998), in good agreement with Hα measurements. In the case of colliding wind binaries, the thermal radio emission can be further enhanced by contributions from the wind-shock zone (Stevens 1995). The radio emission level drops appreciably toward intermediate spectral classes of B, A, and F, probably owing to a steep decrease of the ionized mass loss (Drake & Linsky 1989).

It came as a surprise when several OB stars (Abbott et al. 1984, Bieging et al. 1989, Drake 1990, Contreras et al. 1996) and WR stars (Becker & White 1985, Caillault et al. 1985, Abbott et al. 1986, Churchwell et al. 1992, Chapman et al. 1999) were found to show nonthermal, synchrotron-like radio spectra, some of them associated with short-term variability (Persi et al. 1990). Up to 50% of the WR stars appear to be nonthermal sources, and this fraction is half as large for OB stars (Leitherer et al. 1997, Chapman et al. 1999, Dougherty & Williams 2000). Hot stars are not thought to produce magnetic fields via a dynamo. Moreover, because the wind is optically thick to radio emission out to hundreds of stellar radii, the nonthermal component must originate at such large distances as well. A coronal model therefore should not apply, although highly variable radio emission and a poor correlation with X-ray behavior suggest that magnetic fields play a role in the structuring of the winds (Waldron et al. 1998). Viable alternatives include synchrotron emission from electrons accelerated in shocks of unstable winds of single stars (White 1985, Caillault et al. 1985), in colliding-wind shocks in massive binaries (Eichler & Usov 1993), and in the interaction zone between the thermal wind and a previously ejected shell [luminous blue variable (LBV) phase] (Leitherer et al. 1997). Magnetic fields inferred for the photospheric level are of order 1 Gauss (Bieging et al. 1989, Phillips & Titus 1990), and up to 50 G in synchrotron envelope models (Miralles et al. 1994).

Shocks are attractive features, as they can accelerate electrons by the first-order Fermi mechanism, with a predicted electron energy power-law index $\delta = 2$ (Bell 1978). Skinner et al. (1999), however, found a significant deviation from this model for a colliding-wind binary, but successfully interpreted their radio spectra with an absorbed synchrotron spectrum from a quasi-monoenergetic electron population.

The origin and stability of such a population is unclear. The colliding-wind model found strong support when it became evident that most nonthermal sources are indeed binaries (Dougherty & Williams 2000) and that the nonthermal sources are located between the two stars, separate from the wind (Moran et al. 1989; Churchwell et al. 1992; Dougherty et al. 1996, 2000; Williams et al. 1997; Chapman et al. 1999). Convincing evidence for colliding winds is available for WR 146 (Dougherty et al. 2000), Cygnus OB2 No 5 (Contreras et al. 1997), and WR 147 that all show a thermal source coincident with the WR star and a separate bow-shock–shaped nonthermal source in the wind collision zone close to a lower-mass companion (Figure 10; Contreras & Rodríguez 1999). If the stars are in a

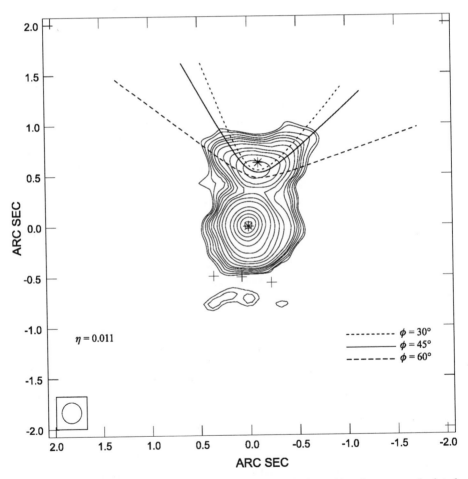

Figure 10   VLA map of the WR 147 system observed at 3.6 cm. Also shown are calculated model curves that follow the shock formed by the two-wind interaction (from Contreras & Rodríguez 1999) (reproduced with the permission of the AAS).

strongly eccentric orbit, the wind-shock zone between two stellar components can enter the thermal-wind radio photosphere and become absorbed, thus producing strong long-term variability of the radio flux (Williams et al. 1990, 1992, 1994; White & Becker 1995; Setia Gunawan et al. 2000). Further long-term synchrotron variability may be caused by long-term modulation in the magnetic field along the orbit, whereas short-term ($\sim$daily) variability may be due to clumps in the wind that arrive at the shock (Setia Gunawan et al. 2000).

The rather inhomogeneous class of Be stars shows evidence for very large ($\sim 100R_*$) high-density outflowing disks that have been probed at radio wavelengths. Steeply increasing radio spectra and some flux variability are characteristic, but there is considerable debate on the source geometry (Taylor et al. 1987, 1990; Drake 1990; Dougherty et al. 1991).

There have been a number of interferometric observations of nonthermal sources among hot stars. Source sizes of order $100R_*$, exceeding the size of the optically thick surface of the thermal wind, have been measured. They imply brightness temperatures up to at least 4×10^7 K and thus further support the nonthermal interpretation (Felli et al. 1989, 1991; Felli & Massi 1991; Phillips & Titus 1990).

STAR FORMATION AND THE SOLAR-STELLAR CONNECTION

Radio Emission from Low-Mass Young Stellar Objects

Large numbers of visible T Tau stars and embedded infrared sources in star-forming regions are strong radio sources (Garay et al. 1987, Felli et al. 1993). Schematically, pre–main sequence evolution is thought to proceed through four consecutive stages with progressive clearing of circumstellar material (Shu et al. 1987): (a) cold condensations of infalling molecular matter, forming a hydrostatic low-luminosity protostellar object, with the bulk mass still accreting (Class 0 source); (b) formation of a deeply embedded protostar (Class I source) through further accretion via a massive accretion disk, associated with strong polar outflows; (c) the classical T Tau (CTT, Class II) phase with an optically visible star accompanied by a thick accretion disk, a weak outflow, and possibly a weakly ionized wind; and (d) the weak-lined T Tau (WTT, Class III) phase, at which disk and circumstellar material have largely been dissipated, and the star approaches the main sequence. The evolutionary sequence is somewhat controversial (see below) and may in fact describe the evolutionary phase of the circumstellar material rather than of the star itself.

At first inspection, and especially in the outer reaches of the molecular clouds such as the ρ Oph cloud (André et al. 1987, 1988; Stine et al. 1988; Magazzù et al. 1999), one encounters predominantly WTTs that appear to have evolved past the CTT phase. But how young can a star be to develop strong radio signatures? Whereas deeper surveys of ρ Oph have accessed several deeply embedded infrared sources with high radio luminosities ($\log L_R > 15$) [some of which are class I sources (André et al. 1987, Brown 1987, Leous et al. 1991, Feigelson et al. 1998)], many of the radio-strong WTTs in the ρ Oph dark cloud are significantly closer to the sites of current star formation and therefore younger than the typical radio-weaker WTTs and CTTs. It seems that radio-strong WTTs evolve directly from embedded protostars. André et al. (1992) speculated that in some cases strong fossil magnetic fields accelerate both dissipation of circumstellar material and formation of large magnetospheric structures on short time scales of $\sim 10^6$ years.

Genuine, embedded class I protostars have most often been detected as thermal sources, and this emission is predominantly due to collimated thermal winds or jets. These jets are probably ionized by neutral winds that collide with the ambient medium at distances of around 10 AU and are aligned with molecular outflows (Bieging & Cohen 1985; Snell et al. 1985; Brown 1987; Curiel et al. 1987, 1989, 1990; Rodríguez et al. 1989, 1990, 1995; Morgan et al. 1990; Martí et al. 1993; Garay et al. 1996; Suters et al. 1996; Torrelles et al. 1997). In the case of more massive stars, the radio emission can also originate from optically thick or thin compact HII regions (e.g., Hughes 1988, Estalella et al. 1991, Gómez et al. 2000), or from ionized winds (Felli et al. 1998); even (thermal or nonthermal) gyrosynchrotron emission has been proposed, given the high brightness temperatures, small sizes, variability, and negative spectral indices of some sources (Hughes 1991, Hughes et al. 1995, Garay et al. 1996). For a review of thermal radio jets driving outflows and Herbig-Haro objects, see, e.g., Anglada (1995, 1996), and Anglada et al. (1998). Ionized circumstellar material and winds easily become optically thick and therefore occult any nonthermal, magnetic emission from close to the star. However, the discovery of polarization in T Tau(S) (Phillips et al. 1993), in IRS 5 (Feigelson et al. 1998), in protostellar jet sources (Yusef-Zadeh et al. 1990), and in the jet outflows themselves (Curiel et al. 1993, Hughes 1997, Ray et al. 1997) (Figure 11), as well as variability and negative spectral indices in T Tau(S) (Skinner & Brown 1994) provided definitive evidence for magnetic fields and particle acceleration in these class I objects.

Moving toward the youngest accreting or class 0 sources, centimetric thermal radio detections probably again relate to jets/collimated winds that drive massive outflows, whereas the bulk of the emitted power leaves the system at mm/submm wavelengths from cold ($\lesssim 20$ K) dust, a defining property of class 0 protostars (André et al. 1993). Several objects have been detected, with ages of the order of only 10^4 years and central stellar masses of only $\approx 0.05\ M_\odot$ (Leous et al. 1991, André et al. 1993, 1999, Bontemps et al. 1995, Yun et al. 1996, Gibb 1999, Reipurth et al. 1999), marking the very beginning of the protostellar accretion phase. Radio emission is thus a sensitive tracer for the presence of an embedded, nascent but already formed protostellar core as opposed to a contracting cloud fragment (see, e.g., Yun et al. 1996).

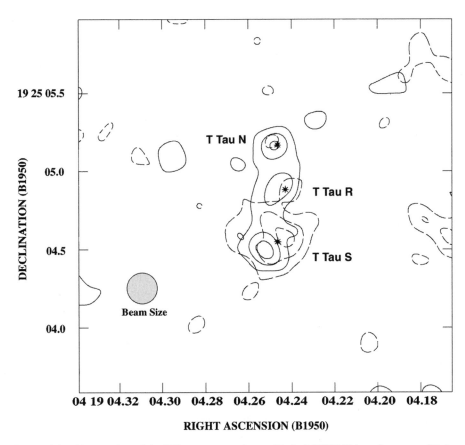

Figure 11   Observation of the T Tau system at 6 cm with the MERLIN interferometer. Right- and left-hand circularly polarized components are shown dashed and solid, respectively. The offset of the polarized flux centroids in T Tau(S) is interpreted in terms of polarized outflows (Ray et al. 1997) (reproduced with the permission of *Nature*).

Evolution to the Main Sequence

Early VLA surveys of CTTs quickly recognized their thermal wind-type emission with rising spectra and large angular sizes (Cohen et al. 1982; Bieging et al. 1984; Cohen & Bieging 1986; Schwartz et al. 1984, 1986). The enormous kinetic wind energy derived under the assumption of a uniform spherical wind suggests anisotropic outflows, whereas structural changes in the radio sources indicate variable outflows, probably along jet-like features (Cohen & Bieging 1986). Mass-loss rates are estimated to be $\leq 10^{-7} M_\odot$ yr^{-1} (André et al. 1987). It is important to note that the thermal radio emission says nothing about the presence or absence of stellar magnetic fields. CTTs do show a number of magnetically induced phenomena, but whatever the possible accompanying radio emission, it is thought to be absorbed by the circumstellar ionized wind, unless huge magnetospheric structures reach beyond the optically thick wind surface (Stine et al. 1988, André et al. 1992).

The CTT stage is pivotal for planetary system formation, as massive accretion disks are present. Their dust emission dominates the systemic luminosity at millimeter wavelengths. Large (>1000 AU) molecular and dust features have been mapped at such wavelengths; however, observations of the solar-system–sized inner disks (100 AU) have been challenging and require high angular resolution. At centimeter wavelengths, thermal collimated outflows may become dominant over dust disks, but both features can be mapped simultaneously in some cases, revealing two orthogonal structures, one a jet and one an edge-on disk (Rodríguez et al. 1992, 1994; Wilner et al. 1996).

At ages of typically $1–20 \times 10^6$ years, most T Tau stars dissipate their accretion disks and circumstellar material and become similar to main-sequence stars, albeit at much higher magnetic activity levels, probably induced by their short rotation periods (O'Neal et al. 1990, White et al. 1992a). The presence of huge flares (Feigelson & Montmerle 1985, Stine et al. 1988, Stine & O'Neal 1998), longer-term variability, and falling spectra clearly point to nonthermal gyrosynchrotron emission (Bieging et al. 1984, Kutner et al. 1986, Bieging & Cohen 1989, White et al. 1992a, Felli et al. 1993, Phillips et al. 1996) analogous to more evolved active stars. Conclusive radio evidence for the presence of solar-like magnetic fields in WTTs came with the detection of weak circular polarization during flare-like modulations but also in quiescence (White et al. 1992b, André et al. 1992, Skinner 1993). Zeeman measurements confirm the presence of kilogauss magnetic fields on the surface of some of these stars (Johns-Krull et al. 1999). Extreme particle energies radiating synchrotron emission may be involved, giving rise to linear polarization in flares on the WTT star HD283447 (Phillips et al. 1996). VLBI observations showing large ($\sim 10R_*$) magnetospheric structures with brightness temperatures of $T_b \approx 10^9$ K fully support the nonthermal picture (Phillips et al. 1991).

As a WTT star ages, its radio emission drops rapidly on time scales of a few million years from luminosities as high as 10^{18} erg s^{-1} Hz^{-1} to values around or below 10^{15} erg s^{-1} Hz^{-1} at ages beyond 10 Myr. The young age of a star is thus favorable for strong radio emission (O'Neal et al. 1990, White et al. 1992a, Chiang et al. 1996), whereas toward the subsequent zero-age main-sequence (ZAMS) stage, it is only the very rapid rotators that keep producing radio emission at a 10^{15} erg s^{-1} Hz^{-1} level (Carkner et al. 1997, Magazzù et al. 1999, Mamajek et al. 1999).

Low-Mass Stars on the Main Sequence

The pace at which radio emission decays toward the ZAMS has prevented systematic radio detections of cool main-sequence stars beyond 10–20 pc. Much of the evolutionary trends known to date have been derived from fairly small and very select samples of extraordinary stars. Although many nearby M dwarfs are probably quite young and located near the ZAMS, their ages are often not well known. An interesting exception is the proper-motion dM4e companion to AB Dor,

Rst137B, detected as a surprisingly luminous steady and flaring radio star (Lim 1993, Beasley & Cram 1993). With an age of $\sim$5–8 $\times$ 10^7 years, it may indicate that stars approaching the main sequence go through a regime of strongly enhanced magnetic activity. The AB Dor pair is a member of the Local Association, also known as the Pleiades Moving Group, a star stream of an age ($\sim$50–100 Myr) corresponding to near-ZAMS for F–K stars. A handful of further stream members with vigorous radio emission are now known in the solar neighborhood, most notably PZ Tel (cited in Lim & White 1995), HD 197890 (K1 V) (Robinson et al. 1994), HD 82558 (K1 V) (Drake et al. 1990), EK Dra (G0 V) (Güdel et al. 1994, 1995c), and 47 Cas (F0 V + G V) (Güdel et al. 1995b, 1998).

These objects are analogs of young open-cluster stars with the observational advantage of being much closer. Clusters, however, are preferred when more precise stellar ages or large statistics are required. Some young clusters house a select group of ultrafast rotators, stars at the extreme of dynamo operation with rotation periods $\lesssim$1 day. The absence of any detections in Bastian et al.'s (1988) Pleiades radio survey suggests that their flare properties are similar to solar neighborhood stars rather than to the much more energetic outbursts occasionally seen in star-forming regions. Flaring or quiescent radio emissions have been detected from G–K-type members of both the Pleiades (Lim & White 1995) and the Hyades (White et al. 1993, Güdel et al. 1996a), although these examples, in part ultrafast rotators at saturated activity levels and in part binary systems, clearly represent only the extreme upper end of magnetic activity. Radio emission of normal, single solar analogs rapidly declines to undetectable levels after a few hundred million years (Gaidos et al. 2000).

Herbig Ae/Be Stars

The evolution of intermediate-mass (3–20 $M_\odot$) stars is quite different from, and much faster than, that of low-mass stars, as they still accrete while already on the main sequence. Given their intermediate spectral range, it is of great interest to know whether Herbig Ae/Be pre–main sequence stars support convective, magnetic dynamos or whether they resemble more massive wind sources. Radio emission is the ideal discriminator. A wind-mass loss interpretation is compatible with the expected mass-loss rates of 10^{-6}–10^{-8} $M_\odot$ yr^{-1} (Güdel et al. 1989b, Skinner et al. 1993). This interpretation is supported by the large radio sizes (order of 1″) and the absence of circular polarization or strong variability. The radio luminosity is also correlated with the stellar temperature and bolometric luminosity. This is expected because wind mass-loss rates increase toward higher stellar masses (Skinner et al. 1993). New radio observations complemented by millimeter measurements further indicate the presence of substantial dust envelopes (Di Francesco et al. 1997).

Nonthermal gyrosynchrotron sources exist as well among Herbig stars, although they are the exception (Skinner et al. 1993). The evidence comes primarily from negative spectral indices, including the extremely X-ray–strong proto-Herbig star EC95 = S68-2 that further supports a coronal model based on its L_X/L_R ratio (Smith et al. 1999).

STARS AT THE INTERFACE BETWEEN
HOT WINDS AND CORONAE

Owing to missing convection, no dynamo-generated magnetic fields are expected on stars earlier than spectral type F, nor should there be massive ionized winds on main-sequence stars of spectral type B and later, given the weak radiation pressure. Indeed, few detections have been reported as early as spectral type F0, although some have been found at quite high luminosities compatible with the gyrosynchrotron mechanism, including main-sequence candidates (Güdel et al. 1995a) and the supergiant α Car (Slee & Budding 1995). Brown et al. (1990) surveyed a number of normal A-type stars and found stringent upper limits to any radio emission and thus to the mass-loss rate—very much in agreement with expectations. However, as it turns out, this spectral range is shared by some of the more provocative radio detections.

Many magnetic chemically peculiar Bp/Ap stars maintain considerable radio emission ($\log L_R \approx 15$–18) (Drake et al. 1987a), including very young objects such as S1 in ρ Oph (André et al. 1988). They all relate to the hotter (O9–A0 spectral type) He-strong and He-weak/Si-strong classes, whereas the cooler (A-type) SrCrEu peculiarity classes remain undetected (Drake et al. 1987a, Willson et al. 1988, Linsky et al. 1992, Leone et al. 1994). Similarly, nonmagnetic Am and HgMn stars remain undetected despite recent claims that these stars may have magnetic fields as well (Drake et al. 1994). The emission mechanism is likely to be gyrosynchrotron, as judged from flat spectra (Linsky et al. 1992, Leone et al. 1996), high brightness temperature (Phillips & Lestrade 1988), variability, and sometimes moderate polarization (Linsky et al. 1992).

The presence of kilogauss magnetic fields on magnetic chemically peculiar stars has been known since 1947 (Babcock 1947). The fields are generally assumed to be of global, dipolar topology. There is little photospheric motion that could stir magnetic footpoints, but weak winds could draw the magnetic field lines into an equatorial current sheet, thus producing a global "van Allen Belt" magnetospheric structure as described in "Magnetospheric Models," above. Estimated nonthermal source radii are a few stellar radii, confirmed by VLBI observations (Phillips & Lestrade 1988, André et al. 1991) that also conclusively established the nonthermal nature of the radio emission, with brightness temperatures of $T_b \approx 10^8$–10^9 K.

The wind-controlled magnetospheric model is further supported qualitatively by a parameter dependence of the form $L_R \propto \dot{M}^{0.38} B_{\mathrm{rms}}^{1.06} P_{\mathrm{rot}}^{-0.32}$, where $\dot{M}$ is the estimated wind mass-loss rate for the spectral type, B_{rms} is the root-mean square value of the longitudinal magnetic field, and P_{rot} is the stellar rotation period (the dependence on the latter is marginal) (Linsky et al. 1992). Evidence for rotational modulation probably owing to field misalignment has been found by Leone (1991), Leone & Umana (1993), and Lim et al. (1996a), the latter authors reporting modulation of the polarization degree and sign. A surprise detection was the phase-dependent 100% polarized radio emission from a Bp star that suggests a strongly beamed, continuously radiating electron cyclotron maser (Trigilio et al. 2000).

A binary class in this spectral range that is thought to be intermediate in evolution between the young semidetached B-type β Lyr system that shows radio evidence for large systemic wind mass loss (Umana et al. 2000) and evolved normal Algol binaries with nonthermal coronal emission is defined by the rather inhomogeneous sample of Serpentid stars or Peculiar Emission Line Algols (PELAs). These typically consist of an A–B-type primary and an F–K-type companion, with strong mass transfer into a geometrically thick accretion disk around the early-type star, and a common thin envelope. They have shown an appreciable level of radio emission, first thought to be gyrosynchrotron emission based on their large luminosities (Elias 1990) but later suspected to be wind sources, given their spectral indices close to the standard wind law (Elias & Güdel 1993).

RADIO EMISSION FROM BROWN DWARFS

Brown dwarf stars have masses below the critical mass of $\approx 0.08\ M_\odot$ required for stable hydrogen burning in the stellar core. Detecting radio emission that shows variability, polarization, or gyrosynchrotron-like spectra from brown dwarfs would provide very strong evidence for the existence of surface magnetic fields. However, after brown dwarfs cease to burn deuterium, the internal convection decreases, probably diminishing the generation of magnetic fields (Neuhäuser et al. 1999). A deep 3.6-cm survey with the VLA, concentrating on older targets, did indeed fail to produce radio detections (Krishnamurthi et al. 1999). One young candidate brown dwarf has been reported with strongly variable radio emission (P. André, cited in Wilking et al. 1999, Neuhäuser et al. 1999). But again, nature offered more riddles: The first radio-detected bona-fide brown dwarf, LP944-20, is in fact quite old (~ 500 Myr), but showed both flaring and quiescent episodes (Berger et al. 2001). Its radio emission is orders of magnitude higher than would be expected from the active-stellar relation (28). In any case, the observed flaring radio emission is a telltale signature of magnetic fields and proves that solar-like coronal activity is present in substellar objects—even old ones.

A promising method for detecting substellar objects with sufficiently strong magnetic fields involves the operation of the cyclotron maser. In fact, Jupiter's decametric radiation can be as strong as solar coherent bursts, although at somewhat longer wavelengths. Searching nearby stars for low-frequency coherent bursts may, by positional analysis, reveal companion planets or brown dwarfs. Such a study was performed by Winglee et al. (1986), albeit with null results. We note in passing that one of their targets, Gl 411/Lalande 21185, has in the meantime been shown by astrometric means to possess at least one planetary companion (Gatewood 1996).

STELLAR ASTROMETRY

VLBI techniques have been used for astrometric purposes, i.e., measurements of positions, proper motions, binary orbital motions, and parallaxes, to an accuracy of fractions of a milliarcsecond (Lestrade et al. 1990, 1993, 1995, 1999).

Whereas one fundamental purpose of such programs is to establish the astrometric link between the radio and the optical coordinate reference frames that is eventually of great importance for numerous dynamical studies, invaluable astrophysical spin-offs were obtained. Astrometry of the Algol triple system (a 2.9-day period A7m + K0 IV close binary with mass exchange, in a wide 680-day orbit around a B8 V star) showed unequivocally that the radio emission is related to the magnetically active K subgiant. The most startling result (confirming earlier, indirect optical polarization measurements) was that the planes of the short and the long orbits are orthogonal to each other (Lestrade et al. 1993, 1999), a challenge for stability theories. Further important highlights include (*a*) precise measurements of distances through parallaxes, including distances to star-forming regions; (*b*) identification of binarity or multiple systems; and (*c*) rapidly moving ejecta after a stellar flare (see Lestrade et al. 1999 for details). A dedicated pre-Hipparcos VLBI study of the AB Doradus system solved for both the accurate distance (moving the star to two thirds its previously assumed distance and placing it on the ZAMS) and a short-period disturbance attributed to the presence of a low-mass object (possibly a brown dwarf) in orbit around this star (Guirado et al. 1997).

SUMMARY AND CONCLUSIONS

Stellar radio astronomy has matured over the past few decades to a science that is indispensable for our understanding of stellar atmospheres. Historical milestones include, among many others, (*a*) the discovery of steady and flaring nonthermal and polarized emission in cool stars, testifying to the importance of highly energetic processes; (*b*) the recognition that these phenomena are ubiquitous in many classes of convective-envelope stars; (*c*) observations of very large, apparently stable magnetospheric structures, unlike anything known from the Sun, around various types of magnetically active stars such as T Tau stars, Bp stars, dMe stars, or RS CVn binaries; (*d*) the discovery of nonthermal emission produced in (wind-collision) shocks of hot-star atmospheres; (*e*) gyromagnetic and flaring emission from deeply embedded protostellar objects, testifying to the importance of magnetic fields back to the earliest moments of a stellar life; and (*f*) flaring radio emission from substellar objects not previously thought to support stellar-like convective outer envelopes. Radio methodology has become a standard to estimate magnetic fields in cool stars, to determine mass loss in stars with ionized winds, to spatially resolve and map structures at the milliarcsecond level, and to simply prove the presence of magnetic fields through polarization measurements.

Far from being an auxiliary science to research at other wavelengths, stellar radio astronomy should prepare to address outstanding problems to which it has unique access, although more sensitive instruments are needed. Questions of particular interest include, Are there relevant high-energy processes and magnetic fields in class 0 protostars? Are accretion processes important for the high-energy mechanisms and the generation of large-scale magnetic fields? Are there magnetic fields in hot stars, and what role do they play in the winds? Are brown dwarfs

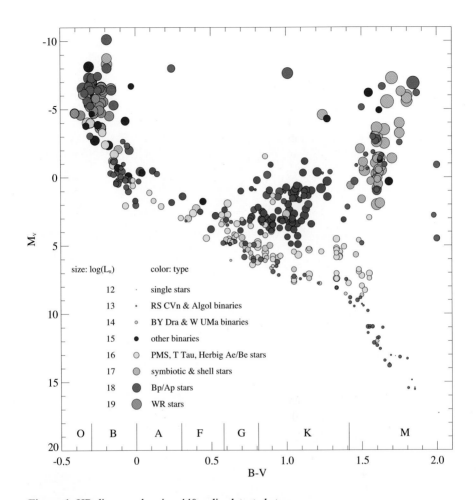

Figure 1 HR diagram showing 440 radio-detected stars.

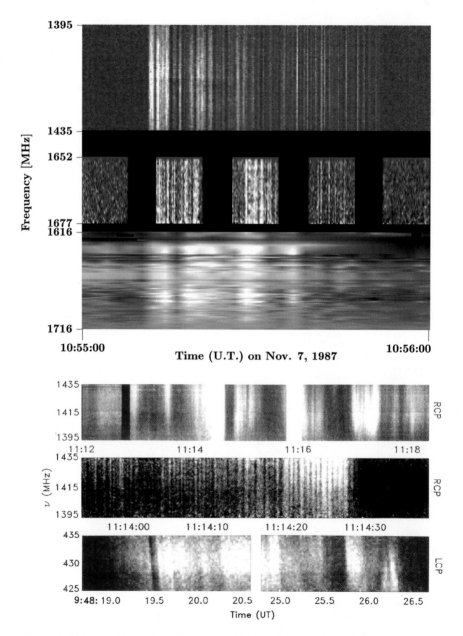

Figure 2 Gallery of radio dynamic spectra of M dwarf flares. Upper three panels show a flare on AD Leo, recorded with the Arecibo (*top*), Effelsberg (*middle*), and Jodrell Bank (*bottom*) telescopes in different wavelength ranges (see also Güdel et al. 1989a). Bottom three panels show flares on AD Leo (*top* and *middle*) and YZ CMi (*bottom*), observed at Arecibo (after Bastian et al. 1990). Reproduced with permission of the AAS.

usually quiescent radio emitters? Do they maintain stable magnetic fields? What is the structure of their coronae? Are there intrabinary magnetic fields in close binary stars? How do large magnetospheres couple to the more compact X-ray coronae? Are quiescent coronae fed by numerous (micro) flares?

ACKNOWLEDGMENTS

It is a pleasure to thank Marc Audard, Arnold Benz, Stephen Skinner, Kester Smith, and Stephen White for helpful discussions, and Philippe André, Tim Bastian, Arnold Benz, Maria Contreras, Jeremy Lim, Robert Mutel, Tom Ray, Luis Rodríguez, and Stephen White for providing figure material. The introductory Einstein quotation is cited from "Einstein sagt," ed. A. Calaprice, A. Ehlers (1997, Munich: Piper) and from "The Expanded Quotable Einstein," ed. A Calaprice (2000, Princeton: Princeton Univ. Press).

The *Annual Review of Astronomy and Astrophysics* is online at
http://astro.annualreviews.org

LITERATURE CITED

Abada-Simon M, Lecacheux A, Aubier M, Bookbinder JA. 1997. *Astron. Astrophys.* 321:841–49

Abada-Simon M, Lecacheux A, Louarn P, Dulk GA, Belkora L, et al. 1994. *Astron. Astrophys.* 288:219–30

Abbott DC, Bieging JH, Churchwell E. 1984. *Ap. J.* 280:671–78

Abbott DC, Bieging JH, Churchwell E, Torres AV. 1986. *Ap. J.* 303:239–61

Airapetian VS, Holman GD. 1998. *Ap. J.* 501:805–12

Alef W, Benz AO, Güdel M. 1997. *Astron. Astrophys.* 317:707–11

André P. 1996. In *Radio Emission from the Stars and the Sun*, ed. AR Taylor, JM Paredes, ASP Conf. Ser. 93, pp. 273–84. San Francisco: ASP

André P, Deeney BD, Phillips RB, Lestrade J-F. 1992. *Ap. J.* 401:667–77

André P, Montmerle T, Feigelson ED. 1987. *Astron. J.* 93:1182–98

André P, Montmerle T, Feigelson ED, Stine PC, Klein K-L. 1988. *Ap. J.* 335:940–52

André P, Motte F, Bacmann A. 1999. *Ap. J. Lett.* 513:L57–60

André P, Phillips RB, Lestrade J-F, Klein K-L. 1991. *Ap. J.* 376:630–35

André P, Ward-Thompson D, Barsony M. 1993. *Ap. J.* 406:122–41

Anglada G. 1995. *Rev. Mex. Astron. Astrofis.* 1:67–76

Anglada G. 1996. See André 1996, pp. 3–14

Anglada G, Villuendas E, Estalella R, Beltrán MT, Rodríguez LF, et al. 1998. *Astron. J.* 116:2953–64

Aschwanden MJ. 1987. *Sol. Phys.* 111:113–36

Ayres TR, Brown A, Osten RA, Huenemoerder DP, Drake JJ, et al. 2001. *Ap. J.* 549:554–77

Babcock HW. 1947. *Ap. J.* 105:105–19

Baliunas SL, Donahue RA, Soon W, Henry GW. 1998. In *10th Cambridge Workshop on Cool Stars, Stellar Systems, and the Sun*, ed. R Donahue, JA Bookbinder, pp. 153–72. San Francisco: ASP

Bastian TS. 1990. *Sol. Phys.* 130:265–94

Bastian TS. 1996. See André 1996, pp. 447–54

Bastian TS, Benz AO, Gary DE. 1998. *Annu. Rev. Astron. Astrophys.* 36:131–88

Bastian TS, Bookbinder JA. 1987. *Nature* 326:678–80

Bastian TS, Bookbinder J, Dulk GA, Davis M. 1990. *Ap. J.* 353:265–73

Bastian TS, Dulk GA, Slee OB. 1988. *Astron. J.* 95:794–803

Beasley AJ, Ball LT, Budding E, Slee OB, Stewart RT. 1993. *Astron. J.* 106:1656–59

Beasley AJ, Cram LE. 1993. *MNRAS* 264:570–78

Beasley AJ, Güdel M. 2000. *Ap. J.* 529:961–67

Becker RH, White RL. 1985. *Ap. J.* 297:649–51

Bell AR. 1978. *MNRAS* 182:147–56

Benz AO. 2002. *Plasma Astrophysics.* Dordrecht, The Netherlands: Kluwer

Benz AO, Alef W. 1991. *Astron. Astrophys. Lett.* 252:L19–22

Benz AO, Alef W, Güdel M. 1995. *Astron. Astrophys.* 298:187–92

Benz AO, Conway J, Güdel M. 1998. *Astron. Astrophys.* 331:596–600

Benz AO, Güdel M. 1994. *Astron. Astrophys.* 285:621–30

Benz AO, Magun A, Stehling W, Hai S. 1992. *Sol. Phys.* 141:335–46

Berger E, Ball S, Becker KM, Clarke M, Frail DA, et al. 2001. *Nature* 410:338–40

Bieging JH, Abbott DC, Churchwell EB. 1982. *Ap. J.* 263:207–14

Bieging JH, Abbott DC, Churchwell EB. 1989. *Ap. J.* 340:518–36

Bieging JH, Cohen M. 1985. *Ap. J. Lett.* 289: L5–8

Bieging JH, Cohen M. 1989. *Astron. J.* 98: 1686–92

Bieging JH, Cohen M, Schwartz PR. 1984. *Ap. J.* 282:699–708

Bontemps S, André P, Ward-Thompson D. 1995. *Astron. Astrophys.* 297:98–102

Bookbinder JA. 1988. In *Activity in Cool Stars Envelopes*, ed. O Havnes, BR Pettersen, JHMM Schmitt, JE Solheim, pp. 257–67. Dordrecht, The Netherlands: Kluwer

Bookbinder JA. 1991. *Mem. Soc. Astron. Ital.* 62:321–36

Borghi S, Chiuderi Drago F. 1985. *Astron. Astrophys.* 143:226–30

Brown A. 1987. *Ap. J. Lett.* 322:L31–34

Brown A, Osten RA, Drake SA, Jones KL, Stern RA. 1998. In *The Hot Universe*, ed. K Koyama, S Kitamoto, M Itoh, IAU Symp 188, pp. 215–16. Dordrecht, The Netherlands: Kluwer

Brown A, Vealé A, Judge P, Bookbinder JA, Hubeny I. 1990. *Ap. J.* 361:220–24

Brown RL, Crane PC. 1978. *Astron. J.* 83: 1504–9

Budding E, Jones KL, Slee OB, Watson L. 1999. *MNRAS* 305:966–76

Bunton JD, Large MI, Slee OB, Stewart RT, Robinson RD, Thatcher JD. 1989. *Proc. Astron. Soc. Aust.* 8:127–31

Caillault J-P, Chanan GA, Helfand DJ, Patterson J, Nousek JA, et al. 1985. *Nature* 313:376–78

Caillault J-P, Drake S, Florkowski D. 1988. *Astron. J.* 95:887–93

Carkner L, Mamajek E, Feigelson E, Neuhäuser R, Wichmann R, Krautter J. 1997. *Ap. J.* 490:735–43

Chapman JM, Leitherer C, Koribalski B, Bouter R, Storey M. 1999. *Ap. J.* 518:890–900

Chiang E, Phillips RB, Lonsdale CJ. 1996. *Astron. J.* 111:355–64

Chiuderi Drago F, Franciosini E. 1993. *Ap. J.* 410:301–8

Churchwell E, Bieging JH, van der Hucht KA, Williams PM, Spoelstra TATh, Abbott DC. 1992. *Ap. J.* 393:329–40

Cohen M, Bieging JH. 1986. *Astron. J.* 92: 1396–402

Cohen M, Bieging JH, Schwartz PR. 1982. *Ap. J.* 253:707–15

Contreras ME, Rodríguez LF. 1999. *Ap. J.* 512:762–66

Contreras ME, Rodríguez LF, Gómez Y, Velázquez A. 1996. *Ap. J.* 469:329–35

Contreras ME, Rodríguez LF, Tapia M, Cardini D, et al. 1997. *Ap. J. Lett.* 488:L153–56

Cox JJ, Gibson DM. 1985. In *Radio Stars*, ed. RM Hjellming, DM Gibson, pp. 233–36. Dordrecht, The Netherlands: Reidel

Curiel S, Cantó J, Rodríguez LF. 1987. *Rev. Mex. Astron. Astrofis.* 14:595–602

Curiel S, Raymond JC, Rodríguez LF, Cantó J, Moran JM. 1990. *Ap. J.* 365:L85–88

Curiel S, Rodríguez LF, Bohigas J, Roth M,

Cantó J. 1989. *Astrophys. Lett. Commun.* 27: 299–309

Curiel S, Rodríguez LF, Moran JM, Cantó J. 1993. *Ap. J.* 415:191–203

Dempsey RC, Neff JE, Lim J. 2001. *Ap. J.* 122: 332–48

Dennis BR, Zarro DM. 1993. *Sol. Phys.* 146: 177–90

Di Francesco J, Evans NJ II, Harvey PM, Mundy LG, Guilloteau S, Chandler CJ. 1997. *Ap. J.* 482:433–41

Doiron DJ, Mutel RL. 1984. *Astron. J.* 89:430–32

Dougherty SM, Taylor AR, Waters LBFM. 1991. *Astron. Astrophys.* 248:175–78

Dougherty SM, Williams PM. 2000. *MNRAS* 319:1005–10

Dougherty SM, Williams PM, Pollacco DL. 2000. *MNRAS* 316:143–51

Dougherty SM, Williams PM, van der Hucht KA, Bode MF, Davis RJ. 1996. *MNRAS* 280:963–70

Doyle JG, Mathioudakis M. 1991. *Astron. Astrophys. Lett.* 241:L41–42

Drake SA. 1990. *Astron. J.* 100:572–78

Drake SA, Abbott DC, Bastian TS, Bieging JH, Churchwell E, et al. 1987a. *Ap. J.* 322:902–8

Drake SA, Johnson HR, Brown A. 1991a. *Astron. J.* 101:1483–88

Drake SA, Linsky JL. 1986. *Astron. J.* 91:602–20

Drake SA, Linsky JL. 1989. *Astron. J.* 98:1831–41

Drake SA, Linsky JL, Bookbinder JA. 1994. *Astron. J.* 108:2203–6

Drake SA, Linsky JL, Elitzur M. 1987b. *Astron. J.* 94:1280–90

Drake SA, Linsky JL, Judge PG, Elitzur M. 1991b. *Astron. J.* 101:230–36

Drake SA, Simon T, Brown A. 1993. *Ap. J.* 406:247–51

Drake SA, Simon T, Linsky JL. 1986. *Astron. J.* 91:1229–32

Drake SA, Simon T, Linsky JL. 1987c. *Astron. J.* 92:163–67

Drake SA, Simon T, Linsky JL. 1989. *Ap. J. Suppl.* 71:905–30

Drake SA, Simon T, Linsky JL. 1992. *Ap. J. Suppl.* 82:311–21

Drake SA, Walter FM, Florkowski DR. 1990. In *6th Cambridge Workshop on Cool Stars, Stellar Systems, and the Sun*, ed. G Wallerstein, pp. 148–51. San Francisco: ASP

Dulk GA. 1985. *Annu. Rev. Astron. Astrophys.* 23:169–224

Dulk GA. 1987. In *5th Cambridge Workshop on Cool Stars, Stellar Systems, and the Sun*, ed. JL Linsky, RE Stencel, pp. 72–82. Berlin: Springer

Dulk GA, Marsh KA. 1982. *Ap. J.* 259:350–58

Eichler D, Usov V. 1993. *Ap. J.* 402:271–79

Elias NM II. 1990. *Ap. J.* 352:300–2

Elias NM II, Güdel M. 1993. *Astron. J.* 106: 337–41

Elias NM II, Quirrenbach A, Witzel A, Naundorf CE, Wegner R, et al. 1995. *Ap. J.* 439: 983–90

Estalella R, Anglada G, Rodríguez LF, Garay G. 1991. *Ap. J.* 371:626–30

Feigelson ED, Carkner L, Wilking BA. 1998. *Ap. J. Lett.* 494:L215–18

Feigelson ED, Montmerle T. 1985. *Ap. J. Lett.* 289:L19–23

Feldman PA, Taylor AR, Gregory PC, Seaquist ER, Balonek TJ, Cohen NL. 1978. *Astron. J.* 83:1471–84

Felli M, Massi M. 1991. *Astron. Astrophys.* 246:503–6

Felli M, Massi M, Catarzi M. 1991. *Astron. Astrophys.* 248:453–57

Felli M, Massi M, Churchwell E. 1989. *Astron. Astrophys.* 217:179–86

Felli M, Taylor GB, Catarzi M, Churchwell E, Kurtz S. 1993. *Astron. Astrophys. Suppl.* 101:127–51

Felli M, Taylor GB, Neckel Th, Staude HJ. 1998. *Astron. Astrophys.* 329:243–48

Fox DC, Linsky JL, Vealé A, Dempsey RC, Brown A, et al. 1994. *Astron. Astrophys.* 284:91–104

Franciosini E, Chiuderi Drago F. 1995. *Astron. Astrophys.* 297:535–42

Franciosini E, Massi M, Paredes JM, Estalella R. 1999. *Astron. Astrophys.* 341:595–601

Gaidos E, Güdel M, Blake GA. 2000. *Geophys. Res. Lett.* 27:501–3

Garay G, Moran JM, Reid MJ. 1987. *Ap. J.* 314:535–50

Garay G, Ramírez S, Rodríguez LF, Curiel S, Torrelles JM. 1996. *Ap. J.* 459:193–208

Gary DE, Linsky JL. 1981. *Ap. J.* 250:284–92

Gary DE, Linsky JL, Dulk GA. 1982. *Ap. J. Lett.* 263:L79–83

Gatewood G. 1996. *Bull. Am. Astron. Soc.* 188:40.11 (Abstr)

Gibb AG. 1999. *MNRAS* 304:1–7

Gibson DM. 1983. In *3rd Cambridge Workshop on Cool Stars, Stellar Systems, and the Sun*, ed. SL Baliunas, L Hartmann, pp. 197–201. Berlin: Springer-Verlag

Gómez Y, Rodríguez LF, Garay G. 2000. *Ap. J.* 531:861–67

Güdel M. 1994. *Ap. J. Suppl.* 90:743–51

Güdel M, Audard M, Smith KW, Behar E, Beasley AJ, Mewe R. 2002. *Ap. J.* In press

Güdel M, Benz AO. 1989. *Astron. Astrophys. Lett.* 211:L5–8

Güdel M, Benz AO. 1993. *Ap. J. Lett.* 405:L63–66

Güdel M, Benz AO, Bastian TS, Fürst E, Simnett GM, Davis RJ. 1989a. *Astron. Astrophys. Lett.* 220:L5–8

Güdel M, Benz AO, Catala C, Praderie F. 1989b. *Astron. Astrophys. Lett.* 217:L9–12

Güdel M, Benz AO, Guinan EF, Schmitt JHMM. 1996a. See André 1996, pp. 306–8

Güdel M, Benz AO, Schmitt JHMM, Skinner SL. 1996b. *Ap. J.* 471:1002–14

Güdel M, Guinan EF, Etzel PB, Mewe R, Kaastra JS, Skinner SL. 1998. See Baliunas et al. 1998, pp. 1247–56

Güdel M, Schmitt JHMM, Benz AO. 1994. *Science* 265:933–35

Güdel M, Schmitt JHMM, Benz AO. 1995a. *Astron. Astrophys.* 302:775–87

Güdel M, Schmitt JHMM, Benz AO. 1995b. *Astron. Astrophys. Lett.* 293:L49–52

Güdel M, Schmitt JHMM, Benz AO, Elias NM II. 1995c. *Astron. Astrophys.* 301:201–12

Güdel M, Schmitt JHMM, Bookbinder JA, Fleming TA. 1993. *Ap. J.* 415:236–39

Guirado JC, Reynolds JE, Lestrade J-F, Preston RA, Jauncey DL, et al. 1997. *Ap. J.* 490:835–39

Gunn AG, Brady PA, Migenes V, Spencer RE, Doyle JG. 1999. *MNRAS* 304:611–21

Gunn AG, Migenes V, Doyle JG, Spencer RE, Mathioudakis M. 1997. *MNRAS* 287:199–210

Havnes O, Goertz CK. 1984. *Astron. Astrophys.* 138:421–30

Hawley SL, Fisher GH, Simon T, Cully SL, Deustua SE, et al. 1995. *Ap. J.* 453:464–79

Helfand DJ, Schnee S, Becker RH, White RL, McMahon RG. 1999. *Astron. J.* 117:1568–77

Hjellming RM. 1988. In *Galactic and Extragalactic Radio Astronomy*, ed. GL Verschur, KI Kellermann, pp. 381–438. Berlin: Springer-Verlag

Hjellming RM, Gibson DM, eds. 1985. *Radio Stars*. Dordrecht: Reidel. 411 pp.

Holman GD. 1985. *Ap. J.* 293:584–94

Holman GD. 1986. In *4th Cambridge Workshop on Cool Stars, Stellar Systems, and the Sun*, ed. M Zeilik, DM Gibson, pp. 271–74. Berlin: Springer-Verlag

Hughes VA. 1988. *Ap. J.* 333:788–800

Hughes VA. 1991. *Ap. J.* 383:280–88

Hughes VA. 1997. *Ap. J.* 481:857–65

Hughes VA, Cohen RJ, Garrington S. 1995. *MNRAS* 272:469–80

Hughes VA, McLean BJ. 1984. *Ap. J.* 278:716–20

Hughes VA, McLean BJ. 1987. *Ap. J.* 313:263–67

Jackson PD, Kundu MR, White SM. 1987. *Ap. J. Lett.* 316:L85–90

Jackson PD, Kundu MR, White SM. 1989. *Astron. Astrophys.* 210:284–94

Johns-Krull CM, Valenti JA, Hatzes AP, Kanaan A. 1999. *Ap. J.* 510:L41–44

Jones KL, Brown A, Stewart RT, Slee OB. 1996. *MNRAS* 283:1331–39

Jones KL, Stewart RT, Nelson GJ. 1995. *MNRAS* 274:711–16

Jones KL, Stewart RT, Nelson GJ, Duncan AR. 1994. *MNRAS* 269:1145–51

Kellermann KI, Pauliny-Toth IIK. 1969. *Ap. J. Lett.* 155:L71–78

Klein K-L. 1987. *Astron. Astrophys.* 183:341–50

Klein K-L, Chiuderi-Drago F. 1987. *Astron. Astrophys.* 175:179–85

Knapp GR, Bowers PF, Young K, Phillips TG. 1995. *Ap. J.* 455:293–99

Krishnamurthi A, Leto G, Linsky JL. 1999. *Astron. J.* 118:1369–72

Kuijpers J. 1985. See Cox & Gibson 1985, pp. 3–31

Kuijpers J. 1989a. *Sol. Phys.* 121:163–85

Kuijpers J. 1989b. In *Plasma Phenomena in the Solar Atmosphere*, ed. MA Dubois, F Bely-Dubau, D Gresillion, pp. 17–31. Les Ulis, France: Les Editions de Physique

Kuijpers J, van der Hulst JM. 1985. *Astron. Astrophys.* 149:343–50

Kundu MR, Jackson PD, White SM, Melozzi M. 1987. *Ap. J.* 312:822–29

Kundu MR, Pallavicini R, White SM, Jackson PD. 1988. *Astron. Astrophys.* 195:159–71

Kundu MR, Shevgaonkar RK. 1985. *Ap. J.* 297:644–48

Kundu MR, Shevgaonkar RK. 1988. *Ap. J.* 334:1001–7

Kutner ML, Rydgren AE, Vrba FJ. 1986. *Astron. J.* 92:895–97

Lang KR. 1990. In *Flare Stars in Star Clusters, Associations and the Solar Vicinity*, ed. LV Mirzoyan, BR Pettersen, MK Tsvetkov, IAU Symp. 137, pp. 125–37. Dordrecht, The Netherlands: Kluwer

Lang KR. 1994. *Ap. J. Suppl.* 90:753–64

Lang KR. 1999. *Astrophysical Formulae*, Vol I. Berlin: Springer Verlag. 614 pp.

Lang KR, Bookbinder J, Golub L, Davis MM. 1983. *Ap. J. Lett.* 272:L15–18

Lang KR, Willson RF. 1986a. *Ap. J. Lett.* 302:L17–21

Lang KR, Willson RF. 1986b. *Ap. J.* 305:363–68

Lang KR, Willson RF. 1988a. *Ap. J.* 326:300–4

Lang KR, Willson RF. 1988b. *Ap. J.* 328:610–16

Large MI, Beasley AJ, Stewart RT, Vaughan AE. 1989. *Proc. Astron. Soc. Aust.* 8:123–26

Lebach DE, Ratner MI, Shapiro II, Ransom RR, Bietenholz MF, et al. 1999. *Ap. J. Lett.* 517:L43–46

Lefèvre E, Klein K-L, Lestrade J-F. 1994. *Astron. Astrophys.* 283:483–92

Leitherer C, Chapman JM, Koribalski B. 1995. *Ap. J.* 450:289–301

Leitherer C, Chapman JM, Koribalski B. 1997. *Ap. J.* 481:898–911

Leone F. 1991. *Astron. Astrophys.* 252:198–202

Leone F, Trigilio C, Umana G. 1994. *Astron. Astrophys.* 283:908–10

Leone F, Umana G. 1993. *Astron. Astrophys.* 268:667–70

Leone F, Umana G, Trigilio C. 1996. *Astron. Astrophys.* 310:271–76

Leous JA, Feigelson ED, André P, Montmerle T. 1991. *Ap. J.* 379:683–88

Lestrade J-F. 1997. In *Stellar Surface Structure*, ed. KG Strassmeier, JL Linsky, pp. 173–80. Dordrecht, The Netherlands: Kluwer

Lestrade J-F, Jones DL, Preston RA, Phillips RB, Titus MA, et al. 1995. *Astron. Astrophys.* 304:182–88

Lestrade J-F, Mutel RL, Phillips RB, Webber JC, Niell AE, Preston RA. 1984a. *Ap. J. Lett.* 282:L23–26

Lestrade J-F, Mutel RL, Preston RA, Phillips RB. 1988. *Ap. J.* 328:232–42

Lestrade J-F, Mutel RL, Preston RA, Scheid JA, Phillips RB. 1984b. *Ap. J.* 279:184–87

Lestrade J-F, Phillips RB, Hodges MW, Preston RA. 1993. *Ap. J.* 410:808–14

Lestrade J-F, Preston RA, Jones DL, Phillips RB, Rogers AEE, et al. 1999. *Astron. Astrophys.* 344:1014–26

Lestrade J-F, Rogers AEE, Whitney AR, Niell AE, Phillips RB, Preston RA. 1990. *Astron. J.* 99:1663–73

Leto G, Pagano I, Linsky JL, Rodonò M, Umana G. 2000. *Astron. Astrophys.* 359:1035–41

Lim J. 1993. *Ap. J. Lett.* 405:L33–37

Lim J, Carilli CL, White SM, Beasley AJ, Marson RG. 1998. *Nature* 392:575–77

Lim J, Drake SA, Linsky JL. 1996a. See André 1996, pp. 324–26

Lim J, Nelson GJ, Castro C, Kilkenny D, van Wyk F. 1992. *Ap. J. Lett.* 388:L27–30

Lim J, White SM. 1995. *Ap. J.* 453:207–13

Lim J, White SM. 1996. *Ap. J. Lett.* 462:L91–94

Lim J, White SM, Cully SL. 1996b. *Ap. J.* 461:1009–15

Lim J, White SM, Nelson GJ, Benz AO. 1994. *Ap. J.* 430:332–41

Lim J, White SM, Slee OB. 1996c. *Ap. J.* 460:976–83

Linsky JL. 1996. See André 1996, pp. 439–46

Linsky JL, Drake SA, Bastian TS. 1992. *Ap. J.* 393:341–56

Linsky JL, Gary DE. 1983. *Ap. J.* 274:776–83

Luttermoser DG, Brown A. 1992. *Ap. J.* 384:634–39

Magazzù A, Umana G, Martín EL. 1999. *Astron. Astrophys.* 346:878–82

Mamajek EE, Lawson WA, Feigelson ED. 1999. *Publ. Astron. Soc. Aust.* 16:257–61

Martí J, Rodríguez LF, Reipurth B. 1993. *Ap. J.* 416:208–17

Massi M, Chiuderi-Drago F. 1992. *Astron. Astrophys.* 253:403–6

Massi M, Felli M, Pallavicini R, Tofani G, Palagi F, Catarzi M. 1988. *Astron. Astrophys.* 197:200–4

Massi M, Neidhöfer J, Torricelli-Ciamponi G, Chiuderi-Drago F. 1998. *Astron. Astrophys.* 332:149–54

Melrose DB. 1987. See Dulk 1987, pp. 83–94

Melrose DB. 1989. *Sol. Phys.* 120:369–81

Melrose DB, Dulk GA. 1982. *Ap. J.* 259:844–58

Melrose DB, Dulk GA. 1984. *Ap. J.* 282:308–15

Miralles MP, Rodríguez LF, Tapia M, Roth M, Persi P, et al. 1994. *Astron. Astrophys.* 282:547–53

Moran JP, Davis RJ, Bode MF, Taylor AR, Spencer RE, et al. 1989. *Nature* 340:449–50

Morgan JA, Snell RL, Strom KM. 1990. *Ap. J.* 362:274–83

Morris DH, Mutel RL. 1988. *Astron. J.* 95:204–14

Morris DH, Mutel RL, Su B. 1990. *Ap. J.* 362:299–307

Mullan DJ. 1985. See Cox & Gibson 1985, pp. 173–84

Mullan DJ. 1989. *Sol. Phys.* 121:239–59

Mullan DJ, Doyle JG, Redman RO, Mathioudakis M. 1992. *Ap. J.* 397:225–31

Mutel RL. 1996. In *Magnetodynamic Phenomena in the Solar Atmosphere. Prototypes of Stellar Magnetic Activity*, ed. Y Uchida, T Kosugi, HS Hudson, Proc. IAU Colloquium 153, pp. 71–80. Dordrecht, The Netherlands: Kluwer

Mutel RL, Doiron DJ, Lestrade J-F, Phillips RB. 1984. *Ap. J.* 278:220–23

Mutel RL, Lestrade J-F. 1985. *Astron. J.* 90:493–98

Mutel RL, Lestrade J-F Preston RA, Phillips RB. 1985. *Ap. J.* 289:262–68

Mutel RL, Molnar LA, Waltman EB, Ghigo FD. 1998. *Ap. J.* 507:371–83

Mutel RL, Morris DH, Doiron DJ, Lestrade J-F. 1987. *Astron. J.* 93:1220–28

Mutel RL, Weisberg JM. 1978. *Astron. J.* 83:1499–503

Neidhöfer J, Massi M, Chiuderi-Drago F. 1993. *Astron. Astrophys. Lett.* 278:L51–53

Neuhäuser R, Briceño C, Comerón F, Hearty T, Martín EL, et al. 1999. *Astron. Astrophys.* 343:883–93

Neupert WM. 1968. *Ap. J. Lett.* 153:L59–64

Olnon FM. 1975. *Astron. Astrophys.* 39:217–23

O'Neal D, Feigelson ED, Mathieu RD, Myers PC. 1990. *Astron. J.* 100:1610–17

Osten RA, Brown A, Ayres TR, Linsky JL, Drake SA, et al. 2000. *Ap. J.* 544:953–76

Pallavicini R, Willson RF, Lang KR. 1985. *Astron. Astrophys.* 149:95–101

Panagia N, Felli M. 1975. *Astron. Astrophys.* 39:1–5

Patterson J, Caillault J-P, Skillman DR. 1993. *Publ. Astron. Soc. Pac.* 105:848–52

Persi P, Tapia M, Rodríguez LF, Ferrari-Toniolo M, Roth M. 1990. *Astron. Astrophys.* 240:93–97

Pestalozzi MR, Benz AO, Conway JE, Güdel M. 2000. *Astron. Astrophys.* 353:569–74

Petrosian V. 1985. *Ap. J.* 299:987–93

Phillips RB. 1991. In *7th Cambridge Workshop on Cool Stars, Stellar Systems, and the Sun*, ed. MS Giampapa, JA Bookbinder, pp. 309–18. San Francisco: ASP

Phillips RB, Lestrade J-F. 1988. *Nature* 334: 329–31

Phillips RB, Lonsdale CJ, Feigelson ED. 1991. *Ap. J.* 382:261–69

Phillips RB, Lonsdale CJ, Feigelson ED. 1993. *Ap. J. Lett.* 403:L43–46

Phillips RB, Lonsdale CJ, Feigelson ED, Deeney BD. 1996. *Astron. J.* 111:918–29

Phillips RB, Titus MA. 1990. *Ap. J. Lett.* 359: L15–18

Ray TP, Muxlow TWB, Axon DJ, Brown A, Corcoran D, et al. 1997. *Nature* 385:415–17

Reid MJ, Menten KM. 1997. *Ap. J.* 476:327–46

Reipurth B, Rodríguez LF, Chini R. 1999. *Astron. J.* 118:983–89

Robinson PA, Melrose DB. 1984. *Aust. J. Phys.* 37:675–704

Robinson RD, Carpenter KG, Slee OB, Nelson GJ, Stewart RT. 1994. *MNRAS* 267:918–26

Rodríguez LF, Anglada G, Raga A. 1995. *Ap. J. Lett.* 454:L149–52

Rodríguez LF, Cantó J, Torrelles JM, Gómez JF, Anglada G, Ho PTP. 1994. *Ap. J. Lett.* 427:L103–6

Rodríguez LF, Cantó J, Torrelles JM, Gómez JF, Ho PTP. 1992. *Ap. J. Lett.* 393:L29–31

Rodríguez LF, Ho PTP, Torrelles JM, Curiel S, Cantó J. 1990. *Ap. J.* 352:645–53

Rodríguez LF, Myers PC, Cruz-González I, Tereby S. 1989. *Ap. J.* 347:461–67

Rucinski SM. 1991. *Astron. J.* 101:2199–206

Rucinski SM. 1992. *Publ. Astron. Soc. Pac.* 104:1177–86

Rucinski SM. 1995. *Astron. J.* 109:2690–97

Rucinski SM, Krogulec M, Seaquist ER. 1993. *Astron. J.* 105:2308–18

Rucinski SM, Seaquist ER. 1988. *Astron. J.* 95:1837–40

Saar SH. 1990. In *Solar Photosphere: Structure, Convection, and Magnetic Fields*, ed. JO Stenflo, IAU Symp. 138, pp. 427–41. Dordrecht, The Netherlands: Kluwer

Schwartz PR, Simon T, Campbell R. 1986. *Ap. J.* 303:233–38

Schwartz PR, Simon T, Zuckerman B, Howell RR. 1984. *Ap. J. Lett.* 280:L23–26

Scuderi S, Panagia N, Stanghellini C, Trigilio C,

Umana G. 1998. *Astron. Astrophys.* 332:251–67

Seaquist ER, 1993. *Rep. Prog. Physics* 56(9): 1145–1208

Seaquist ER, Krogulec M, Taylor AR. 1993. *Ap. J.* 410:260–74

Seaquist ER, Taylor AR. 1990. *Ap. J.* 349:313–27

Seaquist ER, Taylor AR, Button S. 1984. *Ap. J.* 284:202–10

Setia Gunawan DYA, de Bruyn AG, van der Hucht KA, Williams PM. 2000. *Astron. Astrophys.* 356:676–90

Shu FH, Adams FC, Lizano S. 1987. *Annu. Rev. Astron. Astrophys.* 25:23–81

Simon T, Fekel FC Jr, Gibson DM. 1985. *Ap. J.* 295:153–61

Skinner CJ, Dougherty SM, Meixner M, Bode MF, Davis RJ, et al. 1997. *MNRAS* 288:295–306

Skinner SL. 1991. *Ap. J.* 368:272–78

Skinner SL. 1993. *Ap. J.* 408:660–67

Skinner SL, Brown A. 1994. *Astron. J.* 107: 1461–68

Skinner SL, Brown A, Stewart RT. 1993. *Ap. J. Suppl.* 87:217–65

Skinner SL, Itoh M, Nagase F, Zhekov SA. 1999. *Ap. J.* 524:394–405

Slee OB, Budding E. 1995. *MNRAS* 277:1063–70

Slee OB, Nelson GJ, Innis JL, Stewart RT, Vaughan A, Wright AE. 1986. *Proc. Astron. Soc. Aust.* 6:312–15

Slee OB, Nelson GJ, Stewart RT, Wright AE, Innis JL, et al. 1987a. *MNRAS* 229:659–77

Slee OB, Nelson GJ, Stewart RT, Wright AE, Jauncey DL, et al. 1987b. *MNRAS* 227:467–79

Slee OB, Stewart RT, Nelson GJ, Wright AE, Dulk GA, et al. 1988. *Astro. Lett. Comm.* 27:247–56

Slish VI. 1963. *Nature* 199:682

Smith KW, Güdel M, Benz AO. 1999. *Astron. Astrophys.* 349:475–84

Snell RL, Bally J, Strom SE, Strom KM. 1985. *Ap. J.* 290:587–95

Stern RA, Uchida Y, Walter F, Vilhu O, Hannikainen D, et al. 1992. *Ap. J.* 391:760–72

Stevens IR. 1995. *MNRAS* 277:163–72

Stewart RT, Innis JL, Slee OB, Nelson GJ, Wright AE. 1988. *Astron. J.* 96:371–77

Stine PC, Feigelson ED, André P, Montmerle T. 1988. *Astron. J.* 96:1394–406

Stine PC, O'Neal D. 1998. *Astron. J.* 116:890–94

Storey MC, Hewitt RG. 1996. *Publ. Astron. Soc. Aust.* 12:174–79

Suters M, Stewart RT, Brown A, Zealey W. 1996. *Astron. J.* 111:320–26

Taylor AR, Paredes JM, eds. 1996. *Radio Emission from the Stars and the Sun.* San Francisco: ASP. 466 pp.

Taylor AR, Seaquist ER. 1984. *Ap. J.* 286:263–68

Taylor AR, Waters LBFM, Bjorkman KS, Dougherty SM. 1990. *Astron. Astrophys.* 231:453–58

Taylor AR, Waters LBFM, Lamers HJGLM, Persi P, Bjorkman KS. 1987. *MNRAS* 228:811–17

Topka K, Marsh KA. 1982. *Ap. J.* 254:641–45

Torrelles JM, Gómez JF, Rodríguez LF, Ho PTP, Curiel S, Vázquez R. 1997. *Ap. J.* 489:744–52

Torricelli-Ciamponi G, Franciosini E, Massi M, Neidhöfer J. 1998. *Astron. Astrophys.* 333:970–76

Trigilio C, Buemi CS, Umana G, Rodonò M, Leto P, et al. 2001. *Astron. Astrophys.* 373:181–89

Trigilio C, Leto P, Leone F, Umana G, Buemi C. 2000. *Astron. Astrophys.* 362:281–88

Trigilio C, Leto P, Umana G. 1998. *Astron. Astrophys.* 330:1060–66

Trigilio C, Umana G, Migenes V. 1993. *MNRAS* 260:903–7

Umana G, Catalano S, Rodonò M. 1991. *Astron. Astrophys.* 249:217–22

Umana G, Leto P, Trigilio C, Hjellming RM, Catalano S. 1999. *Astron. Astrophys.* 342:709–16

Umana G, Maxted PFL, Trigilio C, Fender RP, Leone F, Yerli SK. 2000. *Astron. Astrophys.* 358:229–32

Umana G, Trigilio C, Catalano S. 1998. *Astron. Astrophys.* 329:1010–18

Umana G, Trigilio C, Hjellming RM, Catalano S, Rodonò M. 1993. *Astron. Astrophys.* 267:126–36

Umana G, Trigilio C, Tumino M, Catalano S, Rodonò M. 1995. *Astron. Astrophys.* 298:143–50

van den Oord GHJ. 1996. See André 1996, pp. 263–72

van den Oord GHJ, de Bruyn AG. 1994. *Astron. Astrophys.* 286:181–93

van den Oord GHJ, Doyle JG. 1997. *Astron. Astrophys.* 319:578–88

van den Oord GHJ, Doyle JG, Rodonò M, Gary DE, Henry GW, et al. 1996. *Astron. Astrophys.* 310:908–22

van den Oord GHJ, Kuijpers J, White NE, van der Hulst JM, Culhane JL. 1989. *Astron. Astrophys.* 209:296–304

Vaughan AE, Large MI. 1986. *MNRAS* 223:399–403

Vilhu O, Caillault J-P, Heise J. 1988. *Ap. J.* 330:922–27

Vilhu O, Rucinski SM. 1983. *Astron. Astrophys.* 127:5–14

Vilhu O, Tsuru T, Collier Cameron A, Budding E, Banks T, et al. 1993. *Astron. Astrophys.* 278:467–77

Waldron WL, Corcoran MF, Drake SA, Smale AP. 1998. *Ap. J. Suppl.* 118:217–38

Wendker HJ. 1995. *Astron. Astrophys. Suppl.* 109:177–79

White RL. 1985. *Ap. J.* 289:698–708

White RL, Becker RH. 1995. *Ap. J.* 451:352–58

White SM. 1996. In *9th Cambridge Workshop on Cool Stars, Stellar Systems, and the Sun,* ed. R Pallavicini, AK Dupree, pp. 21–30. San Francisco: ASP

White SM. 2000. In *Radio Interferometry: The Saga and the Science,* ed. DG Finley, W Miller Goss, NRAO Workshop No. 27, pp. 86–111. Assoc. Univ.

White SM, Franciosini E. 1995. *Ap. J.* 444:342–49

White SM, Jackson PD, Kundu MR. 1989a. *Ap. J. Suppl.* 71:895–904

White SM, Jackson PD, Kundu MR. 1993. *Astron. J.* 105:563–70

White SM, Kundu MR, Jackson PD. 1986. *Ap. J.* 311:814–18

White SM, Kundu MR, Jackson PD. 1989b. *Astron. Astrophys.* 225:112–24

White SM, Kundu MR, Uchida Y, Nitta N. 1990. See Drake et al. 1990, pp. 239–42

White SM, Lim J, Kundu MR. 1994. *Ap. J.* 422:293–303

White SM, Pallavicini R, Kundu MR. 1992a. *Astron. Astrophys.* 257:557–66

White SM, Pallavicini R, Kundu MR. 1992b. *Astron. Astrophys.* 259:149–54

Wilking BA, Greene TP, Meyer MR. 1999. *Astron. J.* 117:469–82

Williams PM, Dougherty SM, Davis RJ, van der Hucht KA, Bode MF, Setia Gunawan DYA. 1997. *MNRAS* 289:10–20

Williams PM, van der Hucht KA, Bouchet P, Spoelstra TATh, Eenens PRJ, et al. 1992. *MNRAS* 258:461–72

Williams PM, van der Hucht KA, Pollock AMT, Florkowski DR, van der Woerd H, Wamsteker WM. 1990. *MNRAS* 243:662–84

Williams PM, van der Hucht KA, Spoelstra TATh. 1994. *Astron. Astrophys.* 291:805–10

Willson RF, Lang KR. 1987. *Ap. J.* 312:278–83

Willson RF, Lang KR, Foster P. 1988. *Astron. Astrophys.* 199:255–61

Wilner DJ, Ho PTP, Rodríguez LF. 1996. *Ap. J. Lett.* 470:L117–21

Winglee RM, Dulk GA, Bastian TS. 1986. *Ap. J. Lett.* 309:L59–62

Wood BE, Linsky JL, Müller H-R, Zank GP. 2001. *Ap. J.* 547:L49–52

Wright AE, Barlow MJ. 1975. *MNRAS* 170:41–51

Yun JL, Moreira MC, Torrelles JM, Afonso JM, Santos NC. 1996. *Astron. J.* 111:841–45

Yusef-Zadeh F, Cornwell TJ, Reipurth B, Roth M. 1990. *Ap. J. Lett.* 348:L61–64

Annu. Rev. Astron. Astrophys. 2002. 40:263–317
doi: 10.1146/annurev.astro.40.060401.093923
Copyright © 2002 by Annual Reviews. All rights reserved

MODIFIED NEWTONIAN DYNAMICS AS AN ALTERNATIVE TO DARK MATTER

Robert H. Sanders

*Kapteyn Astronomical Institute, University of Groningen, Groningen,
The Netherlands; email: sanders@astro.rug.nl*

Stacy S. McGaugh

*Department of Astronomy, University of Maryland, College Park,
Maryland; email: ssm@astro.umd.edu*

Key Words   dark matter, galaxy dynamics, gravitational theory, cosmology

■ **Abstract**   Modified Newtonian dynamics (MOND) is an empirically motivated modification of Newtonian gravity or inertia suggested by Milgrom as an alternative to cosmic dark matter. The basic idea is that at accelerations below $a_o \approx 10^{-8}$ cm/s$^2 \approx cH_o/6$ the effective gravitational attraction approaches $\sqrt{g_n a_o}$, where g_n is the usual Newtonian acceleration. This simple algorithm yields flat rotation curves for spiral galaxies and a mass-rotation velocity relation of the form $M \propto V^4$ that forms the basis for the observed luminosity–rotation velocity relation—the Tully-Fisher law. We review the phenomenological success of MOND on scales ranging from dwarf spheroidal galaxies to superclusters and demonstrate that the evidence for dark matter can be equally well interpreted as evidence for MOND. We discuss the possible physical basis for an acceleration-based modification of Newtonian dynamics as well as the extention of MOND to cosmology and structure formation.

INTRODUCTION

The appearance of discrepancies between the Newtonian dynamical mass and the directly observable mass in large astronomical systems has two possible explanations: either these systems contain large quantities of unseen matter, or gravity (or the response of particles to gravity) on these scales is not described by Newtonian theory. Most attention has focused on the first of these explanations. An intricate paradigm has been developed in which nonbaryonic dark matter plays a central role—not only in accounting for the traditional dynamical mass of bound gravitational systems (Faber & Gallagher 1979, Blumenthal et al. 1984) but also in promoting the formation of structure via gravitational collapse beginning in the highly homogeneous ionized universe (Peebles 1982, Vittorio & Silk 1984). The paradigm of cold dark matter (CDM) is widely purported to be successful in this cosmological context, particularly in predicting the scale-dependence of density

0066-4146/02/0922-0263$14.00

263

fluctuations. Moreover, with the development of cosmic N-body simulations of high precision and resolution, this hypothesis has gained predictive power on the scale of galaxies—a power that considerably restricts the freedom to arrange dark matter as one would wish in order to explain the form and magnitude of the discrepancy in any particular system (Navarro et al. 1996).

It is in this second aspect, the distribution of dark matter in galactic systems, that the CDM paradigm encounters observational difficulties (McGaugh & de Blok 1998a, Sellwood & Kosowsky 2001). It is not the purpose of this review to discuss the possible problems with the CDM hypothesis. We only comment that, as of the date of this review, candidate dark matter particles have not yet been detected by any means independent of their putative global gravitational effect. So long as this is the only evidence for dark matter, its presumed existence is not independent of the assumed law of gravity or inertia on astronomical scales. If a physical law, when extended to a regime in which it has never before been tested, implies the existence of a medium (e.g., an ether) that cannot be detected by any other means, then it would not seem unreasonable to question that law.

Of course, if one chooses to modify Newtonian dynamics or gravity in an ad hoc fashion, then the set of alternative possibilities is large. It is a simple matter to claim that Newton's law of gravity fails on galactic scales and then to cook up a recipe that explains a particular aspect of the observations—such as flat rotation curves of spiral galaxies. To be credible, an empirically based alternative to dark matter should at least provide a more efficient description of the phenomenology. Any viable alternative should account for various aspects of the observations of astronomical systems (such as global scaling relations) with as few additional parameters as possible. A second, but less immediate, requirement is that the suggested alternative should have some basis in sensible physics—it should make contact with familiar physical principles or at least a reasonable extrapolation of those principles.

To date, the only suggestion that goes some way toward meeting these requirements (particularly the former) is Milgrom's modified Newtonian dynamics (MOND) (Milgrom 1983a,b,c). The empirical successes of this hypothesis on scales ranging from dwarf spheroidal galaxies to super-clusters, and its possible physical basis and extension to a cosmological context, is the subject of this review. It may be argued that this is a speculative topic for review in this series. In our opinion the subject of dark matter (Trimble 1987) is, in the absence of its direct detection, no less speculative, particularly considering that the standard model of particle physics does not predict the existence of candidate dark matter particles with the necessary properties. Reasonable extensions of the standard model (e.g., supersymmetry) can, with an appropriate adjustment of parameters, accommodate such particles, but this also requires an extrapolation of known physics (e.g., Griest et al. 1990). Here we demonstrate that the evidence for dark matter can be equally well interpreted as evidence for modified dynamics.

There have been other attempts to modify gravity in order to account for astronomical mass discrepancies without invoking dark matter. We mention some of these efforts, but it is fair to say that none of these alternatives has enjoyed the phenomenological success of MOND or been as extensively discussed in the literature. A considerable lore on MOND has emerged in the past two decades—with contributions not only by advocates reporting phenomenological successes but also by critics pointing out possible problems or inconsistencies. There have also been several contributions attempting to formulate MOND either as a covariant theory in the spirit of General Relativity, or as a modified particle action (modified inertia). Whereas none of these attempts has, so far, led to anything like a satisfactory or complete theory, they provide some insight into the required properties of generalized theories of gravity and inertia.

In the absence of a complete theory, MOND cannot be unambiguously extended to problems of cosmology and structure formation. However, by making certain reasonable assumptions, one may speculate on the form of a MOND cosmology. We discuss the efforts that have been made in this direction. The general expectation is that, because MOND results in effectively stronger gravity for low peculiar accelerations, the rapid growth of structure is possible even in a low-density purely baryonic Universe.

BASICS OF MODIFIED NEWTONIAN DYNAMICS

An Acceleration Scale

The phenomenological basis of MOND consists of two observational facts about spiral galaxies: (*a*) The rotation curves of spiral galaxies are asymptotically flat (Shostak 1973, Roberts & Whitehurst 1975, Bosma 1978, Rubin et al. 1980), and (*b*) there is a well-defined relationship between the rotation velocity in spiral galaxies and the luminosity—the Tully-Fisher (TF) law (Tully & Fisher 1977, Aaronson et al. 1982). This latter implies a mass-velocity relationship of the form $M \propto V^\alpha$, where α is in the neighborhood of 4.

If one wishes to modify gravity in an ad hoc way to explain flat rotation curves, an obvious first choice would be to propose that gravitational attraction becomes more like $1/r$ beyond some length scale that is comparable to the scale of galaxies. So the modified law of attraction about a point mass M would read

$$F = \frac{GM}{r^2} f(r/r_o), \tag{1}$$

where r_o is a new constant of length on the order of a few kpc, and $f(x)$ is a function with the asymptotic behavior: $f(x) = 1$, where $x \ll 1$ and $f(x) = x$, where $x \gg 1$. Finzi (1963), Tohline (1983), Sanders (1984), and Kuhn & Kruglyak (1987) have proposed variants of this idea. In Sanders' (1984) version the Newtonian potential is modified by including a repulsive Yukawa term ($e^{-r/r_o}/r$), which can yield a flat rotation velocity over some range in r. This idea keeps reemerging with various modern justifications (e.g., Eckhardt 1993, Hadjimichef & Kokubun 1997, Drummond 2001, Dvali et al. 2001).

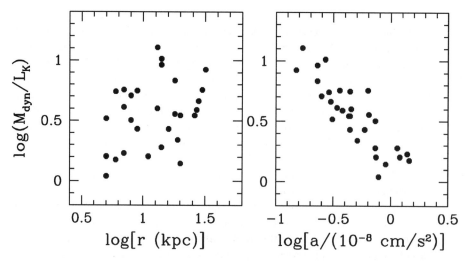

Figure 1   The global Newtonian mass-to-K′-band-luminosity ratio of Ursa Major spirals at the last measured point of the rotation curve plotted first against the radial extent of the rotation curve (*left*) and then against the centripetal acceleration at that point (*right*).

All of these modifications attached to a length scale have one thing in common: Equating the centripetal to the gravitational acceleration in the limit $r > r_o$ would lead to a mass-asymptotic rotation velocity relation of the form $v^2 = GM/r_o$. Milgrom (1983a) realized that this was incompatible with the observed TF law, $L \propto v^4$. Moreover, any modification attached to a length scale would imply that larger galaxies should exhibit a larger discrepancy (Sanders 1986). This is contrary to the observations. There are very small, usually low surface brightness (LSB) galaxies with large discrepancies, and very large high surface brightness (HSB) spiral galaxies with very small discrepancies (McGaugh & de Blok 1998a).

Figure 1 illustrates this. At the left is a log-log plot of the dynamical $M/L_{K'}$ vs. the radius at the last measured point of the rotation curve for a uniform sample of spiral galaxies in the Ursa Major cluster (Tully et al. 1996, Verheijen & Sancisi 2001). The dynamical M/L is calculated simply using the Newtonian formula for the mass v^2r/G (assuming a spherical mass distribution), where r is the radial extent of the rotation curve. Population synthesis studies suggest that $M/L_{K'}$ should be about 1, so anything much above 1 indicates a global discrepancy—a "dark matter problem." It is evident that there is not much of a correlation of M/L with size. On the other hand, the Newtonian M/L plotted against centripetal acceleration (v^2/r) at the last measured point (Figure 1, *right*) looks rather different. There does appear to be a correlation in the sense that $M/L \propto 1/a$ for $a < 10^{-8}$ cm/s². The presence of an acceleration scale in the observations of the discrepancy in spiral galaxies has been pointed out before (Sanders 1990, McGaugh 1998), and as the data have improved, it has become more evident. Any modification of gravity attached to a length scale cannot explain such observations.

Milgrom's insightful deduction was that the only viable sort of modification is one in which a deviation from Newton's law appears at low acceleration. (Data such as that shown in Figure 1 did not exist at the time of Milgrom's initial papers; an acceleration scale was indicated by the slope of the TF relation.) MOND as initially formulated could be viewed as a modification of inertia or of gravity (this dichotomy remains). In the first case the acceleration of a particle with mass m under the influence of an external force would be given by

$$m\mathbf{a}\mu(a/a_o) = \mathbf{F}, \tag{2}$$

where a_o is a new physical parameter with units of acceleration and $\mu(x)$ is a function that is unspecified but must have the asymptotic form $\mu(x) = x$ when $x \ll 1$ and $\mu(x) = 1$ when $x \gg 1$. Viewed as a modification of gravity, the true gravitational acceleration $\mathbf{g}$ is related to the Newtonian gravitational acceleration $\mathbf{g_n}$ as

$$\mathbf{g}\mu(|g|/a_o) = \mathbf{g_n}. \tag{3}$$

Although there are clear differences in principle and practice between these two formulations, the consequence for test particle motion in a gravitational field in the low acceleration regime is the same: The effective gravitational force becomes $g = \sqrt{g_n a_o}$. For a point mass M, if we set g equal to the centripetal acceleration v^2/r, this gives

$$v^4 = GMa_o \tag{4}$$

in the low acceleration regime. Thus, all rotation curves of isolated masses are asymptotically flat, and there is a mass-luminosity relation of the form $M \propto v^4$. These are aspects that are built into MOND, so they cannot rightly be called predictions. However, in the context of MOND, the aspect of an asymptotically flat rotation curve is absolute. MOND leaves rather little room for maneuvering; the idea is in principle falsifiable, or at least it is far more fragile than the dark matter hypothesis. Unambiguous examples of rotation curves of isolated galaxies that decline in a Keplerian fashion at a large distance from the visible object would falsify the idea.

In addition, the mass-rotation velocity relation and implied TF relation is absolute. The TF relation should be the same for different classes of galaxies, independent of surface brightness, and the logarithmic slope (at least of the mass-velocity relation) must be 4.0. Moreover, the relation is essentially one between the total baryonic mass of a galaxy and the asymptotic flat rotational velocity—not the peak rotation velocity but the velocity at large distance. This is the most immediate and most obvious prediction (see McGaugh & de Blok 1998b and McGaugh et al. 2000 for a discussion of these points).

Converting the mass-velocity relation (Equation 4) to the observed luminosity-velocity relation, we find

$$\log(L) = 4\log(v) - \log(Ga_o\langle M/L\rangle). \tag{5}$$

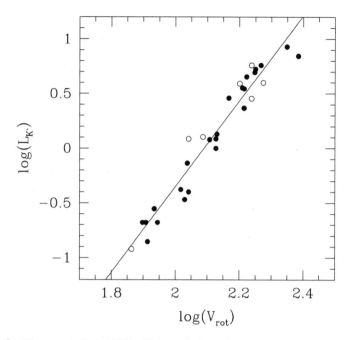

Figure 2 The near-infrared Tully-Fisher relation of Ursa Major spirals (Sanders & Verheijen 1998). The rotation velocity is the asymptotically constant value. The velocity is in units of kilometer/second and luminosity in 10^{10} $L_\odot$. The unshaded points are galaxies with disturbed kinematics. The line is a least-square fit to the data and has a slope of 3.9 ± 0.2.

Figure 2 shows the near-infrared TF relation for Verheijen's UMa sample (Sanders & Verheijen 1998), where the velocity, v, is that of the flat part of the rotation curve. The scatter about the least-square fit line of slope 3.9 ± 0.2 is consistent with observational uncertainties (i.e., no intrinsic scatter). Given the mean M/L in a particular band (≈ 1 in the K' band), this observed TF relation Equation 5 tells us that a_o must be on the order of 10^{-8} cm/s². Milgrom immediately noticed that $a_o \approx cH_o$ to within a factor of 5 or 6. This cosmic coincidence is provocative and suggests that MOND perhaps reflects the effect of cosmology on local particle dynamics.

General Predictions

There are several other direct observational consequences of modified dynamics—all of which Milgrom explored in his original papers—that do fall in the category of predictions in the sense that they are not part of the propositional basis of MOND.

1. There exists a critical value of the surface density

$$\Sigma_m \approx a_o/G. \tag{6}$$

If a system such as a spiral galaxy has a surface density of matter greater than

Σ_m, then the internal accelerations are greater than a_o, so the system is in the Newtonian regime. In systems with $\Sigma \geq \Sigma_m$ (HSB galaxies) there should be a small discrepancy between the visible and classical Newtonian dynamical mass within the optical disk. In the parlance of rotation curve observers, an HSB galaxy should be well represented by the "maximum disk" solution (van Albada & Sancisi 1986), but in LSB galaxies ($\Sigma \ll \Sigma_m$) there is a low internal acceleration, so the discrepancy between the visible and dynamical mass would be large. Milgrom predicted, before the actual discovery of LSB galaxies, that there should be a serious discrepancy between the observable and dynamical mass within the luminous disk of such systems—should they exist. They do exist, and this prediction has been verified, as is evident from the work of McGaugh & de Blok (1998a,b) and de Blok & McGaugh (1998).

2. It is well known since the work of Ostriker & Peebles (1973) that rotationally supported Newtonian systems tend to be unstable to global nonaxisymmetric modes that lead to bar formation and rapid heating of the system. In the context of MOND these systems would be those with $\Sigma > \Sigma_m$, so this would suggest that Σ_m should appear as an upper limit on the surface density of rotationally supported systems. This critical surface density is 0.2 g/cm^2 or 860 M$_\odot$/pc^2. A more appropriate value of the mean surface density within an effective radius would be $\Sigma_m/2\pi$ or 140 M$_\odot/pc^2$, and taking $M/L_b \approx 2$, this would correspond to a surface brightness of about 22 mag/arc sec^2. There is such an observed upper limit on the mean surface brightness of spiral galaxies, and this is known as Freeman's law (Freeman 1970, Allen & Shu 1979). The point is that the existence of such a maximum surface density (McGaugh et al. 1995, McGaugh 1996) is quite natural in the context of MOND but must be put in by hand in dark matter theories (e.g., Dalcanton et al. 1997).

3. Spiral galaxies with a mean surface density near this limit—HSB galaxies— would be, within the optical disk, in the Newtonian regime. Thus, one would expect that the rotation curve would decline in a near Keplerian fashion to the asymptotic constant value. In LSB galaxies, with mean surface density below Σ_m, the prediction is that rotation curves continuously rise to the final asymptotic flat value. Thus, there should be a general difference in rotation curve shapes between LSB and HSB galaxies. Figure 3 shows the rotation curves of two galaxies, an LSB and an HSB, where we see exactly this trend. This general effect in observed rotation curves was first noted by Casertano & van Gorkom (1991).

4. With Newtonian dynamics, pressure-supported systems that are nearly isothermal have infinite mass. However, in the context of MOND it is straightforward to demonstrate that such isothermal systems are of finite mass with the density at large radii falling approximately as r^{-4} (Milgrom 1984). The equation of hydrostatic equilibrium for an isotropic, isothermal system reads

$$\sigma_r^2 \frac{d\rho}{dr} = -\rho g, \tag{7}$$

where, in the limit of low accelerations, $g = \sqrt{GM_r a_o}/r$. σ_r is the radial velocity

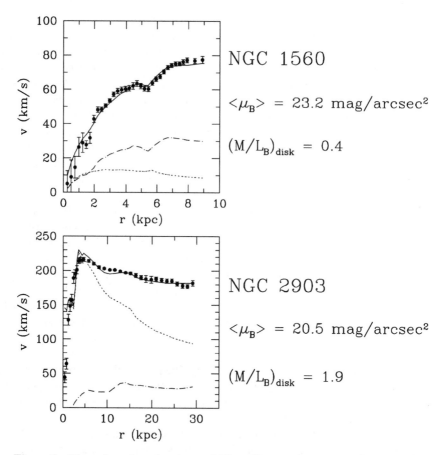

Figure 3   The points show the observed 21-cm line rotation curves of a low surface brightness galaxy, NGC 1560 (Broeils 1992), and a high surface brightness galaxy, NGC 2903 (Begeman 1987). The dotted and dashed lines are the Newtonian rotation curves of the visible and gaseous components of the disk, and the solid line is the MOND rotation curve with $a_o = 1.2 \times 10^{-8}$ cm/s²—the value derived from the rotation curves of 10 nearby galaxies (Begeman et al. 1991). The only free parameter is the mass-to-light ratio of the visible component.

dispersion, ρ is the mass density, and M_r is the mass enclosed within r. It then follows immediately that, in the outer regions, where $M_r = M = $ constant,

$$\sigma_r^4 = GMa_o\left(\frac{d\ln(\rho)}{d\ln(r)}\right)^{-2}. \tag{8}$$

Thus, there exists a mass-velocity dispersion relation of the form

$$(M/10^{11}M_\odot) \approx (\sigma_r/100\,\mathrm{km\,s^{-1}})^4, \tag{9}$$

which is similar to the observed Faber-Jackson relation (luminosity-velocity dispersion relation) for elliptical galaxies (Faber & Jackson 1976). This means that a MOND near-isothermal sphere with a velocity dispersion of 100–300 km/s will always have a galactic mass. Moreover, the same $M - \sigma$ relation (Equation 8) should apply to all pressure-supported, near-isothermal systems, from globular clusters to clusters of galaxies, albeit with considerable scatter owing to deviations from a strictly isotropic, isothermal velocity field (Sanders 2000).

The effective radius of a near-isothermal system is roughly $r_e \approx \sqrt{GM/a_o}$ (at larger radii the system is in the MOND regime and is effectively truncated). This means that a_o appears as a characteristic acceleration in near-isothermal systems and that Σ_m appears as a characteristic surface density—at least as an upper limit (Milgrom 1984). Fish (1964), on the basis of then-existing photometry, pointed out that elliptical galaxies exhibit a constant surface brightness within an effective radius. Although Kormendy (1982) demonstrated that more luminous ellipticals have a systematically lower surface brightness, when the ellipticals are considered along with the bulges of spiral galaxies and globular clusters (Corollo et al. 1997) there does appear to be a characteristic surface brightness on the order of that implied by Σ_m, i.e., the Fish law is recovered for the larger set of pressure-supported objects.

5. The "external field effect" is not a prediction but a phenomenological requirement on MOND that has strong implications for nonisolated systems. In his original papers Milgrom noted that open star clusters in the Galaxy do not show evidence for mass discrepancies, even though the internal accelerations are typically below a_o. He therefore postulated that the external acceleration field of the Galaxy must have an effect upon the internal dynamics of a star cluster—that, in general, the dynamics of any subsystem is affected by the external field in which that system is found. This implies that the theory upon which MOND is based does not respect the equivalence principle in its strong form. (This in no way implies that MOND violates the universality of free fall—the weak version of the equivalence principle—which is the more cherished and experimentally constrained version.) Milgrom suggested that this effect arises owing to the nonlinearity of MOND and can be approximated by including the external acceleration field, $\mathbf{g_e}$, in the MOND equation, i.e.,

$$\mu(|\mathbf{g_e} + \mathbf{g_i}|/a_o)\,\mathbf{g_i} = \mathbf{g_{n_i}}, \qquad (10)$$

where $\mathbf{g_i}$ is the internal gravitational field of the subsystem and $\mathbf{g_{n_i}}$ is the Newtonian field of the subsystem alone. This means that a subsystem with internal accelerations below a_o will exhibit Newtonian dynamics if the external acceleration exceeds a_o ($g_i < a_o < g_e$). If the external and internal accelerations are below a_o and $g_i < g_e < a_o$, then the dynamics of the subsystem will be Newtonian but with a larger effective constant of gravity given by $G/\mu(g_e/a_o)$. If $g_e < g_i < a_o$ the dynamics is MONDian but with a maximum discrepancy given by $[\mu(g_e/a_o)]^{-1}$. In addition the dynamics is anisotropic with dilation along the direction of the external field.

The external field effect would have numerous consequences: It would influence the internal dynamics of globular clusters and the satellite dwarf companions of the Milky Way independently of tides. It may provide an important nontidal mechanism for the maintenance of warps in galaxies owing to the presence of companions. The peculiar accelerations resulting from large-scale structure would be expected to limit the mass discrepancy in any particular galactic system (no object is isolated), and the deceleration (or acceleration) of the Hubble flow may influence the development of large-scale structure.

Dark Halos with an Acceleration Scale

Can the phenomenology of MOND be reproduced by dark halos, specifically the kind of dark halos that emerge from cosmological N-body simulations with an initial fluctuation spectrum given by CDM? This is an important question because the phenomenological success of MOND may be telling us something about the universal distribution of dark matter in galaxies and its relation to the visible component rather than anything about the law of gravity or inertia at low accelerations. This question was first considered by Begeman et al. (1991), who attempted to devise disk-halo coupling rules that could yield a one-parameter fit to rotation curves (M/L of the visible disk) similar to MOND. Without any physical justification, the core radius and asymptotic circular velocity of an isothermal halo were adjusted to the scale length and maximum rotation velocity of the disk to yield a characteristic acceleration. With such coupling rules, the fits to galaxy rotation curves were of lower quality than the MOND fits (particularly for the dwarf systems), and there were numerous ambiguities (e.g., in gas-dominated galaxies what is the proper disk length scale?). Similar ad hoc coupling rules between visible and dark components have also been considered by Giraud (2000).

The idea that the halo might exhibit a characteristic acceleration was carried further by Sanders & Begeman (1994) when the first cosmic N-body calculations with high resolution (Dubinski & Carlberg 1991) indicated that CDM halos were not at all similar to an isothermal sphere with a constant density core. The objects emerging from the simulations exhibited a density law with a r^{-1} cusp that steepened in the outer regions to r^{-4}. Dubinski & Carlberg pointed out that this run of density was well described by the model of Hernquist (1990):

$$\rho(r) = \frac{\Sigma_o}{r}(1 + r/r_o)^{-3}. \tag{11}$$

This has been subsequently confirmed by the extensive calculations of Navarro et al. (1997), who corrected the outer power law to -3 (this is the famous NFW halo). The reality of the cusp, if not the exact power law, seems well established (Moore et al. 1998) and is due to the fact that there are no phase-space constraints upon the density of a collapsed object composed of CDM.

Sanders & Begeman (1994) pointed out that if Σ_o were fixed with only the characteristic length scale varying from halo to halo, this implied that a fixed acceleration scale could be associated with any halo, $a_h = 2\pi G\Sigma_o$. They demonstrated that a halo density law of this form provided a reasonable fit to rotation curves of several HSB galaxies (comparable to that of MOND), where the length scale of the halo was proportional to the mass of the visible disk. This proportionality would follow if the baryonic mass were a fixed fraction (about 0.03) of the dark mass as is usually assumed, so this appeared to be a natural way to explain MOND phenomenology in the context of CDM.

There are two problems with this idea. First, a fixed Σ_o implies that no galaxy could exhibit an acceleration in the inner regions less than a_o; this is not true for a number of LSB galaxies (McGaugh & de Blok 1998b). The problem was already evident in the fits to the LSB galaxies in the sample of Sanders & Begeman in which the one-parameter fitting scheme broke down. Second, the halos that emerge from the cosmic N-body simulations do not have fixed Σ_o, as is evident from the mass-rotation velocity law of $m \propto v^3$ (NFW). No characteristic surface density or acceleration is evident in these objects.

Semianalytic models for the formation of disk galaxies in the context of CDM (van den Bosch & Dalcanton 2000) can be tuned to give rise to a characteristic acceleration. In such models one starts with a specified dark halo density law (NFW); allows some fraction of the halo mass, presumably baryonic, to collapse by a factor determined by the dimensionless spin parameter of the halo; applies a stability criterion to allow some further fraction of this dissipational component to be converted to stars; and removes gas from the system by an appropriate number of supernovae (feedback). In this procedure there are dimensionless parameters that quantify the feedback mechanism. Because these parameters can be adjusted to produce a TF law of the form $L \propto V^4$, it is not surprising that there is a fixed acceleration connected with these models ($a \approx V^4/GL\langle M/L\rangle$). The exercise is essentially that of modeling complicated astrophysical processes by a set of free parameters; these parameters are then tuned in order to achieve a desired result. The fact that this is possible is interesting but not at all compelling.

The possibility of a characteristic acceleration arising from CDM has been revisited by Kaplinghat & Turner (2002), who offered an explanation for why dark matter appears to become dominant beyond an acceleration numerically comparable to cH_o. They argued that halos formed from CDM possess a one-parameter density profile that leads to a characteristic acceleration profile that is only weakly dependent upon the mass (or comoving scale) of the halo. Then with a fixed collapse factor for the baryonic material, the transition from dominance of dark over baryonic occurs at a universal acceleration, which by numerical coincidence, is on the order of cH_o. Milgrom (2002) responded by pointing out that a_o plays several roles in the context of MOND: It not only is a transition acceleration below which the mass discrepancy appears, but it also defines the asymptotic rotation velocity of spiral galaxies (via Equation 4) and thereby determines the zero point of the TF relation (Equation 5). a_o determines the upper limit on the surface density of spirals (i.e., Σ_m); a_o appears as an effective upper limit upon the gravitational acceleration of a halo component in sensible disk-halo fits to observed rotation curves (Brada & Milgrom 1999a); a_o determines the magnitude of the discrepancy within LSB galaxies (where the ratio of missing to visible mass is a_o/g_i); a_o sets the scale of the Faber-Jackson relation via Equation 8; a_o appears as an effective internal acceleration for pressure-supported, quasi-isothermal systems and determines the dynamics of galaxy systems—global and detailed—ranging from small groups to super-clusters. These roles are independent in the context of dark matter and would each require a separate explanation. The explanation of Kaplinghat & Turner applies only to the first of these and by construction prohibits the existance of objects that are dark matter–dominated within their optical radius (such as LSB galaxies).

The basic problem in trying to explain a fixed acceleration scale in galaxies in terms of galaxy formation rather than underlying dynamics is that the process is stochastic: Each galaxy has its own history of formation-evolution-interaction. One would expect these effects to erase any intrinsic acceleration scale, not enhance it. Dark matter may address the general trends but it cannot account for the individual idiosyncrasies of each rotation curve. In the next section we present the evidence that MOND can do this with a_o as the only additional fixed parameter.

ROTATION CURVES OF SPIRAL GALAXIES

Method and Results of Rotation-Curve Fitting

The measured rotation curves of spiral galaxies constitute the ideal body of data to confront ideas such as MOND (Begeman et al. 1991, Sanders 1996, McGaugh & de Blok 1998b, Sanders & Verheijen 1998). That is because in the absence of dark matter the rotation curve is in principle predictable from the observed distribution of stars and gas. Moreover, the rotation curve as measured in the 21-cm line of neutral hydrogen often extends well beyond the optical image of the galaxy where the centripetal acceleration is small and the discrepancy is large. In the particularly critical case of the LSB galaxies, 21-cm line observations can be supplemented by H_α observations (McGaugh et al. 2001, de Blok & Bosma 2002) and compared in detail with the rotation curve predicted from the distribution of detectable matter. The procedure that has usually been followed is outlined below:

1. One assumes that light traces stellar mass, i.e., M/L is constant in a given galaxy. There are color gradients in spiral galaxies, so this cannot be generally true—or at least one must decide which color band is the best tracer of the mass distribution. The general opinion is that the near-infrared emission of spiral galaxies is the optimal tracer of the underlying stellar mass distribution because the old population of low-mass stars contributes to this emission, and the near-infrared is less affected by dust obscuration. Thus, where available, near-infrared surface photometry is preferable.

2. In determining the distribution of detectable matter one must include the observed neutral hydrogen scaled up with an appropriate correction factor (typically 1.3–1.4) to account for the contribution of primordial helium. The gas can make a dominant contribution to the total mass surface density in some (generally low luminosity) galaxies.

3. Given the observed distribution of mass, the Newtonian gravitational force, g_n, is calculated via the classical Poisson equation. Here it is usually assumed that the stellar and gaseous disks are razor thin. It may also be necessary to add a spheroidal bulge if the light distribution indicates the presence of such a component.

4. Given the radial distribution of the Newtonian force, the true gravitational force, g, is calculated from the MOND formula (Equation 3) with a_o fixed. Then the mass of the stellar disk is adjusted until the best fit to the observed rotation curve is achieved. This gives M/L of the disk as the single free parameter of the fit (unless a bulge is present).

In this procedure, one assumes that the motion of the gas is a coplaner rotation about the center of the given galaxy. This is certainly not always the case because there are well-known distortions to the velocity field in spiral galaxies caused by bars and warping of the gas layer. In a fully two-dimensional velocity field these distortions can often be modeled (Bosma 1978, Begeman 1989), but the optimal rotation curves are those in which there is no evidence for the presence of significant deviations from coplanar circular motion. Not all observed rotation curves are perfect tracers of the radial distribution of force. A perfect theory will not fit all rotation curves because of these occasional problems (the same is true of a specified dark-matter halo). The point is that with MOND, usually, there is one adjustable parameter per galaxy and that is the mass or M/L of the stellar disk.

The preferred value of a_o has been derived from a highly selected sample of large galaxies with well-determined rotation curves (Begeman et al. 1991). Assuming a distance scale, $H_o = 75$ km/s Mpc in this case, rotation-curve fits to all galaxies in the sample were made allowing a_o to be a free parameter. The mean value for nine of the galaxies in the sample, excluding NGC 2841 with a distance ambiguity (see below), is $1.2 \pm 0.27 \times 10^{-8}$ cm/s^2. Having fixed a_o in this way, one is no longer free to take this as a fitting parameter.

There is, however, a relation between the derived value of a_o and the assumed distance scale because the implied centripetal acceleration in a galaxy scales as the inverse of the assumed distance. With respect to galaxy rotation curves, this dependence is not straightforward because the relative contributions of the stellar and gaseous components to the total force vary as a function of distance. For a gas-rich sample of galaxies the derived value of a_o scales as H_o^2, and for a sample of HSB galaxies dominated by the stellar component $a_o \propto H_o$. This is related to a more general property of MOND: a_o in its different roles scales differently with H_o. This fact in itself means that MOND cannot live with any distance scale; to be consistent with MOND, H_o must be in the range of 50–80 km/s-Mpc.

Figure 3 shows two examples of MOND fits to rotation curves. The dotted and dashed curves are the Newtonian rotation curves of the stellar and gaseous disks, respectively, and the solid curve is the MOND rotation curve with the standard value of a_o. Not only does MOND predict the general trend for LSB and HSB galaxies, but it also predicts the observed rotation curves in detail from the observed distribution of matter. This procedure has been carried out for about 100 galaxies, 76 of which are listed in Table 1; the results are given in terms of the fitted mass of the stellar disk (in most cases, the only free parameter) and the implied M/Ls.

Rotation curves for the entire UMa sample of Sanders & Verheijen (1998) are shown in Figure 4, where the curves and points have the same meaning as in Figure 3. As noted above, this is a complete and unbiased sample of spiral galaxies all at about the same distance (here taken to be 15.5 Mpc). The sample includes both HSB galaxies (e.g., NGC 3992) and LSB galaxies (e.g., UGC 7089) and covers a factor of 10 in centripetal acceleration at the outermost observed point (Table 1 and Figure 1). The objects denoted by the asterisk in Figure 4 are galaxies previously designated by Verheijen (1997) as being kinematically disturbed (e.g.,

TABLE 1   Rotation-curve fits

Galaxy (1)	Type (2)	L_B (3)	L_r (4)	M_{HI} (5)	V_∞ (6)	M_* (7)	M_*/L_B (8)	M_*/L_r (9)	Ref (10)
UGC 2885	Sbc	21.0	—	5.0	300	30.8	1.5	—	1
NGC 2841[a]	Sb	8.5	17.9	1.7	287	32.3	3.8	1.8	2
NGC 5533	Sab	5.6	—	3.0	250	19.0	3.4	—	1
NGC 6674	SBb	6.8	—	3.9	242	18.0	2.6	—	1
NGC 3992	SBbc	3.1	7.0	0.92	242	15.3	4.9	2.2	3
NGC 7331	Sb	5.4	18.0	1.1	232	13.3	2.5	0.7	2
NGC 3953	SBbc	2.9	8.5	0.27	223	7.9	2.7	0.9	3
NGC 5907	Sc	2.4	4.9	1.1	214	9.7	3.9	2.0	1
NGC 2998	SBc	9.0	—	3.0	213	8.3	1.2	—	1
NGC 801	Sc	7.4	—	2.9	208	10.0	1.4	—	1
NGC 5371	S(B)b	7.4	—	1.0	208	11.5	1.6	—	1
NGC 5033	Sc	1.90	3.90	0.93	195	8.8	4.6	2.3	1
NGC 3893[b]	Sc	2.14	3.98	0.56	188	4.20	2.0	1.1	3
NGC 4157	Sb	2.00	5.75	0.79	185	4.83	2.4	0.8	3
NGC 2903	Sc	1.53	2.15	0.31	185	5.5	3.6	2.6	2
NGC 4217	Sb	1.90	5.29	0.25	178	4.25	2.2	0.8	3
NGC 4013	Sb	1.45	4.96	0.29	177	4.55	3.1	0.9	3
NGC 3521	Sbc	2.40	—	0.63	175	6.5	2.7	—	1
NGC 4088[b]	Sbc	2.83	5.75	0.79	173	3.30	1.1	0.6	3
UGC 6973[b]	Sab	0.62	2.85	0.17	173	1.69	2.7	0.6	3
NGC 3877	Sc	1.94	4.52	0.14	167	3.35	1.7	0.7	3
NGC 4100	Sbc	1.77	3.50	0.30	164	4.32	2.4	1.2	3

(Continued)

TABLE 1   *(Continued)*

Galaxy (1)	Type (2)	L_B (3)	L_r (4)	M_{HI} (5)	V_∞ (6)	M_* (7)	M_*/L_B (8)	M_*/L_r (9)	Ref (10)
NGC 3949	Sbc	1.65	2.33	0.33	164	1.39	0.8	0.6	3
NGC 3726	SBc	2.65	3.56	0.62	162	2.62	1.0	0.7	3
NGC 6946	SABcd	5.30	—	2.7	160	2.7	0.5	—	1
NGC 4051[b]	SBbc	2.58	3.91	0.26	159	3.03	1.2	0.8	3
NGC 3198[c]	Sc	0.90	0.80	0.63	156	2.3	2.6	2.9	2
NGC 2683	Sb	0.60	—	0.05	155	3.5	5.8	—	1
UGC 5999[e]	Im	—	0.13	0.25	155	0.09	—	0.7	4
NGC 4138	Sa	0.82	2.88	0.14	147	2.87	3.5	1.0	3
NGC 3917	Scd	1.12	1.35	0.18	135	1.40	1.3	1.0	3
NGC 4085	Sc	0.81	1.22	0.13	134	1.00	1.2	0.8	3
NGC 2403	Sc	0.79	0.98	0.47	134	1.1	1.4	1.1	2
NGC 3972	Sbc	0.68	1.00	0.12	134	1.00	1.5	1.0	3
UGC 128	Sdm	0.52	0.41	0.91	131	0.57	1.1	1.4	4
NGC 4010	SBd	0.63	1.20	0.27	128	0.86	1.4	0.7	3
F568-V1	Sd	0.22	0.17	0.34	124	0.66	3.0	3.8	4
NGC 3769[b]	SBb	0.68	1.27	0.53	122	0.80	1.2	0.6	3
NGC 6503	Sc	0.48	0.47	0.24	121	0.83	1.7	1.8	2
F568-3	Sd	0.33	0.27	0.39	120	0.44	1.3	1.6	4
F568-1	Sc	0.28	0.21	0.56	119	0.83	3.0	4.0	4
NGC 4183	Scd	0.90	0.73	0.34	112	0.59	0.7	0.8	3
F563-V2	Irr	0.30	—	0.32	111	0.55	1.8	—	4
F563-1	Sm	0.14	0.10	0.39	111	0.40	3.0	4.0	4
NGC 4389[b]	SBbc	0.61	1.22	0.05	110	0.23	0.4	0.2	3
NGC 1003	Scd	1.50	0.45	0.82	110	0.30	0.2	0.7	1
UGC 6917	SBd	0.38	0.42	0.20	110	0.54	1.4	1.3	3
UGC 6930	SBd	0.50	0.40	0.31	110	0.42	0.8	1.0	3
M 33	Sc	0.74	0.43	0.13	107	0.48	0.6	1.1	1
UGC 6983	SBcd	0.34	0.34	0.29	107	0.57	1.7	1.7	3
NGC 247	SBc	0.35	0.22	0.13	107	0.40	1.1	1.8	1
UGC 1230[e]	Sm	0.32	0.22	0.81	102	0.38	1.2	1.7	4
F574-1[d]	Sd	—	0.37	0.49	100	0.26	—	0.7	4
NGC 7793	Scd	0.34	0.17	0.10	100	0.41	1.2	2.4	1
UGC 5005[e]	Im	—	0.15	0.41	99	0.74	—	4.8	4

(Continued)

TABLE 1   (*Continued*)

Galaxy (1)	Type (2)	L_B (3)	L_r (4)	M_{HI} (5)	V_∞ (6)	M_* (7)	M_*/L_B (8)	M_*/L_r (9)	Ref (10)
NGC 300	Sc	0.30	—	0.13	90	0.22	0.7	—	1
NGC 5585	SBcd	0.24	0.14	0.25	90	0.12	0.5	0.9	1
NGC 2915[f]	BCD	0.04	—	0.10	90	0.25	6.9	—	1
UGC 6399	Sm	0.20	0.21	0.07	88	0.21	1.0	1.0	3
NGC 55	SBm	0.43	—	0.13	86	0.10	0.2	—	1
UGC 6667	Scd	0.26	0.28	0.08	86	0.25	1.0	0.9	3
UGC 2259	SBcd	0.10	—	0.05	86	0.22	2.1	—	2
F583-1	Sm	0.06	0.06	0.24	85	0.11	1.7	2.0	4
UGC 6446	Sd	0.25	0.14	0.30	82	0.12	0.5	0.9	3
UGC 6923	Sdm	0.22	0.21	0.08	81	0.16	0.8	0.8	3
UGC 7089	Sdm	0.44	0.21	0.12	79	0.09	0.2	0.4	3
UGC 5750[e]	SBdm	—	0.36	0.14	75	0.32	—	0.9	4
UGC 6818[b]	Sd	0.18	0.12	0.10	73	0.04	0.2	0.3	3
F571-V1[e]	Sdm	0.10	0.80	0.164	73	0.67	0.7	0.8	4
NGC 1560	Sd	0.035	0.063	0.098	72	0.034	1.0	0.5	2
F583-4	Sc	—	0.071	0.077	67	0.022	—	0.3	4
IC 2574	SBm	0.080	0.022	0.067	66	0.010	0.1	0.5	1
DDO 170	Im	0.016	—	0.061	64	0.024	1.5	—	2
NGC 3109	SBm	0.005	—	0.068	62	0.005	0.1	—	2
DDO 154	IB	0.005	—	0.045	56	0.004	0.1	—	2
DDO 168	Irr	0.022	—	0.032	54	0.005	0.2	—	1
F565-V2[e]	Im	0.023	0.019	0.084	51	0.050	2.2	2.7	4

Explanation of columns of Table 1: (1) Galaxy name. (2) Morphological Type. (3) B-band luminosity in units of $10^{10}\,L_\odot$. (4) Red band luminosity in units of $10^{10}\,L_\odot$. The precise band used depends on the reference: Refs. 1 & 2: H-band. Ref. 3: K'-band. Ref. 4: R-band. (5) Mass of neutral hydrogen in units of $10^{10}\,M_\odot$ assuming $M_{gas}=M_{HI}$. (6) Asymptotic flat rotation velocity in km s^{-1}. (7) Stellar mass from MOND fit in units of $10^{10}\,M_\odot$. (8) B-band stellar mass-to-light ratio in units of $M_\odot/L_\odot$. (9) R-band stellar mass-to-light ratio in units of $M_\odot/L_\odot$. Both B- and R-band mass-to-light ratios refer only to the stars (the gas is not included in the mass) and average over disk and bulge where both components are significant. See original references for further details. (10) References: 1. Sanders (1996). 2. Begeman, Broeils, & Sanders (1991). 3. Sanders & Verheijen (1998). 4. de Blok & McGaugh (1998).

Notes for Table 1: [a]The MOND fit for this galaxy is sensitive to its distance, prefering $D \approx 19$ Mpc (Sanders 1996) to the Hubble flow value of ≈ 9 Mpc. Macri et al. (2001) give a Cepheid distance of 14 Mpc, which is marginally tolerable given the uncertainties in this galaxy's warp (R. Bottema, J.L.G. Pestanã, B. Rothberg, R.H. Sanders, unpublished manuscript).

[b]Noted as having disturbed kinematics by Verheijen (1997).

[c]The MOND fit for this galaxy is sensitive to its distance, preferring a smaller value than the Cepheid distance of 13.8 Mpc (R. Bottema, J.L.G. Pestanã, B. Rothberg, R.H. Sanders, unpublished manuscript).

[d]The original MOND fit for this galaxy (de Blok & McGaugh 1998) was not very good. The 21-cm observations of this galaxy were severely affected by beam smearing.

[e]Inclination uncertain (de Blok & McGaugh 1998).

[f]Distance uncertain.

nonaxisymmetric velocity field caused by bars or interactions); the derived rotation curves are less secure for these galaxies. (Note that fits to the UMa rotation curves using a revised cluster distance of 18.5 Mpc from the Cepheid-based recalibrated TF relation of Sakai et al. (2000) would imply that a_o should be reduced to 1×10^{-8} cm/s^2.)

This is a fair selection of MOND fits to rotation curves in which the only free parameter is the M/L of the visible disk (no separate bulge components were fitted in these cases). In HSB objects in which the centripetal acceleration remains large out to the last measure point of the rotation curve, such as NGC 3954, there is a very small difference between the Newtonian curve and the predicted

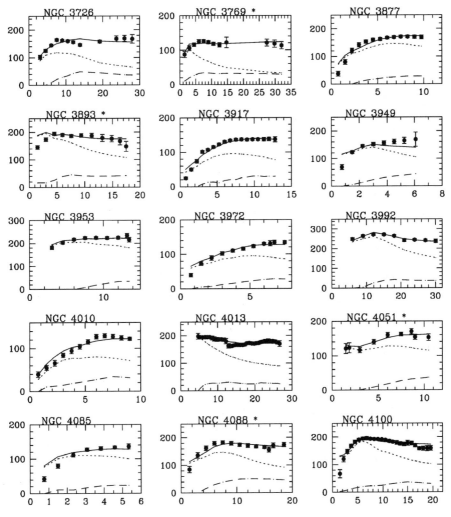

Figure 4 *(Continued)*

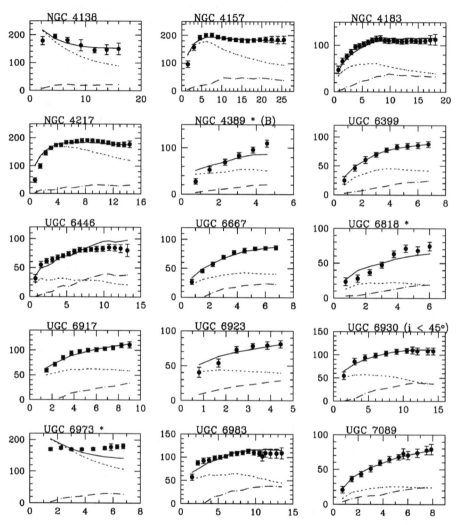

Figure 4   MOND fits to the rotation curves of the Ursa Major galaxies (Sanders & Verheijen 1998). The radius (horizontal axis) is given in kiloparsecs, and in all cases the rotation velocity is in kilometers/second. The points and curves have the same meaning as in Figure 3. The distance to all galaxies is assumed to be 15.5 Mpc, and a_o is the Begeman et al. (1991) value of 1.2×10^{-8} cm/s^2. The free parameter of the fitted curve is the mass of the stellar disk. If the distance to UMa is taken to be 18.6 Mpc, as suggested by the Cepheid-based recalibration of the Tully-Fisher relation (Sakai et al. 2000), then a_o must be reduced to 10^{-8} cm/s^2.

curve; i.e., the observed rotation curve is reasonably well described by Newtonian theory, as expected when accelerations are high. In lower acceleration systems, the discrepancy is larger (e.g., UGC 6667). In gas-rich galaxies, such as UGC 7089, the shape of the rotation curve in the outer regions essentially follows from the shape of the Newtonian rotation curve of the gaseous component, as though the gas surface density were only scaled up by some factor on the order of 10—a property that has been noticed for spiral galaxies (Carignan & Beaulieu 1989, Hoekstra et al. 2001). Empirically, MOND gives the rule that determines the precise scaling.

In general the MOND curves agree well with the observed curves, but there are some cases in which the agreement is less than perfect. Usually these cases have an identifiable problem with the observed curve or its interpretation as a tracer of the radial force distribution. For example NGC 4389 is strongly barred, and the neutral hydrogen is contiguous with the visible disk and bar. Another example is UGC 6818, which is probably interacting with a faint companion at its western edge.

Figure 5 shows a less homogeneous sample of rotation curves. These are curves from the literature based upon observations carried out either at the Westerbork Radio Synthesis Telescope or the Very Large Array (VLA) from Sanders (1996) and McGaugh & de Blok (1998b) and ranked here in order of decreasing circular velocity. These are mostly galaxies with a large angular size, so there are many independent points along the rotation curve. The selection includes HSB and LSB galaxies such as NGC 2403 and UGC 128—two objects with the same asymptotic rotation velocity ($\approx$130 km/s). Here the general trend, predicted by MOND, is evident: The LSB exhibits a large discrepancy throughout the galaxy in contrast to the HSB, where the discrepancy becomes apparent in the outer regions. In several objects, such as NGC 2403, NGC 6503, and M33, the quality of the MOND fit is such that, given the density of points, the fitted curve cannot be distinguished beneath the observations.

The most striking aspect of these studies is the fact that not only general trends but also the details of individual curves are well reproduced by Milgrom's simple formula applied to the observed distribution of matter. In only about 10% of the roughly 100 galaxies considered in the context of MOND does the predicted rotation curve differ significantly from the observed curve.

We have emphasized that the only free parameter in these fits is the M/L of the visible disk, so one may well ask if the inferred values are reasonable. It is useful to consider again the UMa sample because all galaxies are at the same distance and there is K'-band (near infrared) surface photometry of the entire sample. Figure 6 shows the M/L in the B-band required by the MOND fits plotted against B-V color index (*top*) and the same for the K'-band (*bottom*). We see that in the K'-band M/L $\approx$ 0.8 with a 30% scatter. In other words if one were to assume a K'-band M/L of about 1 at the outset, most rotation curves would be quite reasonably predicted from the observed light and gas distribution with no free parameters. In the B-band, on the other hand, the MOND M/L does appear to be a function of color in the sense that redder objects have larger M/L values. This is exactly what is expected from population synthesis models, as is shown by the solid lines in both panels (Bell & de Jong 2001). This is quite interesting because there is

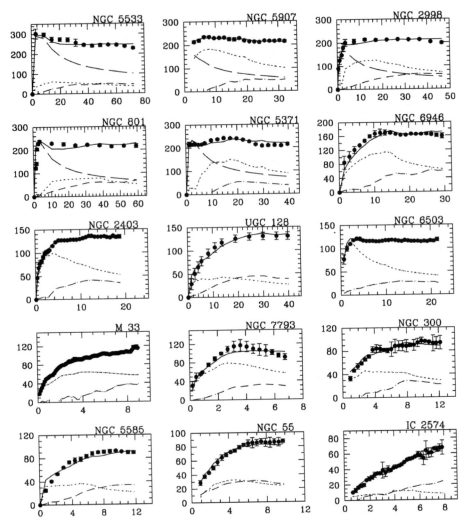

Figure 5 MOND fits to the rotation curves of spiral galaxies with published data, from Sanders (1996) and McGaugh & de Blok (1998). The symbols and curves are as in Figure 4.

nothing built into MOND that would require that redder galaxies should have a higher M/L_b; this simply follows from the rotation-curve fits.

Falsification of Modified Newtonian Dynamics with Rotation Curves

It is sometimes said that MOND is designed to fit rotation curves, so it is not surprising that it does so. This is not only a trivialization of a remarkable phenomenological success, but it is also grossly incorrect. MOND was designed to

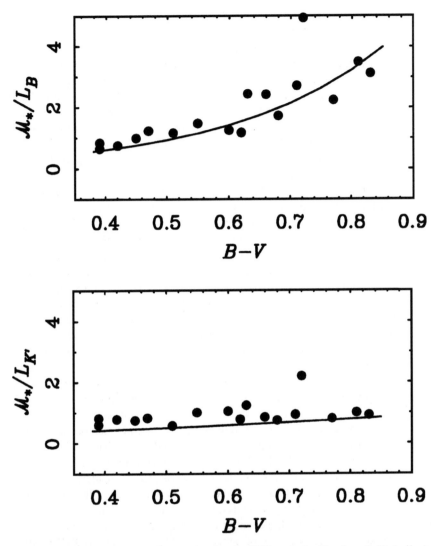

Figure 6   Inferred mass-to-light ratios for the UMa spirals (Sanders & Verheijen) in the B-band (*top*) and the K'-band (*bottom*) plotted against blue-visual (B-V) color index. The solid lines show predictions from population synthesis models by Bell & de Jong (2001).

produce asymptotically flat rotation curves with a given mass-velocity relation (or TF law). It was not designed to fit the details of all rotation curves with a single adjustable parameter—the M/L of the stellar disk (MOND also performs well on galaxies that are gas dominated and have no adjustable parameter). It was certainly not designed to provide a reasonable dependence of fitted M/L on color. Indeed, none of the rotation curves listed in Table 1 were available in 1983; designing a theory to fit data that are not yet taken is called "prediction."

However, MOND is particularly vulnerable to falsification by rotation-curve data. Although there are problems, mentioned above, in the measurement and interpretation of velocity field and photometric data, MOND should not "fail" too often; especially damaging would be a systematic failure for a particular class of objects. In this regard Lake (1989) has claimed that the value of a_o required to fit rotation curves varies with the maximum rotation velocity of the galaxy in the sense that objects with lower rotation velocities (and therefore lower luminosity galaxies) require a systematically lower value of a_o. He supported this claim by rotation-curve fits to six dwarf galaxies with low internal accelerations. If this were true, then it would be quite problematic for MOND, implying at the very least a modification of Milgrom's simple formula. Milgrom (1991) responded to this criticism by pointing out inadequacies in the data used by Lake: uncertainties in the adopted distances and/or inclinations. Much of the rotation-curve data is also of lower quality than the larger galaxies considered in the context of MOND.

R.A. Swaters & R.H. Sanders (2002, unpublished manuscript) reconsidered this issue on the basis of extensive 21-cm line observations of a sample of 35 dwarf galaxies (Swaters 1999). When a_o is taken as an additional free parameter, the effect pointed out by Lake is not seen: There is no systematic variation of a_o with the maximum rotation velocity of a galaxy. There is a large scatter in the fitted a_o, but this is due to the fact that many dwarf galaxies contain large asymmetries or an irregular distribution of neutral hydrogen. Moreover, the galaxies in this sample have large distance uncertainties, the distances in many cases being determined by group membership. The mean a_o determined from the entire sample ($\approx 10^{-8}$ cm/s^2) is consistent with that implied by the revised Cepheid-based distance scale.

Given the well-known uncertainties in the interpretation of astronomical data, it is difficult to claim that MOND is falsified on the basis of a single rotation curve. However, it should generally be possible to identify the cause of failures (i.e., poor resolution, bars, interactions, warps, etc.). An additional uncertainty is the precise distance to a galaxy because, as noted above, the estimated internal acceleration in a galaxy depends upon its assumed distance. For nearby galaxies, such as those of the Begeman et al. (1991) sample, the distances are certainly not known to an accuracy of better than 25%. When the MOND rotation curve is less than a perfect match to the observed curve, it is often possible to adjust the distance, within reasonable limits (i.e., the distance appears as a second free parameter). In principle, precise independent distance determinations place more severe restrictions on this extra degree of freedom and are therefore relevant to rotation-curve tests of MOND.

There are now four galaxies from the original sample of Begeman et al. (1991) with Cepheid-based distances. Three of these (NGC 2403, NGC 3198, and NGC 7331) have been observed as part of the HST Key Project on the extragalactic distance scale (Sakai et al. 2000). For NGC 2403 and NGC 7331 the MOND rotation curve fits precisely the observed curve at the Cepheid-based distances.

For NGC 3198 MOND clearly prefers a distance at least 10% smaller than the Cepheid-based distance of 13.8 Mpc (R. Bottema, J.L.G. Pestanã, B. Rothberg, R.H. Sanders, unpublished manuscript), even with the lower value of a_o implied by the revised distance scale. Given the likely uncertainties in the Cepheid method, and in the conversion of a 21-cm line velocity field to a rotation curve, this cannot be interpreted as problematic for MOND.

A more difficult case is presented by NGC 2841, a large spiral galaxy with a Hubble distance of about 10 Mpc (Begeman et al. 1991). The rotation curve of the galaxy cannot be fit using MOND if the distance is only 10 Mpc; MOND, as well as the TF law, prefers a distance of 19 Mpc (Begeman et al. 1991, Sanders 1996). A Cepheid distance to this galaxy has been determined (Macri et al. 2001) at 14.1 ± 1.5 Mpc. At a distance of 15.6 Mpc the MOND rotation curve of the galaxy still systematically deviates from the observed curve, and the implied M/L is 8.5; thus, this galaxy remains the most difficult case for MOND. It is nonetheless interesting that the Cepheid-calibrated T-F relation (Sakai et al. 2001) implies a distance of about 23 Mpc for NGC 2841, and supernova 1999 by, if a normal type Ia, would imply a distance of 24 Mpc.

Overall, the ability of MOND, as an ad hoc algorithm, to predict galaxy rotation curves with only one free parameter (the M/L of the visible disk) is striking. This implies that galaxy rotation curves are entirely determined by the distribution of observable matter. Regardless of whether or not MOND is correct as a theory, it does constitute an observed phenomenology that demands explanation. Herein lies a real conundrum for the dark matter picture. The natural expectations of dark matter theories for rotation curves do not look like MOND, and hence fail to reproduce a whole set of essential observational facts. The best a dark matter theory can hope to do is contrive to look like MOND and hence reproduce a posteriori the many phenomena that MOND successfully predicts. This gives one genuine pause to consider how science is supposed to proceed.

PRESSURE-SUPPORTED SYSTEMS

General Properties

Figure 7 is a log-log plot of the velocity dispersion versus size for pressure-supported, nearly isothermal astronomical systems. At the bottom left the star-shaped points are globular clusters (Pryor & Meylen 1993, Trager et al. 1993) and the solid points are giant molecular clouds in the Galaxy (Solomon et al. 1987). The group of points (*crosses*) near the center are high–surface brightness elliptical galaxies (Jørgensen et al. 1995a,b, Jørgensen 1999). At the upper right the squares indicate X-ray-emitting clusters of galaxies from the compilation by White et al. (1997). The triangle-shaped points are the dwarf spheroidal systems surrounding the Milky Way (Mateo 1998), and the dashes are compact dwarf ellipticals (Bender et al. 1992). The plotted parameters are taken directly from the relevant observational papers. The measure of size is not homogeneous: For ellipticals and globular

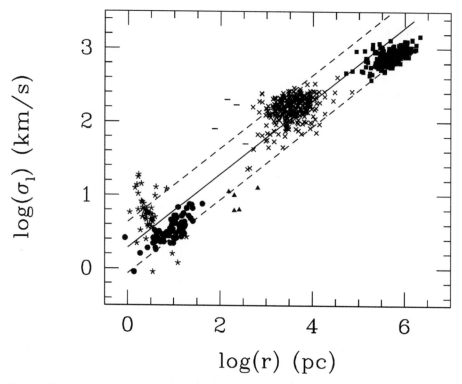

Figure 7 The line-of-sight velocity dispersion vs. characteristic radius for pressure-supported astronomical systems. The star-shaped points are globular clusters (Pryor & Meylen 1993, Trager et al. 1993), the points are massive molecular clouds in the Galaxy (Solomon et al. 1987), the triangles are the dwarf spheroidal satellites of the Galaxy (Mateo 1998), the dashes are compact elliptical galaxies (Bender et al. 1992), the crosses are massive elliptical galaxies (Jørgensen et al. 1995a,b; Jørgensen 1999), and the squares are X-ray-emitting clusters of galaxies (White et al. 1997). The solid line shows the relation $\sigma_l^2/r = a_o$, and the dashed lines a factor of 5 variation about this relation.

clusters it is the well-known effective radius; for the X-ray clusters it is an X-ray intensity isophotal radius; and for the molecular clouds it is an isophotal radius of CO emission. The velocity dispersion refers to the central velocity dispersion for ellipticals and globulars; for the clusters it is the thermal velocity dispersion of the hot gas; for the molecular clouds it is just the typical line width of the CO emission.

The parallel lines represent fixed internal accelerations. The solid line corresponds to $\sigma_l^2/r = 10^{-8}$ cm/s^2, and the parallel dashed lines to accelerations five times larger or smaller than this particular value. It is clear from this diagram that the internal accelerations in most of these systems are within a factor of a few a_o. This also implies that the surface densities in these systems are near the MOND surface density Σ_m.

It is easy to over-interpret such a log-log plot containing different classes of objects covering many orders of magnitude in each coordinate. We do not wish to claim a velocity–dispersion size correlation, although such a relationship has been previously noticed for individual classes of objects—in particular, for molecular clouds (Solomon et al. 1987) and clusters of galaxies (Mohr & Evrard 1997). Probable pressure-supported systems such as super-clusters of galaxies (Eisenstein et al. 1996) and Ly α forest clouds (Schaye 2001) are clearly not on this relation, but there are low-density solutions for MOND isothermal objects (Milgrom 1984) that have internal accelerations far below a_o.

It has been noted above that, with MOND, if certain very general conditions are met, self-gravitating, pressure-supported systems would be expected to have internal accelerations comparable to or less than a_o. The essential condition is that these objects should be approximately isothermal. It is not at all evident how Newtonian theory with dark matter can account for the fact that these different classes of astronomical objects, covering a large range in size and located in very different environments, all appear to have comparable internal accelerations. In the context of MOND the location of an object in this diagram, above or below the $\sigma_l^2/r = a_o$ line, is an indicator of the internal dynamics and the extent to which these dynamics deviate from Newtonian theory.

Now we consider individual classes of objects on Figure 7.

Luminous Elliptical Galaxies

Systems above the solid line in Figure 7, e.g., the luminous elliptical galaxies, are high–surface brightness objects and, in the context of MOND, would not be expected to show a large mass discrepancy within the bright optical object. In other words, if interpreted in terms of Newtonian dynamics, these objects should not exhibit much need for dark matter within an effective radius; this is indicated by analysis of the stellar kinematics in several individual galaxies (e.g., Saglia et al. 1992). MOND isotropic, isothermal spheres have a lower mean internal acceleration within r_e (about one-quarter a_o); i.e., these theoretical objects lie significantly below the solid line in Figure 7. This was noted by Sanders (2000), who pointed out that, to be consistent with their observed distribution in the $r - \sigma_l$ plane, elliptical galaxies cannot be represented by MOND isothermal spheres; these objects must deviate both from being perfectly isothermal (in the sense that the velocity dispersion decreases outward) and from perfect isotropy of the velocity distribution (in the sense that stellar orbits become radial in the outer regions).

The general properties of ellipticals can be matched by representing these objects as high-order, anisotropic polytropic spheres; i.e., objects having a radial velocity dispersion-density relation of the form

$$\sigma_r^2 = A\rho^{1/n}, \tag{12}$$

where A is a constant depending upon n, the polytropic index. In these models the deviation from isotropy toward more radial orbits appears beyond an anisotropy radius, r_a. To reproduce the distribution of ellipticals in the $r_e - \sigma_l$ plane, it must be the case that $12 \leq n \leq 16$ and that $r_a > r_e$ (i.e., the radial orbit anisotropy does not extend within an effective radius). However, the strict homology of models is broken and the mass-velocity relation given above for isotropic, isothermal spheres (Equation 8) is replaced by a more general relation of the form

$$\sigma_l^4 = q(\Sigma/\Sigma_m)\, Ga_o M. \tag{13}$$

That is to say, for these mixed Newtonian-MOND objects, the mean surface density enters as an additional parameter; actual elliptical galaxies would comprise a two-parameter family and not a one-parameter family, as suggested by the simple MOND $M - \sigma_l$ relation for a homologous class of objects.

This is consistent with the fact that elliptical galaxies do seem to comprise a two-parameter family, as indicated by the small scatter about the "fundamental plane"—a relation between the luminosity, effective radius, and central line-of-sight velocity dispersion of the form $L \propto \sigma^a r_e^b$, where $a \approx 1.5$ and $b \approx 0.8$ (Dressler et al. 1987, Djorgovsky & Davis 1987). This has generally been attributed to the traditional virial theorem combined with a systematic dependence of M/L upon luminosity (e.g., van Albada et al. 1995). With MOND high-order polytropes are Newtonian in the inner region and MONDian beyond an effective radius. MOND imposes boundary conditions upon the inner Newtonian solution that restrict these solutions to a dynamical fundamental plane, i.e., $M \propto \sigma^\alpha r_e^\gamma$, where the exponents may differ from the Newtonian expectations. The breaking of homology leads to a considerable dispersion in the $M - \sigma$ relation owing to a factor of 10 dispersion in q in Equation 13. This is shown in Figure 8a, which is the $M - \sigma$ relation for the anisotropic polytropes covering the required range in n and r_a. A least-square fit yields

$$M/(10^{11} M_\odot) = 2 \times 10^{-8} [\sigma_d(\text{km/s})]^{3.47}, \tag{14}$$

where σ_d is the velocity dispersion measured within a circular aperture of fixed linear size as defined by Jørgensen et al. (1995a). The fact that q in Equation 13 is a well-defined function of mean surface brightness (roughly a power-law) results in a tighter fundamental plane relation (Figure 8b) of the form

$$M/(10^{11} M_\odot) = 3 \times 10^{-5} [\sigma_d(\text{km s}^{-1})]^{1.76} [r_e(\text{kpc})]^{0.98} \tag{15}$$

(see Figure 8b). With $M/L \propto M^{0.17}$, the fundamental plane in its observed form is reproduced.

The existence of a fundamental plane, in itself, is not a critical test for MOND because Newtonian theory also predicts such a relation via the virial theorem. However, for MOND a single relation of the form of Equation 15 applies for range of nonhomologous models; this is due to the underlying dynamics and not to the details of galaxy formation or subsequent dynamical evolution. A curious aspect of the Newtonian basis for the fundamental plane is the small scatter in the observed relation given the likely deviations from homology—particularly considering a dynamical history that presumably involves multiple mergers. Moreover, Newtonian

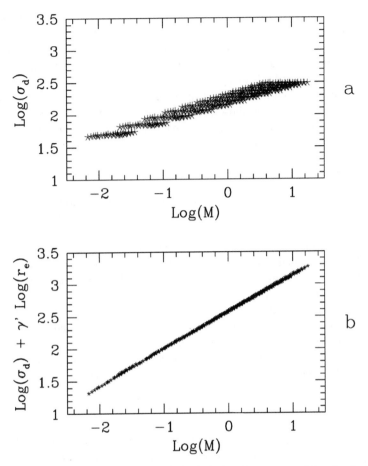

Figure 8   (*a*) The mass-velocity dispersion relation for an ensemble of anisotropic polytropes covering the range necessary to produce the observed properties of elliptical galaxies. Mass is in units of 10^{11} $M_\odot$ and velocity in kilometers/second. (*b*) The result of entering a third parameter; i.e., this is the best-fitting fundamental plane relation. $\log(\sigma_d) + \gamma' \log(r_e)$ is plotted against $\log(M)$ and γ' is chosen to give the lowest scatter (r_e is in kiloparsecs). The resulting slope is about 1.76 with $\gamma' = 0.56$. From Sanders (2000).

theory offers no explanation for the existence of a mass-velocity dispersion relation (even one with large scatter). As noted above, in the context of MOND, a near-isothermal object with a velocity dispersion of a few hundred km/s will always have a galactic mass.

The compact dwarf ellipticals (the dashes in Figure 7) have internal accelerations considerably greater than a_o and a mean surface brightness larger than Σ_m. In the context of MOND this can only be understood if these objects deviate considerably from isothermality. If approximated by polytropic spheres, these objects would have a polytropic index, n, less than about 10 (for $n \leq 5$ Newtonian polytropes no longer have infinite extent and are not necessarily MONDian objects; thus, there are no restrictions upon the internal accelerations or mean surface densities). This leads to a prediction: For such high surface brightness objects the line-of-sight stellar velocity dispersion should fall more dramatically with projected radius than in those systems with $\langle \Sigma \rangle \approx \Sigma_m$. We might also expect the compact dwarfs not to lie on the fundamental plane as defined by the lower surface brightness ellipticals.

Dwarf Spheroidal Systems

With MOND, systems that lie below the solid line in Figure 7, i.e., those systems with low internal accelerations, would be expected to exhibit larger discrepancies. This is particularly true of the dwarf spheroidal systems with internal accelerations ranging down to 0.1 a_o. On the basis of the low surface brightness of these systems Milgrom (1983b) predicted that, when velocity dispersion data became available for the stellar component, these systems should have a dynamical mass 10 or more times larger than that accounted for by the stars. These kinematic data are now available, and, indeed, these systems, when considered in the context of Newtonian dynamics, require a significant dark matter content, as is indicated by M/L values in the range of 10–100 (Mateo 1998).

For a spherically symmetric, isolated, low-density object that is deep in the MOND regime, a general mass estimator is given by

$$M = \frac{81}{4} \frac{\sigma^4}{G a_o}, \qquad (16)$$

where σ is the line-of-sight velocity dispersion (Gerhard & Spergel 1992, Milgrom 1994b). However, in estimating the dynamical mass of dwarf spheroidals with MOND, one must consider the fact that these objects are near the Galaxy, and the external field effect may be important. A measure of the degree of isolation of such an object would be given by

$$\eta = \frac{3\sigma^2/2r_c}{V_\infty^2/R} \approx \frac{g_i}{g_e}, \qquad (17)$$

where r_c is the core radius, V_∞ is the asymptotic rotation velocity of the Galaxy (≈ 200 km/s), and R is the galactocentric distance of the dwarf. For $\eta < 1$ the dwarf spheroidal is dominated by the Galactic acceleration field, and the external field effect must be taken into account. In this case the dynamical mass is simply given by the Newtonian estimate with the effective constant of gravity multiplied by a_o/g_e. In the opposite limit the MOND mass estimator for a system deep in the MOND limit is given by Equation 16.

Gerhard & Spergel (1992) and Gerhard (1994) have argued that MOND M/L values for dwarf spheroidals, based upon these estimators, have a very large range, with, for example, Fornax requiring a global M/L between 0.2 and 0.3, whereas the UMi dwarf requires M/L $\approx$ 10–13; that is to say, although some implied M/L values are unrealistically low, in other cases MOND still seems to require dark matter. Milgrom (1995) has responded to this criticism by pointing out that the kinematic data on dwarf spheroidals is very much in a state of flux and when more recent values for σ are used along with the realistic error estimates, the MOND M/L values for the dwarf spheroidals span a very reasonable range—on the order of one to three. Velocity dispersion data on dwarf spheroidals compiled by Mateo (1998) yield the MOND M/L values shown in Figure 9. In addition, MOND seems to fit the radial variation of velocity dispersion with a plausible amount of anisotropy (comparable to or less than required by dark matter) in the cases for which such data are available (Lokas 2001). It is clear that, for this class of objects, improved data gives MOND estimates for M/L values that are generally consistent with that expected for standard stellar populations.

Globular Clusters and Molecular Clouds

The globular clusters in Figure 7 generally lie well above the solid line; i.e., the internal accelerations are in excess of a_o. This implies that these systems should show no significant mass discrepancy within the half-light radius, as seems to be implied by the very reasonable M/L values based upon Newtonian dynamics. For a set of 56 globular clusters tabulated by Pryor & Meylan (1993) the mean Newtonian M/L$_V$ is 2.4 $\pm$ 1. There are several cases of globular clusters with very low internal accelerations (for example NGC 6366 having $g_i/a_o \approx 0.07$), but these are generally cases in which the external Galactic field dominates (i.e., this object is only 4 kpc from the Galactic Center and $g_e > a_o$). Periodic tidal shocks may also affect the internal dynamics of the systems and result in larger core radii than if the systems were completely isolated.

The massive molecular clouds in the Galaxy are a unique class of objects to be considered in this context, in the sense that they are not generally included in discussions of the dark matter problem or global scaling relations. However, we see from Figure 7 that the internal accelerations within these objects are also roughly comparable to a_o—a fact that emerged from the empirically discovered size-line width relation for molecular clouds in the Galaxy (Solomon et al. 1987). Milgrom (1989b) noticed that this also implies that the surface density of molecular clouds is comparable to Σ_m—a property too striking to be entirely coincidental. The suggested explanation is that molecular clouds with $g_i > a_o \approx g_e$ expand via classical internal two-body evaporation until $g_i \approx a_o$, at which point they encounter a barrier to further evaporation; this can be seen as a consequence of the fact that an isolated system in MOND is always bound. If, alternatively, $g_i \ll a_o \approx g_e$, then there is no barrier to tidal disruption in the Galaxy. Thereby, a_o emerges as a preferred internal acceleration for molecular clouds. Regions in a galaxy where $g_e > a_o$ would, as an additional consequence, lack massive molecular clouds (as in the inner 3 kpc of the Galaxy apart from the exceptional Galactic Center clouds). The

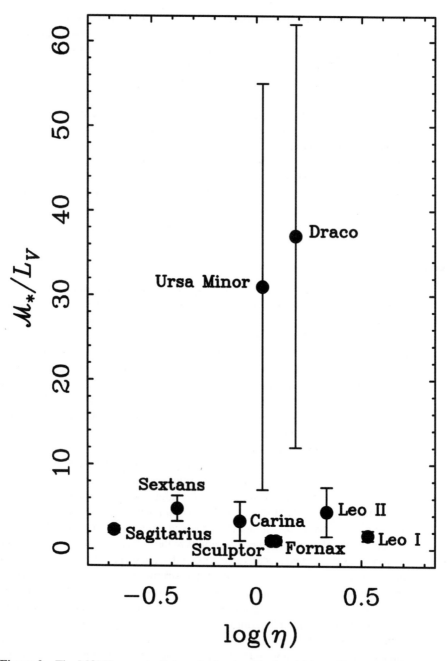

Figure 9 The MOND mass-to-light ratio for dwarf spheroidal satellites of the Galaxy as a function of η, the ratio of the internal to external acceleration. This is the parameter that quantifies the influence of the Galactic acceleration field (the external field effect); when $\eta < 1$ the object is dominated by the external field.

fact that the molecular clouds lie somewhat below the solid line in Figure 7 would also suggest that, viewed in the context of Newtonian dynamics, there should be a dark matter problem for molecular clouds; that is, the classical dynamically inferred mass should be somewhat larger than the mass derived by counting molecules.

Solomon et al. (1987) noted that combining the size–line width relation with the Newtonian virial theorem and an empirical mass-CO luminosity relation for molecular clouds results in a luminosity–line width relation that is analogous to the Faber-Jackson relation for ellipticals. Viewed in terms of MOND, the corresponding mass-velocity dispersion relation is not just analogous: It is the low mass extrapolation of the same relation that applies to all pressure-supported, nearly isothermal systems up to and including clusters of galaxies. If one applies Equation 8 to objects with a velocity dispersion of 4 or 5 km/s (typical of giant molecular clouds), then one deduces a mass of a few times 10^5 $M_\odot$. No explanation of global scaling relations for extragalactic objects in terms of dark matter can accommodate the extension of the relation to such subgalactic objects.

Small Groups of Galaxies

We include small galaxy groups in this section on pressure-supported systems even though this is more properly a small n-body problem. Proceeding from individual galaxies, the next rung on the ladder is binary galaxies, but it is difficult to extract meaningful dynamical information about these systems, primarily because of high contamination by false pairs. The situation improves for small groups because of more secure identification with an increasing number of members. Although uncertainties in the mass determination of individual groups remains large, either in the context of Newtonian dynamics or MOND, it is likely that statistical values deduced for selected samples of groups may be representative of the dynamics.

This problem has been considered by Milgrom (1998), who looked primarily at a catalogue of groups by Tucker et al. (1998) taken from the Las Campanas Redshift Survey. The median orbital acceleration of galaxies in this sample of groups is on the order of a few percent of a_o, so these systems are in the deep MOND regime. Milgrom therefore applied the MOND mass estimator relevant to this limit (Equation 16) and found that median M/L values are reduced from about 100 based on Newtonian dynamics to around 3 with MOND. Given the remaining large uncertainties owing to group identification and unknown geometry, these results are consistent with no dark matter in groups when considered in terms of MOND.

Rich Clusters of Galaxies

Clusters of galaxies lie below the $\sigma_l^2/r = a_o$ line in Figure 7; thus, these objects would be expected to exhibit significant discrepancies. That this is the case has been known for 70 years (Zwicky 1933), although the subsequent discovery of hot X-ray-emitting gas goes some way toward alleviating the original discrepancy. For an isothermal sphere of hot gas at temperature T, the Newtonian dynamical

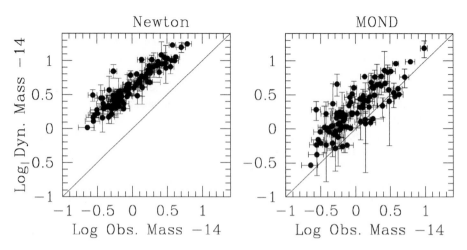

Figure 10 (*Left*) the Newtonian dynamical mass of clusters of galaxies within an observed cutoff radius (r_{out}) vs. the total observable mass in 93 X-ray-emitting clusters of galaxies (White et al. 1997). The solid line corresponds to $M_{dyn} = M_{obs}$ (no discrepancy). (*Right*) the MOND dynamical mass within r_{out} vs. the total observable mass for the same X-ray-emitting clusters. From Sanders (1999).

mass within radius r_o, calculated from the equation of hydrostatic equilibrium, is

$$M_n = \frac{r_o}{G} \frac{kT}{m} \left(\frac{d \ln(\rho)}{d \ln(r)} \right), \tag{18}$$

where m is the mean atomic mass and the logarithmic density gradient is evaluated at r_o. For the X-ray clusters tabulated by White et al. (1997), this Newtonian dynamical mass plotted against the observable mass, primarily in hot gas, is shown in Figure 10 (Sanders 1999), in which we see that the dynamical mass is typically about a factor of 4 or 5 larger than the observed mass in hot gas and in the stellar content of the galaxies. This rather modest discrepancy viewed in terms of dark matter has led to the so-called baryon catastrophe: not enough nonbaryonic dark matter in the context of standard CDM cosmology (White et al. 1993).

With MOND, the dynamical mass [assuming an isothermal gas (Equation 8)] is given by

$$M_m = (Ga_o)^{-1} \left(\frac{kT}{m} \right)^2 \left(\frac{d \ln(\rho)}{d \ln(r)} \right)^2, \tag{19}$$

and this is also shown in Figure 10, again plotted against the observable mass. The larger scatter is due to the fact that the temperature and the logarithmic density gradient enter quadratically in the MOND mass determination. Here we see that, using the same value of a_o determined from nearby galaxy rotation curves, the discrepancy is on average reduced to about a factor of 2. The fact that MOND predicts more mass than is seen in clusters has been pointed out previously in the specific example of the Coma cluster (The & White 1988) and for three clusters with measured temperature gradients for which the problem is most evident in the central regions (Aguirre et al. 2001).

The presence of a central discrepancy is also suggested by strong gravitational lensing in clusters, i.e., the formation of multiple images of background sources in the central regions of some clusters (Sanders 1999). The critical surface density required for strong lensing is

$$\Sigma_c = \frac{1}{4\pi} \frac{cH_o}{G} F(z_l, z_s),$$ (20)

where F is a dimensionless function of the lens and source redshifts that depends upon the cosmological model (Blandford & Narayan 1992); typically for clusters and background sources at cosmological distances $F \approx 10$ (assuming that a MOND cosmology is reasonably standard). It is then evident that $\Sigma_c > \Sigma_m$, which means that strong gravitational lensing always occurs in the Newtonian regime. MOND cannot help with any discrepancy detected by strong gravitational lensing. Because strong gravitational lensing in clusters typically indicates a projected mass within 200–300 kpc between 10^{13} and 10^{14} $M_\odot$, which is not evidently present in the form of hot gas and luminous stars, it is clear that there is a missing mass problem in the central regions of clusters that cannot be repaired by MOND. This remaining discrepancy could be interpreted as a failure, or one could say that MOND predicts that the baryonic mass budget of clusters is not yet complete and that there is more mass to be detected (Sanders 1999). It would have certainly been devastating for MOND had the predicted mass turned out to be typically less than the observed mass in hot gas and stars; this would be a definitive falsification.

There is an additional important aspect of clusters of galaxies regarding global scaling relations. As pressure-supported, near-isothermal objects, clusters should lie roughly upon the same $M - \sigma$ relation defined by the elliptical galaxies. That this is the case was first pointed out by Sanders (1994), using X-ray observations of about 16 clusters that apparently lie upon the extension of the Faber-Jackson relation for elliptical galaxies. From Equation 8 we find that an object having a line-of-sight velocity dispersion of 1000 km/s would have a dynamical mass of about 0.5×10^{14} $M_\odot$, which is comparable to the baryonic mass of a rich cluster of galaxies. The fact that the Faber-Jackson relation—albeit with considerable scatter—extends from molecular clouds to massive clusters of galaxies finds a natural explanation in terms of MOND.

Super-Clusters and Lyα Forest Clouds

The largest coherent astronomical objects with the lowest internal accelerations are super-clusters of galaxies, as exemplified by the Perseus-Pisces filament. If one assumes that this object is virialized in a direction perpendicular to the long axis of the filament, then a linear mass density (μ_o) for the filament may be calculated following the arguments given by Eisenstein et al. (1996); by approximating the filament as an infinitely long, axisymmetric, isothermal cylinder, one finds

$$\mu_o = \frac{2\sigma^2}{G}. \tag{21}$$

Applying this relation to Perseus-Pisces, these authors estimated a global M/L in the super-cluster of 450 h—indicating a serious mass discrepancy.

Milgrom (1997b) has generalized the arguments of Eisenstein et al. and found that a relation similar to Equation 21 holds even if one drops the assumptions of axial symmetry and isothermality. He then derived a MOND estimator for the line density of a filament:

$$\mu_o = Q \frac{\langle \sigma_{\perp}^2 \rangle^2}{a_o G r_h}, \tag{22}$$

where $\sigma_{\perp}$ is the velocity dispersion perpendicular to the filament axis, r_h is the half mass radius, and Q (≈ 2) depends upon the velocity anisotropy factor. Applying this expression to Perseus-Pisces, Milgrom found an M/L value on the order of 10; once again the MOND M/L seems to require little or no dark matter, even on this very large scale. This is significant because the internal acceleration in this object is on the order of 0.03 a_o, which suggests that the MOND formula applies, at least approximately, down to this very low acceleration.

The diffuse intergalactic clouds resulting in the Ly$_\alpha$ forest absorption lines in the spectra of distant quasars are also apparently objects with internal acceleration very much lower than a_o. These have been considered as self-gravitating objects both in the context of dark matter (Rees 1986) and MOND (Milgrom 1988). There is now evidence that the sizes of individual absorbers may be as large as 100 kpc, as indicated by observations of gravitationally lensed quasars and quasar pairs (Schaye 2001). Given that the widths of the absorption lines are on the order of 10 km/s, then the internal accelerations within these systems may be as small as $3 \times 10^{-4} a_o$. Schaye (2001) has argued that the characteristic sizes of the Lyα clouds are most likely to be comparable to the Jeans length. In the context of MOND this would be

$$\lambda_J = \left(\frac{\sigma^4}{G a_o \Sigma} \right)^{\frac{1}{2}}, \tag{23}$$

where Σ is the mean surface density. Because the fractional ionization is likely to be very high (Σ is dominated by protons), one finds that this characteristic size, in terms of MOND should be more on the order of 10 kpc, in contradiction to observations of common lines in quasar pairs. On this basis, Aguirre et al. (2001) have argued that the observed large sizes of the absorbers, perpendicular to the line of sight, are inconsistent with the predictions of MOND.

However, these authors noted that the external field effect provides a possible escape. The implied internal accelerations of the clouds, if they are roughly spherical with sizes of 100 kpc, are likely to be much smaller than the external acceleration field resulting from large-scale structure, which, as we saw above, is on the order of several percent a_o. In this case the Jean's length is given by the traditional Newtonian formula with an effective constant of gravity, which may be 20 or 30 times larger than G, and the sizes can be consistent with the large observed extent.

THE PHYSICAL BASIS OF MOND

The Bekenstein-Milgrom Theory

In spite of its empirical success, MOND remains a largely ad hoc modification of Newtonian gravity or dynamics without connection to a more familiar theoretical framework. This is, at present, the essential weakness of the idea. The original algorithm (Equation 2 or 3) cannot be considered as a theory but as a successful phenomenological scheme for which an underlying theory is necessary. If one attempts to apply Milgrom's original prescription (either as a modification of gravity or inertia) to an N-body system, then immediate physical problems arise, such as nonconservation of linear momentum (Felten 1984).

Bekenstein & Milgrom (1984) recognized this and proposed a nonrelativistic Lagrangian-based theory of MOND as a modification of Newtonian gravity. Given a scalar potential ϕ, the dynamics of the theory is contained in field action

$$S_f = -\int d^3r \left[\rho\phi + (8\pi G)^{-1} a_o^2 F\left(\frac{\nabla\phi^2}{a_o^2}\right) \right]. \tag{24}$$

The particle action takes its standard form. The field equation derived, as usual, under the assumption of stationary action is

$$\nabla \cdot \left[\mu\left(\frac{|\nabla\phi|}{a_o}\right) \nabla\phi \right] = 4\pi G\rho, \tag{25}$$

where the function $\mu(x) = dF/dx^2$ must have the asymptotic behavior required in the simple MOND prescription; i.e., $F(x^2) = (x^2)^{3/2}$ in the MOND limit ($x \ll 1$) and $F(x^2) = x^2$ in the Newtonian limit. The equation of motion for a particle assumes its usual Newtonian form.

Because of the symmetry of the Lagrangian density to space-time translations (and to space rotations), the theory respects the laws of conservation of energy and (angular) momentum. Moreover, Bekenstein & Milgrom demonstrated that, in the context of this theory, the motion of a compound object (e.g., a star or star cluster in the Galaxy) in an external field is independent of its internal structure (or internal accelerations) and may be described in terms of its center-of-mass accelerations; i.e., objects like stars with Newtonian internal accelerations behave like billiard balls in the external field, even in the MOND limit. However, the external acceleration field does affect the internal dynamics of such a subsystem in just the way proposed by Milgrom—the external field effect.

In addition to enjoying the properties of consistency and conservation, this modified Poisson equation has an interesting symmetry property. It is well known that the usual Poisson equation is conformally invariant in two spatial dimensions. Conformal transformations comprise a set of angle-preserving coordinate transformations that represent, in effect, a position-dependent transformation of units of length. Many of the well-known equations of physics (e.g., Maxwell's equations) are invariant under transformations of this form. Milgrom (1997a) discovered that there is a nonlinear generalization of the Poisson equation that is conformally invariant in D spatial dimensions in the presence of a source ρ. This is of the form

$$\nabla \cdot \left\{ [(\nabla \phi)^2]^{D/2 - 1} \nabla \phi \right\} = \alpha_D G \rho. \tag{26}$$

When $D = 2$ Equation 26 becomes the usual Poisson equation, but when $D = 3$ the equation takes the form that is exactly required for MOND phenomenology. In other words, the Bekenstein-Milgrom field equation in the MOND limit is conformally invariant in three spatial dimensions. The full significance of this result is unclear, but it should be recalled that much of modern physics rests upon just such symmetry principles.

The Bekenstein-Milgrom theory is a significant step beyond Milgrom's original prescription. Even though the theory is noncovariant, it demonstrates that MOND can be placed upon a solid theoretical basis and that MOND phenomenology is not necessarily in contradiction with cherished physical principles. Although this is its essential significance, the theory also permits a more rigorous consideration of specific aspects of MOND phenomenology relating to N-body systems—such as the external field effect.

It is, in general, difficult to solve this nonlinear equation except in cases of high symmetry in which the solution reduces to that given by the simple algorithm. Brada & Milgrom (1995) have derived analytic solutions for Kuzmin disks and of their disk-plus-bulge generalizations. The solution can be obtained in the form of a simple algebraic relation between the Bekenstein-Milgrom solution and the Newtonian field of the same mass distribution, and this relation can be extended to a wider class of disk configurations (e.g., exponential disks) where it holds approximately. From this work, it is evident that the simple MOND relation (Equation 3) gives a radial force distribution in a thin disk, which is generally within 10% of that determined by the Bekenstein-Milgrom field equation.

Brada (1997) has developed a numerical method of solution for N-body problems based upon a multigrid technique, and Brada & Milgrom (1999b) used this method to consider the important problem of stability of disk galaxies. They demonstrated that MOND, as anticipated (Milgrom 1989a), has an effect similar to a dark halo in stabilizing a rotationally supported disk against bar-forming modes. However, there is also a significant difference (also anticipated by Milgrom 1989a). In a comparison of MOND and Newtonian truncated exponential disks with identical rotation curves (the extra force in the Newtonian case being provided by a rigid dark halo), Brada & Milgrom found that, as the mean surface density of the disk decreases (the disk sinks deeper into the MOND regime), the growth rate of the bar-forming $m = 2$ mode decreases similarly in the two cases. However, in the limit of very low surface densities the MOND growth rate saturates while

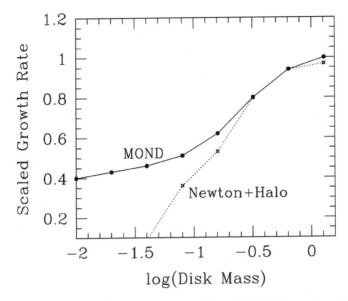

Figure 11   The scaled growth rate of the $m = 2$ instability in Newtonian disks with dark matter (*dashed line*) and MONDian disks as a function of disk mass. In the MOND case as the disk mass decreases, the surface density decreases and the disk sinks deeper into the MOND regime. In the equivalent Newtonian case, the rotation curve is maintained at the MOND level by supplementing the force with an inert dark halo. From Brada & Milgrom 1999b.

the Newtonian growth rate continues to decrease as the halo becomes more dominant. This effect, shown in Figure 11, may provide an important observational test: With MOND, LSB galaxies remain marginally unstable to bar- and spiral-forming modes, whereas in the dark matter case, halo-dominated LSB disks become totally stable. Observed LSB galaxies do have bars and $m = 2$ spirals (McGaugh et al. 1995a). In the context of dark matter, these signatures of self-gravity are difficult to understand in galaxies that are totally halo-dominated (Mihos et al. 1997).

The Brada method has also been applied to calculating various consequences of the external field effect, such as the influence of a satellite in producing a warp in the plane of a parent galaxy (Brada & Milgrom 2000a). The idea is that MOND, via the external field effect, offers a mechanism other than the relatively weak effect of tides in inducing and maintaining warps. As noted above, the external field effect is a nonlinear aspect of MOND, subsumed by the Bekenstein-Milgrom field equation; unlike Newtonian theory, even a constant external acceleration field influences the internal dynamics of a system. Brada & Milgrom (2000a) demonstrated that a satellite at the position and with the mass of the Magellanic clouds can produce a warp in the plane of the Galaxy with about the right amplitude and form.

The response of dwarf satellite galaxies to the acceleration field of a large parent galaxy has also been considered (Brada & Milgrom 2000b). It was found that the satellites become more vulnerable to tidal disruption because of the expansion induced by the external field effect as they approach the parent galaxy (the effective constant of gravity decreases toward its Newtonian value). The distribution of satellite orbits is therefore expected to differ from the case of Newtonian gravity plus dark matter, although it is difficult to make definitive predictions because of the unknown initial distribution of orbital parameters.

Although the Bekenstein-Milgrom theory is an important development for all the reasons outlined above, it must be emphasized that it remains, at best, only a clue to the underlying theoretical basis of MOND. The physical basis of MOND may lie completely in another direction—as modified Newtonian inertia rather than gravity. However, a major advantage of the theory is that it lends itself immediately to a covariant generalization as a nonlinear scalar-tensor theory of gravity.

Modified Newtonian Dynamics as a Modification of General Relativity

As noted above, the near coincidence of a_o with cH_o suggests that MOND may reflect the influence of cosmology upon particle dynamics or $1/r^2$ attraction. However, in the context of general relativity there is no such influence of this order, with or without the maximum permissible cosmological term—a fact that may be deduced from the Birkhoff theorem (Nordtvedt & Will 1972). Therefore, general relativity, in a cosmological context, cannot be the effective theory of MOND, although the theory underlying MOND must effectively reduce to general relativity in the limit of high accelerations [see Will (2001) for current experimental constraints on strong field deviations from general relativity].

The first suggested candidate theory (Bekenstein & Milgrom 1984) is an unconventional scalar-tensor theory that is a covariant extension of the nonrelativistic Bekenstein-Milgrom theory. Here the Lagrangian for the scalar field is given by

$$L_s = \frac{a_o^2}{c^4} F\left[\frac{\phi_{,\alpha}\phi^{,\alpha}c^4}{a_o^2}\right], \tag{27}$$

where F(X) is an arbitrary positive function of its dimensionless argument. This scalar field, as usual, interacts with matter jointly with $g_{\mu\nu}$ via a conformal transformation of the metric, i.e., the form of the interaction Lagrangian is taken to be

$$L_I = L_I[\xi(\phi^2)g_{\mu\nu}\ldots], \tag{28}$$

where ξ is a function of the scalar field (this form preserves weak equivalence where particles follow geodesics of a physical metric $\hat{g}_{\mu\nu} = \xi(\phi^2)g_{\mu\nu}$). The scalar field action $(\int L_s\sqrt{-g}d^4x)$ is combined with the usual Einstein-Hilbert action of general relativity and the particle action formed from L_I to give the complete theory.

Thus, the covariant form of the Bekenstein-Milgrom field equation becomes

$$(\mu \phi^{\cdot \alpha})_{;\alpha} = \frac{4\pi GT}{c^4}, \tag{29}$$

where, again, $\mu = dF/dX$ and T is the contracted energy-momentum tensor. The complete theory includes the Einstein field equation with an additional source term owing to the contribution of the scalar field to the energy-momentum tensor. Again we require that F(X) has the asymptotic behavior $F(X) \to X^{\frac{3}{2}}$ in the limit where $X \ll 1$ (the MOND limit) and $F(X) \to \omega X$ in the limit of $X \gg 1$. Thus, in the limit of large field gradients the theory becomes a standard scalar-tensor theory of Brans-Dicke form (Brans & Dicke 1961); it is necessary that $\omega \gg 1000$ if the theory is to be consistent with local solar system and binary pulsar tests of general relativity (Will 2001). Because of the nonstandard kinetic Lagrangian (Equation 27) Bekenstein (1988) termed this theory the aquadratic Lagrangian or AQUAL theory.

Bekenstein & Milgrom (1984) immediately noticed a physical problem with the theory: Small disturbances in the scalar field propagate at a velocity faster than the speed of light in directions parallel to the field gradient in the MOND regime. This undesirable property appears to be directly related to the aquadratic form of the Lagrangian and is inevitably true in any such theory in which the scalar force decreases less rapidly than $1/r^2$ in the limit of low field gradient. Clearly, the avoidance of causality paradoxes, if only in principle, should be a criterion for physical viability.

The acausal propagation anomaly led Bekenstein (1988a,b) to propose an alternative scalar-tensor theory in which the field is complex

$$\chi = Ae^{i\phi} \tag{30}$$

and the Lagrangian assumes its usual quadratic form

$$L_s = \frac{1}{2}A^2 \phi_{,\alpha}\phi^{\cdot \alpha} + A_{,\alpha}A^{\cdot \alpha} + V(A^2), \tag{31}$$

where $V(A^2)$ is a potential associated with the scalar field. The unique aspect of the theory is that only the phase couples to matter (jointly with $g_{\mu\nu}$ as in Equation 28); hence, it is designated "phase coupling gravitation" (PCG). The field equation for the matter coupling field is then found to be

$$(A^2 \phi^{\cdot \alpha})_{;\alpha} = \frac{4\pi \eta G}{c^4}T, \tag{32}$$

where η is a dimensionless parameter describing the strength of the coupling to matter. Thus, the term A^2 plays the role of the MOND function μ; A^2 is also a function of $(\nabla \phi)^2$, but the relationship is differential instead of algebraic. Bekenstein noted that if $V(A^2) = -kA^6$ precisely, the phenomenology of MOND is recovered by the scalar field ϕ (here a_o is related to the parameters k and η). Because such a potential implies an unstable vacuum, alternative forms were considered by Sanders (1988), who demonstrated that phenomenology similar to MOND is predicted as long as $dV/dA < 0$ over some range of A. Moreover, in a cosmological context (Sanders 1989) PCG, with the properly chosen bare potential, becomes an effective MOND theory where the cosmological $\dot{\phi}$ plays the role of a_o. Romatka (1992) considered PCG as one of a class of two-scalar plus tensor theories in which the scalar fields couple in one of their kinetic terms and demonstrated that, in a certain limit, Bekenstein's sextic potential theory approaches the original AQUAL theory. This suggests that PCG may suffer from a similar physical anomaly as AQUAL, and indeed, Bekenstein (1990) discovered that PCG apparently permits no stable background solution for the field equations—an illness as serious as that of the acausal propagation that the theory was invented to cure.

A far more practical problem with AQUAL, PCG, or, in fact, all scalar-tensor theories in which the scalar field enters as a conformal factor multiplying the Einstein metric (Equation 28) is the failure to predict gravitational lensing at the level observed in rich clusters of galaxies (Bekenstein & Sanders 1994). If one wishes to replace dark matter by a modified theory of gravity of the scalar-tensor type with the standard coupling to matter, then the scalar field produces no enhanced deflection of light. The reason for this is easy to understand: In scalar-tensor theories particles follow geodesics of a physical metric that is conformally related (as in Equation 28) to the usual Einstein metric. But Maxwell's equations are conformally invariant, which means that photons take no notice of the scalar field (null geodesics of the physical and Einstein metrics coincide). In other words, the gravitational lensing mass of an astronomical system should be comparable to that of the detectable mass in stars and gas and thus much less than the traditional virial mass. This is in sharp contrast to the observations (Fort & Mellier 1994).

A possible cure for this ailment is a nonconformal relation between the physical and Einstein metrics; that is, in transforming the Einstein metric to the physical metric, a special direction is picked out for additional squeezing or stretching (Bekenstein 1992, 1993). To preserve the isotropy of space, this direction is usually chosen to be time-like in some preferred cosmological frame as in the classical stratified theories (Ni 1972). In this way one may reproduce the general relativistic relation between the weak-field force on slow particles and the deflection of light (Sanders 1997). However, the Lorentz invariance of gravitational dynamics is broken, and observable preferred frame effects—such as a polarization of the earth-moon orbit (Müller et al. 1996)—are inevitable at some level. It is of interest that an aquadratic Lagrangian for the scalar field (similar to Equation 27) can provide a mechanism for local suppression of these effects; essentially, the scalar force may be suppressed far below the Einstein-Newton force in the limit of solar system accelerations. On this basis, one could speculate that cosmology is described by a preferred frame theory (there is clearly a preferred cosmological frame from an observational point of view). Then it may be argued that the reconciliation of preferred frame cosmology with general relativistic local dynamics (weak local preferred frame effects) requires MOND phenomenology at low accelerations (Sanders 1997). However, any actual theory is highly contrived at this point.

In summary, it is fair to say that, at present, there is no satisfactory covariant generalization of MOND as a modification of general relativity. But this does not imply that MOND is wrong any more than the absence of a viable theory of quantum gravity implies that general relativity is wrong. It is simply a statement that the theory remains incomplete and that perhaps the tinkering with general relativity is not the ideal way to proceed.

Modified Newtonian Dynamics as a Modification of Newtonian Inertia

A different approach has been taken by Milgrom (1994a, 1999), who considers the possibility that MOND may be viewed as a modification of particle inertia. In such theories, at a nonrelativistic level, one replaces the standard particle action $(\int v^2/2\, dt)$ by a more complicated object, $A_m S[\mathbf{r}(t), a_o]$, where A_m depends upon the body and can be identified with the particle mass, and S is a functional of the particles trajectory, $\mathbf{r}(t)$, characterized by the parameter a_o. This form ensures weak equivalence. Milgrom (1994a) proved that if such an action is to be Galilei invariant and have the correct limiting behavior (Newtonian as $a_o \to 0$ and MONDian as $a_o \to \infty$), then it must be strongly nonlocal; i.e., the motion of a particle at a point in space depends upon its entire past trajectory. This nonlocality has certain advantages in a dynamical theory: For example, because a particle's motion depends upon an infinite number of time-derivatives of the particle's position, the theory does not suffer from the instabilities typical of higher derivative (weakly nonlocal) theories. Moreover, because of the nonlocality, the acceleration of the center of mass of a composite body emerges as the relevant factor in determining its dynamics (Newtonian or MONDian) rather than the acceleration of its individual components. Milgrom further demonstrated that, in the context of such theories, the simple MOND relation (Equation 2) is exact for circular orbits in an axisymmetric potential (although not for general orbits).

Although these results on the nature of generalized particle actions are of considerable interest, this, as Milgrom stresses, is not yet a theory of MOND as modified inertia. The near coincidence of a_o with cH_o suggests that MOND is, in some sense, an effective theory; that is to say, MOND phenomenology only arises when the theory is considered in a cosmological background (the same may also be true if MOND is due to a modified theory of gravity). However, the cosmology does not necessarily directly affect particle motion; the same agent—a cosmological constant—may affect both cosmology and dynamics. Suppose, for example, that inertia results from the interaction of an accelerating particle with the vacuum. Suppose further that there is a nonzero cosmological constant (which is consistent with a range of observations). Then because a cosmological constant is an attribute of the vacuum, we might expect that it has a nontrivial effect upon particle inertia at accelerations corresponding to $\approx c\sqrt{\Lambda}$.

The phenomenon of Unruh radiation provides a hint of how this might happen (Unruh 1975). An observer uniformly accelerating through Minkowski space sees a nontrivial manifestation of vacuum fields as a thermal bath at temperature

$$kT = \frac{\hbar}{2\pi c} a, \tag{33}$$

where a is the acceleration (this is exactly analogous to Hawking radiation, where a is identified with the gravitational acceleration at the event horizon). In other words, an observer can gain information about his state of motion by using a quantum detector. However, the same observer accelerating through de Sitter space sees a modified thermal bath now characterized by a temperature

$$kT_\Lambda = \frac{\hbar}{2\pi c} \sqrt{a^2 + \frac{c^2 \Lambda}{3}} \tag{34}$$

(Narnhofer et al. 1996, Deser & Levin 1997). The presence of a cosmological constant changes the accelerating observer's perception of the vacuum through the introduction of a new parameter with units of acceleration ($c\sqrt{\Lambda}$) and a magnitude comparable to a_o. If the observer did not know about the cosmological constant this would also change his perception of the state of motion.

The Unruh radiation itself is too miniscule to be directly implicated as the field providing inertia: It may be, in effect, a tracer of the particle's inertia. Milgrom (1999) has suggested that inertia may be what drives a noninertial body back to some nearby inertial state—attempting to reduce the vacuum radiation to its minimum value. If that were so then the relevant quantity with which to identify inertia would be $\Delta T = T_\Lambda - T$. In that case one could write

$$\frac{2\pi c}{\hbar} k\Delta T = a\mu(a/a_o), \tag{35}$$

where

$$\mu(x) = [1 + (2x)^{-2}]^{1/2} - (2x)^{-1} \tag{36}$$

with $a_o = 2c(\Lambda/3)^{1/2}$. Inertia defined in this way would have precisely the two limiting behaviors of MOND.

Again, this is not a theory of MOND as modified inertia but only a suggestive line of argument. To proceed further along this line, a theory of inertia derived from interaction with vacuum fields is necessary—something analogous to induced gravity (Sakharov 1968), in which the curvature of space-time modifies the behavior of vacuum fields producing an associated action for the metric field. If this approach is correct, the free action of a particle must be derived from the interactions with vacuum fields.

Gravitational Lensing and No-Go Theorems

We have noted above that the phenomenon of gravitational lensing places strong constraints upon scalar-tensor theories of modified dynamics, specifically upon the relation between the physical metric and the gravitational metric (Bekenstein & Sanders 1994, Sanders 1997). Here we wish to discuss gravitational lensing in a more general sense because it is the only measureable relativistic effect that exists on the scale of galaxies and clusters and therefore is generally relevant to proposed modifications of general relativity that may be only effective on this scale. Indeed, several authors have attempted to formulate no-go theorems for modified gravity on the basis of the observed gravitational lensing.

The first of these was Walker (1994), who considered general metrics of the standard Schwarzschild form $ds^2 = -B(r)dt^2 + A(r)dr^2 + r^2d\Omega^2$ with the condition that $A = B^{-1}$ and $B - 1 = 2\phi(r)$, where $\phi(r)$ is a general weak field potential of the form $\phi \propto r^n$. With these assumptions he demonstrated that gravity is actually repulsive for photons if $0 < n < 2$—i.e., gravitational lenses would be divergent. He used this argument to rule out the Mannheim & Kazanas (1989) spherical vacuum solution for Weyl conformal gravity in which $\phi(r)$ contains such a linear term. But then Walker went on to consider the mean convergence $\langle \kappa \rangle$ and the variance of the shear σ_γ in the context of $\phi(r) \propto \log r$ which might be relevant to MOND. Again with the condition that $A = B^{-1}$, this form of the potential would imply a mean convergence of $\langle \kappa \rangle \approx 10^7$ while the observations constrain $\kappa < 1$; that is to say, with such an effective potential the optical properties of the Universe would be dramatically different than observed.

The second no-go theorem is by Edery (1999) and is even more sweeping. Basically the claim is that, again assuming a metric in the standard Schwarzschild form with $A = B^{-1}$, any potential that yields flat rotation curves (i.e., ϕ falls less rapidly than $1/r$) is repulsive for photons even though it may be attractive for nonrelativistic particles. This was disputed by Bekenstein et al. (2000), who pointed out that solar system tests do not constrain the form of A and B on galactic scale, except in the context of a specific gravity theory. An alternative theory may exhibit $AB = 1$ to high accuracy on a solar system scale but $AB \neq 1$ on a galactic scale; indeed, this is a property of the stratified scalar-tensor theory of Sanders (1997), which predicts enhanced deflection in extragalactic sources but is also consistent with solar system gravity tests at present levels of precision.

Walker's objection would actually seem to be more problematic for MOND; here the estimate of enormous mean convergence due to galactic lenses is independent of the assumption of $AB = 1$. This problem has also emerged in a different guise in the galaxy-galaxy lensing results of Hoekstra et al. (2002). These results imply that galaxy halos have a maximum extent; if represented by isothermal spheres, the halos do not extend beyond 470 h^{-1} on average. With MOND the equivalent halo for an isolated galaxy would be infinite. Walker also noted that for MOND to be consistent with a low mean convergence, the modified force law could only extend to several Mpc at most, beyond which there must be a return to r^{-2} attraction.

The external field effect provides a likely escape from such objections in the context of pure (unmodified) MOND. Basically, no galaxy is isolated. For an L^* galaxy the acceleration at a radius of 470 kpc is about $0.02a_o$. This is at the level of the external accelerations expected from large-scale structure. For lower accelerations one would expect a return to a $1/r^2$ law with a larger effective constant of gravity. For this reason, such objections cannot be considered as a falsification of MOND. It might also be that at very low accelerations the attraction really does return to $1/r^2$, as speculated by Sanders (1986), although there is not yet a compelling reason to modify MOND in this extreme low acceleration regime.

MONDian gravitational lensing in a qualitative sense is also considered by Mortlock & Turner (2001), who after making the reasonable assumption that the relation between the weak field force and deflection in MOND is the same as in general relativity (specifically including the factor 2 over the Newtonian deflection) considered the consequences when the force is calculated from the MOND equation. The first result of interest is that the thin lens approximation (i.e., the deflection in an extended source depends only upon the surface density distribution) cannot be made with MOND; this means that the deflection depends, in general, upon the density distribution along the line of sight. They further pointed out that observations of galaxy-galaxy lensing (taking the galaxies to be point masses) is consistent with MOND, at least within the truncation noted by Hoekstra et al. (2002). Mortlock & Turner proposed that a test discriminating between MOND and dark halos would be provided by azimuthal symmetry of the galaxy-galaxy lensing signal: MOND would be consistent with such symmetry, whereas halos would not. They further noted that gravitational microlensing in the context of MOND would produce a different signature in the light curves of lensed objects (particularly in the wings) and that this could be observable in cosmological microlensing (however, this effect may be limited by the external field of the galaxy containing the microlensing objects).

In general, in extragalactic lenses such as galaxy clusters distribution of shear in background sources (and hence apparent dark mass distribution) should be calculable from the distribution of observable matter; i.e., there should be a strong correlation between the visible and, in terms of general relativity, the dark mass distribution. The theme of correlation between the observable (visible) structure of a lens and the implied shape of the dark matter distribution was taken up by Sellwood & Kosowsky (2002), who emphasized that the observed correlation in position angles between the elongated light distribution and implied mass distribution; Kochanek (2002) argues strongly in favor of some form of modified gravity.

It is clear from these discussions that gravitational lensing may provide generic tests of the MOND hypothesis vs. the dark matter hypothesis and that any more basic theory must produce lensing at a level comparable to that of general relativity with dark matter. This already strongly constrains the sort of theory that may underpin MOND.

COSMOLOGY AND THE FORMATION OF STRUCTURE

Considerations of cosmology in the context of MOND might appear to be premature in the absence of a complete theory. However, there are some very general statements that can be made about a possible MOND universe independently of any specific underlying theory. First of all, the success of the hot Big Bang with respect to predicting the thermal spectrum and isotropy of the cosmic microwave background as well as the observed abundances of the light isotopes (e.g., Tytler et al. 2000) strongly implies that a theory of MOND should preserve the standard model—at least with respect to the evolution of the early hot Universe. This, in fact, may be considered as a requirement on an underlying theory. Second, it would be contrary to the spirit of MOND if there were cosmologically significant quantities of nonbaryonic CDM ($\Omega_{cdm} \ll 1$); i.e., dark matter that clusters on the scale of galaxies. This is not to say dark matter is nonexistent; the fact that Ω_V (luminous matter) is substantially less than the Ω_b (baryons) implied by primordial nucleosynthesis (Fukugita et al. 1998) means that there are certainly as-yet-undetected baryons. Moreover, particle dark matter also exists in the form of neutrinos. It is now clear from the detection of neutrino oscillations (Fukuda et al. 1998) that at least some flavors of neutrinos have mass: The constraints are $.004h^{-2} < \Omega_v < 0.1h^{-2}$ (Turner 1999) with the upper limit imposed by the experimental limit on the electron neutrino mass (3 ev). However, it would be entirely inconsistent with MOND if dark matter, baryonic or nonbaryonic, contributed substantially to the mass budget of galaxies. Neutrinos near the upper limit of 3 ev cannot accumulate in galaxies owing to the well-known phase space constraints (Tremaine & Gunn 1979), but they could collect within and contribute to the mass budget of rich clusters of galaxies (which would not be inconsistent with MOND, as noted above). Apart from this possibility it is reasonable to assume that MOND is most consistent with a purely baryonic universe.

This possibility has been considered by McGaugh (1998), who first pointed out that, in the absence of CDM, oscillations should exist in the present power spectrum of large-scale density fluctuations—at least in the linear regime. These oscillations are the relic of the sound waves frozen into the plasma at the epoch of recombination and are suppressed in models in which CDM makes a dominant contribution to the mass density of the Universe (Eisenstein & Hu 1998). McGaugh (1999b) further considered whether or not a cosmology with $\Omega_m \approx \Omega_b \approx 0.02h^{-2}$ would be consistent with observations of anisotropies in the cosmic microwave background, particularly the pattern of acoustic oscillations predicted in the angular power spectrum (e.g., Hu et al. 1997). McGaugh, using the standard CMB-FAST program (Seljak & Zaldarriaga 1996), pointed out that in a purely baryonic Universe with vacuum energy density being the dominant constituent the second acoustic peak would be much reduced with respect to the a priori expectations of the concordance ΛCDM model (Ostriker & Steinhardt 1995). The reason for this low amplitude is Silk damping (Silk 1968) in a low Ω_m, pure baryonic universe— the shorter wavelength fluctuations are exponentially suppressed by photon diffusion. When the Boomerang and Maxima results first appeared (Hanany et al. 2000, Lange et al. 2001), much of the initial excitement was generated by the unexpected low amplitude of the second peak. With $\Omega_{total} = 1.01$ and $\Omega_m = \Omega_b$ (no CDM or nonbaryonic matter of any sort), McGaugh (2000) produced a good match to these initial Boomerang results. A further prediction is that the third acoustic peak should be even further suppressed. There are indications from the more complete analysis of Boomerang and Maxima data (Netterfield et al. 2001,

Lee et al. 2001) that this may not be the case, but the systematic uncertainties remain large.

The SNIa results on the accelerated expansion of the Universe (Perlmutter et al. 1999) as well as the statistics of gravitational lensing (Falco et al. 1998) seem to exclude a pure baryonic, vacuum energy–dominated universe, although it is unclear that all of the systematic effects are well understood. It is also possible that a MOND cosmology differs from a standard Friedmann cosmology in the low-z Universe, particularly with regard to the angular size distance–redshift relation. At this point it remains unclear whether these observations require CDM. It is evident that such generic cosmological tests for CDM relate directly to the viability of MOND. Nonetheless, cosmological evidence for dark matter, in the absence of its direct detection, is still not definitive, particularly considering that a MOND universe may be non-Friedmannian.

Can we then, in the absence of a theory, reasonably guess what form a MOND cosmology might take? When a theory is incomplete, the way to proceed is to make several assumptions (*Ansätze*) in the spirit of the theory as it stands and determine the consequences. This has been done by Felten (1984) and by Sanders (1998), who following the example of Newtonian cosmology, considered the dynamics of a finite expanding sphere. Here it is assumed that the MOND acceleration parameter a_o does not vary with cosmic time. The second critical assumption is that, in the absence of a relativistic theory, the scale factor of the sphere is also the scale factor of the Universe, but then an immediate contradiction emerges. MONDian dynamics of a sphere permits no dimensionless scale factor. Uniform expansion of a spherical region is not possible, and any such region will eventually recollapse regardless of its initial density and expansion velocity (Felten 1984). In the low acceleration regime the dynamical equation for the evolution of the sphere, the MONDian equivalent of the Friedmann equation, is given by

$$\dot{r} = u_i^2 - \left[2\Omega_m H_o^2 r_o^3 a_o\right]^{1/2} \ln(r/r_i), \tag{37}$$

where r_i is the initial radius of the sphere, r_o is a comoving radius, and u_i is the initial expansion velocity. From the form of Equation 37 it is obvious that the sphere will eventually recollapse. It would also appear that a MONDian universe is inconsistent with the cosmological principle.

However, looking at the Newtonian equations for the dynamics of a spherical region, one finds that, at any given epoch, the acceleration increases linearly with radial distance from the center of the sphere. This means that there exists a critical radius given by

$$r_c = \sqrt{GM_{r_c}/a_o}, \tag{38}$$

where M_{r_c} is the active gravitational mass within r_c. Beyond r_c (which is epoch dependent) the acceleration exceeds a_o; therefore, on larger scales the dynamics of any spherical region is Newtonian and the expansion may be described by a dimensionless scale factor. During this Newtonian expansion, it would appear possible to make the standard assumption of Newtonian cosmology that the scale factor of the sphere is identical to the scale factor of the universe, at least on comoving scales corresponding to M_{r_c} or larger.

Making use of the Friedmann equations we find that

$$r_c = \frac{2a_o}{\Omega_m H_o^2} x^3 \tag{39}$$

during the matter-dominated evolution of the Universe (Sanders 1998). Here x is the dimensionless scale factor with $x = 1$ at present. Therefore, larger and larger comoving regions become MONDian as the Universe evolves. Because in the matter-dominated period the horizon increases as $r_h \propto x^{1.5}$, it is obvious that at some point in the past the scale over which MOND applies was smaller than the horizon scale. This would suggest that, in the past, whereas small regions may have been dominated by modified dynamics, the evolution of the Universe at large is described by the usual Friedmann models. In particular, in the early radiation-dominated universe, r_c is very much smaller than the MONDian Jeans length, $\lambda_j \approx (c/H_o)x^{4/3}$ (the Hubble deceleration is very large at early times), so the expectation is that the dynamical history of the early MOND universe would be identical to the standard Big Bang.

After recombination, the Jeans length of the baryonic component falls very much below r_c, and by a redshift of 3 or 4, r_c approaches the horizon scale; the entire Universe becomes MONDian. Thus, we might expect that the evolution of a postrecombination MOND universe might differ in significant ways from the standard Friedmann-Lemaitre models, particularly with respect to structure formation. One could assume that when the critical radius (Equation 39) grows beyond a particular comoving scale, then the MOND relation (Equation 37) applies for the subsequent evolution of regions on that scale, and recollapse will occur on a time scale comparable to the Hubble time at that epoch. For a galaxy mass object ($10^{11} M_\odot$, $r_c \approx 14$ kpc) this happens at a redshift of about 140, and recollapse would occur on time scale of several hundred million years. Therefore, we might expect galaxies to be in place as virialized objects by a redshift of 10. It is evident that larger scale structure forms later, with the present turn around radius being at about 30 Mpc.

There are two problems with such a scenario for structure formation. The first is conceptual: In a homogeneous Universe what determines the point or points about which such recollapse occurs? Basically in this picture small density fluctuations play no role,whereas we might expect that in the real world structure develops from the field of small density fluctuations as in the standard picture. The second problem is observational: This picture predicts inflow out to scales of tens of megaparsecs; it would have been quite difficult for Hubble to have discovered his law if this were true.

These problems may be overcome in the context of a more physically consistent, albeit nonrelativistic, theory of MOND cosmology (Sanders 2001). Following Bekenstein & Milgrom (1984), one begins with a two-field Lagrangian-based theory of MOND (nonrelativistic) in which one field is to be identified with the usual Newtonian field and the second field describes an additional MOND force that dominates in the limit of low accelerations. The theory embodies two important properties: MOND plays no role in the absence of fluctuations (the MOND field couples only to density inhomogeneities). This means the basic Hubble flow is left intact. Second, although the Hubble flow is not influenced by MOND, it enters as an external field that influences the internal dynamics of a finite-size region. Basically, if the Hubble deceleration (or acceleration) over some scale exceeds a_o, the evolution of fluctuations on that scale is Newtonian.

Figure 12 shows the growth of fluctuations of different comoving scales compared with the usual Newtonian growth. It is evident that MOND provides a considerable boost, particularly at the epoch during which the cosmological constant begins to dominate. This is due to the fact that the external acceleration field vanishes at this point and therefore plays no role in suppressing the modified dynamics. This adds a new aspect to an anthropic argument originally given by

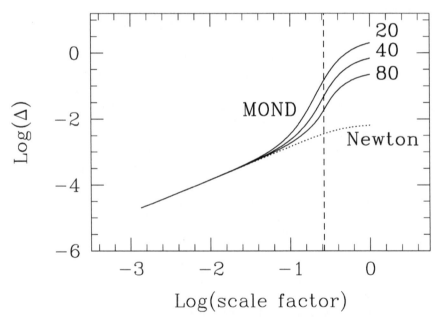

Figure 12   The growth of spherically symmetric over-densities in a low-density baryonic universe as a function of scale factor in the context of a two-field Lagrangian theory of MOND. The solid curves correspond to regions with comoving radii of 20, 40, and 80 Mpc. The dotted line is the corresponding Newtonian growth. With MOND, smaller regions enter the low-acceleration regime sooner and grow to larger final amplitude. The vertical dashed line indicates the epoch at which the cosmological constant begins to dominate the Hubble expansion.

Milgrom (1989c): We are observing the universe at an epoch during which Λ has only recently emerged as the dominant term in the Friedmann equation because that is when structure formation proceeds rapidly. The predicted MOND power spectrum (Sanders 2001) is rather similar in form to the CDM power spectrum in the concordance model but contains the baryonic oscillations proposed by McGaugh (1998). These would be telling but are difficult to resolve in large-scale structure surveys.

It is the external field effect owing to Hubble deceleration that tames the very rapid growth of structure in this scenario. In the context of the two-field theory this effect may be turned off, and then it is only the peculiar accelerations that enter the MOND equation. This case has been considered by Nusser (2002), who finds extremely rapid growth to the nonlinear regime and notes that the final MOND power spectrum is proportional to k^{-1}, independent of its original form. He found that to be consistent with the present amplitude of large-scale fluctuations a_o must be reduced by about a factor of 10 over the value determined from rotation curve-fitting. He has confirmed this with N-body simulations, again applying the MOND equation only in determining the peculiar accelerations. This suggests that the Hubble deceleration should come into play in a viable theory of MOND structure formation.

All of these conclusions are tentative; their validity depends upon the validity of the original assumptions. Nonetheless, it is evident that MOND is likely to promote the formation of cosmic structure from very small initial fluctuations; this, after all, was one of the primary motivations for nonbaryonic cosmic dark matter.

CONCLUSIONS

It is noteworthy that MOND, as an ad hoc algorithm, can explain many systematic aspects of the observed properties of bound gravitating systems: (a) the presence of a preferred surface density in spiral galaxies and ellipticals (the Freeman and Fish laws); (b) the fact that pressure-supported, nearly isothermal systems ranging from molecular clouds to clusters of galaxies are characterized by specific internal acceleration ($\approx a_o$); (c) the existence of a tight rotation velocity-luminosity relation for spiral galaxies (the TF law)—specifically revealed as a correlation between the total baryonic mass and the asymptotically flat rotation velocity of the form $M \propto V^4$; (d) the existence of a luminosity-velocity dispersion relation in elliptical galaxies (Faber-Jackson)—a relation that extends to clusters of galaxies as a baryonic mass-temperature relation; and (e) the existence of a well-defined two-parameter family of observed properties, the fundamental plane, of elliptical galaxies—objects that have varied formation and evolutionary histories and nonhomologous structure. Moreover, this is all accomplished in a theory with a single new parameter with units of acceleration, a_o, that must be within an order of magnitude of the cosmologically interesting value of cH_o. Further, many of these systematic aspects of bound systems do not have any obvious connection to what has been traditionally called the "dark matter problem." This capacity to connect seemingly unrelated points is the hallmark of a good theory.

Impressive as these predictions (or explanations) of systematics may be, it is the aspect of spiral galaxy rotation curves that is most remarkable. The dark matter hypothesis may, in principle, explain trends, but the peculiarities of an individual rotation curve must result from the unique formation and evolutionary history of that particular galaxy. The fact that there is an algorithm—MOND—that allows the form of individual rotation curves to be successfully predicted from the observed distribution of detectable matter—stars and gas—must surely be seen, at the very least, as a severe challenge for the dark matter hypothesis. This challenge would appear to be independent of whether or not the algorithm has a firm foundation in theoretical physics because science is, after all, based upon experiment and observation. Nonetheless, if MOND is, in some sense, correct then the simple algorithm carries with it revolutionary implications about the nature of gravity and/or inertia—implications that must be understood in a theoretical sense before the idea can be unambiguously extended to problems of cosmology and structure formation.

Does MOND reflect the influence of cosmology on local particle dynamics at low accelerations? The coincidence between a_o and cH_o would suggest a connection. Does inertia result from interaction of an acceleration object with the vacuum as some have suggested? If so, then one would expect a cosmological vacuum energy density to influence this interaction. Are there long-range scalar fields in addition to gravity, which in the manner anticipated by the Bekenstein-Milgrom theory become more effective in the limit of low field gradients? Additional fields with gravitational strength coupling are more or less required by string theory, but their influence must be suppressed on the scale of the solar system (high accelerations); otherwise, they would have revealed themselves as deviations from the precise predictions of general relativity at a fundamental level—violations of the equivalence principle or preferred frame effects. Such suppression can be achieved via the Bekenstein-Milgrom field equation.

Ideally, a proper theory of MOND would make predictions on a scale other than extragalactic; this would provide the possibility of a more definitive test. An example of this is the stratified aquadratic scalar-tensor theory, which predicts local preferred frame effects at a level that should soon be detectable in the lunar laser ranging experiment (Sanders 1997). An additional prediction that is generic of viable scalar-tensor theories of MOND is the presence of an anomalous acceleration, on the order of a_o, in the outer solar system. The reported anomalous acceleration detected by the Pioneer spacecrafts beyond the orbit of Jupiter (Anderson et al. 1998) is most provocative in this regard, but the magnitude (8×10^{-8} cm/s^2) is somewhat larger than would be naively expected if there is a connection with MOND. However, this is an example of the kind of test that, if confirmed, would establish a breakdown of Newtonian gravity or dynamics at low acceleration.

In a 1990 review on dark matter and alternatives Sanders (1990) wrote, "overwhelming support for dark matter would be provided by the laboratory detection of candidate particles with the required properties, detection of faint emission from low mass stars well beyond the bright optical image of galaxies, or the definite observation of 'micro-lensing' by condensed objects in the dark outskirts of galaxies." Now, more than a decade later, a significant baryonic contribution to the halos of galaxies in the form of "machos" or low mass stars seems to have been ruled out (Alcock et al. 2000). Particle dark matter has been detected in the form of neutrinos, but of such low mass—certainly less than 3 ev and probably comparable to 0.15 ev (Turner 1999)—that they cannot possibly constitute a significant component of the dark matter—either cosmologically or on the scale of galaxies. At the same time, the inferred contribution of CDM to the mass budget of the Universe has dropped from 95% to perhaps 30%, and both observational and theoretical problems have arisen with the predicted form of halos (Sellwood & Kosowsky 2001). However, all of this has not deterred imaginative theorists from speculative extrapolations of the standard model to conjure particles having the properties desired to solve perceived problems with dark matter halos. It is surely time to apply Occam's sharp razor and seriously consider the suggestion that Newtonian dynamics may break down in the heretofore unobserved regime of low accelerations.

ACKNOWLEDGMENTS

We are very grateful to Moti Milgrom for sharing his deep insight with us over a number of years and for many comments and constant encouragement. We also thank Jacob Bekenstein for his ongoing interest in this problem and for many hours of stimulating discussions. We wish to express our profound gratitude to the observers and interpreters of the distribution of neutral hydrogen in nearby galaxies—to the radio astronomers of the Kapteyn Institute in Groningen. In particular we thank Renzo Sancisi and Tjeerd van Albada and "generations" of Groningen students, especially, Albert Bosma, Kor Begeman, Adrick Broeils, Roelof Bottema, Marc Verheijen, Erwin de Blok, and Rob Swaters.

The *Annual Review of Astronomy and Astrophysics* is online at
http://astro.annualreviews.org

LITERATURE CITED

Aaronson M, Huchra J, Mould J, Tully RB, Fisher JR, et al. 1982. *Ap. J. Suppl.* 50:241–62

Aguirre A, Schaye J, Quataert E. 2002. *Ap. J.* 561:550–58

Alcock C, Allsman RA, Alves DR, Axelrod T, Becker AC, et al. 2000. *Ap. J.* 542:281–307

Allen RJ, Shu FH. 1979. *Ap. J.* 227:67–72

Anderson JD, et al. 1998. *Phys. Rev. Lett.* 81: 2858–61

Begeman KG. 1989. *Astron. Astrophys.* 223: 47–60

Begeman KG, Broeils AH, Sanders RH. 1991. *MNRAS* 249:523–37

Bekenstein JD. 1988a. In *Second Canadian Conference on General Relativity and Relativistic Astrophysics*, ed. A Coley, C Dyer, T Tupper, pp. 68–104. Singapore: World Sci.

Bekenstein JD. 1988b. *Phys. Lett. B* 202:497–99

Bekenstein JD. 1990. In *Developments in General Relativity, Astrophysics and Quantum Theory: A Jubilee in Honour of Nathan Rosen*, ed. FI Cooperstock, LP Horwitz,

J Rosen, pp. 155–74. Bristol, UK: IOP

Bekenstein JD. 1992. In *Proceedings of the Sixth Marcel Grossman Meeting on General Relativity*, ed. H Sato, T Nakamura, pp. 905–24. Singapore: World Sci.

Bekenstein JD. 1993. *Phys. Rev. D* 48:3641–47

Bekenstein JD, Milgrom M. 1984. *Ap. J.* 286:7–14

Bekenstein JD, Sanders RH. 1994. *Ap. J.* 429: 480–90

Bekenstein JD, Milgrom M, Sanders RH. 2000. *Phys. Rev. Lett.* 85:1346

Bell EF, de Jong RS. 2001. *Ap. J.* 550:212–29

Bender R, Burstein D, Faber SM. 1992. *Ap. J.* 399:462–77

Blandford RD, Narayan R. 1992. *Annu. Rev. Astron. Astrophys.* 30:311–58

Blumenthal GR, Faber SM, Primack JR, Rees MJ. 1984. *Nature* 311:517–25

Bosma A. 1978. *The distribution and kinematics of neutral hydrogen in spiral galaxies of various morphological types*. PhD thesis. Univ. Groningen, The Netherlands

Brada R. 1997. *Problems in galactic dynamics using the modified dynamics (MOND)*. PhD thesis. Rehovot, Israel: The Weizmann Inst.

Brada R, Milgrom M. 1995. *MNRAS* 276:453–59

Brada R, Milgrom M. 1999a. *Ap. J.* 512:L17–18

Brada R, Milgrom M. 1999b. *Ap. J.* 519:590–98

Brada R, Milgrom M. 2000a. *Ap. J.* 531:L21–24

Brada R, Milgrom M. 2000b. *Ap. J.* 541:556–64

Brans C, Dicke RH. 1961. *Phys. Rev.* 124:925–35

Broeils AH. 1992. *Dark and visible matter in spiral galaxies*. PhD thesis. Univ. Groningen, The Netherlands

Carignan C, Beaulieu S. 1989. *Ap. J.* 347:760–70

Casertano S, van Gorkom JH. 1991. *Astron. J.* 101:1231–41

Corollo CM, Stiavelli M, de Zeeuw PT, Mack J. 1997. *Astron. J.* 114:2366–80

Dalcanton JJ, Spergel DN, Summers FJ. 1997. *Ap. J.* 482:659–76

de Bernardis P, Ade PAR, Bock JJ, Bond JR, Borrill J, et al. 2000. *Nature* 404:955–57

de Blok WJG, Bosma A. 2002. *Astron. Astrophys.* In press. astro-ph/0201276

de Blok WJG, McGaugh SS. 1998. *Ap. J.* 508: 132–40

de Blok WJG, McGaugh SS, Bosma A, Rubin VC. 2001. *Ap. J. Lett.* 552:L23–26

Deser S, Levin O. 1997. *Class. Quant. Grav.* 14:L163–68

Djorgovski S, Davis M. 1987. *Ap. J.* 313:59–68

Dressler A, et al. 1987. *Ap. J.* 313:42–58

Drummond IT. 2001. *Phys. Rev. D* 63:043503-1–13

Dubinski J, Carlberg RG. 1991. *Ap. J.* 378:495–503

Dvali G, Gabadadze G, Shifman M. 2001. *Mod. Phys. Lett.* A16:513–30

Eckhardt DH. 1993. *Phys. Rev. D* 48:3762–67

Edery A. 1999. *Phys. Rev. Lett.* 83:3990–92

Eisenstein DJ, Loeb A, Turner EL. 1996. *Ap. J.* 475:421–28

Eisenstein DJ, Hu W. 1998. *Ap. J.* 496:605–14

Faber SM, Jackson RE. 1976. *Ap. J.* 204:668–83

Faber SM, Gallagher JS. 1979. *Annu. Rev. Astron. Astrophys.* 17:135–87

Falco EE, Kochanek CS, Munoz JA. 1998. *Ap. J.* 494:47–59

Felten JE. 1984. *Ap. J.* 286:3–6

Finzi A. 1963. *MNRAS* 127:21–30

Fish RA. 1964. *Ap. J.* 139:284–305

Fort B, Mellier Y. 1994. *Astron. Astrophys. Rev.* 5:239–92

Freeman KC. 1970. *Ap. J.* 160:811–30

Fukuda Y, Hayakawa T, Ichihara E, Inoie K, Ishihara K, et al. 1998. *Phys. Rev. D* 81:1562–66

Fukugita M, Hogan CJ, Peebles PJE. 1998. *Ap. J.* 503:518–30

Gerhard OE. 1994. In *Proc. ESO/OHP Workshop, Dwarf Galaxies*, ed. G Meylen, P Prugniel, pp. 335–49. Garching, Germany: ESO

Gerhard OE, Spergel DN. 1992. *Ap. J.* 397:38–43

Giraud E. 2000. *Ap. J.* 531:701–15

Griest K, Kamionkowski M, Turner MS 1990. *Phys. Rev. D* 41:3565–82

Hadjimichef D, Kokubun F. 1997. *Phys. Rev. D* 55:733–38

Hanany S, Ade P, Balbi A, Bock J, Borrill J, et al. 2000. *Ap. J. Lett.* 545:L5–9

Hernquist L. 1990. *Ap. J.* 356:359–64

Hoekstra H, van Albada TS, Sancisi R. 2001. *MNRAS* 323:453–59

Hoekstra H, Yee HKC, Gladders MD. 2002. In *Where's the Matter? Tracing Dark and Bright Matter with the New Generation of Large Scale Surveys*, ed. M Treyer, L Tresse. Marseille, France. In press. astro-ph/0109514

Hu W, Sugiyama N, Silk J. 1997. *Nature* 386: 37–43

Jørgensen I. 1999. *MNRAS* 306:607–36

Jørgensen I, Franx M, Kærgard P. 1995a. *MNRAS* 273:1097–128

Jørgensen I, Franx M, Kærgard P. 1995b. *MNRAS* 276:1341–64

Kaplinghat M, Turner M. 2001. *Ap. J. Lett.* In press. astro-ph/0107284

Kochanek CS. 2002. In *Shapes of Galaxies and Their Halos*. New Haven, CT: Yale Cosmol. Workshop. In press. astro-ph/0106495

Kormendy J. 1982. In *Morphology and Dynamics of Galaxies*, ed. L Martinet, M Mayor, pp. 115–270. Sauverny: Geneva Observatory

Kuhn JR, Kruglyak L. 1987. *Ap. J.* 313:1–12

Lake G. 1989. *Ap. J.* 345:L17–19

Lange AE, Ade PAR, Bock JJ, Bond JR, Borrill J, et al. 2001. *Phys. Rev. D* 63:042001-1–8

Lee AT, Ade P, Albi A, Bock J, Borrill J, et al. 2001. *Ap. J. Lett.* 561:L1–5

Lokas EL. 2001. *MNRAS* 327:L21–26

Macri LM, Stetson PB, Bothun GD, Freedman WL, Garnavich PM, et al. 2001. *Ap. J.* 559: 243–59

Mannheim PD, Kazanas D. 1989. *Ap. J.* 342: 635–38

Mateo M. 1998. *Annu. Rev. Astron. Astrophys.* 36:435–506

McGaugh SS. 1996. *MNRAS* 280:337–54

McGaugh SS. 1998. In *After the Dark Ages: When Galaxies Were Young*, ed. SS Holt, EP Smith, pp. 72–75. Woodbury, VT: AIP. astro-ph/9812328

McGaugh SS. 1999a. In *Galaxy Dynamics*, ed. D Merritt, JA Sellwood, M Valluri, pp. 528–

38. San Francisco: Astron. Soc. Pac. astro-ph/9812327

McGaugh SS. 1999b. *Ap. J. Lett.* 523:L99–102

McGaugh SS. 2000. *Ap. J. Lett.* 541:L33–36

McGaugh SS, de Blok WJG. 1998a. *Ap. J.* 499: 41–65

McGaugh SS, de Blok WJG. 1998b. *Ap. J.* 499: 66–81

McGaugh SS, Rubin VC, de Blok WJG. 2001. *Astron. J.* 122:2381–95

McGaugh SS, Schombert JM, Bothun GD. 1995a. *Astron. J.* 109:2019–33

McGaugh SS, Schombert JM, Bothun GD. 1995b. *Astron. J.* 110:573–80

McGaugh SS, Schombert JM, Bothun GD, de Blok WJG. 2000. *Ap. J. Lett.* 533:L99–102

Mihos JC, McGaugh SS, de Blok WJG. 1997. *Ap. J.* 477:L79–83

Milgrom M. 1983a. *Ap. J.* 270:365–70

Milgrom M. 1983b. *Ap. J.* 270:371–83

Milgrom M. 1983c. *Ap. J.* 270:384–89

Milgrom M. 1984. *Ap. J.* 287:571–76

Milgrom M. 1988. *Astron. Astrophys.* 202:L9–12

Milgrom M. 1989a. *Ap. J.* 338:121–27

Milgrom M. 1989b. *Astron. Astrophys.* 211:37–40

Milgrom M. 1989c. *Comments Astrophys.* 13: 215–30

Milgrom M. 1991. *Ap. J.* 367:490–92

Milgrom M. 1994a. *Ann. Phys.* 229:384–415

Milgrom M. 1994b. *Ap. J.* 429:540–44

Milgrom M. 1995. *Ap. J.* 455:439–42

Milgrom M. 1997a. *Phys. Rev. E* 56:1148–59

Milgrom M. 1997b. *Ap. J.* 478:7–12

Milgrom M. 1998. *Ap. J. Lett.* 496:L89–91

Milgrom M. 1999. *Phys. Lett. A* 253:273–79

Milgrom M. 2002. astro-ph/0110362

Mohr JJ, Evrard AE. 1997. *Ap. J.* 491:38–44

Moore B, Governato F, Quinn T, Stadel J, Lake G. 1998. *Ap. J. Lett.* 499:L5–8

Mortlock DJ, Turner EL. 2002. *MNRAS.* 327: 557–66

Múller J, Nordtvedt K, Vokrouhlický D. 1996. *Phys. Rev. D* 54:5927–30

Narnhofer H, Peter I, Thirring W. 1996. *Int. J. Mod. Phys. B* 10:1507–20

Navarro JF, Frenk CS, White SDM. 1996. *Ap. J.* 462:563–75

Netterfield CB, et al. 2001. astro-ph/104460

Ni W-T. 1972. *Ap. J.* 176:769–96

Nordvedt K, Will CM. 1972. *Ap. J.* 177:775–92

Nusser A. 2002. *MNRAS.* In press. astro-ph/0109016

Ostriker JP, Peebles PJE. 1973. *Ap. J.* 186:467–80

Ostriker JP, Steinhardt PJ. 1995. *Nature* 377:600–2

Peebles PJE. 1982. *Ap. J.* 263:L1–5

Perlmutter S. et al. 1999. *Ap. J.* 517:565–86

Pryor C, Meylan G. 1993. In *Structure and Dynamics of Globular Clusters*, APS Conf. Ser. Vol. 50, ed. SG Djorgovski, G Meylan, pp. 357–71. San Francisco: Astron. Soc. Pac.

Rees MJ. 1986. *MNRAS* 218:25P–30P

Roberts MS, Whitehurst RN. 1975. *Ap. J.* 201:327–46

Romatka R. 1992. *Alternativen zur "dunklen Materie."* PhD thesis. Münich, Germany: Max-Plank-Inst. Phys.

Rubin VC, Ford WK, Thonnard N. 1980. *Ap. J.* 238:471–87

Saglia RP, Bertin G, Stiavelli M. 1992. *Ap. J.* 384:433–47

Sakai S, et al. 2000. *Ap. J.* 529:698–722

Sakharov AD. 1968. *Sov. Phys. Dok.* 12:1040–42

Sanders RH. 1984. *Astron. Astrophys.* 136:L21–23

Sanders RH. 1986. *MNRAS* 223:539–55

Sanders RH. 1988. *MNRAS* 235:105–21

Sanders RH. 1989. *MNRAS* 241:135–51

Sanders RH. 1990. *Astron. Astrophys. Rev.* 2:1–28

Sanders RH. 1994. *Astron. Astrophys.* 284:L31–34

Sanders RH. 1996. *Ap. J.* 473:117–29

Sanders RH. 1997. *Ap. J.* 480:492–502

Sanders RH. 1998. *MNRAS* 296:1009–18

Sanders RH. 1999. *Ap. J.* 512:L23–26

Sanders RH. 2000. *MNRAS* 313:767–74

Sanders RH. 2001. *Ap. J.* 560:1–6

Sanders RH, Begeman KG. 1994. *MNRAS* 266:360–66

Sanders RH, Verheijen MAW. 1998. *Ap. J.* 503:97–108

Schaye J. 2002. *Ap. J.* 559:507–15

Seljak U, Zeldarriaga M. 1996. *Ap. J.* 469:437–44

Sellwood JA, Kosowsky A. 2001. In *Gas & Galaxy Evolution*, ASP Conf. Ser., ed. JE Hibbard, MP Rupen, JH van Gorkom, pp. 311–18. San Francisco: Astron. Soc. Pac.

Sellwood JA, Kosowsky A. 2002. In *The Dynamics, Structure and History of Galaxies*, ed. GS Da Costa, EM Sadler. astro-ph/0109555. In press

Shostak GS. 1973. *Astron. Astrophys.* 24:411–19

Silk J. 1968. *Ap. J.* 151:459–72

Solomon PM, Rivolo AR, Barrett J, Yahil A. 1987. *Ap. J.* 319:730–41

Swaters RA. 1999. *Dark matter in late-type dwarf galaxies.* PhD thesis. Univ. Groningen, The Netherlands

The LS, White SDM. 1988. *Astron. J.* 95:1642–46

Tohline JE. 1983. In *The Internal Kinematics and Dynamics of Galaxies*, IAU Symp. 100, ed. E Athanassoula, pp. 205–6. Dordrecht, The Netherlands: Reidel

Trager SC, Djorgovski S, King IR. 1993. In *Structure and Dynamics of Globular Clusters*, APS Ser. V. 50, pp. 347–55. San Francisco: Astron. Soc. Pac.

Tremaine S, Gunn JE. 1979. *Phys. Rev. Lett.* 42:407–10

Trimble V. 1987. *Annu. Rev. Astron. Astrophys.* 25:425–72

Tucker DL, et al. 1998. In *Large Scale Structure: Tracks and Traces. Proc. 12th Potsdam Cosmology Workshop*, ed. V Mueller, S Gottloeber, JP Muechet, J Wambsganss, pp. 105–6. Singapore: World Sci.

Tully RB, Fisher JR. 1977. *Astron. Astrophys.* 54:661–73

Tully RB, Verheijen MAW, Pierce MJ, Hang J-S, Wainscot R. 1996, *Astron. J.* 112:2471–99

Turner MS. 1999. *Publ. Astron. Soc. Pac.* 111:264–73

Tytler D, O'Meara JM, Suzuki N, Lubin D. 2000. *Physica Scripta:* T85:12–31

Unruh WG. 1975, *Phys. Rev. D* 14:870–92

van Albada TS, Sancisi R. 1986. *Philos. Trans. R. Soc. Lond. Ser. A* 320:447–64

van Albada TS, Bertin G, Stiavelli M. 1995. *MNRAS* 276:1255–61

van den Bosch FC, Dalcanton JJ. 2000. *Ap. J.* 534:146–64

Verheijen MAW. 1997. *The Ursa Major Cluster of Galaxies*. PhD thesis. Univ. Groningen, The Netherlands

Verheijen MAW, Sancisi R. 2001. *Astron. Astrophys.* 370:765–867

Vittorio N, Silk J. 1984. *Ap. J.* 285:L39–43

Walker MA. 1994. *Ap. J.* 430:463–66

White DA, Jones C, Forman W. 1997. *MNRAS* 292:419–67

White SDM, Navarro JF, Evrard AE, Frenk CS. 1993. *Nature* 336:429–31

Will CM. 2001. *Living Rev. Rel.* 4:4–104

Zwicky F. 1933, *Helv. Phys. Acta* 6:110–27

Annu. Rev. Astron. Astrophys. 2002. 40:319–48
doi: 10.1146/annurev.astro.40.060401.093852
Copyright © 2002 by Annual Reviews. All rights reserved

CLUSTER MAGNETIC FIELDS

C. L. Carilli and G. B. Taylor

National Radio Astronomy Observatory, Socorro, New Mexico 87801;
email: ccarilli@nrao.edu, gtaylor@nrao.edu

Key Words   galaxy clusters, intergalactic medium, intracluster medium, cosmic
rays, observations: radio, x-ray

■ **Abstract**   Magnetic fields in the intercluster medium have been measured using
a variety of techniques, including studies of synchrotron relic and halo radio sources
within clusters, studies of inverse Compton X-ray emission from clusters, surveys of
Faraday rotation measures of polarized radio sources both within and behind clusters,
and studies of cluster cold fronts in X-ray images. These measurements imply that most
cluster atmospheres are substantially magnetized, with typical field strengths of order 1
μGauss with high areal filling factors out to Mpc radii. There is likely to be considerable
variation in field strengths and topologies both within and between clusters, especially
when comparing dynamically relaxed clusters to those that have recently undergone
a merger. In some locations, such as the cores of cooling flow clusters, the magnetic
fields reach levels of 10–40 μG and may be dynamically important. In all clusters the
magnetic fields have a significant effect on energy transport in the intracluster medium.
We also review current theories on the origin of cluster magnetic fields.

INTRODUCTION

Magnetic fields play an important role in virtually all astrophysical phenomena.
Close to home, the Earth has a bipolar magnetic field with a strength of 0.3 G at the
equator and 0.6 G at the poles. This field is thought to originate in a dynamo owing to
fluid motions within the liquid core (Soward 1983). With its faster angular rotation,
Jupiter leads the planets with an equatorial field strength of ~4 G (Warwick 1963,
Smith et al. 1974). A similar mechanism produces the solar magnetic fields that give
rise to spectacular sunspots, arches, and flares (Parker 1979). Within the interstellar
medium, magnetic fields are thought to regulate star formation via the ambipolar
diffusion mechanism (Spitzer 1978). Our own Galaxy has a typical interstellar
magnetic field strength of ~2 μG in a regular ordered component on kiloparsec
scales, and a similar value in a smaller scale, random component (Beck et al.
1996, Kulsrud 1999). Other spiral galaxies have been estimated to have magnetic
field strengths of 5 to 10 μG, with fields strengths up to 50 μG found in starburst
galaxy nuclei (Beck et al. 1996). Magnetic fields are fundamental to the observed
properties of jets and lobes in radio galaxies (Bridle & Perley 1984), and they
may be primary elements in the generation of relativistic outflows from accreting,

0066-4146/02/0922-0319$14.00

319

massive black holes (Begelman et al. 1984). Assuming equipartition conditions apply, magnetic field strengths range from a few μG in kpc-scale extended radio lobes, to mG in pc-scale jets.

The newest area of study of cosmic magnetic fields is on even larger scales still, those of clusters of galaxies. Galaxy clusters are the largest virialized structures in the universe. The first spatially resolving X-ray observations of clusters in the early 1970s (e.g., Forman et al. 1972) revealed atmospheres of hot gas (10^7 to 10^8 K) that extend to Mpc radii and that dominate the baryonic mass of the systems (10^{13} to 10^{14} M$_\odot$). Soon thereafter came the first attempts to measure magnetic field strengths in the intercluster medium (ICM) (Jaffe 1977). Only in the last decade has it become clear that magnetic fields are ubiquitous in cluster atmospheres, certainly playing a critical role in determining the energy balance in cluster gas through their effect on heat conduction, and in some cases, perhaps even becoming dynamically important.

Cluster magnetic fields have been treated as secondary topics in reviews of cluster atmospheres (Sarazin 1988, Fabian 1994) and in general reviews of cosmic magnetic fields (Kronberg 1996, Ruzmaikin et al. 1987). To date there has been no dedicated review of cluster magnetic fields.

The focus of this review is primarily observational. We summarize and critique various methods used for measuring cluster magnetic fields. In the course of the review, we consider important effects of magnetic fields in clusters, such as their effect on heat conduction and gas dynamics, and other issues such as the lifetimes of relativistic particles in the ICM. We then attempt to synthesize the various measurements and develop a general picture for cluster magnetic fields, with the caveat that there may be significant differences between clusters, and even within a given cluster atmosphere. We conclude with a section on the possible origin of cluster magnetic fields.

We assume $H_0 = 75$ km s^{-1} Mpc^{-1} and $q_0 = 0.5$, unless stated otherwise. Spectral index, α, is defined as $S_\nu \propto \nu^\alpha$.

SYNCHROTRON RADIATION

Radio Halos

Over 40 years ago, Large (1959) discovered a radio source in the Coma cluster that was extended even when observed with a 45′ beam. This source (Coma C) was studied by Willson (1970) who found that it had a steep spectral index and could not be made up of discrete sources, but instead was a smooth "radio halo" with no structure on scales less than 30′. Willson further surmised that the emission mechanism was likely to be synchrotron, and if in equipartition required a magnetic field strength of 2 μG. In Figure 1, we show the best image yet obtained of the radio halo in the Coma cluster. Other radio halos were subsequently discovered, although the number known remained under a dozen until the mid-1990s (Hanisch 1982).

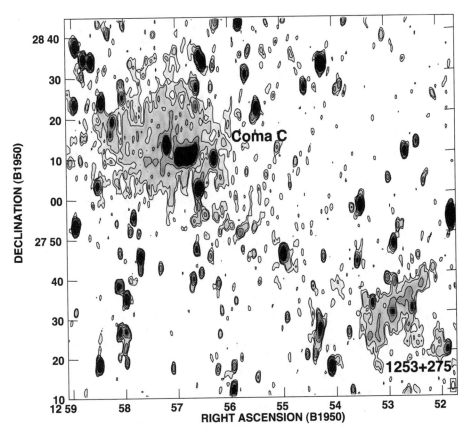

Figure 1   Westerbork Synthesis Radio Telescope (WSRT) radio image of the Coma cluster region at 90 cm, with angular resolution of $55'' \times 125''$ (HPBW, RA $\times$ DEC) for the radio telescope from Feretti & Giovannini (1998). Labels refer to the halo source Coma C and the relic source $1253 + 275$. The gray-scale range displays total intensity emission from 2 to 30 mJy/beam, whereas contour levels are drawn at 3, 5, 10, 30, and 50 mJy/beam. The bridge of radio emission connecting Coma C to $1253 + 275$ is resolved and visible only as a region with an apparent higher positive noise. The Coma cluster is at a redshift of 0.023, such that $1' = 26$ kpc for $H_0 = 75$.

Using the Northern VLA Sky Survey [NVSS (Condon et al. 1998)] and X-ray selected samples as starting points, Giovannini & Feretti (2000) and Giovannini et al. (1999) have performed moderately deep VLA observations (integrations of a few hours) that have more than doubled the number of known radio halo sources. Several new radio halos have also been identified from the Westerbork Northern Sky Survey (Kemper & Sarazin 2001). These radio halos typically have sizes ~ 1 Mpc, steep spectral indices ($\alpha < -1$), low fractional polarizations (<5%), low surface brightnesses ($\sim 10^{-6}$ Jy arcsec^{-2} at 1.4 GHz), and centroids close to the cluster center defined by the X-ray emission.

A steep correlation between cluster X-ray and radio halo luminosity has been found, as well as a correlation between radio and X-ray surface brightnesses in clusters (Liang et al. 2000, Feretti et al. 2001, Govoni et al. 2001a). A complete (flux limited) sample of X-ray clusters shows only 5% to 9% of the sources are detected at the surface brightness limits of the NVSS of 2.3 mJy beam^{-1}, where the beam has FWHM $= 45''$ (Giovannini & Feretti 2000, Feretti et al. 2001). But this sample contains mostly clusters with X-ray luminosities $<10^{45}$ erg s^{-1}. If one selects for clusters with X-ray luminosities $>10^{45}$ erg s^{-1}, the radio detection rate increases to 35% (Feretti et al. 2001, Owen et al. 1999). Likewise, there may be a correlation between the existence of a cluster radio halo and the existence of substructure in X-ray images of the hot cluster atmosphere, indicative of merging clusters, and a corresponding anticorrelation between cluster radio halos and clusters with relaxed morphologies, e.g., cooling flows (Govoni et al. 2001a), although these correlations are just beginning to be quantified (Buote 2001).

Magnetic fields in cluster radio halos can be derived, assuming a minimum energy configuration for the summed energy in relativistic particles and magnetic fields (Burbidge 1959), corresponding roughly to energy equipartition between fields and particles. The equations for deriving minimum energy fields from radio observations are given in Miley (1980). Estimates for minimum energy magnetic field strengths in cluster halos range from 0.1 to 1 μG (Feretti 1999). One of the best studied halos is that in Coma, for which Giovannini et al. (1993) report a minimum energy magnetic field of 0.4 μG. These calculations typically assume $k = 1, \eta = 1, \nu_{low} = 10$ MHz, and $\nu_{high} = 10$ GHz, where k is the ratio of energy densities in relativistic protons to that in electrons, η is the volume filling factor, ν_{low} is the low frequency cut-off for the integral, and ν_{high} is the high frequency cut-off. All of these parameters are poorly constrained, although the magnetic field strength only behaves as these parameters raised to the $\frac{2}{7}$ power. For example, using a value of $k \sim 50$, as observed for Galactic cosmic rays (Meyer 1969), increases the fields by a factor of three.

Brunetti et al. (2001a) present a method for estimating magnetic fields in the Coma cluster radio halo independent of minimum energy assumptions. They base their analysis on considerations of the observed radio and X-ray spectra, the electron inverse Compton and synchrotron radiative lifetimes, and reasonable mechanisms for particle reacceleration. They conclude that the fields vary smoothly from 2 ± 1 μG in the cluster center, to 0.3 ± 0.1 μG at 1 Mpc radius.

Radio Relics

A possibly related phenomena to radio halos is a class of sources found in the outskirts of clusters known as radio relics. Like the radio halos, these are very extended sources without an identifiable host galaxy (Figure 1). Unlike radio halos, radio relics are often elongated or irregular in shape, are located at the cluster periphery (by definition), and are strongly polarized [up to 50% in the case of the relic 0917 + 75 (Harris et al. 1993)]. As the name implies, one of the first explanations put forth to explain these objects was that these are the remnants of a radio jet associated with an active galactic nucleus (AGN) that has since turned off and moved on. A problem with this model is that, once the energy source is removed, the radio source is expected to fade on a timescale $\ll 10^8$ years due to adiabatic expansion, inverse Compton, and synchrotron losses (see "Electron Lifetimes" below). This short timescale precludes significant motion of the host galaxy from the vicinity of the radio source.

A more compelling explanation is that the relics are the result of first order Fermi acceleration (Fermi I) of relativistic particles in shocks produced during cluster mergers (Ensslin et al. 1998), or are fossil radio sources revived by compression associated with cluster mergers (Ensslin & Gopal-Krishna 2001, Röttgering et al. 1994). Equipartition field strengths for relics range from 0.4–3.0 μG (Ensslin et al. 1998). If the relics are produced by shocks or compression during a cluster merger, then Ensslin et al. (1998) calculate a pre-shock cluster magnetic field strength in the range 0.2–0.5 μG.

FARADAY ROTATION

Cluster Center Sources

The presence of a magnetic field in an ionized plasma sets a preferential direction for the gyration of electrons, leading to a difference in the index of refraction for left versus right circularly polarized radiation. Linearly polarized light propagating through a magnetized plasma experiences a phase shift of the left versus right circularly polarized components of the wavefront, leading to a rotation of the plane of polarization, $\Delta\chi = \text{RM}\,\lambda^2$, where $\Delta\chi$ is the change in the position angle of polarization, λ is the wavelength of the radiation, and RM is the Faraday rotation measure. The RM is related to the thermal electron density, n_e, and the magnetic field, **B**, as:

$$\text{RM} = 812 \int_0^L n_e \mathbf{B} \cdot d\mathbf{l}\,\text{radians m}^{-2}, \qquad (1)$$

where **B** is measured in μGauss, n_e in cm^{-3} and $d\mathbf{l}$ in kpc, and the boldface symbols represent the vector product between the magnetic field and the direction of propagation. This phenomenon can also be understood qualitatively by considering the forces on the electrons.

Synchrotron radiation from cosmic radio sources is well known to be linearly polarized, with fractional polarizations up to 70% in some cases (Pacholczyk 1970). Rotation measures (RM) can be derived from multifrequency polarimetric observations of these sources by measuring the position angle of the polarized radiation as a function of frequency. The RM values can then be combined with measurements of n_e to estimate the magnetic fields. Due to the vector product in Equation 1, only the magnetic field component along the line-of-sight is measured, so the results depend on the assumed magnetic field topology.

Most extragalactic radio sources exhibit Faraday rotation measures (RMs) of the order of tens of rad m^{-2} due to propagation of the emission through the interstellar medium of our galaxy (Simard-Normandin et al. 1981). Sources at Galactic latitudes $\leq 5°$ can exhibit $\sim$300 rad m^{-2}. For the past 30 years, however, a small number of extragalactic sources were known to have far higher RMs than could be readily explained as Galactic in origin. Large intrinsic RMs were suspected, but the mechanism(s) producing them were unclear.

Mitton (1971) discovered that the powerful radio galaxy Cygnus A had large and very different RMs (35 versus -1350 rad m^{-2}), in its two lobes (see also Alexander et al. 1984). Whereas its low galactic latitude (5.8°) could possibly be invoked to explain the high RMs, the large difference in RMs over just 2′ was difficult to reproduce in the context of Galactic models. This "RM anomaly" was clarified when Dreher et al. (1987) performed the first high resolution RM studies with the VLA and found complex structure in the RM distribution on arcsec scales (Figure 2, see color insert), with gradients as large as 600 rad m^{-2} arcsec^{-1}. These large gradients conclusively ruled out a Galactic origin for the large RMs.

Perhaps just as important as the observed RM structure across the lobes of Cygnus A was the discovery that the observed position angles behave quadratically with wavelength to within very small errors over a wide range in wavelengths (Dreher et al. 1987). Examples of this phenomenon are shown in (Figure 5). Moreover, the change in position angle from short to long wavelengths is much larger than π radians in many cases, whereas the fractional polarization remains constant. This result is critical for interpreting the large RMs for cluster center radio sources, providing proof that the large RMs cannot be due to thermal gas mixed with the radio emitting plasma (Dreher et al. 1987). Such mixing would lead to rapid depolarization with increasing wavelength and departures from a quadratic behavior of χ with wavelength (Burn 1966).

The Cygnus A observations were the first to show that the large RMs must arise in an external screen of magnetized, ionized plasma, but cannot be Galactic in origin. Dreher et al. (1987) considered a number of locations for the Faraday screen toward Cygnus A, and concluded that the most likely site was the X-ray emitting cluster atmosphere enveloping the radio source (Fabbiano et al. 1979). They found that magnetic fields in the cluster gas of 2–10 μG could produce the observed RMs.

Since the ground-breaking observations of Cygnus A, RM studies of cluster center radio sources have become a standard tool for measuring cluster fields. RM studies of radio galaxies in clusters can be divided into studies of cooling-flow and noncooling-flow clusters. Cooling-flow clusters are those in which the X-ray emission is strongly peaked at the center, leading to high densities, and cooling times of the hot ICM in the inner $\sim$100 kpc of much less than the Hubble time. To maintain hydrostatic equilibrium, an inward flow may be required (Fabian et al. 1991). Typical mass cooling flow rates are 100 M$_\odot$ yr^{-1}. The actual presence

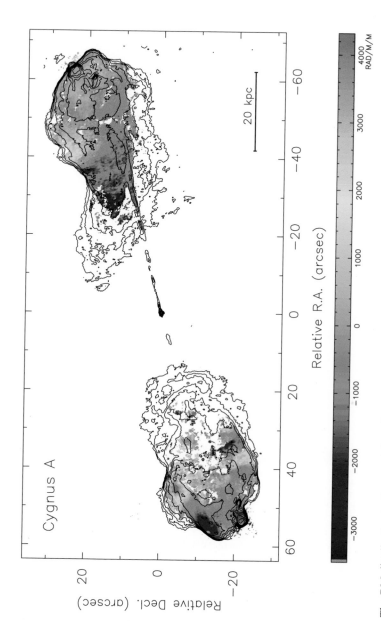

Figure 2 The RM distribution in Cygnus A based on multifrequency, multiconfiguration VLA observations. The resolution is 0.35″ (Dreher, Carilli & Perley 1987). The colorbar indicates the range in RMs from -3400 to +4300 rad m^{-2}. Note the undulations in RM on scales of 10–30 kpc. Contours are overlaid from a 5 GHz total intensity image. The RM was solved for by fitting for the change in polarization angle with frequency on a pixel-by-pixel basis (see Figure 4).

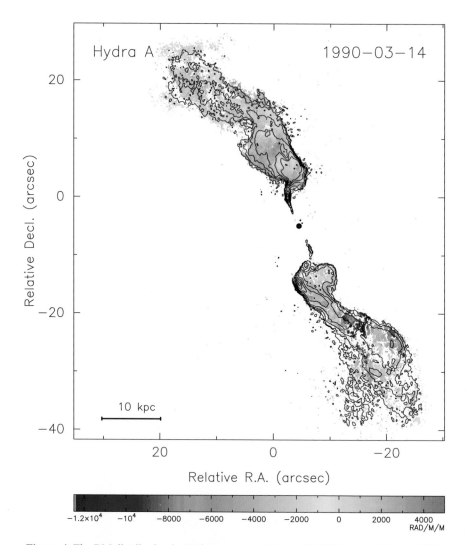

Figure 4 The RM distribution in Hydra A at a resolution of 0.3″ (Taylor & Perley 1993) with total intensity contours overlaid. Multi-configuration VLA observations were taken at four widely spaced frequencies within the 8.4 GHz band, and a single frequency in the 15 GHz band. The colorbar indicates the range in RMs from -12000 to +5000 rad m⁻². The dramatic difference in color between the lobes is due to a reversal in the sign of the magnetic field.

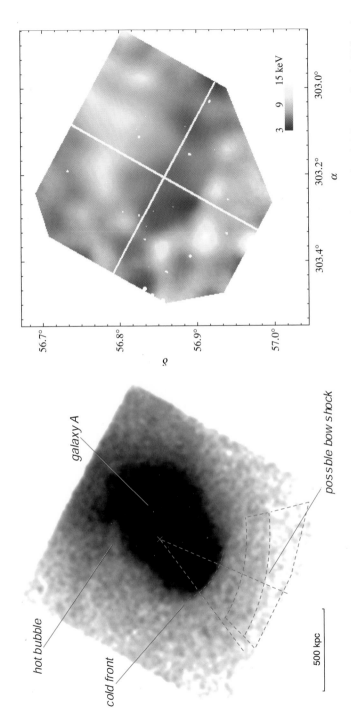

Figure 9 (*left*) A smoothed 0.5–4 keV Chandra image of Abell 3667. The most prominent feature is the sharp surface brightness edge (cold front). The front shape is nearly circular as indicated by the arc. (*right*) Temperature map. The typical statistical error in this image is ±1 keV. The cold, ~4 keV, region near the center of the map coincides with the inside of the cold front (Vikhlinin et al. 2001a).

of material "cooling" and "flowing" is a topic that is hotly debated at present (Binney 2002). What is more agreed upon is that cooling-flow clusters are more dynamically relaxed than noncooling-flow clusters, which often show evidence of cluster mergers (T. Markovic, J.A. Eilek, unpublished manuscript).

Radio galaxies in cooling-flow clusters attracted some of the first detailed RM studies by virtue of their anomalously high RMs [e.g., A1795 (Ge & Owen 1993), Hydra A (Taylor & Perley 1993)]. Out of a sample of 14 cooling-flow clusters with strong embedded radio sources, Taylor et al. (1994, 2001b) found that 10/14 display RMs in excess of 800 rad m^{-2}, two (PKS 0745-191 and 3C 84 in Abell 426) could not be measured due to a lack of polarized flux, and two [Abell 119 (Feretti et al. 1999) and 3C 129 (Taylor et al. 2001a)] that have lower RMs, but with better X-ray observations, turn out not to be in cooling-flow clusters. Hence, current data are consistent with all radio galaxies at the center of cooling-flow clusters having extreme RMs, with the magnitude of the RMs roughly proportional to the cooling-flow rate (see Figure 3).

The RM distributions for radio sources found at the centers of cooling-flow clusters tend to be patchy with coherence lengths of 5–10 kpc (Figure 4, see color insert). Larger "patches" up to 30 kpc are seen, for example, in Cygnus A (Figure 2, see color insert). In both Cygnus A and Hydra A one can find "bands" of alternating high and low RM (see Figures 2 and 4, color inserts). Such bands are also found in the noncooling-flow cluster sources (Eilek & Owen 2002), along with slightly larger coherence lengths of 15–30 kpc. In Hydra A, there is a strong trend for all the RMs to the north of the nucleus to be positive and, to the south, negative. To explain this requires a field reversal and implies a large-scale (100 kpc) ordered component to the cluster magnetic fields in Hydra A. Taylor & Perley (1993) found the large-scale field strength to be ~7 μG and more tangled fields to have a strength of ~40 μG. A similar RM sign reversal across the nucleus is seen in A1795, although in this case the radio source is only 11 kpc in extent.

Minimum cluster magnetic field strengths can be estimated by assuming a constant magnetic field along the line-of-sight through the cluster. Such estimates usually lead to magnetic field strengths of 5 to 10 μG in cooling-flow clusters, and a bit less (factor ~2) in the noncooling-flow clusters.

From the patchiness of the RM distributions it is clear that the magnetic fields are not regularly ordered on cluster (Mpc) scales, but have coherence scales between 5 and 10 kpc. Beyond measuring a mean line-of-site field, the next level of RM modeling entails cells of constant size and magnetic field strength, but random magnetic field direction, uniformly filling the cluster. The RM produced by such a screen will be built up in a random-walk fashion and will thus have an average value of 0 rad m^{-2}, but a dispersion in the RM, σ_{RM}, that is proportional to the square root of the number of cells along the line-of-sight through the cluster. The most commonly fit form to the X-ray observations to obtain the radial electron density distribution, $n_e(r)$, through a cluster is the modified King model (Cavaliere & Fusco-Femiano 1976):

$$n_e(r) = n_0\left(1 + r^2/r_c^2\right)^{-3\beta/2}, \tag{2}$$

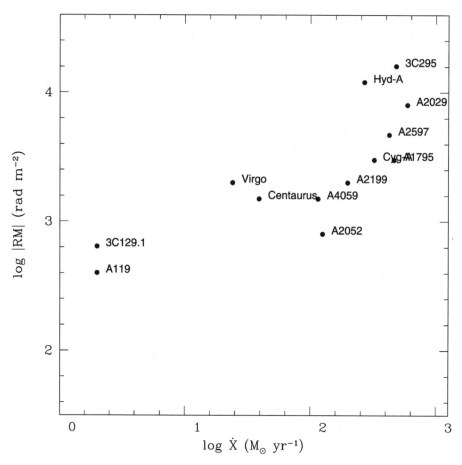

Figure 3 The maximum absolute RM plotted as a function of the estimated cooling flow rate, $\dot{X}$, for a sample of X-ray luminous clusters with measured RMs from Taylor et al. (2002). Both RM and $\dot{X}$ are expected to depend on density to a positive power, so in that sense, the correlation is expected.

where n_0 is the central density, r_c is the core radius, and β is a free parameter in the fit. Typical values for these parameters are $r_c \sim 200$ kpc, $\beta \sim \frac{2}{3}$, and $n_o \sim 0.01$ cm^{-3}.

For this density profile and cells of constant magnetic strength but random orientation, Felten (1996) and Feretti et al. (1995) derived the following relation for the RM dispersion:

$$\sigma_{RM} = \frac{KB\, n_0\, r_c^{1/2} l^{1/2}}{\left(1 + r^2/r_c^2\right)^{(6\beta - 1)/4}} \sqrt{\frac{\Gamma(3\beta - 0.5)}{\Gamma(3\beta)}}, \qquad (3)$$

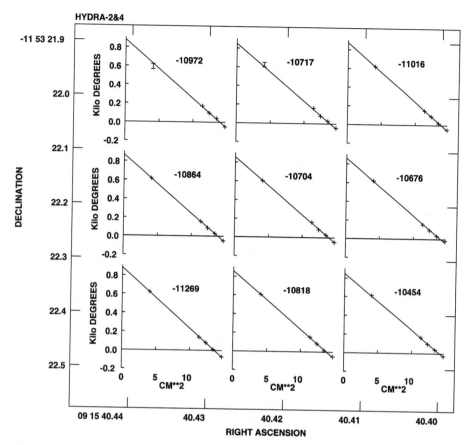

Figure 5   The observed position angles, χ, of the linearly polarized radio emission as a function of the square of the observing wavelength, λ^2, for a number of positions in the southern lobe of Hydra A at a resolution of 0.3″ (Taylor & Perley 1993). The points plotted are each separated by approximately one beamwidth and thus are independent of each other. This illustrates the consistency of the RMs within a coherence length of $\sim$7 kpc. Notice also the excellent agreement to a λ^2-law for $\Delta\chi = 600$ degrees, nearly two complete turns.

where l is the cell size in kpc, r is the distance of the radio source from the cluster center, also in kpc, Γ is the Gamma function, and K is a factor that depends on the location of the radio source along the line-of-sight through the cluster: $K = 624$ if the source is beyond the cluster, and $K = 441$ if the source is halfway through the cluster. Note that Equation 3 assumes that the magnetic field strength, B, is related to the component along the line of sight, $\langle B_{\|}\rangle$, by $B = \sqrt{3}B_{\|}$. The cell size, l, can be estimated to first order from the observed coherence lengths of the RM distributions. Both cooling-flow and noncooling-flow clusters yield typical estimates of 5 to 10 kpc. Magnetic field strength estimates, however are two to three times higher in the cooling-flow clusters—19 μG in the 3C 295 cluster (Perley & Taylor 1991, Allen et al. 2001a) compared to 6 μG in the 3C 129 cluster (Taylor et al. 2001a) using the methodology described above.

Most radio sources found embedded in clusters are located at the center and identified with a cD galaxy. This relatively high pressure environment has been found in many cases to confine or distort the radio galaxy (Taylor et al. 1994), as well as giving rise to extreme RMs. For this same reason, the extended radio sources in Hydra A and Cygnus A are unique in that they sample regions over 100 kpc in linear extent. There are, however, a few clusters containing more than one strong, polarized radio source. The cluster Abell 119 (Feretti et al. 1999) contains three radio galaxies. Using an analysis based on Equation 3 above, Feretti et al. (1999) find that a magnetic field strength of 6–12 μG extending over 3 Mpc could explain the RM distributions for all 3 sources, although they note that such a field would exceed the thermal pressure in the outer parts of the cluster. In a reanalysis of the Abell 119 measurements, Dolag et al. (2001) find that the field scales as $n_e^{0.9}$. This power-law behavior is marginally steeper than that expected assuming flux conservation, for which the tangled field scales as $n_e^{2/3}$, and significantly steeper than that expected assuming a constant ratio between magnetic and thermal energy density, for which the tangled field scales as $n_e^{1/2}$ for an isothermal atmosphere. In the 3C 129 cluster, there are two extended radio galaxies whose RM observations can be fit by a field strength of 6 μG. Finally, in A514, Govoni et al. (2001a) has measured the RM distributions of five embedded (and background) radio sources and found cluster magnetic field strengths of 4–9 μG spread over the central 1.4 Mpc of the cluster. If the magnetic field scales with the density raised to a positive power, then the product of B and n_e in Equation 1 implies that the observed rotation measures are heavily weighted by the innermost cells in the cluster (Dreher et al. 1987).

It has been suggested that high RMs may result from an interaction between the radio galaxy and the ICM, such that the RMs are generated locally and are not indicative of cluster magnetic fields. Bicknell et al. (1990) proposed a model in which the RM screen is due to a boundary layer surrounding the radio source in which the large magnetic fields within the radio source are mixed with the large thermal densities outside the radio source by Kelvin-Helmholtz waves along the contact discontinuity. This model predicts a Faraday depolarized region of a few kpc extent surrounding the radio source, where the synchrotron emitting material has mixed with the thermal gas. Such a depolarized shell has not been observed to date.

In general, extreme RMs have been observed in sources of very different morphologies, from edge-brightened [Fanaroff-Riley Class II (FR) (Fanaroff & Riley 1974)], to edge-darkened (FR I) sources. The models for the hydrodynamic evolution of these different classes of sources are thought to be very different, with the FR II sources expanding supersonically, whereas the FR I sources expand subsonically (Begelman et al. 1984). The different dynamics of FR I and FR II sources argues that the high RMs are not solely a phenomenon arising from a local interaction between the radio source and its environment, but are more likely to be a property of the large-scale (i.e., cluster) environment.

Although we feel that large RMs for cluster center radio sources most likely arise in the large-scale cluster atmosphere, we should point out that there are some cases in which the radio source does appear to compress the gas and fields in the ICM to produce local RM enhancements. For example, there is evidence for an RM enhancement at the bow shock preceding the radio hot spots in Cygnus A and 3C 194 (Carilli et al. 1988, Taylor et al. 1992). However, even in these cases, the implied external (i.e., unperturbed) ICM fields are a few μG.

Background and Embedded Sources

The first successful demonstration of Faraday rotation of the polarized emission from background radio sources seen through a cluster atmosphere was presented by Vallee et al. (1986) for A2319. Vallee et al. (1987) combined the RM excess in A2319 with density estimates from X-ray observations by Jones et al. (1979) to estimate a field strength of 2 μG if the field is organized in 20 kpc-sized cells. Hennessy et al. (1989) studied a sample of 16 sources located behind Abell clusters and found no significant RM excess compared to a control sample of field sources. Kim et al. (1991) considered a larger sample of 161 sources and found that those sources projected within one third of an Abell radius of the cluster center had a significant RM excess over sources with larger impact parameters. Kim et al. (1990) found that the RMs toward sources within 20' of the Coma cluster center had an enhanced RM dispersion (by 38 ± 6 rad m^{-2}). From this excess they derived a magnetic field strength of 2.0 ± 1.1 μG assuming a cell size in the range 7–26 kpc. Feretti et al. (1995) found evidence from the RM distribution of the embedded cluster source NGC 4869 for smaller cell sizes ($\sim$1 kpc), and subsequently estimated the field strength in Coma to be 6.2 μG.

The most significant work in this area is the recent VLA survey by Clarke et al. (2001), in which they observed radio sources in and behind a representative sample of 16 Abell clusters at $z < 0.1$. They found enhanced rotation measures on the large majority of the lines of sight within 0.5 Mpc of the cluster centers (Figure 6), from which they derived an areal filling of the magnetic fields of 95%. Their modeling results in magnetic fields of $\sim$5 μG, assuming a cell size of 10 kpc. These clusters were chosen for their lack of cooling flows, but are otherwise unremarkable in their properties. With approximately one half their sources behind the clusters, these observations demonstrated an embedded powerful radio galaxy is not required to produce significant RMs. Another advantage of this technique is that it permits estimation of the spatial extent of the magnetic fields within the cluster ($\sim$0.5 Mpc). The areal filling factor of 95% (assuming constant magnetic fields in and among all clusters) suggests a relatively large volume-filling factor for the fields, with a formal (extreme) lower limit being about 8% for 10-kpc cell sizes.

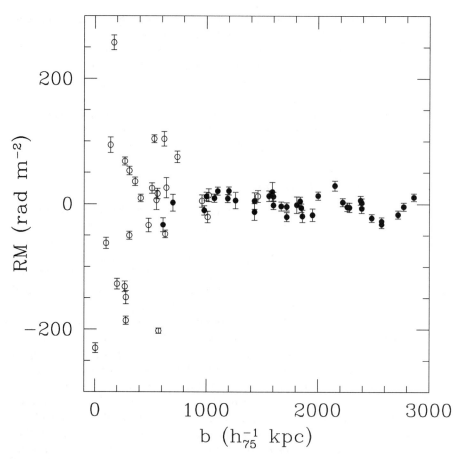

Figure 6   The integrated RM plotted as a function of source impact parameter in kiloparsecs for the sample of 16 Abell clusters described in Clarke et al. (2001). The *open symbols* represent sources viewed through the cluster, whereas the *closed symbols* represent the control sample of field sources.

High Redshift Sources

Radio galaxies and radio loud quasars have been detected to $z = 5.2$ (van Breugel 2000). The extended polarized emission from these sources provides an ideal probe of their environments through Faraday rotation observations. Extensive radio imaging surveys of $z > 2$ radio galaxies and quasars have shown large rotation measures, and Faraday depolarization, in at least 30% of the sources, indicating that the sources are situated behind dense Faraday screens of magnetized, ionized plasma (Chambers et al. 1990; Garrington et al. 1988; Carilli et al. 1994, 1997; Pentericci et al. 2000; Lonsdale et al. 1993; Athreya et al. 1998), with a possible increase in this fraction with redshift (Pentericci et al. 2000). Drawing the analogy to lower z radio galaxies, these authors proposed that the high z sources may be embedded in magnetized (proto-) cluster atmospheres, with μG field strengths.

A difficulty with the study of high redshift sources is that the sources are typically small (<few arcseconds), requiring higher frequency observations (5 to 8 GHz) in order to properly resolve the source structure. This leads to two problems. First, the rest frame frequencies are then ≥ 20 GHz, such that only extreme values of Faraday rotation can be measured (RM ≥ 1000 rad m^{-2}). Second, only the flatter spectrum, higher surface brightness radio emitting structures in the sources are detected, thereby allowing for only a few lines-of-site through the ICM as RM probes. Imaging at frequencies of 1.4 GHz or lower with subarcsecond resolution is required to address this interesting issue.

INVERSE COMPTON X-RAY EMISSION

Cosmic magnetic fields can be derived by comparing inverse Compton X-ray emission and radio synchrotron radiation (Harris & Grindlay 1979, Rephaeli et al. 1987). Inverse Compton (IC) emission is the relativisitic extrapolation of the Sunyaev-Zel'dovich effect (Rephaeli 1995), involving up-scattering of the ambient photon field by the relativisitic particle population. The IC process involves two Lorentz transforms (to and from the rest frame of the electron), plus Thompson scattering in the rest frame of the electron, leading to $\nu_{IC} \sim \frac{4}{3}\gamma^2\nu_{bg}$, where ν_{IC} is the emergent frequency of the scattered radiation, γ is the electron Lorentz factor, and ν_{bg} is the incident photon frequency (Bagchi et al. 1998). From a quantum mechanical perspective, synchrotron radiation is directly analogous to IC emission, with synchrotron radiation being the up-scattering of the virtual photons that constitute the static magnetic field. Given a relativistic electron population, the IC emissivity is directly proportional to the energy density in the photon field, U_{bg}, whereas the synchrotron emissivity is proportional to the energy density in the magnetic field, U_B, leading to a simple proportionality between synchrotron and IC luminosity: $\frac{L_{syn}}{L_{IC}} \propto \frac{U_B}{U_{bg}}$. Given that they originate from the same (assumed power-law) relativistic electron population, IC X-rays and synchrotron radio emission share the same spectral index, α. The spectral index relates to the index for the power-law electron energy distribution, Γ, as $\Gamma = 2\alpha - 1$, and to the photon index as $\alpha - 1$.

In most astrophysical circumstances, U_{bg} is dominated by the cosmic microwave background (CMB), except in the immediate vicinity of active–star-forming regions and AGN (Brunetti et al. 2001a, Carilli et al. 2001). The Planck function at T = 2.73 K peaks near a frequency of $\nu_{bg} \sim 1.6 \times 10^{11}$ Hz, hence IC X-rays observed at 20 keV ($\nu_{IC} = 4.8 \times 10^{18}$ Hz), are emitted predominantly by electrons at $\gamma \sim 5000$, independent of redshift.[1] The corresponding radio synchrotron

---

[1] γ is independent of redshift because ν_{bg} increases as $1 + z$.

emission from $\gamma = 5000$ electrons peaks at a (rest frame) frequency of $\nu_{\mathrm{syn}} \sim 4.2 \, (\frac{B}{1\,\mu G}) \gamma^2$ Hz $= 100$ MHz (Bagchi et al. 1998).

Many authors have considered the problem of deriving magnetic fields by comparing synchrotron radio and inverse Compton X-ray emission (Blumenthal & Gould 1970, Harris & Grindlay 1979, Rephaeli et al. 1987, Feigelson et al. 1995, Bagchi et al. 1998). Assuming $\alpha = -1$, the magnetic field is given by:

$$B = 1.7(1 + z)^2 \left(\frac{S_r \nu_r}{S_x \nu_x} \right)^{0.5} \mu G, \qquad (4)$$

where S_r and S_x are the radio and X-ray flux densities at observed frequencies ν_r and ν_x, respectively. Note that, unlike Faraday rotation measurements, the geometry of the field does not play a critical role in this calculation, except in the context of the electron pitch angle distribution (see "Reconciling IC- and RM-Derived Fields" below).

The principle difficulty in studying IC emission from clusters of galaxies is confusion by the thermal emission from the cluster atmosphere. One means of separating the two emission mechanisms is through spectroscopic X-ray observations at high energy. The IC emission will have a harder, power-law spectrum relative to thermal brehmstrahlung emission. Recent high-energy X-ray satellites such as Beppo/Sax and RXTE have revolutionized this field by allowing for sensitive observations to be made at energies well above 10 keV (Rephaeli 2002). Prior to these instruments, most studies of IC emission from clusters with radio halos only provided lower limits to the magnetic fields of about 0.1 μG (Rephaeli et al. 1987).

Recent observations of four clusters with radio halos with Beppo/Sax and RXTE have revealed hard X-ray tails that dominate the integrated emission above 20 keV (Rephaeli et al. 1999; Fusco-Femiano et al. 2001, 2000). In Figure 7, we reproduce the RXTE observations of the Coma cluster. The hard X-ray emission in these sources has a spectral index $\alpha = -1.3 \pm 0.3$, roughly consistent with the radio spectral index. Comparing the IC X-ray and radio synchrotron emission in these sources leads to a volume-averaged cluster magnetic field of 0.2 to 0.4 μG, with a relativistic electron energy density $\sim 10^{-13}$ erg cm^{-3}.

Spatially resolving X-ray observations can also be used to separate nonthermal and thermal X-ray emission in clusters. This technique has been used recently in the study of the steep spectrum radio relic source in Abell 85 (Bagchi et al. 1998). An X-ray excess relative to that expected for the cluster atmosphere is seen with the ROSAT PSPC detector at the position of the diffuse radio source in Abell 85 (see Figure 8). Bagchi et al. (1998) subtract a model of the thermal cluster X-ray emission in order to derive the IC contribution, from which they derive a magnetic field of 1.0 $\pm$ 0.1 μG.

Emission above that expected from the hot cluster atmosphere has also been detected in the extreme ultraviolet (EUV = 0.1 to 0.4 keV) in a few clusters (Berghöfer et al. 2000, Bowyer et al. 2001, Bonamente et al. 2001). It has been suggested that this emission may also be IC in origin, corresponding to relativistic electrons with $\gamma \sim 400$ (Atoyan & Völk 2000). However, the emission spectrum

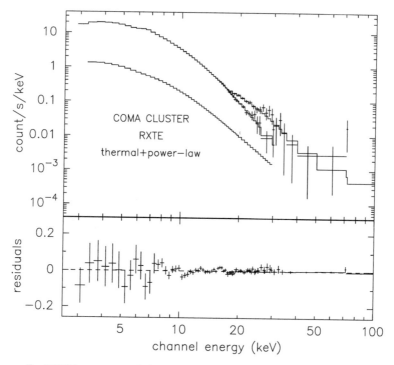

Figure 7 RXTE spectrum of the Coma cluster. Data and folded Raymond-Smith ($kT \simeq 7.51$ keV), and power-law (photon index = 2.34) models are shown in the *upper frame*; the latter component is also shown separately in the *lower line*. Residuals of the fit are shown in the *lower frame* (Rephaeli 2002).

is steep ($\alpha \leq -2$), and the EUV emitting regions are less extended than the radio regions. Neither of these properties are consistent with a simple extrapolation of the radio halo properties to low frequency (Bowyer et al. 2001). Also, the pressure in this low γ relativistic component would exceed that in the thermal gas by at least a factor of three (Bonamente et al. 2001).

Electron Lifetimes

An important point concerning IC and synchrotron emission from clusters is that of particle lifetimes. The lifetime of a relativistic electron is limited by IC losses off the microwave background to $t_{IC} = 7.7 \times 10^9 (\frac{300}{\gamma})(1+z)^{-4}$ years (Sarazin 2002).[2] Relativistic electrons emitting in the hard X-ray band via IC scattering

---

[2]For $\gamma > 300$, IC losses dominate [or synchrotron losses for B $> 2.3 \, (1+z)^2 \, \mu$G], whereas for lower γ electrons Brehmstrahlung losses dominate in cluster environments (Sarazin 2001b).

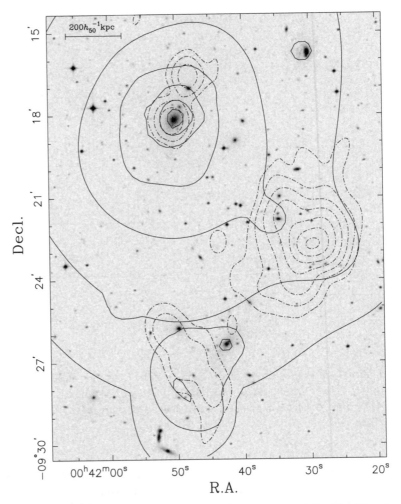

Figure 8   The cluster Abell 85 central region at different wavelengths. The photographic R-filter image (UK Schmidt Telescope and the Digitized Sky Survey) is shown in gray scale. Full contour lines show the multiscale wavelet reconstruction of the ROSAT PSPC X-ray data. The Ooty Synthesis Radio Telescope (OSRT) 326.5 MHz radio surface-brightness image is depicted using dot-dashed contour lines. All contours are spaced logarithmically (Bagchi et al. 1998).

of the CMB have lifetimes of about 10^9 years, whereas the lifetimes for 1.4 GHz synchrotron emitting electrons are a factor of four or so shorter in μG fields. Diffusion timescales (set by streaming along the tangled magnetic field) for cluster relativistic electrons are thought to be longer than the Hubble time (Sarazin 2002, Colafrancesco & Blasi 1998, Casse et al. 2001), making cluster atmospheres efficient traps of relativistic electrons, much like galaxy disks. The fact that the diffusion timescales are much longer than the energy loss timescales for $\gamma > 10^4$ electrons requires in situ acceleration in order to explain radio halo sources (Schlikeiser et al. 1987).

Cluster merger shock fronts are obvious sites for first-order Fermi acceleration, whereas subsequent turbulence may lead to second-order (stochastic) Fermi acceleration (Brunetti et al. 2001b, Eilek 1999, Ensslin et al. 1998, Markevitch & Vikhlinin 2001). Active particle acceleration during cluster mergers provides a natural explanation for the observed correlation between cluster radio halos and substructure in cluster atmospheres (Govoni et al. 2001a), and between cluster radio luminosity and cluster atmosphere temperature, assuming that the gas temperature increases during a merger (Liang et al. 2000). Brunetti et al. (2001a) develop a two-phase model in which initial relativistic particle injection into the ICM occurs early in the cluster lifetime by starburst-driven winds from cluster galaxies, and/or by shocks in early subcluster mergers. The second phase involves re-acceleration of the radiatively aged particle population via more recent cluster mergers. Their detailed application of this model to the Coma cluster suggests a merger has occurred within the last 10^9 years.

Another mechanism proposed for in situ relativistic particle injection is secondary electron production via the decay of π-mesons generated in collisions between cosmic ray ions (mostly protons) and the thermal ICM (Dennison 1980, Dolag & Ensslin 2000). The important point, in this case, is that the energetic protons have radiative lifetimes orders of magnitude longer than the lower mass electrons. The problem with this hypothesis is that the predicted γ-ray fluxes exceed limits set by EGRET by a factor of 2 to 7 (Blasi & Colafrancesco 1999).

Reconciling IC- and RM-Derived Fields

The IC-estimated cluster magnetic fields are typically 0.2 to 1 μG, whereas those obtained using RM observations are a factor four to ten higher. Petrosian (2001) has considered this discrepancy in detail, and finds that the different magnetic field estimates can be reconciled in most cases by making a few reasonable assumptions concerning the electron energy spectrum and pitch-angle distribution.

First, an anisotropic pitch-angle distribution biased toward low angles would clearly weaken the radio synchrotron radiation relative to the IC X-ray emission. Such a distribution will occur naturally given that electrons at large pitch angles have greater synchrotron losses. A potential problem with this argument is that pitch-angle scattering of the relativistic electrons by Alfven waves self-induced by particles streaming along field lines is thought to be an efficient process in the ISM and ICM (Wentzel 1974), such that re-isotropization of the particle distribution will occur on a timescale short compared to radiative timescales. Petrosian (2001) points out that most derivations of magnetic fields from IC emission assume the electrons are gyrating perpendicular to the magnetic field lines. Just assuming an isotropic relativistic electron pitch-angle distribution raises the IC-estimated magnetic field by a factor of two or so.

Second, the IC hard X-ray emission is from relativistic electrons with $\gamma \sim 5000$. This corresponds to radio continuum emission at 100 MHz for μG magnetic fields. Most surveys of cluster radio halos have been done at 1.4 GHz (Giovanni et al. 1999, Govoni et al. 2001a), corresponding to electron Lorentz factors of $\gamma \sim 18000$. A steepening in the electron energy spectrum at Lorentz factors around 10^4 will reduce the 1.4 GHz radio luminosities, but retain the IC hard X-ray emission. For example, Petrosian (2001) finds that a change in the power-law index for the particle energy distribution from $\Gamma = -3$ to $\Gamma = -5$ (corresponding to $\alpha = -1$ to -2) at $\gamma \sim 10^4$ raises the IC-estimated fields to ~ 1 μG. Such a steepening of the electron energy spectrum at $\gamma \sim 10^4$ will arise naturally if no relativistic particle injection occurs over a timescale $\sim 10^8$ years (see "Electron Lifetimes" above). The problem in this case is the fine tuning required to achieve the break in the relevant energy range for a number of clusters. In general, a negatively curved (in log-space) electron energy distribution will inevitably lead to IC-estimated fields being lower than those estimated from 1.4 GHz radio observations, unless a correction is made for the spectral curvature.

Others have pointed out that magnetic substructure, or filamentation, can lead to a significant difference between fields estimated using the different techniques. A large relativistic electron population can be hidden from radio observations by putting them in weak-field regions (Feretti et al. 2001, Rephaeli et al. 1987, Rephaeli 2002, Goldshmidt & Rephaeli 1993, Rudnick 2000). A simple example of this is if the relativistic particles have a larger radial scale-length than the magnetic fields in the cluster. In this case, most of the IC emission will come from the weak-field regions in the outer parts of the cluster, whereas most of the Faraday rotation and synchrotron emission occurs in the strong-field regions in the inner parts of the cluster.

Another explanation for the discrepancy between IC- and RM-derived magnetic fields is to assume that the hard X-rays are not IC in origin, in which case the IC estimates become lower limits. A number of authors have considered high energy X-ray emission by nonthermal Brehmstrahlung, i.e., Brehmstrahlung radiation from a suprathermal tail of electrons arising via stochastic acceleration in a turbulent medium (Fermi II acceleration) (Blasi 2000, Sarazin 2002, Ensslin et al. 1999, Dogiel 2000). The problem with this hypothesis is the energetics: Brehmstrahlung is an inefficient radiation mechanism, with most of the collisional energy going into heat. Assuming an energy source is available to maintain the suprathermal tail, Petrosian (2001) shows that the collisional energy input by the suprathermal particles would be adequate to evaporate the cluster atmosphere on a timescale of $\leq 10^9$ years. Hence the mechanism maintaining the suprathermal tail must be short lived (Blasi 2000).

The current hard X-ray spectroscopic observations are limited to very low spatial resolution ($\sim 1°$), whereas X-ray imaging instruments have high energy cut-offs at around 10 keV. Likewise, sensitive arcminute resolution radio images for a large number of clusters are available only at 1.4 GHz, corresponding to electrons with Lorentz factors 3 to 4 times higher than those emitting hard X-rays. Both of these limitations will be overcome in the coming years with the launch of hard-X-ray imaging satellites such as Constellation-X, and improvements in radio imaging capabilities at 300 MHz and below at the Very Large Array and the Giant Meter wave Radio Telescope.

COLD FRONTS

In order to maintain temperature gradients in X-ray clusters, the thermal conduction must be suppressed by two orders of magnitude relative to the classical Spitzer value (Mckee & Begelman 1990, Fabian 1994, Spitzer 1962). If not, cooler structures on scales of $\sim$0.1 Mpc will be evaporated by thermal conduction from the hot surrounding medium on timescales of $\sim 10^7$ years. Examples of such cooler structures in clusters include cooling-flow cluster cores and X-ray corona surrounding large galaxies (Fabian 1994, Vikhlinin et al. 2001a), with temperature differences ranging from a factor of 2 to 5 relative to the hot ICM.

Cowie & McKee (1977) show that the conductivity can be suppressed by almost an order of magnitude below the Spitzer value in cases where the Coulomb mean free path (mfp) is comparable to the scale of thermal gradients owing to the development of electrostatic fields. For large-scale structure in cluster atmospheres, this reduction is not adequate because the mfp $\sim 1.2 \times 10^{22} \left(\frac{T}{5 \times 10^7 \, K}\right)^2 \left(\frac{n}{0.001 \, cm^{-3}}\right)^{-1}$ cm, or just a few kpc for a typical cluster.

The presence of magnetic fields will reduce the conductivity in a thermal plasma (Field 1965, Parker 1979, Binney & Cowie 1981, Chandran et al. 1999). The simple point is that the gyro radius for thermal electrons in the ICM is $\sim 2 \times 10^8$ $\left(\frac{B^{-1}}{1 \, \mu G}\right)\left(\frac{T}{5 \times 10^7 \, K}\right)$ cm, many orders of magnitude below the collisional mfp. Tribble (1993) shows that the presence of a cluster magnetic field will lead naturally to the development of a multiphase ICM, with thermally isolated regions on scales set by the magnetic structures [but compare Rosner & Tucker 1989).

The idea of magnetic suppression of thermal conductivity in cluster gas has been verified with the recent discovery of cold fronts in clusters of galaxies (Markevitch et al. 2000; Vikhlinin et al. 2001a,b). These fronts manifest themselves as sharp discontinuities in X-ray surface brightness (Figure 9, see color insert). They are not shocks, because the increase in density is accompanied by a decrease in temperature such that there is no dramatic change in the pressure and entropy across the front (Markevitch et al. 2000, Ettori & Fabian 2000). These structures are interpreted as resulting from cluster mergers, where a cooler subcluster core falls into a hot ICM at sub- or trans-sonic velocities ($\sim 10^3$ km s^{-1}). A discontinuity is formed where the internal pressure of the core equals the combined ram and thermal pressure of the external medium. Gas beyond this radius is stripped from the merging subcluster, and the core is not penetrated by shocks owing to its high internal pressure.

The best example of a cluster cold front is that seen in Abell 3667 (see Figure 9) (Vikhlinin et al. 2001a,b). In this case, the temperature discontinuity occurs over a scale of ∼5 kpc, comparable to the collisional mfp, thereby requiring thermal isolation. Magnetic fields play a fundamental role in allowing for such structures in two ways: (a) by suppressing thermal conduction and (b) by suppressing Kelvin-Helmholtz mixing along the contact discontinuity. Vikhlinin et al. (2001a) present a model in which the field is tangentially sheared by fluid motions along the contact discontinuity. They invoke magnetic tension to suppress mixing, and show that the required magnetic pressure is between 10% and 20% of the thermal pressure. The implied fields are between 7 and 16 μG. They also argue that the fields cannot be much stronger than this, because dynamically dominant fields would suppress mixing along the entire front, which does not appear to be the case.

The existence of cold fronts provides strong evidence for cluster magnetic fields. However, the field strengths derived correspond to those in the tangentially sheared boundary region around the front. Relating these back to the unperturbed cluster field probably requires a factor of a few reduction in field strength, implying unperturbed field strengths between 1 and 10 μG, although the exact scale factor remains uncertain (Vikhlinin et al. 2001b).

MAGNETIC SUPPORT OF CLUSTER GAS AND THE BARYON CRISIS

Two interesting issues have arisen in the study of cluster gas and gravitational masses. First is the fact that total gravitating masses derived from weak gravitational lensing are a factor of a few higher than those derived from X-ray observations of cluster atmospheres, assuming hydrostatic equilibrium of isothermal atmospheres (Loeb & Mao 1994, Miralda-Escude & Babul 1995). Second is the baryon crisis, in which the baryonic mass of a cluster, which is dominated by the mass of gas in the hot cluster atmosphere, corresponds to roughly 5% of the gravitational mass derived assuming hydrostatic equilibrium for an isothermal cluster atmosphere. This baryon fraction is a factor of three to five larger than the baryon fraction dictated by big bang nucleosynthesis in inflationary world models (White et al. 1993).

A possible solution to both these problems is to invoke nonthermal pressure support for the cluster atmosphere, thereby allowing for larger gravitating masses relative to those derived assuming hydrostatic equilibrium. A number of authors have investigated the possibility of magnetic pressure support for cluster atmospheres (Loeb & Mao 1994, Miralda-Escude & Babul 1995, Dolag & Schindler 2000). The required fields are about 50 μG, which is an order of magnitude, or more, larger than the measured fields in most cases, except perhaps in the inner 10s of kpc of cooling-flow clusters. For most relaxed clusters, Dolag & Schindler (2000) find that magnetic pressure affects hydrostatic mass estimates by at most 15%.

Other mechanisms for nonthermal pressure support of cluster atmospheres involve motions of the cluster gas other than thermal, such as turbulent or bulk motions owing to a recent cluster merger (Mazzotta et al. 2001, Wu 2000). For relaxed clusters, a number of groups have shown that the lensing and X-ray mass estimates can be reconciled by using nonisothermal models for the cluster atmospheres, i.e., by allowing for radial temperature gradients (Allen et al. 2001b, Markevitch et al. 1999).

GZK LIMIT

Greisen (1966) and Zatsepin & Kuzmin (1966) pointed out that cosmic rays with energies >10^{19} eV lose energy due to photo-pion production through interaction with the CMB. These losses limit the propagation distance for 10^{20} eV particles to about 60 Mpc. Yet, no clear correlation has been found between the arrival direction of high energy cosmic rays and the most likely sites of origin for EeV particles, namely, AGN at distances less than 60 Mpc (Elbert & Sommers 1995, Biermann 1999). One possible solution to the GZK paradox is to assume the energetic particles are isotropized in the IGM by tangled magnetic fields, effectively randomizing their observed arrival direction. Such isotropization requires fields in the local super-cluster ≥ 0.3 μG (Farrar & Piran 2000; but compare Isola et al. 2000).

SYNTHESIS

In Table 1, we summarize the cluster magnetic field measurements. Given the limitations of the current instrumentation, the limited number of sources studied thus far, and the myriad physical assumptions involved with each method, we are encouraged by the order-of-magnitude agreement between cluster field strengths derived from these different methods. Overall, the data are consistent with cluster atmospheres containing $\sim\mu$G fields, with perhaps an order-of-magnitude scatter in field strength between clusters, or within a given cluster.

TABLE 1   Cluster magnetic fields

Method	Strength μG	Model parameters
Synchrotron halos	0.4–1	Minimum energy, $k = \eta = 1$, $\nu_{low} = 10$ MHz, $\nu_{high} = 10$ GHz
Faraday rotation (embedded)	3–40	Cell size = 10 kpc
Faraday rotation (background)	1–10	Cell size = 10 kpc
Inverse Compton	0.2–1	$\alpha = -1$, $\gamma_{radio} \sim 18000$, $\gamma_{xray} \sim 5000$
Cold fronts	1–10	Amplification factor ~ 3
GZK	>0.3	AGN = site of origin for EeV CRs

The RM observations of embedded and background radio sources suggest that μG magnetic fields with high areal filling factors are a standard feature in clusters, and that the fields extend to large radii (0.5 Mpc or more). The RM observations of extended radio galaxies embedded in clusters impose order on the fields, with coherence scales of order 10 kpc, although larger scale coherence in overall RM sign can be seen in some sources. Observations of inverse Compton emission from a few clusters with radio halos provide evidence against much stronger, pervasive, and highly tangled fields.

In most clusters the fields are not dynamically important, with magnetic pressures one to two orders of magnitude below thermal gas pressures. But the fields play a fundamental role in the energy budget of the ICM through their effect on heat conduction, as is dramatically evident in high-resolution X-ray observations of cluster cold fronts.

If most clusters contain μG magnetic fields, then why don't most clusters have radio halos? The answer may be the short lifetimes of the relativistic electrons responsible for the synchrotron radio emission (see "Electron Lifetimes" above). Without re-acceleration or injection of relativistic electrons, a synchrotron halo emitting at 1.4 GHz will fade in about 10^8 years due to synchrotron and inverse Compton losses. This may explain the correlation between radio halos and cluster mergers, and the anticorrelation between radio halos and clusters with relaxed X-ray morphologies. In this case, the fraction of clusters with radio halos should increase with decreasing survey frequency.

The existence of μG-level fields in cluster atmospheres appears well established. The challenge for observers now becomes one of determining the detailed properties of the fields, and how they relate to other cluster properties. Are the fields filamentary, and to what extent do the thermal and nonthermal plasma mix in cluster atmospheres? What is the radial dependence of the field strength? How do the fields depend on cluster atmospheric parameters, such as gas temperature, metalicity, mass, substructure, or density profile? How do the fields evolve with cosmic time? And do the fields extend to even larger radii, perhaps filling the IGM? The challenge to the theorists is simpler: How were these fields generated? This topic is considered briefly in the next section.

FIELD ORIGIN

When attempting to understand the behavior of cosmic magnetic fields, a critical characteristic to keep in mind is their longevity. The Spitzer conductivity (Spitzer 1962) of the ICM is $\sigma \sim 3 \times 10^{18}$ sec^{-1} (for comparison, the conductivity of liquid mercury at room temperature is 10^{16} sec^{-1}). The timescale for magnetic diffusion in the ICM is then $\tau_{\mathrm{diff}} = 4\pi\sigma(\frac{L}{c})^2 \sim 10^{36}(\frac{L}{10\,\mathrm{kpc}})^2$ years, where L is the spatial scale for magnetic fluctuations. The magnetic Reynold's number is $R_m = \frac{\tau_{\mathrm{diff}}}{\tau_{\mathrm{conv}}} \sim 10^{29}(\frac{L}{10\,\mathrm{kpc}})(\frac{V}{1000\,\mathrm{km\,s^{-1}}})$, where $\tau_{\mathrm{conv}} = $ the convective timescale $= \frac{L}{V}$, and V is the bulk fluid velocity. The essentially infinite diffusion timescale for the fields implies that once a field is generated within the ICM, it will remain extant unless some anomalous resistive process occurs, e.g., reconnection via plasma-wave generation in shocks.

Perhaps the simplest origin for cluster magnetic fields is compression of an intergalactic field. Clusters have present-day overdensities $\delta \sim 10^3$. In order to get $B_{ICM} > 10^{-7}$ G by adiabatic compression ($B \propto \delta^{\frac{2}{3}}$) then requires IGM fields $B_{IGM} > 10^{-9}$ μG.

Of course, this solution merely pushes the field origin problem from the ICM into the IGM. An upper limit to IGM fields of 10^{-9} G is set by Faraday rotation measurements of high z radio loud QSOs, assuming a cell size of order 1 Mpc (Kronberg 1996, Blasi et al. 1999). A limit to IGM magnetic fields at the time of recombination can also be set by considering their affect on the CMB. Dynamically significant magnetic fields will exert an anisotropic pressure on the gas, which must be balanced by gravity. Detailed studies of this phenomenon in the context of recent measurements of the CMB anisotropies shows that the comoving IGM fields[3] must be less than a few $\times$ 10^{-9} G (Barrow et al. 1997, Barrow & Subramanian 1998, Clarkson & Coley 2001, Adams et al. 1996). A comoving field of 10^{-9} G at recombination would lead to Faraday rotation of the polarized CMB emission by $1°$ at an observing frequency of 30 GHz, a measurement that is within reach of future instrumentation (Kosowsky & Loeb 1996, Grasso & Rubinstein 2001). Considerations of primordial nucleosynthesis and the affect of magnetic fields on weak interactions and electron densities imply upper limits to comoving IGM fields of 10^{-7} G (Grasso & Rubinstein 1995).

The origin of IGM magnetic fields has been considered by many authors. One class of models involves large-scale field generation prior to recombination. An excellent review of pre-recombination magnetic field generation is presented by Grasso & Rubinstein (2001). Early models for pre-recombination field generation involved the hydrodynamical Biermann battery effect (Biermann 1950). In general, the hydrodynamic battery involves charge separation arising from the fact that electrons and protons have the same charge, but very different masses. For instance, protons will have larger Coulomb stopping distances than electrons, and be less affected by photon drag. Harrison (1970) suggested that photon drag on protons relative to electrons in vortical turbulence during the radiation era could lead to charge separation, and hence, magnetic field generation by electric currents. Subsequent authors have argued strongly against vortical density perturbations just prior to recombination, because vortical (and higher order) density perturbations decay rapidly with the expansion of the universe (Rees 1987). This idea has been revisited recently in the context of vortical turbulence generated by moving cosmic strings (Vachaspati & Vilenkin 1991, Avelino & Shellard 1995). Other mechanisms for field generation prior to recombination include battery affects during the quark-hadron (QCD) phase transition (Quashnock et al. 1989), dynamo

---

[3]Comoving fields correspond to equivalent present-epoch field strengths, i.e., corrected for cosmic expansion assuming flux freezing.

mechanisms during the electro-weak (QED) phase transition (Baym et al. 1996), and mechanisms relating to the basic physics of inflation (Turner & Widrow 1988).

A problem with all these mechanisms is the survivability of the fields on relevant scales during the radiation era. Battaner & Lesch (2000) argue that magnetic and photon diffusion will destroy fields on comoving scales ≤ few Mpc during this epoch, thereby requiring generation of the fields in the post-recombination universe by normal plasma processes during proto-galactic evolution (see also Lesch & Birk 1998).

Models for post-recombination IGM magnetic field generation typically involve ejection of the fields from normal or active galaxies (Kronberg 1996). A simple but cogent argument for this case is that the metalicity of the ICM is typically about 30% solar, implying that cluster atmospheres have been polluted by outflows from galaxies (Aguirre et al. 2001). A natural extension of this idea would be to include magnetic fields in the outflows (Goldshmidt & Rephaeli 1993). It has also been suggested that IGM fields could be generated through turbulent dynamo processes and/or shocks occurring during structure formation (Zweibel 1988, Kulsrud et al. 1997, Waxman & Loeb 2000), or by battery effects during the epoch of reionization (Gnedin et al. 2000).

Seed magnetic fields will arise in the earliest stars via the normal gas kinematical Biermann battery mechanism. These fields are amplified by the $\alpha - \Omega$ dynamo operating in convective stellar atmospheres (Parker 1979), and then are ejected into the ISM by stellar outflows and supernova explosions. The ISM fields can then be injected into the IGM by winds from active star-forming galaxies (Heckman 2001). Kronberg et al. (1999) consider this problem in detail and show that a population of dwarf starburst galaxies at $z \geq 6$ could magnetize almost 50% of the universe, but that at lower redshifts the IGM volume is too large for galaxy outflows to affect a significant fraction of the volume.

De Young (1992) and Rephaeli (1988) show that galaxy outflows, and/or gas stripping by the ICM, in present-day clusters are insufficient to be solely responsible for cluster fields ∼1 μG without invoking subsequent dynamo amplification of the field strength by about an order of magnitude in the cluster atmosphere. A simple argument in this case is that the mean density ratio of the ICM versus the ISM, $\delta \sim 0.01$, such that ICM fields would be weaker than ISM fields by $\delta^{\frac{2}{3}} \sim$ 0.05, corresponding to maximum ICM fields of 0.2 to 0.5 μG.

Fields can be ejected from Active Galactic Nuclei (AGN) by relativistic outflows (radio jets) and Broad Absorption Line outflows (BALs) (Rees & Setti 1968, Daly & Loeb 1990). The ultimate origin of the fields in this case may be a seed field generated by a gas kinematic battery operating in the dense accretion disk around the massive black hole, plus subsequent amplification by an $\alpha - \Omega$ dynamo in the rotating disk (Colgate & Li 2000). Detailed consideration of this problem (Furlanetto & Loeb 2001, Kronberg et al. 2001) using the statistics for high z QSO populations shows that by $z \sim 3$, between 5% and 20% of the IGM may be permeated by fields with energy densities corresponding to ≥10% the thermal energy density of the photoionized IGM at 10^4 K, corresponding to comoving field strengths of order 10^{-9} μG.

Kronberg et al. (2001) point out that powerful double radio sources such as Cygnus A (radio luminosities $\sim 10^{45}$ erg s^{-1}) typically have total magnetic energies of about 10% of that of the ICM as a whole. Hence, about ten powerful double radio sources over a cluster lifetime would be adequate to magnetize the cluster at the μG level.

Galaxy turbulent wakes have been proposed as a means of amplifying cluster magnetic fields (Jaffe 1980, Tribble 1993, Ruzmaikin et al. 1989). The problem in this case is that the energy appears to be insufficient, with expected field strengths of at most ~ 0.1 μG. Also, the size scale of the dominant magnetic structures is predicted to be significantly smaller than the 5 to 10 kpc scale sizes observed (Goldshmidt & Rephaeli 1993, De Young 1992).

Cluster mergers are the most energetic events in the universe since the big bang, releasing of order 10^{64} ergs in gravitational binding energy (Sarazin 2002). For comparison, the total thermal energy in the cluster atmosphere is $\sim 10^{63}$ $(\frac{M_{gas}}{10^{14} M_\odot})(\frac{T}{5 \times 10^7 K})$ ergs, and the total energy contained in the cluster magnetic fields is $\sim 10^{60}(\frac{B}{1 \mu G})^2$ ergs. Hence, only a fraction of a percent of the cluster merger energy need be converted into magnetic fields. One possibility for merger-generated magnetic fields is a rotational dynamo associated with helical turbulence driven by off-center cluster mergers. This mechanism requires net cluster rotation—a phenomenon that has yet to be seen in cluster galaxy velocity fields (compare Dupke & Bregman 2001). The lack of observed rotation for clusters suggests low-impact parameters for mergers (≤ 100 kpc) on average (Sarazin 2002), as might arise if most mergers occur along filamentary large-scale structure (Evrard & Gioia 2002). The energetics of even slightly off-center cluster mergers is adequate to generate magnetic fields at the level observed, but the slow cluster rotation velocities (≤ 100 km s^{-1}) imply only one or two rotations in a Hubble time (Colgate & Li 2000), which is insufficient for mean field generation via the inverse cascade of the $\alpha - \Omega$ dynamo (Parker 1979).

A general treatment of the problem of magnetic field evolution during cluster formation comes from numerical studies of heirarchical merging of large-scale structure including an initial intergalactic field $\sim 10^{-9}$ G (Dolag & Schindler 2000, Roettiger et al. 1999). These studies show that a combination of adiabatic compression and nonlinear amplification in shocks during cluster mergers may lead to ICM mean fields of order 1 μG.

A related phenomenon is field amplification by (possible) cooling flows. Soker & Sarazin (1990) have considered this mechanism in detail, and show that the amplification could be a factor of 10 or larger in the inner 10s of kpc. They predict a strong increase in RMs with radius ($\propto r^2$), with centrally peaked radio halos. Such an increase may explain the extreme RM values seen in powerful radio sources at the centers of cooling flow clusters (see "Cluster Center Sources" above), although the existence of gas inflow in these systems remains a topic of debate (Binney 2002).

Overall, there are a number of plausible methods for generating cluster magnetic fields, ranging from injection of fields into the IGM (or early ICM) by active star-forming galaxies and/or radio jets at high redshift, to field amplification by cluster mergers. It is likely that a combination of these phenomena give rise to the μG fields observed in nearby cluster atmospheres. Tests of these mechanisms will require observations of (proto-) cluster atmospheres at high redshift, and a better understanding of the general IGM field.

ACKNOWLEDGMENTS

We thank Juan Uson and Ken Kellermann for suggesting this review topic. We are grateful to P. Blasi, J. Barrow, Stirling Colgate, Steve Cowley, Torsten Ensslin, Luigina Feretti, Bill Forman, Gabriele Giovannini, Federica Govoni, Avi Loeb, Hui Li, Vahe Petrosian, and Robert Zavala for insightful corrections and comments on initial drafts of this manuscript. We also thank Rick Perley, John Dreher, and Frazer Owen for fostering our initial studies of cluster magnetic fields. And we thank G. Giovannini, T. Clarke, J. Bachi, Y. Rephaeli, and A. Vikhlinin for allowing us to reproduce their figures.

The National Radio Astronomy Observatory is a facility of the National Science Foundation operated under a cooperative agreement by Associated Universities, Inc. This research has made use of the NASA/IPAC Extragalactic Database (NED), which is operated by the Jet Propulsion Laboratory, Caltech, under contract with NASA, and use of NASA's Astrophysics Data System Bibliographic Services.

The *Annual Review of Astronomy and Astrophysics* is online at
http://astro.annualreviews.org

LITERATURE CITED

Adams J, Danielsson UH, Grasso D, Rubinstein H. 1996. *Phys. Rev. D* 388:253–58

Aguirre A, Hernquist L, Schaye J, Katz N, Weinberg D, Gardner J. 2001. *Ap. J.* 561: 521–49

Alexander P, Brown MT, Scott PF. 1984. *MNRAS* 209:851–68

Allen SW, Ettori S, Fabian AC. 2001b. *MNRAS* 324:877–90

Allen SW, Taylor GB, Nulsen PEJ, Johnstone RM, David LP, et al. 2001a. *MNRAS* 324:842–58

Athreya RM, Kapahi VK, McCarthy PJ, van Breugel W. 1998. *Astron. Astrophys.* 329:809–20

Atoyan AM, Völk HJ. 2000. *Ap. J.* 535:45–52

Avelino PP, Shellard EPS. 1995. *Phys. Rev. D* 51:5946–49

Bagchi J, Pislar V, Neto GBL. 1998. *MNRAS* 296:L23–28

Barrow JD, Ferreira PG, Silk J. 1997. *Phys. Rev. Lett.* 78:3610–13

Barrow JD, Subramanian K. 1998. *Phys. Rev. Lett.* 81:3575–79

Battaner E, Lesch H. 2000. *An. Fis.* In press astro-ph/0003370

Baym G, Bödeker D, McLerran L. 1996. *Phys. Rev. D* 53:662–67

Beck R, Brandenburg A, Moss D, Shukurov A, Sokoloff D. 1996. *Annu. Rev. Astron. Astrophys.* 34:155–206

Begelman MC, Blandford RD, Rees MJ. 1984. *Rev. Mod. Phys.* 56:255–351

Berghöfer TW, Bowyer S, Korpela E. 2000. *Ap. J.* 545:695–700

Bicknell GV, Cameron RA, Gingold RA. 1990. *Ap. J.* 357:373–87

Biermann L. 1950. *Z. Naturforsch. Tiel A* 5: 65

Biermann PL. 1999. *Astrophys. Space Sci.* 264:423–35

Binney J. 2002. In *Particles and Fields in Radio Galaxies*, ed. RA Laing, KM Blundell. San Francisco: Astron. Soc. Pac. In press

Binney J, Cowie LL. 1981. *Ap. J.* 247:464–72

Blasi P. 2000. *Ap. J. Lett.* 532:L9–12

Blasi P, Burles S, Olinto AV. 1999. *Ap. J. Lett.* 514:L79–82

Blasi P, Colafrancesco S. 1999. *Astropart. Phys.* 12:169–83

Blumenthal GR, Gould RJ. 1970. *Rev. Mod. Phys.* 42:237–70

Böhringer H, Feretti L, Schuecker P, eds. 1999. *Diffuse Thermal and Relativistic Plasma in Galaxy Clusters.* MPE Rep. 271

Bonamente M, Lieu R, Nevalainen J, Kaastra JS. 2001. *Ap. J. Lett.* In press

Bowyer S, Korpela E, Berghöfer T. 2001. *Ap. J. Lett.* 548:L135–38

Bridle AH, Perley RA. 1984. *Annu. Rev. Astron. Astrophys.* 22:319–58

Brunetti G, Setti G, Feretti L, Giovannini G. 2001a. *MNRAS* 320:365–78

Brunetti G, Setti G, Feretti L, Giovannini G. 2001b. *New Astron.* 6:1–15

Buote D. 2001. *Ap. J.* 553:L15–18

Burbidge GR. 1959. *Ap. J.* 129:849–51

Burn BJ. 1966. *MNRAS* 133:67–83

Carilli CL, et al. 1997. In *Extragalactic Radio Sources*, ed. C Fanti, R Ekers, p. 159. Dordrecht: Kluwer

Carilli CL, et al. 2001. In *Starburst Galaxies Near and Far*, ed. L Tacconi, D Lutz, pp. 309–17. Berlin: Springer-Verlag. In press

Carilli CL, Owen FN, Harris DE. 1994. *Astron. J.* 107:480–93

Carilli CL, Perley RA, Dreher JW. 1988. *Ap. J.* 334:L73–76

Carilli CL, Roettgering HJA, van Ojik R, Miley GK, van Breugel WJM. 1997. *Astrophys. J. Suppl. Ser.* 109:1

Casse F, Lemoine M, Pelletier G. 2001. *Phys. Rev. D.* 65:3002–17

Cavaliere A, Fusco-Femiano R. 1976. *Astron. Astrophys.* 49:137–44

Chambers KC, Miley GK, van Breugel WJM. 1990. *Ap. J.* 363:21–39

Chandran BDG, Cowley SC, Albright B. 1999. See Böhringer et al. 1999, pp. 242–46

Clarke TE, Kronberg PP, Böhringer H. 2001. *Ap. J.* 547:L111–14

Clarkson CA, Coley AA. 2001. *Class. Quantum Gravity* 18:1305–10

Colafrancesco S, Blasi P. 1998. *Astropart. Phys.* 9:227–46

Colgate SA, Li H. 2000. In *Highly Energetic Physical Processes and Mechanisms for Emission from Astrophysical Plasmas. Proc. IAU Symp. 195, San Francisco*, pp. 255–65. San Francisco: Astron. Soc. Pac.

Condon JJ, Cotton WD, Greisen EW, Yin QF, Perley RA, et al. 1998. *Astron. J.* 115:1693–716

Cowie LL, McKee CF. 1977. *Ap. J.* 211:135–46

Daly RA, Loeb A. 1990. *Ap. J.* 364:451–55

Dennison B. 1980. *Ap. J.* 239:L93–96

De Young DS. 1992. *Ap. J.* 386:464–72

Dogiel VA. 2000. *Astron. Astrophys.* 357:66–74

Dolag K, Ensslin TA. 2000. *Astron. Astrophys.* 362:151–57

Dolag K, Schindler S. 2000. *Astron. Astrophys.* 364:491–96

Dolag K, Schindler S, Govoni F, Feretti L. 2001. *Astron. Astrophys.* 378:777–86

Dreher JW, Carilli CL, Perley RA. 1987. *Ap. J.* 315:611–25

Dupke RA, Bregman JN. 2001. *Ap. J.* 547:705–13

Eilek JA. 1999. See Böhringer et al. 1999, pp. 71–76

Eilek JA, Owen FN. 2002. *Ap. J.* In press

Elbert JW, Sommers P. 1995. *Ap. J.* 441:151–61

Ensslin TA, Biermann PL, Klein U, Kohle S. 1998. *Astron. Astrophys.* 332:395–409

Ensslin TA, Gopal-Krishna XX. 2001. *Astron. Astrophys.* 366:26–34

Ensslin TA, Lieu R, Biermann PL. 1999. *Astron. Astrophys.* 344:409–20

Ettori S, Fabian AC. 2000. *MNRAS* 317:L57–59

Evrard AE, Gioia IM. 2002. In *Merging Processes in Clusters of Galaxies*, ed. L Feretti, IM Gioia, G Giovannini. Dordrecht: Kluwer. In press

Fabbiano G, Schwartz DA, Schwarz J, Doxsey RE, Johnston M. 1979. *Ap. J.* 230:L67–71

Fabian AC. 1994. *Annu. Rev. Astron. Astrophys.* 32:277–318

Fabian AC, Nulsen PEJ, Canizares CR. 1991. *Astron. Astrophys. Rev.* 2:191–226

Fanaroff BL, Riley JM. 1974. *MNRAS* 167:L31–35

Farrar GR, Piran T. 2000. *Phys. Rev. Lett.* 84:3527–30

Feigelson ED, Laurent-Muehleisen SA, Kollgaard RI, Fomalont EB. 1995. *Ap. J. Lett.* 449:L149–52

Felten JB. 1996. In *Clusters, Lensing and the Future of the Universe*, ed. V Trimble, A Reisenegger, 88:271–73. San Francisco: Astron. Soc. Pac. Conf. Ser.

Feretti L. 1999. See Böhringer et al. 1999, pp. 3–8

Feretti L, Dallacasa D, Giovannini G, Tagliani A. 1995. *Astron. Astrophys.* 302:680–90

Feretti L, Dallacasa D, Govoni F, Giovannini G, Taylor GB, Klein U. 1999. *Astron. Astrophys.* 344:472–82

Feretti L, Fusco-Femiano R, Giovannini G, Govoni F. 2001. *Astron. Astrophys.* 373:106–12

Feretti L, Giovannini G. 1998. In *A New View of An Old Cluster: Untangling Coma Berenices*, ed. A Mazure, F Casoli, F Durret, D Gerbal, pp. 123. Singapore: World Sci.

Field GB. 1965. *Ap. J.* 142:531–67

Forman W, Kellogg E, Gursky H, Tananbaum H, Giacconi R. 1972. *Ap. J.* 178:309

Furlanetto S, Loeb A. 2001. *Ap. J.* 556:619–34

Fusco-Femiano R, Dal Fiume D, De Grandi S, Feretti L, Giovannini G, et al. 2000. *Ap. J. Lett.* 534:L7–10

Fusco-Femiano R, Dal Fiume D, Orlandini M, Brunetti G, Feretti L, Giovannini G. 2001. *Ap. J. Lett.* 552:L97–100

Garrington ST, Leahy JP, Conway RG, Laing RA. 1988. *Nature* 331:147–49

Ge JP, Owen FN. 1993. *Astron. J.* 105:778–87

Giovannini G, Feretti L. 2000. *New Astron.* 5:335–47

Giovannini G, Feretti L, Venturi T, Kim K-T, Kronberg PP. 1993. *Ap. J.* 406:399–406

Giovannini G, Tordi M, Feretti L. 1999. *New Astron.* 4:141–55

Gnedin NY, Ferrara A, Zweibel EG. 2000. *Ap. J.* 539:505–16

Goldshmidt O, Rephaeli Y. 1993. *Ap. J.* 411:518–28

Govoni F, Feretti L, Giovannini G, Böhringer H, Reiprich TH, Murgia M. 2001a. *Astron. Astrophys.* 376:803–69

Govoni F, Taylor GB, Dallacasa D, Feretti L, Giovannini G. 2001b. *Astron. Astrophys.* 379:807–22

Grasso D, Rubinstein HR. 1995. *Nucl. Phys. B* 43:303–7

Grasso D, Rubinstein HR. 2001. *Phys. Rep.* 348:163–266

Greisen K. 1966. *Phys. Rev. Lett.* 16:748–58

Hanisch RJ. 1982. *Astron. Astrophys.* 116:137–46

Harris DE, Grindlay JE. 1979. *MNRAS* 188:25–37

Harris DE, Stern CP, Willis AG, Dewdney PE. 1993. *Astron. J.* 105:769–77

Harrison ER. 1970. *MNRAS* 147:279

Heckman TM. 2001. In *Extragalactic Gas at Low Redshift*, ed. J Mulchaey, J Stocke, San Francisco: Astron. Soc. Pac.

Hennessy GS, Owen FN, Eilek JA. 1989. *Ap. J.* 347:144–51

Isola C, Lemoine M, Sigl G. 2002. *Phys. Rev. D* 65:023004 astro-ph/0104289

Jaffe WJ. 1977. *Radio Astronomy and Cosmology. IAU Symp. 74*, pp. 305

Jaffe WJ. 1980. *Ap. J.* 241:925–27

Jones C, Mandel E, Schwarz J, Forman W, Murray SS, Harnden FR. 1979. *Ap. J. Lett.* 234:L21–24

Kempner JC, Sarazin CL. 2001. *Ap. J.* 548:639–51

Kim KT, Kronberg PP, Dewdney PE, Landecker TL. 1990. *Ap. J.* 355:29–37

Kim KT, Kronberg PP, Tribble PC. 1991. *Ap. J.* 379:80–88

Kosowsky A, Loeb A. 1996. *Ap. J.* 469:1–6

Kronberg PP. 1996. *Space Sci. Rev.* 75:387–99

Kronberg PP, Dufton QW, Li H, Colgate SA. 2001. *Ap. J.* 560:178–86

Kronberg PP, Lesch H, Hopp U. 1999. *Ap. J.* 511:56–64

Kulsrud RM. 1999. *Annu. Rev. Astron. Astrophys.* 37:37–64

Kulsrud RM, Cen R, Ostriker JP, Ryu D. 1997. *Ap. J.* 480:481–91

Large MI. 1959. *Nature* 183:1663–64

Lesch H, Birk GT. 1998. *Phys. Plasmas* 5: 2773–76

Liang H, Hunstead RW, Birkinshaw M, Andreani P. 2000. *Ap. J.* 544:686–701

Loeb A, Mao S. 1994. *Ap. J. Lett.* 435:L109–12

Lonsdale CJ, Barthel PD, Miley GK. 1993. *Astrophys. J. Suppl. Ser.* 87:63

Markevitch M, Ponman TJ, Nulsen PEJ, Bautz MW, Burke DJ, et al. 2000. *Ap. J.* 541:542–49

Markevitch M, Vikhlinin A. 2001. *Ap. J.* 563: 95–102

Markevitch M, Vikhlinin A, Forman WR, Sarazin CL. 1999. *Ap. J.* 527:545–53

Mazzotta P, Markevitch M, Vikhlinin A, Forman WR, David LP, VanSpeybroeck L. 2001. *Ap. J.* 555:205–14

McKee CF, Begelman MC. 1990. *Ap. J.* 358: 392–98

Meyer P. 1969. *Annu. Rev. Astron. Astrophys.* 7:1–38

Miley G. 1980. *Annu. Rev. Astron. Astrophys.* 18:165–218

Miralda-Escude J, Babul A. 1995. *Ap. J.* 449: 18–27

Mitton S. 1971. *MNRAS* 153:133–43

Owen F, Morrison G, Vogues W. 1999. Radio halos in luminous RASS clusters. See Böhringer, et al. 1999, pp. 9–11

Pacholczyk AG. 1970. *Radio Astrophysics.* San Francisco: Freeman

Parker EN. 1979. *Cosmical Magnetic Fields, Their Origin and Their Activity.* Oxford: Clarendon

Pentericci L, Van Reeven W, Carilli CL, Röttgering HJA, Miley GK. 2000. *Astron. Astrophys. Suppl. Ser.* 145:121–59

Perley RA, Taylor GB. 1991. *Astron. J.* 101:1623–31

Petrosian V. 2001. *Ap. J.* 557:560–72

Quashnock JM, Loeb A, Spergel DN. 1989. *Ap. J. Lett.* 344:L49–51

Rees MJ. 1987. *Q. J. R. Astron. Soc.* 28:197–206

Rees MJ, Setti G. 1968. *Nature* 219:127–31

Rephaeli Y. 1988. *Comm. Mod. Phys.* 12:265–79

Rephaeli Y. 1995. *Annu. Rev. Astron. Astrophys.* 33:541–80

Rephaeli Y. 2001. In *High Energy Gamma-Ray Astronomy.* In press astro-ph/0101363

Rephaeli Y, Gruber DE, Blanco P. 1999 *Ap. J. Lett.* 511:L21–24

Rephaeli Y, Gruber DE, Rothschild RE. 1987. *Ap. J.* 320:139–44

Roettiger K, Stone JM, Burns JO. 1999. *Ap. J.* 518:594–602

Rosner R, Tucker WH. 1989. *Ap. J.* 338:761–69

Rottgering HJ, Snellen I, Miley G, De Jong JP, Hanish R, Perley R. 1994. *Ap. J.* 435:654–88

Rudnick L. 2000. In *Cluster Mergers and Their Connection to Radio Sources, 24th Meet. IAU,* JD10:E22

Ruzmaikin A, Shukurov A, Sokolov D. 1987. *Magnetic Fields of Galaxies*

Ruzmaikin A, Sokolov D, Shukurov A. 1989. *MNRAS* 241:1–14

Sarazin CL. 1988. *X-Ray Emission from Clusters of Galaxies.* Cambridge, UK: Cambridge Univ. Press

Sarazin CL. 2002. In *Galaxy Clusters and the High Redshift Universe Observed in X-Rays,* ed. D Neumann, F Durret, J Tran Thanh Van. In press astro-ph/0105458

Sarazin CL. 2001b. In *Merging Processes in Clusters of Galaxies,* ed. L Feretti, IM Gioia, G Giovannini. Dordrecht: Kluwer

Schlikeiser R, Sievers A, Thiemanns H. 1987. *Astron. Astrophys.* 182:21–35

Simard-Normandin M, Kronberg PP, Button S. 1981. *Astrophys. J. Suppl. Ser.* 45:97–111

Smith E, Jones DE, Coleman PJ, Colburn DS, Dyal P, et al. 1974. *J. Geophys. Res.* 79: 3501–11

Soker N, Sarazin CL. 1990. *Ap. J.* 348:73–84

Soward AM. 1983. *Stellar and Planetary Magnetism.* New York: Gordon & Breach

Spitzer L. 1962. *The Physics of Fully Ionized Gases.* New York: Intersci.

Spitzer L. 1978. *Physical Processes in the Interstellar Medium.* New York: Wiley

Taylor GB, Allen SW, Fabian AC. 2002. *MNRAS* Submitted astro-ph/0109337

Taylor GB, Barton EJ, Ge JP. 1994. *Astron. J.* 107:1942–52

Taylor GB, Govoni F, Allen SW, Fabian AC. 2001a. *MNRAS.* 326:2–10

Taylor GB, Inoue M, Tabara H. 1992. *Astron. Astrophys.* 264:421–27

Taylor GB, Perley RA. 1993. *Ap. J.* 416:554–62

Tribble PC. 1993. *MNRAS* 263:31–36

Turner MS, Widrow LM. 1988. *Phys. Rev. D* 37:2743–54

Vachaspati T, Vilenkin A. 1991. *Phys. Rev. D* 43:3846–55

Vallee JP, MacLeod JM, Broten NW. 1986. *Astron. Astrophys.* 156:386–90

Vallee JP, MacLeod JM, Broten NW. 1987. *Ap. J. Lett.* 25:L181–86

van Breugel WJ. 2000. *Proc. Soc. Photo-Opt. Instrum. Eng.* 4005:83–94

Vikhlinin A, Markevitch M, Forman W, Jones C. 2001a. *Ap. J. Lett.* 555:L87–90

Vikhlinin A, Markevitch M, Murray SS. 2001b. *Ap. J. Lett.* 549:L47–50

Warwick JA. 1963. *Ap. J.* 137:41–60

Waxman E, Loeb A. 2000. *Ap. J. Lett.* 545:11–14

Wentzel DG. 1974. *Annu. Rev. Astron. Astrophys.* 12:71–96

White SDM, Navarro JF, Evrard AE, Frenk CS. 1993. *Nature* 366:429–31

Willson MAG. 1970. *MNRAS* 151:1–44

Wu X-P. 2000. *MNRAS* 316:299–306

Zatsepin GT, Kuzmin VA. 1966. *Phys. JETP Lett.* 4:78–83

Zweibel EG. 1988. *Ap. J. Lett.* 329:L1–4

Annu. Rev. Astron. Astrophys. 2002. 40:349–85
doi: 10.1146/annurev.astro.40.060401.093810
Copyright © 2002 by Annual Reviews. All rights reserved

THE ORIGIN OF BINARY STARS

Joel E. Tohline
*Department of Physics and Astronomy, Louisiana State University, Baton Rouge,
Louisiana 70803; email: tohline@rouge.phys.lsu.edu*

Key Words star formation, binaries

■ **Abstract** Although we have a general understanding of the manner in which individual stars form, our understanding of how binary stars form is far from complete. This is in large part due to the fact that the star formation process happens very quickly and in regions of the Galaxy that are difficult to study observationally. We review the theoretical models that have been developed in an effort to explain how binaries form. Several proposed mechanisms appear to be quite promising, but none is completely satisfactory.

INTRODUCTION

This review of the ideas that have been put forward to explain the origin of binary stars builds upon a foundation that has been laid by two earlier articles in this Annual Review series: The first, published 15 years ago by Shu, Adams & Lizano (1987), focused on a discussion of the origin of single stars; the second, published 8 years ago by Mathieu (1994), reviewed the observational evidence of binary systems in the pre–main-sequence (PMS) stellar population. Because the ideas presented in these two earlier reviews are essential ingredients to any discussion of the origin of binary stars, I spend a bit of time recapitulating them, but lack of space requires me to refer the reader to these articles for details. The task of reviewing this topic has been considerably simplified because of the conference that was held in April, 2000, on the topic of "The Formation of Binary Stars" (Zinnecker & Mathieu 2001).

Background

Shu et al. (1987) have provided the following outline of the various stages of the birth of single stars. The process begins with a molecular cloud, after its basic constituent material has been assembled somehow and somewhere in a galaxy. The cloud is gravitationally bound, but as a whole it is supported against gravitational collapse by the presence of a magnetic field. Stage I: Via ambipolar diffusion, the magnetic field slowly leaks out of relatively over-dense regions of the cloud, allowing these regions to become more and more dense, relative to their surroundings.

0066-4146/02/0922-0349$14.00

In this way, observationally identifiable "molecular cloud cores" are formed. Stage II: The star formation process begins in earnest when a condensing cloud core "passes the brink of instability" (Shu et al. 1987) and collapses dynamically toward stellar densities. This leads to the formation of a central protostar embedded within an infalling envelope of dust and gas. A disk almost always surrounds the embedded protostar, reflecting the fact that molecular cloud cores are almost always rotating. The protostar accretes matter, largely from the infalling cloud in which it is embedded, but to some degree also from its surrounding disk. As it contracts toward the main sequence, the protostar develops a stellar wind, which initially is unobservable because "the ram pressure from material falling directly onto the [proto]stellar surface suppresses breakout" (Shu et al. 1987). Stage III: Gradually, direct infall onto the protostar's surface weakens as incoming material with relatively high specific angular momentum falls preferentially onto the disk. The stellar wind is then able to break out in the direction of the system's rotational poles, creating an observable bipolar flow. Stage IV: The amount of material added to the protostar via direct infall continues to decrease, and the opening angle of the wind steadily widens until the young central star (along with its surrounding nebular disk) is revealed as a bonafide PMS star—for example, a T Tauri star. Stage V: Over time, the nebular disk finally disappears.

As Mathieu (1994) has reviewed, it is clear from an observational perspective that the PMS stellar population is rich in multiple systems. The overall frequency of occurrence of binary stars among the PMS population is at least as large as has been documented for main-sequence stars (Duquennoy & Mayor 1991), that is, certainly greater than 50%. Young, stellar-mass binary systems have been found with semimajor axes ranging from 0.02 to 10^3 AU (orbital periods ranging from a couple of days to 10^4 years), with a binary frequency distribution as a function of semimajor axis that is qualitatively consistent with the log-normal–like distribution found for main-sequence stars. With this evidence in hand, Mathieu (1994) concluded that "binary formation is the *primary* branch of the star-formation process." Also, it seems clear that the process (or processes) responsible for creating binary stars generally exerts its influence before stage IV. These realizations have led to an increased effort over the past decade to understand from a theoretical perspective how binary star systems form.

In the eight years since Mathieu compiled his review, significant advances have been made in the techniques and instrumentation that are available to identify and study binary stars that have a wide range of orbital periods, mass ratios, and ages. These include HST (WFPC2 & NICMOS) imaging (Padgett et al. 1997, 1999; Reid et al. 2001), submillimeter imaging (Smith et al. 2000), optical and infrared long-baseline interferometry (Quirrenback 2001a,b), millimeter and submillimeter interferometry (Launhardt et al. 2000, Launhardt 2001, Guilloteau 2001), adaptive optics (Simon et al. 1999, Close 2001), Hipparcos astrometric observations (Söderhjelm 1999; Quist & Lindegren 2000, 2001), and microlensing (Alcock et al. 2001, Di Stefano 2001). A fair fraction of the increased observational activity in this arena certainly has been stimulated by the discovery of extrasolar planetary systems (Mayor & Queloz 1995; Marcy & Butler 1998, 2000). Although all of this work is important in establishing a database that can, in the long run, be used to critically evaluate theoretical models proposed to explain how binary (and other multiple) star systems form, it is fair to say that, at present, our understanding of process(es) by which binary systems form is far too crude to take full advantage of the detailed information that is housed in such a database. For the purpose of this review, it is sufficient to build our discussion on the two broad conclusions drawn by Mathieu (1994): Stars preferentially form in pairs, and binary formation occurs prior to the PMS phase of a star's evolution.

Basic Physical Principles

Throughout this review, there is very little discussion of the role that magnetic fields play in the star formation process. This stands in stark contrast to the earlier review by Shu et al. (1987). The reasons for downplaying the role of magnetic fields here are twofold. First, although there is a considerable body of evidence [see the discussion by Shu et al. (1987)] supporting the conclusion that the interstellar magnetic field significantly influences the onset of gravitational collapse in molecular clouds, the general consensus is that the field will largely decouple from a contracting cloud at number densities $\gtrsim 10^{10}$ cm^{-3} because, at such high densities, the fractional ionization of the gas becomes extraordinarily small. Because, as is emphasized below, the processes likely to be responsible for transforming single gas clouds into binary protostellar systems largely operate at densities higher than this limit, neglect of the magnetic field is justified. Second—and largely justified by the first—researchers who have focused their modeling efforts on the binary star formation problem generally have ignored the effects of magnetic fields, so a review of this body of work must naturally downplay the influence of magnetic fields as well. We return to the issue of the influence that magnetic fields have on the onset of collapse in "Summary and Conclusions," below.

By ignoring the effects of magnetic fields, we in practice (although not in spirit) depart somewhat from the storyline presented by Shu et al. (1987) and summarized above. Specifically, when the text of the earlier review refers to a cloud core that has "passed the brink of instability" and thereby entered stage II of the star formation process, it is referring to a critical physical condition in the cloud core that is established, in part, by the strength of the cloud's magnetic field. We instead use this phrase to refer to the classic thermal Jeans instability (see the relevant definition in "The Jeans Instability," below). Even in the case in which the effects of magnetic fields are considered, of course, the Jeans instability is relevant, but it may establish only a necessary rather than sufficient condition for collapse.

PHYSICAL PARAMETERS   We characterize a protostellar gas cloud or protostar by its total mass M and radius R, its mean temperature T and mean molecular weight

μ, and its mean angular velocity ω. From M, R, and μ we can determine the cloud's mean mass density,

$$\bar{\rho} \equiv \frac{3M}{4\pi R^3},$$ (1)

and its corresponding number density,

$$n = \frac{\bar{\rho}}{\mu m_{\mathrm{p}}},$$ (2)

where m_{p} is the mass of a proton. We also specify that, upon compression, the cloud's temperature varies with its density as

$$T \propto \bar{\rho}^{\gamma-1},$$ (3)

where γ is the effective adiabatic exponent of the gas. It is understood that γ itself is likely to be a function of the cloud's density (see the discussion below in association with Figure 2). From T/μ and γ, the mean sound speed of the gas can be determined via the expression,

$$c_{\mathrm{s}} = \left[\gamma \frac{\Re T}{\mu}\right]^{1/2},$$ (4)

where $\Re$ is the gas constant.

As Shu et al. (1987) have reminded us, star formation is a complex process that spans many orders of magnitude in mass and linear scale. If, following Shu et al., we concentrate on aspects of the problem that occur on scales ranging from that of a giant molecular cloud to the shortest period PMS binaries (e.g., Table A2 of Mathieu 1994), then we have to deal with systems having masses in the range of $10^6 \gtrsim M/M_\odot \gtrsim 1$ and linear scales in the range of 10^{20} cm $\gtrsim l \gtrsim 10^{12}$ cm (that is, 30 pc $\gtrsim l \gtrsim 10$ R$_\odot$). The bottom diagram in Figure 1 illustrates this range of length scales, with the letters GMC drawn on the left to mark the size of a giant molecular cloud and the letter B drawn near the right to mark the scale (orbital separation) of a PMS binary star with an orbital period of only a couple of days. The letter P in this diagram marks the linear scale of our own planetary system (40 AU, as set by the orbit of Pluto), which lies between the two extremes. Rotationally flattened disks of this size or larger are now almost always found in association with the youngest binary or single PMS stars. As a point of reference, the top diagram in Figure 1 illustrates that a comparable range of length scales takes us from the present scale of the universe (marked by U on the diagram), through the size of a typical galaxy (marked by G), to the size of a typical globular cluster (marked by GC). This comparison of scales serves to emphasize the complexity of the problem at hand. That is, we should not be surprised to find that the process by which binary stars form from molecular clouds in our Galaxy is at least as difficult to understand and to model uniquely as the process by which globular clusters form from an homogeneous and isotropic universe.

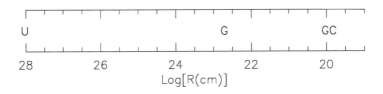

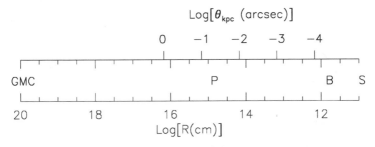

Figure 1   Two scales that, together, cover over 40 orders of magnitude in length. They stretch from the scale of the universe (U at the left-hand edge of top scale), through our Galaxy (G) to the scale of an individual globular cluster (GC); and from the scale of a giant molecular cloud (GMC), through our planetary system (P) down to the scale of the shortest period pre–main-sequence binaries (B) and the radius of the Sun (S). Along the bottom scale we have also indicated the angular resolution θ_{kpc} (in arcsec) that is required to resolve a system of radius R at a distance of 1 kpc. Notice that there is roughly the same separation between the scale of the universe and the size of individual GCs as there is between the scale of a GMC and the shortest period binary systems.

At a distance d_{kpc} (measured in kiloparsecs) from the solar system, an object of linear scale R_{AU} (measured in astronomical units) will subtend an angle θ_{kpc} (measured in seconds of arc) given by the expression,

$$\theta_{kpc} = 10^{-3} \frac{R_{AU}}{d_{kpc}}. \tag{5}$$

Hence, as the top scale in the bottom diagram of Figure 1 indicates, at a distance of 1 kpc a 40-AU disk will subtend an angle θ_{kpc} of 40 milli-arcseconds on the sky; at this same distance a resolution of a tenth of a milli-arcsecond or better would be required to spatially resolve the shortest period PMS binaries.

Figure 2 illustrates the range of densities and temperatures that a stellar-mass gas cloud must traverse as it contracts from a molecular cloud state toward a structure (on the zero-age main sequence) that is hot enough and dense enough to fuse hydrogen. For purposes of illustration, the solid curve identifies the approximate $\rho - T$ path that is expected to be followed by the central-most region of a nonrotating (spherically symmetric) protostellar cloud containing a total mass $M \sim 1\text{--}3\,M_\odot$ (drawn from Figure 3 of Tohline 1982 and Figure 27.3 of Kippenhahn & Weigert

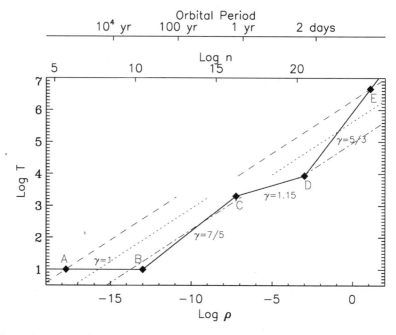

Figure 2 The evolutionary trajectory (*solid curve*) of the central region of a proto-stellar gas cloud is shown in the temperature-density plane. (Patterned after Figure 3 of Tohline 1982.) The slope of each segment of the curve is indicated by the value of the effective adiabatic exponent γ, as defined in Equation 3. The density is shown both in g cm^{-3} (*bottom horizontal axis*) and in cm^{-3} (*top horizontal axis*); the temperature is given in degrees Kelvin. Also shown along the top of the plot is the orbital period of a binary system that has the equivalent mean density, as determined through Equations 10 and 6. For reference, lines of constant M_{equil}, as defined by Equation 22, have been drawn at values of 1 $M_\odot$ (*dashed*), 10^{-1} $M_\odot$ (*dotted*), and 10^{-2} $M_\odot$ (*dash-dot*); the vertical gap between temperatures of 2,000 K and 10,000 K signifies that the molecular gas is being dissociated and ionized; hence, μ changes from 2 (*lower section of each line*) to 1/2 (*upper section of each line*).

1990; see also Figure 2 of Winkler & Newman 1980, Figure 3 of Boss 1984, and Figure 1 of Bate 1998). Various segments of this curve have been labeled with the value of the effective adiabatic exponent γ that governs the illustrated $\rho - T$ relationship, as defined above by Equation 3.

KEY TIMESCALES From $\bar{\rho}$, c_s, ω, and R, we can derive three key timescales: The free-fall time,

$$\tau_{\text{ff}} = \left[\frac{3\pi}{32G\bar{\rho}} \right]^{1/2},$$

(6)

where, G is the gravitational constant; the sound-crossing time,

$$\tau_s = \frac{R}{c_s};$$ (7)

and the rotation period of the cloud,

$$\tau_{rot} = \frac{2\pi}{\omega}.$$ (8)

When discussing binary star (or binary protostellar) systems, a fourth relevant timescale is the binary's orbital period,

$$P = \left[\frac{4\pi^2 a^3}{GM_{tot}}\right]^{1/2},$$ (9)

where M_{tot} is the total mass of the system, and a is the system's semimajor axis [for circular orbits, a is the distance between the two stars (or protostars)]. Notice, however, that the orbital period is not much different from the free-fall time, in the following sense. If a gas cloud of radius R and mass M is transformed (by some, as yet, unspecified mechanism) into a binary star system of mass $M_{tot} = M$ and separation $a = R$, then by Equations 9 and 6 the binary will have an orbital period,

$$P = 32^{1/2}\tau_{ff} \approx 5.7\tau_{ff}.$$ (10)

Because, as shown by Equation 6, the free-fall time of a gas cloud (or protostar) depends only on the cloud's mean mass density, Equation 10 suggests that the measured orbital period of a binary system tells us something directly about the density of the gas cloud from which the binary system formed. With this in mind, the top horizontal axis in Figure 2 has been labeled with the binary orbital period that corresponds to the mean mass density that is given along the bottom horizontal axis of the figure.

Also by way of illustration, Table 1 lists the values of some relevant physical parameters at a variety of different scales that should be of interest to researchers

TABLE 1   Scales of interest

P_{orbit} [years]	n_{H_2} [cm^{-3}]	$\Delta v/\Delta\theta_{kpc}$ [km s^{-1}/arcsec]	a [AU] for 1 M$_\odot$
380000	3×10^5	0.04	5200
20000	1×10^8	0.75	740
3800	3×10^9	3.7	250
1000	4×10^{10}	15.	100
300	5×10^{11}	50.	45
100	4×10^{12}	150.	20

searching for evidence of protostellar binary systems in molecular clouds. For example, a binary system that forms from an environment in which the mean cloud density $n_{H_2} \sim 3 \times 10^5$ cm^{-3} should have an orbital period $P \sim 4 \times 10^5$ years and should exhibit a velocity gradient on the sky $\Delta v / \Delta \theta_{kpc} \sim 0.04$ km s^{-1}/arcsec [e.g., the properties of the Bok Globule CB230 described by Launhardt et al. (2000)]. If this system contains 1 M$_\odot$ of material, the separation of its binary components will be ~ 5200 AU, which at a distance of 1 kpc will subtend an angle of $\theta_{kpc} \sim 5.2$ arcsec. A bound system with an orbital period of ~ 1000 years, however, must form in an environment that has a much higher mean density, $n_{H_2} \sim 4 \times 10^{10}$ cm^{-3}. It should exhibit a significantly higher velocity gradient on the sky, $\Delta v / \Delta \theta_{kpc} \sim 15$ km s^{-1}/arcsec, but will subtend a much smaller angle on the sky: At 1 kpc, a 1 M$_\odot$ system should have $\theta_{kpc} \sim 0.1$ arcsec.

Finally, we should mention the accretion timescale,

$$\tau_{accrete} = \frac{m_0}{\dot{M}}, \tag{11}$$

which gives the time that it takes an equilibrium structure (e.g., the central core of a collapsing cloud) of mass m_0 to double in mass as it accretes material from a surrounding cloud or disk at a rate $\dot{M}$. Over intervals of time that are short compared to $\tau_{accrete}$, the mass of the central structure remains relatively unchanged, so the central structure can be considered dynamically isolated from its surroundings. The accretion rate $\dot{M}$ varies from situation to situation, but in discussions of the free-fall collapse of protostellar gas clouds, two rates are of particular interest. The first comes simply from the ratio of a cloud's total mass to its initial free-fall time,

$$\dot{M}_{ff} = \frac{M}{\tau_{ff}}. \tag{12}$$

The second is the accretion rate $\dot{M}_{sis}$ highlighted by Shu (1977) that arises from the collapse of the so-called singular isothermal sphere,

$$\dot{M}_{sis} = \frac{c_s^3}{G}. \tag{13}$$

IMPLICATIONS OF THE VIRIAL THEOREM   From M and R we obtain the following estimate of the configuration's total gravitational potential energy,

$$E_{grav} \sim -\frac{3}{5} \frac{GM^2}{R}. \tag{14}$$

From M and T/μ we obtain an estimate of its total thermal energy via the expression,

$$E_{therm} \sim \frac{3}{2} \frac{\Re}{\mu} MT. \tag{15}$$

The configuration's rotational kinetic energy is

$$E_{\text{rot}} = \frac{1}{2}I\omega^2 = \frac{1}{2}\frac{J^2}{I} \sim \frac{1}{5}MR^2\omega^2, \tag{16}$$

where $I \approx \frac{2}{5}MR^2$ is the configuration's principal moment of inertia and $J = I\omega$ is its total angular momentum. From the virial theorem (e.g., Hansen & Kawaler 1994), we know that the following relationship between these three global energy reservoirs must hold if a protostellar gas cloud (or protostar) is in equilibrium:

$$2(E_{\text{therm}} + E_{\text{rot}}) + E_{\text{grav}} = 0 \tag{17}$$

or

$$\alpha + \beta = \frac{1}{2}, \tag{18}$$

where

$$\alpha \equiv \frac{E_{\text{therm}}}{|E_{\text{grav}}|} \sim \frac{5}{2}\frac{\Re}{\mu}T\frac{R}{GM}, \tag{19}$$

$$\beta \equiv \frac{E_{\text{rot}}}{|E_{\text{grav}}|} \sim \frac{1}{3}\frac{R^3\omega^2}{GM}. \tag{20}$$

If a cloud (or protostar) is in equilibrium but is not rapidly rotating ($\beta \ll 1/2$), according to Equations 18 and 19, $\alpha \approx 1/2$ and its mass must be related to its radius and mean temperature via the expression

$$M_{\text{equil}} \sim 5\frac{\Re}{\mu}T\frac{R}{G}. \tag{21}$$

Using Equation 1, we can alternatively express this equilibrium mass in terms of the cloud's mean density and temperature as follows:

$$M_{\text{equil}} \sim 5.5\left[\frac{\Re}{\mu}\frac{T}{G}\right]^{3/2}\bar{\rho}^{-1/2}. \tag{22}$$

For reference, lines of constant M_{equil} have been drawn in Figure 2 at values of 1 $M_\odot$ (*dashed*), 10^{-1} $M_\odot$ (*dotted*), and 10^{-2} $M_\odot$ (*dash-dot*); the vertical gap between temperatures of 2000°K and 10,000°K signifies that the molecular gas is being dissociated and ionized, hence μ changes from 2 (lower section of each line) to 1/2 (upper section of each line). (The lines have a slope of 1/3 in this log-log plot.) Combining Equation 21 with Equations 1, 4, 6, and 7, this statement of equilibrium also means that the sound-crossing time and the free-fall time are approximately equal to one another in the cloud (or protostar), that is,

$$\tau_s \approx \tau_{\text{ff}}. \tag{23}$$

On the other hand, if a cloud's equilibrium state is balanced against gravity largely by rotation ($\alpha \ll 1/2$ and $\beta \approx 1/2$), then the virial theorem states that the cloud's mean angular velocity will be near its maximum allowable value,

$$\omega \approx \omega_{max} \approx [2\pi G \bar{\rho}]^{1/2}; \qquad (24)$$

that is, the cloud's rotation period will be

$$\tau_{rot} = \frac{2\pi}{\omega} \approx 4.6\tau_{ff}. \qquad (25)$$

Not surprisingly, this is essentially equal to the orbital period of a binary system that has a separation $a = R$ and the same total mass, as discussed above in connection with Equation 10.

THE JEANS INSTABILITY   If, for a given $M, R, T/\mu$, and ω, one finds that $\alpha + \beta < 1/2$ in a protostellar gas cloud, then the cloud will collapse on a free-fall timescale. In the absence of rotation, this condition ($\alpha < 1/2$) is simply a statement that

$$M > M_{equil} \qquad (26)$$

and, hence, that the cloud has encountered the classic Jeans instability (Jeans 1919). In this context, Equations 21 and 22 given above for M_{equil} also serve to define the familiar Jeans mass, M_J; that is to say,

$$M_J \sim 5\frac{\Re}{\mu}T\frac{R}{G} \sim 5.5\left[\frac{\Re}{\mu}\frac{T}{G}\right]^{3/2}\bar{\rho}^{-1/2}. \qquad (27)$$

When the mass of a molecular cloud or cloud core exceeds this critical mass, it has "passed the brink of instability" and entered stage II of the star formation process (Shu et al. 1987. For example, at $10°K$ a uniform-density, 1 $M_\odot$ molecular gas cloud will encounter this Jeans instability at the point marked A in Figure 2 ($\bar{\rho} = 1.8 \times 10^{-18}$ g cm^{-3}) (see Table 2). As it collapses, it will evolve (to the right in Figure 2) to configurations of higher density and smaller radius.

TABLE 2   Typical conditions in collapsing cloud core (see Figure 2)

Case	ρ [g/cm^3]	T [°K]	γ	μ	$\frac{1}{\gamma}c_s^2$ [cm^2/s^2]	M_{equil} [M$_\odot$]	j_{max} [cm^2/s]
A	1.8×10^{-18}	10	1	2	4.2×10^8	1	3.6×10^{21}
B	1.0×10^{-13}	10	$\frac{7}{5}$	2	4.2×10^8	0.004	1.5×10^{19}
C	5.7×10^{-8}	2000	1.15	1	1.7×10^{11}	0.05	8.3×10^{18}
D	1.0×10^{-3}	8.7×10^3	$\frac{5}{3}$	$\frac{1}{2}$	1.4×10^{12}	0.008	5.1×10^{17}
E	1.2×10^1	4.6×10^6	$\frac{5}{3}$	$\frac{1}{2}$	7.7×10^{14}	1	2.6×10^{18}

Notice that the mass of a cloud can be expressed very simply in terms of the Jeans mass and the dimensionless energy parameter α as follows:

$$M = \frac{M_J}{2\alpha}. \tag{28}$$

Therefore, if a protostellar gas cloud collapses from a state in which α is only slightly less than $1/2$, that is equivalent to saying that the cloud's mass is only slightly greater than one Jeans mass initially. In this case, one can show (see, for example, the discussion associated with Equation 23 of Shu et al. 1987) that the relevant accretion timescale τ_{accrete} can be derived from the accretion rate defined by either Equation 12 or 13. That is to say, the model of a collapsing singular isothermal sphere is relevant (Larson 1969, Penston 1969, Shu 1977) and $\dot{M}_{\text{ff}} \approx \dot{M}_{\text{sis}}$. If, however, $\alpha \ll 1/2$ initially, then the cloud initially encloses many Jeans masses and is well past the brink of instability. (One might fairly ask how the cloud was brought to such a drastic state in the first place. See "Summary and Conclusions," below.) In this case, the singular isothermal sphere solution becomes irrelevant and the appropriate accretion timescale must be estimated from Equation 12.

THE IMPORTANCE OF THE EFFECTIVE ADIABATIC EXPONENT γ Once a cloud encounters the Jeans instability, it will evolve dynamically until it acquires a new configuration in which virial equilibrium is achieved, that is, until the sum $(\alpha + \beta)$ climbs up to the value $1/2$. We can therefore estimate how far a given cloud will collapse before it settles into an equilibrium state by examining how the two energy ratios α and β scale with the cloud's radius or mean density. Assuming that the cloud's mass M and angular momentum J are conserved during its dynamical collapse, Equations 1, 16, and 20 give

$$\beta \propto R^{-1} \propto \bar{\rho}^{1/3}, \tag{29}$$

and Equations 3 and 19 give,

$$\alpha \propto TR \propto \bar{\rho}^{\gamma - 4/3}. \tag{30}$$

Equation 30 identifies the critically important role that the effective adiabatic exponent plays in star formation. According to this expression, if $\gamma < 4/3$, the energy ratio α actually decreases during a collapse. Therefore it is impossible for thermal pressure alone to stop the cloud's free-fall collapse as long as the cloud evolves through a density-temperature regime where the effective adiabatic exponent $\gamma < 4/3$, such as the isothermal ($\gamma = 1$) regime illustrated in the left-hand portion of Figure 2. This is why in models of spherically symmetric collapse (e.g., Larson 1969, Winkler & Newman 1980) the cloud's free-fall is not slowed until its central-most region reaches a density $\sim 10^{-13}$ g cm^{-3} (the point marked B in Figure 2) and starts to become opaque to the cloud's primary cooling radiation. Quoting directly from Larson (1969), "...the heat generated by the collapse in this region is then no longer freely radiated away, and the compression becomes approximately adiabatic" with an effective adiabatic exponent ($\gamma \approx 7/5$) that is greater than $4/3$.

Thus, in a spherically symmetric cloud, a minimum of two conditions must be met before a stable equilibrium configuration can be achieved: $\alpha = 1/2$, that is, $M = M_{\text{equil}}$; and $\gamma > 4/3$. In Figure 2, the segments of the (*solid*) evolutionary track for which a stable equilibrium is possible are the segments for which $\gamma > 4/3$. On each of these segments the mass required to achieve an equilibrium ($\alpha = 1/2$) is given by the value of the (*dashed, dotted,* or *dash-dot*) M_{equil} line that intersects the evolutionary track. For example, at the point marked B in Figure 2, equilibrium will be achieved with a mass $\sim 4 \times 10^{-3}$ $M_{\odot}$. This is why the first core that forms in the collapse models of Larson (1969) and Winkler & Newman (1980) contains only a very small fraction of the cloud's total mass—only a few Jupiter masses!

Notice that a spherically symmetric cloud that contains an entire solar mass of material can achieve virial equilibrium in only two places along the evolutionary track shown in Figure 2: at the point marked A [at a very low density and temperature ($\bar{\rho} \sim 1.8 \times 10^{-18}$ g cm^{-3}, $T \sim 10$ K)] or at the point marked E [at a very high density and temperature ($\bar{\rho} \sim 12$ g cm^{-3}, $T \sim 5 \times 10^{6}$ K)]. The only stable configuration is the high-density one because it resides on a portion of the evolutionary track for which $\gamma > 4/3$. Thus, once the Jeans instability is encountered (at point A), the dynamical collapse must proceed on a free-fall timescale—through almost 19 orders of magnitude in density—to a star-like configuration. A central, low-mass core will form along the way (as in the models of Larson 1969 and Winkler & Newman 1980) only if the collapse proceeds in a nonhomologous fashion (see "Nonhomologous Collapses," below).

The free-fall collapse of a rotating cloud can be halted at much lower densities because, as shown by Equation 29, β always increases as the cloud contracts. Hence, even during the isothermal phase of a cloud's contraction, virial balance can be achieved when β grows to a value of $(1/2 - \alpha)$. Hence, there will be a tendency for β to climb up to a value $\sim 1/2$ during the phase of isothermal contraction because, for the cloud as a whole, α will drop to a very small value.

Once again, though, virial equilibrium alone does not guarantee a configuration that is stable against further collapse. As in the nonrotating case just described, an additional condition involving the effective adiabatic exponent must be met for stability. According to the detailed stability analysis of nearly spherical systems presented by Ledoux (1945; see also the discussion associated with Figure 3 in Tohline 1984), the condition for stability is,

$$\gamma > \gamma_{\text{crit}} \approx \frac{2}{3} \frac{(2 - 5\beta)}{(1 - 2\beta)}. \tag{31}$$

When $\beta \to 0$, this expression produces the value of $\gamma_{\text{crit}} = 4/3$ that is familiar for spherically symmetric, nonrotating gas clouds. When $\beta \neq 0$, this expression shows that γ_{crit} is somewhat less than $4/3$. However, according to Equation 31, a rotating, isothermal ($\gamma_{\text{crit}} = 1$) gas will not be stable against further radial collapse unless $\beta > 1/4$. Complementary analyses of equilibrium structures by Hayashi et al. (1982), Tohline (1984), and Hachisu & Eriguchi (1985) have confirmed that rotationally flattened, axisymmetric, isothermal gas clouds are stable against dynamical collapse (or expansion) only if $\beta \gtrsim 0.25$–0.3.

Possible Formation Mechanisms

The mechanisms proposed for forming binary stars can be divided into three broad categories. First, it is possible that Jeans-unstable gas clouds preferentially collapse to form single stars; then, after formation, the stars become bound together in pairs via a process usually referred to as "capture." Second, either during or immediately after its free-fall collapse, an individual rotating gas cloud may spontaneously break into two pieces that are in orbit about one another. In this process, which we refer to as "prompt fragmentation," the cloud's original spin angular momentum is converted fairly directly into orbital angular momentum of the binary system. Third, the central-most regions of a rotating gas cloud may collapse to form an equilibrium configuration that is initially stable against fragmentation. Then, as this relatively dense core contracts toward the main sequence while accreting relatively high specific angular momentum material from the outer regions of the cloud, the core (or its surrounding accretion disk) may encounter an instability that leads to the formation of a binary system. We refer to this process as "delayed breakup."

CAPTURE

It is possible that binary stars form by the relatively simple mechanism of capture. That is to say, it is possible that stars preferentially form as single objects along the lines described by Shu et al. (1987), then after formation become grouped together in bound pairs through dynamical encounters. As Clarke (1992) has reviewed, though, the formation of a binary from two initially unbound stars requires the dissipation of some fraction of the energy of their relative orbit. In favorable three-body encounters, the energy lost from the relative orbit of the two stars can be transferred as kinetic energy to a third star. Although rare, such encounters can significantly influence the evolution of the central-most regions of globular star clusters (e.g., Portegies et al. 1997). But this cannot be the mechanism responsible for the formation of most binary systems in large clusters or in the field because the frequency of such favorable encounters is extremely low. Also, in large virialized clusters, the typical velocity of approach of unpaired stars is unfavorably hyperbolic. Whereas this is less of a problem in smaller N-body systems, simulations show that purely gravitational encounters yield relatively few binaries per cluster, and these binaries tend to form from the most massive stellar components of a cluster (Clarke 1992, Valtonen & Mikkola 1991).

Two-body encounters can result in the formation of a binary system from an initially unbound pair if the interaction is not purely gravitational. For example, orbital energy can be converted into heat through the excitation of tides in one or both stars. However, significant tides are raised only during very close encounters—which require relatively special initial conditions—and in the absence of strong tides many encounters are required to dissipate a significant amount of energy. Hence, this traditional tidal capture mechanism is unlikely to explain how young, PMS binaries are formed. In protostellar environments, however, the material that resides in the extended disks around young stars (or protostars) can also be tidally disturbed during an encounter and thereby absorb a portion of the orbital energy. Larson (1990) estimated that a large fraction of all stars might be incorporated into binaries through this mechanism during the PMS phase. Subsequent detailed investigations have shown, however, that even when disks are included to enhance the cross-section for collisions, tidal capture still does not work effectively enough. Typically the velocities of encounter are sufficiently large that they disrupt the disk (Clarke & Pringle 1993). It appears, therefore, that in all but the smallest virialized clusters, star-disk capture cannot be responsible for the formation of most binaries.

PROMPT FRAGMENTATION

Prompt fragmentation is the binary formation process that has received by far the most attention over the past two decades. This is in large part due to the relative capabilities and limitations of the numerical tools that have been employed to study binary star formation. Binary fragmentation is, by definition, a nonlinear process that exhibits no simple geometric symmetries. To model such a process in the midst of a free-fall collapse therefore requires a fully three-dimensional, nonlinear hydrodynamical simulation with adequate spatial resolution. Smoothed-particle hydrodynamic (e.g., Lucy 1977, Benz 1991, Monaghan 1992) and finite-difference hydrodynamic (e.g., Boss & Mayhill 1992, Truelove et al. 1998) techniques have both been successfully employed to study various aspects of this problem. With either of these techniques the system is advanced forward in time via an explicit (rather than implicit) integration of the time-dependent equations that govern the evolution of self-gravitating fluids. Hence, both techniques are constrained by the Courant-Freidrichs-Lewey condition (Courant et al. 1928; see also p. 45 of Roache 1976) to take time steps that are very small compared with the physical system's sound-crossing time, free-fall time, and rotation period. This means many integration time steps are required to model fragmentation, even for systems that fragment promptly. Then when you consider how many grid cells (in finite-difference hydrodynamic techniques) or particles (in smoothed-particle hydrodynamic techniques) are required to adequately resolve a fragmentation event (Truelove et al. 1997, 1998; Whitworth 1998), it becomes clear that each modeled event requires a very large amount of computing resources. With these available simulation tools it is usually impractical to examine mechanisms that might lead to the breakup of a cloud only after a significant evolutionary delay (see discussion in "Delayed Breakup," below).

Many different groups have utilized three-dimensional hydrodynamical techniques to study the early, approximately free-fall phase of the collapse of rotating,

Jeans-unstable gas clouds in an effort to see whether prompt fragmentation occurs. Generally an isothermal ($\gamma = 1$) equation of state has been assumed because, as illustrated in Figure 2, that is what appears to be appropriate for the early phases of collapse (Boss & Bodenheimer 1979, Tohline 1980, Boss 1980, Bodenheimer et al. 1980, Różyczka et al. 1980, Gingold & Monaghan 1981, Wood 1982, Miyama et al. 1984, Monaghan & Lattanzio 1986, Bonnell et al. 1991, Burkert & Bodenheimer 1993, Sigalotti 1997, Truelove et al. 1998, Tsuribe & Inutsuka 1999). However, some attempts have been made to include the effects of adiabatic compression and heating at intermediate densities (Boss 1986, Myhill & Kaula 1992, Bonnell & Bate 1994b, Bate 1998, Tsuribe & Inutsuka 2000) or to focus just on the likelihood of prompt fragmentation in adiabatic collapse regimes (Boss 1981, Arcoragi et al. 1991). Before reviewing what has been learned about prompt fragmentation from these investigations, it is useful to summarize the general behavior of a cloud's collapse based on some of the general principles outlined above in "Basic Physical Principles." For clarity, we use the subscript cl to identify global properties of the cloud that do not change with time, such as the cloud's total mass M_{cl} and total angular momentum J_{cl} and the subscript (or sometimes superscript) i to denote the initial properties of the cloud at the onset of its collapse.

Nearly Homologous Collapses

If a rotating cloud begins to collapse from a spherical or spheroidal configuration that is uniform in density and whose mass M_{cl} is significantly larger than the local Jeans mass M_J^i—that is, from a configuration in which $\alpha_i \ll 1/2$—then the cloud collapses fairly homologously. On a free-fall timescale τ_{ff}^i governed by its initial mean density $\bar{\rho}_i$, the cloud evolves through a sequence of flatter and flatter configurations, not unlike the collapsing pressure-free spheroids modeled some time ago by Lynden-Bell (1962, 1964), Lin et al. (1965), and Hutchins (1976). Indeed, because the ratio of thermal to gravitational energy α drops during the isothermal phase of collapse (see Equation 30), the local Jeans mass also steadily decreases and the cloud more and more closely approximates a pressure-free spheroid.

As the cloud's evolutionary time t approaches one initial free-fall time τ_{ff}^i and its degree of flattening becomes most extreme, pressure gradients build to the point at which they are able to decelerate the collapse—at least in the vertical direction. If the cloud's angular momentum J_{cl} is sufficiently large, this deceleration of the collapse will occur while the cloud is still in the isothermal phase of its contraction. As a result (see the discussion in "The Importance of the Effective Adiabatic Exponent γ," above), the cloud will necessarily be very flat and β will have grown to a value close to $1/2$. If the cloud's total angular momentum is relatively small, the cloud can enter an adiabatic phase of its contraction before the collapse begins to decelerate. In this case as well, however, the homologously collapsing cloud is destined to be very flat when the deceleration occurs because α will have dropped to a very small value during the cloud's earlier, isothermal phase of contraction.

Using Equation 29, one can estimate the radial size (and the mean density) of such a flattened configuration (Tohline 1981, Hachisu & Eriguchi 1985).

Note that in a homologously contracting cloud, virtually all of the cloud's mass reaches the final, flattened configuration at approximately the same time. That is, at approximately one free-fall time, the accretion rate $\dot{M}$ becomes very high, but there is not an extended phase of accretion thereafter. The early, spherically symmetric simulation by Narita et al. (1970) presents a nonrotating analog to this type of evolution; the collapse started from a configuration in which α_i was relatively small, and after a central equilibrium core formed, the core experienced only a brief period of relatively rapid accretion (and a correspondingly high accretion luminosity).

Nonhomologous Collapses

If, on the other hand, a cloud begins to collapse from a configuration that is fairly centrally condensed and/or the cloud initially is only marginally Jeans unstable (that is, $\alpha_i \lesssim 1/2$), then the collapse proceeds in a nonhomologous fashion. The central region of the cloud collapses ahead of the rest of the cloud, producing a steep central density gradient. If the cloud is initially centrally condensed, this happens because, at every position in the cloud the timescale for collapse is governed by the "local" free-fall time; regions of higher mean density have shorter free-fall times, so the central region runs away from the rest (e.g., Section 3.2 of Tohline 1982). If the cloud is only marginally Jeans unstable initially, this happens because $\tau_s^i \approx \tau_{ff}^i$ (see the discussion associated with Equation 23). A rarefaction wave propagating in from the edge of the cloud reaches the cloud center in approximately one free-fall time and establishes a relatively steep density gradient in its wake (Bodenheimer & Sweigart 1968, Larson 1969, Penston 1969).

As a result, a relatively small volume of material at the core of the cloud will have an opportunity to find an equilibrium configuration first (locally, at least, $\alpha + \beta \approx 1/2$), then that configuration will change with time as material (and angular momentum) from the rest of the cloud accretes onto the core. However, as explained above in "The Importance of the Effective Adiabatic Exponent γ," as long as the core remains isothermal it will be unable to settle into an equilibrium state without the aid of rotation. If there is enough angular momentum in the cloud's core, the core's collapse can be stopped by rotation in the isothermal phase; in this case the first equilibrium core will necessarily exhibit a large value of β. Otherwise, the cloud's core must contract to a density regime (see Figure 2) in which $\gamma > 4/3$ before it finds its initial equilibrium configuration. In either case, owing to the nonhomologous nature of the collapse, the first equilibrium core will contain a relatively small fraction of the entire cloud's mass, and an extended period of mass accretion will follow the core's formation. The early spherically symmetric simulation by Larson (1969) presents a nonrotating analog to this type of evolution.

It is important to note that the key timescales associated with the first equilibrium core (τ_{ff}^{ec}, τ_s^{ec}, and τ_{rot}^{ec}) all generally will be very short compared with the initial free-fall time $\tau_{\rm ff}^i$ of the cloud and, hence, short compared with the core's accretion timescale $\tau_{\rm accrete}^{ec}$. This is because, as shown in "Key Timescales" and "Implications of the Virial Theorem," above, all three of these key timescales are shorter in higher density configurations, and the density of the collapsed core $\bar{\rho}_{ec}$ is necessarily higher (usually much higher) than the cloud's original mean density $\bar{\rho}_i$. (See Table 1 and the uppermost axis of Figure 2 for values of the dynamical timescale—specifically, the orbital period—that correspond to various density regimes.) For example, the first core that formed in Larson's (1969) simulations had a mean density that was approximately 5 orders of magnitude higher than the cloud's initial density, so the core's sound-crossing time and local free-fall time were roughly 2.5 orders of magnitude shorter than the cloud's initial free-fall time. Therefore, in a dynamical sense, the core will be relatively isolated from its surroundings. If the core is found to be unstable toward the development of a dynamical instability (characterized by an e-folding time that is of order the local free-fall time $\tau_{\rm ff}^{ec}$), this instability has an opportunity to manifest itself and even grow to nonlinear amplitude before the core is significantly influenced by further accretion from the surrounding cloud material.

Does Prompt Fragmentation Occur?

All of the three-dimensional hydrodynamical simulations that have been conducted to date to investigate prompt fragmentation in collapsing protostellar gas clouds support the following conclusion: Fragmentation does not occur during a phase of free-fall collapse. If fragmentation occurs at all, it happens only after one initial free-fall time, that is, after the cloud—or at least the core of the cloud—has collapsed into a rotationally flattened, quasi-equilibrium configuration (see discussion below). This finding argues strongly against the idea of "hierarchical" fragmentation that was proposed by Hoyle (1953) to explain how clusters of stars might form in less than a free-fall time. At first this result was somewhat surprising because linear stability analyses of the pressure-free collapse of spherical (Hunter 1962, Mestel 1965) and spheroidal (Silk 1982) gas clouds lent some support to Hoyle's idea. However, as numerical simulations have become more and more refined (e.g., Truelove et al. 1998; Bate 1998; Tsuribe & Inutsuka 1999, 2000), this result has been repeatedly reaffirmed. The presence of finite gas pressure and, more to the point, nonzero gas pressure gradients in a collapsing protostellar gas cloud is apparently what makes all the difference. (See Section 7 of Tohline 1982 for a more exhaustive discussion of this issue.)

On the other hand, multidimensional collapse simulations have shown that fragmentation of a rotating gas cloud can occur immediately after the cloud (or at least the cloud's core) has settled into its first rotationally flattened, quasi-equilibrium configuration. As Tsuribe & Inutsuka (1999, 2000) have summarized, fragmentation seems to occur relatively easily in clouds that have collapsed in a nearly homologous fashion; in contrast to this, simulations of nonhomologous collapses generally have not resulted in prompt fragmentation. Attempts have been made to develop a formula (in terms of the cloud's initial values of α_i and β_i, for example) that readily predicts under what specific circumstances prompt fragmentation will or will not occur (Tohline 1981, Hachisu & Eriguchi 1984, Miyama et al. 1984, Tsuribe & Inutsuka 2000). Generally speaking, these attempts have met with only marginal success. This is, perhaps, not surprising considering that hydrodynamical flows can be extremely sensitive to initial conditions and that the available parameter space for initial cloud conditions is huge. For a given choice of α_i and β_i, for example, one can select different initial radial density profiles, distributions of angular momentum, geometries for the cloud (e.g., spherical, oblate, prolate), amplitude and shape of nonaxisymmetric perturbations, degrees of internal turbulent motions, and influences from an external environment. In addition, it has become clear that the end results can depend sensitively on numerical resolution—to the extent that many of the simulations of strictly isothermal collapse carried out prior to 1997 may have produced misleading results (Truelove et al. 1997, 1998; Boss et al. 2000).

All in all, though, the general conclusions drawn by Tsuribe & Inutsuka (1999, 2000) are likely to remain intact. They make sense on simple physical grounds. When the initial conditions are chosen in such a way that a cloud is able to collapse in a nearly homologous fashion, fluctuations that contain more than one M_J of material can amplify to some degree during the collapse (Hunter 1962). Furthermore, the cloud as a whole will evolve to an equilibrium configuration that is highly susceptible to the further growth of these, if not other, nonaxisymmetric perturbations: The parameter α is very small, so the configuration contains many Jeans masses; β is so high that the configuration is likely to be unstable to one or more global, nonaxisymmetric instabilities, akin to the unstable modes that are present in Maclaurin spheriods (see discussion, below), and the Toomre Q parameter (Toomre 1964) is likely to be very small, making the rotationally flattened configuration susceptible to the same types of instabilities observed in galaxies.

On the other hand, a nonhomologous collapse discourages fragmentation. Even during a phase of isothermal collapse the development of radial pressure gradients retards the growth of fluctuations (Hunter 1962). Because the first equilibrium core that forms will contain only a small fraction of the total cloud mass, α will be relatively high in the core, so it will be much less susceptible to the growth of nonaxisymmetric perturbations of the types just mentioned.

Are Binaries the Result?

Although numerical simulations have demonstrated that the free-fall collapse of a rotating protostellar gas cloud can produce a flattened configuration that is susceptible to prompt fragmentation, it is not yet clear how often this process will directly produce a binary system (as opposed to producing a larger number of fragments, for example). The outcome may be sensitive to the spectrum of fluctuations present in the initial cloud state. Even when it seemed likely from early calculations that some specific sets of initial conditions would directly produce binaries, the results are being called into question (Tsuribe & Inutsuka 1999, Boss et al. 2000). More importantly, owing to the constraints imposed by present numerical schemes, few simulations have been able to follow an individual cloud's evolution very far beyond the initial instant of fragmentation. For example, published results that purport to show the formation of a binary virtually never follow the binary through even one full orbit. It is therefore impossible to deduce (*a*) whether either (or both) of the components will undergo an additional fragmentation event, (*b*) whether or not the components will subsequently merge, or (*c*) how any subsequent phase of accretion will affect the parameters of the binary (e.g., period, eccentricity, mass ratio). We clearly still have a lot to learn about the relationship between prompt fragmentation and the formation of binary stars.

DELAYED BREAKUP

As discussed above, the dynamical collapse of a Jeans-unstable, rotating gas cloud may often end with the formation of an equilibrium configuration that is stable against fragmentation. This is especially true of collapses that occur in a non-homologous fashion. Even when prompt fragmentation occurs, the individual fragments may initially be found to be stable against further fragmentation. We should then investigate whether, through its subsequent "slow" evolution toward the main sequence, such a configuration will become susceptible to breakup and thereby produce a binary system.

This equilibrium configuration (hereafter, referred to as a "core" or referenced by the subscript "ec") will be rotating, although not necessarily rapidly; it generally will contain only a small fraction of the cloud's total mass (remember that the first core that formed in Larson's 1969 nonrotating models contained only a few Jupiter masses of gas); and it may or may not initially be surrounded by a rotationally supported (accretion) disk. For the purpose of discussion, we will assign to this core a mass $m_{ec} \ll M_{cl}$, an equatorial radius R_{ec}, a mean temperature T_{ec}, and a mean angular velocity ω_{ec}. From these quantities we can, in turn, determine many other key properties of the core—such as ρ_{ec} and β_{ec}—via the expressions given in "Basic Physical Principles," above.

In the context of this discussion of delayed breakup, the phrase "slow evolution" is intended to convey the idea that the core is no longer in free-fall collapse and that its overall structure is not changing significantly on a free-fall time τ_{ff}^{ec}, as measured by the core's mean density. Keep in mind, however, that this evolution may still be fast compared with the cloud's initial free-fall time τ_{ff}^{i} because $\bar{\rho}_{ec} \gg \bar{\rho}_i$. In the absence of significant nonaxisymmetric distortions, the core's evolution should be driven primarily by the same two processes that have been found to be important in spherically symmetric models of cloud collapse: radiation loses and accretion (see, for example, the detailed description provided by Winkler & Newman 1980). Both processes cause the mean density and the mean temperature of the core to steadily rise, along an evolutionary trajectory like the one displayed in Figure 2.

In the case of a rotating core it is clear by Equation 29 that as the core's density increases, β_{ec} increases as well.

The core's mass m_{ec} will steadily increase through accretion of material that is either falling directly in from the cloud or migrating in from a surrounding disk. The core's total angular momentum will steadily increase as a result of accretion as well. Normally, the accretion process will add relatively high specific angular momentum material to the core because there is a natural tendency for the cloud's low specific angular momentum material to accumulate in the core first. Accretion will therefore tend to increase the ratio of the core's rotational to gravitational energy, β_{ec}, faster than one would have expected from the core's contraction alone. And if the core is not initially surrounded by a disk, one will almost certainly develop as a result of direct infall from the cloud (see Cassen & Moosman 1981, Terebey et al. 1984, and Section 4.1.3 of Shu et al. 1987).

In the following subsections, we examine the stability of an equilibrium core as it undergoes slow evolution of the type described above. Then we examine the stability of the accretion disk that surrounds such a core. More specifically, we examine whether nonaxisymmetric instabilities might arise in either the core or its accompanying disk that would lead to the "delayed breakup" of the protostellar cloud into a binary star system.

Nonaxisymmetric Instabilities in Rapidly Rotating, Equilibrium Cores

As a result of accretion and radiation losses, a rotating equilibrium core will contract and become more rapidly rotating, in the sense that β_{ec} will steadily increase. When discussing the stability of such a system, it is useful to refer to a diagram, such as the one presented here in Figure 3 (see also Figure 15 of Chandrasekhar 1969 and Figure 1 of Durisen & Tohline 1985), in which one can identify rotating ellipsoidal configurations of any shape. In particular, for an ellipsoid with principal axes (a_1, a_2, a_3) and rotation about its a_3-axis, the ordinate of Figure 3 (a_3/a_1) specifies the object's degree of rotational flattening, and the abscissa (a_2/a_1) specifies the degree of the configuration's equatorial ellipticity. For example, the point in the upper right-hand corner (marked by an S) identifies a sphere; points along the right-hand, vertical axis (passing from S through the points marked M_2 and O_2) identify flattened, axisymmetric (oblate spheroidal) configurations; the curves connecting points S, M_2, and O_2 to the origin identify sequences of more and more distorted ellipsoids.

Strictly speaking, Figure 3 can be used to accurately identify the properties of ellipsoidal figures of equilibrium that arise only for uniform-density objects with velocities that are linear functions of the coordinates. In this context, the right-hand vertical axis represents the sequence of uniformly rotating, Maclaurin spheroids, and the three curves connecting the Maclaurin sequence to the origin specify certain subsets of the general class of Riemann ellipsoids. We have adopted the conventions used by Chandrasekhar (1969) and Lebovitz (1974) in labeling key bifurcation

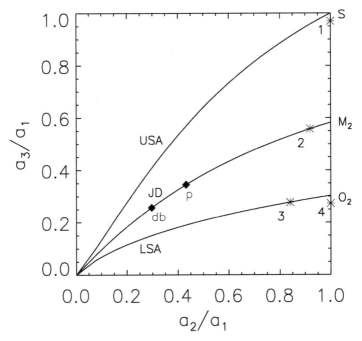

Figure 3 A phase-space diagram for ellipsoidal configurations with principal axis lengths (a_1, a_2, a_3). Figures are rotating about the a_3-axis. The right-hand vertical axis represents the sequence of rotationally flattened Maclaurin spheroids $(a_2 = a_1)$. The three curves connecting this vertical axis to the origin represent three separate sequences of Riemann S-type ellipsoids: USA, the upper self-adjoint $(x = -1)$ sequence; JD, the Jacobi-Dedekind sequence; LSA, the lower self-adjoint $(x = +1)$ sequence. Key bifurcation points along the Maclaurin sequence are labeled by M_2 and O_2. Along the Jacobi ellipsoid sequence, bifurcation points to the pear-shaped (p) and dumbbell (db) sequences are also marked. The data used to construct this plot has been drawn directly from Chandrasekhar (1969).

points along the Maclaurin sequence and in labeling the key ellipsoidal sequences: *USA* is the upper $(x = -1)$ self-adjoint series of Riemann S-type ellipsoids; it branches off of the Maclaurin sequence at $\beta = 0$. *JD* refers to the Jacobi-Dedekind sequence; it bifurcates from the Maclaurin sequence at $\beta = 0.1375$. *LSA* is the lower $(x = +1)$ self-adjoint series of Riemann S-type ellipsoids; it bifurcates from the Maclaurin sequence at $\beta = 0.2738$. In a less formal way, however, it also is useful to refer to Figure 3 when discussing the evolution of less idealized (e.g., centrally condensed and differentially rotating) configurations that more closely resemble real protostellar cloud cores.

In the context of Figure 3, a slowly contracting cloud core will move down along the right-hand vertical axis from, e.g., point 1 toward the bifurcation point marked

M_2. However, the cloud core will be unable to become arbitrarily flat and maintain an axisymmetric structure. Stability analyses (e.g., Lyttleton 1953, Chandrasekhar 1969, Tassoul 1978, Durisen & Tohline 1985) tell us that after reaching a critical degree of flattening (that is, a critical value of β_{ec}) the cloud will begin to distort into a rotating bar-like (ellipsoidal) configuration that has a lower total energy.

CLASSICAL VIEW OF FISSION   If the equilibrium core's contraction occurs on a timescale that is long compared with the viscous dissipation timescale of the gas, then the core will begin to deform into a triaxial configuration when it acquires the degree of flattening identified by the point identified as M_2 in Figure 3, that is, once the rotational energy of the core climbs to a value of $\beta_{ec} \approx 0.14$. Thereafter, further contraction should drive the evolution along the JD sequence toward, for example, point 2 in Figure 3. As the cloud evolves beyond this point on the JD sequence and becomes even more elongated, stability analyses tell us that it will become susceptible toward even higher-order figure deformations. In particular, as Lyttleton (1953) has reviewed, from the nineteenth and early twentieth century works of Poincaré, Darwin, Liapounoff, Jeans, and Cartan, it has been known for over 100 years that ellipsoids along the JD sequence become susceptible to a pear-shaped deformation at the point marked p in Figure 3; and these ellipsoids are susceptible to a dumbbell-shaped deformation at the point marked db. More recently, Eriguchi et al. (1982) have explicitly demonstrated that a sequence of pear-shaped configurations and a sequence of dumbbell-shaped configurations bifurcate at the points p ($a_2/a_1 = 0.432$; $a_3/a_1 = 0.345$; $\beta = 0.1268$) and db ($a_2/a_1 = 0.297$; $a_3/a_1 = 0.258$; $\beta = 0.1863$), respectively. Very early on, the realization that a pear-shaped sequence branches off of the Jacobi sequence "gave rise to the notion that if the mass were stable and evolved by equilibrium forms along this [pear-shaped] series, with the furrow continually deepening as the figure elongated, the final result would be two detached masses rotating in circular orbital motion about each other" (Lyttleton 1953, p. 3). This is the classical formulation of the "fission" theory of binary star formation.

It is discouraging to realize that, more than 100 years after its formulation, the full, nonlinear evolutionary scenario that underpins this classical theory of fission has never been fully tested. That is, to date, nobody has evolved a uniform-density, uniformly rotating ellipsoid slowly along the JD sequence to see whether it spontaneously deforms into a pear-shaped configuration and then whether the slow contraction of the "pear" leads to its breakup. There are, however, a number of reasons why the astrophysical community has decided that this classical fission scenario will not work.

First, it has been determined by stability analyses that ellipsoids are dynamically (rather than secularly) unstable toward the pear-shaped deformation at all points beyond point p on the Jacobi (JD) sequence (Lyttleton 1953). Hence, any attempt to slowly evolve a configuration beyond this critical point of bifurcation will almost certainly cause the configuration to depart rapidly from the desired sequence of uniformly rotating pears. Second, the sequence of pear-shaped configurations discovered by Eriguchi et al. (1982) shows no sign of developing a "deepening furrow," but rather, terminates by equatorial mass-shedding. Third, it is now generally accepted that the viscosity of protostars is too low for viscous dissipation to drive evolution along the JD sequence (Lynden-Bell 1964, Lebovitz 1974) in the first place. That is, the contraction of protostellar cloud cores almost certainly occurs on a timescale that is short compared with the viscous dissipation timescale. For these reasons the classical theory of fission has been abandoned as an explanation of how binary stars form.

LEBOVITZ'S REVISED VERSION OF THE FISSION THEORY   With a thorough understanding of the classical theory's shortcomings, Lebovitz (1974, 1984) has proposed a revised formulation of the classical fission hypothesis. It parallels the classical theory in that the rotating cloud core slowly contracts along a sequence of axisymmetric (or at least approximately axisymmetric) spheroids; then, at a critical degree of flattening, the evolution shifts to a family of more and more elongated ellipsoids. The difference is that Lebovitz considers the evolution of inviscid configurations (acknowledging the idea that contraction times are short compared with viscous times). As a result, the relevant bifurcation point off of the axisymmetric spheroidal sequence is the point marked O_2 in Figure 3, and the relevant ellipsoidal sequence is the one labelled LSA. Instead of following an evolutionary trajectory that progresses from point 1 in Figure 3 to a point on the JD sequence, the core will evolve as an axisymmetric structure through the location labeled M_2, continuing to flatten until it reaches the bifurcation point marked O_2. At this point ($\beta_{ec} \approx 0.27$) the equilibrium core will deform into an ellipsoidal structure and, upon further contraction, it will evolve along the sequence labeled LSA toward, for example, point 3 in Figure 3.

As explained by Lebovitz (1974, 1984, 1989), this evolutionary scenario potentially avoids all of the objections that have been raised in the context of the classical fission hypothesis. First, it does not matter that the viscous timescale is long compared with the contraction timescale in protostellar clouds because deformation into an ellipsoidal configuration occurs in the absence of viscosity. Second, on the direct LSA sequence (see Lebovitz 1974) the order of the points p and db are reversed from their positions on the JD sequence, so by evolving along this sequence the cloud can avoid altogether problems that might be associated with third-harmonic (pear-shaped) instability. Because Eriguchi & Hachisu (1985) have shown that dumbbell-binary sequences do exist and that they bifurcate smoothly from sequences of Riemann ellipsoids, it is tempting to suggest that further slow contraction of the equilibrium core will drive it smoothly along a dumbbell sequence to a binary configuration. This is the essence of Lebovitz's revised version of the fission theory.

A number of questions remain to be answered before the viability of Lebovitz's revised theory can be properly assessed. Most importantly, it has not been demonstrated whether bifurcation from the relevant ellipsoidal sequence to a dumbbell sequence occurs stably and, if it does, whether continued slow contraction of the configuration will actually proceed smoothly to a binary configuration, as envisioned by Eriguchi & Hachisu (1985).

Direct Breakup from an Axisymmetric State

Both the original and revised version of the fission hypothesis of binary star formation have grown out of the expectation that a slowly contracting, equilibrium cloud core will deform from an axisymmetric structure into a triaxial (bar-like) configuration that is dynamically stable. Fission is hypothesized to occur only after this triaxial configuration has undergone further slow contraction. These ideas have not been fully tested in the context of realistic protostellar clouds largely because it has not been possible to construct equilibrium models of rapidly rotating, triaxial configurations with realistic (compressible) equations of state. In the absence of such equilibrium models, we cannot even begin to examine the critical questions of stability that are at the root of these two fission hypotheses. (As discussed in "Slow Contraction of a Rapidly Rotating Ellipsoid," below, some relief to this situation may be forthcoming.)

However, we do have the tools in hand to construct equilibrium models of rapidly rotating, axisymmetric configurations with realistic equations of state and a wide variety of different distributions of angular momentum (e.g., Hachisu 1986). Over the past 15 years a considerable amount of work has been directed toward understanding the stability properties of these axisymmetric systems (Tohline et al. 1985, Williams & Tohline 1987, Luyten 1990, Pickett et al. 1996, Toman et al. 1998, Shibata et al. 2000). Through these studies, for example, it has been determined that virtually all axisymmetric configurations are dynamically unstable toward the development of a "barmode" structure if $\beta_{ec} \geq 0.27$. This finding overlaps well with the classical stability analysis of Maclaurin spheroids, wherein a bifurcation to the LSA sequence of Riemann ellipsoids occurs at $\beta = 0.2738$ (the point marked O_2 in Figure 3). It is in this sense that Figure 3 provides a useful context in which to discuss the evolutionary trajectories of slowly contracting, protostellar gas clouds even though it was originally constructed from studies of much simpler (e.g., uniform density, incompressible) equilibrium structures.

In addition, hydrodynamical codes (like the ones developed to study the prompt fragmentation problem) have been used extensively to determine whether or not the nonlinear development of this barmode instability results directly in the breakup of an equilibrium cloud core (Tohline et al. 1985, Durisen et al. 1986, Williams & Tohline 1988, Pickett et al. 1996, New et al. 2000, Cazes & Tohline 2000, Brown 2000). The initial model used in these investigations generally has been a structure that sits on the axisymmetric sequence and just beyond the point (marked O_2 in Figure 3), where inviscid systems are expected to evolve dynamically away from an axisymmetric structure—that is, starting from a point like point 4 in Figure 3. From all of these investigations, the unanimous consensus is that breakup does not occur (see Tohline & Durisen 2001 for a recent overview). Instead, within a few rotation periods τ_{rot} the cloud core generally deforms into a bar-like structure with a slight two-armed spiral character; gravitational torques quickly facilitate a local redistribution of angular momentum within the core (Imamura et al. 2000); a relatively small amount of high specific angular momentum material is shed in the equatorial plane; and the core settles down into a dynamically stable, roughly ellipsoidal bar that is spinning about its shortest axis. As Cazes & Tohline (2000) discussed in detail, when viewed from a frame rotating with this bar, it is clear that the structure is very robust and supports nontrivial internal motions. In a qualitative sense, the bar resembles a Riemann ellipsoid with a structure similar to the one identified by point 3 on the LSA sequence in Figure 3. Hence, Cazes & Tohline (2000) referred to the configuration as a "compressible analog of a Riemann Ellipsoid."

These nonlinear investigations of the development of the bar-mode instability from initially axisymmetric configurations also have been categorized as "fission studies" (e.g., Durisen & Tohline 1985). Because there is unanimous agreement that such evolutions do not lead directly to the formation of a binary system, there has been a broad pronouncement that the fission hypothesis is dead (Boss 1988, Bodenheimer et al. 1993, Bonnell 2001). This pronouncement seems premature, however. Neither the classical fission theory nor the revised one proposed by Lebovitz predicted that fission would occur directly from an instability that arises in axisymmetric configurations. Instead, as summarized above, the expectation was that a slowly contracting, equilibrium cloud core would deform first into a triaxial (bar-like) configuration. Fission should then occur only after the triaxial configuration undergoes further slow contraction past a point at which the structure becomes unstable to a higher-order surface distortion.

Slow Contraction of a Rapidly Rotating Ellipsoid

It has proven difficult to critically evaluate this last step of the classical or revised fission hypothesis because few models of equilibrium, triaxial configurations exist for systems with a physically relevant, compressible equation of state. A limited set of rigidly rotating, polytropic bars have been constructed for very slightly compressible fluids (Vandervoort & Welty 1981, Ipser & Managan 1981, Hachisu & Eriguchi 1982), but focusing on Lebovitz's scenario in particular, no one has yet figured out how to routinely construct compressible models of rotating ellipsoids that are spinning fast enough to serve as analogues to the Riemann ellipsoids that lie along the LSA sequence in Figure 3.

As just summarized, however, groups that have modeled the nonlinear development of the bar-mode instability in rapidly rotating, axisymmetric gas clouds have discovered that, after shedding a bit of material in the equatorial plane, the system usually settles down into a steady-state bar. Also, as Cazes & Tohline (2000) have pointed out, this bar in many respects serves as a compressible analog of the Riemann ellipsoids (CARE). With Lebovitz's revised fission hypothesis in mind, Cazes (1999; see also Tohline & Durisen 2001) slowly cooled one of these CAREs, following its cooling evolution in a self-consistent fashion with a finite-difference hydrodynamic code. Because the gas was being modeled as a polytrope, cooling was accomplished by slowly decreasing the system's polytropic constant, K, according to the relation $K = K_0(1 - t/\tau_{cool})$ with $\tau_{cool} = 4P_{pattern}$, where $P_{pattern}$ was the initial pattern rotation period of the bar. The configuration slowly contracted and began to spin somewhat faster, as expected, and it also became somewhat more elongated. Roughly speaking, from a configuration like point 3 in Figure 3, the system slowly evolved to the left along the LSA sequence. Then, after K had decreased to approximately half of its original value, the model began to oscillate dynamically between a centrally condensed, bar-like state and a distinctly dumbbell shape. In its dumbbell state the model presented a pair of clearly defined off-axis density maxima and a velocity field that showed circulation about each density maximum. Evidently the bar had reached a point in its evolution along the LSA sequence where there were two equally plausible equilibrium configurations into which it was permitted to settle. Cazes (1999) hypothesized that upon further cooling the system was likely to evolve along a dumbbell-binary sequence analogous to the one depicted by Eriguchi & Hachisu (1985) and through a common-envelope binary state like the one discussed by Tohline et al. (1999).

Unfortunately, owing to computational constraints, Cazes was unable to follow his cooling evolution further, and when cooling was stopped the model settled into a centrally condensed, rather than a binary, configuration. Thus, it is debatable whether Cazes' model was actually progressing along a route to fission, but his simulation provides tantalizing evidence that Lebovitz's revised fission hypothesis may be correct under certain circumstances.

Stability of Accretion Disks

After an equilibrium core forms at the center of a collapsing gas cloud, it will continue to grow in mass through the direct accretion of infalling cloud material, but the core will be able to directly accrete only the cloud material that has relatively low specific angular momentum. Specifically, only material that arrives at the surface of the core with an angular velocity $\lesssim \omega_{\max}^{ec}$, as given by Equation 24, will be gravitationally bound to the core. For a core of radius R_{ec} this means that only material with a specific angular momentum $j < j_{\max}^{ec} \approx R_{ec}^2 \omega_{\max}^{ec}$ will fall directly onto the core. The rest of the infalling material must form or become part of a centrifugally supported disk that surrounds the core. For purposes of illustration we have listed in Table 2 the value of $j_{\max}^{ec}$ that is associated with the equilibrium core that will form during a spherically symmetric collapse at the points marked B, C, and D along the evolutionary path shown in Figure 2. (In each case, R_{ec} has been estimated by treating the core as a sphere of mass M_{equil} and mean density ρ as given in Table 2.)

In practice, unless the molecular gas cloud is initially very slowly rotating, only a small fraction of the cloud's gas will have $j < j_{\max}^{ec}$. Consider, for example, the canonical uniform-density, spherical gas cloud that is initially rotating with an angular velocity ω_i and, hence, initially has a ratio of rotational to gravitational potential energy $\beta_i = \frac{1}{2}(\omega_i/\omega_{\max}^{cl})^2$. [Cameron (1978) and Cassen & Summers (1983) examined models of this type in the context of the formation of the solar nebula.] The specific angular momentum of the material in such a cloud increases as one moves out from the center of the cloud. Specifically, in terms of the fractional mass

m_ϖ that is enclosed inside each cylindrical radius ϖ,

$$j = \frac{5}{2}\frac{J}{M}[1 - (1 - m_\varpi)^{2/3}] \tag{32}$$

(Hachisu et al. 1987). By inverting Equation 32, we obtain the following expression for the fraction of the cloud's material that has a specific angular momentum $j \le j_{max}^{ec}$:

$$m_\varpi = 1 - [1 - \eta(2\beta_i)^{-1/2}]^{3/2} \approx \eta\beta_i^{-1/2}, \tag{33}$$

where $\eta \equiv (j_{max}^{ec}/j_{max}^{cl})$ and j_{max}^{cl} is the maximum specific angular momentum that material can have in the initial, marginally Jeans-unstable gas cloud. Examining the values of j_{max}^{ec} listed in Table 2 (and realizing that the value given for Case A supplies $j_{max}^{cl} = 3.6 \times 10^{21}$), we see that η ranges between 4×10^{-3} (Case B) and 1×10^{-4} (Case D). Hence, unless β_i is extremely small, m_ϖ will be very small, indicating that the majority of the cloud's mass $(1 - m_\varpi)$ will be unable to fall directly onto the central equilibrium core. One arrives at this conclusion even if the uniformly rotating cloud is initially as centrally condensed as a singular isothermal sphere (Cassen & Summers 1983, Terebey et al. 1984).

We must therefore ask, as Cameron (1978) and Cassen & Moosman (1981) did in the context of the primitive solar nebula, whether the surrounding disk of gaseous material will remain stable as an axisymmetric configuration when it accumulates as much or even more mass than the central core. Cameron was interested in the possibility that a giant gaseous proto-planet might form from a gravitational instability in such a disk. Here we are of course interested in whether an object (or more than one object) with a mass comparable to the central core might form from such a disk. With this in mind, we focus on recent studies that have examined whether or not protostellar disks are susceptible to global, nonaxisymmetric instabilities with relatively long wavelengths (low azimuthal mode numbers m) so that their development incorporates a sizeable fraction of the gas that is in the disk.

As Shu et al. (1990) have summarized, there is a long history of research into the stability of disks in the context of galaxies, evolved stars, and planetary rings. Some of this work is relevant to discussions of the stability of protostellar accretion disks [such as the classic works of Toomre (1964) and Goldreich & Tremaine (1978)], but protostellar disks are sufficiently different from planetary rings and galaxy disks that issues related to their stability have to be addressed separately (Adams et al. 1989). Despite continuing improvements in computing resources, it is still very difficult to model with adequate spatial and time resolution the full three-dimensional evolution of protostellar disks. This is primarily because they have a significant radial extent, so the dynamical timescale governing events near the inner edge of the disk (near the surface of the protostar) can be many orders of magnitude shorter than the relevant timescale at the outer edge of the disk. The vertical thickness of the disk is also generally expected to be very small compared with the radial extent of the disk, which puts extra demands on a model's spatial resolution. As a result, the models that have been developed have either (*a*) assumed that the disk is infinitesimally thin (has no vertical extent) and examined the evolution only of two-dimensional (ϖ, θ) systems or (*b*) assumed that the disk has some vertical thickness but does not have a large radial extent relative to its inner edge, in which case the structure resembles a torus. The work of Papaloizou & Lin (1989) and Adams et al. (1989) have led the way in the former category of investigations; Papaloizou & Pringle (1984) and Goodman & Narayan (1988) have led the way in the latter category.

Studies of infinitesimally thin, self-gravitating disks generally have shown that, under a wide range of conditions the disk will become unstable toward the development of long-wavelength, spiral shaped instabilities if the disk-to-central-object mass ratio, M_d/M_c, is sufficiently large (Papaloizou & Lin 1989; Shu et al. 1990; Heemskerk et al. 1992; Noh et al. 1991, 1992; Laughlin & Korchagin 1996; Taga & Iye 1998). Most significantly, Adams et al. (1989) discovered that an $m = 1$ "eccentric mode" instability arises in disks with $M_d/M_c \gtrsim 1$ even in situations in which an evaluation of the Toomre (1964) Q parameter indicates that the growth of many other modes is suppressed. It appears as though any of the spiral modes whose relative stability is governed by the importance of self-gravity in the disk will be effective at redistributing angular momentum within the disk and, thereby, can drive accretion of disk material onto the central protostar (e.g., Laughlin & Różyczka 1996). However, the "eccentric mode" appears to be special in the sense that, through its development, a "clump" of material preferentially accumulates on one side of the disk, suggesting that a separate protostellar core may be able to form from material in the disk and remain in orbit about the original, central object. This is dynamically allowed because, as explained by Adams et al. (1989), as the $m = 1$ distortion grows in the disk, the central protostar wanders away from the center of mass of the system along a spiral trajectory that keeps it and the clump of material in the disk on opposite sides of the system's center of mass. With finite-difference hydrodynamical techniques, it has been difficult to follow this proposed disk fragmentation event to completion. As Laughlin & Różyczka (1996) explained following one of their simulations, before the clump became fully developed, "the central star had spiraled out and crashed into the inner disk edge, effectively terminating the computation due to numerical difficulties." Smoothed-particle hydrodynamic simulations have been more successful at illustrating how this full fragmentation event may occur (Adams & Benz 1992, Bonnell 1994).

It is significant that all of the research on infinitesimally thin disks conducted after the insightful work of Adams et al. (1989) has confirmed that protostellar disks are susceptible to an $m = 1$ instability if $M_d/M_c \gtrsim 1$; and almost all agree that this is the first unstable mode that is likely to spontaneously develop as the mass ratio M_d/M_c grows through the accretion of infalling cloud material (see Laughlin & Różyczka 1996 for a counter-example). At present, however, there is little agreement regarding the precise physical mechanism responsible for exciting this mode—indeed, five explanations have been proposed (Shu et al. 1990, Heemskerk et al. 1992, Noh et al. 1992, Laughlin & Korchagin 1996, Taga & Iye 1998). As a result, there is little agreement regarding the precise mass ratio at which the mode first becomes unstable, although it seems in most cases to be within a factor of a few of the value $3/4\pi \approx 0.24$ suggested by the derivation of Shu et al. (1990).

Self-gravitating, geometrically thick disks (tori) that orbit a central point mass also appear to be unstable toward the development of long-wavelength, nonaxisymmetric modes. As Goodman & Narayan (1988) first outlined, two different types of modes appear to be excitable in addition to the "sonic" mode identified by Papaloizou & Pringle (1984) in massless accretion tori. Analogs of all three modes appear as well in "annuli," that is, in differentially rotating systems that have limited radial extent but extend vertically to infinity (see Goodman & Narayan 1988, Christodoulou & Narayan 1992, Christodoulou 1993). Nonlinear dynamical simulations have demonstrated that radially slender tori and annuli readily distort into m clumps, where m is the mode number of the fastest growing, unstable mode (Hawley 1990, Tohline & Hachisu 1990, Christodoulou 1993, Woodward et al. 1994). For a given disk-to-central-object mass ratio M_d/M_c the relevant mode number decreases as the radial extent (thickness) of the torus increases; for a given radial extent the mode number decreases as the mass ratio decreases.

In the simulation that seems most relevant to our discussion of binary star formation, Woodward et al. (1994) evolved to nonlinear amplitude a toroidal system with $M_d/M_c = 1$ that was unstable only toward the development of an $m = 1$ spiral disturbance. As it grew, the distortion produced a single, high-density clump that was orbiting "about the center of mass of the system along with (but roughly oppositely positioned from) the central object," in accordance with the behavior of the "eccentric instability" that was first discussed by Adams et al. (1989) in the context of infinitesimally thin disks. Also, as predicted, the central object moved along a spiral trajectory, progressively farther from the center of the system. Unfortunately [and reminiscent of the account given by Laughlin & Różyczka (1996)], "before the high-density blob in the disk . . . developed to the point where it clearly could be identified as a compact entity distinct from the rest of the disk material, the central object . . . impacted the inner edge of the disk." The simulation was terminated, so it never became clear whether a binary protostellar system would be the outcome.

Andalib et al. (1997) have attempted to summarize and provide a unified discussion of these results. Generally speaking, the picture that emerges from these studies of geometrically thick disks is consistent with the one that has arisen from the studies of infinitesimally thin disks: As protostellar disks grow in mass and radius by accreting infalling cloud material, they are likely to become unstable toward the development of long-wavelength, nonaxisymmetric structure, with the so-called "eccentric mode" generally being dominant. It is therefore conceivable that some protostellar disks will break up into one or more pieces containing a sizeable fraction of the disk's entire mass.

SUMMARY AND CONCLUSIONS

The theory of star formation would be tremendously simplified if we could point to one physical mechanism that is primarily responsible for the transformation of single clumps of gas into binary systems. Boss (1988) has argued rather persuasively that prompt fragmentation is the preferred mechanism and, given only meager evidence to the contrary, others have been inclined to agree (Clarke & Pringle 1993, Bodenheimer et al. 1993, Bonnell 2001). From an extensive review of the literature, I must conclude that, to date, none of the ideas that has been put forth to explain the origin of binary star systems is completely satisfactory. At the same time, I would argue that several of the ideas appear to be quite promising, but a significant amount of additional effort must be devoted toward the development of each of these ideas before any of them will become fully convincing. Before outlining the areas that require more work, let us summarize the various points on which there seems to be broad agreement.

Issues on Which There is Broad Agreement

1. Capture holds little promise as a binary formation process. It seems clear from observations that young stars generally have paired themselves up into bound systems before they have reached stage IV in the evolutionary scenario presented by Shu et al. (1987), that is, well before they have reached the main sequence. This provides very little time for the capture process to operate because it is inherently inefficient.

2. Clouds do not fragment during a phase of free-fall collapse. This runs contrary to earliest expectations, but it seems to be borne out repeatedly by modern numerical simulations of cloud collapse that have included realistic effects of gas pressure and rotation. Instead, clouds tend to collapse to a rotationally flattened, quasi-equilibrium configuration before any fragmentation occurs, if at all.

3. Prompt fragmentation generally works immediately following a phase of free-fall collapse if a significant fraction of the cloud's mass falls into a rotationally flattened configuration approximately in unison, especially if this configuration forms while the cloud is still in the isothermal phase of its contraction (that is, while it maintains a mean density $\lesssim 10^{-13}\,\mathrm{g\,cm^{-3}}$). Clouds are therefore susceptible to prompt fragmentation if they begin to collapse from a configuration that is relatively uniform in density and contains more than a few Jeans masses of material (i.e., α_i is well below $1/2$).

4. Prompt fragmentation usually does not work if a cloud collapses in a nonhomologous fashion. Instead, a central core of relatively small mass forms first, followed by an extended phase of accretion. This core will necessarily become rapidly rotating as it accretes infalling material from the surrounding, free-falling cloud.

5. Rapidly rotating, axisymmetric cloud cores do not break up when they encounter the dynamical bar-mode instability, like the one that spontaneously arises at the bifurcation point labeled O_2 in Figure 3 ($\beta \gtrsim 0.27$). Instead, gravitational torques are effective at redistributing angular momentum within the core, a bit of high specific angular momentum material gets shed in the

equatorial plane, and the core settles down into a new, dynamically stable—although still rapidly rotating—configuration.

6. The core generally settles into a spinning ellipsoidal or bar-like structure after it encounters the dynamical bar-mode instability. By many accounts, the configuration appears to be a compressible analog of the Riemann ellipsoids that lie, for example, along the model sequence labeled LSA in Figure 3.

7. A substantial disk will almost certainly form around the central core through the additional infall of high specific angular momentum cloud material, and it will likely grow to a mass that is comparable to or larger than the mass of the core.

8. Protostellar disks become dynamically and globally unstable toward the development of long-wavelength, nonaxisymmetric structure when the mass contained in the disk becomes comparable to or greater than the mass of the core that it surrounds.

Out of this list, items 3, 6, and 8 provide the most promising leads in connection with our search to find the process or processes by which binary stars form. Further study is required, however, before we will be able to specify with confidence the degree to which any or all three of them is, in practice, relevant.

Issues that Require Further Investigation

By all accounts, prompt fragmentation along the lines described in item 3, above, is a strong candidate for explaining how binaries form. Numerous simulations by many groups have demonstrated that, under the conditions described, a collapsed cloud configuration will undergo rapid, nonaxisymmetric deformation. I remain concerned about several aspects of this proposed process, however; some of these concerns are alluded to in "Does Prompt Fragmentation Occur?" and "Are Binaries the Result?". First, to be effective, the process requires that at the onset of collapse the cloud must contain more than a few Jeans masses of material. How is nature able to construct such an artificially unstable initial configuration? Second, the outcome of the fragmentation process seems to be relatively sensitive to initial conditions. While it is true that many published simulations show the development of two fragments (or at least a strong $m = 2$ azimuthal mode deformation), it is usually also true that some type of two-fold deformation was imposed as a form of perturbation in the initial, unstable cloud. Why should nature preferentially introduce two-fold deformations into significantly Jeans-unstable initial states? Third, few simulations have been able to follow an individual cloud's evolution very far beyond the initial instant of fragmentation to determine whether a binary protostellar state is actually the outcome. Usually simulations are terminated before even one full orbit (and one rotation period of the initial cloud) has been completed. Fourth, collapsed cloud configurations are much less susceptible to prompt fragmentation after the gas has entered a phase of its evolution in which it heats up upon further compression, that is, at densities $\bar{\rho} > 10^{-13}$ g cm^{-3} (see Figure 2).

Hence, it is difficult to understand how prompt fragmentation can explain the formation of binary systems with periods shorter than about 1000 years (see the top-most horizontal axis of Figure 2).

The model developed recently by Shu et al. (2000) and Galli et al. (2001) may very well provide an answer to the first two of these concerns. It relies on the introduction of magnetic fields to support the initial cloud state and ambipolar diffusion to slowly decouple the cloud from the field so that the cloud is brought gradually to the "brink of instability." While studying the properties of rotating gas clouds that are supported against collapse by a magnetic field, this group has found that multiple-lobed ($m = 2, 3, 4, \ldots$) configurations naturally bifurcate from an underlying axisymmetric sequence. Thus, it is reasonable to expect rotating clouds to already exhibit appreciable nonaxisymmetric distortions before the field decouples from the gas. Furthermore, if the field rather quickly decouples from the gas, the cloud will begin its free-fall collapse from a configuration that has a relatively small α_i. Shadmehri & Ghanbari (2001) and Nakamura & Li (2001) have presented related investigations of the structure and evolution of nonaxisymmetric distortions in magnetically supported clouds.

In connection with the third concern that has been raised in the context of models of prompt fragmentation, it is clear why most numerical simulations have been terminated prematurely. Integration time steps are forced to be very small because they are linked to the free-fall (and sound-crossing) timescale associated with the densest region of a fragment, whereas the relevant orbital period is tied to the mean density of the configuration and, particularly when dealing with isothermal flows, these two timescales can become separated by many orders of magnitude. This unfortunate situation will be overcome only when we develop numerical tools that permit us to model hydrodynamical flows in three dimensions for times that are very long compared with the Courant-Freidrichs-Lewey constraint ("Prompt Fragmentation"). Until we overcome this practical limitation of our modeling techniques, we will be unable to determine (*a*) whether prompt fragmentation preferentially forms binary sytems, (*b*) whether any initial fragments survive subsequent phases of accretion or merge back together, (*c*) what the final mass ratio of the binary system will be, and (*d*) whether the circum-stellar and/or circum-binary disks that form from the infalling gas indeed become massive relative to the mass of the newly formed binary cores. By building certain simplifying assumptions into their smoothed-particle hydrodynamic scheme, Bate et al. (1996) and Bate & Bonnell (1997) have begun to address this challenging but important numerical techniques problem.

In connection with item 6, above, it appears as though the results of our modeling to date are consistent with the revised fission hypothesis put forward by Lebovitz (1974, 1984) (see "Lebovitz's Revised Version of the Fission Theory," above). That is, rapidly rotating axisymmetric configurations that become susceptible to the dynamical bar-mode instability evolve to a configuration that, for all practical purposes, resembles a Riemann ellipsoid. In order to complete the test of this revised fission hypothesis, a method will have to be found to slowly evolve such configurations along a sequence of more and more distorted ellipsoids (like the LSA sequence illustrated in Figure 3) to see whether the sequence eventually bifurcates to a dumbbell-binary sequence and whether slow evolution along this new sequence indeed leads to fission. As described in "Slow Contraction of a Rapidly Rotating Ellipsoid," above, Cazes (1999) made an early attempt to model such an evolution, but his work needs to be extended; his results need to be confirmed by other research groups, and similar evolutions starting from different initial ellipsoidal states need to be investigated. The challenge will be to develop a technique by which a variety of rapidly rotating, ellipsoidal configurations can readily be constructed out of differentially rotating, compressible fluids. Numerical models of fully three-dimensional structures can be constructed one at a time by following through to completion the dynamical bar-mode instability, but this is an extremely inefficient way to proceed. More efficient methods recently have been devised to construct two-dimensional analogs of these rapidly rotating ellipsoidal configurations under certain restrictions (Syer & Tremaine 1996, Andalib 1998). Extending such techniques to three-dimensional systems would be extremely desirable.

Finally, we address the shortcomings associated with item 8, above. With few exceptions, simulations that have attempted to follow the nonlinear development of nonaxisymmetric instabilities in massive protostellar disks have run into difficulties following the evolutions to completion. This has especially been true in the case of the $m = 1$ "eccentric instability," which appears to be most promising in the context of the problem of binary star formation (see "Stability of Accretion Disks," above for details). Again, dynamical range appears to be the problem. For simplicity, the protostellar core that is initially positioned at the center of the disk usually has been treated as a point mass. When the core migrates away from the center of mass of the system and impacts the disk, the simulation grinds to a halt as it tries to follow flow of the disk material into the point-mass singularity. A method needs to be devised to overcome this problem so we can determine with confidence what the outcome of this instability is; in particular, whether the disk material clumps up into a binary companion to the core. In this context, Pickett et al. (1998, 2000) have been studying the stability of systems in which the central protostar is fully resolved along with the disk. Their models have included disks with relatively modest radial extent (and therefore a relatively modest range of dynamical timescales), so they have only been able to test the viability of the eccentric mode instability in a limited way.

When examining gravitational instabilities in a protostellar disk, we must also keep in mind that viscous processes—or indeed the gravitational instability itself—acting within the disk may prevent it from ever holding a significant amount of material. Viscosity (or gravitational torques) can facilitate the radial redistribution of angular momentum within the disk and thereby allow material to move radially inward and fall onto the central protostar. If the rate of accretion of material through the disk onto the central core ever exceeds the rate of accretion of material from the surrounding gas cloud onto the disk, then the disk will decrease in mass, significantly reducing the likelihood that a binary star will form through the process of disk fragmentation. Our understanding of the processes that lead to the formation of binary stars is therefore tightly coupled to our understanding of viscous transport processes in protostellar disks.

In conclusion, we reflect back on Mathieu's (1994) statement—supported through observations and not through theoretical models—that binary formation is the primary branch of the star formation process. Obviously nature knows how to form binary star systems. Hopefully, in the coming decade, our numerical models will find one or more fully convincing ways to do so as well.

The *Annual Review of Astronomy and Astrophysics* is online at
http://astro.annualreviews.org

LITERATURE CITED

Adams FC, Benz W. 1992. In *Gravitational Instabilities in Circumstellar Disks & the Formation of Binary Companions*, pp. 185–94. San Francisco: Astron. Soc. Pac.

Adams FC, Ruden SP, Shu FH. 1989. *Astrophys. J.* 347:959–76

Alcock C, Allsman RA, Alves DR, Axelrod TS, Becker AC, et al. 2001. *Astrophys. J.* 552:259–67

Andalib SW. 1998. *The structure and stability of selected 2-D self-gravitating systems*. PhD thesis. Baton Rouge: Louisiana State Univ.

Andalib SW, Tohline JE, Christodoulou DM. 1997. *Astrophys. J. Suppl.* 108:471–87

Arcoragi J-P, Bonnell I, Martel H, Benz W, Bastien P. 1991. *Astrophys. J.* 380:476–83

Bate MR. 1998. *Astrophys. J. Lett.* 508:L95–98

Bate MR, Bonnell IA. 1997. *MNRAS* 285:33–48

Bate MR, Bonnell IA, Price NM. 1995. *MNRAS* 277:362–76

Benz W. 1991. In *Late Stages of Stellar Evolution Computational Methods in Astrophysical Hydrodynamics*, pp. 259–312. New York: Springer-Verlag

Bodenheimer P, Ruzmajkina T, Mathieu RD. 1993. In *Protostars and Planets III*, pp. 367–404. Tucson: Univ. Arizona Press

Bodenheimer P, Sweigart A. 1968. *Astrophys. J.* 152:515–22

Bodenheimer P, Tohline JE, Black DC. 1980. *Astrophys. J.* 242:209–18

Bonnell IA. 1994. *MNRAS* 269:837–48

Bonnell IA. 2001. See Zinnecker & Mathieu 2001, pp. 23–32

Bonnell IA, Bate MR. 1994. *MNRAS* 271:999–1004

Bonnell IA, Martel H, Bastien P, Arcoragi J-P, Benz W. 1991. *Astrophys. J.* 377:553–58

Boss AP. 1980. *Astrophys. J.* 237:866–76

Boss AP. 1981. *Astrophys. J.* 250:636–44

Boss AP. 1984. *Astrophys. J.* 277:768–82

Boss AP. 1986. *Astrophys. J. Suppl.* 62:519–52

Boss AP. 1988. *Comments Astrophys.* 12:169–90

Boss AP, Bodenheimer P. 1979. *Astrophys. J.* 234:289–95

Boss AP, Fisher RT, Klein RI, McKee CF. 2000. *Astrophys. J.* 528:325–35

Boss AP, Mayhill EA. 1992. *Astrophys. J. Suppl.* 83:311–27

Brown JD. 2000. *Phys Rev. D* 62:084024

Burkert A, Bodenheimer P. 1993. *MNRAS* 264:798–806

Cameron AGW. 1978. *Moon Planets* 18:5–40

Cassen P, Moosman A. 1981. *Icarus* 48:353–76

Cassen P, Summers A. 1983. *Icarus* 53:26–40

Cazes JE. 1999. *The formation of short period binary star systems from stable, self-gravitating, gaseous bars*. PhD thesis, Baton Rouge: Louisiana State Univ.

Cazes JE, Tohline JE. 2000. *Astrophys. J.* 532:1051–68

Chandrasekhar S. 1969. *Equilibrium Figures of Equilibrium.* New Haven, CT: Yale Univ. Press

Christodoulou DM. 1993. *Astrophys. J.* 412: 696–719

Christodoulou DM, Narayan R. 1992. *Astrophys. J.* 388:451–66

Clarke C. 1992. In *Complementary Approaches to Double and Multiple Star Research*, pp. 176–84. San Francisco: Astron. Soc. Pac.

Clarke CJ, Pringle JE. 1993. *MNRAS* 261:190–202

Close LM. 2001. See Zinnecker & Mathieu 2001, pp. 555–58

Courant R, Friedrichs KO, Lewy H. 1928. *Math. Ann.* 100:32–38

Di Stefano R. 2001. See Zinnecker & Mathieu 2001, pp. 529–38

Duquennoy A, Mayor M. 1991. *Astron. Astrophys.* 248:485–524

Durisen RH, Gingold RA, Tohline JE, Boss AP. 1986. *Astrophys. J.* 305:281–308

Durisen RH, Tohline JE. 1985. In *Protostars and Planets II*, pp. 534–75. Tucson: Univ. Arizona Press

Eriguchi Y, Hachisu I. 1985. *Astron. Astrophys.* 142:256–62

Eriguchi Y, Hachisu I, Sugimoto D. 1982. *Prog. Theor. Phys.* 67:1068–75

Galli D, Shu FH, Laughlin G, Lizano S. 2001. *Astrophys. J.* 551:367–86

Gingold RA, Monaghan JJ. 1981. *MNRAS* 197:461–75

Goldreich P, Tremaine S. 1978. *Astrophys. J.* 222:850–58

Goodman J, Narayan R. 1988. *MNRAS* 231:97–114

Guilloteau S. 2001. See Zinnecker & Mathieu 2001, pp. 547–54

Hachisu I. 1986. *Astrophys. J. Suppl.* 61:479–507

Hachisu I, Eriguchi Y. 1982. *Prog. Theor. Phys.* 68:206–21

Hachisu I, Eriguchi Y. 1984. *Astron. Astrophys.* 140:259–64

Hachisu I, Eriguchi Y. 1985. *Astron. Astrophys* 143:355–64

Hachisu I, Tohline JE, Eriguchi Y. 1987. *Astrophys. J.* 323:592–613

Hansen CJ, Kawaler SD. 1994. *Stellar Interiors: Physical Principles, Structure, and Evolution.* New York: Springer

Hayashi C, Narita S, Miyama SM. 1982. *Prog. Theor. Phys.* 68:1949–66

Hawley JF. 1990. *Astrophys. J.* 356:580–90

Heemskerk MHM, Papaloizou JC, Savonije GJ. 1992. *Astron. Astrophys.* 260:161–74

Hoyle F. 1953. *Astrophys. J.* 118:513–28

Hunter C. 1962. *Astrophys. J.* 136:594–608

Hutchins JB. 1976. *Astrophys. J.* 205:103–121

Imamura JN, Durisen RH, Pickett BK. 2000. *Astrophys. J.* 528:946–64

Ipser JR, Managan RA. 1981. *Astrophys. J.* 250:362–72

Jeans J. 1919. *Problems of Cosmogony and Stellar Dynamics.* Cambridge: Cambridge Univ. Press

Kippenhahn R, Weigert A. 1990. *Stellar Structure and Evolution.* New York: Springer-Verlag

Larson RB. 1969. *MNRAS* 145:271–95

Larson RB. 1990. In *Physical Processes in Fragmentation and Star Formation*, pp. 389–99. Dordrecht, The Netherlands: Kluwer

Laughlin G, Korchagin V. 1996. *Astrophys. J.* 460:855–68

Laughlin G, Różyczka M. 1996. *Astrophys. J.* 456:279–91

Launhardt R. 2001. See Zinnecker & Mathieu 2001, pp. 117–21

Launhardt R, Sargent AI, Henning Th, Zylka R, Zinnecker H. 2000. In *Birth and Evolution of Binary Stars, IAU Symp. No. 200*, ed. B. Reipurth, H. Zinnecker, pp. 103–5. San Francisco: Astron. Soc. Pac.

Lebovitz NR. 1974. *Astrophys. J.* 190:121–30

Lebovitz NR. 1984. *Astrophys. J.* 284:364–80

Lebovitz NR. 1989. *Highlights Astron.* 8:129–31

Ledoux P. 1945. *Astrophys. J.* 102:143–53

Lin CC, Mestel L, Shu FH. 1965. *Astrophys. J.* 142:1431–46

Lucy LB. 1977. *Astron. J.* 82:1013–24

Luyten PJ. 1990. *MNRAS* 245:614–22

Lynden-Bell D. 1962. *Proc. Cambridge Philos. Soc.* 58:709–11

Lynden-Bell D. 1964. *Astrophys. J.* 139:1195–216

Lyttleton RA. 1953. *The Stability of Rotating Liquid Masses.* Cambridge: Cambridge Univ. Press

Marcy GW, Butler RP. 1998. *Annu. Rev. Astron. Astrophys.* 36:57–98

Marcy GW, Butler RP. 2000. *Publ. Astron. Soc. Pac.* 112:137–40

Mathieu R. 1994. *Annu. Rev. Astron. Astrophys.* 32:465–530

Mayor M, Queloz D. 1995. *Nature* 378:355–59

Mestel L. 1965. *Q. J. Roy. Astron. Soc.* 6:161–98

Miyama SM, Hayashi C, Narita S. 1984. *Astrophys. J.* 279:621–32

Monaghan JJ. 1992. *Annu. Rev. Astron. Astrophys.* 30:543–74

Monaghan JJ, Lattanzio JC. 1986. *Astron. Astrophys.* 158:207–11

Myhill EA, Kaula WM. 1992. *Astrophys. J.* 386:578–86

Nakamura F, Li Z-Y. 2001. *Astrophys. J. Lett.* 566:L101–4

Narita S, Nakano J, Hayashi C. 1970. *Prog. Theor. Phys.* 43:942–64

New KCB, Centrella JM, Tohline JE. 2000. *Phys. Rev.* 62:064019

Noh H, Vishniac ET, Cochran WD. 1991. *Astrophys. J.* 383:372–79

Noh H, Vishniac ET, Cochran WD. 1992. *Astrophys. J.* 397:347–52

Padgett DL, Brandner W, Stapelfeldt R, Strom SE, Terebey S, Koerner D. 1999. *Astron. J.* 117:1490–1504

Padgett DL, Strom SE, Ghez A. 1997. *Astrophys. J.* 477:705–10

Papaloizou JCB, Lin DNC. 1989. *Astrophys. J.* 344:645–68

Papaloizou JCB, Pringle JE. 1984. *MNRAS* 208:721–50

Penston MV. 1969. *MNRAS* 144:425–48

Pickett BK, Cassen P, Durisen RH, Link R. 1998. *Astrophys. J.* 504:468–91

Pickett BK, Durisen RH, Cassen P, Mejia AC. 2000. *Astrophys. J. Lett.* 540:L95–98

Picket BK, Durisen RH, Davis GA. 1996. *Astrophys. J.* 458:714–38

Portegies Z, Simon F, Hut P, McMillan SLW, Verbunt F. 1997. *Astron. Astrophys.* 328:143–57

Quirrenbach A. 2001a. See Zinnecker & Mathieu 2001, pp. 539–46

Quirrenbach A. 2001b. *Annu. Rev. Astron. Astrophys.* 39:353–401

Quist CF, Lindegren L. 2000. *Astron. Astrophys.* 361:770–80

Quist CF, Lindegren L. 2001. See Zinnecker & Mathieu 2001, pp. 64–68

Reid IN, Gizis JE, Kirkpatrick JD, Koerner DW. 2001. *Astron. J.* 121:489–502

Roache PJ. 1976. *Computational Fluid Dynamics.* Albuquerque, NM: Hermosa

Różyczka M, Tscharnuter WM, Winkler K-H, Yorke HW. 1980. *Astron. Astrophys.* 83:118–28

Shadmehri M, Ghanbari J. 2001. *Astrophys. J.* 557:1028–34

Shibata M, Baumgarte TW, Shapiro SL. 2000. *Astrophys. J.* 542:453–63

Shu FH. 1977. *Astrophys. J.* 214:488–97

Shu FH, Adams FC, Lizano S. 1987. *Annu. Rev. Astron. Astrophys.* 25:23–81

Shu FH, Laughlin G, Lizano S, Galli D. 2000. *Astrophys. J.* 535:190–210

Shu FH, Tremaine S, Adams F, Ruden SP. 1990. *Astrophys. J.* 358:495–514

Sigalotti LDiG. 1997. *Astron. Astrophys.* 328:586–94

Silk J. 1982. *Astrophys. J.* 256:514–22

Simon M, Close LM, Beck TL. 1999. *Astron. J.* 117:1375–86

Smith KW, Bonnell IA, Emerson JP, Jenness T. 2000. *MNRAS* 319:991–1000

Söderhjelm S. 1999. *Astron. Astrophys.* 341:121–40

Syer D, Tremaine S. 1996. *MNRAS* 281:925–36

Taga M, Iye M. 1998. *MNRAS* 299:1132–38

Tassoul JL. 1978. *Theory of Rotating Stars.* Princeton, NJ: Princeton Univ. Press

Terebey S, Shu FH, Cassen P. 1984. *Astrophys. J.* 286:529–51

Tohline JE. 1980. *Astrophys. J.* 235:866–81

Tohline JE. 1981. *Astrophys. J.* 248:717–26

Tohline JE. 1982. *Fundam. Cosmic Phys.* 8:1–81

Tohline JE. 1984. *Astrophys. J.* 285:721–28

Tohline JE, Cazes JE, Cohl HS. 1999. *Numerical Astrophysics*, pp. 155–58. Boston: Kluwer

Tohline JE, Durisen RH. 2001. See Zinnecker & Mathieu 2001, pp. 40–43

Tohline JE, Durisen RH, McCollough M. 1985. *Astrophys. J.* 298:220–34

Tohline JE, Hachisu I. 1990. *Astrophys. J.* 361:394–407

Toman J, Imamura JN, Pickett BK, Durisen RH. 1998. *Astrophys. J.* 497:370–87

Toomre A. 1964. *Astrophys. J.* 139:1217–38

Truelove JK, Klein RI, McKee CF, Holliman JH II, Howell LH, Greenough JA. 1997. *Astrophys. J.* 489:L179–83

Truelove JK, Klein RI, McKee CF, Holliman JH II, Howell LH, et al. 1998. *Astrophys. J.* 495:821–52

Tsuribe T, Inutsuka S. 1999. *Astrophys. J. Lett.* 523:L155–58

Tsuribe T, Inutsuka S. 2000. See Launhardt et al. 2000, pp. 184–86

Valtonen M, Mikkola S. 1991. *Annu. Rev. Astron. Astrophys.* 29:9–29

Vandervoort PO, Welty DE. 1981. *Astrophys. J.* 248:504–15

Whitworth A. 1998. *MNRAS* 296:442–44

Williams HA, Tohline JE. 1987. *Astrophys. J.* 315:594–601

Williams HA, Tohline JE. 1988. *Astrophys. J.* 334:449–64

Winkler K-H, Newman MJ. 1980. *Astrophys. J.* 236:201–11

Wood D. 1982. *MNRAS* 199:331–43

Woodward JW, Tohline JE, Hachisu I. 1994. *Astrophys. J.* 420:247–67

Zinnecker H, Mathieu RD, eds. 2001. *The Formation of Binary Stars*. San Francisco: Astron. Soc. Pac.

Annu. Rev. Astron. Astrophys. 2002. 40:387–438
doi: 10.1146/annurev.astro.40.060401.093744
Copyright © 2002 by Annual Reviews. All rights reserved

RADIO EMISSION FROM SUPERNOVAE AND GAMMA-RAY BURSTERS

Kurt W. Weiler
Naval Research Laboratory, Code 7213, Washington, DC 20375-5320;
email: Kurt.Weiler@nrl.navy.mil

Nino Panagia
Space Telescope Science Institute, 3700 San Martin Drive, Baltimore, Maryland 21218
and Astrophysics Division, Space Science Department of European Space Agency;
email: panagia@stsci.edu

Marcos J. Montes
Naval Research Laboratory, Code 7212, Washington, DC 20375-5320;
email: Marcos.Montes@nrl.navy.mil

Richard A. Sramek
P.O. Box 0, National Radio Astronomy Observatory, Socorro, New Mexico 87801;
email: dsramek@nrao.edu

Key Words   SN1998bw, GRB980425, afterglows, GRB970508, GRB980329,
GRB980519, GRB991208, GRB991216, GRB000301C

■ **Abstract**   Study of radio supernovae over the past 20 years includes two dozen
detected objects and more than 100 upper limits. From this work it is possible to
identify classes of radio properties, demonstrate conformance to and deviations from
existing models, estimate the density and structure of the circumstellar material and,
by inference, the evolution of the presupernova stellar wind, and reveal the last stages
of stellar evolution before explosion. It is also possible to detect ionized hydrogen
along the line of sight, to demonstrate binary properties of the stellar system, and to
show clumpiness of the circumstellar material. More speculatively, it may be possible
to provide distance estimates to radio supernovae.

   Over the past four years the afterglow of gamma-ray bursters has occasionally
been detected in the radio, as well in other wavelengths bands. In particular, the
interesting and unusual gamma-ray burst GRB980425, thought to be related to
SN1998bw, is a possible link between supernovae and gamma-ray bursters. Analyzing the extensive radio emission data avaiable for SN1998bw, one can describe
its time evolution within the well-established framework available for the analysis of radio emission from supernovae. This allows relatively detailed description
of a number of physical properties of the object. The radio emission can best be
explained as the interaction of a mildly relativistic ($\Gamma \sim 1.6$) shock with a dense

0066-4146/02/0922-0387$14.00

preexplosion stellar wind–established circumstellar medium that is highly structured both azimuthally, in clumps or filaments, and radially, with observed density enhancements. Because of its unusual characteristics for a Type Ib/c supernova, the relation of SN1998bw to GRB980425 is strengthened and suggests that at least some classes of GRBs originate in massive star explosions. Thus, employing the formalism for describing the radio emission from supernovae and following the link through SN1998bw/GRB980425, it is possible to model the gross properties of the radio and optical/infrared emission from the half-dozen GRBs with extensive radio observations. From this we conclude that at least some members of the "slow-soft" class of GRBs can be attributed to the explosion of a massive star in a dense, highly structured circumstellar medium that was presumably established by the preexplosion stellar system.

RADIO SUPERNOVAE

Introduction

A series of papers published over the past 20 years on radio supernovae (RSNe) has established the radio detection and, in a number of cases, radio evolution for approximately two dozen objects: three Type Ib supernovae, five Type Ic supernovae (because the differences between these two supernova optical classes are slight, Type Ib show strong He I absorption, whereas Type Ic show weak He I absorption, and there are no obvious radio differences, we often refer to the classes as Type Ib/c), and the rest Type II SNe. A much larger list of more than 100 additional SNe have low radio upper limits (See Table 1 and http://rsd-www.nrl.navy.mil/7214/weiler/kwdata/rsnhead.html).

In this extensive study of the radio emission from SNe, several effects have been noted: (a) Type Ia SNe are not radio emitters to the detection limit of the Very Large Array (VLA[1]); (b) Type Ib/c SNe are radio luminous with steep spectral indices (generally $\alpha < -1$; $S \propto \nu^{+\alpha}$) and a fast turn-on/turn-off, usually peaking at 6-cm near or before optical maximum; and (c) Type II SNe show a range of radio luminosities with flatter spectral indices (generally $\alpha > -1$) and a relatively slow turn-on/turn-off, usually peaking at 6-cm significantly after optical maximum. Type Ib/c may be fairly homogeneous in some of their radio properties, whereas Type II, as in the optical, are quite diverse.

A large number of physical properties of SNe can be determined from radio observations. Very long baseline interferometry (VLBI) imaging shows the symmetry of the blastwave and the local circumstellar medium (CSM), estimates the speed and deceleration of the supernova blastwave propagating outward from the explosion, and with assumptions of symmetry and optical

---

[1]The VLA telescope of the National Radio Astronomy Observatory is operated by Associated Universities, Inc. under a cooperative agreement with the National Science Foundation.

TABLE 1 Observed supernovae[a]

SN	Type	Radio	SN	Type	Radio	SN	Type	Radio	SN	Type	Radio
1895B	I		1974G	Ia		1986I	IIP		1994W	IIn	
1901B	I		1975N	Ia		1986J	IIn	LC	1994Y	IIn	
1909A	II		1977B	?		1986O	Ia		1994ai	Ib/c	
1914A	?		1978B	II		1987A	IIpec	LC	1994aj	II	
1917A	I?		1978G	II		1987B	IIn		1994ak	IIn	
1921B	II		1978K	II	LC	1987D	Ia		1995G	IIn	
1921C	I		1979B	Ia		1987F	IIpec		1995N	IIn	DT
1923A	IIP	DT	1979C	IIL	LC	1987K	IIb		1995X	IIP	
1937C	Ia		1980D	IIP		1987M	Ib/c		1995ad	IIP	
1937F	II		1980I	Ia		1987N	Ia		1995ag	II	
1939C	II		1980K	IIL	LC	1988I	IIn		1995al	Ia	
1940A	IIL		1980L	?		1988Z	IIn	LC	1996L	IIL	
1945B	?		1980N	Ia		1989B	Ia		1996N	Ib/c	DT
1948B	II?		1980O	II		1989C	IIn		1996W	IIpec	
1950B	II?	DT	1981A	II		1989L	IIL		1996X	Ia	
1954A	I		1981B	Ia		1989M	Ia		1996ae	IIn	
1954J	II		1981K	II	LC	1989R	IIn		1996an	II	
1957D	II?	DT	1982E	Ia?		1990B	Ib/c	LC	1996aq	Ib/c	
1959D	II		1982R	Ib/c?		1990K	IIL		1996bu	IIn	
1959E	I		1982aa	?	LC	1990M	Ia		1996cb	IIb?	DT
1963J	I		1983G	I		1991T	Ia		1997X	Ib/c	DT
1966B	II		1983K	IIP/n		1991ae	IIn		1997ab	IIn	
1968D	II	DT	1983N	Ib/c	LC	1991ar	Ib/c		1997db	II	
1968L	IIP		1984A	I		1991av	IIn		1997dq	Ib/c	
1969L	IIP		1984E	IIL/n		1991bg	Ia		1997ei	Ib/c	
1969P	?		1984L	Ib/c	LC	1992A	Ia		1997eg	IIn	LC
1970A	II?		1984R	?		1992H	IIP		1998S	IIn	DT
1970G	IIL	LC	1985A	Ia		1992ad	IIP?	DT	1998bu	Ia	
1970L	I?		1985B	Ia		1992bd	II		1998bw	Ib/c	LC
1970O	?		1985F	Ib/c		1993G	II		1999D	II	
1971G	I?		1985G	IIP		1993J	IIb	LC	1999E	IIn	
1971I	Ia		1985H	II		1993N	IIn		1999em	IIP	DT
1971L	Ia		1985L	IIL	DT	1993X	II		2000P	IIn	
1972E	Ia		1986A	Ia		1994D	Ia				
1973R	IIP		1986E	IIL	LC	1994I	Ib/c	LC			
1974E	?		1986G	Ia		1994P	II				

[a]DT, detection; LC, light curve available.

line/radio-sphere velocities, allows independent distance estimates to be made (see, e.g., Marcaide et al. 1997, Bartel et al. 1985).

Measurements of the multifrequency radio light curves and their evolution with time show the density and structure of the CSM, evidence for possible binary companions, clumpiness or filamentation in the presupernova wind, mass-loss rates and changes therein for the presupernova stellar system, and through stellar evolution models, estimates of the zero age main sequence (ZAMS) presupernova stellar mass and the stages through which the star passed on its way to explosion. It has also been proposed (Weiler et al. 1998) that the time from explosion to 6-cm radio maximum may be an indicator of the radio luminosity, and thus an independent distance indicator for Type II SNe and that Type Ib/c SNe may be approximate radio standard candles at 6-cm radio peak flux density.

A summary of the radio information on SNe can be found at http://rsd-www.nrl. navy.mil/7214/weiler/sne-home.html.

Models

GENERAL PROPERTIES   All known RSNe appear to share common properties of (a) nonthermal synchrotron emission with high brightness temperature; (b) a decrease in absorption with time, resulting in a smooth, rapid turn-on, first at shorter wavelengths and later at longer wavelengths; (c) a power-law decline of the flux density with time at each wavelength after maximum flux density (optical depth ~1) is reached at that wavelength; and (d) a final, asymptotic approach of spectral index α($S \propto \nu^{+\alpha}$) to an optically thin, nonthermal, constant negative value (Weiler et al. 1986, 1990).

The characteristic RSN ratio light curves such as those shown in Figure 2 arise from the competing effects of slowly declining nonthermal radio emission and more rapidly declining thermal or nonthermal absorption yielding a rapid turn-on and slower turn-off of the radio emission at any single frequency. This characteristic light curve shape is also illustrated schematically in Figure 1 of Weiler et al. (1998). Because absorption processes are greater at lower frequencies, transition from optically thick to optically thin (turn-on) occurs first at higher frequencies and later at lower frequencies. After the radiation is completely optically thin and showing the ongoing decline of the underlying emission process (turn-off), the nonthermal spectrum causes lower frequencies to have higher flux density. These two effects cause the displacement in time and flux density of the light curves at different frequencies also seen in Figure 2.

Chevalier (1982a,b) has proposed that the relativistic electrons and enhanced magnetic field necessary for synchrotron emission arise from the supernova blastwave interacting with a relatively high-density CSM that has been ionized and heated by the initial UV/X-ray flash. This CSM density (ρ), which decreases as the inverse square of the radius (r) from the star is presumed to have been established by a constant mass-loss ($\dot{M}$) rate, constant velocity (w_{wind}) wind (i.e., $\rho \propto \frac{\dot{M}}{w_{wind} r^2}$) from a massive stellar progenitor or companion. This ionized CSM is the source of some or all of the initial thermal gas absorption, although Chevalier (1998) has proposed that synchrotron self-absorption (SSA) may play a role in some objects.

A rapid rise in the observed radio flux density results from a decrease in these absorption processes as the radio-emitting region expands and the absorption processes, either internal or along the line of sight, decrease. Weiler et al. (1990) have suggested that this CSM can be "clumpy" or "filamentary," leading to a slower radio turn-on, and Montes et al. (1997) have proposed the possible presence of an additional ionized medium along the line of sight that is sufficiently distant from the explosion that it is unaffected by the blastwave and can cause a spectral turn-over at low radio frequencies. In addition to clumps or filaments, the CSM may be radially structured with significant density irregularities such as rings, disks, shells, or gradients.

PARAMETERIZED RADIO LIGHT CURVES Weiler et al. (1986, 1990) and Montes et al. (1997) adopted a parameterized model for which the notation is extended and rationalized here from previous publications. However, the "old" notation of τ, τ', and τ'', which has been used previously, is noted, where appropriate, for continuity.

$$S(\text{mJy}) = K_1 \left(\frac{\nu}{5\,\text{GHz}} \right)^\alpha \left(\frac{t - t_0}{1\,\text{day}} \right)^\beta e^{-\tau_{\text{external}}} \left(\frac{1 - e^{-\tau_{\text{CSM}_{\text{clumps}}}}}{\tau_{\text{CSM}_{\text{clumps}}}} \right) \left(\frac{1 - e^{-\tau_{\text{internal}}}}{\tau_{\text{internal}}} \right)$$

(1)

with

$$\tau_{\text{external}} = \tau_{\text{CSM}_{\text{uniform}}} + \tau_{\text{distant}} = \tau + \tau'', \tag{2}$$

where

$$\tau_{\text{CSM}_{\text{uniform}}} = \tau = K_2 \left(\frac{\nu}{5\,\text{GHz}} \right)^{-2.1} \left(\frac{t - t_0}{1\,\text{day}} \right)^\delta, \tag{3}$$

$$\tau_{\text{distant}} = \tau'' = K_4 \left(\frac{\nu}{5\,\text{GHz}} \right)^{-2.1}, \tag{4}$$

and

$$\tau_{\text{CSM}_{\text{clumps}}} = \tau' = K_3 \left(\frac{\nu}{5\,\text{GHz}} \right)^{-2.1} \left(\frac{t - t_0}{1\,\text{day}} \right)^{\delta'}, \tag{5}$$

with K_1, K_2, K_3, and K_4 determined from fits to the data and corresponding, formally, to the flux density (K_1), uniform (K_2, K_4), and clumpy or filamentary (K_3) absorption at 5 GHz one day after the explosion date t_0. The terms $\tau_{\text{CSM}_{\text{uniform}}}$ and $\tau_{\text{CSM}_{\text{clumps}}}$ describe the attenuation of local, uniform CSM and clumpy CSM that are near enough to the supernova progenitor that they are altered by the rapidly expanding supernova blastwave. The $\tau_{\text{CSM}_{\text{uniform}}}$ absorption is produced by an ionized medium that uniformly covers the emitting source ("uniform external absorption"), and the $(1 - e^{-\tau_{CSM_{clumps}}})\tau_{CSM_{clumps}}^{-1}$ term describes the attenuation produced by an inhomogeneous medium ("clumpy absorption") (see Natta & Panagia 1984 for a more detailed discussion of attenuation in inhomogeneous media). The $\tau_{distant}$ term describes the attenuation produced by a homogeneous medium that uniformly covers the source but is so far from the supernova progenitor that it is not affected by the expanding supernova blastwave and is constant in time. All external and clumpy absorbing media are assumed to be purely thermal, singly ionized gas that absorbs via free-free (f-f) transitions with frequency dependence $\nu^{-2.1}$ in the radio. The parameters δ and δ' describe the time dependence of the optical depths for the local uniform and clumpy or filamentary media, respectively.

The f-f optical depth outside the emitting region is proportional to the integral of the square of the CSM density over the radius. Because in the simple Chevalier model the CSM density decreases as r^{-2}, the external optical depth is proportional to r^{-3}, and because the blastwave radius increases as a power of time, $r \propto t^m$ with $m \leq 1$ (i.e., m = 1 for undecelerated blastwave expansion), it follows that the deceleration parameter, m, is

$$m = -\delta/3. \tag{6}$$

Chevalier's model (1982a,b) relates β and δ to the energy spectrum of the relativistic particles $\gamma (\gamma = 2\alpha - 1)$ by $\delta = \alpha - \beta - 3$ so that, for cases in which $K_2 = 0$ and δ is, therefore, indeterminate, one can use

$$m = -(\alpha - \beta - 3)/3. \tag{7}$$

Because it is physically realistic and may be needed in some RSNe in which radio observations have been obtained at early times and high frequencies, Equation 1 also includes the possibility of an internal absorption term[2]. This internal absorption ($\tau_{internal}$) term may consist of two parts—synchrotron self-absorption (SSA) ($\tau_{internal_{SSA}}$), and mixed, thermal f-f absorption/nonthermal emission ($\tau_{internal_{ff}}$).

$$\tau_{internal} = \tau_{internal_{SSA}} + \tau_{internal_{ff}} \tag{8}$$

where

$$\tau_{internal_{SSA}} = K_5 \left(\frac{\nu}{5\,\text{GHz}}\right)^{\alpha-2.5} \left(\frac{t-t_0}{1\,\text{day}}\right)^{\delta''}, \tag{9}$$

---

[2]Note that for simplicity an internal absorber attenuation of the form ($\frac{1-e^{-\tau_{CSM_{internal}}}}{\tau_{CSM_{internal}}}$), which is appropriate for a plane-parallel geometry, is used instead of the more complicated expression (e.g., Osterbrock 1974) valid for the spherical case. The assumption does not affect the quality of the analysis because, to within 5% accuracy, the optical depth obtained with the spherical case formula is simply three fourths of that obtained with the plane-parallel slab formula.

and

$$\tau_{\text{internal}_{\text{ff}}} = K_6 \left(\frac{\nu}{5\,\text{GHz}} \right)^{-2.1} \left(\frac{t - t_0}{1\,\text{day}} \right)^{\delta'''}, \tag{10}$$

with K_5 corresponding formally to the internal, nonthermal ($\nu^{\alpha - 2.5}$) SSA and K_6 corresponding formally to the internal thermal ($\nu^{-2.1}$) f-f absorption mixed with nonthermal emission, at 5 GHz one day after the explosion date t_0. The parameters δ'' and δ''' describe the time dependence of the optical depths for the SSA and f-f internal absorption components, respectively.

A cartoon of the expected structure of a supernova and its surrounding media is presented in Figure 1 (see also Lozinskaya 1992). The radio emission is expected to arise near the blastwave (Chevalier & Fransson 1994).

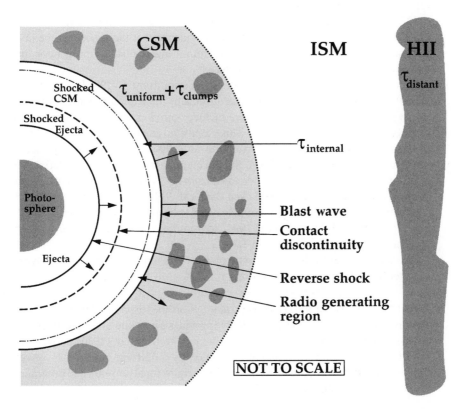

Figure 1 Cartoon, not to scale, of the supernova and its shocks, along with the stellar wind–established circumstellar medium (CSM), the interstellar medium (ISM), and more distant ionized hydrogen (H II) absorbing gas. The radio emission is thought to arise near the blastwave front. The expected locations of the several absorbing terms in Equations 1–10 are illustrated.

Results

The success of the basic parameterization and modeling has been shown in the good correspondence between the model fits and the data for all subtypes of RSNe: e.g., Type Ib SN1983N (Figure 2a) (Sramek et al. 1984), Type Ic SN1990B (Figure 2b) (Van Dyk et al. 1993a), Type II SN1979C (Figure 3a) (Weiler et al. 1991, 1992b; Montes et al. 2000) and SN1980K (Figure 3b) (Weiler et al. 1992a, Montes et al. 1998), and Type IIn SN1988Z (Figure 4) (Van Dyk et al. 1993b, Williams et al. 2002). [Note that after day ~4000 (a) the evolution of the radio emission from both SN1979C and SN1980K deviates from the expected model evolution; (b) SN1979C shows a sinusoidal modulation in its flux density prior to day ~4000; (c) and SN1988Z changes its evolution characteristics after day ~1750.]

Thus, the radio emission from SNe appears to be relatively well understood in terms of blastwave interaction with a structured CSM as described by the Chevalier (1982a,b) model and its extensions by Weiler et al. (1986, 1990), and Montes et al. (1997). For instance, the fact that the uniform external absorption exponent δ is ~−3, or somewhat less, for most RSNe is evidence that the absorbing medium is generally a $\rho \propto r^{-2}$ wind as expected from a massive stellar progenitor that explodes in the red supergiant (RSG) phase.

Additionally, in their study of the radio emission from SN1986J, Weiler et al. (1990) found that the simple Chevalier (1982a,b) model could not describe the relatively slow turn-on. They therefore included terms described mathematically by $\tau_{CSM_{clumps}}$ in Equations 1 and 5. This extension greatly improved the quality of the fit and was interpreted by Weiler et al. (1990) to represent the presence of filaments or clumps in the CSM. Such a clumpiness in the wind material was again required for modeling the radio data from SN1988Z (Van Dyk et al. 1993b, Williams et al. 2002) and SN1993J (Van Dyk et al. 1994). Since that time, evidence for filamentation in the envelopes of SNe has also been found from optical and UV observations (e.g., Filippenko et al. 1994, Spyromilio 1994). The best fit parameters for a number of RSNe are listed in Table 2.

From this modeling there are several physical properties of supernovae that can be determined from radio observations.

MASS-LOSS RATE FROM RADIO ABSORPTION   From the Chevalier (1982a,b) model, the turn-on of the radio emission for RSNe provides a measure of the presupernova mass-loss rate to wind velocity ratio ($\dot{M}/w_{wind}$). Weiler et al. (1986) derived this ratio for the case of pure, external absorption by a uniform medium. However, Weiler et al. (2001) proposed several possible origins for absorption and generalize Equation 16 of Weiler et al. (1986) to

$$\frac{\dot{M}(M_{\odot}\ \mathrm{yr}^{-1})}{(w_{wind}/10\ \mathrm{km\ s}^{-1})} = 3.0 \times 10^{-6} \left\langle \tau_{eff}^{0.5} \right\rangle m^{-1.5} \left(\frac{v_i}{10^4\ \mathrm{km\ s}^{-1}}\right)^{1.5}$$

$$\times \left(\frac{t_i}{45\ \mathrm{days}}\right)^{1.5} \left(\frac{t}{t_i}\right)^{1.5m} \left(\frac{T}{10^4\ \mathrm{K}}\right)^{0.68}. \qquad (11)$$

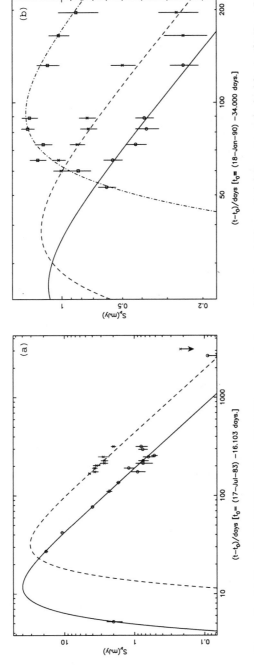

Figure 2 (*a*) Type Ib SN1983N at 6 cm (4.9 GHz; *open circles, solid line*) and 20 cm (1.5 GHz; *stars, dashed line*). (*b*) Type Ic SN1990B at 3.6 cm (8.4 GHz; *open circles, solid line*), 6 cm (4.9 GHz; *stars, dashed line*), and 20 cm (1.5 GHz; *open squares, dash-dot line*).

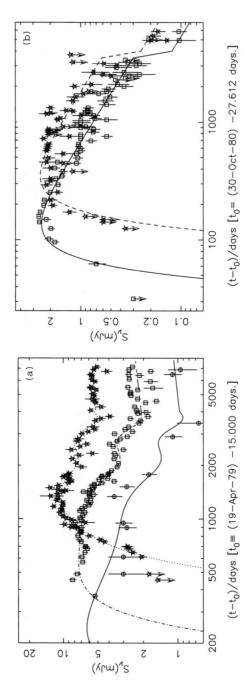

Figure 3 (*a*) Type II SN1979C at 2 cm (14.9 GHz; *crossed circles, solid line*), 6 cm (4.9 GHz; *open squares, dash-dot line*), and 20 cm (1.5 GHz; *open stars, dotted line*) (For discussion of the increase in the radio flux density after day ∼4000 and the sinusoidal modulation of the radio emission before day ∼4000, see Weiler et al. 1991, 1992a and Montes et al. 2000.) (*b*) Type II SN1980K at 6 cm (4.9 GHz; *open squares, solid line*), and 20 cm (1.5 GHz; *open stars, dashed line*). (For discussion of the sharp drop in flux density after day ∼4000, see Montes et al. 1998.)

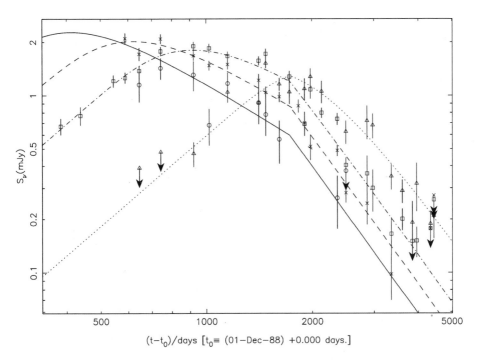

Figure 4 Radio "light curves" for SN1988Z in MCG +03-28-022. The four wavelengths [2 cm (14.9 GHz, *open circles, solid curve*), 3.6 cm (8.4 GHz, *crosses, dashed curve*), 6 cm (4.9 GHz, *open squares, dot-dash curve*), and 20 cm (1.5 GHz, *open triangles, dotted curve*)] are shown together with their best fit light curves. The age of the supernova is measured in days from the adopted (Stathakis & Sadler 1991) explosion date of December 1, 1988. A break in the model radio light curves can be seen around day 1750.

Because the appearance of optical lines for measuring supernova ejecta velocities is often delayed a bit relative to the time of the explosion, Weiler et al. (1986) arbitrarily took $t_i = 45$ days. Because observations have shown that, generally, $0.8 \leq m \leq 1.0$ and from Equation 11 $\dot{M} \propto t_i^{1.5(1-m)}$, the dependence of the calculated mass-loss rate on the date t_i of the initial ejecta velocity v_i measurement is weak, $\dot{M} \propto t_i^{<0.3}$, so that the best optical or VLBI velocity measurements available can be used without worrying about the deviation of the exact measurement epoch from the assumed 45 days after explosion. For convenience, and because many supernova measurements indicate velocities of $\sim$10,000 km s^{-1}, one usually assumes $v_i = v_{\text{blastwave}} = 10,000$ km s^{-1} and takes values of $T = 20,000$ K, $w_{\text{wind}} = 10$ km s^{-1} (which is appropriate for an RSG wind), $t = (t_{6\,\text{cm peak}} - t_0)$ days from best fits to the radio data for each radio supernova (RSN), and m from Equation 6 or 7, as appropriate.

The optical depth term $\langle \tau_{\text{eff}}^{0.5} \rangle$ used by Weiler et al. (1986) was extended by Weiler et al. (2001), who identified at least three possible absorption regimes:

TABLE 2   Fitting parameters for RSNe[a]

SN	Type	α	β	K_1	K_2	δ	K_3	δ'	Explosion date	References
Type Ib/c										
SN1983N	Ib	−1.08	−1.55	3.30×10^3	3.01×10^2	−2.53			30-Jun-83	1, 2
SN1984L	Ib	−1.15	−1.56	3.52×10^2	3.01×10^2	−2.59			13-Aug-84	2, 3
SN1990B	Ic	−1.07	−1.24	1.77×10^2	1.24×10^4	−2.83			15-Dec-89	4
SN1994I[b]	Ic	−1.16	−1.57	8.70×10^3	3.43×10^1	−1.64	2.52×10^4	−2.70	30-Mar-94	5
SN1998bw[c]	Ib/c	−0.71	−1.38	2.37×10^3			1.73×10^3	−2.80	25-Apr-98	6
Type II										
SN1970G	IIL	−0.55	−1.87	1.77×10^6	1.80×10^7	−3.00			25-Jun-70	7
SN1978K[d]	II	−0.77	−1.41	1.14×10^7			3.34×10^8	−2.91	22-May-78	8
SN1979C[e]	IIL	−0.75	−0.80	1.72×10^3	3.38×10^7	−2.94			04-Apr-79	9, 10, 11
SN1980K	IIL	−0.60	−0.73	1.15×10^2	1.42×10^5	−2.69			02-Oct-80	12
SN1981K	II	−0.74	−0.70	7.61×10^1	1.00×10^4	−3.04			16-Jul-81	2, 13
SN1982aa	II?	−0.73	−1.22	5.28×10^4	3.55×10^7	−2.96			28-Apr-79	7
SN1986J	IIn	−0.66	−1.65	1.19×10^7			3.06×10^9	−3.33	01-Jan-84	14
SN1988Z	IIn	−0.69	−1.25	1.47×10^4			5.39×10^8	−2.95	01-Dec-88	15, 16
SN1993J	IIb	−1.07	−0.93	1.86×10^4	1.45×10^3	−2.02	6.31×10^4	−2.14	27-Mar-93	17

References: 1. Sramek et al. 1984; 2. Weiler et al. 1986; 3. Panagia et al. 1986; 4. Van Dyk et al. 1993a; 5. Rupen, private communication; 6. Weiler et al. 2001; 7. K. Weiler, private communication; 8. Montes et al. 1997; 9. Weiler et al. 1991; 10. Weiler et al. 1992a; 11. Montes et al. 2000; 12. Weiler et al. 1992b; 13. Van Dyk et al. 1992; 14. Weiler et al. 1990; 15. Van Dyk et al. 1993b; 16. Williams et al. 2002; 17. Van Dyk et al. 1994.

[a]In some cases the fitting parameters are determined with various input assumptions and/or from very limited data. Thus, all reference should be to the published literature.

[b]SN1994I fit includes $K_5 = 2.21 \times 10^4$, $\delta'' = -3.07$, $K_6 = 2.65 \times 10^2$, $\delta''' = -2.28$.

[c]SN1998bw fit includes $K_4 = 1.24 \times 10^{-2}$.

[d]SN1978K fit includes $K_4 = 1.18 \times 10^{-2}$.

[e]SN1979C is improved by inclusion of a sinusoidal component. See Weiler et al. (1992a) for discussion and parameter values.

(a) absorption by a uniform external medium, (b) absorption by a clumpy or filamentary external medium with a statistically large number of clumps, and (c) absorption by a clumpy or filamentary medium with a statistically small number of clumps. These three cases have different formulations for $\langle \tau_{\text{eff}}^{0.5} \rangle$.

Case 1: Absorption by a Uniform External Medium is the simplest case and has been treated by Weiler et al. (1986). Their result is obtained by substituting

$$\langle \tau_{\text{eff}}^{0.5} \rangle = \tau_{\text{CSM}_{\text{uniform}}}^{0.5}, \tag{12}$$

which is the uniform absorption described in Equation 3.

Case 2: Absorption by a Statistically Large Number of Clumps or Filaments is applicable if the number density and the geometric cross section of clumps is large enough so that any line of sight from the emitting region intersects many clumps. Then one can use a statistical approach in a scenario that has numerous clumps immersed in a uniform medium. For the case of $\delta = \delta'$, it is clear that the fraction of clumpy material remains constant throughout the whole wind–established CSM and, therefore, that the radio signal from the supernova suffers an absorption $\tau_{CSM_{uniform}}$ from the uniform component of the CSM plus an additional absorption, with an even probability distribution between 0 and $\tau_{CSM_{clumps}}$ from the clumpy or filamentary component of the CSM. In such a case, the appropriate average over the possible extremes of the optical depth is taken as

$$\langle \tau_{eff}^{0.5} \rangle = 0.67 \left[(\tau_{CSM_{uniform}} + \tau_{CSM_{clumps}})^{1.5} - \tau_{CSM_{uniform}}^{1.5} \right] \tau_{CSM_{clumps}}^{-1}, \tag{13}$$

with $\tau_{CSM_{uniform}}$ and $\tau_{CSM_{clumps}}$ described in Equations 3 and 5. Note that in the limit of $\tau_{CSM_{clumps}} \to 0$ then $\langle \tau_{eff}^{0.5} \rangle \to \tau_{CSM_{uniform}}$ and in the limit of $\tau_{CSM_{uniform}} \to 0$ then $\langle \tau_{eff}^{0.5} \rangle \to 0.67 \tau_{CSM_{clumps}}^{0.5}$.

Case 3: Absorption by a Statistically Small Number of Clumps or Filaments is appropriate when the number density of clumps or filaments is small and the probability that the line of sight from a given clump intersects another clump is low. Then both the emission and the absorption will occur effectively within each clump. One still expects a situation with a range of optical depths from zero for clumps on the far side of the blastwave-CSM interaction region to a maximum corresponding to the optical depth through a clump for clumps on the near side of the blastwave-CSM interaction region. One also expects an attenuation of the form $(1 - e^{-\tau_{CSM_{clumps}}}) \tau_{CSM_{clumps}}^{-1}$, but now $\tau_{CSM_{clumps}}$ represents the optical depth along a clump diameter. Moreover, in this case the clumps occupy only a small fraction of the volume and have volume-filling factor $\phi \ll 1$. Because the probability that the line of sight from a given clump intersects another clump is low, a condition between the size of a clump, the number density of clumps, and the radial coordinate can be written as

$$\eta \pi r^2 R \approx N < 1, \tag{14}$$

where η is the volume number density of clumps, r is the radius of a clump, R is the distance from the center of the supernova to the blastwave-CSM interaction region, and N is the average number of clumps along the line of sight, with N appreciably lower than unity by definition. It is easy to verify that there is a relation between the volume-filling factor ϕ, r, R, and N, of the form

$$\phi = \frac{4}{3} \frac{r}{R} N. \tag{15}$$

One can then express the effective optical depth $\langle \tau_{eff}^{0.5} \rangle$ as

$$\langle \tau_{eff}^{0.5} \rangle = 0.47 \, \tau_{CSM_{clumps}}^{0.5} \phi^{0.5} N^{0.5}, \tag{16}$$

where for initial estimates one can take $N \sim 0.5$ and a constant ratio $rR^{-1} \sim 0.33$ so that, from Equation 15, $\phi \sim 0.22$.

Although intermediate cases between these three yield results with larger errors, it is felt, considering other uncertainties in the assumptions, that Equation (11) with the relations for $\langle \tau_{eff}^{0.5} \rangle$ given in Equations 12, 13, and 16 yields reasonable estimates of the mass-loss rates of the presupernova star. Mass-loss rate estimates from radio

TABLE 3   Estimated mass-loss rates for RSNe[a]

SN	Type	Explosion to 6 cm peak (days)	Flux density at 6 cm peak (mJy)	Peak 6 cm radio luminosity (erg s⁻¹ Hz⁻¹)	Absorption mass-loss rate (M$_\odot$ yr⁻¹)
Type Ib/c					
SN1983N	Ib	11.6	40.10	1.41×10^{27}	8.74×10^{-7}
SN1984L	Ib	11.0	4.59	2.57×10^{27}	7.45×10^{-7}
SN1990B	Ic	37.5	1.26	5.64×10^{26}	2.69×10^{-6}
SN1994I	Ic	38.1	14.3	1.35×10^{27}	8.80×10^{-6}
SN1998bw	Ib/c	13.3	37.4	6.70×10^{28}	2.60×10^{-5}
Type II					
SN1970G	IIL	307.0	21.50	1.40×10^{27}	6.76×10^{-5}
SN1978K	II	802.0	518.0	1.25×10^{28}	1.52×10^{-4}
SN1979C	IIL	556.0	7.32	2.55×10^{27}	1.06×10^{-4}
SN1980K	IIL	134.0	2.45	1.18×10^{26}	1.28×10^{-5}
SN1981K	II	33.7	5.15	2.14×10^{26}	1.46×10^{-6}
SN1982aa[b]	II?	476.0	19.10	1.27×10^{29}	1.03×10^{-4}
SN1986J	IIn	1210.0	135.0	1.97×10^{28}	4.28×10^{-5}
SN1988Z	IIn	898.0	1.85	2.32×10^{28}	1.14×10^{-4}
SN1993J	IIb	180.0	95.20	1.50×10^{27}	2.41×10^{-5}

[a]See Table 2 for references.

[b]SN1982aa is not optically identified but behaves like an unusually radio luminous Type II.

absorption obtained in this manner tend to be $\sim 10^{-6}$ M$_\odot$ yr⁻¹ for Type Ib/c SNe and $\sim 10^{-4}$–10^{-5} M$_\odot$ yr⁻¹ for Type II SNe. Estimates for particular RSNe are listed in Table 3.

MASS-LOSS RATE FROM RADIO EMISSION   For comparison purposes, one can also try to estimate the presupernova mass-loss rate that established the circumstellar medium (CSM) by considering the radio emission directly. From the Chevalier model and Weiler et al. (1989) one can write the mass-loss rate to wind velocity ratio for an RSN in the form:

$$\frac{\dot{M}(\text{M}_\odot \text{ yr}^{-1})}{(w_{\text{wind}}/10 \text{ km s}^{-1})} = 8.6 \times 10^{-9} \left(\frac{L_{6\text{ cm peak}}}{10^{26} \text{ ergs s}^{-1} \text{ Hz}^{-1}} \right)^{0.71} \left(\frac{t_{6\text{ cm peak}} - t_0}{\text{(days)}} \right)^{1.14} \tag{17}$$

for Type Ib/c supernovae and

$$\frac{\dot{M}(\text{M}_\odot \text{ yr}^{-1})}{(w_{\text{wind}}/10 \text{ km s}^{-1})} = 1.0 \times 10^{-6} \left(\frac{L_{6\text{cm peak}}}{10^{26} \text{ ergs s}^{-1} \text{ Hz}^{-1}} \right)^{0.54} \left(\frac{t_{6\text{cm peak}} - t_0}{\text{(days)}} \right)^{0.38} \tag{18}$$

for Type II supernovae, assuming that the absorption $\tau_{6\,cm}$ peak ~ 1, from whatever origin, at the time of observed peak in the 6-cm flux density.

The coefficients in Equations 17 and 18 depend on the amount of kinetic energy that is transferred to accelerate relativistic electrons and on the details of the acceleration mechanism. Although Fermi acceleration is generally accepted as the relativistic electron acceleration process and it is usually assumed that some fixed fraction of the explosion kinetic energy is transformed into relativistic synchrotron electrons (often assumed $\sim 1\%$), the physics of these two aspects is not known in detail a priori. Therefore, Equations 17 and 18 can only be "calibrated" by using the values of well-studied RSNe and the assumption that all RSNe of the same type have similar characteristics. The constants in Equations 17 and 18 have thus been determined from averages within the two RSN subtypes (Type Ib/c and Type II) of those RSNe that have pure, uniform absorption (i.e., $K_3 = 0$), with the omission of SN1987A because of its blue supergiant (BSG) rather than RSG progenitor.

It should be kept in mind that, because the detailed mechanism of radio emission from supernovae is not well understood and the estimates have to rely on ad hoc calibrations from the few RSNe that have well-measured light curves, the mass-loss rate estimated from Equations 17 and 18 can only complement, and perhaps support, the more accurate determinations done from radio absorption.

CHANGES IN MASS-LOSS RATE   A particularly interesting case of mass-loss from an RSN is SN1993J, for which detailed radio observations are available starting only a few days after explosion (Figure 5a). Van Dyk et al. (1994) found evidence for a changing mass-loss rate (Figure 5b) for the presupernova star that was as high as $\sim 10^{-4} M_\odot$ yr^{-1} approximately 1000 years before explosion and decreased to $\sim 10^{-5} M_\odot$ yr^{-1} just before explosion, resulting in a relatively flat density profile of $\rho \propto r^{-1.5}$.

Fransson & Björgsson (1998) have suggested that the observed behavior of the f-f absorption for SN1993J could alternatively be explained in terms of a systematic decrease of the electron temperature in the circumstellar material as the supernova expands. It is not clear, however, what the physical process is that determines why such a cooling might occur efficiently in SN1993J but not in supernovae such as SN1979C and SN1980K, where no such behavior is required to explain the observed radio turn-on characteristics. Also, recent X-ray observations with ROSAT of SN1993J indicate a non-r^{-2} CSM density surrounding the supernova progenitor (Immler et al. 2001), with a density gradient of $\rho \propto r^{-1.6}$.

Moreover, changes in presupernova mass-loss rates are not unusual. Montes et al. (2000) found that Type II SN1979C had a slow increase in its radio light curve after day ~ 4300 (see Figure 3a), which implied by day 7100 an excess in flux density by a factor of ~ 1.7 with respect to the standard model, or a density enhancement of $\sim 30\%$ over the expected density at that radius. This may be understood as a change of the average CSM density profile from the r^{-2} law, which was applicable until day ~ 4300, to an appreciably flatter behavior of $\sim r^{-1.4}$ (Montes et al. 2000).

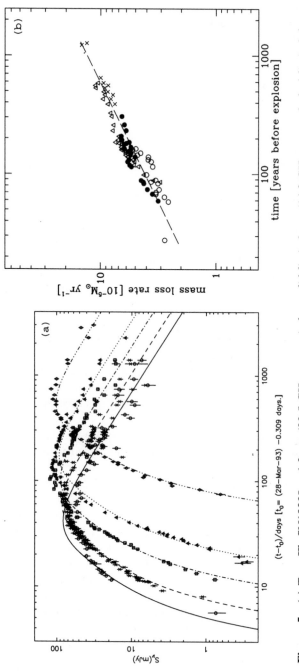

Figure 5   (*a*) Type IIb SN1993J at 1.3 cm (22.5 GHz; *open circles, solid line*), 2 cm (14.9 GHz; *stars, dashed line*), 3.6 cm (8.4 GHz; *open squares, dash–dot line*), 6 cm (4.9 GHz; *open triangles, dotted line*), and 20 cm (1.5 GHz; *open diamonds, dash–triple dot line*). (*b*) Changing mass-loss rate of the presumed red supergiant progenitor to SN1993J versus time before the explosion.

On the other hand, Type II SN1980K showed a steep decline in flux density at all wavelengths (see Figure 3b) by a factor of $\sim$2 occurring between day $\sim$3700 and day $\sim$4900 (Montes et al. 1998). Such a sharp decline in flux density implies a decrease in ρ_{CSM} by a factor of $\sim$1.6 below that expected for a r^{-2} CSM density profile. If one assumes the radio emission arises from an $\sim$10^4 km s^{-1} blastwave traveling through a CSM established by a $\sim$10 km s^{-1} preexplosion stellar wind, this implies a significant change in the stellar mass-loss rate, for a constant speed wind, at $\sim$12,000 years before explosion for both supernovae.

BINARY SYSTEMS   In the process of analyzing a full decade of radio measurements from SN1979C, Weiler et al. (1991, 1992b) found evidence for a significant, quasi-periodic variation in the amplitude of the radio emission at all wavelengths of $\sim$15% with a period of 1575 days or $\sim$4.3 years (see Figure 3a at age <4000 days). They interpreted the variation as due to a minor ($\sim$8%) density modulation, with a period of $\sim$4000 years, on the larger, relatively constant presupernova stellar mass-loss rate. Because such a long period is inconsistent with most models for stellar pulsations, they concluded that the modulation may be produced by interaction of a binary companion in an eccentric orbit with the stellar wind from the presupernova star.

This concept was strengthened by more detailed calculations for a binary model from Schwarz & Pringle (1996). Since that time, the presence of binary companions has been suggested for the progenitors of SN1987A (Podsiadlowski 1992), SN1993J (Podsiadlowski et al. 1993), and SN1994I (Nomoto et al. 1994), indicating that binaries may be common in presupernova systems.

IONIZED HYDROGEN (HII) ALONG THE LINE OF SIGHT   A reanalysis of the radio data for SN1978K from Ryder et al. (1993) clearly shows flux density evolution characteristic of normal Type II supernovae. Additionally, the data indicate the need for a time-independent, f-f absorption component. Montes et al. (1997) interpreted this constant absorption term as indicative of the presence of HII along the line of sight to SN1978K, perhaps as part of an HII region or a distant circumstellar shell associated with the supernova progenitor. SN1978K had already been noted for its lack of optical emission lines broader than a few thousand km s^{-1} since its discovery in 1990 (Ryder et al. 1993), suggesting the presence of slowly moving circumstellar material.

To determine the nature of this absorbing region, Chu et al. (1999) obtained a high-dispersion spectrum of SN1978K at the wavelength range 6530–6610 Å. The spectrum shows not only the moderately broad Hα emission of the supernova ejecta, but also narrow nebular Hα and [N II] emission. The high [N II] 6583/Hα ratio of 0.8–1.3 suggests that this radio absorbing region is a stellar ejecta nebula. The expansion velocity and emission measure of the nebula are consistent with those seen in ejecta nebulae of luminous blue variables. Previous low-dispersion spectra have detected a strong [N II] 5755 Å line, indicating an electron density of $(3–12) \times 10^5$ cm^{-3}. These data suggest that the ejecta nebula detected toward

SN1978K is probably part of a large, dense, structured circumstellar envelope of SN1978K.

RAPID PRESUPERNOVA STELLAR EVOLUTION   Supernova radio emission that preserves its spectral index while deviating from the standard model is taken to be evidence for a change of the circumstellar density behavior from the canonical r^{-2} law expected for a presupernova wind with a constant mass-loss rate, $\dot{M}$, and a constant wind velocity, w_{wind}. Because the radio luminosity of a supernova is proportional to $(\dot{M}/w_{\text{wind}})^{(\gamma-7+12m)/4}$ (Chevalier 1982a) or, equivalently, to the same power of the circumstellar density (because $\rho \propto \frac{\dot{M}}{w_{\text{wind}}\, r^2}$), a measure of the deviation from the standard model provides an indication of deviation of the circumstellar density from the r^{-2} law. Monitoring the radio light curves of RSNe also provides a rough estimate of the time scale of deviations in the presupernova stellar wind density. Because the supernova blastwave travels through the CSM roughly 1000 times faster than the stellar wind velocity that established the CSM ($v_{\text{blastwave}} \sim 10,000$ km s^{-1} versus $w_{\text{wind}} \sim 10$ km s^{-1}), one year of radio light curve monitoring samples roughly 1000 years of stellar wind mass-loss history.

In addition to the changes in the radio light curves of SN1979C and SN1980K discussed in "Changes in Mass-Loss Rate," above, the Type IIn SN1988Z can best be described by two evolution phases—an "early" phase that extends roughly from explosion through day 1479 and a "late" phase that extends roughly from day 2129 through the end of the data set. Figure 4 shows a clear steepening of the light curves sometime between these two measurement epochs, but the actual "break" date is somewhat arbitrary owing to uncertainties in the flux density measurements for this relatively faint source and the likely smoothness of any transition region. Williams et al. (2002) chose to describe the flux density evolution separately for these two time intervals with the period from day $\sim$1500 to day $\sim$2000 as a transition. With the explosion date assumed to be December 1, 1998, from optical estimates (Stathakis & Sadler 1991), applying the fitting procedures separately to the early and late periods, with the data points between day 1500 and day 2000 included in both fits to provide a smooth transition, yields a spectral index (α) and clumpy absorption parameters (K_3 and δ') that are the same in the two time intervals within the fitting errors. However, the emission decay rate parameter β steepens significantly from $\sim$$-1.3$ for the early period to $\sim$$-2.8$ for the late period.

For a purely clumpy CSM ($K_2 = 0$), the sharp steepening of β around day 1750, with constant K_3 and δ', implies (a) that the number density of clumps per unit volume (η) starts decreasing more rapidly with radius by approximately $R^{-1.5/m}$ (i.e., $\eta_{\text{after day 1750}} = \eta_{\text{before day 1750}}(\frac{R}{R_{\text{day 1750}}})^{-1.5/m}$), with the average characteristics of the individual clumps remaining unchanged, and (b) that most of the absorption occurs within the emitting clumps themselves. In other words, the spatial distribution is so sparse that the average number of clumps along the line of sight is less than one ($N < 1$). This second condition is consistent with Case 3 from Weiler et al. (2001) and discussed above in "Mass-Loss Rate from Radio Absorption." It is interesting to note that Weiler et al. (2001) also found that Case 3 applies to the radio light curves for the unusual SN1998bw/GRB980425.

The best-observed example of rapid presupernova evolution is the Type II SN1987A, whose proximity makes it easily detectable even at very low radio luminosity. The progenitor to SN1987A was in a blue supergiant (BSG) phase at the time of explosion and had ended an RSG phase some 10,000 years earlier. After an initial, very rapidly evolving radio outburst (Turtle et al. 1987), which reached a peak flux density at 6 cm, $\sim$3 orders of magnitude fainter than other known Type II RSNe (possibly owing to sensitivity-limited selection effects), the radio emission declined to a low radio brightness within a year. However, at an age of $\sim$3 years the radio emission started increasing again and continues to increase (Ball et al. 1995, 2001; Gaensler et al. 1997).

Although its extremely rapid development resulted in the early radio data at higher frequencies being very sparse, the evolution of the initial radio outburst is roughly consistent with the models described above in Equations 1–10 (i.e., a blastwave expanding into a spherically symmetric circumstellar envelope). The density implied by such modeling is appropriate to a presupernova mass-loss rate of a few $\times 10^{-6}$ $M_\odot$ yr^{-1} for a wind velocity of $w_{\text{wind}} = 1,000$ km s^{-1} (more appropriate to a BSG progenitor), a blastwave velocity of $v_{\text{blastwave}} = v_i = 35,000$ km s^{-1}, and a CSM temperature of $T = 20,000$ K.

Because the Hubble Space Telescope can actually image the denser regions of the CSM around SN1987A, we know that the current rise in radio flux density is caused by the interaction of the SN blastwave with the diffuse material at the inner edge of the well-known inner circumstellar ring (Gaensler et al. 1997). Because the density increases as the supernova blastwave interaction region moves deeper into the main body of the optical ring, the flux density is expected to continue to increase steadily at all wavelengths. Recently, increases at optical and X-ray have also been reported (Garnavich et al. 1997, Hasinger et al. 1996). Best estimates are that the blastwave-CSM interaction will reach a maximum by $\sim$2003.

DISCUSSION OF CIRCUMSTELLAR MEDIUM STRUCTURE   For at least four Type II supernovae, SN1979C, SN1980K, SN1987A, and SN1988Z, there are sharp deviations from smooth modeling of the radio flux density occurring a few years after the explosion. (The smooth change of mass-loss rate for SN1993J is discussed above.) Because the supernova blastwave is moving about 1000 times faster than the wind material of the RSG progenitor (i.e., $\sim$10,000 km s^{-1} versus $\sim$10 km s^{-1}), such a time interval implies a significant change in the presupernova stellar wind properties several thousand years before the explosion. Such an interval is short compared with the lifetimes of typical RSN progenitors (e.g., 10–30 Myrs) but is a sizeable fraction of their RSG phase ($t_{\text{RSG}} \sim$2–5 $\times 10^5$ yrs), suggesting that a significant transition occurs in the evolution of presupernova stars just before the final explosive event.

SN1987A is an unusual case in that its BSG wind was almost certainly of a higher velocity ($\sim 10^3$ km s^{-1}) than the usually assumed RSG wind velocity (~ 10 km s^{-1}). However, it also clearly underwent significant evolution in the past few thousand ($\sim 10^4$) years before explosion.

Additional evidence for altered mass-loss rates from Type II supernova progenitors over time intervals of several thousand years is provided by the detection of relatively narrow emission lines with typical widths of several 100 km s^{-1} in some of the optical spectra (e.g., SN1978K: Ryder et al. 1993, Chugai et al. 1995, Chu et al. 1999; SN1997ab: Salamanca et al. 1998; SN1996L: Benetti et al. 1999). This indicates the presence of dense circumstellar shells ejected by the supernova progenitors in addition to more diffuse, steady wind activity.

Because the radio emission is determined by the circumstellar density that is proportional to the mass-loss rate to stellar-wind velocity ratio ($\dot{M}/w$), one of these quantities, or both, is required to change by as much as a factor of two over the last few thousand years before the supernova explosion. However, $\sim 10^4$ years is considerably shorter than the H and He burning phases but much longer than any of the successive nuclear burning phases that a massive star goes through before core collapse (see, e.g., Chieffi et al. 1998), so it is unlikely that the stellar luminosity (which determines the mass-loss rate, $\dot{M} \propto L^{1-1.5}$), can vary on a time scale needed to account for the observed changes.

However, the wind velocity, w_{wind}, is roughly proportional to the square of the effective temperature ($w_{\text{wind}} \propto T_{\text{eff}}^2$) (see e.g., Panagia & Macchetto 1982), so a change of a factor of ~ 2 in w_{wind} requires a change of a factor of only ~ 1.4 in T_{eff}, (e.g., from $\sim 3,500$ K to $\sim 5,000$ K) or, correspondingly, a change from an early M to an early K supergiant spectrum. Such a transition would define a loop in the Hertzsprung-Russell (HR) diagram reminiscent of the blue loops characteristic of the evolution of moderately massive stars (see, e.g., Brocato & Castellani 1993, Langer & Maeder 1995), but classical blue loops are much slower and more extreme processes, occurring several $\times 10^5$ years before the terminal stages of an RSG and involving temperature excursions from $\sim 3,500$ K to $> 10,000$ K.

The smaller temperature changes inferred from the radio data require a star to change only from a very red to a moderately red spectrum, and back, corresponding to a transition in the HR diagram, which is more appropriately dubbed a "pink loop". A possibility for explaining the implied CSM density changes derives from a recent study by Panagia & Bono (2001), who found that the pulsational instability of stars in the mass range 10–20 M$_\odot$ appears, in some cases, to be of suitable period and magnitude to account for the observed radio light curve changes.

Another mechanism that may in some cases account for sudden changes of CSM density is the presence of a companion in a wide binary system. If the companion star has a sufficiently strong wind at the time of the primary star explosion, then the wind pressure could partially confine and compress the supernova progenitor wind, increasing the average density at distances comparable to or larger than the binary system separation (Boffi & Panagia 1996). This mechanism could explain the case of SN1979C, whose recent slow increase of radio flux density occurred roughly when the supernova blastwave is believed to have traveled a distance comparable to the binary system separation estimated from the sinusoidal modulation of the radio light curves. However, the sudden declines in the radio emission observed for SN1980K and SN1988Z cannot be explained by this mechanism.

DECELERATION OF BLASTWAVE EXPANSION   Radio studies also offer the only possibility for measuring the deceleration of the blastwave from the supernova explosion. So far, this has been directly possible for two objects, SN1987A and SN1993J, although in some cases the deceleration can be estimated from model fitting and Equation 6 or 7. Manchester et al. (2002) have shown that the blastwave from the explosion of SN1987A traveled through the tenuous medium of the bubble created by the high-speed wind of its BSG progenitor at an average speed of $\sim$10% of the speed of light ($\sim$35,000 km s^{-1}) but has decelerated dramatically to only $\sim$3,000 km s^{-1} since it has reached the inner edge of the prominent optical ring (Figure 6a).

Marcaide et al. (1997), through the use of VLBI techniques, have followed the expansion of the radio shell of SN1993J in detail and have found that it is starting to decelerate. While the expansion speed of SN1993J is quite high at $\sim$15,000 km s^{-1} (Marcaide et al. 1997), the deceleration is much more gradual (Figure 6b) than that of SN1987A.

PEAK RADIO LUMINOSITIES AND DISTANCES   Weiler et al. (1998) found from their long-term monitoring of the radio emission from supernovae that the radio light curves evolve in a systematic fashion with a distinct peak flux density (and thus, in combination with a distance, a peak spectral luminosity) at each frequency and a well-defined time from explosion to that peak. Studying these two quantities at 6-cm wavelength, peak spectral luminosity ($L_{6\,cm\,peak}$) and time after explosion date (t_0) to reach that peak ($t_{6\,cm\,peak} - t_0$), they found that the two appear related. In particular, based on four objects, Weiler et al. (1998) suggested that Type Ib/c supernovae may be approximate radio standard candles with a peak 6-cm spectral luminosity of

$$L_{6\,cm\,peak} \approx 1.3 \times 10^{27} \text{ erg s}^{-1} \text{ Hz}^{-1}, \tag{19}$$

with no error estimated because of the small size and poor quality of their data set and, based on 12 objects, Type II supernovae appear to obey a relation

$$L_{6\,cm\,peak} \simeq 5.5^{+8.7}_{-3.4} \times 10^{23} (t_{6\,cm\,peak} - t_0)^{1.4\pm0.2} \text{ erg s}^{-1} \text{ Hz}^{-1}, \tag{20}$$

with time measured in days.

If these relations are supported by further observations, they may provide a means for determining distances to supernovae, and thus to their host galaxies, from purely radio continuum observations.

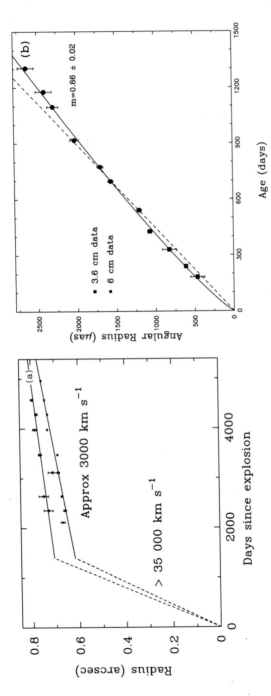

Figure 6 The deceleration of the radio generating blastwaves from SN1987A (*a*) and SN1993J. (*b*) The average expansion speed of SN1987A is ~0.1c until the blastwave reaches the inner edge of the optical ring when it slows abruptly to only ~3,000 km s⁻¹. The expansion speed of SN1993J is slower, but still quite high at ~15,000 km s⁻¹, and the deceleration is much more gradual.

SN1998bw/GRB980425

The suggestion of an association of the Type Ib/c SN1998bw with the gamma-ray burst GRB980425 may provide evidence for another new phenomenon generated by supernovae: At least some types of gamma-ray bursters (GRBs) may originate in some types of supernova explosions. Because SN1998bw/GRB980425 is by far the nearest and best studied of the gamma-ray bursters, it is worthwhile to examine its radio emission in detail before proceeding to the discussion of the radio emission from other GRBs.

Background

Although generally accepted that most GRBs are extremely distant and energetic (see, e.g., Paczynski 1986, Goodman 1986), the discovery of GRB980425 (Soffitta et al. 1998) in April 1998 25.90915 and its possible association with a bright supernova [SN1998bw at RA(J2000) $= 19^h 35^m 03^s\!.31$, Dec(J2000) $= -52°50'44''\!.7$ (Tinney et al. 1998)] in the relatively nearby spiral galaxy ESO 184-G82 at $z = 0.0085$ (distance $\sim$40 Mpc for $H_0 = 65$ km s^{-1} Mpc^{-1}) (Galama et al. 1998a,b,c, 1999; Lidman et al. 1998; Tinney et al. 1998; Sadler et al. 1998; Woosley et al. 1999), has introduced the possibility of a supernova origin for at least some types of GRBs. The estimated explosion date of SN1998bw in 1998 between April 21–27 (Sadler et al. 1998) corresponds rather well with the time of the GRB980425 outburst. Iwamoto et al. (1998) felt that they could restrict the core collapse date for SN1998bw even more from hydrodynamical modeling of exploding C + O stars and, assuming that the SN1998bw optical light curve is energized by ^{56}Ni decay as in Type Ia supernovae, they placed the coincidence between the core collapse of SN1998bw to within $+0.7/-2$ days of the outburst detection of GRB980425.

Classified initially as a supernova optical Type Ib (Sadler et al. 1998), then Type Ic (Patat & Piemonte 1998), then peculiar Type Ic (Kay et al. 1998, Filippenko 1998), then later, at an age of 300–400 days, again as a Type Ib (Patat et al. 1999), SN1998bw presents a number of optical spectral peculiarities that strengthen the suspicion that it may be the counterpart of the gamma-ray burst.

When the more precise BeppoSAX Narrow Field Instruments (NFI) were pointed at the BeppoSAX error box 10 h after the detection of GRB980425, two X-ray sources were present (Pian et al. 1999). One, named S1 by Pian et al. (1999), was coincident with the position of SN1998bw and declined slowly between April 1998 and November 1998. The second X-ray source, S2, which was $\sim$4' from the position of SN1998bw, was not (or at best only marginally with a $<3\sigma$ possible detection 6 days after the initial detection) detectable again in follow-up observations in April, May, and November 1998 (Pian et al. 1999, 2000).

However, although concern remains that the Pian et al. (1999, 2000) X-ray source, S2, might have been the brief afterglow from GRB980425 rather than the Pian source S1 associated with SN1998bw, Pian et al. (2000) concluded that S1 has a "high probability" of being associated with GRB980425 and that S2 is more likely a variable field source.

Radio Emission

The radio emission from SN1998bw reached an unusually high 6-cm spectral luminosity at a peak of $\sim 6.7 \times 10^{28}$ erg s^{-1} Hz^{-1}, i.e., ~ 3 times higher than either of the well-studied, very radio luminous Type IIn supernovae SN1986J and SN1988Z, and ~ 40 times higher than the average peak of 6-cm spectral luminosity of Type Ib/c supernovae. It also reached this 6-cm peak rather quickly, only ~ 13 days after explosion.

SN1998bw is unusual in its radio emission, but not extreme. For example, the time from explosion to peak 6-cm luminosity for both SN1987A and SN1983N was shorter, and in spite of the fact that SN1998bw has been called "the most luminous radio supernova ever observed," its 6-cm spectral luminosity at peak is exceeded by that of SN1982aa (Yin 1994). However, SN1998bw is certainly the most radio luminous Type Ib/c RSN observed so far by a factor of ~ 25, and it reached this higher radio luminosity very early.

Expansion Velocity

Although unique in neither the speed of radio light curve evolution nor in peak 6-cm radio luminosity, SN1998bw is certainly unusual in the combination of these two factors—very radio luminous very soon after explosion. Kulkarni et al. (1998) have used these observed qualities, together with the lack of interstellar scintillation (ISS) at early times, brightness temperature estimates, and physical arguments to conclude that the blastwave from SN1998bw that gives rise to the radio emission must have been expanding relativistically. On the other hand, Waxman & Loeb (1999) argued that a subrelativistic blastwave can generate the observed radio emission. However, both sets of authors agree that a very high expansion velocity ($\gtrsim 0.3c$) is required for the radio-emitting region under a spherical geometry.

Simple arguments confirm this high velocity. To avoid the well-known Compton Catastrophe, Kellermann & Pauliny-Toth (1969) have shown that the brightness temperature $T_B < 10^{12}$ K must hold, and Readhead (1994) has better defined this limit to $T_B < 10^{11.5}$ K. From geometrical arguments, such a limit requires the radio-sphere of SN1998bw to have expanded at an apparent speed of $\gtrsim 230,000$ km s^{-1}, at least during the first few days after explosion. Although such a value is only mildly relativistic ($\Gamma \sim 1.6$; $\Gamma = (1 - \frac{v^2}{c^2})^{-\frac{1}{2}}$), it is unusually high. However, measurements by Gaensler et al. (1997) and Manchester et al. (2002) have demonstrated that the radio-emitting regions of the Type II SN1987A expanded at an average velocity of $\sim 35,000$ km s^{-1} ($\sim 0.1c$) over the 3 years from February 1987 to mid-1990 so that, in a very low-density environment such as one finds around Type Ib/c supernovae, very high blastwave velocities appear to be possible.

Radio Light Curves

An obvious comparison of SN1998bw with other RSNe is the evolution of its radio flux density at multiple frequencies and its description by known RSN models. The radio data available at http://www.narrabri.atnf.csiro.au/~mwiering/grb/grb980425/ are plotted in Figure 7. SN1998bw shows an early peak that reaches a maximum as early as day 10–12 at 8.64 GHz, a minimum almost simultaneously for the higher frequencies ($v \geq 2.5$ GHz) at day ~20–24, then a secondary, somewhat lower peak after the first dip. An interesting characteristic of this "double humped" structure is that it dies out at lower frequencies and is relatively inconspicuous in the 1.38 GHz radio measurements (see Figure 7).

Such a double humped structure of the radio light curves can be reproduced by a single energy blastwave's encountering differing CSM density regimes as it travels rapidly outward. This is a reasonable assumption because complex density

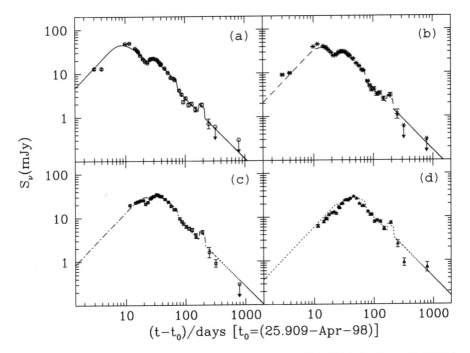

Figure 7   The radio light curves of SN1998bw at (*a*) 3.5 cm (8.6 GHz), (*b*) 6.3 cm (4.8 GHz), (*c*) 12 cm (2.5 GHz) and (*d*) 21 cm (1.4 GHz). The curves are derived from a best fit model described by Equations 1–10 and the parameters and assumptions listed in Table 4. During the 50-day intervals from day 25–75 and from day 165–215, the emission and absorption terms (K_1 and K_3) increase by factors of 1.6 and 2.0, respectively, corresponding to a density increase of 40% with a 6 day boxcar-smoothed turn-on and turn-off of the enhanced emission/absorption.

structure in the CSM surrounding supernovae, giving rise to structure in the radio light curves, is very well known in such objects as SN1979C (Weiler et al. 1991, 1992b; Montes et al. 2000), SN1980K (Weiler et al. 1992b, Montes et al. 1998) and, particularly, SN1987A (Jakobsen et al. 1991).

Weiler et al. (2001) pointed out what has not been previously recognized, that there is a sharp drop in the radio emission near day ∼75 and a single measurement epoch at day 192 that is significantly (∼60%) higher at all frequencies than expected from the preceeding data on day 149 and the following data on day 249.

Weiler et al. were able to explain both of these temporary increases in radio emission by the supernova blastwave's encountering physically similar shells of enhanced density. The first enhancement or "bump" after the initial outburst peak is estimated to start on day 25 and end on day 75, i.e., having a duration of ∼50 days and turn-on and turn-off times of about 12 days, where the radio emission (K_1) increased by a factor of 1.6 and absorption (K_3) increased by a factor of 2.0 implying a density enhancement of ∼40% for no change in clump size. Exactly the same density enhancement factor and length of enhancement is compatible with the bump observed in the radio emission at day 192 (i.e., the single measurement within the 100-day gap between measurements on day 149 and day 249), even though the logarithmic time scale of Figure 7 makes the time interval look much shorter. The decreased sampling interval has only one set of measurements altered by the proposed day 192 enhancement, so Weiler et al. (2001) could not determine its length more precisely than <100 days.

Li & Chevalier (1999) proposed an initially synchrotron self-absorbed, rapidly expanding blastwave in a $\rho \propto r^{-2}$ circumstellar wind model to describe the radio light curve for SN1998bw. This is in many ways similar to the Chevalier (1998) model for Type Ib/c supernovae, which also included synchrotron self-absorption (SSA). To produce the first bump in the radio light curves of SN1998bw, Li & Chevalier (1999) postulated a boost of blastwave energy by a factor of ∼2.8 on day ∼22 in the observer's time frame. They did not discuss the second bump.

Modeling of the radio data for SN1998bw with the well-established formalism for RSNe presented above shows that such an energy boost is not needed. A fast blastwave interacting with a dense, slow, stellar wind–established, ionized CSM, which is modulated in density over time scales similar to those seen for RSNe, can produce a superior fit to the data. No blastwave reacceleration is required. The parameters of the best-fit model are given in Table 4 and shown as the curves in Figure 7. A visual comparison of the curves in Figure 7 with those of Li & Chevalier (1999) Figure 9, shows that the purely thermal absorption model with structured CSM provides a superior fit.

One should note that the fit listed in Table 4 and shown as the curves in Figure 7 requires no "uniform" absorption $(K_2 = 0)$, so all of the free-free (f-f) absorption is due to a clumpy medium as described in Equations 1 and 5. These results, combined with the estimate of a high blastwave velocity, suggest that the CSM around SN1998bw is highly structured with little, if any, interclump gas. The clump filling factor must be high enough to intercept a considerable fraction of the

TABLE 4   SN1998bw/GRB980425 modeling results

Parameter	Value
α	-0.71
β	-1.38
K_1^a	2.4×10^3
K_2	0
δ	—
K_3^a	1.7×10^3
δ'	-2.80
K_4	1.2×10^{-2}
t_0 (Explosion Date)	1998 Apr. 25.90915
$(t_{6\,\text{cm peak}} - t_0)$ (days)	13.3
$S_{6\,\text{cm peak}}$ (mJy)	37.4
d(Mpc)	38.9
$L_{6\,\text{cm peak}}$ (ergs s^{-1} Hz^{-1})	6.7×10^{28}
$\dot{M}$ (M$_\odot$ yr^{-1})^b	2.6×10^{-5}

^{a}Enhanced by a factors of 1.6 (K_1) and 2.0 (K_3), corresponding to a density increase of 40%, over the intervals day 25–75 and day 165–215, although the latter interval could be as long as 100 days and still be compatible with the available data.

^{b}Assuming $t_i = 23$ days, $t = (t_{6\,\text{cm peak}} - t_0) = 13.3$ days, $m = -(\alpha - \beta - 3)/3 = 0.78$, $w_{\text{wind}} = 10$ km s^{-1}, $v_i = v_{\text{blastwave}} = 230{,}000$ km s^{-1}, $T = 20{,}000$ K, average number of clumps along the line-of-sight $N = 0.5$ and volume filling factor $\phi = 0.22$.

blastwave energy and low enough to let radiation escape from any given clump without being appreciably absorbed by any other clump, which is the Case 3 discussed in "Mass-Loss Rate from Radio Absorption," above. The blastwave can then easily move at a speed that is a significant fraction of the speed of light, because it is moving in a very low-density medium, but still cause strong energy dissipation and relativistic electron acceleration at the clump surfaces facing the supernova explosion center.

Weiler et al. (2001) also noted from the fit given in Table 4 that the presence of a K_4 factor implies there is a more distant, uniform screen of ionized gas surrounding the exploding system, which is too far to be affected by the rapidly expanding blastwave and provides a time-independent absorption.

Physical Parameter Estimates

Using the fitting parameters from Table 4 and Equations 11 and 16, Weiler et al. (2001) estimated a mass-loss rate from the preexplosion star. The proper parameter assumptions are rather uncertain for these enigmatic objects, but for a preliminary estimate, they assumed $t_i = 23$ days, $t = (t_{6\,cm\,peak} - t_0) = 13.3$ days, $m = -(\alpha - \beta - 3)/3 = 0.78$ (Equation 7), $w_{wind} = 10$ km s^{-1} (for an assumed RSG progenitor), $v_i = v_{blastwave} = 230{,}000$ km s^{-1}, and $T = 20{,}000$ K. They also assumed, because the radio emission implies that the CSM is highly clumped (i.e., $K_2 = 0$), that the CSM volume is only sparsely occupied ($N = 0.5$, $\phi = 0.22$, see "Mass-Loss Rate from Radio Absorption," Case 3, above). Within these rather uncertain assumptions, Equations 11 and 16 yield an estimated mass-loss rate of $\dot{M} \sim 2.6 \times 10^{-5}$ M$_\odot$ yr^{-1} with density enhancements of $\sim 40\%$ during the two known, extended bump periods.

Assuming that the blastwave is traveling at a constant speed of $\sim 230{,}000$ km s^{-1}, the first bump initializing on day 25 and terminating on day 75 implies that the density enhancement extends from $\sim 5.0 \times 10^{16}$ cm to $\sim 1.5 \times 10^{17}$ cm from the star. Correspondingly, if it was established by a 10 km s^{-1} RSG wind, the 50 days of enhanced mass-loss ended $\sim 1{,}600$ years and started $\sim 4{,}700$ years before the explosion. The earlier high mass-loss rate epoch indicated by the enhanced emission on day 192 in the measurement gap between day 149 and day 249 implies, with the same assumptions, that the enhanced mass-loss rate occurred in the interval between $\sim 9{,}400$ years and $\sim 15{,}700$ years before explosion. It is interesting to note that the time between the presumed centers of the first and second increased mass-loss episodes of $\sim 9{,}400$ years is comparable to the $\sim 12{,}000$ years before explosion at which SN1979C had a significant mass-loss rate increase (Montes et al. 2000) and SN1980K had a significant mass-loss rate decrease (Montes et al. 1998), thus establishing a possible characteristic time scale of $\sim 10^4$ years for significant changes in mass-loss rate for preexplosion massive stars.

Radio Emission from SN1998bw/GRB980425 and Other Gamma-Ray Bursters

Since the suggestion of a possible relation between SN1998bw and GRB980425, it has remained a tantalizing possibility that the origin of at least some gamma-ray bursters (GRBs) is in the better-known Type Ib/c supernova phenomenon. First, of course, one must keep in mind that there may be (and probably is) more than one origin for GRBs, a situation that is true for most other classes of objects. For example, supernovae, after having been identified as a new phenomenon in the early part of the last century, were quickly split into several subgroups such as Zwicky's Types I–V, then coalesced back into just two subgroupings based on Hα absent (Type I) or Hα present (Type II) in their optical spectra (see, e.g., Minkowski 1964). This simplification has not withstood the test of time, however, and subgroupings of Type Ia, Ib, Ic, II, IIb, IIn, and others have come into use over the past 20 years.

GRBs, although at a much earlier stage of understanding, have similarly started to split into subgroupings. The two currently accepted groupings are referred to as "fast-hard" and "slow-soft," from the tendency of the gamma-ray emission for some to evolve more rapidly (mean duration ~ 0.2 s) and to have a somewhat harder spectrum than for others that evolve more slowly (mean duration $\sim$20 s) with a somewhat softer spectrum (Fishman & Meegan 1995).

Because we are only concerned with the radio afterglows of GRBs here, all of our examples fall into the slow-soft classification, at least partly because the fast-hard GRBs fade too quickly for follow-up observations to obtain the precise positional information needed for identification at longer wavelengths. It is therefore uncertain whether fast-hard GRBs have radio afterglows or even whether the slow-soft GRBs represent a single phenomenon. If, however, we assume that at least some types of slow-soft GRBs have a similar origin and that GRB980425/SN1998bw is a key to this puzzle, telling us that slow-soft GRBs have their origin in at least some types of supernovae, we can investigate relations between the two observational phenomena.

GAMMA-RAY BURSTERS

Afterglows

Gamma-ray bursters (GRBs) produce mysterious flashes of high-energy radiation that appear from random directions in space and typically last a few seconds. They were first discovered by U.S. Air Force Vela satellites in the 1960s, and since then numerous theories of their origin have been proposed. NASA's Compton Gamma-Ray Observatory satellite detected several thousand bursts, with an occurrence rate of approximately one per day. The uniform distribution of the bursts on the sky led theoreticians to initially suggest that their sources were either very near, and thus uniformly distributed around the solar system, in an unexpectedly large halo around the Galaxy, or at cosmological distances—not very restrictive proposals.

Only after the launch of the Italian/Dutch satellite BeppoSAX in 1996 was it possible to couple a quick-response pointing system with relatively high precision position-sensitive detectors for gamma-rays and hard X-rays. This quick response, coupled with high-accuracy position information, finally permitted rapid and accurate follow-up observations at other wavelengths with ground- and space-based telescopes and led to the discovery of long-lived afterglows of the bursts in soft X-rays, visible and infrared light, and radio waves. Although the gamma-ray bursts generally last only seconds, their afterglows have in a few cases been studied for minutes, hours, days, or even weeks after discovery. These longer wavelength observations have allowed observers to probe the immediate environment of GRBs and to assemble clues as to their nature.

The first GRB-related optical transient was identified for GRB970228 by Groot et al. (1997) with follow-up by the Hubble Space Telescope (HST) (Sahu et al. 1997). It showed that the GRB was associated with a faint (thus probably distant) late-type galaxy. A few months later, Fruchter & Bergeron (1997) (see also Pian et al. 1998) imaged the afterglow of GRB970508 with the HST-WFPC2, finding this source to be associated with a late-type galaxy at a redshift of $z = 0.835$ and finally demonstrating conclusively that GRBs are at cosmological distances.

GRB970508 was also the first GRB to be detected in its radio afterglow (Frail & Kulkarni 1997).

More than a dozen GRBs have now been associated with afterglows in one or more longer wavelength bands. As with the RSNe, we concentrate on the GRBs that have been detected in their radio afterglow. Fortunately, radio observations are a particularly useful technique for studying GRB afterglows. Frail et al. (2000a) pointed out that radio observations are particularly useful because (*a*) radio observations are relatively immune to the geometry of the relativistic fireball (presently the preferred model for the GRB phenomenon); (*b*) the radio afterglow is much slower to develop than at optical or X-ray wavelengths, which permits, within the logistics of discovering, pinpointing, and following up on GRB reports, observation of the critical early phases of the source evolution; and (*c*) interstellar scintillation (ISS), which is only observable at radio wavelengths provides a possibility for placing observational size limits on the emitting region.

Radio Detections

Beyond GRB980425, there are relatively few GRBs with detected radio afterglows, and only six have sufficient radio light curve information to permit approximate model fits. Additionally, because the optical/infrared (OIR) data appear consistent with a synchrotron origin similar to that of the radio emission, we have collected the available OIR data and included it in our model fitting to better constrain the source parameters. Although detailed modeling is beyond the scope of this review (the general characteristics of available models are summarized below in "Theoretical Models"), we apply the parameterization of Equations 1 to 10 to the available radio and OIR data in an attempt to highlight some of the gross properties of the GRB afterglow processes. Because the OIR data suffer an extinction that is absent in the radio, a zero redshift color excess [E(B−V)] was also obtained from the fitting by adopting the galactic extinction law of Savage & Mathis (1979).

GRB970508 was discovered by the BeppoSAX team on May 8.904, 1997 Universal Time (UT) (Costa et al. 1997). These results showed detection of an afterglow in all wavelength bands including X-ray (Piro et al. 1997), optical (Bond 1997), and radio (Frail & Kulkarni 1997). Frail & Kulkarni (1997) derived a position from their 8.46 GHz VLA observations of RA(J2000) = $06^h53^m49^s.45$, Dec(J2000) = $+79°16'19''.5$ with an error of $\sim0''.1$ in both coordinates. Metzger et al. (1997) found a redshift of z = 0.835, which Bloom et al. (1998) confirmed for the host galaxy.

The radio data were obtained from Bremer et al. (1998), Shepherd et al. (1998), Galama et al. (1998c), Smith et al. (1999), and Frail et al. (2000a). The OIR data were obtained from Castro-Tirado et al. (1998), Sahu et al. (1997), Djorgovski et al. (1997), Galama et al. (1998c,d), Sokolov et al. (1998), Schaefer et al. (1997), Chary et al. (1998), and Garcia et al. (1998). Representative data for GRB970508 are plotted, along with curves from the best fit model, in Figure 8 and the parameters of the fit are listed in Table 5.

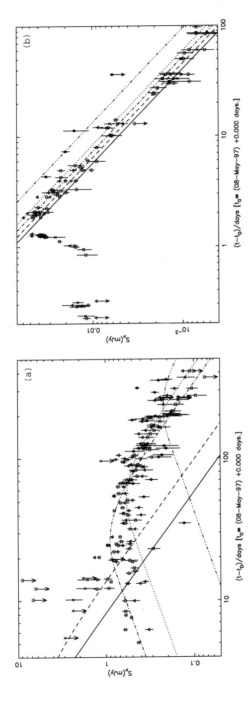

Figure 8 GRB970508 at (*a*) radio wavelengths of 1.3 mm (232 GHz; *open circles, solid line*), 3.5 mm (86.7 GHz; *crosses, dashed line*), 3.5 cm (8.5 GHz; *open squares, dash-dot line*), 6 cm (4.9 GHz; *open triangles, dotted line*), and 20 cm (1.5 GHz; *filled diamonds, dash-triple dot line*), (*b*) optical/IR bands of B (681 THz; *circles, solid line*), V (545 THz; *crosses, dashed line*), R (428 THz; *open squares, dash-dot line*), I (333 THz; *open triangles, dotted line*), and K (138 THz; *filled diamonds, dash-triple dot line*). (To enhance clarity, the scales are different between the two figures even though the fitting parameters are the same and not all available bands are plotted even though they were used in the fitting.)

TABLE 5 Fitting parameters for GRB radio afterglows[a]

GRB	$E(B-V)$ Milky Way[b] (mag)	$E(B-V)$ Host galaxy[c] (mag)	Redshift[d] (z)	α	β	K_1^e	K_3^e	δ'	Peak 6 cm radio luminosity[e] (erg s^{-1} Hz^{-1})	Explosion date (observed)
GRB970508	0.08	0.00	0.835	−0.63	−1.18	1.35×10^2	1.81×10^3	−1.75	1.39×10^{31}	1997 May 8.904
GRB980329	0.07	0.28	≡1.000	−1.33	−1.09	1.54×10^4	1.33×10^5	−1.16	7.69×10^{30}	1998 Mar 29.1159
GRB980425[f]	—	—	0.0085	−0.71	−1.38	2.37×10^3	1.73×10^3	−2.80	6.70×10^{28}	1998 April 25.9092
GRB980519	0.27	0.00	≡1.000	−0.75	−2.08	8.45×10^1	1.37×10^4	−3.57	7.52×10^{30}	1998 May 19.51410
GRB991208	0.02	0.07	0.707	−0.58	−2.27	5.11×10^2	3.53×10^3	−3.29	2.07×10^{31}	1999 Dec 08.1923
GRB991216	0.63	0.11	1.020	−0.28	−1.38	7.07×10^0	2.50×10^1	−1.40	1.05×10^{31}	1999 Dec 16.6715
GRB000301C[g]	0.05	0.05	2.034	−0.60	≡−1.75	2.31×10^2	2.72×10^3	−2.06	3.77×10^{31}	2000 March 1.4108

[a]The fits do not generally require use of K_2, δ, K_4, K_5, δ'', K_6, δ''' so that these parameters are not included in Table 5.

[b]Galactic extinction in the direction of the gamma-ray burster (GRB) was obtained from Schlegel et al. (1998), Reichart et al. (1999), and Halpern et al. (2000).

[c]Additional extinction, at the host galaxy redshift (see "Radio Detections"), is needed in some cases to provide a good fit to the optical/infrared (OIR) data.

[d]Where unknown, the redshift is defined to be z = 1.000.

[e]Derived for 6 cm in the rest frame of the observer, not the emitter.

[f]The best fit includes a $K_4 = 1.24 \times 10^{-2}$ term.

[g]The OIR data and radio data appear to have different rates of decline ($\beta_{OIR} \sim -2.0$, $\beta_{radio} \sim -1.5$) implying that there may be a break between the two regimes. For the purposes of this review the average of $\beta \equiv -1.75$ has been adopted.

Examination of Figure 8 shows that the parameterization listed in Table 5 describes the data well in spite of the very large frequency and time range. In Figure 8*a* the 232-GHz upper limits and the 86.7-GHz limits and detections are in rough agreement with the model fitting; the 8.5-GHz and 4.9-GHz measurements are described very well, and even the 1.5-GHz data are consistent with the parameterization if significant ISS is present (see "Interstellar Scintillation," below). Waxman et al. (1998) have already ascribed the large fluctuations in the flux density at both 8.5 and 4.9 GHz to ISS, and one expects such ISS to also be present at 1.4 GHz.

In Figure 8*b*, although the OIR data show significant scatter at individual frequencies, the data are consistent with a nonthermal, synchrotron origin that has the same decline rate (β) and spectral index (α) as in the radio regime, indicating that no spectral breaks have occurred between the two observing ranges. The color excess of $E(B - V) = 0.08$ mag from Schlegel et al. (1998) is consistent with the best fit to the data, implying little extinction in the host galaxy.

The most obvious characteristic of the OIR data is that in the initial interval between the time of the GRB and day 1.75 the data are not well fitted by the parameterization. This is not surprising because the modeling contains no parameters to describe any turn-on effects for the synchrotron emission. Thus, the initial OIR data prior to day 1.75 are particularly suitable for constraining explosion models such as those described in "Theoretical Models," below.

GRB980329 was discovered by the BeppoSAX team (BATSE Trigger # 6665) (Briggs et al. 1998) on March 29.1559 UT, 1998 (Frontera et al. 1998). The afterglow was detected in all wavelength bands including X-ray (Zand et al. 1998), optical (Klose et al. 1998), and radio (Taylor et al. 1998a). Taylor et al. (1998a) derived a position from their VLA observations of RA(J2000) = $07^h 02^m 38^s.022$, Dec(J2000) = $+38°50'44''.02$ with an uncertainty of $\pm 0''.05$ in each coordinate. Unfortunately, no redshift has been obtained for the optical afterglow of GRB980329 or its inferred host galaxy so that, except for arguments that are quite distant with, perhaps, $z \sim 5$, no reliable distance estimate is available.

The radio data were obtained from Taylor et al. (1998b), Smith & Tilanus (1998), and Smith et al. (1999). The OIR data were obtained from Gorosabel et al. (1999) and Reichart et al. (1999). Representative data for GRB980329 are plotted, along with curves from the best fit model, in Figure 9 and the parameters of the fit are listed in Table 5.

Examination of Figure 9 shows that the parameterization listed in Table 5 describes the data rather well over the broad parameter space in time and frequency. In Figure 9*a* the 352-GHz detections and upper limits are in good agreement with the parameterization; the 8.5-GHz and 4.9-GHz measurements are described very well, although there may be some ISS present ("Interstellar Scintillation," below). The 1.4-GHz upper limits are consistent with the parameterization.

In Figure 9*b* the data are mostly upper limits, with only a few detections at K-band (136 THz) and R-band (428 THz). However, these are surprisingly well described by the parameterization if a zero redshift color excess of $E(B - V) = 0.75$ mag is assumed. This value for color excess is much higher than the value of

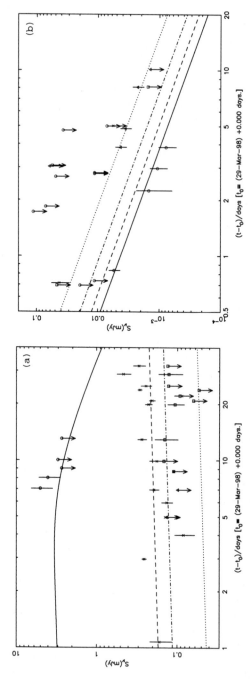

Figure 9 GRB980329 at (*a*) radio wavelengths of 0.9 mm (352 GHz; *open circles, solid line*), 3.5 cm (8.5 GHz; *crosses, dashed line*), 6 cm (4.9 GHz; *open squares, dash-dot line*), and 20 cm (1.4 GHz; *open triangles, dotted line*), (*b*) optical/IR bands of R (428 THz; *open circles, solid line*), I (333 THz; *crosses, dashed line*), J (240 THz; *open squares, dash-dot line*), and K (136 THz; *open triangles, dotted line*). (To enhance clarity, the scales are different between the two figures even though the fitting parameters are the same and not all available bands are plotted even though they were used in the fitting.)

$E(B - V) = 0.073$ from Reichart et al. (1999) for Galactic color excess in that direction. However, if one assumes that the additional color excess arises in the host galaxy and that the color excess scales with redshift as $(1 + z)^{-1.25}$, as is appropriate for an adopted Small Magellanic Cloud extinction law (Prevot et al. 1984), then the host galaxy contributes $E(B - V) = 0.28$ mag (Table 5). Such a color excess is fairly typical for the disk of a late-type galaxy.

Taylor et al. (1998b) have also modeled the radio data and invoked a somewhat steeper, inverted spectrum with $\alpha = -1.7$ between 4.9 and 8.3 GHz, flattening to $\alpha = -0.8$ between 15 and 90 GHz caused by an SSA component with a turnover frequency near 13 GHz. Extrapolating to higher frequencies, their model predicts a rather low 350-GHz flux density of only ~ 1.7 mJy, which is incompatible with the James Clerk Maxwell Telescope (JCMT) measurements.

Such complexity is not needed, however. Our parameterization listed in Table 5 and shown in Figure 9a yields a good description of the data and predicts a 350-GHz flux density of ~ 3.0 mJy, in much better agreement with the observations.

Our model fit to the available first 30 days of radio data indicates that GRB980329 should be detectable with the VLA at centimeter wavelengths, with little decline, for an extended period. Although prediction of exact flux densities at later times is not reliable because the decline phase of the model is not well constrained by the few available data, it is interesting to note that GRB980329 has apparently been detected by the VLA at 8.46 GHz, 4.86 GHz, and 1.43 GHz for up to 500 days after the outburst (Young et al. 1999). Unfortunately, our request for the later data was refused.

GRB980519 was discovered by the BeppoSAX team (BATSE Trigger # 6764) on May 19.51410 UT, 1998 (Muller et al. 1998). The afterglow was detected in all wavelength bands including X-ray (Nicastro et al. 1998), optical (Jaunsen et al. 1998), and radio (Frail et al. 1998). Frail et al. (1998) derived a position from their VLA observations of RA(J2000) $= 23^h 22^m 21^s.49$, Dec(J2000) $= +77° 15' 43''.2$ with an uncertainty of $\pm 0''.1$ in each coordinate. Unfortunately, no redshift has been obtained for the optical afterglow, so no distance estimate is available.

The radio data were obtained from Frail et al. (2000b) and the OIR data from Halpern et al. (1999), Vrba et al. (2000), and Jaunsen et al. (2001). Representative data for GRB980519 are plotted, along with curves from the best fit model, in Figure 10 and the parameters of the fit are listed in Table 5.

Examination of Figure 10 shows that the parameterization listed in Table 5 describes the data reasonably well, even though the radio data shown in Figure 10a are very limited and of relatively poor quality. For example, the reported detections at 4.9-GHz after day 50 are barely more than 3σ and thus of limited reliability, so the fit is not well constrained and the significance of their deviation from the best fit curve is unknown. The data at both 8.5-GHz and 4.9-GHz have significant fluctuations yielding detections and 3σ upper limits at relatively small time separations, so GRB980519 may be undergoing ISS (see "Interstellar Scintillation," below). Only upper limits are available at 1.4 GHz, but they are consistent with the best fit model.

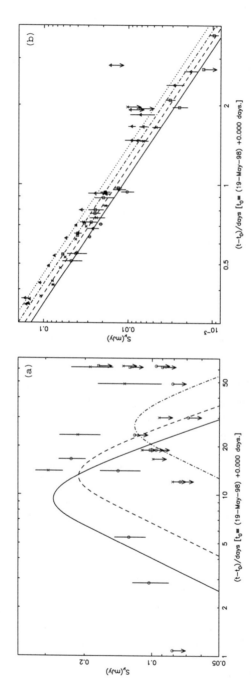

Figure 10   GRB980519 at (*a*) radio wavelengths of 3.5 cm (8.5 GHz; *open circles, solid line*), 6 cm (4.9 GHz; *crosses, dashed line*), and 20 cm (1.4 GHz; *open squares, dash-dot line*), (*b*) optical/IR bands of B (681 THz; *open circles, solid line*), V (545 THz; *crosses, dashed line*), R (428 THz; *open squares, dash-dot line*), and I (333 THz; *open triangles, dotted line*). (To enhance clarity, the scales are different between the two figures even though the fitting parameters are the same and not all available bands are plotted even though they were used in the fitting.)

In Figure 10b there is good coverage in both frequency and time. Although there is some fluctuation, the data are rather well described by the parameterization if a color excess of $E(B - V) = 0.25$ mag is assumed. This is rather close to the value of $E(B - V) = 0.267$ mag suggested by Schlegel et al. (1998) in that direction from their Infrared Astronomical Satellite (IRAS) 100 μm maps implies that there is little extinction in the host galaxy.

GRB991208 was discovered by the Ulysses, KONUS, and NEAR teams on Dec. 08.1923 UT, 1999 (Hurley & Cline 1999). The afterglow was detected in the optical (Castro-Triado 1999) and radio (Frail & Kulkarni 1999) wavelength bands. There does not appear to have been an X-ray detection. Frail & Kulkarni (1999) derived a position for the radio transient of RA(J2000) = $16^h 33^m 53\overset{s}{.}50$, Dec(J2000) = $+46°27'20\overset{''}{.}9$ with no error given, but normally for the VLA positional errors are $<0\overset{''}{.}1$. Dodonov et al. (1999) found a redshift for the host galaxy of $z = 0.707 \pm 0.002$.

The radio data were obtained from Galama et al. (2000) and the OIR data from Castro-Tirado et al. (2001). Representative data for GRB991208 are plotted, along with curves from the best fit model, in Figure 11 and the parameters of the fit are listed in Table 5.

Examination of Figure 11 shows that the parameterization listed in Table 5 describes the data reasonably well. In Figure 11a the radio data are quite limited and have relatively large scatter. For example, the reported detection at 15 GHz on day 7.4 is preceeded by a 3σ upper limit of comparable magnitude on day 5.4 and followed by an upper limit on day 12.5, so the reality of the day 7.4 detection must be called into question. In any case, although the fit is not well constrained by the radio data, the model curves describe its evolution well. The data are too sparse to judge if ISS is present.

In Figure 11b there is good coverage in both frequency and time. There is some fluctuation, particularly for the four R-band observations on day 4. However, examination of the data reveals that the three high R-band measurements on day 4.026, 4.058, and 4.096 have significantly larger errors ($\sim$0.4 mag) than the R-band datum on day 4.068 (error = 0.050 mag), which is consistent with the fitted curves. In general, the data are rather well described by the parameterization if a zero redshift color excess of $E(B - V) = 0.15$ mag is assumed. This is appreciably higher than the value of $E(B - V) = 0.016$ mag suggested by Schlegel et al. (1998) in that direction from their IRAS 100 μm maps, implying, under the same assumptions as were used for GRB980329 above, a $E(B - V) = 0.07$ at the redshift of the host galaxy (Table 5).

GRB991216 was discovered by BATSE on Dec. 16.671544 UT, 1999 (BATSE trigger # 7906) (Kippen et al. 1999). The afterglow was detected in all wavelength bands including X-ray (Takeshima et al. 1999), optical (Uglesich et al. 1999), and radio (Taylor & Berger 1999). Taylor & Berger (1999) dervied a position from their 8.5-GHz VLA observations of RA(J2000) = $05^h 09^m 31\overset{s}{.}297$, Dec(J2000) = $+11°17'07\overset{''}{.}25$, with an error of $\pm 0\overset{''}{.}1$ in each coordinate. Vreeswijk et al. (1999) suggested a redshift of $z \geq 1.02$ based on the highest redshift of three possible absorption systems seen in the optical afterglow.

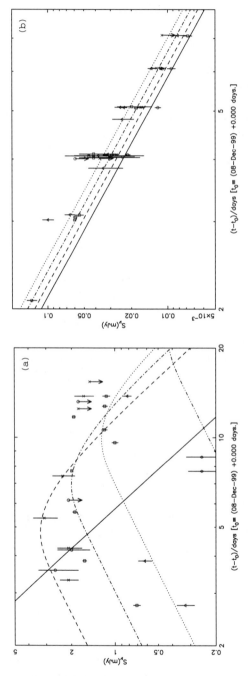

Figure 11 GRB991208 at (*a*) radio wavelengths of 1.2 mm (250 GHz; *open circles, solid line*), 2.0 cm (15.0 GHz; *crosses, dashed line*), 3.5 cm (8.5 GHz; *open squares, dash-dot line*), 6 cm (4.9 GHz; *open triangles, dotted line*), and 20 cm (1.4 GHz; *open diamonds, dash–triple dot line*), (*b*) optical/IR bands of B (681 THz; *open circles, solid line*), V (545 THz; *crosses, dashed line*), R (428 THz; *open squares, dash-dot line*), and I (333 THz; *open triangles, dotted line*). (To enhance clarity, the scales are different between the two figures even though the fitting parameters are the same and not all available bands are plotted and even though they were used in the fitting.)

The radio data were obtained from Frail et al. (2000a,c) and the OIR data from Halpern et al. (2000), Frail et al. (2000c), and Garnavich et al. (2000). Representative data for GRB991216 are plotted, along with curves from the best fit model, in Figure 12 and the parameters of the fit are listed in Table 5.

Examination of Figure 12 shows that the parameterization listed in Table 5 fits the data rather well, even with the radio data (Figure 12a) being quite limited and with relatively large scatter. The single detection at 8.5 GHz on day 42.49 surrounded by much lower 3σ upper limits on days 38.28 and 50.51, if correct, is not well described. The data are too sparse to judge if ISS is observed.

In Figure 12b there is reasonably good coverage in both frequency and time, particularly at R-band. A satisfactory fit to the data requires a rather high zero redshift color excess of $E(B - V) = 0.90$ mag. This is significantly higher than the value of $E(B - V) = 0.63$ mag suggested by Schlegel et al. (1998) in that direction from their IRAS 100 μm maps or the value of $E(B - V) = 0.40$ mag derived by Halpern et al. (2000) from the Galactic 21-cm column density in that direction. If, as above, one assumes that the additional extinction arises in the host galaxy, then with the same assumptions as were used for GRB980329, one obtains a color excess of $E(B - V) = 0.11$ at the redshift of the host galaxy (Table 5).

GRB000301C was discovered by the Ulysses and NEAR teams on Mar. 01.4108 UT, 2000 (Smith et al. 2000). The afterglow was detected in the optical (Fynbo et al. 2000) and radio (Bertoldi et al. 2000) wavelength bands. There does not appear to have been an X-ray detection. Fynbo et al. (2000) derived a position for the optical transient of RA(J2000) $= 16^h20^m18.''6$, Dec(J2000) $= +29°26'36''$, with an error of $\sim 1''$ in both coordinates. Castro et al. (2000) (see also Smette et al. 2000) found a redshift for the host galaxy of $z = 2.0335 \pm 0.003$.

The radio data were obtained from Berger et al. (2000) and the OIR data from Sagar et al. (2000), Bhargavi & Cowsik (2000), Masetti et al. (2000), Rhoads & Fruchter (2001), and Jensen et al. (2001). Representative data for GRB000301C are plotted, along with curves from the best fit model, in Figure 13 and the parameters of the fit are listed in Table 5.

Examination of Figure 13 shows that the parameterization listed in Table 5 is successful in that it describes the data rather well over the large frequency and time range. In Figure 13a the 350-GHz upper limits and the 250-GHz detections and limits are in rough agreement with the parameterization; the 15-GHz and 8.5-GHz measurements are also consistent with the modeling, although the model drops off a bit faster at late times than the data, and the fit would be improved by a somewhat flatter decline rate of $\beta \sim -1.5$. The 1.4-GHz data are generally consistent with the model, although ISS (see "Interstellar Scintillation," below) may be present in the first measurement.

In Figure 13b although the OIR data show significant scatter at individual frequencies, they are consistent with a nonthermal, synchrotron origin. A decline rate of $\beta \sim -2.0$, somewhat steeper than in the radio, would provide an improved fit. The fit indicates a best value for the zero redshift color excess of $E(B - V) = 0.25$ mag, significantly higher than the value of $E(B - V) = 0.052$ mag suggested by Schlegel et al. (1998) in that direction from their IRAS 100 μm maps. As above,

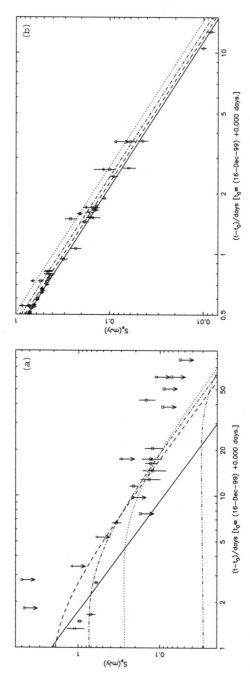

Figure 12 GRB991216 at (*a*) radio wavelengths of 0.9 mm (350 GHz; *open circles, solid line*), 2.0 cm (15.0 GHz; *crosses, dashed line*), 3.5 cm (8.5 GHz; *open squares, dash-dot line*), 6 cm (4.9 GHz; *open triangles, dotted line*), and 20 cm (1.4 GHz; *open diamonds, dash–triple dot line*), (*b*) optical/IR bands of R (428 THz; *crosses, solid line*), I (333 THz; *crosses, dashed line*), J (239 THz; *open squares, dash-dot line*), and K (136 THz; *open triangles, dotted line*). (To enhance clarity, the scales are different between the two figures even though the fitting parameters are the same and not all available bands are plotted and even though they were used in the fitting.)

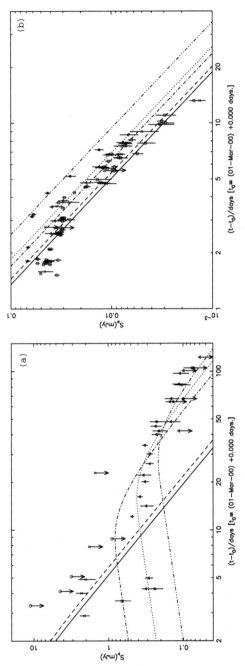

Figure 13 000301C at (*a*) radio wavelengths of 0.9 mm (350 GHz; *open circles, solid line*), 1.2 mm (250 GHz; *crosses, dashed line*), 2.0 cm (15.0 GHz; *open squares, dash-dot line*), 3.5 cm (8.5 GHz; *open triangles, dotted line*), and 6.0 cm (4.9 GHz; *filled diamonds, dash-triple dot line*), (*b*) optical/IR bands of U (832 THz; *open circles, solid line*), B (681 THz; *crosses, dashed line*), R (428 THz; *open squares, dash-dot line*), I (333 THz; *open triangles, dotted line*), and K (136 THz; *filled diamonds, dash-triple dot line*). (To enhance clarity, the scales are different between the two figures even though the fitting parameters are the same and not all available bands are plotted even though they were used in the fitting.)

this implies a color excess of $E(B - V) = 0.05$ mag at the redshift of the host galaxy (Table 5).

The most obvious characteristic of the OIR data is that the initial interval between the time of the GRB and day 3.5 is not well described by the parameterization. This is not surprising, because the modeling contains no parameters to describe any turn-on effects for the synchrotron emission. A similar turn-on effect was seen for the early OIR data for GRB970508, although not for the other GRBs discussed here.

It should be noted that the possible steepening of the flux density decline rate from $\beta \sim -1.5$ (radio) to $\beta \sim -2.0$ (OIR) may indicate a break in the decline rate somewhere between the two observing ranges, possibly owing to synchrotron losses of the high energy electrons.

Parameterization

Even though the study of the GRB phenomenon is still very much in its early stages and, as has been pointed out above, the GRBs for which we have any afterglow information are all of the slow-soft category, with those about which we have significant radio information, only a small subset, the parameterization studies (see Table 5), allows us to draw some tentative conclusions:

1. Of the seven relatively well-observed radio GRBs (RGRBs) (including GRB980425/SN1998bw) five appear to have similar relativistic electron acceleration processes that generate a spectral index $\alpha \sim -0.6$ to ~ -0.7. The two possible exceptions are GRB980329 ($\alpha \sim -1.3$) and GRB991216 ($\alpha \sim -0.3$), which interestingly enough, straddle the average spectral index of the other objects. The average spectral index value is very similar to the canonical value for extragalactic radio sources.

2. The RGRBs can be described rather well from radio through OIR by the same formalism of Equations 1 to 10 that is successful in describing the radio emission from RSNe as arising from a blastwave/CSM interaction. The obvious conclusion is that at least this subset of GRBs arises from the explosion of a compact, presumably stellar-sized object embedded in a dense CSM.

3. Only two of the objects, GRB970508 and GRB000301C, appear to have been observed in a phase for which they show the turn-on of their OIR radiation. Such a turn-on is expected to be related to the unmodeled relativistic electron acceleration at very early times and provides a test of explosion models.

4. None of the objects requires the inclusion of a uniformly distributed absorbing component to the early time radio absorption (K_2 in Equation 1), which implies a highly filamentary or clumpy structure in the CSM with which the blastwave is interacting.

5. Estimates for the color excess, $E(B - V)$, obtained from the modeling imply that many of the objects suffer appreciable extinction at the redshift of their host galaxies because fitted zero redshift values significantly exceed estimates of Galactic extinction in their directions. The values for the excess

extinction at the redshifts of the host galaxies noted above and in Table 5 are rather typical for late-type galaxies.

6. All of the RGRBs except for GRB980425/SN1998bw have similar observed peak 6-cm radio spectral luminosities of $\sim 10^{31}$ ergs s^{-1} Hz^{-1}. This implies that GRBs are rare objects for which one must survey very large volumes of space in order to detect examples. Such examples are, therefore, generally at very large distances, and we detect only the brightest members of the luminosity distribution. GRB980425/SN1998bw is the only known nearby, lower luminosity exception.

7. The prevalence of interstellar scintillation (ISS) in the early radio emission from many of these objects implies that the RGRBs are initially very small in angular size, although they expand very rapidly.

Discussion

Although many useful conclusions can be drawn from the light curve parameterizations, one should keep in mind that a more detailed study of radio GRBs will have to take into account several physical effects that are not seen in RSNe and that have not been included in this brief overview but need further consideration.

INTERSTELLAR SCINTILLATION   Because their high radio luminosity and low absorption allows detection at great distance and quite early, when they are still of very small angular size, GRB radio afterglows appear to be so compact as to exhibit ISS during the first few days or weeks of detectability. This possibility was proposed by (Goodman 1997) for GRB970508 based on earlier work by Rickett (1970) for pulsars. After consideration of the several regimes of strong, weak, refractive, and diffractive scattering, Goodman concluded that both diffractive and refractive scintillation are possible for radio afterglows and that observation of the effects of scintillation can place limits on their μas (micro-arcsecond) sizes at levels far too small to be resolved with VLBI.

GRB970508 shows strong flux density fluctuations at both 8.46 and 4.86 GHz until age ~ 4 weeks after explosion, which Waxman et al. (1998) attributed to diffractive scintillation. After ~ 1 month, they found that the modulation amplitude decreased, which is consistent with the diffractive scintillation being quenched by the increased size of the radio emitting region. They also took this increasing source size to be consistent with, and supportive of, the "fireball" model predictions (see "Relativistic Fireball," below). From their 4.86- and 8.46-GHz results, Waxman et al. (1998) concluded that the quenching of diffractive scintillation at ~ 4 weeks implies a size at that epoch of $\sim 10^{17}$ cm and an expansion speed comparable to that of light.

In contrast to the conclusion of Waxman et al. (1998) that GRB970508 is undergoing strong diffractive scintillation during the first 30 days after explosion, Smirnova & Shishov (2000) concluded that the radio afterglow is, in fact, undergoing only weak scintillation at 4.86 and 8.46 GHz at early times, with refractive scintillation dominating at 1.43 GHz.

GRB980329 shows rapid flux density fluctuations at 4.9 and 8.3 GHz, which are extinguished by age ~3 weeks. Although they did not analyze the scintillation data in detail, Taylor et al. (1998b) pointed out a similarity to the better-studied scintillations of GRB970508 and suggested that the early-time angular size of GRB980329 may be even smaller than the ~3 μas inferred for GRB970508 by Goodman (1997) and Waxman et al. (1998) because GRB980329 is at a lower Galactic latitude, which should more quickly quench ISS.

GRB980519 shows strong modulation of the flux density at 4.86 and 8.46 GHz during the first ~20 days after the GRB burst, which Frail et al. (2000b) interpreted as being due to diffractive ISS. They derived a resulting maximum radio source size of <0.4 μas, an extremely compact object. As was seen for GRB970508, the 1.4-GHz emission from GRB980519 seems to be suppressed, in this case below their detectability limit because their three measurements at 1.4 GHz are all upper limits. Frail et al. (2000b) attributed this suppression to synchrotron self-absorption (SSA) with a turn-over frequency between 1.43 and 4.86 GHz.

COSMOLOGICAL EFFECTS   Because the GRBs are at cosmological distances ($z \sim 1$), there are two effects that must be taken into account:

1. There is a time dilation that slows the light curve evolution in the observer's frame with respect to the time evolution in the emission frame. This is a straightforward correction and has been elaborated by a number of authors (see, e.g., Colgate 1979, Leibundgut 1990, Hamuy et al. 1993). The time dilation results in a true emitted time to 6-cm peak flux density $[(t_{6\,cm\,peak} - t_0)_{emit}]$ shorter than that actually observed $[(t_{6\,cm\,peak} - t_0)_{obs}]$. The correction takes the form

$$(t_{6\,cm\,peak} - t_0)_{emit} = (t_{6\,cm\,peak} - t_0)_{obs} (1 + z)^{-1} \qquad (21)$$

and must be applied to the measured times from explosion to 6-cm peak flux density to obtain true times.

2. There is a correction of the observed flux density, $(S_{6\,cm})_{obs}$, owing to the redshift's making the observed frequency different from that emitted, $(S_{6\,cm})_{emit}$, for sources with nonzero spectral indices. The correction normally takes the form, for $S \propto \nu^{+\alpha}$,

$$(S_{6\,cm})_{emit} = (S_{6\,cm})_{obs} (1 + z)^{-\alpha} \qquad (22)$$

However, Chevalier & Li (2000) have proposed an "equality of peaks" on theoretical grounds, so such a correction may be less important than expected.

RELATIVISTIC EFFECTS   Essentially all researchers agree that the GRB phenomenon involves relativistic motion of the emitting region. Whether the motion involves a spherical, relativistic fireball or a directed relativistic jet is probably not of concern from the radio observer's standpoint because the emission from the CSM interaction is probably not highly directed.

For relativistic corrections there are two factors: the relativistic motion factor Γ [$\Gamma = (1 - \frac{v^2}{c^2})^{-\frac{1}{2}}$] and, if the motion is directed, the viewing angle θ. If, as expected, the radio emission is not highly directed, then taking $\theta = 0$ leaves only the relativistic motion factor Γ to be determined.

The factor Γ is difficult to estimate from observations. Theoretical modeling predicts Γ of several hundred very early in the expansion phase (Waxman 1997) declining to subrelativistic motion after only a few days to a few weeks (Huang et al. 1998, Dai et al. 1999). However, although highly speculative, there is perhaps one method for estimating Γ. In "Expansion Velocity," above we discussed how, by assuming a relation between Type Ib SN1998bw and GRB980425, Weiler et al. (2001) obtained estimates of $\Gamma \sim 1.6$–2.0 for the radio afterglow. They obtained one estimate by postulating that SN1998bw was expanding at $\sim 230{,}000$ km s^{-1} if its brightness temperature at 6-cm peak flux density was $10^{11.5}$ K, as required by the inverse Compton limit. The second estimate was obtained by assuming that the true 6-cm peak spectral luminosity was the same as that for the average of all known Type Ib/c RSNe. If one assumes (a) all GRB radio afterglows arise in Type Ib/c supernovae, (b) the postulate of Weiler et al. (1998) that all Type Ib/c supernovae may be approximate standard radio candles with an $L_{6\,\mathrm{cm\,peak}} \sim 1.3 \times 10^{27}$ erg s^{-1} Hz^{-1}, and (c) any increase in the observed flux density is solely due to relativistic boosting, then one can estimate a relativistic Γ factor for each GRB radio afterglow.

Applying these assumptions to the peak 6-cm radio luminosity ($L_{6\,\mathrm{cm\,peak}}$) values obtained for the GRBs and listed in Table 5, we obtain Γ values ~ 10. Such Γs are similar to those found for VLBI "superluminal" sources (see e.g., Scheuer 1984, Kellermann 1985) and for the Galactic micro-QSO superluminal sources (Levinson & Blandford 1996, Mirabel & Rodriguez 1994, Tingay et al. 1995, Hjellming & Rupen 1995).

Theoretical Models

The origin of gamma-ray bursts is a long-standing problem dating back to the 1960s. With little actual data to constrain them, many theories were developed. Ruderman (1975) first summarized these in the mid-1970s. Cavallo & Rees (1978) were probably the first to use the term "fireball" in an article discussing several possibilities for the conversion of massive amounts of injection energy into gamma-rays, including the possibility of relativistic expansion and the conversion of kinetic energy into gamma-ray luminosity through the impact on an external medium.

The 1980s brought the more detailed calculations of Paczynski (1986) and Goodman (1986) suggesting that the GRBs might be at cosmological distances of $z \sim 1$ or more and that the energies involved were at least comparable to that expected from a supernova ($\sim 10^{51}$ ergs). In the 1990s, of course, the field finally came of age with the Compton Gamma-Ray Observatory to provide almost daily detections of GRBs and BeppoSAX to provide rapid follow-up for many bursts and accurate position determination for X-ray afterglows. This increased data flow refined the model studies for the origin of GRBs so that there are basically only two models presently under serious consideration: relativistic fireball and relativistic jet. Because of the observational (particularly radio) orientation of this review, we do not discuss the mechanisms of the source of the energy release, whether neutron star or black hole formation in supernovae, coalescence of compact objects, or more exotic phenomena. We are more concerned here with the observable consequences of this great energy release and, where possible, similarities to known phenomena.

RELATIVISTIC FIREBALL   Many workers have concluded that the relativistic fireball model is the most likely scenario for GRB creation and most prefer conversion of the blastwave energy into emission through blastwave interaction with an external medium. The basic outlines of the fireball model are probably best summarized by Waxman (1997, pp. L33–34):

A compact $r_0 \sim 10^7$ cm source releases an energy E comparable to that observed in gamma-rays, $E \sim 10^{51}$ ergs, over a time t < 100 s. The large energy density in the source results in an optically thick plasma that expands and accelerates to relativistic velocity. After an initial acceleration phase, the fireball energy is converted to proton kinetic energy. A cold shell of thickness ct is formed and continues to expand with time-independent Lorentz factor $\Gamma \sim 300$. The GRB is produced once the kinetic energy is dissipated at large radius, $r > 10^{13}$ cm, due to internal collisions within the ejecta and radiated as gamma-rays through synchrotron and possibly inverse-Compton emission of blastwave accelerated electrons. Following internal collisions, which convert part of the energy to radiation and which result from variations in Γ across the expanding shell, the fireball rapidly cools and continues to expand with approximately uniform Lorentz factor Γ. As the cold shell expands, it drives a relativistic blastwave into the surrounding gas.

Huang et al. (1998, p. 459) described this interaction at larger radius as

After producing the main GRB, the cooling fireball is expected to expand as a thin shell into the interstellar medium (ISM) and generate a relativistic blastwave, although whether the expansion is highly radiative (Vietri 1997) or adiabatic is still controversial. Afterglows at longer wavelengths are produced by the shocked ISM.

Rees & Mészáros (1992) described the possibility of generation of GRBs from the interaction of a highly relativistic ($\Gamma \sim 10^3$) blastwave interacting with a tenuous ISM or CSM to generate a $\sim 10^{51}$ erg gamma-ray burst. Mészáros et al. (1993) worked this out in more detail (see also Mészáros 1995 for a review of fireball models). When the first optical afterglow of a GRB was finally found for GRB970228, although the data were very limited, Wijers et al. (1997) found support for their model of an expanding, relativistic fireball of total energy $\sim 10^{51}$ ergs plowing into a surrounding medium.

As early as 1993, Paczynski & Rhoads (1993) pointed out the similarity of GRB relativistic fireball interaction with CSM or ISM material to relativistic jets interacting with intergalactic material, Supernova Remnants (SNRs) interacting with the ISM, or RSNe interacting with CSM and predicted the presence of delayed radio emission from the GRB process. Mészáros & Rees (1997) and Vietri (1997) also predicted late-time radio, optical, and soft X-ray emission based on the fireball model.

Huang et al. (1998) proposed that the extremely large Γ factors of the GRB-producing relativistic fireball decay quickly, within 3–4 days, so that the blast-wave/CSM interaction generating the synchrotron emission is only slightly relativistic with $2 < \Gamma < 5$. Even more rapidly evolving is the fireball model of Dai et al. (1999), who proposed that the fireball is only mildly relativistic after $\sim$3 h and nonrelativistic after a few days.

Waxman (1997) studied the optical and radio afterglow from GRB970508 and found agreement with the cosmological fireball model for a fireball energy of $\sim$10^{52} ergs and the blastwave expanding into an ambient medium with density $\sim$1 cm^{-3}. Waxman et al. (1998) also compared the radio light curves of GRB970508 to the predictions of the Waxman (1997) fireball model and, although their fit is extremely poor after $\sim$4 weeks of age, they ascribed the deviation to transition of the fireball from relativistic to subrelativistic expansion with a Lorentz factor $\Gamma < 2$ or, possibly, to a lower energy "jet" structure for the fireball at early epochs. Their model also predicts transition to a more uniformly emitting fireball structure by age $\sim$5 weeks, becoming a spherical, nonrelativistic emitter by age $\sim$12 weeks.

RELATIVISTIC JETS AND HYPERNOVAE   Support for the relativistic fireball model is not universal, however. Dar (1998) proposed an origin for GRBs and their longer wavelength afterglows in relativistic jets produced by mergers and/or accretion-induced collapse of compact stellar objects. He felt that such a scenario can better solve the "energy problem" (where some estimates of relativistic fireball energies can range up to a rather high $> 10^{52}$ ergs), the short time variability of many GRBs, and other apparent deviations from the relativistic fireball models. A somewhat different variant on the fireball model has been proposed by Paczynski (1998), who suggested that a "hypernova" explosion produces a "dirty" fireball from the violent death of a massive star.

MULTIPLE ORIGINS   Livio & Waxman (2000) took the broader stance that the GRBs with afterglows, although probably not originating in neutron star-neutron star (ns-ns) mergers, could well have several possible origins including the collapse of massive stars as in supernovae (GRB980519 and GRB980326) or could be produced by Black Hole–He star mergers (GRB990123 and GRB990510), with a third group such as GRB970228 and GRB970508 being possibly produced by Black Hole–He star mergers but without collimated jets being formed by the collapse. They disagreed with other workers that the transition from relativistic expansion to nonrelativistic expansion of a spherical fireball will produce a detectable break in the afterglow decline rate. They also felt that the GRBs creating detectable afterglows are diverse, with only some producing highly collimated jets but with all originating from massive stars and all interacting with a stellar wind environment.

Perhaps the broadest summary and comparison of models with radio data for several GRBs is that given by Chevalier & Li (2000) (see also, Chevalier & Li 1999), who were able to describe rather well the radio light curves of several GRB radio afterglows as arising from blastwave interaction with an external medium. They also suggested that there is evidence for two groupings: (*a*) GRBs arising from the explosion of a massive star, possibly a Wolf-Rayet star, embedded in a dense $\rho \propto r^{-2}$ stellar wind–formed CSM and accompanied by a supernova, and (*b*) GRBs arising from the explosion caused by a compact binary merger that occur in a constant density ISM and are not accompanied by a supernova.

SUMMARY

We have assembled an overview of the radio emission from supernovae and have shown that there are many similarities. In particular, the radio emission from supernovae can be modeled in great detail and used to describe the physical properties of the CSM and the nature and final stages of evolution of the presupernova system.

This same formalism can also be applied to the radio and OIR emission from a number of GRBs and provides a satisfactory description of their light curves over extremely broad frequency and time ranges. These results then imply that at least some GRBs probably arise in the explosions of massive stars in dense, highly structured circumstellar media. Such media are presumed to have been established by mass-loss from the preexplosion stellar systems. We hope that the continued study of GRB afterglows and the slowly increasing statistics of radio observations of supernovae will soon establish the true nature of both phenomena and strengthen their probable relations.

ACKNOWLEDGMENTS

KWW & MJM thank the Office of Naval Research (ONR) for the 6.1 funding supporting this research. Additional information and data on radio emission from supernovae and GRBs can be found on http://rsd-www.nrl.navy.mil/7214/weiler/ and linked pages.

The *Annual Review of Astronomy and Astrophysics* is online at
http://astro.annualreviews.org

LITERATURE CITED

Ball L, Campbell-Wilson D, Crawford DF, Turtle AJ. 1995. *Ap. J.* 453:864–72

Ball L, Crawford DF, Hunstead RW, Kalmer I, McIntyre VJ. 2001. *Ap. J.* 549:599–607

Bartel N, Shapiro II, Gorenstein MV, Gwinn CR, Rogers AEE. 1985. *Nature* 318:25–30

Benetti S, Turatto M, Cappellaro E, Danziger IJ, Mazzali P. 1999. *MNRAS* 305:811–19

Berger E, Sari R, Frail DA, Kulkarni SR, Bertoldi F, et al. 2000. *Ap. J.* 545:56–62

Bertoldi F. 2000. *GCN* 580

Bhargavi SG, Cowsik R. 2000. *Ap. J. Lett.* 545:L77–80

Bloom JS, Djorgovski SG, Kulkarni SR, Frail DA. 1998. *Ap. J. Lett.* 507:L25–28

Boffi FR, Panagia N. 1996. *Radio Emission from the Stars and the Sun*, ed. AR Taylor, JM Paredes, *Astr. Soc. Pac. Conf. Series* 93:153–57

Bond HE. 1997. *IAU Circ.* 6654

Bremer M, Kirchbaum TP, Galama TJ, Castro-Tirado AJ, Frontera F, et al. 1998. *Astron. Astrophys.* 332:L13–16

Briggs MS, Richardson G, Kippen RM, Woods PM. 1998. *IAU Circ.* 6856

Brocato E, Castellani V. 1993. *Ap. J.* 410:99–109

Castro SM, Diercks A, Djorgovski SG, Kulkarni SR, Galama TJ, et al. 2000. *GCN* 605

Castro-Tirado AJ, Gorosabel J, Benitez N, Wolf C, Fockenbrock R, et al. 1998. *Science* 279:1011–21

Castro-Tirado AJ, Gorosabel J, Greiner J, Pedersen H, Pian E, et al. 1999. *GCN* 452

Castro-Tirado AJ, Sokolov VV, Gorosabel J, Castro-Cern JM, Greiner J, et al. 2001. *Astron. Astrophys.* 370:398–406

Cavallo G, Rees MJ. 1978. *MNRAS* 183:359–65

Chary R, Neugebauer G, Morris M, Becklin EE, Matthews K, et al. 1998. *Ap. J. Lett.* 498:L9–11

Chevalier RA. 1982a. *Ap. J.* 259:302–10

Chevalier RA. 1982b. *Ap. J. Lett.* 259:L85–89

Chevalier RA. 1998. *Ap. J.* 499:810–19

Chevalier RA, Fransson C. 1994. *Ap. J.* 420:268–85

Chevalier RA, Li Z-Y. 1999. *Ap. J. Lett.* 520:L29–32

Chevalier RA, Li Z-Y. 2000. *Ap. J.* 536:195–212

Chieffi A, Limongi M, Straniero O. 1998. *Ap. J.* 502:737–62

Chu Y-H, Caulet A, Montes MJ, Panagia N, Van Dyk SD, Weiler KW. 1999. *Ap. J. Lett.* 512:L51–54

Chugai NN, Danziger IJ, della Valle M. 1995. *MNRAS* 276:530–36

Colgate SA. 1979. *Ap. J.* 232:404–8

Costa E, Feroci M, Piro L, Soffitta P, Amati L, et al. 1997. *IAU Circ.* 6649

Dai ZG, Huang YF, Lu T. 1999. *Ap. J.* 520:634–40

Dar A. 1998. *Ap. J. Lett.* 500:L93–96

Djorgovski SG, Metzger MR, Kulkarni SR, Odewahn SC, Gal RR, et al. 1997. *Nature* 387:876–78

Dodonov SN, Afanasiev VL, Sokolov VV, Moiseev AV, Castro-Tirado AJ. 1999. *GCN* 465

Filippenko A. 1998. *IAU Circ.* 6969

Filippenko A, Matheson T, Barth A. 1994. *Astron. J.* 108:2220–25

Fishman GJ, Meegan CA. 1995. *Annu. Rev. Astron. Astrophys.* 33:415–58

Frail DA, Kulkarni SR. 1997. *IAU Circ.* 6662

Frail DA, Kulkarni SR. 1999. *GCN* 451

Frail DA, Taylor GB. 2000. *GCN* 574

Frail DA, Taylor GB, Kulkarni SR. 1998. *GCN* 89

Frail DA, Berger E, Galama T, Kulkarni SR, Moriarty-Schieven GH, et al. 2000a. *Ap. J. Lett.* 538:L129–32

Frail DA, Kulkarni SR, Sari R, Taylor GB, Shepherd DS, et al. 2000b. *Ap. J.* 534:559–64

Frail DA, Waxman E, Kulkarni SR. 2000c. *Ap. J.* 537:191–204

Fransson C, Björgsson C-I. 1998. *Ap. J.* 509:861–78

Frontera F, Costa E, Piro L, Soffitta P, Zand J. in 't, et al. 1998. *IAU Circ.* 6853

Fruchter A, Bergeron L. 1997. *IAU Circ.* 6674

Fynbo JU, Jensen BL, Hjorth J, Pedersen H, Gorosabel J. 2000. *GCN* 570

Gaensler BM, Manchester RN, Staveley-Smith L, Tzioumis AK, Reynolds JE, Kesteven MJ. 1997. *Ap. J.* 479:845–58

Galama TJ, Bremer M, Bertoldi F, Menten KM, Lisenfeld U, et al. 2000. *Ap. J. Lett.* 541:L45–49

Galama TJ, Groot PJ, van Paradijs J, Kouveliotou C, Strom RG, et al. 1998a. *Ap. J. Lett.* 497:L13–16

Galama TJ, Vreeswijk PM, Pian E, Frontera F, Doublier V, et al. 1998b. *IAU Circ.* 6895

Galama TJ, Vreeswijk PM, van Paradijs J, Kouveliotou C, Augusteijn T, et al. 1998c. *Nature* 395:670–72

Galama TJ, Wijers RAMJ, Bremer M, Groot PJ, Strom RG, et al. 1998d. *Ap. J. Lett.* 500: L101–9

Galama TJ, Vreeswijk PM, van Paradijs J, Kouveliotou C, Augusteijn T, et al. 1999. *Astron. Astrophys. Suppl. Ser.* 138:465–66

Garcia MR, Callanan PJ, Moraru D, McClintock JE, Tollestrup E, et al. 1998. *Ap. J. Lett.* 500:L105–9

Garnavich P, Kirshner R, Challis P. 1997. *IAU Circ.* 6710

Garnavich PM, Jha S, Pahre MA, Stanek KZ, Kirshner RP, et al. 2000. *Ap. J.* 543:61–65

Goodman J. 1986. *Ap. J. Lett.* 308:L47–50

Goodman J. 1997. *New Astr.* 2:449–60

Gorosabel J, Castro-Tirado AJ, Pedrosa A, Zapatero-Osorio MR, Fernades AJL, et al. 1999. *Astron. Astrophys. Lett.* 347:L31–34

Groot PJ, Galama TJ, van Paradijs J, Strom R, Telting J, et al. 1997. *IAU Circ.* 6584

Halpern JP, Kemp J, Piran T, Bershady MA. 1999. *Ap. J. Lett.* 517:L105–8

Halpern JP, Uglesich R, Mirabal N, Kassin S, Thorstensen J, et al. 2000. *Ap. J.* 543:697–703

Hamuy M, Phillips MM, Wells LA, Maza J. 1993. *Publ. Astron. Soc. Pac.* 105:787–93

Hasinger G, Aschenbach B, Truemper J. 1996. *Astron. Astrophys.* 312:L9–12

Hjellming RM, Rupen MP. 1995. *Nature* 375:464–66

Huang YF, Dai ZG, Wei DM, Lu T. 1998. *MNRAS* 298:459–63

Hurley K, Cline T. 1999. *GCN* 450

Immler S, Aschenbach B, Wang QD. 2001. *Ap. J. Lett.* 561:L107–10

Iwamoto K, Mazzali PA, Nomoto K, Umeda H, Nakamura T, et al. 1998. *Nature* 395:672–74

Jakobsen P, Albrecht R, Barbieri C, Blades JC, Boksenberg A, et al. 1991. *Ap. J. Lett.* 369:L63–66

Jaunsen AO, Hjorth J, Andersen MI, Kjernsmo K, Pedersen H, Palazzi E. 1998. *GCN* 78

Jaunsen AO, Hjorth J, Bjrnsson G, Andersen MI, Pedersen H, et al. 2001. *Ap. J.* 546:127–33

Jensen BL, Fynbo JU, Gorosabel J, Hjorth J,

Holland S, et al. 2001. *Astron. Astrophys.* 370:909–22

Kay LE, Halpern JP, Leighly KM. 1998. *IAU Circ.* 6969

Kellermann KI. 1985. *Comm. Astrophys.* 11: 69–81

Kellermann KI, Pauliny-Toth IIK. 1969. *Ap. J. Lett.* 155:L71–78

Kippen MR, Preece RD, Giblin T. 1999. *GCN* 463

Klose S, Meusinger H, Lehmann H. 1998. *IAU Circ.* 6864

Kulkarni SR, Bloom JS, Frail DA, Ekers R, Wieringa M, et al. 1998. *IAU Circ.* 6903

Langer N, Maeder A. 1995. *Astron. Astrophys.* 295:685–92

Leibundgut B. 1990. *Astron. Astrophys.* 229:1–6

Levinson A, Blandford R. 1996. *Astron. Astrophys. Suppl. Ser.* 120:129–32

Li Z-Y, Chevalier RA. 1999. *Ap. J.* 526:716–26

Lidman C, Doublier V, Gonzalez J-F, Augusteijn T, Hainaut OR, et al. 1998. *IAU Circ.* 6895

Livio M, Waxman E. 2000. *Ap. J.* 538:187–91

Lozinskaya TA. 1992. *Supernovae and Stellar Wind in the Interstellar Medium.* New York: Am. Inst. Phys. p. 190

Manchester RN, Gaensler BM, Wheaton VC, Staveley-Smith L, Tzioumis AK, et al. 2002. *Pub. Astron. Soc. Austr.* 19:207–21

Marcaide JM, Alberdi A, Ros E, Diamond P, Shapiro II, et al. 1997. *Ap. J. Lett.* 486:L31–34

Masetti N, Bartolini C, Bernabei S, Guarnieri A, Palazzi E, et al. 2000. *Astron. Astrophys. Lett.* 359:L23–26

Mészáros P. 1995. *NY Acad. Sci. Ann.* 759:440–45

Mészáros P, Laguna P, Rees MJ. 1993. *Ap. J.* 415:181–90

Mészáros P, Rees MJ. 1997. *Ap. J.* 476:232–37

Metzger MR, Djorgovski SG, Steidel CC, Kulkarni SR, Adelberger KL, Frail DA. 1997. *IAU Circ.* 6655

Minkowski R. 1964. *Annu. Rev. Astron. Astrophys.* 2:247–66

Mirabel IF, Rodriguez LF. 1994. *Nature* 371: 46–48

Montes MJ, Van Dyk SD, Weiler KW, Sramek RA, Panagia N. 1998. *Ap. J.* 506:874–79

Montes MJ, Weiler KW, Panagia N. 1997. *Ap. J.* 488:792–98

Montes MJ, Weiler KW, Van Dyk SD, Sramek RA, Panagia N, Park R. 2000. *Ap. J.* 532: 1124–31

Muller JM, Heise J, Butler C, Frontera F, Di Ciolo L, et al. 1998. *IAU Circ.* 6910

Natta A, Panagia N. 1984. *Ap. J.* 287:228–37

Nicastro L, Antonelli LA, Celidonio G, Daniele MR, De Libero C, et al. 1998. *IAU Circ.* 6912

Nomoto K, Yamaoka H, Pols OR, van den Heuvel E, Iwamoto K, et al. 1994. *Nature* 371:227–31

Osterbrock DE. 1974. *Astrophysics of Gaseous Nebulae.* San Francisco: Freeman. p. 82

Paczynski B. 1986. *Ap. J. Lett.* 308:L43–46

Paczynski B. 1998. *Ap. J. Lett.* 494:L45–48

Paczynski B, Rhoads JE. 1993. *Ap. J. Lett.* 418: L5–8

Panagia N, Bono G. 2001. *STScI 1999 May Symposium: The Largest Explosions Since the Big Bang, Supernovae and Gamma Ray Bursts,* ed. M Livio, N Panagia, K Sahu, pp. 184–97. Cambridge, UK: Cambridge Univ. Press

Panagia N, Macchetto F. 1982. *Astron. Astrophys.* 106:266–73

Panagia N, Sramek RA, Weiler KW. 1986. *Ap. J.* 300:L55–58

Patat F, Cappellaro E, Rizzi L, Turatto M, Benetti S. 1999. *IAU Circ.* 7215

Patat F, Piemonte A. 1998. *IAU Circ.* 6918

Pian E, Amati L, Antonelli LA, Butler RC, Costa E, et al. 1999. *Astron. Astrophys. Suppl. Ser.* 138:463–64

Pian E, Amati L, Antonelli LA, Butler RC, Costa E, et al. 2000. *Ap. J.* 536:778–87

Pian E, Fruchter AS, Bergeron LE, Thorsett SE, Frontera F, et al. 1998. *Ap. J. Lett.* 492: L103–6

Piro L, Costa E, Feroci M, Soffitta P, Antonelli LA, et al. 1997. *IAU Circ.* 6656

Podsiadlowski Ph. 1992. *Publ. Astron. Soc. Pac.* 104:717–29

Podsiadlowski Ph, Hsu J, Joss P, Ross R. 1993. *Nature* 364:509–11

Prevot ML, Lequeux J, Prevot L, Maurice E, Rocca-Volmerange B. 1984. *Astron. Astrophys.* 132:389–92

Readhead ACS. 1994. *Ap. J.* 426:51–59

Rees MJ, Mészáros P. 1992. *MNRAS* 258:41–43

Reichart DE, Lamb DQ, Metzger MR, Quashnock JM, Cole DM, et al. 1999. *Ap. J.* 517: 692–99

Rhoads JE, Fruchter AS. 2001. *Ap. J.* 546:117–26

Rickett BJ. 1970. *MNRAS* 150:67–91

Ruderman M. 1975. *NY Acad. Sci. Ann.* 262:164–80

Ryder S, Staveley-Smith L, Dopita M, Petre R, Colbert E, et al. 1993. *Ap. J.* 416:167–81

Sadler EM, Stathakis RA, Boyle BJ, Ekers RD. 1998. *IAU Circ.* 6901

Sagar R, Mohan V, Pandey SB, Pandey AK, Stalin CS, Castro-Tirado AJ. 2000. *Bull. Astron. Soc. India.* 28:499–513

Sahu KC, Livio M, Petro L, Bond HE, Macchetto F, et al. 1997. *Ap. J. Lett.* 489:L127–31

Salamanca I, Cid-Fernandes R, Tenorio-Tagle G, Telles E, Terlevich RJ, et al. 1998. *MNRAS* 300:L17–21

Savage BD, Mathis JS. 1979. *Annu. Rev. Astron. Astrophys.* 17:73–111

Schaefer B, Schaefer M, Smith P, Mackey C, Wilcots E, et al. 1997. *IAU Circ.* 6658

Scheuer PAG. 1984. *IAU Symp. 110: VLBI and Compact Radio Sources* 110:197–207

Schlegel DJ, Finkbeiner DP, Davis M. 1998. *Ap. J.* 500:525–53

Schwarz DH, Pringle JE. 1996. *MNRAS* 282: 1018–26

Shepherd DS, Frail DA, Kulkarni SR, Metzger MR. 1998. *Ap. J.* 497:859–64

Smette A, Fruchter A, Gull T, Sahu K, Ferguson H, et al. 2000. *GCN* 603

Smirnova TV, Shishov VI. 2000. *Astron. Rep.* 44:421–25

Smith DA, Hurley K, Cline T. 2000. *GCN* 568

Smith IA, Tilanus RPJ. 1998. *GCN* 50

Smith IA, Tilanus RPJ, van Paradijs J, Galama

TJ, Groot PJ, et al. 1999. *Astron. Astrophys.* 347:92–98

Soffitta P, Feroci M, Piro L, Zand J in 't, Heise J, et al. 1998. *IAU Circ.* 6884

Sokolov VV, Kopylov AI, Zharikov SV, Feroci M, Nicastro L, Palazzi E. 1998. *Astron. Astrophys.* 334:117–23

Spyromilio J. 1994. *MNRAS* 266:L61–64

Sramek RA, Panagia N, Weiler KW. 1984. *Ap. J. Lett.* 285:L59–62

Stathakis RA, Sadler EM. 1991. *MNRAS* 250:786–95

Takeshima T, Markwardt C, Marshall F, Giblin T, Kippen RM. 1999. *GCN* 478

Taylor GB, Berger E. 1999. *GCN* 483

Taylor GB, Frail DA, Kulkarni SR. 1998a. *GCN* 40

Taylor GB, Frail DA, Kulkarni SR, Shepherd DS, Feroci M, Frontera F. 1998b. *Ap. J. Lett.* 502:L115–18

Tingay SJ, Jauncey DL, Preston RA, Reynolds JE, Meier DL, et al. 1995. *Nature* 374:141–44

Tinney C, Stathakis R, Cannon R, Galama T. 1998. *IAU Circ.* 6896

Turtle AJ, Campbell-Wilson D, Bunton JD, Jauncey DL, Kesteven MJ. 1987. *Nature* 327:38–40

Uglesich R, Mirabal N, Halpern J, Kassin S, Novati S. 1999. *GCN* 472

Van Dyk SD, Sramek RA, Weiler KW, Panagia N. 1993a. *Ap. J.* 409:162–69

Van Dyk S, Sramek RA, Weiler K, Panagia N. 1993b. *Ap. J. Lett.* 419:L69–72

Van Dyk S, Weiler KW, Sramek RA, Panagia N. 1992. *Ap. J.* 396:195–200

Van Dyk S, Weiler K, Sramek R, Rupen M, Panagia N. 1994. *Ap. J. Lett.* 432:L115–18

Vietri M. 1997. *Ap. J. Lett.* 478:L9–12

Vrba FJ, Henden AA, Canzian B. Levine SE, Luginbuhl CB, et al. 2000. *Ap. J.* 528:254–59

Vreeswijk PM, Rol E, Hjorth J, Kouveliotou C, Pian E, et al. 1999. *GCN* 496

Waxman E. 1997. *Ap. J. Lett.* 489:L33–36

Waxman E, Kulkarni SR, Frail DA. 1998. *Ap. J.* 497:288–93

Waxman E, Loeb A. 1999. *Ap. J.* 515:721–25

Weiler KW, Panagia N, Montes MJ. 2001. *Ap. J.* 562:670–78

Weiler K, Panagia N, Sramek R. 1990. *Ap. J.* 364:611–25

Weiler KW, Panagia N, Sramek RA, van der Hulst JM, Roberts MS, Nguyen L. 1989. *Ap. J.* 336:421–28

Weiler K, Sramek R, Panagia N, van der Hulst J, Salvati M. 1986. *Ap. J.* 301:790–812

Weiler KW, Van Dyk SD, Montes MJ, Panagia N, Sramek RA. 1998. *Ap. J.* 500:51–58

Weiler K, Van Dyk S, Panagia N, Sramek R, Discenna J. 1991. *Ap. J.* 380:161–66

Weiler K, Van Dyk S, Panagia N, Sramek R. 1992a. *Ap. J.* 398:248–53

Weiler K, Van Dyk S, Pringle J, Panagia N. 1992b. *Ap. J.* 399:672–79

Wijers RAMJ, Rees MJ, Mészáros P. 1997. *MNRAS* 288:L51–56

Williams CL, Panagia N, Lacey CK, Weiler KW, Sramek RA, Van Dyk SD. 2002. *Ap. J.* In press

Woosley SE, Eastman RG, Schmidt BP. 1999. *Ap. J.* 516:788–96

Yin QF. 1994. *Ap. J.* 420:152–58

Young CH, Frail DA, Kulkarni SR. 1999. *Bull. Am. Astron. Soc.* 31:1474

Zand J in 't, Heise J, Piro L, Antonelli LA, Daniele M, et al. 1998. *IAU Circ.* 6854

Annu. Rev. Astron. Astrophys. 2002. 40:439–86
doi: 10.1146/annurev.astro.40.060401.093849
Copyright © 2002 by Annual Reviews. All rights reserved

SHAPES AND SHAPING OF PLANETARY NEBULAE

Bruce Balick
Department of Astronomy, University of Washington, Seattle, Washington 98195-1580;
email: balick@astro.washington.edu

Adam Frank
Department of Physics and Astronomy, University of Rochester, Rochester,
New York 14627-0171; email: afrank@pas.rochester.edu

Key Words   stellar evolution, mass loss, protoplanetary nebulae, hydrodynamics,
magnetohydrodynamics

■ **Abstract**   We review the state of observational and theoretical studies of the shaping of planetary nebulae (PNe) and protoplanetary nebulae (pPNe). In the past decade,
high-resolution studies of PNe have revealed a bewildering array of morphologies with
elaborate symmetries. Recent imaging studies of pPNe exhibit an even richer array of
shapes. The variety of shapes, sometimes multiaxial symmetries, carefully arranged
systems of low-ionization knots and jets, and the often Hubble-flow kinematics of
PNe and pPNe indicate that there remains much to understand about the last stages
of stellar evolution. In many cases, the basic symmetries and shapes of these objects
develop on extremely short timescales, seemingly at the end of AGB evolution when
the mode of mass loss abruptly and radically changes. No single explanation fits all
of the observations. The shaping process may be related to external torques of a close
or merging binary companion or the emergence of magnetic fields embedded in dense
outflowing stellar winds. We suspect that a number of shaping processes may operate
with different strengths and at different stages of the evolution of any individual object.

INTRODUCTION

PN Morphologies: The Smirk of the Cheshire Cat

Perhaps the first and youngest of the many "standard models" in astronomy to
fall victim to the penetrating spatial resolution and dynamic range of the Hubble
Space Telescope (HST) was that for planetary nebulae (PNe). Historically, in 1993,
Frank et al. confidently claimed that the morphologies of nearly all PNe could be
understood as the evolving hydrodynamic interaction between fast winds from a
central star and the nozzle formed by a dense torus of material presumably ejected
earlier in the life of the central star. In 1994, the now-famous HST image of the
Cat's Eye Nebula (Harrington & Borkowski 1994) mocked Frank et al.'s simple
paradigm in several ways. First, no signs of dense tori were seen in close association

with either member of an odd pair of orthogonal ellipsoidal features in the nebular core. Second, the HST image showed an incredible array of meticulously organized knotty or jet-like features that extant hydro models simply cannot explain with any credible set of presumed initial and boundary conditions. Insofar as our comprehension of the shapes of PNe is concerned, the HST image of NGC 6543 is redolent of the frustrating ambiguity of the Cheshire Cat.

HST images of other PNe with very different but equally spectacular structures and symmetries followed quickly. Each new image was greeted with a combination of aesthetic delight and interpretive apprehension, if not terror, by nebular dynamicists. Taken together, and divided into morphological classes, the complex symmetries of PNe raise embarrassing issues for our understanding of stellar evolution as well as difficult questions about the physics of gas dynamical processes. The underlying issue raised by modern images of PNe is how gasping, dying stars with huge, fluffy, marginally bound atmospheres can produce such complex but highly organized outflows. This paper 1. reviews the morphologies and their first derivative, the kinematics of both planetary nebulae and their progenitors, protoplanetary nebulae (pPNe); 2. summarizes the astronomical and physical challenges revealed by the data; and 3. probes the physical shaping mechanisms of PNe.

Studies of nebular dynamics in PNe are probably extensible to other more elusive types of outflows. The resemblance of some PNe to H-H objects, YSOs, η Car, some axisymmetric supernovae, and bipolar AGNS is striking. For example, compare the pairs of rings near SN1987A (e.g., Burrows et al. 1995) with the similar rings in MyCn18 (Sahai et al. 1999b) and He2-104 (Corradi et al. 2001) that are inscribed in their bipolar lobes. Obvious counterparts to the "homunculus" of η Car (Morse et al. 1998), a rapidly expanding bilobed LBV nebula, are Menzel 3 (Redman et al. 2000) and Hubble 5 (Riera et al. 2000). He2-90 (Sahai & Nyman 2000, Guerrero et al. 2001) is a dead ringer for an H-H object, other than it has no connection whatsoever to a star-forming region. Therefore some of the mechanisms that shape PNe may have broad application in related objects such as YSOs in which extinction or distance often occlude the dynamically important regions.

PNe research is entering another renaissance. This is a time of intellectual puzzlement, debate, and play in which the imaginations of observers and theoreticians are crossing disciplinary boundaries in order to explain the morphologies and kinematics of PNe and pPNe. PNe are the testing grounds for many of these ideas thanks to their brightness, the full range of their shapes, the paucity of local extinction, and their large numbers. The expanding constellation of interpretive ideas that has emerged to explain PN morphologies are ready to be organized and reviewed.

The Historical Setting

The modern field of PNe mass ejection started when Deutsch (1956) found displaced absorption lines in the spectrum of α Her and concluded that late-type AGB stars and supergiants were shedding their outer layers. Shklovskii (1956a,b) and Abell & Goldreich (1966) showed how luminous AGBs might eject their envelopes via structural instabilities. Within a decade the idea that subsurface shell flashes and/or surface pulsations could drive loosely bound matter from AGB stars had gained popularity, though the effectiveness of each mechanism remained unclear.

It is now well established that AGB and post-AGB stars lose mass at rates as high as 10^{-5} $M_\odot$ yr^{-1} or more at super-escape velocities of about 10 km s^{-1}. The general idea is that during the AGB phase IR photons accelerate dust particles that then drag the gas with them to form a nebula. Acoustic waves and surface pulsations might provide an additional boost (Pijpers & Hearn 1989; Pijpers & Habing 1989). Mass loss mechanisms from cool giant stars have been reviewed thoroughly by Willson (2000).

Until 1978 it was believed that mass loss ceased at this point and the ejected envelope expanded isotropically. A paradigm shift occurred with the opening of the ultraviolet window. Kwok et al. (1978), Cerruti-Sola & Perinotto (1989), Patriarchi & Perinotto (1991), and others found that as the star evolves from AGB to PN nucleus, its mass loss rate decreases precipitously (10^{-8} $M_\odot$ yr^{-1}) but does not cease. At the same time, the speed of the winds, V_w, rises almost reciprocally, from about 10 to 10^3 km s^{-1}. The material ejected earlier becomes ionized once the temperature at the exposed surface of the central "nucleus" exceeds 25,000 K. Eventually, hard uv photons (10–50 eV) couple strongly to blanketed lines of multiply ionized metals (Lamers & Cassinelli 1999). Multiple scattering of the uv photons in the dense acceleration zone could enhance the radiation pressure and drive the wind with a momentum flux $\tau_{scat} \cdot L_*/c$, where $\tau_{scat} \leq 10$ is the scattering opacity.

In general, all winds driven by these mechanisms are inherently isotropic; however, the fruits of their labors, pPNe and PNe, conspicuously display a variety of symmetries. The present controversy is how axisymmetric and more complex types of pPNe and PNe form and evolve. Astronomically, how do highly evolved stars in isolated environments collimate their outflows and orchestrate the grand design of the nebula? Physically, what is the nature of the shaping process(es)?

Wind-Shaping 101

Here we describe the basic concept behind the now-challenged standard model of PN morphologies (a fuller description appears in "Variations on a Theme: Theoretical Models"). Kwok et al. (1978) introduced the interacting stellar wind (ISW) concept into the field of PNe. Fast stellar winds quickly overtake the slower, denser material ejected as or just after the AGB phase and interact hydrodynamically with it. In steady state, fast winds do not reach the "slow winds" directly. Most of the kinetic energy in fast winds is first converted to thermal energy at a generally radiationless shock where the ram pressure of the diluted wind matches the pressure in the world upstream. If the fast wind speed exceeds 150 km s^{-1}, then a "hot bubble" (10^{7-8} K) with a large adiabatic sound speed forms just ahead of this shock. Its high internal pressure pushes the bubble supersonically into the older and denser gas upstream. Hence, the expanding bubble displaces and sweeps the slow wind into a thin, compressed, efficiently cooling dense rim at its leading edge.

Consequently, if we assume isotropy of the old slow and young fast winds, then all PNe should consist of a hot, nearly invisible central cavity separated from a smooth mantle of slow winds by a bright rim of plowed-up gas. The potential relevance of ISW models is nicely confirmed by most round and mildly elliptical PNe, such as IC 3568 and NGC 3132 (Figure 1, see color insert). Frank et al. (1990) demonstrated detailed agreement between predictions of ISW models and the observed density distribution, or shells, of round PNe. (Hereafter we avoid the use of the term "shell" that has become extremely ambiguous over the years.)

Bipolar PNe—those with two lobes on opposite sides of the nuclear region—form a greater challenge to explain. Calvet & Peimbert (1983) were the first to suggest that a dense circumstellar torus can deflect the winds toward a polar axis and form a pair of expanding bubbles. Kahn & West (1985) used analytical calculations to demonstrate the plausibility of the mechanism. Balick (1987), Balick et al. (1987), and others observationally confirmed that dense equatorial disks gird the bright rims of nearly all axisymmetric PNe. Analytic and numerical studies of fast winds by disks and tori by Icke (1988), Icke et al. (1989, 1992), Soker & Livio (1989), Frank (1994), Mellema (1994, 1995), and Mellema & Frank (1995) followed shortly. Their calculations became known as generalized ISW (GISW) models discussed in detail in "Variations on a Theme: Theoretical Models."

The origin of the putative collimating disk or torus remains a matter of conjecture to this day. Historically, Morris (1981, 1987), Soker & Livio (1989), and many others suggested that the disks could be the remnants of binary mergers and tidal interactions. As observations have improved, PN morphologies with far too much high-order structure (e.g., multiple pairs of lobes each with their own axes of symmetry, and jets of very thin cross section) have clearly demonstrated that GISW models are, by themselves, inadequate. Meanwhile, additional paradigms such as magnetic collimation have emerged as well. The pursuit of the paradigms and the data that constrain them are the subject of this review.

Scope of this Review

Samuel Clemens once quipped that horse races are less interesting than the people who bet on them. The same is true for PNe research. In "Classical PNe" and "ProtoPNe" we'll examine the horses, PNe and pPNe, and organize their properties. In "Observational Results and Questions: Summary" we develop a list of issues that need insight and understanding. In "Variations on a Theme: Theoretical Models" below we shall look at the bettors: the mechanisms proposed to account for the observed morphologies and kinematics of PNe and pPNe. A brief summary and assessment of the field is given in "Conclusions and Future Directions." Regretfully, we cannot consider the growing body of research regarding the central stars, chemical abundances, luminosities, masses, distances, radiative wind-driving mechanisms, ionization structures, numbers, or galactic distribution of PNe.

CLASSICAL PNe

Organizing PNe: Morphological Types

Like the taxonomy of moths, nebular classification is important as it can uncover recurrent patterns and themes that reveal common outcomes of processes shaping PNe and other nebulae. A useful system must be based on clear class definitions, be easy to apply, and incorporate the vast majority of nebulae. Here we describe two complementary systems, both of which are widely used.

The most common description of global morphology, originally assembled from ideas in the literature and evaluated by Balick (1987), is based on the outline of the spatially resolved nebula. The four basic nebular types are Round (R, prototype IC 3568), Elliptical (E, NGC 3132 and 6826), Bipolar (Bp, a pair of lobes), and irregular (relatively rare). [See Figure 1 (color insert) for images of prototypes.] The first three classes are not completely distinct: Mildly elliptical nebulae form the transition from R to E, and peanut-shaped objects divide the E and Bp classifications. It is very important to note the Galactic scale height of Bp PNe, 130 pc, is that of ≥ 1.5 M$_\odot$ stars, whereas E and R PNe have the same scale height of ≤ 1.1 M$_\odot$ stars (Zuckerman & Aller 1986, Corradi & Schwarz 1995). Therefore, Bp nebulae almost certainly evolve from higher-mass progenitors than other types.

In this paper we subdivide Bp into "butterfly," in which the waist is pinched into the center, and "bilobed" PNe, in which a pair of larger outer lobes connects to a central and generally smaller R or E nebula. M2-9 (Doyle et al. 2000) and He2-104 (Corradi et al. 2001) are examples of the former and NGC 6886 and 7026 are characteristic of the latter. [See Figure 1 (color insert) and Corradi & Schwarz (1995), Balick (2000) and López (2000) for image galleries of these and other bilobed and butterfly PNe.] The bipolar class has recently been generalized to quadrupolar and multipolar (Manchado et al. 1996, Guerrero & Manchado 1998, Muthu & Anandarao 2001).

GISW models predict that bilobed PNe will form as the interior hot bubble punctures the outer edge of the confining mantle on the symmetry axis (Balick 1987, Icke et al. 1989). The lobes grow rapidly at first as the escaping hot gas plows outwards into sparse gas outside the mantle and cools adiabatically. The growth of the lobes is decelerated as ambient material of relatively low specific momentum is accreted. This compressed material will cool and may partially recombine in time.

The structure and kinematic patterns of butterfly PNe are much more complicated and diverse than for bilobed PNe. We consider the properties of butterfly PNe later in this section. For now we note that several of them have luminous and relatively cool central stars sometimes showing spectral features of symbiotics or

Miras. Thus, they may be bone fide PNe based on morphological selection criteria; nonetheless, their central stars will not reside in the H-R diagram on the evolutionary tracks where most other PN nuclei are found, and the history of their evolution may not be consonant with true PNe.

All morphological classes are blurred by projection effects (Frank et al. 1993). For example, ellipticals and bipolars degenerate into round PNe when we view the nebulae along their symmetry axes. In addition, the morphological classes can depend somewhat on the emission line used for making an image. It is not unusual for images obtained in [NII] and [OI] lines to appear edge-brightened, whereas [OIII] and Hα emission seems to arise in the nebular interior. Any taxonomic system that is based on outlines is best defined from the former, low-ionization images in which the nebular edges are generally exaggerated.

A second classification scheme favored by Manchado, Stanghellini, Corradi, and others is based on the highest degree of overall symmetry of the nebular interior. The basic types are axisymmetric, reflection symmetric, point symmetric, and asymmetric. The first requires reflection symmetry around the major and minor axes, the second requires reflection symmetry about the minor axis only, the third requires symmetry only through the nebular center, and the final category has no symmetry about an axis or point. Most PNe are very nearly axisymmetric. Point-symmetric nebulae tend to be S-shaped, corkscrew-like, or multipolar (showing more than one axis of symmetry). Reflection-symmetric PNe are rare; M2–9 is an example since its lobe edges are brighter on one side than the other.

Some PNe have striking symmetries that combine or transcend standard taxonomical schemes above. The crossed ellipses of the Cat's Eye Nebula were already mentioned. Hu 2–1 is the PN equivalent of a Russian doll, with nested nebulae of highly elliptical outlines at various position angles (Miranda et al. 2001b). Sahai & Trauger (1998) published images of several very compact, low-ionization (young) PNe that have striking multi-axis symmetries, some of which look like textbook hydrogen wave functions! Any process that forms such high-order symmetries requires a high degree of global coordination or internal communication. We note also that in some cases certain emission characteristics may single out morphological types as in H_2 emission for bipolar nebula (Kastner et al. 1994, 1996).

Knots and Jets

In addition to the global structures, so-called "fine structures," such as knots and jets, are found in about half of PNe (see review by Gonçalves et al. 2001). It is common for the knots and jets to appear primarily or exclusively in emission lines of low ionization, so Gonçalves et al. suggested the acronym "LIS" (low-ionization structures) for them. Many knots have directly associated tails that point radially outwards (O'Dell & Ball 1985, O'Dell & Handron 1996, Balick et al. 1998). Such a morphology indicates the tails develop when winds from the central star sweep into gas ablating or evaporating from dense neutral knots.

A special subclass of LIS are FLIERs, or Fast Low-Ionization Emission Regions, in which symmetric pairs of low-ionization knots exhibit distinctly opposite, supersonic Doppler shifts (≥ 20 km s^{-1}). These high Doppler shifts imply that FLIERs have a smaller kinematic age than the gas that surrounds them. They appear to be enriched in nitrogen/oxygen (N/O) (Balick et al. 1994), so they may have been ejected directly from the central star after the slow wind ejection ended. But this picture is incomplete. For example, ejected material should produce bow shocks whose heads point outwards; however, the reverse is found (Balick et al. 1998). [See López (2000) and "Variations on a Theme: Theoretical Models" below for a more extensive summary of FLIERs.]

The word "jets" is generally restricted to thin and often radial features with no visible sign of widening. (Jets and thin tails are functionally the same morphology.) Some jets have corkscrew shapes, called Bipolar Episodic Rotating jeTs (BRETs) (López et al. 1993; see reviews by López 1997, 2000). Unlike H-H objects/jets, the base of nearly all visible jets in PNe is widely separated from their central star.

Knots in the Helix Nebula (appearing by the hundreds) have been studied intensively by O'Dell and collaborators in a series of papers. Knots like these, many with long radial tails, are also found in the Eskimo, the Dumbbell, Abell 30, and other PNe where their properties are fairly similar. These knots exhibit surface ionization that is strongest where the knots directly face the star. O'Dell concluded that the knots are being slowly photoevaporated, and that the evaporated gas is swept back by gentle outflowing winds. Meaburn et al. (1998) found the knots to be moving radially more slowly than the outflowing winds in which they are immersed. They assert that the knots are primarily eroded by the wind. However, O'Dell et al. (2000) show that the emission is extremely well modeled by simple photoionization alone. The aggregate gas dispersed from these knots may very well determine the large-scale morphology and line flux of some of their parent nebulae, such as the Helix, at least at the present time. It is extremely unclear how or why the knots form in such large numbers and highly organized patterns.

The kinematics of small features—usually LISs—show exciting but puzzling results. [See the excellent review and synthesis of the kinematics of LISs by Gonçalves et al. (2001).] We have already mentioned that FLIERs come in pairs with equal but opposite supersonic velocities. Extremely high speeds in remote knots along the symmetry axis of some butterfly PNe—up to 630 km s^{-1} for MyCn18 (O'Connor et al. 2000), 500 km s^{-1} for Menzel 3 (Redman et al. 2000), 350 km s^{-1} for He2–111 (Meaburn & Walsh 1989), 164 km s^{-1} for M2–9 (Schwarz et al. 1997), and 300 km s^{-1} for M1–16 (Corradi & Schwarz 1993b)—are more the rule than the exception. These speeds pose huge challenges for GISW models since the transfer of momentum from stellar winds ($V_w \approx 10^3$ km s^{-1}) to nearby dense gas is not very efficient.

Halos

A halo often surrounds the core of a far brighter round or elliptical PN core at the halo's center. Halos are almost exclusively round in shape, very faint in surface brightness, smooth or slightly mottled, and limb-brightened (an exception is the outermost halo of the Cat's Eye Nebula). Most halos seem to be spherical bubbles seen in projection. Some of the deepest images are by Balick et al. (1992) and Hajian et al. (1997). The properties of halos are summarized by Terzian & Hajian (2000).

Frank et al. (1990) used hydrodynamic modeling to show that the density distributions of single ring-like halos are consistent with an earlier and highly evolved episode of mass ejection expanding into an isotropic medium. The most complete observational investigations of individual halos are by Bryce et al. (1992a,b) and Bryce et al. (1994). Their kinematical observations do not always confirm this picture.

Recent observations of halos in both PNe and pPNe have revealed a second type of halo morphology: more or less evenly spaced concentric rings or arcs (e.g., Sahai et al. 1998, Balick et al. 2001, Hrivnak et al. 2001) that lie outside of but in close contact with their brighter cores. Again, these nested features are interpreted to be the edges of concentric bubbles or bubbles segments seen in projection. These are found in pPNe and PNe of sundry morphological types including bipolars.

For halos composed of concentric rings or arcs, the intra-bubble time scales are close to 1000 years. The repetition periods of surface pulsations are much shorter (≤ 10 years), and core flash timescales are much longer ($\sim 10^5$ years). Simis et al. (2001) have found a hydrodynamic instability in dusty outflows with the appropriate 1000-year timescale. García-Segura et al. (2001) show that regularly spaced bubbles can be created by pressure waves associated with periodic reversals of the stellar magnetic field.

One implication of the both single and concentric halos surrounding much more structured inner PNe is that the mode of stellar mass loss changes abruptly from isotropic to something far more complex, dense, massive, and structured. Judging from the images, the mode change is abrupt and permanent. As we show later, the change is likely to occur near the end of the AGB phase of evolution as the PN phase commences.

Large-Scale Kinematics and Proper Motions

Speaking crudely, kinematics and proper motions trace the first derivative of the nebular structure. Hence kinematic studies bear directly on our understanding of nebular shaping processes. The primary technique of probing nebular kinematics has been Doppler shift mapping using an imaging Fabry-Perot interferometer or a multiple long-slit, high-dispersion (Echelle) spectrograph.

The most systematic studies of the kinematics of individual PNe are by Meaburn, Bryce, López, and other collaborators (since 1980); Miranda, Solf and several collaborators over the past 20 years; Balick et al. (1987) and Icke et al. (1989); Corradi and various collaborators (starting in 1993); Cuesta, Phillips, Mampaso and their collaborators (from 1993); Guerrero et al. (1998). Compilations of, and a search for, statistical trends in kinematic data from many PNe by Weinberger (1989) and Sabbadin (1984), Sabbadin et al. (1984), and Sabbadin et al. (1986) are still very useful.

In summary, R and E nebulae expand almost uniformly and homologously. Smooth mantles expand in such a way that velocity scales with distance. The bright rims on the inside edges of the mantles are generally expanding faster than the inner mantle, as expected where the bubble is overtaking and plowing up the material upstream from it. Gas motions reflect the symmetries of the structure with which they are associated. R and E PNe expand radially and tend (with minor exceptions) to grow faster along their projected symmetry axis than in their equatorial plane. This is as might be expected if they are to maintain their overall shapes as they evolve. Bilobed (as distinct from butterfly nebulae) follow the trend of ellipticals.

The kinematic patterns for butterfly PNe often reveal a more complex story. M1–16, Hubble 5, NGC 6537 (Corradi & Schwarz 1993a,b; Huggins et al. 2000) and He2–104, (Corradi et al. 2001) exhibit patterns of "Hubble Flows"; i.e., radial outflows whose speeds increase linearly with distance from the nucleus. Corradi et al. surmised that the expansion of three very distinct morphological components of He2-104 all have the same expansion age, suggesting that these three parts of the nebulae were formed in one event and then flung ballistically outward. In contrast, two butterfly nebulae exhibit fairly constant velocities with radial offset in their inner regions (e.g., M2–9, Solf 2000; He2–90, Guerrero et al. 2001). This kinematic pattern is much as expected for a constant, sustained outflow from the wind source, much as seen in most H-H systems. Why this kinematic signature is common in YSOs and rare in PNe is not clear.

Another method of charting the internal motions of PNe is to make a time series of images. The resulting movies provide dynamical information that nicely complements Doppler-velocity mapping. Multi-epoch images spanning two to five years with sufficient spatial resolution to monitor the changes in the structure of over a dozen PNe are being obtained by Hajian and his collaborators. Reed et al. (1999) found that NGC 6543 has expanded fairly uniformly (by a magnification factor of 1.00275) in three years. Schwarz et al. (1997) found a similar pattern in the outer regions of M2–9. Similar self-similar patterns have been found in some shocked pPNe such as OH231 ("ProtoPNe). Distances to these nebulae can be found using expansion parallax.

Nebular Momenta

All of the wind's momentum is transferred into the snowplowed gas in the rim of a model wind-blown bubble. Note, however, that only 20% of the wind's kinetic energy is deposited (Dyson & Williams 1980). It is therefore useful to compare the momenta of the nebula to that supplied by stellar radiation since the AGB phase to see if radiatively driven stellar winds are likely to be the driving mechanism behind PNe.

The cumulative momentum of a radiatively driven wind P_w is $(L_*/c)\Delta t$, where L_* is the luminosity of the star during its nebular expansion lifetime Δt. A low-mass star will sustain a luminosity of $\sim 10^3 \, L_\odot$ for about 10^4 year, providing a cumulative wind momentum P_w of about $10^{37.5}$ g cm s^{-1}. By comparison, a typical round or elliptical PN consists of about ~ 0.2 M$_\odot$ with an expansion velocity of 10 km s^{-1} ahead of the expanding hot bubble in its interior. The corresponding momentum of the nebula, P_{neb}, is 10^{39} g cm s^{-1}. Multiple scattering of photons at the base of the wind is believed to account for the momentum excess $P_{neb}/P_w \sim \tau_{scat} \sim 10$–20. Hence the nebular momentum agrees with expectations of ISW models.

Higher-mass central stars have higher luminosities and somewhat offsetting shorter evolution times, or about the same P_w as for their lower-mass counterparts. The masses of their bipolar nebulae (up to 1 M$_\odot$ in ionized and molecular gas) and outflow speeds (20–50 km s^{-1}) are several times larger than for round and elliptical PNe. Bujarrabal et al. (2001) find that the momentum budgets of Bp nebulae are in accord with the momentum yield from their luminosity-driven winds.

Extended X rays in PNe

The Chandra X-ray telescope was the first to map extended X rays in the elliptical PNe BD $+ 30°3639$ (Kastner et al. 2000), NGC 7027 (Kastner et al. 2001) and NGC 6543 (Chu et al. 2001a,b), and more efforts are under way.

So far, X rays are closely associated with the innermost optically dark hot bubbles of these three nebulae. The observed X-ray emission temperatures are 10^6 K whereas ISW models predict temperatures as much as 100 times hotter. Mixing of hot gas with much cooler, denser gas across the contact discontinuity is ruled out because each region is observed to be chemically differentiated. Thus the observations suggest that additional physics, such as thermal conduction, must be added to ISW models to properly describe the growth of the interior structure of the hot, expanding bubble and its impact on the slower gas upstream.

PROTOPNe

Several pPNe had been identified prior to 1990, most all of them owing to their unusual IR or molecular properties. The best known of these are CRL 618, CRL 2688 (the Egg), IRC $+$ 10216, HR 44179 (the Red Rectangle), and IRAS09371 $+$ 1212 (Frosty Leo). These and a few others were legacies of opportunistic discoveries— some from observations of radio and IR molecular lines, and others from early IR surveys done during the 1970s and 1980s. These early pPNe were the tip of the iceberg; many more, most of them heavily obscured, were found starting in the 1980s as IR measurements improved. Today, hundreds of IR color-selected candidate objects and OH/IR stars have been identified, all of which can be loosely described as pPNe.

AGB and post-AGB stars are differentiated from each other and YSOs in two ways: van der Veen & Habing (1988) and van der Veen et al. (1989) showed how the IR colors of pPNe systematically distinguish them their state of evolution as the central star evolves from AGB to PN nucleus. In addition, Kwok (1993) and his collaborators showed that the IR spectral energy distributions (SEDs) of pPNe undergo a qualitative change as the star enters the post-AGB phase of its evolution. Kwok attributes the changes of the SEDs to the ejection and detachment of a cool shell from the core as the star evolves from AGB to post-AGB states.

pPNe are just beginning their expansions, so they tend to be intrinsically small. pPNe lifetimes are short, so they tend to be rarer and, hence, located at larger distances than typical PNe. Most pPNe are (or have recently been) forming dust particles, which heavily obscure them. In spite of all of these difficulties, optical images of pPNe in dust-scattered starlight and IR images in thermal dust continua, along with emission-line images of extremely youthful PNe, show that pPNe are just as important to understanding mass outflow properties of AGB and post-AGB stars as are PNe, their much brighter, less obscured, and more mature siblings.

Morphology: Optical and IR Continuum Observations

The first systematic imaging survey of pPNe by Kwok et al. (1996) uncovered about a dozen resolved pPNe. They found that the morphologies of these pPNe had about the same distribution of morphological types as ordinary PNe and surmised that PNe originally obtain their underlying symmetries during the slow-wind phase of mass loss. This work, which was limited by a small sample size and poor spatial resolution, was prescient.

HST images have been obtained for several pPNe with post-AGB nuclei, and the results are as spectacular as they are unanticipated. These images, made in reflected starlight and emission lines arising in shocks, show striking bipolar and multipolar symmetries and in several cases, large multiple outer halos. Two such HST images appear at the bottom of Figure 1. Many more observations can be found in Kwok et al. (1998, 2000), Su et al. (1998), Hrivnak et al. (1999, 2001), Skinner et al. (1998), Ueta et al. (2000), Sahai et al. (1998, 1999a,c, 2000); Trammell & Goodrich (1996) and others.

Sahai & Trauger (1998) published a gallery of some of the most stunning of all HST imaging observations. Their targets are very young PNe that not surprisingly are small in size, low in overall ionization, and high in IR excess flux (Zhang & Kwok 1993). These criteria select objects in transition from pPNe to classical PNe in which the central star is just starting to ionize the gas. A few evolved low-ionization bipolars fell into their net; we ignore these here. The remaining objects show a panoply of morphologies, all of them exhibiting startlingly high-order symmetries—more so than mature PNe. Many have several axes of symmetry in different orientations. Names such as "starfish" (He2-47 and He2-339; see Figure 1) tell the story. Similar starfish morphologies have been found in outflows of various traditional pPNe (CRL 2688—Cox et al. 2000; CRL 618—Trammell 2000, Ueta et al. 2001; IRAS 16594-4656—Hrivnak et al. 1999; Roberts 22—Sahai et al. 1999c; and Frosty Leo—Sahai et al. 2000). Other young PNe such as He2–131 and He2–138, have serpentine inner boundaries that resemble atomic wave functions. For a recent compilation of observations of these young objects see Section 1 of Sahai (2000).

Although the many odd morphologies of pPNe and very young PNe are unexplained, several straightforward conclusions emerge: 1. as suggested by Sahai & Trauger, GISW models simply do not have the complexity necessary to account for the common high-order symmetries; 2. the winds that form these complex structures may not be isotropic; and 3. normal PNe only rarely show such high-order symmetries. The last conclusion suggests that fast winds, hot bubbles, and ionization fronts burnish or eradicate many structural details and, consequently, invite an unrealistically simplistic approach to constructing hydrodynamic models. Sahai & Trauger proposed that the multipolar lobes form as a flailing nozzle ejects occasional jets. Several authors have proposed that such complex symmetries mandate magnetohydrodynamic models ("Variations on a Theme: Theoretical Models").

Morphologies of pPNe with AGB Nuclei

Imaging studies specifically directed at stars in their AGB phase, though relatively few in number, show that mass loss is generally isotropic. IRC + 10216 is a spectacular example of a halo of a pPNe in its AGB phase that consists of multiple concentric dust shells—actually, more like arc segments with irregular spacings of 200–800 years (Mauron & Huggins 1999, 2000). Dust rings and asymmetries have been mapped with lower resolution in several other post-AGB stars: (AC Her—Jura et al. 2000, HD 56126—Jura et al. 2000) and other papers cited therein. Jura & Kahane (1999) discuss orbiting molecular reservoirs around many evolved stars. A nice review is by Jura (1999).

CO imaging observations generally support the isotropy of AGB-star outflows. TT Cygni, an AGB carbon star, shows a nearly perfect spherical (albeit clumpy) thin bubble of CO expanding at 12.6 km s^{-1} with a mass of 0.007 M$_{\odot}$ and a expansion age of 7000 y if no ambient gas of low specific momentum has been accreted (Olofsson et al. 2000). U Cam has a similar but younger (800 y) bubble-like shell (Lindqvist et al. 1999). IRC + 10420 may be similar (Castro-Carrizo et al. 2001). An exception to isotropic outflow is the carbon star V Hydrae. Kahane et al. (1996) and Knapp et al. (1997) argue that its observed CO profile shapes are best fitted with a tilted equatorial disk expanding at 15 km s^{-1} and collimated outflows of 45 and 200 km s^{-1} along its symmetry axis.

Interestingly, although the vast majority of scattered starlight from IRC + 10216 comes from the many concentric shells in its extended halo, a short exposure HST image shows that the nuclear region has bipolar symmetry (Kwok 2000). Apparently the mode of mass loss in this object has recently changed.

Morphologies of pPNe with Post-AGB Nuclei

Meixner et al. (1997, 1999) have been systematically imaging pPNe in the mid-IR where extinction is probably not significant (but the contribution from cold dust—probably containing a majority of the dust's mass—is minimal). [See the comprehensive review of these surveys by Meixner (2000).] Several very significant results are these: 33 of 73 pPN candidates were spatially resolved, and all of them are axisymmetric. The mid-IR morphologies of pPNe split into two distinct groups, SOLE and DUPLEX, which are possibly the antecedents of elliptical and bipolar PNe, respectively, though there is disagreement on whether these groups are truly distinct or just the result of inclination effects.

Some spectacular individual cases of highly collimated outflows, many showing shock-excited optical emission lines, have been studied in the papers listed earlier. The lobes of He3-401 show magnificent pencil-like collimation very clearly traceable to a central disk (Sahai et al. 1999a). Kwok et al. (2000) uncovered a spectacular example, IRAS 17106-3046, which clearly shows a bipolar outflow along the symmetry axis of a disk seen clearly in reflected light. IRAS 04296 + 3429 may be a very similar structure (Sahai (1999). [See the bottom row of Figure 1 for selected images.]

Kinematical studies of post-AGB nebulae are generally made in the bright low-lying transitions of CO. Early single dish observations showed that most pPNe have an integrated CO profile that has a Doppler half width of 15–30 km s^{-1} (Huggins & Healy 1989). Recently CO interferometer maps have revealed unexpected details (e.g., Alcolea et al. 2000). Deeper, higher-transition CO lines occasionally show additional broad (≥ 100 km s^{-1}) faint wings (Bujarrabal et al. 2001). Of those that have been imaged in CO, virtually all are butterfly pPNe. The most notable are CRL 2688, CRL 618, OH231.8 + 4.2 (Figure 1), M1–92, and He3–1475 (Figure 1), in no particular order. Each is a highly individualistic case. Some show optical emission-line radiation, generally from shocks, which allows their morphology and kinematics to be traced in great detail.

In summary, although pPNe associated with AGB stars are normally round, those of post-AGB objects are largely axisymmetric and those of high-luminosity systems are often highly bipolar. It is easy to conjecture that the halos of ordinary PNe form first and then the mode of mass loss changes to axisymmetric. Indeed, the images of young PNe by Sahai & Trauger suggest that the new mode of mass loss is one of very carefully orchestrated outflow patterns that soften considerably as photoheating and ionization fronts alter the nebula.

Outflow Momenta of pPNe

We highlight a very important kinematic result that has not received adequate attention: Radiatively accelerated winds cannot account for the momenta implied by the broad wings of the CO profiles of pPNe with relatively high-mass nuclei. This problem was identified first by Knapp (1986), and most recently and comprehensively investigated by Bujarrabal et al. (2001). The latter summarize the CO data for 16 objects that they studied and another 21 objects compiled from the literature.

Almost 90% of 32 objects with excellent data show optically thin, broad wings containing at least 0.1 M$_\odot$ of gas. By taking the first moment of the line profile, Bujarrabal et al. 2001 computed the component of momentum along the observer's line of sight (a fraction of the total momentum, of course). As discussed earlier, this scalar momentum P can be compared to the momentum in the stellar radiation,

$(L_*/c)\Delta t$, where here L_* is the stellar luminosity emitted during the pPNe outflow expansion lifetime, Δt. They find that after conservatively culling objects for which data are dubious, 21 of 23 CO-emitting ppNe with high-mass central stars show highly excessive (factor of >100) outflow momenta. The excess momentum was not seen in any of the four low-mass stars in their survey. In light of these results, both the launching and collimation of winds in ppNe becomes problematic.

Two Case Studies

A detailed understanding of exceptional objects may be useful for uncovering and evaluating the key physical processes involved in a class of objects. Therefore we present the observational results of two case studies of phenomenological extrema in which the shaping processes act with outstanding vigor. We elect to consider OH231.8 + 4.2 (a.k.a., the Calabash or Juggler or Rotten Egg Nebula, OH07399–1435, QX Pup, CRL 5237) and He3–1475 (a.k.a. Henize 3–1475, IRAS 17423-1755). Both show strong IR excesses, shock-excited optical emission line nebulosity, very fast outflows (300 km s^{-1} for OH231 and 2000 km s^{-1} for He3–1475), and highly bipolar (butterfly) structures. Images of both nebulae are found in Figure 1.

OH231.8 + 4.2   The nucleus of OH231 is QX Pup, a 700-day variable Mira of spectral type M9III (Cohen 1981). A slowly expanding SO disk seems to surround the nucleus (Sánchez Contreras et al. 2000b). Evidence of binarity was found by Gómez & Rodríguez (2001), who suggest that the lobes originate $\sim 1''$ from QX Pup at a putative invisible companion in a torus-like region of very high extinction. Obviously the nucleus may be a symbiotic star.

OH231 exhibits two closed, edge-brightened, seemingly hollow lobes seen in Hα, [N II], and other low-ionization optical lines (Sánchez Contreras et al. 2000a and Reipurth 1987). As seen in Figure 1, the southern (receding) lobe is twice as large but fainter in surface brightness than its northern counterpart. Two finger-shaped columns seen in dust-scattered starlight and molecular line emission span the lobes from the nucleus to the outermost lobe tips along the nebular symmetry axis. Optical emission lines and HCO$^+$ peak at the outer tips of the lobes where these lines are probably excited by local shocks, as outflowing gas flowing along the fingers splashes into the lobe edges (Sánchez Contreras 2000a).

Alcolea et al. (2001) mapped the distribution and kinematics of CO in the fingers of OH231. One third of the total CO mass of 1 M$_\odot$ is associated with the fingers, and the rest is concentrated close to the nucleus. The CO Doppler shift increases linearly with distance from the nucleus along the fingers to -210 km s^{-1} at the tip of the northern lobe and $+430$ km s^{-1} in the southern, suggesting a single expansion age of 780 y for both lobes. Thus, the collimated fingers may have been produced in a brief but highly organized event. (We hasten to add the pPN M1–92 and the bipolar PN He2–104, each with a symbiotic/Mira nucleus, share many morphological and kinematic similarities with OH231.) Another key point

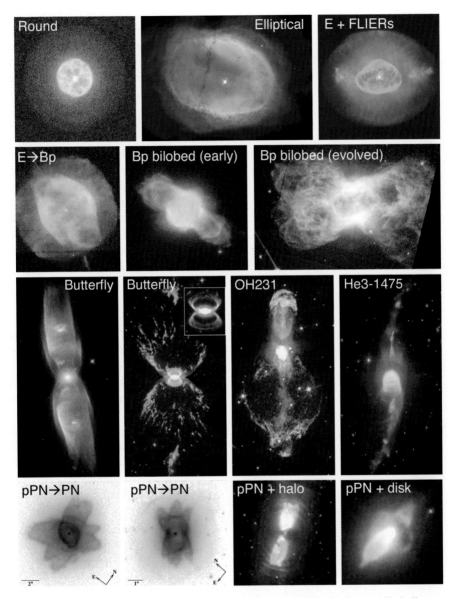

Figure 1 Illustrations of morphological types of PNe and pPNe. Red generally indicates regions of low ionization (colors of NGC 3132 and OH231 have been altered for this purpose). See papers by author for details. (*Top row*) IC 3568 (Bond & NASA), NC 3132 (Sahai & NASA), NC 6826 with FLIERs in red (Balick & NASA); (*Second row*) NC 7354, NC 6886, and NC 7026 (all Hajian & NASA); (*Third row*) M2-9 (Balick & NASA), He2-104 (Corradi & NASA), OH231 (Bujarrabal & ESA), He3-1475 (Borkowski & NASA); (*Bottom row*) He2-47 (Sahai & NASA), He2-339 (Sahai & NASA), IRAS 17150-3224 (Kwok & NASA), IRAS 16594-4656 (Hrivnak & NASA).

raised by Alcolea et al. is that the excess of scalar momentum observed in the CO outflows is far too high—by a factor of 1000—to be explained by the momentum carried by radiation pressure.

He3–1475   Riera et al. (1995) (hereafter RGMPR95) first called attention to this remarkable Pop II object that lies in the bulge 800 pc above the Galactic nucleus where its luminosity is $2 \cdot 10^4$ $L_\odot$ (Borkowski & Harrington 2001; BH01). RGMPR95 found that P-Cyg features and strong permitted lines such as FeII and CaII arise in the nucleus, suggesting a B[e] or B[q]-type nuclear spectrum. They also found near-nuclear knots with outflow speeds, which, after an uncertain correction for inclination, imply outflow speeds of 2000 km s^{-1}. BH01 and Sánchez Contreras & Sahai (2001) (SS01), using HST spectra with ~0.05″ spatial resolution, resolved individual knots of high-speed outflows, one characterized by 1200 km s^{-1} within $5 \cdot 10^{16}$ cm of the nucleus, and the other by 2300 km s^{-1} within 10^{16} cm.

RGMPR95 were also the first to note that much of the forbidden line emission (largely [NII]) arises from a series of N-rich knots emanating along an S-shaped (point symmetric) axis through the nucleus. The knots have decreasing densities, very high electron temperatures (~18,000 K) and line splittings consistent with fast-moving and expanding blobs excited by local shock heating. The Doppler shifts of the knots steadily decrease with radius, which is highly unusual among bipolar PNe and pPNe. RGMPR95 suggested that the knots are ejected from a precessing "cannon" anchored to the nucleus. If so they must decelerate as they propagate in order to prevent merging.

HST images by Borkowski et al. (1997 hereafter BBH97) spatially resolve the knots and some of the bow shocks along with a thick dusty torus associated with the nucleus. One of the innermost knots is seen to have an outward-facing arrow-like shape. However, SS01 assert that the streamlines of 2300 km s^{-1} outflow appear to be pristine stellar winds that emerge fully collimated from the unresolved nuclear region.

Single-dish CO observations of He3–1475 were reported by Bujarrabal et al. They show relatively broad (~50 km s^{-1}) but weak emission wings. Bujarrabal et al. find a CO mass of about 0.6 $M_\odot$ and a large excess (1000) of scalar momentum over that expected from the cumulative pressure of stellar radiation.

OBSERVATIONAL RESULTS AND QUESTIONS: SUMMARY

A summary list of results and unresolved issues posed by observations makes a nice transition from the observations reviewed in "Classical PNe" and "ProtoPNe" to theories and paradigms starting in "Variations on a Theme: Theoretical Models".

- The morphologies of PNe and pPNe consist of three major types: round, elliptical, and bipolar, all of which normally show reflection symmetry about their major and minor axes. We further distinguish butterfly bipolars with strongly pinched waists from bilobed bipolars containing an elliptical core. A small fraction of PNe show point reflection through their nucleus, and some bipolars exhibit multiple axes of axisymmetry. Is there a single unified model/paradigm that can explain the large-scale shapes of pPNe and PNe, or is more than one set of shaping processes universally at work?

- Disks and tori are nearly ubiquitous among PNe and pPNe (other than round ones). What process or processes produces these structures?

- The outflows in butterfly PNe and pPNe can sometimes be highly collimated. Does the inertial pressure of a dense equatorial disk or torus serve as a collimation nozzle? Or are the flows collimated by magnetic forces or by accretion from a companion star?

- Many PNe and pPNe exhibit kinematic patterns that are readily assignable to Hubble Flows (expansion speed $\propto$ radius). What does this infer about the ejection and/or expansion of the nebular material? Are the motions ballistic (i.e., nonhydrodynamic) or self-similar (in which hydrodynamic shaping establishes a constant shape which simply magnifies in time)?

- The GISW paradigm has been successful for explaining the large-scale shapes and kinematics of round, elliptical, and bilobed PNe. In stark contrast, most butterfly PNe—which are generally associated with symbiotic stars and/or massive nuclei—attain their shapes in a brief event and then seemingly expand ballistically. What is the nature of the event that drives these outflows? Do some PNe avoid hydrodynamic shaping altogether?

- pPNe surrounding AGB stars tend to be isotropic, whereas those around the nuclei of post-AGB pPNe are highly axisymmetric. The change in morphology may be associated with a qualitative change in their IR colors and SEDs during the transition from AGB to post-AGB states. Abrupt changes in the mode of mass loss appear at about this state in the evolution of PNe. All of these results provide circumstantial evidence for the formation of a cool, dense disk/torus at the AGB–post-AGB transition. What process or processes drive the mode change?

- Outflows are generally believed to be the results of radiation-driven stellar winds. Yet many pPNe with relatively high-mass nuclei exhibit momenta orders of magnitude greater than stellar radiation pressure can provide during the expansion lifetime of the outflow. Do the nebular momentum excesses imply that bipolar pPNe are formed by highly disruptive events?

- A small but significant subsample of PNe and pPNe exhibit S-shaped (point-symmetric) morphologies or multiple lobe symmetry axes. It has been suggested repeatedly that the flow collimator precesses or becomes unstable in these systems. How is the collimator destabilized? Is it related to tidal disruption or merger of a companion star? Are magnetic or disk instabilities potentially important?

■ YSOs, some LBV nebulae, and some AGN outflows share some striking morphological similarities with their brighter, less obscured counterparts, PNe and pPne. What shaping processes do they share in common?

Observational Support for Collimation Mechanisms

Producing any preferred axis of mass ejection and symmetry is a nontrivial problem for isolated AGB stars. The two most obvious ways to impose a preferred axis are rotation and/or surface magnetic fields. However, the angular momentum needed to distort a star's surface or form a stable disk are at least two orders of magnitude larger than the angular momentum in the Sun (Wood 1997). Models that require rotation to near-breakup speeds are not plausible without accounting for special sources of angular momentum (see below.)

The recent literature is replete with suggestions that either interacting binary systems or magnetic fields must be the agent(s) that collimate outflows. Are these suggestions backed by direct evidence, circumstantial arguments, or merely speculation? We end this section by addressing the observational evidence for closely orbiting companions and magnetic fields. The putative shaping effects of fields and companions are discussed later.

BINARITY AND RAPID ROTATION IN THE NUCLEI OF PNe   Livio, Soker, and many others have long pointed out that the loosely bound envelope of an AGB star is particularly susceptible to external torques or tidal forces exerted by a nearby orbiting companion. Extremely close companions can lead to mass transfers and the formation of an accretion disk on the secondary. Mergers can spin up the AGB star to breakup, rapidly initiating various disk-forming processes in its newly bloated outer layers. Companion stars potentially solve a host of awkward problems—but only if they actually exist.

Let us consider the evidence of frequent binarity in AGB stars that might account for the statistical properties of PN asymmetries. Only 16 Galactic PNe are observed to have close binary nuclei (Bond 2000). On the other hand, companions may quickly merge as the energy and angular momenta of their orbits are transferred to the AGB star, leaving no remaining trace of binarity by the time asymmetries are visible in the nebula. Indirect evidence of recent mergers would be rapid rotation in AGB stars. However, Barnbaum et al. (1995), in a study of 74 carbon stars, found only one, V Hydrae, with detectable rotation (which they attribute to mass transfer from an unseen companion). Interestingly, V Hya and perhaps IRC + 10216 are unusual AGB stars that show strong signs of collimated outflows.

The strongest arguments for companions are by Yungelson et al. (1993) who were the first to show theoretically that the expected frequency of binarity in the nuclei of PNe is consistent with the histograms of PN morphological types. Their results were corroborated and extended by Soker & Rappaport (2001). Observationally, Wood et al. (1999) found that 25% of the AGB stars in the LMC bar are variable. They ascribe the variability to mass transfer from the AGB star to an invisible companion in a semidetached binary. These results agree well with predictions by Han et al. (1995): 30–40% of stars could experience mergers on the AGB.

Soker (1998a) made a critical observation: the outflow speeds of most highly collimated PNe, often hundreds of km s^{-1}, is far more than a factor of ten larger than the escape velocity from the surface of an AGB or post-AGB star. He suggests that these speeds can only be obtained in close binary systems with high orbital speeds. As we have seen, the outflow momenta of luminous pPNe are far too great to have been produced by stellar radiation pressure, perhaps also suggesting that the system must extract this excess momentum from a closely orbiting companion.

The bottom line is that although binarity is a popular mechanism for forming axisymmetric structures on PNe, direct evidence to support the efficacy of the process is not strong. However, the indirect evidence for binarity in highly bipolar PNe is rather suggestive but still far from compelling.

MAGNETIC FIELDS IN PNe   The role of magnetic fields in shaping PNe is under increasing consideration, as described in later sections (García-Segura et al. 1999, Frank 2000). In order to be effective, the energy density of the fields must be equal to or surpass that of thermal pressures and bulk motions in the outflowing winds. Models described below require at least a magnetic field strength of a gauss at or near the stellar surface. Are such fields common? Do fields play their role close to the star where they would be much stronger or do magnetized winds sweep into the nebula where the tension of the fields can deflect streamlines?

The evidence in favor of fields is sparse, particularly in ionized gas where thermal broadening renders any search for Zeeman splitting hopeless. Even in neutral gas such fields are very difficult to detect because detectable synchrotron and cyclotron emission is not expected, and any polarized line radiation is swamped by brighter foreground and background gas. So far, the evidence for magnetic fields is limited to the high degree of circular polarization found in OH masers. Palen & Fix (2001) found 100% circular polarization in some OH-IR stars, requiring the Zeeman splitting associated with at least a milligauss field in the maser path. Recently, Miranda et al. (2001a) examined a young PN K3-35 and found partial circular polarization of OH masers in its disk and at the tips of its butterfly lobes that they also ascribe to magnetic fields.

In short, observations weakly support the notion that orbiting companions and/or magnetic fields can strongly influence the shaping of PNe. The biggest problem is the extreme difficulty of direct and decisive measurements. The paucity of evidence provides theoreticians with an unconstrained opportunity to model how companions and fields—or any other mechanisms—might serve as shaping tools without the fear and useful feedback of embarrassment.

VARIATIONS ON A THEME: THEORETICAL MODELS

In the second half of this paper, we review theoretical models that attempt to explain the shapes and shaping of PNe and pPNe. It is the close coupling between rapid variations in the evolution of the star and the processes of nebular shaping that makes these objects so interesting from a theoretical standpoint. Whereas radiation-(magneto)hydrodynamic models of nebular evolution may give shapes and kinematics that match observations, a truly successful model must also account for the model's initial conditions that are ultimately associated with the evolution of the star. At present, this is just a hope. This lack of unification in stellar and nebular evolution represents one of the greatest challenges facing the field.

Before launching into the theoretical details of this section, of which there are many, it is important to provide a sense of structure for the discussion. The primary goal is to identify workable, effective physical mechanisms that are grounded in well-understood astronomical contexts. These must explain deviations from isotropic outflows during the formation and evolution of PNe. Thus, we build a foundation based on simple physical models and the results of their numerical derivatives. At the end we present and evaluate specific collimation concepts. Along the way, however, we address several of the questions and challenges posed in "Observational Results and Questions: Summary," such as the formation of knots and jets, the effects of magnetic fields, close companion stars (including those that form common envelopes), ionization fronts, rapid stellar rotation, etc.

Basic Models: Physics, Initial Conditions, and Equations

The published theoretical work on ppN/PN shaping falls into two broad categories: inertial confinement and self-confinement.

INERTIAL CONFINEMENT MODELS   In GISW models a fast wind from the central star of a ppNe or PNe expands into an environment created by an older slow wind whose density distribution is assumed to be toroidal. The essence of this mechanism is basically gas bouncing off a pre-existing equatorial wall of some kind. Note that most GISW models are incomplete because they do not specify how the aspherical environment was created. Popular models for creating the torus are discussed elsewhere in this section.

SELF-CONFINEMENT   The variety of nonaxisymmetric features observed in PNe over the past decade, such as point-symmetric jets, have led to the emergence of a different class of theoretical models in which the wind from the central star is self-collimated. The published models to date have all relied on magnetic fields embedded in the wind to achieve this collimation. The use of MHD effects to create various classes of features begs the question of how the fields are created. This is similar to the way GISW models ignore the origin of toroidal ambient density distributions. Thus, once again, there is a division between models that explain the shapes of ppNe and PNe and those that attempt to explain the origin of the initial conditions that allow the nebular MHD models to operate. A schematic overview of various ppNe and PNe shaping mechanisms is shown in Figure 2 and described in its caption.

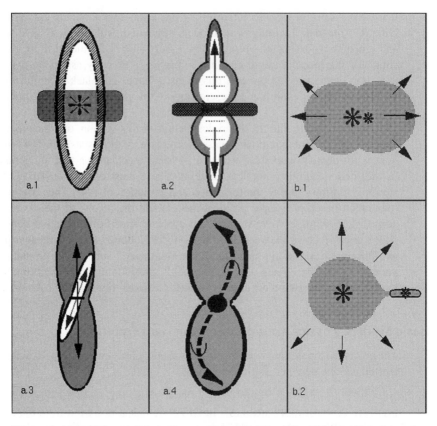

Figure 2   Set of six schematic diagrams showing possible ppNe and PNe formation mechanisms. Schematic a.1 represents the Generalized Interacting Stellar Winds model (GISW) in which an isotropic fast wind from the star expands into a previously ejected toroidally shaped slow wind (Icke et al. 1989, Mellema & Frank 1995). Schematic a.2 represents the Magnetized Wind Blown Bubble (MWBB) in which a weakly magnetized fast wind expands into an aspherical density distribution (Chevalier & Luo 1994; García-Segura et al. 1999). Toroidal magnetic fields (*dashed horizontal lines*) strengthened after passage through the inner shock constrain the outflow and produce jets (*arrows*) along the axis. Schematic a.3 represents disk/star magneto-centrifugal models. A rapidly rotating central object (a disk and/or a star) launches a wind that collimates via hoop stresses. Misalignment of disk and star rotation axis produces multipolar lobes (Blackman et al. 2001a). Schematic a.4 represents outflows driven by episodic jets (collimated on unresolved scales). Precession of the episodic jet can create point symmetric nebulae with interior bow shocks (Cliffe et al. 1996, Steffen & López 1998, Soker & Rapport 2001). Schematic b.1 and b.2 represent processes that can create the toriodal slow wind. Schematic b.1 represents common envelope evolution for short period binaries (Soker 1997, Rasio & Livio 1996). Schematic b.2 represents accretion disk formation via Bondi accretion and Roche lobe effects (Mastrodemos & Morris 1999).

BASIC PHYSICAL ELEMENTS   In studying the shaping of PNe the theorist's task is to track an outflow driven by a star embedded in some form of ambient medium. The structure of the theory is built on the equations of mass, momentum, and energy conservation. The first task is to identify the physical processes that need to be included in the equations. As already noted, magnetic fields may play a dynamically significant role. Thus we need to model a multidimensional, time-dependent MHD flow. We should also include heating and cooling of the gas due to radiation. A proper treatment of energy source terms requires calculation of the microphysical state (ionization, chemistry, level populations). Radiation transfer should be treated explicitly.

The hydrodynamic equations written with the correct microphysical terms and radiative transfer equations can be found in (Frank & Mellema 1994a, Marten & Szczerba 1997). The relevant form of the ideal MHD equations can be found in many references (Priest 1984, Choudhuri 1998). In solving the hydro/MHD equations for pPNe/PNe, many authors have neglected dissipative processes such as molecular and turbulent viscosity, and heat conduction and resistivity. In some settings these processes, particularly heat conduction, may become important (Weaver et al. 1977, Soker 1994a, Kastner et al. 2000). Resistive processes can drive reconnection events in the flow and may be important since they convert magnetic into thermal energy while altering magnetic field topologies.

Hydrodynamic Models of Inertial Confinement

SPHERICAL INTERACTING WINDS MODELS   The majority of hydrodynamic theories of PN shaping are based on a generalization of the interacting stellar winds model for PNe that was first applied to PNe by Kwok et al. (1978). We first consider the theory for spherical, hydrodynamic ($\mathbf{B} = 0$) wind-blown bubbles to clarify basic dynamical issues. In what follows we borrow liberally from a number of excellent treatments of spherical wind-blown bubble theory (Dyson & de Vries 1972, Weaver et al. 1977, Dyson 1977, Kwok et al. 1978, Dyson & Williams 1980, Koo & McKee 1992, Kahn 1983). A more complete version of this review of hydrodynamic shaping appears in Frank (1999).

When a stellar wind is initiated it expands ballistically from the source until enough ambient material is swept up for significant momentum to be exchanged between the two fluids. A triplet of hydrodynamic discontinuities forms defining an interaction region bounded internally (externally) by undisturbed wind (ambient) gas. At the outer boundary of this region is an outward-facing shock called the ambient shock at position R_{as}. The inner boundary is defined by an inward-facing shock called the wind shock at position R_{ws}. A contact discontinuity (CD), R_{cd}, separates the shocked wind and shocked ambient material. In a spherical bubble [one-dimensional, (1-D)] these discontinuities form a sequence in radius: $R_{ws} < R_{cd} < R_{as}$.

The compressed gas behind either or both shocks will emit strongly in optical, UV, and IR wavelengths producing a bright nebular rim that defines the observable bubble. The dynamics of the bubble will be determined by the strength of post-shock cooling. We can define a cooling timescale $t_c = E_t/\dot{E}_t$ for each shock, where E_t is the thermal energy density of the gas. $\dot{E}_t \propto n^2 \Lambda(T)$ when $\Lambda(T)$ is an appropriate cooling curve. Thus, $t_c \propto T/n\Lambda$. We also define a dynamical timescale $t_d = \frac{R_{as}}{V_{as}}$, where V_{as} is the speed of the ambient shock. Comparison of t_c and t_d then separates wind-blown bubbles into two classes: radiative (also known as momentum conserving) and adiabatic (a.k.a. energy conserving).

In PNe and pPNe, the densities in the ambient medium (the slow AGB wind, a circumbinary torus, etc.) are high enough to ensure a radiative ambient shock. Thus, $R_{as} \approx R_{cd}$ and the bubble will have a thin outer rim. The cooling properties of inner shock, however, can vary, and these changes can determine the bubble's dynamics. If $t_c > t_d$, the cooling is weak. The gas retains thermal energy gained after passing through the shock transition, and a hot bubble forms. This type of shock is sometimes referred to as adiabatic. High pressure behind an adiabatic shock limits gas compression ($\rho_{post} \leq 4\rho_{pre}$). In the co-moving frame of the CD, the wind shock is pushed back toward the star, $R_{ws} \ll R_{cd}$. In these adiabatic or energy-conserving bubbles, it is the thermal energy (pressure) of the shocked wind that drives the expansion of the bubble as a whole.

If $t_c < t_d$—which generally applies to pPNe for which the speed of the fast wind is under 150 km s^{-1}—then the gas cools quickly relative to the bubble's growth. The loss of thermal energy behind a strongly cooling radiative shock means the loss of pressure support as well. Because its internal energy is quickly converted into escaping radiation, the shock collapses back toward the contact discontinuity (in a frame moving with the shock). A thin dense shell forms with $R_{ws} \approx R_{cd}$. The expansion is now driven directly by the ram pressure of the stellar wind $\rho_w V_w^2$, and the bubble is often referred to as wind driven, momentum conserving, or radiative. At higher fast-wind speeds, the shocked gas becomes too hot to cool efficiently, so a hot bubble grows between R_{ws} and R_{cd}.

The expansion speed of pressure-driven and wind-driven nebulae can be determined through either exact solutions (Koo & McKee 1992) or through dimensionless arguments. Beginning with a generic spherically symmetric environment $\rho(R) = \rho_0 R^{-l}$, and a wind with mechanical luminosity $L_w = .5\dot{M}_w V_w^2$ (or momentum input $\dot{\Pi} = \dot{M}_w V_w$), one finds the following expressions for the radius of energy- and momentum-conserving systems:

$$R_{as}^E(t) \propto \left(\frac{L_w}{\rho_0}\right)^{\left(\frac{1}{5-l}\right)} t^{\left(\frac{3}{5-l}\right)}, \quad R_{as}^M(t) \propto \left(\frac{\dot{\Pi}}{\rho_0}\right)^{\left(\frac{1}{4-l}\right)} t^{\left(\frac{2}{4-l}\right)}. \tag{1}$$

The first derivative of these expressions gives the expansion speed of the ambient shock. An important point to note here is that the velocity of the ambient shock and nebular rim will be constant for an environment created by a previously deposited wind $l = 2$. Thus wind-wind interactions, which characterize much of PNe and pPNe dynamics, will naturally produce flows that mimic Hubble-law kinematics.

ASPHERICAL NEBULAE—THE GISW MODEL   The ISW paradigm can be extended
to embrace elliptical and bipolar nebulae by generalizing the model to include
aspherical environments. Most work has focused on toroidal density distributions,
$\rho(R, \theta)$, with an equator-to-pole density contrast, $q = \rho(0°)/\rho(90°) = \rho_e/\rho_p$, and
$q \geq 1$. An isotropic stellar wind encountering inertial gradients in the gaseous toroid
will drive an ambient shock that expands more rapidly along the poles than the
equator. In his study of energy-conserving GISW dynamics, Icke (1988) showed
that

$$\frac{\partial R_{as}}{\partial t} \propto \left\{ \frac{\gamma + 1}{2} \frac{P_{hb}}{\rho_0(\theta)} \right\}^{\frac{1}{2}}. \tag{2}$$

Although the pressure in the hot bubble P_{hb} is nearly constant, Equation 2 shows
that the angular variation of $\rho_0(\theta)$ determines the asphericity of the nebula. It
is, therefore, the inertia in the ambient medium that determines the shape of the
bubble. We note that in these models, the ambient thermal pressure plays no role
in the collimation (Reipurth & Bally 2001).

   Analytical determination of aspherical wind-blown bubble evolution involves
solving partial differential equations in two spatial dimensions and time, and it
remains a difficult problem. Dyson (1977) and Kahn & West (1985) provided
analytical solutions for GISW bipolar energy-conserving nebula. Icke's (1988)
investigation explored a wider range of solutions demonstrating that strongly collimated solutions for $R_{as}(\theta, t)$ could be generated if the torus had a narrow opening
angle. Icke et al. (1989) also provided synthetic observations of such models. In
a study of R Aquarii, Henny & Dyson (1992) used a model based on momentum
conservation in the shell and included emission characteristics of the bubble for
both shock-excited and photoionized emission. Working off a novel method proposed by Giuliani (1982), Dwarkadas et al. (1996) produced similarity solutions
for bipolar wind-blown bubbles. These solutions predicted not only the shape of
the bubble, but also the mass motions along the shell of swept up ambient gas that
can lead to substantial modifications of shell density.

   Any study of the GISW formalism must include specification of the ambient
density distribution. The majority of GISW models of PNe assumed ad hoc forms
for the ambient density: $\rho(r, \theta) \propto R^{-l} F(\theta)$ (Dwarkadas et al. 1996). More recent
studies (García-Segura 1997, García-Segura et al. 1999, Collins et al. 1999) have
attempted to incorporate specific physical models for the creation of the toroidal
environment into GISW calculations. This appears to be one of the most important
paths for future research.

NUMERICAL MODELS   The inherent complexity of the GISW problem make simulations a necessary tool for explicating the true range of hydrodynamic flow
patterns. One-dimensional spherically symmetric numerical models have reached
a fairly high degree of sophistication in terms of their treatments of both hydrodynamics, radiation transfer, and microphysical processes (Frank 1994, Mellema
1994, Arthur et al. 1996, Steffen & Schönberner 2001). Recent 1-D models have

made significant steps linking stellar and nebular evolution models (Schönberner & Steffen 2001). Multidimensional studies have been more limited in their ability to link source physics to the nebular evolution.

Soker & Livio (1989) were the first to explore time-dependent GISW simulations of bipolar PNe. Using higher-order methods and higher-resolution grids, Mellema et al. (1991) and Icke et al. (1992) explored the hydrodynamic flow pattern in bipolar wind-blown bubbles in more detail. They provided an extensive mapping of parameter space and articulated the dynamics of the ambient/wind shocks in greater detail. The formation of highly collimated supersonic jets inside the hot bubble was one of the most surprising results of these simulations.

Radiative losses were included in a series of studies by Frank & Mellema (Frank & Mellema 1994a,b; Mellema & Frank 1995; Mellema 1995) who developed a 2-D numerical code including hydrodynamics, microphysics, and radiation transfer from the hot central star. These models tracked cooling and emission behind the ambient shock and explored the limits, in terms of pole-to-equator density contrast q, of the nebular shaping. For low values of q, ($q < 2$), the results showed bubbles become mildly elliptical. Intermediate values of q, ($2 < q < 5$), produce distinct equatorial and polar regions of the ambient shock; i.e., bipolar lobes develop. Larger values of q produce bubbles that are highly collimated. The inclusion of radiative loss terms allowed confrontation with observations of both morphology and kinematics. The models were successful at recovering many of the observed morphological, kinematic, and ionization structures seen in a considerable range of elliptical and some bipolar PNe. Figure 3 compares the results of a GISW model along with synthetic observations with real data for NGC 3242.

Dwarkadas et al. (1996) explored the effect of slow wind speeds on bipolar PNe evolution. Their results demonstrated that the effects of a density contrast in the ambient medium can be washed out when ambient shock and slow wind velocities are comparable. This occurs because the bubble shell is always catching up to the slow wind and no inertial constraint can be imposed.

Dwarkadas et al. also provided the most detailed comparison between numerical simulations and self-similar analytical models available to date. Their results confirm that the two approaches yield similar bubble shapes when the hot bubble remains isobaric ($P_{HB} \approx Const.$). This isobaric condition then determines when energy-conserving bubbles will achieve Hubble-Flow kinematics.

THEORETICAL MODELS AND HUBBLE FLOWS   In "ProtoPNe" we emphasized that many, if not most PNe and pPNe are characterized by uniform, homologous expansions that we described as Hubble Flows. We noted that such kinematic patterns could be the result of ballistic motions or self-similar hydrodynamic growth patterns. Hydro models of radial outflow from a central source have a natural tendency to settle into self-similar growth. As noted above, the underlying reason is that when the fast and slow winds are largely radial, geometric dilution in consort with conservation of mass and momentum densities tend to relax nebular shapes to configurations where $v(r)$ (at each latitude) is constant. This behavior may only be modified by aspherical shock effects.

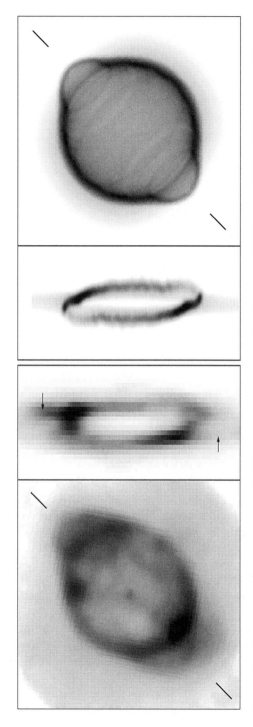

Figure 3 Morphologies and kinematics of GISW model: comparison with reality. *Left:* [OIII] image of NGC 3242 (Balick 1987) along with long slit echelle spectra along major axis (Y. Chu & Jacoby, personal communication). Solid lines mark the location of the Echelle slit. Arrows point to features related to the FLIERs of NGC 3242, which are not a part of the model shown to the right. *Right:* Corresponding synthetic [OIII] image and kinematic from a purely hydrodynamic model of a fast wind expanding into a toroidal slow wind with moderate density contrast. [For more details see Frank & Mellema (1994b)]. Some panels have been modified with permission of the authors.

Which Hubble Flows are ballistic or self-similar? The primary difference between the two types of Hubble flows will be the presence of large-scale working surfaces, such as prolate elliptical or peanut-like shocks, which will indicate whether hydrodynamic shaping is globally active. Observationally, we might expect to find dense filaments of atypical ionization or regions of abrupt changes in kinematics where pressures across these surfaces are actively shaping the nebula.

THE ROLE OF IONIZATION   The toroidal circumstellar environment it is likely to form before the star produces significant UV flux. Thus, nebular shaping begins while the torus is still neutral and cold. Mellema (1997) included evolution of stellar ionizing flux in bipolar PNe simulations and found that rapidly evolving, higher-mass stars ionize the entire environment on timescales $<t_d$. Lower-mass stars tend to have ionization fronts (IFs) in their nebulae for longer periods of time. In addition, ionization of high q environments lead to latitudinal pressure gradients. In these systems, the effective q drops with time leading to more elliptical PNe.

Trapped ionization fronts can also lead to instabilities of various types including Rayleigh-Taylor modes. Analytical models including time-evolution of both the wind and the ionization flux were calculated by Breitschwerdt & Kahn (1990) and Kahn & Breitschwerdt (1990) who showed that a brief period of ionization-front trapping could drive instabilities. García-Segura et al. (1999) have published results showing instabilities in the nebular rim that appear to occur directly from local variations in ionization fractions. Thus, stellar UV photons can affect both the global properties of the flows as well as the development of micro-features. These ionization models may help us to understand how ripples and fingers grow in very young PNe, and may also provide insight into the formation of the serpentine and starfish morphologies of very young PNe ("ProtoPNe").

MHD Models of PN Formation: Self-Confinement

The long lever arm offered by magnetic fields makes them a natural candidate for imposing ordered nonaxisymmetric structures on expanding plasma systems. Delemarter (2000) explored numerical GISW models for CRL 2688 (The Egg Nebula). Numerical simulations (with H_2 and scattered light synthetic observations), showed that a jet, collimated on small scales ($<10^{15}$ cm), was needed to explain the observations. These results indicate that for at least one well-studied object, the GISW model could not capture the proper dynamics.

In their groundbreaking works on jets in PNe, Morris (1987) and Soker & Livio (1994) mapped out scenarios in which accretion disks form around binary PNe progenitors. Each study equated the existence of disks with the existence of jets. The details of the jet launching and collimation mechanism were not, however, specified. Recent work by Soker & Rapport (2001), Soker (2001), and Soker & Livio (2001) have put collimated winds from disks to good use but these works also do not specify how such winds are launched or collimated. As we will see, the twin concerns of creating the wind and focusing it into a narrow outflow determines much of the current debate on MHD PN models.

A number of MHD models of PNe exist in the literature. Gurzadyan (1997 and references therein) assumed a dipolar field in the nebulae as a shaping agent. In a more recent series of papers, Pascoli (1985, 1992, 1997) explored models where toroidal fields embedded in the previously ejected AGB wind produce the bipolar morphologies. Whereas the Pascoli models have been successful at articulating certain limits of MHD shaping, they have yet to find empirical grounding as an explanatory framework. Currently, two flavors of MHD model appear particularly promising.

THE MAGNETIZED WIND-BLOWN BUBBLE (MWBB) MODEL   Motivated to produce bipolar outflows without the need of an inertially confining torus, Chevalier & Luo (1994) developed the MWBB model that relies on hydromagnetic forces in the shocked stellar wind itself. This model assumed a dynamically weak field embedded in a spherically expanding fast wind with foot-points tied to a rotating stellar surface (similar to the Parker solar wind solution). Applying flux conservation gives a poloidal magnetic field B_p that scales as $1/r^2$ and a toroidal field B_ϕ that (for $r \gg R_*$) scales as $1/r$. Here, R_* is the stellar radius. The growth of the toroidal field is directly related to the rotation rate of the star and takes the form $B_\phi = -B_r(\frac{r\Omega \sin(\theta)}{v_r})(1 - \frac{R_*}{r})$.

At large radii the field becomes dominated by the toroidal component $B \sim B_\phi$. When $v_w = Const.$ and $\rho \propto 1/r^2$ the ratio of magnetic to kinetic energy $\sigma = B_\phi^2/(4\pi \rho v_w^2)$ is constant. σ then becomes the principle parameter determining the behavior of the wind driven bubble and can be expressed as,

$$\sigma = \frac{B_*^2 R_*^2}{\dot{M}_w V_w}\left(\frac{V_{rot}}{V_w}\right)^2, \tag{3}$$

where B_*, V_{rot}, and v_w are the stellar radius, stellar rotation rate, and stellar wind speed, respectively. We note that the form $B_\phi \propto r^{-1}$ is a generic feature of toroidally dominated winds even when the field is weak as in this case (Matzner & McKee 1999).

In Chevalier & Luo's MWBB model, the field remains dynamically insignificant until the fast wind passes through the inner shock. Compression then strengthens the field and latitudinal variations in the total pressure $P = P_g + B^2/8\pi$ drive the evolution of globally aspherical morphologies.

Using an analytical formulation with an adiabatic wind shock, Chevalier & Luo demonstrated that magnetic forces in the hot bubble could produce significant departures from spherical morphologies. For a strong bipolar morphology to develop they found that $\sigma > 10^{-4}$.

NUMERICAL MWBB MODELS   The first numerical simulations of the MWBB model were carried out by Różyczka & Franco (1996). Whereas these models showed magnetic collimation in the hot bubble, globally it was restricted to a mild lengthening of the bubble's long axis. More dramatic, however, was the formation of dense, higher velocity features on the poles giving the computed nebula a distinct lemon shape. To achieve this level of magnetic collimation, Różyczka & Franco found $\sigma > 0.05$, i.e., far stronger fields than required for analytical models.

In the case of adiabatic wind shocks and low σ, the long timescales for magnetic collimation are to be expected. In the absence of strong cooling, the plasma beta parameter, $\beta = 8\pi P_{gas}/B^2$, can only increase across a shock (Priest 1984). Globally, magnetic forces in the hot bubble will act on Alfvénic crossing times R_{hb}/v_a, which will be longer than sonic crossing times. Thus, the overall morphology of the bubble will only change after many bubble expansion times. As Różyczka & Franco point out, Chevalier & Luo used a self-similar (scale-free) formulation of the problem, and collimation may simply occur on a larger scale than most observed nebulae when $\sigma < 0.05$.

When cooling is efficient, however, the field is further compressed (along with the gas), decreasing $\beta = 8\pi P_{gas}/B^2$. Once material passes through the wind shock and cools, Lorentz forces are strongly out of balance with gas pressure gradients. If radiative cooling makes $\beta \leq 1$, magnetic forces (hoop stresses) dominate and plasma will be drawn toward the axis in $t < t_d$.

In order to facilitate higher levels of collimation, the MWBB formalism has also been generalized to include a confining gaseous torus (as in the GISW model). García-Segura (1997), García-Segura et al. (1999), and García-Segura & López (2000) have included gaseous tori in their simulations (as well as a simplified treatment of ionization). In general, these models used relatively low fast-wind velocities ($v_w \sim 100$ km s^{-1}) leading to short post-shock cooling times. It is noteworthy that these speeds are more appropriate to the pPN stage than the fully developed PN stage. The explicit form of the gaseous torus was derived via the wind compressed disk model of (Bjorkman & Cassinelli 1993) by assuming high stellar rotation rates and near-Eddington luminosities.

The combination low β in the shocked wind and an inertially confining environment allowed the García-Segura et al. studies to produce a wide variety of nebular morphologies (confirming the feasibility of the Chevalier & Luo scenario). Note that under these conditions the limiting value of the field was $\sigma > 0.01$.

Based on their results, García-Segura et al. (1999) proposed a scheme for the evolution of PNe shapes based on the values of the stellar rotation rate Ω_*. and the magnetic field strength σ. In general, high Ω_*, high B_* central stars produced the highest collimation. This implies higher mass stars are the progenitors of bipolar nebulae (a result that appears to find support in observations).

One of the more striking aspects of the MWB studies has been the results of fully 3-D models (García-Segura 1997, García-Segura & López 2000), which include precession of the star's magnetic axis. The magnetic field at the source provides the needed lever to drive precession in the bubble as a whole. Most importantly, jets form within the bubble and these are observed to precess with the magnetic axis. We take up the issue of jets in a later section noting here that the resulting shapes in the simulations are similar to point-symmetric morphologies observed in some PNe.

Collimation, Initial Conditions, and Challenges to MWBB Models

Early criticism of the MWB models hinged on both the strength and topology of the field (Soker 1998b, Frank 1999). This criticism holds for all MHD models. It is unclear as yet if field strengths of the order required for the models can be generated in AGB or pPNe or PNe stars. Soker (1998b) raised the issue of field topology because MWBB models require fields to circle the star in the planes parallel to the equator. This is not the case for the solar wind that is composed of distinct field-reversing sectors. The field direction should also change polarity across the equatorial plane making $B_\phi(\theta = 90°) = 0$ where it may be needed most for collimation. Such polarity reversals could also allow reconnection at the equator providing a sink for magnetic energy (Frank 1999). The severity of the above issues, however, remains to be determined.

More important to the explicit application of the model is the issue of collimation in the wind before it passes through the inner shock. This has been investigated by Gardiner & Frank (2001). The MWBB model assumes that the field is weak and can have no dynamical effect on the flow material until it passes through the inner shock. Given typical wind shock and launching scales of $R \geq 10^{16}$ and $R \approx 10^{11}$, respectively, the magnetic field does not affect the flow over at least five orders of magnitude of expansion. A freely expanding magnetized wind must, however, experience an unbalanced Lorentz force associated with the helical field. This force has a component perpendicular to the wind velocity and can be written as

$$\frac{\rho v_r}{r}\frac{\partial}{\partial r}(r v_\theta) = \hat{e}_\theta \cdot \frac{1}{c}(\mathbf{J} \times \mathbf{B}) = \frac{-B_\phi^2 \cot \theta}{2\pi r}. \tag{4}$$

For some length scale and for some range of σ, the fast wind may collimate on its own. Using a perturbation analysis, Gardiner & Frank (2001) demonstrated that significant redirection of streamlines occurs even for weak fields. Their calculation showed deflections of streamlines of $\delta\theta \approx 10°$ to $23°$ for $\sigma = 0.02, 0.05$, respectively. Thus, the initial conditions used in the MWBB simulations are not likely to be correct.

Finally, the MWBB model addresses only the issue of outflow collimation and not wind launching. Because the field is explicitly assumed to be weak, it cannot help drive the wind (Lamers & Cassinelli 1999). Because many bipolar pPNe may suffer from a momentum excess ("ProtoPNe"), this must be a concern for MWBB models.

Magneto-Centrifugal Models

Momentum excesses have been encountered before in the YSO community (Lada 1985) in studies of molecular outflows. In those environments, magneto-centrifugal launching from a rapidly rotating magnetized object has became an attractive solution to the problem of launching and collimating the flow. Such models could be applied to pPNe and PNe as well.

Explicit models of magneto-centrifugal launching in PNe systems were made by Blackman et al. (2001a,b) that attempted to explain launching, collimation, and multipolarity. In the Blackman et al. (2001a) scenario, an accretion disk forms around the recently exposed core of an AGB star. The disk forms from companion disruption after the common envelope phase (Soker & Livio 1994, Reyes-Ruiz & López 1999). Blackman et al. (2001b) modeled magneto-centrifugal winds from a central star. These works focused on the magnetic luminosity L_m as the driver for a mechanical wind luminosity L_w in pPNe/PNe systems,

$$L_w \approx \dot{M}_w v_w^2 \sim L_m = \int (\mathbf{E} \times \mathbf{B}) \cdot d\mathbf{S} \sim \int_{R_i}^{R_o} (\Omega R) B_p B_\phi R dR, \tag{5}$$

where B_ϕ and B_p are the toroidal and poloidal field, respectively, Ω and R are the rotational frequency and radius of the magnetized wind source (star or disk). R_i and R_o define the radial boundaries of the wind-producing region.

Using published models for pPNe accretion disk parameters (Reyes-Ruiz & López 1999), Blackman et al. (2001a) calculated $L_m(t)$ for both the disk and the star separately while also accounting for their coupling. The power of each wind evolved along with the disk and the spin-down of the star. Whereas the results showed that either the stellar or disk wind would dominate at any point in time, the energy and momentum flux from the combined wind system was always sufficient to drive observed pPNe and PNe bipolar outflows and perhaps solve the energy and momentum problems (i.e., $L_m \sim 10^{36}$ erg/s). Note also that the star will spin-down as angular momentum is transferred, via the field, to the stellar wind. Thus, magneto-centrifugal winds in PNe may offer a natural explanation for the slow rotation of white dwarfs.

Using a dynamo model based on conservation of mass on shells, Blackman et al. (2001) further concluded that once a magnetized pPNe core is exposed via mass loss or binary effects, it will be spinning fast enough to produce a collimated wind regardless of the existence of an accretion disk.

The degree to which a magnetized stellar rotator will collimate a wind driven off its surface has been explored by numerous authors (Weber & Davis 1967, Sakurai 1985). The efficacy of the magnetic rotator can be expressed via a rotation parameter Q (Tsinganos & Bogovalov 2000) given by

$$Q/Q_\odot \simeq 4(\psi_c/5 \times 10^{26} \text{ G cm}^2)(\Omega_c/2 \times 10^{-5} \text{ s}^{-1})$$
$$\times (\dot{M}/6 \times 10^{21} \text{ g s}^{-1})^{-1/2}(V/400 \text{ km s}^{-1})^{-3/2}, \tag{6}$$

where ψ_c is the magnetic flux at large distances ($\propto BR^2$), V is the outflow speed, and $Q_\odot \sim 0.12$ is the value for the solar wind. In the expression above, the scales have been set using values that may be appropriate to pPNe. For $Q \geq 1$ the system is classified as a fast magnetic rotator and is expected launch a magneto-centrifugal wind. Note the MWBB model utilizes winds with $Q < 1$. In the $Q > 1$ regime, larger Q values imply more strongly self-collimated outflows (Lery et al. 1998, 1999). In the single star model of Blackman et al. (2001), values of up to $Q/Q_\odot \sim 15$ were found, leading to the conclusion that significant MHD launching and collimation is possible in ppNe/PNe systems.

Some fraction of this magnetic energy may also be available to power X-ray emission from the PNe central stars via reconnection-driven flares. X rays have been reported in a number of PNe (Kastner et al. 2000, 2001), and recent observations of X rays in the NGC 6543 (Guerrero et al. 2001, Chu et al. 2001) have provided direct evidence for hard emission from the central star.

Dynamos and the Challenge to the Magneto-Centrifugal Models

Currently, theories of PNe shaping via fast magnetic rotators remain in their infancy. Whereas these models can make quantitative predictions in terms of the evolution, energy, and momentum of the winds, they have yet to produce detailed predictions of nebular shapes either analytically or via numerical models.

The largest uncertainty facing these models is the presence of a strong magnetic field that, most likely, will require some form of effective dynamo. It is encouraging that milligauss magnetic fields have been observed in both AGB and young PNe. But the MHD significance of these fields, so far found only in scattered masers ("Observational Results and Questions: Summary"), is yet to be understood.

Pascoli (1997) published the first dynamo calculation for an AGB star demonstrating the potential efficacy of the classic differential rotation/convection based $\alpha - \Omega$ formalism (Priest 1984) for producing extended fields in evolved stars. An analysis of the dynamo number N_D, (the ratio of field amplification rate to the dissipation timescale), led Soker (2000a) to the conclusion that $N_D < 1$ and that $\alpha - \Omega$ dynamos could not operate in AGB stars. He focused instead on turbulent-based $\alpha^2\Omega$ dynamos and their ability to produce weak fields.

In Blackman et al. (2001b), detailed models of AGB stars (S. Kawaler, personal communication) were used as input to radially averaged $\alpha - \Omega$ studies. These were first calibrated to the solar dynamo. An interface dynamo was found at the base of the convection zone just above the inert CO core with field strength of $1G$ estimated at the AGB surface. These models assumed momentum conservation on shells in the AGB star allowing core contraction and envelope expansion to create the necessary gradients in $\Omega(R)$. Rapid rotation of the core is also needed to drive a magnetized wind once the outer layers are removed. Note that the Blackman et al. (2001) model did not use the field to drive AGB winds. These were assumed to be radiatively driven.

Soker & Zoabi (2002) recently criticized the assumption of zero-angular momentum transfer, arguing that any field extending into the convection zones would spin down (up) the core (envelope). Thus, $\Omega(R)$ would be too flat for $\alpha - \Omega$ dynamos (see also Hager & Langer 1998). This conclusion is unlikely to hold as strong field lines will collect into flux-tubes that do not provide significant drag in the envelope. In addition, reconnection in the turbulent convective zone will tend to disconnect the flux tubes from their original foot points. In either case, there remains great uncertainty as to which, if any, dynamo model operates in AGB/pPNe/PNe stars. Further work in these evolved star dynamos is one of the more important future areas of theoretical work in the field.

Theories of Nebular Features: Jets, BRETS, FLIERS

It remains unclear the degree to which the linear structures seen in PNe and pPNe are physically similar to continuous hypersonic plasma beams so prominent in YSOs and other astronomical environments (Gonçalves et al. 2001, Livio 2001). The similarities have, however, led some authors to assume the existence of jets and then model bipolar outflows through jet interactions with the surrounding material (Steffen & López 1998, Miranda 1999, Soker 2002). A number of theoretical efforts have also focused the formation of jets in evolved star systems. We consider formation scenarios first.

HYDRODYNAMIC JET COLLIMATION MECHANISMS   GISW simulations showed jets could be effectively generated within bipolar nebulae. The collimation relies on the development of an aspherical wind shock to act as a hydrodynamic lens and focus post-shock streamlines toward the axis (Icke 1988). The process, termed Shock Focused Inertial Confinement (SFIC), is robust (Icke et al. 1992) and has been explored in a number of contexts including low- and high-mass YSOs (Frank & Mellema 1996, Mellema & Frank 1997, Yorke & Welz 1996) and relativistic jets from AGN (Eulderink & Mellema 1994). Icke (1988) and Frank & Mellema (1996) presented analytical models of the action of the oblique inner shock demonstrating that even mildly aspherical wind shocks could produce significant flow focusing and supersonic post-shock speeds without a delaval nozzle.

For radiative wind shocks, high compression forms a thin prolate shell of material that streams along the contact discontinuity. This becomes a converging conical flow near the poles redirecting gas into an axial jet (Cantó et al. 1988, Tenorio-Tagle et al. 1988). Mellema & Frank (1997) placed the development of these flows in the context of global bipolar wind-blown bubble evolution. Their simulations demonstrated that such flows may be a natural and robust consequence of wind/ambient material interactions. Borkowski et al. (1997) also performed simulations of winds expanding into pressure stratified environment. Their intent was to model the jets in pPNe He3-1475. Their results, well matched with analytical models, showed the formation of an empty cavity at the base with a narrow jet emanating from the converging flow at the cavity's tip.

In the GISW model, the fast wind encounters a decreasing ambient density with $\rho(R) \propto R^{-2}$. At some finite radius, the wind shock will change from momentum conserving to energy conserving. Jet collimation via converging conical flows may then be followed by an energy-conserving SFIC phase (Mellema & Frank

1997, Frank et al. 1996). Simulations with evolving winds, however, (Dwarkadas & Balick 1998) showed converging flows may be disrupted by instabilities along the walls of the bubble. Thus, the formation of converging flows in wind-blown bubbles remains an open issue. In a fully 3-D situation, the convergence point could be unstable (García-Segura et al. 1997; J. Blondin, personal communication 1998). Recent laboratory experiments using high-energy density plasma machines have, however, created scaled versions of radiative conical flows (Remington et al. 2000, Lebedev 2001). These studies indicate that the convergence point is remarkably stable.

MHD MODELS OF JET FORMATION   The natural tendency for toroidal fields to produce collimation flows make them attractive for producing jets (Lynden-Bell 1996). Simulations of the MWB model (García-Segura 1997, García-Segura et al. 1999) have been quite successful at producing collimated pPNe and PN flows where jets form within the hot bubble far from the star. Magnetic pressure then drives a linear expansion of the jet away from and back toward the star. Such a magnetic spring effect may give velocity segregation seen in some pPNe and PNe. These simulations give very encouraging results in terms of reproducing both morphologies and aspects of the kinematics.

The exact nature of these jets has yet to be explicitly studied. Their formation may be similar to plasma gun effects (Contopoulos 1995). Gardiner & Frank (T.A. Gardiner, A. Frank, unpublished manuscript) recently studied MWBB jets, concluding their formation occurs as axial currents in the hot bubble attempt to achieve cylindrical equilibrium, i.e.,

$$q \frac{1}{\varpi} \frac{d\varpi B_\phi}{d\varpi} = \frac{4\pi}{c} j(\varpi).$$ (7)

Here ϖ is the cylindrical radius and j is the axial current. Gardiner & Frank (2001) proposed that flow in the hot bubble can be described as a jet with a non-zero current density in its core surrounded by a current free region where the field has relaxed to a force-free configuration with $B_\phi \propto 1/\varpi$ (Matzner & McKee 1999).

The consensus in other fields however holds that jets are launched and collimated via magnetized accretion disks (Konigl & Pudritz 2000). As discussed before, in their papers on jets in pPNe/PNe, Morris (1987) and Soker & Livio (1994) proposed theoretical models not of the jets themselves but of the formation of accretion disks via binary interactions. The assumption, based on the success of disk-based magneto-centrifugal models in other contexts, was that disks equal jets. Livio (2000, 2001) has extended this conclusion, articulating the view that all jet-bearing systems (YSOs, AGN, microquasars) require both accretion disks and an additional source of energy (Ogilvie & Livio 1998). In the case of PNe, the extra energy may be associated with the luminosity of the central star (Livio 2000). Whereas the models of Blackman et al. (2001a,b) provide more explicit arguments for magneto-centrifugally driven jets in pPNe/PNe, detailed calculations of jet structures and properties must still be carried out. Of course, any model requiring accretion disks in PNe will likely require a binary companion to act as a mass reservoir (Soker & Rapport 2001, Soker 2001).

BIPOLAR ROTATING EMISSION LINE JETS (BRETS)   The objects known as BRETS (López 1997, Guerrero 2000) provide intriguing problems for evolved-star theorists. The immediate interpretation of BRETS as precessing jets has been strengthened by numerical simulations (Cliffe et al. 1995).

It appears difficult to reconcile point symmetry with hydrodynamic collimation mechanisms due to problems of getting the entire torus to precess rigidly on the correct timescales (Frank 1999, Guerrero 2000). García-Segura (1997) demonstrated that precession of stellar magnetic axes (driven by a binary) can produce point symmetric morphologies. García-Segura & López (2001) extended the idea to multipolar nebula via slow axial precession; however, no physical justification for such a configuration was given. A binary companion may also be required for precession in the Blackman et al. (2001) models where only a rapidly rotating star drives the outflows.

The precession of an accretion disk offers a natural means for inducing precession in magneto-centrifugal jet models. Pringle (1996) demonstrated that radiation shadowing from the central star can drive a warping instability in disks surrounding luminous objects. In Livio & Pringle (1996, 1997), these models were applied to PNe where it was shown that the disk precession timescale is comparable to the observed jet precession periods.

ANASE AND FLIERS   The nature of bright knots seen along the axis on many elliptical and bipolar PPNe and PNe remains a mystery. Whereas these features may be second-order effects attributable to intrinsically statistical variations in the initial conditions of a particular PN (i.e., "weather" rather than "climate"), the fact that many are oriented along the major axis of the outflows means they might be interpreted as remnants of early phases of outflow collimation.

A number of authors have invoked jets to explain polar FLIERs (Icke 1988, Icke et al. 1992, Frank et al. 1996). These models fare poorly when compared with the many observed morphologies of FLIERs that show bow shocks pointing back toward the central star. Steffen & López (2001) have proposed a stagnation flow model for polar FLIERs. Their scenario employs a time-dependent collimated jet driving into circumstellar material (e.g., the planetary nebulae KjPne 8, Steffen & López 1998). When the jet shuts off, thrust is lost from behind the rapidly cooling plug that forms at the head of the jet. The plug is decelerated as it sweeps up the material ahead of it. This model has promising features because it combines strong collimation and time-dependence in the flow from the star to explain both the presence of short knots and their orientation.

FLIERs and anase have also been modeled as clumps of ambient material subject to photoionization, ablation, and acceleration (the Oort-Spitzer rocket effect). Both analytical and numerical studies of this process have been carried out (Mellema et al. 1998, López-Martín et al. 2001). These models hold some promise in terms of recovering the kinematics and ionization structures seen in real FLIERs. Unfortunately, they also fail to account for observed morphologies while begging the question of why polar-orientated clumps should exist in the ambient medium to begin with.

Some mix of the jet and photoionized clumps models may be needed to explain FLIERs. Redman & Dyson (1999) have proposed a model in which a clumpy jet experiences mass loading as clump material is ablated and driven downstream. It is noteworthy that the general effect of clumpy flows on PNe (and other astrophysical systems) remains an open one. Dyson, Hartquist, and collaborators have pioneered the study of how clumpy or inhomogeneous flows differ from the smooth flows assumed in the majority of PNe hydro/MHD studies (Dyson et al. 2001). Recent numerical simulations (Poludnenko et al. 2001) and high-energy density laboratory studies (Dannenberg et al. 2001) have also taken up the issue. We agree with Dyson that some fraction of PNe theory may need to be revised in light of clumpy flow dynamics. Thus the issue deserves further theoretical consideration.

Stellar Origins of Disks and Tori

Much of PNe and pPNe shaping theory relies on circumstellar material in either an expanding toroidal density distribution or an accretion disk. Whereas the latter clearly requires a secondary star, it is still not clear if the former can be obtained by a single star. Thus the presence, frequency, and effect of binary companions on PNe evolution remains one of the most contentious issues in the field. Here we briefly review the subject of binary effects (for a review see Iben & Livio 1993, Iben 2000).

SINGLE-STAR MECHANISMS FOR PRODUCING TORI   A number models exist for creating circumstellar torii from single stars. All of these mechanisms rely on stellar rotation in some way (Asida & Tuchman 1995) and may require spin-up via a companion. Based on a suggestion by Tom Harrison, Frank (1995), Soker (1998a), and Soker (2000b) have proposed a model where large cool spots on the surface of an AGB star produce increased dust production and an equatorial enhancement of wind density. Dorfi & Höfner (1996) examined the direct coupling of rotation, dust formation, and mass loss, finding that temperature variations in a rapidly spinning AGB star could produce enhanced equatorial mass loss. Matt et al. (2000) found that dipole magnetic fields in an AGB star could guide mass loss along the equator forming a torus. This model clearly requires the presence of a dynamo.

The Wind Compressed Disk (WCD) mechanism of Bjorkman & Cassinelli (1993) has been one of the most promising single-star models explored to date (Owocki et al. 1994). The WCD disk model predicts outflowing excretion disks will form around rapidly rotating stars when streamline are deflected via Coriolis forces. Though the model was originally intended for B[e] stars, it has been applied to other classes of objects including AGB stars (Ignace et al. 1996). In most cases, the rotation rate of the star needs to be a significant fraction of the break-up rate for wind compression to occur. García-Segura (1997) have noted that the critical rotation rate for break-up depends on the luminosity of the star via

$$\Omega_c = \sqrt{\frac{GM_*(1-\Gamma)}{R_*^3}}, \tag{8}$$

where $\Gamma = L_*/L_{Ed}$ and L_{Ed} is the Eddington luminosity. Ω_c may approach the actual rotation rate of the star as its luminosity increases.

The WCD model links the properties of the stellar progenitor directly to the bipolar bubble producing density and velocity distributions that can be fed directly into hydro or MHD models. Some studies have cast doubt on the effectiveness of the WCD mechanism in line-driven winds (Owocki et al. 1996), and its fate remains a important area of research.

EFFECTS OF BINARY STARS   Soker (1997) attempted to distinguish between different classes of binary interaction scenarios affecting PNe shaping. Careful in his use of the term binary because substellar mass companions (brown dwarfs and Jupiter mass planets: Soker 1994b, 1996, 1999; Harpaz & Soker 1994) could also affect shaping, Soker defined nonsingle star evolution as at least one property of the primary being determined by an orbiting object. Soker concluded that the majority of aspherical PNe had undergone a binary interaction. Binaries could be separated into extremely wide binaries with $T_{pn} < T_{orb}$ and $a > 5000$ AU ($T_{orb} =$ orbital period, $T_{pn} =$ PNe lifetime $a =$ orbital separation); wide binaries with $T_{pn} \approx T_{orb}$ and $100 < a < 1000$ AU; close binaries that avoid common envelope phase; or close binaries experiencing common envelope evolution. It is the last two categories that are likely to have the most dramatic effect on PNe evolution. Other promising mechanisms that exist, such as the radiative effects of a secondary on the primary's temperature distribution (Frankowski & Tylenda 2001), deserve further exploration.

Whether close companion(s) ultimately and sufficiently influence the environment to account for the shaping of PNe remains a topic of considerable controversy (Bujarrabal et al. 2000). This is particularly true because both inertial- and self-confinement–based models rely on initial conditions (gaseous torii, magnetic fields) that could be supplied directly from mergers, accretion disks, or tidal torquing resulting from interactions with a binary companion. Thus, regardless of the details of the specific model, binary companions often do something good for nebular shaping (M. Livio, personal communication 2001).

COMMON ENVELOPE (CE) EVOLUTION   CE evolution is a consequence of dynamical mass transfer in a close binary system where the more massive star has a deep convective envelope (Rasio & Livio 1996). CE evolution occurs when the more massive, mass-losing star is unable to contract as rapidly as its Roche lobe and envelope material is rapidly transferred to the secondary. The secondary is driven out of thermal equilibrium, and the final configuration is a common envelope shared by both stars, with the cores rotating about the center of mass. Differential rotation of the binary cores and CE gas leads to energy and angular momentum transfer potentially liberating the entire envelope. Binaries that contain a compact object in a close orbit around the secondary (i.e., whose orbital period is a few days) are thought to have undergone CE evolution. The field has received considerable attention, and a number of excellent reviews on the subject exist (Iben & Livio 1993, Taam & Sandquist 2000).

In the domain of PNe studies, CE evolution has been one of the principal mechanisms believed to produce the expanding slow-wind torus needed in GISW models. One of the principal uncertainties in CE models has been the effectiveness of envelope liberation. The quantity $\alpha_{CE} = \Delta E_{bind}/\Delta E_{orb}$ serves as a measure of CE ejection efficiency. Using a primary mass of $M_p = 5\,M_\odot$ and and a variety of secondary masses, Livio & Soker (1988) presented results with low values of α_{CE}. More accurate SPH simulations by Rasio & Livio (1996) examined the early phases of CE evolution with a system of similar masses and realistic initial conditions for the primary star. Whereas the simulations tracked only 100 dynamical times of evolution, they did show rapid in-spiral of the secondary and an ejection of approximately 10% of envelope mass. Extrapolating the energy transfer rate, Rasio & Livio concluded that envelope ejection ($\alpha_{CE} = 1$) would occur on timescales of a year. Terman & Taam (1996) also found envelope ejection in systems with extended primaries and more massive companions. Sandquist et al. (1998) followed the later stages of CE evolution of an AGB star with low mass companion. These simulations showed fairly high efficiency of envelope ejection (40%) as the orbit decayed from $P = 200$ days to 1 day. Most importantly, the Sandquist et al. (1998) models showed mass ejection concentrated toward the equatorial plane with $q \approx 5$. The expanding torus had a form similar to those assumed in GISW models. Note that the highest velocities in the ejected material occur near the equator. Thus, the issues raised by Dwarkadas et al. (1996) concerning pole-to-equator velocity contrasts may limit the ability of CE ejection to provide constraining torii.

FORMATION OF ACCRETION DISKS   It is unlikely that an accretion disk could survive the long main-sequence lifetime of a PN central star. Thus, accretion disks in PNe systems must form via binary interactions. Disks may form around secondaries via Roche lobe overflow or Bondi accretion of AGB winds (Morris 1981, Morris 1987, Mastrodemos & Morris 1998, 1999). Such systems would be similar to symbiotic stars. Accretion disks could also form around the primary after CE evolution and disruption of the secondary star (Soker & Livio 1994, Miranda 1999). Mastrodemos & Morris (1998), and Soker (1998c) carried out detailed simulations of the first scenario. Using an SPH method they modeled the gravitational interaction of a dense AGB wind with lower mass companion (Figure 4). The models were noteworthy for their inclusion of a host of physical processes including gas-dust interactions in the wind and molecular cooling. Typical parameters for the models were $M_p = 1.5$, $M_s = 1$, and binary separations between 3 and 6 R_p (orbital periods of 4 to 14 years). The orbital separation was close enough for both Roche flow and wind accretion to play a role. Accretion disk formation was seen in all the simulations. The disks were stable throughout the life of the simulations (3–5 orbits or

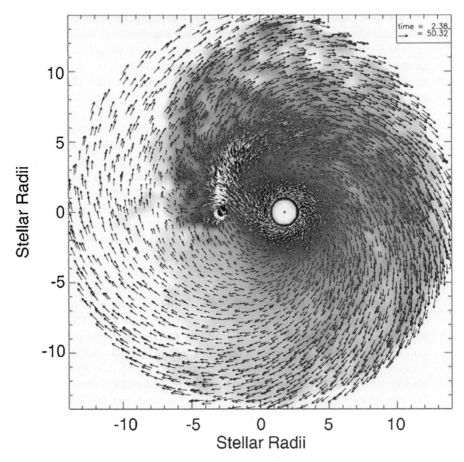

Figure 4   The formation of an accretion disk and a trailing wake associated with a compact companion (*Left*) in a binary system in which the secondary (*Right*) has evolved to an AGB star. The figure shows the results of SPH simulations by Mastrodemos & Morris (1998) in which the density and velocity vector maps of their model M3 are overlaid. The disk and its wake consist of mass accreted from the winds and the Roche-lobe overflows from the AGB star.

>10 accretion flow timescales R_{acc}/v_{acc}) and were geometrically thin owing to strong cooling.

These properties make them amenable to the so-called α disk formalism used in many MHD launching models (Hartmann 1998). The disks also exhibited warping modes that are promising for the development of point-symmetry should jets arise. The size of the disks was shown to vary with the ratio of the orbital to wind speed v_o/v_w. As this ratio increases the secondary becomes more effective at capturing passing wind material. Extrapolations from their models allowed Mastrodemos &

Morris (1998) to set a limiting orbital separation for disk accretion of $\approx$20 AU. In a second paper, Mastrodemos & Morris (1999) examined a wider range of initial conditions for close binaries that would avoid CE evolution. Beginning with binaries whose companion mass spanned a range 0.25–2 $M_\odot$ and orbital separations of 3.6–50 AU, Mastrodemos & Morris (1999) found a continuous variation in behavior from accretion disk formation through the development of outflowing toroidal density distributions. The latter configurations are of great interest in that density contrasts of $5 < q < 10$ were created. As with the CE models, this is enough to account for shaping via GISW mechanisms. Based on these solutions, they estimated that $\approx$34% to 40% of all detached binaries will lead to bipolar morphologies.

The formation of accretion disks via disruption of the secondary after CE evolution is an intriguing possibility (Soker & Livio 1994). This model implies a finite lifetime for the disk as the mass reservoir of the distrusted companion is slowly drained onto the primary. Such a mechanism may be attractive in explaining certain transient properties of ppNe outflows (Blackman et al. 2001a). Reyes-Ruiz & López (1999) carried out more detailed calculations disk formation and evolution in this scenario (using α disk models). Their results showed the disk accretion rate decreased as a power-law in time from an initial value of $\dot{M}_a = 10^{-4}$ $M_\odot$/years. The outer disk radius increased with time as a power-law as angular momentum was transferred to outer disk annuli.

CONCLUSIONS AND FUTURE DIRECTIONS

In the early 1990s, confidence was building that interacting winds scenarios might successfully account for the morphologies of PNe. Since 1994 that confidence has been upended by stunning new images from the Hubble Space Telescope, followed shortly thereafter by puzzling results from the Chandra X-ray satellite and various ground-based studies. Simply put, long-standing theories of stellar evolution, mass transfer in binaries, and the effects of putative magnetic fields have been challenged to supply a paradigm in accord with the observations. This challenge has spurred intense research activity in these and related fields which rooted first within the context of mass loss by evolved stars. The outcomes should be applicable to similar outflow phenomena across the H-R diagram, from YSOs and η Carinae to asymmetric supernova detonations and their remnants.

The observational highlights of the past decade are these:

- PNe and their younger neutral counterparts ppNe have complex, fascinating structures that can be systematically classified by common characteristics of their shapes and symmetries.

- Mass loss is isotropic, sometimes percussive and sometimes pulsed on time scales of 1000 years, as stars make their final ascent to the tip of asymptotic giant branch.

- The mode of mass ejection changes in amplitude $\dot{M}$ and symmetry at or shortly after attaining the tip of the AGB, called the post-AGB phase of evolution. Very high-order symmetries are possible, and squid, jellyfish, and starfish shapes are not uncommon.

- Luminous post-AGB stars can eject mass at very high velocities of hundreds of km s^{-1}, up to 100 times in excess of the surface escape velocity and expected surface wind speeds. Additionally, CO observations show that the momentum of the collimated outflows in collimated, high-luminosity pPNe is as much as 1000 times larger than radiatively driven stellar winds can explain.

- The collimation can be extremely effective, with opening angles from a few to tens of degrees not uncommon. In many cases, though not all, the outflows follow a Hubble Flow form in which Doppler shift increases linearly with stellar radius out to the visible end of the nebulosity. Superficially, the ejection of mass seems better described as an event than an ongoing process.

- The onset of ionization in the pPNe-to-PNe transition probably both amplifies and disrupts the complex symmetries owing to the disruptive passage of an ionization front through the nebula. The sudden heating of the gas from 10s to 10,000 degrees probably reshapes and softens the nebular shape.

- Low-ionization knots and jets are found in many highly ionized PNe, generally in pairs, whose nuclear offsets and Doppler speeds are equal and opposite. The speeds of these knots are both highly supersonic and much larger than the material around them, which suggests that the knots and jets were ejected later than the larger, more amorphous gas that surrounds them.

- A direct link between highly collimated outflows and stellar properties has shown that the most collimated PNe are associated with relatively massive ($M > 3$ M$_\odot$) progenitor stars. However, any direct physical connection between PN morphologies and other nuclear properties that might account for strong flow collimation, e.g., binarity and/or nearby magnetic fields, is weak.

Whereas YSOs have an obvious external reservoir of mass that is rich in angular momentum (and gravitational potential energy) and threaded with magnetic fields upon which to draw, isolated AGB stars do not. The search for an explanation for PN morphologies and kinematics is presently leading in the direction of new ways to look at the evolution of old stars. For example, the formation of bipolar and bilobed PNe requires some sort of fundamentally important axis of symmetry in the ejection system.

One path forward is to assume that external torque from a companion star exerts a substantial influence on the evolution and mass loss of an AGB star, or that mass lost from the AGB star forms an accretion disk on its companion and/or generates a disk from mass overflow. Another is to suppose that AGB stars, particularly the massive ones, harbor deeply hidden magnetized flywheels whose fields once raised to the surface by deep convection, or exposed when the envelope is lost, can define a symmetry axis for the collimated outflows. Although direct observational support for interacting binary systems and magnetic fields is circumstantial at best, these concepts for accounting for outflow collimation are very popular. ppNe and PNe are ideal laboratories for testing these models owing to their low internal extinctions and high surface brightnesses.

Interacting binary systems were the first of these ideas to receive advocacy by Morris, Livio, Soker, and their many colleagues. They suggest that accretion disks or strong tidal forces from companion stars and/or large planets in orbits near the AGB surface supply large amounts of angular momentum to the highly bloated AGB star with a loosely bound and perhaps pulsationally unstable envelope. Several ingenious mechanisms for generating a disk from this angular momentum have been proposed, including wind streamlines from a star rotating near breakup that converge to the midplane and form a disk; spin-induced levitation and cooling of the equatorial regions forming a torus of dust; and external tidal forces that create a toroidal accretion flow around the entire system. However, whereas disks and tori may be assembled in this way, the high degree of outflow collimation, particularly in high-mass AGB stars, is still difficult to understand. The effects of companion stars stand as fundamental challenges to the community. More work providing explicit links between nebular shapes and specific orbital scenarios will be forthcoming.

As early as 1958 Aller had speculated that magnetic fields played a key dynamical role in the shaping of PNe. Magnetic collimation via weak fields have been applied to PNe by García-Segura, Franco, López, Matt, and their colleagues. Magneto-centrifugal models similar to those used in YSOs have been applied to PNe by Blackman, Frank, and collaborators. Simulations are showing how magnetic fields in rotating and nonrotating environments build disks and, sometimes, collimated polar flows. This work is still in its infancy, the initial results (García-Segura et al. 1999) look very promising (Figure 5), and the outcomes will be of interest to several astrophysics communities. A successful effort requires hard work on both observational and theoretical fronts. The recent detection of milligauss fields embedded in the torus of K3–35 will motivate more work on magnetic collimation processes that are already applicable to other types of collimated astrophysical outflows, including the solar wind. However, the source of sufficiently strong fields to channel the winds of AGB stars is as yet not explained.

Not long ago we believed that the evolution of a planetary nebula could be predicted by one number or perhaps two parameters: progenitor stellar mass and the chemical properties of its ejecta accelerated to winds by surface radiation pressure. If nothing else, the Hubble images warn us that other parameters—perhaps several of them—may need to be added to the prescription for forming PNe. These can include the presence of a companion, the orbital attributes of the companion, the angular momentum of the star at birth, and details of the dredge ups that occur at the tip of the AGB. On top of this, we need to hone our physics of how flows are collimated.

An astrophysically interesting aspect of the study of PNe shaping at this time may be the transition from a single monolithic paradigm (GISW) to a multiplicity of processes whose relative importances will depend on the where, when, and how a star enters its final stages of evolution. A number of authors have noted

Figure 5   A MWBB Model for the Ant Nebula, Menzel 3. This figure compares an image of Menzel 3, the Ant Nebula (*left*) with results of axisymmetric MHD calculations of the magnetized wind-blown bubble model (*right*, García-Segura et al. 1999). Note the presence of a jet in the two-dimensional density map of the MWBB model (*top right*). Ionization effects produce the filaments seen in the tilted and projected three-dimensional synthetic emission line image (*lower right*).

the similarity of some PNe shapes to the classic images of wave functions for the hydrogen atom. This playful metaphor may provide some inspiration in the coming years. Whereas a new dominant paradigm may emerge, it is also possible that researchers may have to treat the variety of potential shaping processes as a kind set of basis vectors. The projection onto an individual nebula (i.e., which processes actually act on that nebulae) will then have to be read off the object via models that link evolutionary processes to those that yield specific modes of shaping.

Whether we will be led to a new dominant paradigm or a Hilbert space of many coexisting shaping models is unclear. If the past decade has taught us anything, however, ten years from now we can expect, once again, to find ourselves surprised. That's life at the endless frontier.

ACKNOWLEDGMENTS

There are so many people who have contributed to this article that we would have to add many pages to list them all. We are particularly indebted to the following people for the discussions and help: Mario Livio, Eric Blackman, Jack Thomas, Hugh van Horn, Joel Kastner, Garrelt Mellema, Vincent Icke, Tom Gardiner, Alexei Polundenko. Support to AF was provided at the University of Rochester by NSF grant AST-9702484, NASA grant NAG5-8428, and the Laboratory for Laser Energetics. BB benefitted from the inexhaustible patience of his wife and the serenity and hospitality of the Helen Riaboff Whiteley Center at the Friday Harbor Laboratories of Marine Biology of the University of Washington. AF also wishes to thank his family for their understanding.

The *Annual Review of Astronomy and Astrophysics* is online at
http://astro.annualreviews.org

LITERATURE CITED

Abell GO, Goldreich P. 1966. *Publ. Astron. Soc. Pac.* 78:232

Alcolea J, Bujarrabal V, Castro-Carrizo A, Sanchez Contreras C, Neri R, Zweigle J. 2000. See Kastner et al. 2000, p. 347

Alcolea J, Bujarrabal V, Sánchez Contreras C, Neri R, Zweigle J. 2001. *Astron. Astrophys.* 373:932

Aller LH. 1958. *Astron. J.* 63:47

Arthur SJ, Henney WJ, Dyson JE. 1996. *Astron. Astrophys.* 313:897

Asida SM, Tuchman Y. 1995. *Ap. J.* 455:286

Balick B. 1987. *Astron. J.* 94:671

Balick B. 2000. See Kastner et al. 2000, p. 41

Balick B, Alexander J, Hajian AR, Terzian Y, Perinotto M, Patriarchi P. 1998. *Astron. J.* 116:360

Balick B, Perinotto M, Maccioni A, Terzian Y, Hajian A. 1994. *Ap. J.* 424:800

Balick B, Preston HL, Icke V. 1987. *Astron. J.* 94:1641

Balick B, Wilson J, Hajian AR. 2001. *Astron. J.* 121:354

Barnbaum C, Morris M, Kahane C. 1995. *Ap. J.* 450:862

Bjorkman J, Cassinelli J. 1993. *Ap. J.* 409:429

Blackman EG, Frank A, Welch C. 2001a. *Ap. J.* 546:288

Blackman EG, Frank A, Markiel JA, Thomas JH, Van Horn HM. 2001b. *Nature* 409:485

Bond HE. 2000. See Kastner et al. 2000, p. 115

Borkowski KJ, Blondin JM, Harrington JP. 1997. *Ap. J. Lett.* 482:L97

Borkowski KJ, Harrington JP. 2001. *Ap. J.* 550:778

Breitschwerdt D, Kahn FD. 1990. *MNRAS* 244:521

Bryce M, Balick B, Meaburn J. 1994. *MNRAS* 266:721

Bryce M, Meaburn J, Walsh JR. 1992a. *MNRAS* 259:629

Bryce M, Meaburn J, Walsh JR, Clegg RES. 1992b. *MNRAS* 254:477

Bujarrabal V, Castro-Carrizo A, Alcolea J, Sánchez Contreras C. 2001. *Astron. Astrophys.* 377:868

Bujarrabal V, García-Segura G, Morris M, Soker N, Terzian Y. 2000. See Kastner et al. 2000, p. 201

Burrows CJ, Krist J, Hester JJ, Sahai R, Trauger JT, et al. 1995. *Ap. J.* 452:680

Calvet N, Peimbert M. 1983. *Rev. Mex. Astron. Astrofis.* 5:319

Cantó J, Tenorio-Tagle G, Różyczka M. 1988. *Astron. Astrophys.* 192:227

Castro-Carrizo A, Lucas R, Bujarrabal V, Colomer F, Alcolea J. 2001. *Astron. Astrophys.* 368:L34

Cerruti-Sola M, Perinotto M. 1989. *Ap. J.* 345:339

Chevalier RA, Luo D. 1994. *Ap. J.* 435:815

Choudhuri AR. 1998. *The Physics of Fluids and Plasmas: An Introduction for Astrophysicists*, p. 359. New York: Cambridge Univ. Press

Chu Y, Guerrero M, Gruendl RA, Williams RM, Kaler JB. 2001a. *Ap. J. Lett.* 553:L69

Chu Y, Guerrero M, Gruendl RA, Williams RM, Kaler JB. 2001b. *Ap. J. Lett.* 554: L233

Cliffe JA, Frank A, Livio M, Jones J. 1995. *Ap. J. Lett.* 44:L49

Cohen M. 1981. *Publ. Astron. Soc. Pac.* 93:288

Collins TJB, Frank A, Bjorkman JE, Livio M. 1999. *Ap. J.* 512:322

Contopoulos J. 1995. *Ap. J.* 450:616

Corradi RLM, Livio M, Balick B, Munari U, Schwarz HE. 2001. *Ap. J.* 553:211

Corradi RLM, Schwarz HE. 1993a. *Astron. Astrophys.* 269:462

Corradi RLM, Schwarz HE. 1993b. *Astron. Astrophys.* 278:247

Corradi RLM, Schwarz HE. 1995. *Astron. Astrophys.* 293:87

Cox P, Lucas R, Huggins PJ, Forveille T, Bachiller R, et al. 2000. *Astron. Astrophys.* 35:L25

Dannenberg KK, Drake RP, Furnish MD, Knudson JD, Asay JR, et al. 2001. *Bull. Am. Astron. Soc.* 198:6402

Delamarter G. 2000. *The hydrodynamic shaping of planetary nebulae and young stellar objects.* PhD thesis. Univ. Rochester. 224 pp.

Deutsch AJ. 1956. *Ap. J.* 123:210

Dorfi EA, Hoefner S. 1996. *Astron. Astrophys.* 313:605

Doyle S, Balick B, Corradi RLM, Schwarz HE. 2000. *Astron. J.* 119:1339

Dwarkadas VV, Balick B. 1998. *Ap. J.* 497: 267

Dwarkadas VV, Chevalier RA, Blondin JM. 1996. *Ap. J.* 457:773

Dyson JE. 1977. *Ap. Space Sci.* 51:197

Dyson JE, de Vries J. 1972. *Astron. Astrophys.* 20:223

Dyson JE, Williams DA. 1980. *The Physics of the Interstellar Medium.* New York: Halsted

Dyson JE, Hartquist TW, Redman MP, Williams RJR. 2000. *Ap. Space Sci.* 272:197

Frank A. 1994. *Astron. J.* 107:261

Frank A. 1995. *Astron. J.* 110:2457

Frank A. 1999. *New Astron. Rev.* 43:31

Frank A. 2000. See Kastner et al. 2000, p. 225

Frank A, Balick B, Icke V, Mellema G. 1993. *Ap. J. Lett.* 404:L25

Frank A, Balick B, Livio M. 1996. *Ap. J. Lett.* 471:L53

Frank A, Balick B, Riley J. 1990. *Astron. J.* 100:1903

Frank A, Mellema G. 1994a. *Astron. Astrophys.* 289:937

Frank A, Mellema G. 1994b. *Ap. J.* 430:800

Frank A, Mellema G. 1996. *Ap. J.* 472:684

Frankowski A, Tylenda R. 2001. *Astron. Astrophys.* 367:513

García-Segura G. 1997. *Ap. J. Lett.* 489:L189

García-Segura G, Langer N, Różyczka M, Franco J. 1999. *Ap. J.* 517:767

García-Segura G, López JA. 2000. *Ap. J.* 544: 336

García-Segura G, López JA, Franco J. 2001. *Ap. J.* 560:928

Gardiner TA, Frank A. 2001. *Ap. J.* 557:250

Giuliani JL. 1982. *Ap. J.* 256:624

Gómez Y, Rodríguez LF. 2001. *Ap. J. Lett.* 557:L109

Gonçalves DR, Corradi RLM, Mampaso A. 2001. *Ap. J.* 547:302

Guerrero MA. 2000. See Kastner et al. 2000, p. 371

Guerrero MA, Chut Y-H, Gruendl RA, Williams RM, Kaler JB. 2001. *Ap. J. Lett.* 553: L55

Guerrero MA, Manchado A. 1998. *Ap. J.* 508: 262

Guerrero MA, Miranda LF, Chu Y-H, Rodríguez M, Williams RM. 2001. *Ap. J.* 563: 883

Guerrero MA, Villaver E, Manchado A. 1998. *Ap. J.* 507:889

Gurzadyan GA. 1997. *The Physics and Dynamics of Planetary Nebulae.* Berlin/Heidelberg/ New York: Springer-Verlag

Hajian AR, Frank A, Balick B, Terzian Y. 1997. *Ap. J.* 477:22

Han Z, Podsiadlowski P, Eggleton PP. 1995. *MNRAS* 272:800

Harpaz A, Soker N. 1994. *MNRAS* 270:734

Harrington JP, Borkowski KJ. 1994. *Bull. Am. Astron. Soc.* 26:1469

Hartmann L. 1998. *Accretion Processes in Star Formation.* Cambridge, UK/New York: Cambridge Univ. Press

Heger A, Langer N. 1998. *Astron. Astrophys.* 334:210

Henney WJ, Dyson JE. 1992. *Astron. Astrophys.* 261:301

Hrivnak BJ, Kwok S, Su KYL. 1999. *Ap. J.* 524:849

Hrivnak BJ, Kwok S, Su KYL. 2001. *Astron. J.* 121:2775

Hrivnak BJ, Langill PP, Su KYL, Kwok S. 1999. *Ap. J.* 513:421

Huggins PJ, Forveille T, Bachiller R, Cox P. 2000. *Ap. J.* 544:889

Huggins PJ, Healy AP. 1989. *Ap. J.* 346:201

Iben IJ, Livio M. 1993. *Ap. J. Lett.* 406: L15

Iben IJ. 2000. See Kastner et al. 2000, p. 107

Icke V. 1988. *Astron. Astrophys.* 202:177

Icke V, Balick B, Frank A. 1992. *Astron. Astrophys.* 253:224

Icke V, Mellema G, Balick B, Eulderink F, Frank A. 1992. *Nature* 355:524

Icke V, Preston HL, Balick B. 1989. *Astron. J.* 97:462

Ignace R, Cassinelli JP, Bjorkman JE. 1996. *Ap. J.* 459:671

Jura M. 1999. In *IAU Symp. 191: Asymptotic Giant Branch Stars*, ed. T Le Bertr, A Lebre, C Waelkens, p. 603. Dordrecht: Kluwer Acad.

Jura M, Chen C, Werner MW. 2000. *Ap. J.* 541:264

Jura M, Chen C, Werner MW. 2000. *Ap. J. Lett.* 544:L141

Jura M, Kahane C. 1999. *Ap. J.* 521:302

Kahane C, Audinos P, Barnbaum C, Morris M. 1996. *Astron. Astrophys.* 314:871

Kahn FD. 1983. In *IAU Symp. 103: Planetary Nebulae*, ed. DR Flower, p. 305. Dordrecht: Reidel

Kahn FD, Breitschwerdt D. 1990. *MNRAS* 242:505

Kahn FD, Breitschwerdt D. 1990. *MNRAS* 242:209

Kahn FD, West KA. 1985. *MNRAS* 212:837

Kastner JH, Soker N, Rappaport S, eds. 2000. *Asymmetrical Planetary Nebulae II, ASP Conf. Ser. Vol. 199*, San Francisco: Publ. Astron. Soc. Pac.

Kastner JH, Gatley I, Merrill K, Probst R, Weintraub D. 1994. *Ap. J.* 421:600

Kastner JH, Weintraub D, Gatley I, Merrill K, Probst R. 1996. *Ap. J.* 462:777

Kastner JH, Soker N, Vrtilek SD, Dgani R. 2000. *Ap. J. Lett.* 545:L57

Kastner JH, Vrtilek SD, Soker N. 2001. *Ap. J. Lett.* 550:L189

Knapp GR. 1986. *Ap. J.* 311:731

Knapp GR, Jorissen A, Young K. 1997. *Astron. Astrophys.* 326:318

Konigl A, Pudritz RE. 2000. In *Protostars and Planets IV*, ed. V Manning, AP Boss, SS Russell, p. 759. Tucson: Univ. Ariz. Press

Koo BC, McKee CF. 1992. *Ap. J.* 388:93

Kwok S. 1993. See Weinberger & Acker 1993, p. 263

Kwok S. 2000. See Kastner et al. 2000, p. 33

Kwok S, Hrivnak BJ, Su KYL. 2000. *Ap. J. Lett.* 544:L149

Kwok S, Hrivnak BJ, Zhang CY, Langill PL. 1996. *Ap. J.* 472:287

Kwok S, Purton CR, Fitzgerald PM. 1978. *Ap. J. Lett.* 219:L125

Kwok S, Su KYL, Hrivnak BJ. 1998. *Ap. J. Lett.* 501:L11

Lada CJ. 1985. *Annu. Rev. Astron. Astrophys.* 23:267

Lamers HJGLM, Cassinelli JP. 1999. *Introduction to Stellar Winds*. Cambridge, UK: Cambridge Univ. Press

Lebedev SV, Chittenden JP, Beg FN, Bland SN, Ciard A, et al. 2002. *Ap. J.* 564:113

Lery T, Heyvaerts J, Appl S, Norman CA. 1998. *Astron. Astrophys.* 337:603

Lery T, Heyvaerts J, Appl S, Norman CA. 1999. *Astron. Astrophys.* 347:1055

Lindqvist M, Olofsson H, Lucas R, Schöier FL, Neri R, et al. 1999. *Astron. Astrophys.* 351:L1

Livio M. 2000. See Kastner et al. 2000, p. 243

Livio M. 2001. In *Probing the Physics of Active Galactic Nuclei, ASP Conf. Ser.* ed. BM Peterson, RW Pogge, RS Polidan, 224:225. San Francisco: Publ. Astron. Soc. Pac.

Livio M, Pringle JE. 1996. *Ap. J. Lett.* 465:L55

Livio M, Pringle JE. 1997. *Ap. J.* 486:835

Livio M, Soker N. 1988. *Ap. J.* 329:764

Livio M, Soker N. 2001. *Ap. J.* 552:685

López JA. 1997. See Habing & Lamers 1997, p. 180

López JA. 2000. *Rev. Mex. Astron. Astrofis. Conf. Ser.* 9:201

López JA, Meaburn J, Rodríguez LF, Vázque R, Steffen W, Bryce M. 2000. *Ap. J.* 538:233

López JA, Roth M, Tapi M. 1993. *Rev. Mex. Astron. Astrofis.* 26:110

López-Martín L, Raga AC, Mellema G, Henney WJ, Cantó J. 2001. *Ap. J.* 548:288

Lynden-Bell D. 1996. *MNRAS* 279:389

Manchado A, Stanghellini L, Guerrero MA. 1996. *Ap. J. Lett.* 466:L95

Marten H, Szczerba R. 1997. *Astron. Astrophys.* 325:1132

Mastrodemos N, Morris M. 1998. *Ap. J.* 497: 303

Mastrodemos N, Morris M. 1999. *Ap. J.* 523: 357

Matt S, Balick B, Winglee R, Goodson A. 2000. *Ap. J.* 545:965

Matzner CD, McKee CF. 1999. *Ap. J. Lett.* 526:L109

Mauron N, Huggins PJ. 1999. *Astron. Astrophys.* 349:203

Mauron N, Huggins PJ. 2000. *Astron. Astrophys.* 359:707

Meaburn J, Clayton CA, Bryce M, Walsh JR, Holloway AJ, Steffen W. 1998. *MNRAS* 294:201

Meaburn J, Walsh JR. 1989. *Astron. Astrophys.* 223:277

Meixner M. 2000. See Kastner et al. 2000, p. 135

Meixner M, Skinner CJ, Graham JR, Keto E, Jernigan JG, Arens JF. 1997. *Ap. J.* 482:897

Meixner M, Ueta T, Dayal A, Hora JL, Fazio G, et al. 1999. *Ap. J. Suppl.* 122:221

Mellema G, Eulderink F, Icke V. 1991. *Astron. Astrophys.* 252:718

Mellema G. 1994. *Astron. Astrophys.* 290:915

Mellema G. 1995. *MNRAS* 277:173

Mellema G. 1997. *Astron. Astrophys.* 321:L29

Mellema G, Eulderink F, Icke V. 1991. *Astron. Astrophys.* 252:718

Mellema G, Frank A. 1995. *MNRAS* 273:401

Mellema G, Frank A. 1997. *MNRAS* 292:795.

Mellema G, Raga AC, Canto J, Lundqvist P,

Balick B, et al. 1998. *Astron. Astrophys.* 331:335

Miranda LF, Gómez Y, Anglada G, Torrelles JM. 2001a. *Nature* 414:284

Miranda LF, Torrelles J, Guerrero M, Vázquez R, Gómez Y. 2001b. *MNRAS* 321:487

Miranda LF, Vázquez R, Corradi RLM, Guerrero M, López JA, Torrelles J. 1999. *Ap. J.* 520:714

Morris M. 1981. *Ap. J.* 249:572

Morris M. 1987. *Publ. Astron. Soc. Pac.* 99: 1115

Morse JA, Davidson K, Bally J, Ebbets D, Balick B, Frank A. 1998. *Astron. J.* 116:2443

Muthu C, Anandarao BG. 2001. *Astron. J.* 121:2106

O'Connor JA, Redman MP, Holloway AJ, Bryce M, López JA, Meaburn J. 2000. *Ap. J.* 531:336

O'Dell CR, Henney WJ, Burkert A. 2000. *Astron. J.* 119:2910

O'dell CR, Ball ME. 1985. *Ap. J.* 289:526.

O'Dell CR, Handron KD. 1996. *Astron. J.* 111:1630

Ogilvie GI, Livio M. 1998. *Ap. J.* 499:329

Olofsson H, Bergman P, Lucas R, Eriksson K, Gustafsson B, Bieging JH. 2000. *Astron. Astrophys.* 353:583

Owocki S, Cranmer S, Blondin J. 1994. *Ap. J.* 424:887

Owocki S, Cranmer S, Gayley KG. 1996. *Ap. J.* 472:L115

Palen S, Fix JD. 2000. *Ap. J.* 531:391

Pascoli G. 1985. *Astron. Astrophys.* 147:257

Pascoli G. 1992. *Publ. Astron. Soc. Pac.* 104: 350

Pascoli G. 1997. *Ap. J.* 489:946

Patriarchi P, Perinotto M. 1991. *Astron. Astrophys. Suppl.* 91:325

Pijpers FP, Habing HJ. 1989. *Astron. Astrophys.* 215:334

Pijpers FP, Hearn AG. 1989. *Astron. Astrophys.* 209, 198

Poludnenko A, Frank A, Blackman E. 2001. *Ap. J.* In press

Priest ER. 1984. *Solar Magnetohydrodynamics, Geophys. Astrophys. Monogr.* Dordrecht: Reidel

Pringle JE. 1996. *MNRAS* 281:357

Rasio FA, Livio M. 1996. *Ap. J.* 471:366

Redman MP, Dyson JE. 1999. *MNRAS* 302:L17

Redman MP, O'Connor JA, Holloway AJ, Bryce M, Meaburn J. 2000. *MNRAS* 312:L23

Reed DS, Balick B, Hajian AR, Klayton TL, Giovanardi S. 1999. *Astron. J.* 118:2430

Reipurth B. 1987. *Nature* 325:787

Reipurth B, Bally J. 2001. *Annu. Rev. Astron. Astrophys.* 39:403

Remington BA, Arnett D, Drake RP, Takabe H. 1999. *Science* 284:1488

Reyes-Ruiz M, López JA. 1999. *Ap. J.* 524:952

Riera A, Balick B, Mellema G, Xilouri K, Terzian Y. 2000. See Kastner et al. 2000, p. 297

Riera A, Garcia-Lario P, Manchado A, Pottasch SR, Raga AC. 1995. *Astron. Astrophys.* 302:137

Różyczka M, Franco J. 1996. *Ap. J. Lett.* 469:L127

Sabbadin F. 1984. *Astron. Astrophys. Suppl.* 58:273

Sabbadin F, Bianchini A, Hamzaoglu E. 1984. *Astron. Astrophys.* 136:200

Sabbadin F, Bianchini A, Strafella F. 1986. *Astron. Astrophys. Suppl.* 65:259

Sahai R. 1999. *Ap. J. Lett.* 524:L125

Sahai R. 2000. *Ap. J. Lett.* 537:L43

Sahai R, Bujarrabal V, Castro-Carrizo A, Zijlstra A. 2000. *Astron. Astrophys.* 360:L9

Sahai R, Bujarrabal V, Zijlstra A. 1999a. *Ap. J. Lett.* 518:L115

Sahai R, Dayal A, Watson AM, Trauger JT, Stapelfeldt KR, et al. 1999b. *Astron. J.* 118:468

Sahai R, Nyman L. 2000. *Ap. J. Lett.* 538:L145

Sahai R, Trauger JT. 1998. *Astron. J.* 116:1357

Sahai R, Trauger JT, Watson AM, Stapelfeldt KR, Hester JJ, et al. 1998. *Ap. J.* 493:301

Sahai R, Zijlstra A, Bujarrabal V, Te Lintel Hekkert P. 1999c. *Astron. J.* 117:1408

Sakurai T. 1985. *Astron. Astrophys.* 152:121

Sánchez Contreras C, Bujarrabal V, Miranda LF, Fernández-Figueroa MJ. 2000a. *Astron. Astrophys.* 355:1103

Sánchez Contreras C, Bujarrabal V, Neri R, Alcolea J. 2000b. *Astron. Astrophys.* 357:651

Sánchez Contreras C, Sahai R. 2001. *Ap. J. Lett.* 553:L173

Sandquist EL, Taam RE, Chen X, Bodenheimer P, Burkert A. 1998. *Ap. J.* 500:909

Steffen M, Schönberner D. 2001. In *Post-AGB Objects as a Phase of Stellar Evolution,* ed. R Szczerba, SK Goacuterny, p. 131. Boston/Dordrecht/London: Kluwer

Schwarz HE, Aspin C, Corradi RLM, Reipurth B. 1997. *Astron. Astrophys.* 319:267

Shklovskii I. 1956a. *Astron. Zh.* 33:222

Shklovskii I. 1956b. *Astron. Zh.* 33:315

Simis YJW, Icke V, Dominik C. 2001. *Astron. Astrophys.* 371:205

Skinner CJ, Meixner M, Bobrowsky M. 1998. *MNRAS* 300:L29

Soker N. 1994a. *Astron. J.* 107:276

Soker N. 1994b. *Publ. Astron. Soc. Pac.* 106:59

Soker N. 1996. *Ap. J.* 468:774

Soker N. 1997. *Ap. J. Suppl.* 112:487

Soker N. 1998a. *Ap. J.* 496:833

Soker N. 1998b. *MNRAS* 299:1242

Soker N. 1998c. In *Cool Stars, Stellar Systems, and the Sun, 10th, ASP Conf. Ser. 154,* ed. RA Donahue, JA Bookbinder, p. 1901. San Francisco: Astron. Soc. Pac.

Soker N. 1999. *MNRAS* 306:806

Soker N. 2000a. *Ap. J.* 540:436

Soker N. 2000b. *MNRAS* 312:217

Soker N. 2001. *Ap. J.* 558:157

Soker N. 2002. *Ap. J.* 568:726

Soker N, Livio M. 1989. *Ap. J.* 339:268

Soker N, Livio M. 1994. *Ap. J.* 421:219

Soker N, Rappaport S. 2001. *Ap. J.* 557:256

Soker N, Zoabi E. 2002. *MNRAS* 329:204

Solf J. 2000. *Astron. Astrophys.* 354:674

Steffen W, López JA. 1998. *Ap. J.* 508:696

Steffen M, Schönberner D. 2000. *Astron. Astrophys.* 357:180

Steffen W, López JA, Lim A. 2001. *Ap. J.* 556:823

Su KYL, Volk K, Kwok S, Hrivnak BJ. 1998. *Ap. J.* 508:744

Taam RE, Sandquist EL. 2000. *Annu. Rev. Astron. Astrophys.* 38:113

Tenorio-Tagle G, Cantó J, Różyczka M. 1988. *Astron. Astrophys.* 202:256

Terman JL, Taam RE. 1996. *Ap. J.* 458:692

Terzian Y, Hajian AR. 2000. See Kastner et al. 2000, p. 33

Trammell SR. 2000. See Kastner et al. 2000, p. 147

Trammell SR, Goodrich RW. 1996. *Ap. J. Lett.* 468:L107

Tsinganos K, Bogovalov S. 2000. *Astron. Astrophys.* 356:989

Ueta T, Fong D, Meixner M. 2001. *Ap. J. Lett.* 557:L117

Ueta T, Meixner M, Bobrowsky M. 2000. *Ap. J.* 528:861

van der Veen WECJ, Habing HJ. 1988. *Astron. Astrophys.* 194:125

van der Veen WECJ, Habing HJ, Geballe TR. 1989. *Astron. Astrophys.* 226:108

Weaver R, McCray R, Castor J, Shapiro P, Moore R. 1977. *Ap. J.* 218:377

Weber EJ, Davis L Jr. 1967. *Ap. J.* 148:217

Weinberger R. 1989. *Astron. Astrophys. Suppl.* 78:301

Weinberger R, Acker A, eds. 1993. *IAU Symp. 155: Planetary Nebulae.* Dordrecht: Kluwer

Willson LA. 2000. *Annu. Rev. Astron. Astrophys.* 38:573

Wood PR. 1997. See Habing & Lamers 1997, p. 297

Wood PR, Alcock C, Allsman RA, Alves D, Axelrod TS, et al. 1999. In *Asymptotic Giant Branch Stars, IAU Symp. 191*, ed. T Le Bertre, A Lebre, C Waelkens, p. 151. Dordrecht: Kluwer

Yorke H, Welz A. 1996. *Astron. Astrophys.* 315:555

Yungelson LR, Tutukov AV, Livio M. 1993. *Ap. J.* 418:794

Zhang CY, Kwok S. 1993. *Ap. J. Suppl.* 88:137

Zuckerman B, Aller LH. 1986. *Ap. J.* 301:772

Annu. Rev. Astron. Astrophys. 2002. 40:487–537
doi: 10.1146/annurev.astro.40.060401.093840
Copyright © 2002 by Annual Reviews. All rights reserved

THE NEW GALAXY: Signatures of Its Formation

Ken Freeman
*Mount Stromlo Observatory, Australia National University, Weston Creek, ACT 2611,
Australia; email: kcf@mso.anu.edu.au*

Joss Bland-Hawthorn
*Anglo-Australian Observatory, 167 Vimiera Road, Eastwood, NSW 2122, Australia;
email: jbh@aao.gov.au*

Key Words   cosmology, local group, stellar populations, stellar kinematics

■ **Abstract**   The formation and evolution of galaxies is one of the great outstanding problems of astrophysics. Within the broad context of hierachical structure formation, we have only a crude picture of how galaxies like our own came into existence. A detailed physical picture where individual stellar populations can be associated with (tagged to) elements of the protocloud is far beyond our current understanding. Important clues have begun to emerge from both the Galaxy (near-field cosmology) and the high redshift universe (far-field cosmology). Here we focus on the fossil evidence provided by the Galaxy. Detailed studies of the Galaxy lie at the core of understanding the complex processes involved in baryon dissipation. This is a necessary first step toward achieving a successful theory of galaxy formation.

PROLOGUE

The New Galaxy

Weinberg (1977) observed that "the theory of the formation of galaxies is one of the great outstanding problems of astrophysics, a problem that today seems far from solution." Although the past two decades have seen considerable progress, Weinberg's assessment remains largely true.

Eggen, Lynden-Bell and Sandage (1962; ELS) were the first to show that it is possible to study galactic archaeology using stellar abundances and stellar dynamics; this is probably the most influential paper on the subject of galaxy formation. ELS studied the motions of high velocity stars and discovered that, as the metal abundance decreases, the orbit energies and eccentricities of the stars increased while their orbital angular momenta decreased. They inferred that the metal-poor stars reside in a halo that was created during the rapid collapse of a relatively uniform, isolated protogalactic cloud shortly after it decoupled from the universal expansion. ELS are widely viewed as advocating a smooth monolithic collapse of the protocloud with a timescale of order 10^8 years. But Sandage (1990) stresses that this is an over-interpretation; a smooth collapse was not one of the inferences they drew from the stellar kinematics.

In 1977, the ELS picture was challenged by Searle (see also Searle & Zinn 1978) who noted that Galactic globular clusters have a wide range of metal abundances essentially independent of radius from the Galactic Center. They suggested that this could be explained by a halo built up over an extended period from *independent* fragments with masses of $\sim 10^8 \, M_\odot$. In contrast, in the ELS picture, the halo formed in a rapid free-fall collapse. But halo field stars, as well as globular clusters, are now believed to show an age spread of 2–3 Ga (Marquez & Schuster 1994); for an alternative view, see Sandage & Cacciari (1990). The current paradigm, that the observations argue for a halo that has built up over a long period from infalling debris, has developed after many years of intense debate.

This debate parallelled the changes that were taking place in theoretical studies of cosmology (e.g., Peebles 1971, Press & Schechter 1974). The ideas of galaxy formation via hierarchical aggregation of smaller elements from the early universe fit in readily with the Searle & Zinn view of the formation of the galactic halo from small fragments. The possibility of identifying debris from these small fragments was already around in Eggen's early studies of moving groups, and this is now an active field of research in theoretical and observational stellar dynamics. It offers the possibility to reconstruct at least some properties of the protogalaxy and so to improve our basic understanding of the galaxy formation process.

We can extend this approach to other components of the Galaxy. We will argue the importance of understanding the formation of the galactic disk, because this is where most of the baryons reside. Although much of the information about the properties of the protogalactic baryons has been lost in the dissipation that led to the galactic disk, a similar dynamical probing of the early properties of the disk can illuminate the formation of the disk, at least back to the epoch of last significant dissipation. It is also clear that we do not need to restrict this probing to stellar dynamical techniques. A vast amount of fossil information is locked up in the detailed stellar distribution of chemical elements in the various components of the Galaxy, and we will discuss the opportunities that this offers.

We are coming into a new era of galactic investigation, in which one can study the fossil remnants of the early days of the Galaxy in a broader and more focussed way, not only in the halo but throughout the major luminous components of the Galaxy. This is what we mean by *The New Galaxy*. The goal of these studies is to reconstruct as much as possible of the early galactic history. We review what has been achieved so far, and point to the future.

Near-Field and Far-Field Cosmology

What do we mean by the reconstruction of early galactic history? We seek a detailed physical understanding of the sequence of events which led to the Milky Way. Ideally, we would want to tag (i.e., associate) components of the Galaxy to elements of the protocloud—the baryon reservoir which fueled the stars in the Galaxy.

From theory, our prevailing view of structure formation relies on a hierarchical process driven by the gravitational forces of the large-scale distribution of cold, dark matter (CDM). The CDM paradigm provides simple models of galaxy formation within a cosmological context (Peebles 1974, White & Rees 1978, Blumenthal et al. 1984). N-body and semi-analytic simulations of the growth of structures in the early universe have been successful at reproducing some of the properties of galaxies. Current models include gas pressure, metal production, radiative cooling and heating, and prescriptions for star formation.

The number density, properties and spatial distribution of dark matter halos are well understood within CDM (Sheth & Tormen 1999, Jenkins et al. 2001). However, computer codes are far from producing realistic simulations of how baryons produce observable galaxies within a complex hierarchy of dark matter. This a necessary first step towards a viable theory or a working model of galaxy formation.

In this review, our approach is anchored to observations of the Galaxy, interpreted within the broad scope of the CDM hierarchy. Many of the observables in the Galaxy relate to events which occurred long ago, at high redshift. Figure 1 shows the relationship between look-back time and redshift in the context of the ΛCDM model: The redshift range ($z \lesssim 6$) of discrete sources in contemporary observational cosmology matches closely the known ages of the oldest components in the Galaxy. The Galaxy (near-field cosmology) provides a link to the distant universe (far-field cosmology).

Before we embark on a detailed overview of the relevant data, we give a descriptive working picture of the sequence of events involved in galaxy formation. For continuity, the relevant references are given in the main body of the review where these issues are discussed in more detail.

A Working Model of Galaxy Formation

Shortly after the Big Bang, cold dark matter began to drive baryons towards local density enhancements. The first stars formed after the collapse of the first primordial molecular clouds; these stars produced the epoch of reionization. The earliest recognizable protocloud may have begun to assemble at about this time.

Within the context of CDM, the dark halo of the Galaxy assembled first, although it is likely that its growth continues to the present time. In some galaxies, the first episodes of gas accretion established the stellar bulge, the central black hole, the first halo stars and the globular clusters. In the Galaxy and similar systems, the small stellar bulge may have formed later from stars in the inner disk.

The early stages of the Galaxy's evolution were marked by violent gas dynamics and accretion events, leading to the high internal densities of the first globular clusters, and perhaps to the well-known black hole mass–stellar bulge dispersion relation. The stellar bulge and massive black hole may have grown up together during this active time. We associate this era with the Golden Age, the phase before $z \sim 1$ when star formation activity and accretion disk activity were at their peak.

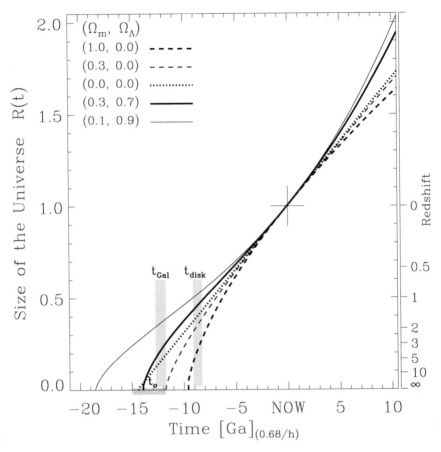

Figure 1   Look-back time as a function of redshift and the size of the Universe (Lineweaver 1999) for five different world models. The approximate ages of the Galactic halo and disk are indicated by hatched regions.

At that time, there was a strong metal gradient from the bulge to the outer halo. The metal enrichment was rapid in the core of the Galaxy such that, by $z \sim 1$, the mean metallicities were as high as [Fe/H] ~ -1 or even higher. In these terms, we can understand why the inner stellar bulge that we observe today is both old and moderately metal rich. The first halo stars ([Fe/H] ≈ -5 to -2.5) formed over a more extended volume and presumably date back to the earliest phase of the protocloud. The first globular clusters formed over a similar volume from violent gas interactions ([Fe/H] ≈ -2.5 to -1.5). We believe now that many of the halo stars and globulars are remnants of early satellite galaxies which experienced independent chemical evolution before being accreted by the Galaxy.

The spread in [Fe/H], and the relative distribution of the chemical elements, is a major diagnostic of the evolution of each galactic component. If the initial mass function is constant, the mean abundances of the different components give a rough indication of the number of SN II enrichments which preceded their formation, although we note that as time passes, an increasing fraction of Fe is produced by SN Ia events. For a given parcel of gas in a closed system, only a few SN II events are required to reach [Fe/H] ≈ -3, 30 to 100 events to get to [Fe/H] ≈ -1.5, and maybe a thousand events to reach solar metallicities. We wish to stress that [Fe/H] is not a clock: Rather it is a measure of supernova occurrences and the depth of the different potential wells that a given parcel of gas has explored.

During the latter stages of the Golden Age, most of the baryons began to settle to a disk for the first time. Two key observations emphasize what we consider to be the mystery of the main epoch of baryon dissipation. First, there are no stars with [Fe/H] < -2.2 that rotate with the disk. Second, despite all the activity associated with the Golden Age, at least 80% of the baryons appear to have settled gradually to the disk over many Ga; this fraction could be as high as 95% if the bulge formed after the disk.

About 10% of the baryons reside in a thick disk which has [Fe/H] ≈ -2.2 to -0.5, compared to the younger thin disk with [Fe/H] ≈ -0.5 to $+0.3$. It is striking how the globular clusters and the thick disk have similar abundance ranges, although the detailed abundance distributions are different. There is also a similarity in age: Globular clusters show an age range of 12 to 14 Ga, and the thick disk appears to be at least 12 Ga old. Both the thick disk and globulars apparently date back to the epoch of baryon dissipation during $z \sim 1-5$.

Figure 2 (see color insert) summarizes our present understanding of the complex age-metallicity distribution for the various components of the Galaxy.

It is a mystery that the thick disk and the globulars should have formed so early and over such a large volume from material that was already enriched to [Fe/H] ~ -2. Could powerful winds from the central starburst in the evolving core have distributed metals throughout the inner protocloud at about that time?

Finally, we emphasize again that 90% of the disk baryons have settled quiescently to the thin disk since $z \sim 1$.

Timescales and Fossils

The oldest stars in our Galaxy are of an age similar to the look-back time of the most distant galaxies in the Hubble Deep Field (Figure 1). For the galaxies, the cosmological redshift measured from galaxy spectra presently takes us to within 5% of the origin of cosmic time. For the stars, their upper atmospheres provide fossil evidence of the available metals at the time of formation. The old Galactic stars and the distant galaxies provide a record of conditions at early times in cosmic history, and both harbor clues to the sequence of events that led to the formation of galaxies like the Milky Way.

The key timescale provided by far-field cosmology is the look-back time with the prospect of seeing galaxies at an earlier stage in their evolution. However, this does not imply that these high-redshift objects are unevolved. We know that the stellar cores of galaxies at the highest redshifts ($z \sim 5$) observed to date exhibit solar metallicities, and therefore appear to have undergone many cycles of star formation (Hamann & Ferland 1999). Much of the light we detect from the early universe probably arises from the chemically and dynamically evolved cores of galaxies.

Near-field cosmology provides a dynamical timescale, $\tau_D \sim (G\rho)^{-\frac{1}{2}}$, where ρ is the mean density of the medium. The dynamical timescale at a radial distance of 100 kpc is of order several Ga, so the mixing times are very long. Therefore, on larger scales, we can expect to find dynamical and chemical traces of past events, even where small dynamical systems have long since merged with the Galaxy.

We note that the CDM hierarchy reflects a wide range of dynamical timescales, such that different parts of the hierarchy may reveal galaxies in different stages of evolution. In this sense, the hierarchy relates the large-scale density to the morphology and evolution of its individual galaxies; this is the so-called morphology-density relation (Dressler 1980, Hermit et al. 1996, Norberg et al. 2001). Over a large enough ensemble of galaxies, taken from different regions of the hierarchy, we expect different light-weighted age distributions because one part of the hierarchy is more evolved than another. In other words, the evolution of small-scale structure (individual galaxies) must at some level relate to the environment on scales of 10 Mpc or more.

The near field also provides important evolutionary timescales for individual stars and groups of stars (see "Stellar Age Dating" below). Individual stars can be dated with astero-seismology (Christensen-Dalsgaard 1986, Gough 2001) and nucleo-cosmochronology (Fowler & Hoyle 1960, Cowan et al. 1997). Strictly speaking, nucleo-cosmochronology dates the elements rather than the stars. Coeval groups of stars can be aged from the main-sequence turn-off or from the He-burning stars in older populations (Chaboyer 1998). Furthermore, the faint end cut-off of the white dwarf luminosity function provides an important age constraint for older populations (Oswalt et al. 1996). Presently, the aging methods are model dependent.

Goals of Near-Field Cosmology

We believe that the major goal of near-field cosmology is to tag individual stars with elements of the protocloud. Some integrals of motion are likely to be preserved while others are scrambled by dissipation and violent relaxation. We suspect that complete tagging is impossible. However, some stars today may have some integrals of motion that relate to the protocloud at the epoch of last dissipation (see "Zero Order Signatures—Information Preserved Since Dark Matter Virialized" below).

As we review, different parts of the Galaxy have experienced dissipation and phase mixing to varying degrees. The disk, in contrast to the stellar halo, is a highly dissipated structure. The bulge may be only partly dissipated. To what extent can we unravel the events that produced the Galaxy as we see it today? Could some of the residual inhomogeneities from prehistory have escaped the dissipative process at an early stage?

Far-field cosmology currently takes us back to the epoch of last scattering as seen in the microwave background. Cosmologists would like to think that some vestige of information has survived from earlier times (compare Peebles et al. 2000). In the same spirit, we can hope that fossils remain from the epoch of last dissipation, i.e., the main epoch of baryon dissipation that occurred as the disk was being assembled.

To make a comprehensive inventory of surviving inhomogeneities would require a vast catalog of stellar properties that is presently out of reach (Bland-Hawthorn 2002). The Gaia space astrometry mission (Perryman et al. 2001), set to launch at the end of the decade, will acquire detailed phase space coordinates for about one billion stars, within a sphere of diameter 20 kpc (the Gaiasphere). In "The Gaiasphere and The Limits of Knowledge" below, we look forward to a time when all stars within the Gaiasphere have complete chemical abundance measurements (including all heavy metals). Even with such a vast increase in information, there may exist fundamental—but unproven—limits to unravelling the observed complexity.

The huge increase in data rates from ground-based and space-based observatories has led to an explosion of information. Much of this information from the near field is often dismissed as weather or unimportant detail. But in fact fundamental clues are already beginning to emerge. In what is now a famous discovery, a large photometric and kinematic survey of bulge stars revealed the presence of the disrupting Sgr dwarf galaxy (Ibata et al. 1994), now seen over a large region of sky and in a variety of populations (see "Structures in Phase Space" below). Perhaps the most important example arises from the chemical signatures seen in echelle spectroscopy of bulge, thick disk and halo stars. In "Epilogue: Challenges for the Future," we envisage a time when the analysis of thousands of spectral lines for a vast number of stars will reveal crucial insights into the sequence of events early in the formation of the Galaxy.

In this review, we discuss fossil signatures in the Galaxy. A key aspect of fossil studies is a reliable time sequence. In "Stellar Age Dating," we discuss methods for age-dating individual stars and coeval groups of stars. In "Structure of the Galaxy," we describe the main components of the Galaxy. In "Signatures of Galaxy Formation," we divide the fossil signatures of galaxy formation into three parts: zero order signatures that preserve information since dark matter virialized; first order signatures that preserve information since the main epoch of baryon dissipation; and second order signatures that arise from major processes involved in subsequent evolution. In "The Gaiasphere and The Limits of Knowledge," we look forward to a time when it is possible to measure ages, phase space coordinates and chemical properties for a vast number of stars in the Galaxy. Even then, what are the prospects for unravelling the sequence of events that gave rise to the Milky Way? We conclude with some experimental challenges for the future.

STELLAR AGE DATING

Nucleo-cosmochronology (or cosmochronometry), or the aging of the elements through radioactive decay, has a long history (Rutherford 1904, Fowler & Hoyle 1960, Butcher 1987). A related technique is widely used in solar system geophysics. Independent schemes have aged the oldest meteorites at 4.53 ± 0.04 Ga (Guenther & Demarque 1997, Manuel 2000). The small uncertainties reflect that the age dating is direct. Element pairs like Rb and Sr are chemically distinct and freeze out during solidification into different crystalline grains. The isotope ^{87}Rb decays into ^{87}Sr, which can be compared to ^{86}Sr, a nonradiogenic isotope, measured from a control sample of Sr-rich grains. This provides a direct measure of the fraction of a ^{87}Rb half-life ($\tau_{1/2} = 47.5$ Ga) since the meteorite solidified.

It appears that, until we have a precise understanding of BBNS and the early chemical evolution history of the Galaxy, geophysical precision will not be possible for stellar ages. The major problem is that, as far as we know, there is no chemical differentiation that requires that we know precisely how much of each isotope was originally present. Modern nucleo-cosmochronology compares radioactive isotope strengths to a stable r-process element (e.g. Nd, Eu, La, Pt). The thorium method (^{232}Th, $\tau_{1/2} = 14.0$ Ga) was first applied by Butcher (1987) and refined by Pagel (1989). Other radioactive chronometers include ^{235}U ($\tau_{1/2} = 0.70$ Ga) and ^{238}U ($\tau_{1/2} = 4.47$ Ga), although Yokoi et al. (1983) have expressed concerns about their use (compare Cayrel et al. 2001). Arnould & Goriely (2001) propose that the isotope pair ^{187}Re$-^{187}$Os ($\tau_{1/2} = 43.5$ Ga) may be better suited for future work.

With the above caveats, we point out that several groups are now obtaining exquisite high-resolution data on stars with enhanced r-process elements (Cayrel et al. 2001, Sneden et al. 2000, Burris et al. 2000, Westin et al. 2000, Johnson & Bolte 2001, Cohen et al. 2002, Hill et al. 2002). For a subset of these stars, radioactive ages have been derived (Truran et al. 2001) normalized to the heavy element abundances observed in meteorites.

There are few other direct methods for deriving ages of individual stars. A promising field is astero-seismology that relies on the evolving mean molecular weight in stellar cores (Christensen-Dalsgaard 1986, Ulrich 1986, Gough 1987, Guenther 1989). Gough (2001) has determined 4.57 ± 0.12 Ga for the Sun, which should be compared with the age of meteorites quoted above. The Eddington satellite under consideration by ESA for launch at the end of the decade proposes to use stellar oscillations to age 50,000 main sequence stars (Gimenez & Favata 2001).

It has long been known that disk stars span a wide range of ages from the diversity of main-sequence stars. Edvardsson et al. (1993) derived precise stellar evolution ages for nearby individual post-main-sequence F stars using Strömgren photometry, and showed that the stars in the Galactic disk exhibit a large age spread with ages up to roughly 10 Ga (Figure 2). Using the inverse age-luminosity relation for RR Lyrae stars, Chaboyer et al. (1996) found that the oldest globular clusters are older than 12 Ga with 95% confidence, with a best estimate of 14.6 ± 1.7 Ga (Chaboyer 1998). But Hipparcos appears to show that the RR Lyr distances are underestimated leading to a downward revision of the cluster ages: 8.5–13.3 Ga (Gratton et al. 1997); 11–13 Ga (Reid 1998); 10.2–12.8 Ga (Chaboyer et al. 1998). For a coeval population (e.g. open and globular clusters), isochrone fitting is widely used. The ages of the galactic halo and globular clusters, when averaged over eight independent surveys, lead to 12.2 ± 0.5 Ga (Lineweaver 1999).

Other traditional methods rely on aging a population of stars that are representative of a particular component of the Galaxy. For example, Gilmore et al. (1989) use the envelope of the distribution in a color-abundance plane to show that all stars more metal poor than [Fe/H] $= -0.8$ are as old as the globular clusters. Similarly, the faint end of the white dwarf luminosity function is associated with the coolest, and therefore the oldest, stars (Oswalt et al. 1996). The present estimate for the age of the old thin disk population when averaged over five independent surveys is 8.7 ± 0.4 Ga (Lineweaver 1999), although Oswalt et al. argue for $9.5^{+1.1}_{-0.8}$ Ga.

For a world model with ($\Omega_\Lambda = 0.7$, $\Omega_m = 0.3$), the Big Bang occurred 14 Ga ago (Efstathiou et al. 2002)—in our view, there is no compelling evidence for an age crisis from a comparison of estimates in the near and far field. But the inaccuracy of age dating relative to an absolute scale does cause problems. At present, the absolute ages of the oldest stars cannot be tied down to better than about 2 Ga, the time elapsed between $z = 6$ and $z = 2$. This is a particular handicap to identifying specific events in the early Universe from the stellar record.

STRUCTURE OF THE GALAXY

Like most spiral galaxies, our Galaxy has several recognizable structural components that probably appeared at different stages in the galaxy formation process. These components will retain different kinds of signatures of their formation. We will describe these components in the context of other disk galaxies, and use images of other galaxies in Figure 3 (see color insert) to illustrate the components.

The Bulge

First, compare images of M104 and IC 5249 (Figures 3c,g): These are extreme examples of galaxies with a large bulge and with no bulge. Large bulges like that of M104 are structurally and chemically rather similar to elliptical galaxies: Their surface brightness distribution follows an $r^{1/4}$ law (e.g., Pritchet & van den Bergh 1994) and they show similar relations of [Fe/H] and [Mg/Fe] with absolute magnitude (e.g., Jablonka et al. 1996). These properties lead to the view that the large bulges formed rapidly. The smaller bulges are often boxy in shape, with a more exponential surface brightness distribution (e.g., Courteau et al. 1996). The current belief is that they may have arisen from the stellar disk through bending mode instabilities.

Spiral bulges are usually assumed to be old, but this is poorly known, even for the Galaxy. The presence of bulge RR Lyrae stars indicates that at least some fraction of the galactic bulge is old (Rich 2001). Furthermore, the color-magnitude diagrams for galactic bulge stars show that the bulge is predominantly old. McWilliam & Rich (1994) measured [Fe/H] abundances for red giant stars in the bulge of the Milky Way. They found that, while there is a wide spread, the abundances ([Fe/H] $\approx$ -0.25) are closer to the older stars of the metal rich disk than to the very old metal poor stars in the halo and in globular clusters, in agreement with the abundances of planetary nebulae in the Galactic bulge (e.g., Exter et al. 2001).

The COBE image of the Milky Way (Figure 3b) shows a modest somewhat boxy bulge, typical of an Sb to Sc spiral. Figure 3d shows a more extreme example of a boxy/peanut bulge. If such bulges do arise via instabilities of the stellar disk, then much of the information that we seek about the state of the early galaxy would have been lost in the processes of disk formation and subsequent bulge formation. Although most of the more luminous disk galaxies have bulges, many of the fainter disk galaxies do not. Bulge formation is not an essential element of the formation processes of disk galaxies.

The Disk

Now look at the disks of these galaxies. The exponential thin disk, with a vertical scale height of about 300 pc, is the most conspicuous component in edge-on disk galaxies like NGC 4762 and IC 5249 (Figures 3e,g). The thin disk is believed to be the end product of the quiescent dissipation of most of the baryons and contains almost all of the baryonic angular momentum. For the galactic disk, which is clearly seen in the COBE image in Figure 3b, we know from radioactive dating, white dwarf cooling and isochrone estimates for individual evolved stars and open clusters that the oldest disk stars have ages in the range 10 to 12 Ga (see "Stellar Age Dating" above).

The disk is the defining stellar component of disk galaxies, and understanding its formation is in our view the most important goal of galaxy formation theory. Although much of the information about the pre-disk state of the baryons has been lost in the dissipative process, some tracers remain, and we will discuss them in the next section.

Many disk galaxies show a second fainter disk component with a larger scale height (typically about 1 kpc); this is known as the thick disk. Deep surface photometry of IC 5249 shows only a very faint thick disk enveloping the bright thin disk (Abe et al. 1999): Compare Figures 3g,h. In the edge-on S0 galaxy NGC 4762, we see a much brighter thick disk around its very bright thin disk (Tsikoudi 1980): The thick disk is easily seen by comparing Figures 3e,f. The Milky Way has a significant thick disk (Gilmore & Reid 1983): Its scale height ($\sim$1 kpc) is about three times larger than the scale height of the thin disk, its surface brightness is about 10% of the thin disk's surface brightness, its stellar population appears to be older than about 12 Ga, and its stars are significantly more metal poor than the stars of the thin disk. The galactic thick disk is currently believed to arise from heating of the early stellar disk by accretion events or minor mergers (see "Disk Heating by Accretion" below).

The thick disk may be one of the most significant components for studying signatures of galaxy formation because it presents a snap frozen relic of the state of the (heated) early disk. Although some apparently pure-disk galaxies like IC 5249 do have faint thick disks, others do not (Fry et al. 1999): These pure-disk galaxies show no visible components other than the thin disk. As for the bulge, formation of a thick disk is not an essential element of the galaxy formation process. In some galaxies the dissipative formation of the disk is clearly a very quiescent process.

The Stellar Halo

There are two further components that are not readily seen in other galaxies and are shown schematically in Figure 3a. The first is the metal-poor stellar halo, well-known in the Galaxy as the population containing the metal-poor globular clusters and field stars. Its mass is only about 1% of the total stellar mass (about 10^9 $M_\odot$: e.g., Morrison 1993). The surface brightness of the galactic halo, if observed in other galaxies, would be too low to detect from its diffuse light. It can be seen in other galaxies of the Local Group in which it is possible to detect the individual evolved halo stars. The metal-poor halo of the Galaxy is very interesting for galaxy formation studies because it is so old: Most of its stars are probably older than 12 Ga and were probably among the first galactic objects to form. The galactic halo has a power law density distribution $\rho \propto r^{-3.5}$ although this appears to depend on the stellar population (Vivas et al. 2001, Chiba & Beers 2000). Unlike the disk and bulge, the angular momentum of the halo is close to zero (e.g., Freeman 1987), and it is supported almost entirely by its velocity dispersion; Some of its stars are very energetic, reaching out to at least 100 kpc from the galactic center (e.g., Carney et al. 1990).

The current view is that the galactic halo formed at least partly through the accretion of small metal-poor satellite galaxies that underwent some independent chemical evolution before being accreted by the Galaxy (Searle & Zinn 1978, Freeman 1987). Although we do still see such accretion events taking place now, in the apparent tidal disruption of the Sgr dwarf (Ibata et al. 1995), most of them must have occurred long ago. Accretion of satellites would dynamically heat the thin disk, so the presence of a dominant thin disk in the Galaxy means that most of this halo-building accretion probably predated the epoch of thick disk formation ~12 Ga ago. We can expect to see dynamically unmixed residues or fossils of at least some of these accretion events (e.g., Helmi & White 1999).

Of all the galactic components, the stellar halo offers the best opportunity for probing the details of its formation. There is a real possibility to identify groups of halo stars that originate from common progenitor satellites (Eggen 1977, Helmi & White 1999, Harding et al. 2001, Majewski et al. 2000). However, if the accretion picture is correct, then the halo is just the stellar debris of small objects accreted by the Galaxy early in its life. Although it may be possible to unravel this debris and associate individual halo stars with particular progenitors, this may tell us more about the early chemical evolution of dwarf galaxies than about the basic issues of galaxy formation. We would argue that the thin disk and thick disk of our Galaxy retain the most information about how the Galaxy formed. On the other hand, we note that current hierarchical CDM simulations predict many more satellites than are currently observed. It would therefore be very interesting to determine the number of satellites that have already been accreted to form the galactic stellar halo.

We should keep in mind that the stellar halo accounts for only a tiny fraction of the galactic baryons and is dynamically distinct from the rest of the stellar baryons. We should also note that the stellar halo of the Galaxy may not be typical: The halos of disk galaxies are quite diverse. The halo of M31, for example, follows the $r^{1/4}$ law (Pritchet & van den Bergh 1994) and is much more metal rich in the mean than the halo of our Galaxy (Durrell et al. 2001) although it does have stars that are very metal weak. It should probably be regarded more as the outer parts of a large bulge than as a distinct halo component. For some other disk galaxies, like the LMC, a metal-poor population is clearly present but may lie in the disk rather than in a spheroidal halo.

The Dark Halo

The second inconspicuous component is the dark halo, which is detected only by its gravitational field. The dark halo contributes at least 90% of the total galactic mass and its $\rho \sim r^{-2}$ density distribution extends to at least 100 kpc (e.g., Kochanek 1996). It is believed to be spheroidal rather than disklike (Crézé et al. 1998, Ibata et al. 2001b; see Pfenniger et al. 1994 for a contrary view). In the current picture of galaxy formation, the dark halo plays a very significant role. The disk is believed to form dissipatively within the potential of the virialized spheroidal halo, which itself formed through the fairly rapid aggregation of smaller bodies.

CDM simulations suggest that the halo may still be strongly substructured (see "Signatures of the CDM Hierarchy" below). If this is correct, then the lumpy halo would continue to influence the evolution of the galactic disk, and the residual substructure of the halo is a fossil of its formation. If the dark matter is grainy, it may be possible to study the dynamics of this substructure through pixel lensing of the light of background galaxies (Widrow & Dubinski 1998, Lewis et al. 2000). Another possibility is to look for the signatures of substructure around external galaxies in gravitationally lensed images of background quasars (Metcalf 2001, Chiba 2001). The dispersal of tidal tails from globular clusters appears to be sensitive to halo substructure (Ibata et al. 2001b), although this is not the case for dwarf galaxies (Johnston et al. 2001).

Within the limitations mentioned above, each of these distinct components of the Galaxy preserves signatures of its past and so gives insights into the galaxy formation process. We now discuss these signatures.

SIGNATURES OF GALAXY FORMATION

Our framework is that the Galaxy formed through hierarchical aggregation. We identify three major epochs in which information about the proto-hierarchy is lost:

1. the dark matter virializes—this could be a time of intense star formation but need not be, as evidenced by the existence of very thin pure disk galaxies;
2. the baryons dissipate to form the disk and bulge;

3. an ongoing epoch of formation of objects within the disk and accretion of objects from the environment of the galaxy, both leaving some long-lived relic.

We classify signatures relative to these three epochs. The role of the environment is presently difficult to categorize in this way. Environmental influences must be operating across all of our signature classes.

Zero Order Signatures—Information Preserved Since Dark Matter Virialized

INTRODUCTION   During the virialization phase, a lot of information about the local hierarchy is lost: This era is dominated by merging and violent relaxation. The total dark and baryon mass are probably roughly conserved, as is the angular momentum of the region of the hierarchy that went into the halo. The typical density of the environment is also roughly conserved: Although the structure has evolved through merging and relaxation, a low density environment remains a low density environment (see White 1996 for an overview).

SIGNATURES OF THE ENVIRONMENT   The local density of galaxies (and particularly the number of small satellite systems present at this epoch) affects the incidence of later interactions. For the Local Group, the satellite numbers appear to be lower than expected from CDM (Moore et al. 1999, Klypin et al. 1999). However there is plenty of evidence for past and ongoing accretion of small objects by the Milky Way and M31 (Ibata et al. 1995, 2001b).

The thin disk component of disk galaxies settles dissipatively in the potential of the virialized dark halo (e.g., Fall & Efstathiou 1980). The present morphology of the thin disk depends on the numbers of small galaxies available to be accreted: A very thin disk is an indication of few accretion events (dark or luminous) after the epoch of disk dissipation and star formation (e.g., Freeman 1987, Quinn et al. 1993, Walker et al. 1996). The formation of the thick disk is believed to be associated with a discrete event that occurred very soon after the disk began to settle, at a time when about 10% of the stars of the disk had already formed. In a low density environment, without such events, thick disk formation may not occur. Since the time of thick disk formation, the disk of the Galaxy appears to have been relatively undisturbed by accretion events. This is consistent with the observation that less than 10% of the metal-poor halo comes from recent accretion of star forming satellites (Unavane et al. 1996).

The existence and structure of the metal-poor stellar halo of the Galaxy may depend on accretion of small objects. This accretion probably took place after the gaseous disk had more or less settled—the disk acts as a resonator for the orbit decay of the small objects. So again the environment of our proto-galaxy may have a strong signature in the very existence of the stellar halo, and certainly in its observed substructure. We would not expect to find a stellar halo encompassing pure disk galaxies, consistent with the limited evidence now available (Freeman et al. 1983, Schommer et al. 1992).

SIGNATURES OF GLOBAL QUANTITIES   During the process of galaxy formation, some baryons are lost to ram pressure stripping and galactic winds. Most of the remaining baryons become the luminous components of the galaxy. The total angular momentum J of the dark halo may contribute to its shape, which in turn may affect the structure of the disks. For example, warps may be associated with misalignment of the angular momentum of the dark and baryonic components. The dark halo may have a rotating triaxial figure; the effect of a rotating triaxial dark halo on the self-gravitating disk has not yet been seriously investigated (see Bureau et al. 1999).

The binding energy E at the epoch of halo virialization affects the depth of the potential well and hence the characteristic velocities in the galaxy. It also affects the parameter $\lambda = J|E|^{\frac{1}{2}} G^{-1} M^{-\frac{5}{2}}$, where M is the total mass: λ, which is critical for determining the gross nature of the galactic disk as a high or low surface brightness system (e.g., Dalcanton et al. 1997).

The relation between the specific angular momentum J/M and the total mass M (Fall 1983) of disk galaxies is well reproduced by simulations (Zurek et al. 1988). Until recently, ellipticals and disk galaxies appeared to be segregated in the Fall diagram: From the slow rotation of their inner regions, estimates of the J/M ratios for ellipticals were about 1 dex below those for the spirals. More recent work (e.g., Arnaboldi et al. 1994) shows that much of the angular momentum in ellipticals appears to reside in their outer regions, so ellipticals and spirals do have similar locations in the Fall diagram. Internal redistribution of angular momentum has clearly occurred in the ellipticals (Quinn & Zurek 1988).

The remarkable Tully-Fisher law (1977) is a correlation between the HI line-width and the optical luminosity of disk galaxies. It appears to relate the depth of the potential well and the baryonic mass (McGaugh et al. 2000). Both of these quantities are probably roughly conserved after the halo virializes, so the Tully-Fisher law should be regarded as a zero-order signature of galaxy formation. The likely connecting links between the (dark) potential well and the baryonic mass are (*a*) a similar baryon/dark matter ratio from galaxy to galaxy, and (*b*) an observed Faber-Jackson law for the dark halos, of the form $M \propto \sigma^4$: i.e., surface density independent of mass for the dark halos (J. Kormendy & K.C. Freeman, in preparation).

SIGNATURES OF THE INTERNAL DISTRIBUTION OF SPECIFIC ANGULAR MOMENTUM
The internal distribution of specific angular momentum $\mathcal{M}(h)$ of the baryons (i.e., the mass with specific angular momentum $< h$) largely determines the shape of the surface brightness distribution of the disk rotating in the potential well of the dark matter. Together with $\mathcal{M}(h)$, the total angular momentum and mass of the baryons determine the scale length and scale surface density of the disk. Therefore the distribution of total angular momentum and mass for protodisk galaxies determines the observed distribution of the scale length and scale surface density for disk galaxies (Freeman 1970, de Jong & Lacey 2000).

Many studies have assumed that $\mathcal{M}(h)$ is conserved through the galaxy formation process, most notably Fall & Efstathiou (1980). It is not yet clear if this

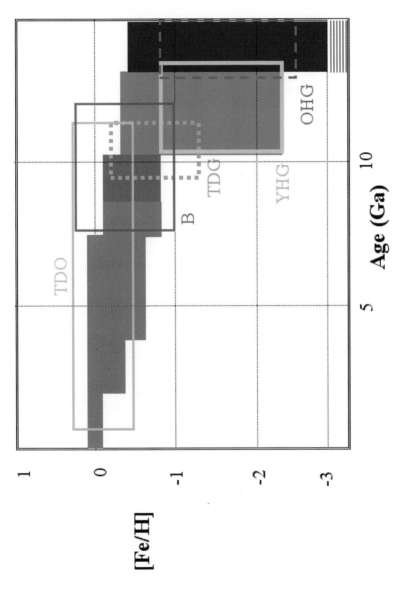

Figure 2 The age-metallicity relation of the Galaxy for the different components (see text): TDO—thin disk open clusters; TDG—thick disk globular clusters; B—bulge; YHG—young halo globular clusters; OHG—old halo globular clusters. The blue corresponds to thin disk field stars, the green to thick disk field stars, and the black shows the distribution of halo field stars extending down to [Fe/H] = -5.

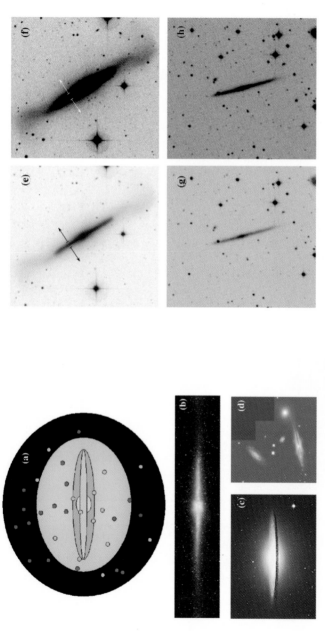

Figure 3 (*a*) Sketch of Milky Way showing the stellar disk (*light blue*), thick disk (*dark blue*), stellar bulge (*yellow*), stellar halo (*mustard yellow*), dark halo (*black*) and globular cluster system (*filled circles*). The radius of the stellar disk is roughly 15 kpc. The baryon and dark halos extend to a radius of at least 100 kpc. (*b*) Infrared image of the Milky Way taken by the DIRBE instrument on board the Cosmic Background Explorer (COBE) Satellite. [We acknowledge the NASA Goddard Space Flight Center and the COBE Science Working Group for this image.] (*c*) M104, a normal disk galaxy with a large stellar bulge (from AAO). (*d*) Hubble Heritage image of the compact group Hickson 87; one galaxy has a peanut-shaped stellar bulge due to dynamical interaction with other group members. (*e*) Image of the S0 galaxy NGC 4762 (Digital Sky Survey) shows its thin disk and stellar bulge. (*f*) A deeper image of NGC 4762 (DSS) shows its more extended thick disk. The base of the arrows in (*e*) and (*f*) shows the height above the plane at which the thick disk becomes brighter than the thin disk (Tsikoudi 1980). (*g*) Image of the pure disk galaxy IC 5249 (DSS) shows its thin disk and no stellar bulge. (*h*) A deeper image of IC 5249 (DSS) shows no visible thick disk, although a very faint thick disk has been detected in deep surface photometry.

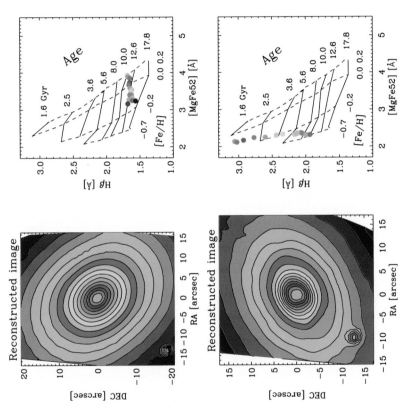

Figure 6 Sauron integral field observations of NGC 4365 (*top panels*) and NGC 4150 (*bottom panels*). Left panels: reconstructed surface brightness maps. Right panels: Hβ versus [MgFe5270] diagram. The points were derived from the Sauron datacubes by averaging along the corresponding color-coded isophotes (Bacon et al. 2001, Davies et al. 2001). The [Fe/H] vs. age grid is derived from Vazdekis (1999). [We acknowledge Harald Kuntschner and the Sauron team for these images.]

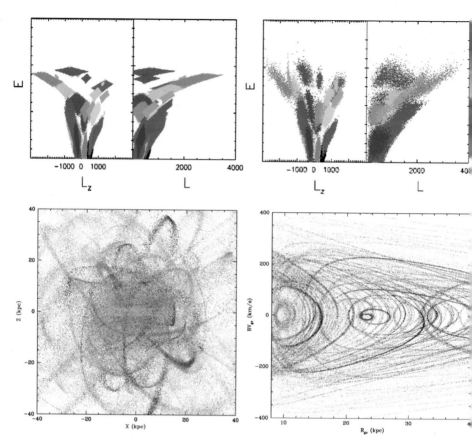

Figure 10 (*a*) Initial distribution of particles for 33 systems falling into the Galactic halo in integral of motion space. (*b*) The final distribution of particles in (*a*) after 12 Ga; the data points have been convolved with the errors expected for Gaia. [We acknowledge A. Helmi for these images.] (*c*) A simulation of the baryon halo built up through accretion of 100 satellite galaxies. The different colors show the disrupted remnants of individual satellites. (*d*) This is the same simulation shown in a different coordinate frame, i.e., the orbit radius (horizontal) plotted against the observed radial velocity (vertical) of the star. [We acknowledge P. Harding and H. Morrison for these images.]

assumption is correct. Conservation of the internal distribution of specific angular momentum $\mathcal{M}(h)$ is a much stronger requirement than the conservation of the *total* specific angular momentum J/M. Many processes can cause the internal angular momentum to be redistributed, while leaving the J/M ratio unchanged. Examples include the effects of bars, spiral structure (Lynden-Bell & Kalnajs 1972) and internal viscosity (Lin & Pringle 1987).

The maximum specific angular momentum h_{max} of the baryons may be associated with the truncation of the optical disk observed at about four scale lengths (de Grijs et al. 2001, Pohlen et al. 2000). This needs more investigation. The truncation of disks could be an important signature of the angular momentum properties of the early protocloud, but it may have more to do with the critical density for star formation or the dynamical evolution of the disk. Similarly, in galaxies with very extended HI, the edge of the HI distribution may give some measure of h_{max} in the protocloud. On the other hand, it may be that the outer HI was accreted subsequent to the formation of the stellar disk (van der Kruit 2001), or that the HI edge may just represent the transition to an ionized disk (Maloney 1993).

This last item emphasizes the importance of understanding what is going on in the outer disk. The outer disk offers some potentially important diagnostics of the properties of the protogalaxy. At present there are too many uncertainties about the significance of (*a*) the various cutoffs in the light and HI distributions, (*b*) the age gradient seen by Bell & de Jong (2000) from integrated light of disks but not by Friel (1995) for open clusters in the disk of the Galaxy, and (*c*) the outermost disk being maybe younger but not "zero age," which means that there is no real evidence that the disk is continuing to grow radially. It is possible that the edge of the disk has something to do with angular momentum of baryons in the protocloud or with disk formation process, so it may be a useful zero-order or first-order signature.

SIGNATURES OF THE CDM HIERARCHY   CDM predicts a high level of substructure that is in apparent conflict with observation. Within galaxies, the early N-body simulations appeared to show that substructure with characteristic velocities in the range $10 < V_c < 30$ km s^{-1} would be destroyed by merging and virialization of low mass structures (Peebles 1970, White 1976, White & Rees 1978). It turned out that the lack of substructure was an artefact of the inadequate spatial and mass resolution (Moore et al. 1996). Current simulations reveal 500 or more low mass structures within 300 kpc of an L$_*$ galaxy's sphere of influence (Moore et al. 1999, Klypin et al. 1999). This is an order of magnitude larger than the number of low mass satellites in the Local Group. Mateo (1998) catalogues about 40 such objects and suggests that, at most, we are missing a further 15–20 satellites at low galactic latitude. Kauffmann et al. (1993) were the first to point out the satellite problem and suggested that the efficiency of dynamical friction might be higher than usually quoted. However, without recourse to fine tuning, this would remove essentially all of the observed satellites in the Local Group today.

Since the emergence of the CDM paradigm, an inevitable question is whether a basic building block can be recognized in the near field. Moore et al. emphasize the self-similar nature of CDM sub-clustering and point to the evidence provided by the mass spectrum of objects in rich clusters, independent of the N-body simulations. The lure of finding a primordial building block in the near field has prompted a number of tests. If the dark mini-halos comprise discrete sources, it should be possible to detect microlensing towards a background galaxy (see "The Dark Halo" above).

The satellite problem appears to be a fundamental prediction of CDM in the nonlinear regime. Alternative cosmologies have been suggested involving the reduction of small-scale power in the initial mass power spectrum (Kamionkowski & Liddle 2000), warm dark matter (Hogan & Dalcanton 2001, White & Croft 2000, Colin et al. 2000), or strongly self-interacting dark matter (Spergel & Steinhardt 2000). Several authors have pointed out that some of the direct dark-matter detection experiments are sensitive to the details of the dark matter in the solar neighborhood. Helmi et al. (2001) estimate that there may be several hundred kinematically cold dark streams passing through the solar neighborhood.

If CDM is correct in detail, then we have simply failed to detect or to recognize the many hundreds of missing objects throughout the Local Group. For example, the satellites may be dark simply because baryons were removed long ago through supernova-driven winds (Dekel & Silk 1986, Mac Low & Ferrara 1999). In support of this idea, X-ray halos of groups and clusters are almost always substantially enriched in metals ($[Fe/H] \geq -0.5$; Renzini 2000, Mushotzky 1999). In fact, we note that up to 70% of the mass fraction in metals is likely to reside in the hot intracluster and intragroup medium (Renzini 2000). Another explanation may be that the absence of baryons in hundreds of dark satellites was set in place long ago during the reionization epoch. Many authors note that the accretion of gas on to low-mass halos and subsequent star formation is heavily suppressed in the presence of a strong photoionizing background (Ikeuchi 1986, Rees 1986, Babul & Rees 1992, Bullock et al. 2000). This effect appears to have a cut-off at low galactic mass at a characteristic circular velocity close to 30 km s^{-1} (Thoul & Weinberg 1996, Quinn et al. 1996), such that the small number of visible Local Group dwarfs are those that exceed this cutoff or acquired most of their neutral hydrogen before the reionization epoch.

Blitz et al. (1999) suggested that the high-velocity HI gas cloud (HVC) population is associated with dark mini-halos on megaparsec scales within the Local Group. This model was refined by Braun & Burton (1999) to include only the compact HVCs. The HVCs have long been the subject of wide-ranging speculation. Oort (1966) realized that distances derived from the virial theorem and the HI flux would place many clouds at Mpc distances if they are self-gravitating. If the clouds lie at about a Mpc and are associated with dark matter clumps, then they could represent the primordial building blocks. However, Hα distances (Bland-Hawthorn et al. 1998) suggest that most HVCs lie within 50 kpc and are unlikely to be associated with dark matter halos (Bland-Hawthorn & Maloney 2001, Weiner et al. 2001). We note that several teams have searched for but failed to detect a faint stellar population in HVCs.

Moore et al. (1999; see also Bland-Hawthorn & Freeman 2000) suggested that ultrathin disks in spirals are a challenge to the CDM picture in that disks are easily heated by orbiting masses. However, Font et al. (2001) found that in their CDM simulations, very few of the CDM sub-halos come close to the optical disk.

At present there are real problems in reconciling the predictions of CDM simulations with observations on scales of the Local Group.

First Order Signatures—Information Preserved
Since the Main Epoch of Baryon Dissipation

THE STRUCTURE OF THE DISK   At what stage in the evolution of the disk are its global properties defined? In part, we have already discussed this question in "Signatures of the Internal Distribution of Specific Angular Momentum" above. The answer depends on how the internal angular momentum distribution $\mathcal{M}(h)$ has evolved as the disk dissipated and various nonaxisymmetric features like bars and spiral structure came and went. Viscous processes associated with star formation, as suggested by Lin & Pringle (1987), may also contribute to the evolution of the $\mathcal{M}(h)$ distribution.

The global structure of disks is defined by the central surface brightnes I_o and the radial scalelength h of the disk. de Jong & Lacey (2000) evaluated the present distribution of galaxies in the (I_o, h) plane (Figure 4). If $\mathcal{M}(h)$ has indeed remained roughly constant, as is often assumed for discussions of disk formation (e.g., Fall & Efstathiou 1980, Fall 1983), then the global parameters of the disk—the scale length, central surface brightness and the Tully-Fisher relation—are relics of the main epoch of baryon dissipation.

CAN DISKS PRESERVE FOSSIL INFORMATION?   Here, we consider radial and vertical fossil gradients in the disk, in particular of abundance and age. Our expectation is that much of the information will be diluted through the dynamical evolution and radial mixing of the disk.

For spirals, different mechanisms may be at work to establish gradients (Molla et al. 1996): (a) a radial variation of the yield due either to the stellar metal production or to the initial mass function, (b) a radial variation of the timescale for star formation, (c) a radial variation for the timescale of infall of gas from outside the disk. Once gradients are established, these can be amplified or washed out by radial mixing (Edmunds 1990, Goetz & Koeppen 1992).

Most stars are born in large clusters numbering hundreds or even thousands of stars. Some clusters stay together for billions of years, whereas others become unbound shortly after the initial starburst, depending on the star formation efficiency. When a cloud disperses, each star suffers a random kick superimposed on the cloud's mean motion. Thereafter, stars are scattered by transient spiral arm perturbations and star-cloud encounters.

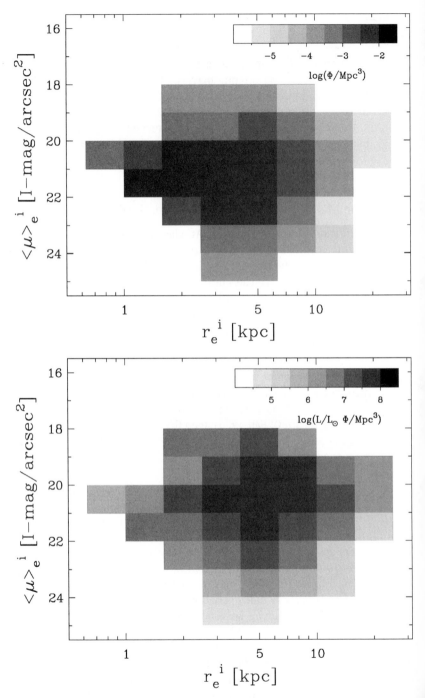

Figure 4 The density distribution of Sa to Sm galaxies over effective radius r_e and effective surface brightness μ_e. The top panel shows the raw (unweighted) distribution and the bottom panel shows the luminosity weighted distribution (de Jong & Lacey 2000).

These perturbations allow the star to migrate in integral space. During inter-action with a single spiral event of pattern speed Ω_p, a star's energy and angular momentum change while it conserves its Jacobi integral: In the (E, J) plane, stars move along lines of constant $I_J = E - \Omega_p J$. The star undertakes a random walk in the (E, J) plane, perturbed by a series of spiral arm events (Sellwood 1999, Dehnen 2000). N-body models of disk evolution indicate that radial mixing is strong (Sellwood 2001, Lynden-Bell & Kalnajs 1972). This is believed to be driven by *transient* spiral waves that heat the in-plane motions, although the process is not yet well understood. Long-term spiral arms produce no net effect. Remarkably, a *single* spiral wave near co-rotation can perturb the angular momentum of a star by $\sim$20% without significant heating: The star is simply moved from one circular or-bit to another, inwards or outwards, by up to 2 kpc (Sellwood & Kosowsky 2000). Substantial variations in the angular momentum of a star are possible over its lifetime.

In addition to radial heating, stars experience vertical disk heating: Their ver-tical velocity dispersion increases as they age. This is believed to occur through a combination of in-plane spiral-arm heating and scattering off giant molecular clouds (e.g., Spitzer & Schwarzschild 1953, Carlberg & Sellwood 1985). The in-plane heating is most effective at the inner and outer Lindblad resonances and vanishes at corotation. In the vertical direction, an age-velocity dispersion relation is observed for stars younger than about 3 Ga, but older disk stars show a velocity dispersion that is independent of age (Figure 5). Thus, the vertical structure does depend on the mean age of the population for $\tau < 3$ Ga (Edvardsson et al. 1993, confirmed from Hipparcos data by Gomez et al. 1997).

As the amplitude of the random motions increases, the star becomes less vul-nerable to heating by transient spiral waves, and the heating process is expected to saturate. This probably happens after about 3 Ga (Binney & Lacey 1988, Jenkins & Binney 1990), consistent with observation. This is important for our purpose here. It means that dynamical information is preserved about the state of the thin disk at an early epoch, or roughly $\tau_L - 3 \approx 7$ Ga ago, for which τ_L is the look-back time when the disk first began to form.

The survival of old open clusters like NGC 6791, Berkeley 21 and Berkeley 17 (Friel 1995, van den Bergh 2000) is of interest here. The oldest open clusters exceed 10 Ga in age and constitute important fossils (Phelps & Janes 1996). Both old and young open clusters are part of the thin disk. If the heating perturbations occur over a lengthscale that significantly exceeds the size of an open cluster, it seems likely that the cluster will survive. A large spiral-arm heating event will heat many stars along their I_J trajectories. The trace of the heating event is likely to survive for a very long time but be visible only in integral space (Sellwood 2001). We note that vertical abundance gradients have not been seen among the open clusters (Friel & Janes 1993).

About 4% of disk stars are super metal-rich (SMR) relative to the Hyades (Castro et al. 1997). SMR stars of intermediate age appear to have formed a few kpc inside of the Solar circle from enriched gas. The oldest SMR stars appear to

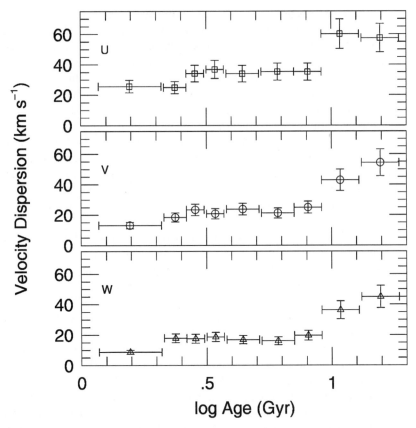

Figure 5 The relation between the three components of the velocity dispersion and the stellar age, as derived by Quillen & Garnett (2001) for stars from the sample of Edvardsson et al. (1993). Stars with ages between 2 and 10 Ga belong to the old thin disk: Their velocity dispersion is independent of age. The younger stars show a smaller velocity dispersion. The velocity dispersion doubles abruptly at an age of about 10 Ga; these older stars belong to the thick disk.

come from the Galactic Center: Their peculiar kinematics and outward migration may be associated with the central bar (Carraro et al. 1998, Grenon 1999).

In summary, our expectation is that fossil gradients within the disk are likely to be weak. This is borne out by observations of both the stars and the gas (Chiappini et al. 2001).

The vertical structure of the disk preserves another fossil—the thick disk—which we discuss in the next section. Like the open clusters, this component also does not show a vertical abundance gradient (Gilmore et al. 1995). In later sections, we argue that this may be the most important fossil to have survived the early stages of galaxy formation.

DISK HEATING BY ACCRETION: THE THICK DISK   Heating from discrete accretion events also imposes vertical structure on the disk (Quinn & Goodman 1986, Walker et al. 1996). Such events can radically alter the structure of the inner disk and the bulge (see Figure 3d for an example) and are currently believed to have generated the thick disk of the Galaxy.

The galactic thick disk was first recognized by Gilmore & Reid (1983). It includes stars with a wide range of metallicity, from $-2.2 \le$ [Fe/H] ≤ -0.5 (Chiba & Beers 2000): Most of the thick disk stars are in the more metal-rich end of this range. The velocity ellipsoid of the thick disk is observed to be $(\sigma_R, \sigma_\phi, \sigma_z) =$ $(46 \pm 4, 50 \pm 4, 35 \pm 3)$ km s^{-1} near the sun, with an asymmetric drift of about 30 km s^{-1}. For comparison, the nearby halo has a velocity ellipsoid $(\sigma_R, \sigma_\phi, \sigma_z) =$ $(141 \pm 11, 106 \pm 9, 94 \pm 8)$ km s^{-1} and its asymmetric drift is about 200 km s^{-1}.

The mean age of the thick disk is not known. From photometric age-dating of individual stars, the thick disk appears to be as old as the globular clusters. Indeed, the globular cluster 47 Tuc (age 12.5 ± 1.5 Ga; Liu & Chaboyer 2000) is often associated with the thick disk.

After Quinn & Goodman (1986), Walker et al. (1996) showed in detail that a low mass satellite could substantially heat the disk as it sinks rapidly within the potential well of a galaxy with a live halo. The conversion of satellite orbital energy to disk thermal energy is achieved through resonant scattering. Simulations of satellite accretion are important for understanding the survival of the thin disk and the origin of the thick disk. This is particularly relevant within the context of CDM. The satellites which do the damage are those that are dense enough to survive tidal disruption by the Galaxy. We note that even dwarf spheroidals which appear fluffy are in fact rather dense objects dominated by their dark matter (J. Kormendy & K. Freeman, in preparation).

It is fortuitous that the Galaxy has a thick disk, since this is not a generic phenomenon. The disk structure may be vertically stepped as a consequence of past discrete accretion events. The Edvardsson et al. (1993) data (Figure 5) appears to show an abrupt increase in the vertical component of the stellar velocity dispersion at an age of 10 Ga; see also Strömgren (1987). Freeman (1991) argued that the age–velocity dispersion relation shows three regimes: stars younger than 3 Ga with $\sigma_z \sim 10$ km s^{-1}, stars between 3 and 10 Ga with $\sigma_z \sim 20$ km s^{-1}, and stars older than 10 Ga with $\sigma_z \sim 40$ km s^{-1}. The first regime probably arises from the disk heating process due to transient spiral arms which we described in the previous section. The last regime is the thick disk, presumably excited by an ancient discrete event.

Can we still identify the disrupting event that led to the thick disk? There is increasing evidence now that the globular cluster ω Cen is the stripped core of a dwarf elliptical (see "Globular Clusters" below). It is possible that the associated accretion event or an event like it was the event that triggered the thick disk to form.

In summary, it seems likely that the thick disk may provide a snap-frozen view of conditions in the disk shortly after the main epoch of dissipation. Any low level chemical or age gradients would be of great interest in the context of dissipation models. In this regard, Hartkopf & Yoss (1982) argued for the presence of a vertical abundance gradient in the thick disk, although Gilmore, Wyse & Jones (1995) found no such effect. Because stars of the thick disk spend relatively little time near the galactic plane, where the spiral arm heating and scattering by giant molecular clouds is most vigorous, radial mixing within the thick disk is unlikely to remove all vestiges of a gradient. If our earlier suggestions are right (see "Signatures of the Internal Distribution of Specific Angular Momentum" above), we might expect to see a different truncation radius for the thick disk compared to the thin disk.

IS THERE AN AGE-METALLICITY RELATION? Some fossil information has likely been preserved since the main epoch of baryon dissipation. The inner stellar bulge is a striking example. It is characterized by old, metal-rich stars, which seems to be at odds with the classical picture where metals accumulate with time (Tinsley 1980). However, the dynamical timescales in the inner bulge are very short compared to the outer disk and would have allowed for rapid enrichment at early times. This is consistent with the frequent occurrence of metal-rich cores of galaxies observed at high redshift (Hamann & Ferland 1999). The dynamical complexity of the Galactic bulge may not allow us to determine the sequence of events that gave rise to it. We anticipate that this will come about from far-field cosmology (Ellis et al. 2000).

The existence of an age-metallicity relation (AMR) in stars is a very important issue, about which there has long been disagreement. Twarog (1980) and Meusinger et al. (1991) provide evidence for the presence of an AMR, while Carlberg et al. (1985) find that the metallicity of nearby F stars is approximately constant for stars older than about 4 Ga. More recently it has become clear that an AMR is apparent only in the solar neighborhood and is strictly true only for stars younger than 2 Ga and hotter than $\log T_{\mathrm{eff}} = 3.8$ (Feltzing et al. 2001). Edvardsson et al. (1993) demonstrate that there is no such relation for field stars in the old disk. Similarly, Friel (1995) shows that there is no AMR for open clusters (see "Open Clusters" below): she goes on to note that

> Apparently, over the entire age of the disk, at any position in the disk, the oldest clusters form with compositions as enriched as those of much younger objects.

In fact, it has been recognized for a long time (e.g., Arp 1962, Eggen & Sandage 1969, Hirshfeld et al. 1978) that old, metal-rich stars permeate the galaxy, throughout the disk, the bulge and the halo. We regard the presence of old metal-rich stars as a first-order signature. An age-metallicity relation which applies to all stars would have been an important second-order signature, but we see no evidence for such a relation, except among the young stars.

EFFECTS OF ENVIRONMENT AND INTERNAL EVOLUTION Environmental influences are operating on all scales of the hierarchy and across all stages of our signature classification, so our attempts to classify signatures are partly artificial. Within CDM, environmental effects persist throughout the life of the galaxy.

The parameters that govern the evolution of galaxies are among the key unknowns of modern astrophysics. Are the dominant influences internal (e.g., depth of potential) or external (e.g., environment) to galaxies? We consider here the effects of environment and internal evolution on the validity of the first-order signatures of galaxy formation (i.e., the properties that may have been conserved since the main epoch of baryon dissipation).

The well-known G dwarf problem indicates that external influences are important. A simple closed box model of chemical evolution predicts far too many metal-poor stars in the solar neighborhood (Tinsley 1980). This problem is easily remedied by allowing gas to flow into the region (Lacey & Fall 1983, 1985; Clayton 1987, 1988; Wyse & Silk 1989; Matteuci & Francois 1989; Worthey et al. 1996). In the context of CDM, this is believed to arise from the continued accretion of gas-rich dwarfs (e.g., Cole et al. 1994, Kauffmann & Charlot 1998).

Environment is clearly a key factor. Early type galaxies are highly clustered compared to late type galaxies (Hubble & Humason 1931, Dressler 1980). Trager et al. (2000) find that for a sample of early-type galaxies in low-density environments, there is a large spread in the $H\beta$ index (i.e., age), but little variation in metallicity. For galaxies in the Fornax cluster, Kuntschner (2000) finds the opposite effect: A large spread in metallicity is present with little variation in age. This probably reflects strong differences in environment between the field and the cluster.

Another likely environmental effect is the fraction of S0 galaxies in clusters, which shows a rising trend with redshift since $z \approx 0.4$ (Jones et al. 2000). Furthermore, S0 galaxies in the Ursa Major cluster show age gradients that are inverted compared to field spirals, in the sense that the cores are young and metal-rich (Tully et al. 1996, Kuntschner & Davies 1998). Both of these effects involve more recent phenomena and would be properly classified as second-order signatures.

Internal influences are also at work. A manifestation is the color-magnitude relation (CMR) in early-type (Sandage & Visvanathan 1978) and late-type (Peletier & de Grijs 1998) galaxies. The CMR does not arise from dust effects (Bell & de Jong 2000) and must reflect systematic variations in age and/or metallicity with luminosity. In the case of ellipticals, the CMR is believed to reflect a mass-metallicity dependence (Faber 1973, Bower et al. 1998). The relation is naturally explained by supernova-driven wind models in which more massive galaxies retain supernova ejecta and thus become more metal rich and redder (Larson 1974, Arimoto & Yoshii 1987). The CMR is presumably established during the main phase of baryon dissipation and is a genuine first-order signature.

Concannon et al. (2000) analyzed a sample of 100 early-type galaxies over a large range in mass. They found that lower-mass galaxies exhibit a larger range in age than higher-mass galaxies. This appears to show that smaller galaxies have had a more varied star formation history, which is at odds with the naive CDM picture of low-mass galaxies being older than high-mass galaxies (Baugh et al. 1996, Kauffman 1996). The work of Concannon et al. (2000) shows the presence of a real cosmic scatter in the star formation history. It is tempting to suggest that this cosmic scatter relates to different stages of evolution within the hierarchy. In this sense, we would regard the Concannon et al. result as a first-order manifestation of galaxy formation (see "Timescales and Fossils" above).

Spiral galaxies commonly show color gradients that presumably reflect gradients in age and metallicity (Peletier & de Grijs 1998). Faint spiral galaxies have younger ages and lower metallicities relative to bright spirals. In a study of 120 low-inclination spirals, Bell & de Jong (2000) found that the local surface density within galaxies is the most important parameter in shaping their star formation and chemical history. However, they find that metal-rich galaxies occur over the full range of surface density. This fact has a remarkable resonance with the distribution of the metal-rich open clusters that are found at any position in the Galactic disk (see "Is There an Age-Metallicity Relation?" above). Bell & de Jong argue that the total mass is a secondary factor that modulates the star formation history. Once again, these authors demonstrate the existence of cosmic scatter that may well arise from variations in environment.

Second Order Signatures—Major Processes Involved in Subsequent Evolution

INTRODUCTION   Here we consider relics of processes that have taken place in the Galaxy since most of the baryonic mass settled to the disk. There are several manifestations of these processes, probably the most significant of which is the star formation history of the disk, for which the open clusters are particularly important probes.

There is a wealth of detail relating to anomalous populations throughout the Galaxy, discussed at length by Majewski (1993). Examples include an excess of stars on extreme retrograde orbits (Norris & Ryan 1989, Carney et al. 1996), metal-poor halo stars of intermediate age (Preston et al. 1994) and metal-rich halo A stars (Rodgers et al. 1981).

In an earlier section, we discussed observational signatures of the CDM hierarchy in the Galactic context. In fact, detailed observations in velocity space are proving to be particularly useful in identifying structures that have long since dispersed in configuration space. In external galaxies, related structures are showing up as low surface brightness features. We do not know what role globular clusters play in the galaxy formation picture, but we include them here because at least one of them appears now to be the nucleus of a disrupted dwarf galaxy.

STAR FORMATION HISTORY   The star formation history (SFH) of our Galaxy has been very difficult to unravel. Derived star formation histories range from a roughly uniform star formation rate over the history of the disk to a SFH that was highly peaked at early times (e.g., Twarog 1980, Rocha-Pinto et al. 2000, Just 2001). Galaxies of the Local Group show a great diversity in SFH (Grebel 2001), although the average history over the Local Group appears consistent with the mean cosmic history (Hopkins et al. 2001). The present emphasis is on star formation studies that make use of the integrated properties of external galaxies, but it should be noted that this is necessarily weighted towards the most luminous populations. Key results for external galaxies are reviewed in "Effects of Environment and Internal Evolution" above. It was concluded that environmental effects are very significant in determining the SFH for individual galaxies.

The conventional approach to the study of chemical evolution in galaxy disks is to consider the solar neighborhood a closed box, and to assume that it is representative of all disks. Simple mathematical formulations have developed over the past 40 years (van den Bergh 1962, Schmidt 1963, Pagel & Patchett 1975, Talbot & Arnett 1971, Tinsley 1980, Twarog 1980, Pitts & Tayler 1989). Most observations are interpreted within this framework. The SFH is quantified in terms of stellar age, stellar (+gas) metallicity and, to a lesser extent, the existing gas fraction.

The use of broadband photometry coupled with stellar population synthesis is a well-established technique for probing the SFH of galaxy populations from integrated light. The power of the method is its simplicity, although it cannot uniquely disentangle the age-metallicity degeneracy (Bica et al. 1990, Charlot & Silk 1994).

Another widely used technique is the Lick index system (Burstein et al. 1984) further refined in Worthey et al. (1994) and Trager et al. (1998). In this system, the $H\beta$ index is the primary age-sensitive spectral indicator, whereas the Mg and Fe indices are the primary metallicity indicators. The Lick indices have well-known limitations: They correspond to low spectroscopic resolution (8–9 Å), require difficult corrections for internal galaxy motions, and are not calibrated onto a photometric scale. Furthermore, two of the most prominent Lick indices—Mg_2 $\lambda5176$ and Fe $\lambda5270$—are now known to be susceptible to contamination from other elements, in particular Ca and C (Tripicco & Bell 1995).

How best to measure galaxy ages is a subject with a long history. The most reliable methods to date involve the low order transitions ($n < 4$) of the Balmer series. Ages derived from the $H\gamma$ equivalent width have been used by Jones & Worthey (1995). Rose (1994) and Caldwell & Rose (1998) have pioneered the use of even higher-order Balmer lines to break the age-metallicity degeneracy (Worthey 1994). These higher-order lines are less affected by Balmer line emission from the interstellar medium. They develop a line ratio index Hn/Fe which is a sum over $H\gamma$, $H\delta$ and H8 lines with respect to local Fe lines. The most recent demonstration of the power of this index can be found in Concannon et al. (2000).

Ultimately, full spectrum fitting matched to spectral synthesis models holds the most promise (Vazdekis 1999). The new models, which have a fourfold increase in spectroscopic resolution compared to the Lick system, show that the isochrone or isochemical grid lines overlaid on a plot of two Lick indices are more orthogonal than the Worthey models. Thus, galaxies like NGC 4365 that exhibit no age gradient in the Vazdekis models (Davies et al. 2001; see Figure 6a, color insert) appear to show an age spread in the Worthey models. Interestingly, NGC 4150 exhibits an abundance spread with constant age (Figure 6b).

LOW SURFACE BRIGHTNESS STRUCTURES IN GALAXIES   Dynamical interaction between galaxies led to a range of structures including stellar shells (Malin & Carter 1980, Quinn 1984), fans (Weil et al. 1997), and tidal streamers (Gregg & West 1998, Calcaneo-Roldan et al. 2000, Zheng et al. 1999). Some excellent examples are shown in Figure 7. We see evidence of multiple nuclei, counter-rotating cores, and gas in polar orbits. At low light levels, the outermost stellar contours of spiral disks appear frequently to exhibit departures from axisymmetry (Rix & Zaritsky 1995). The same is true for spiral arms in all Hubble types (Schoenmakers et al. 1997, Cianci 2002).

The stellar streamers are particularly interesting, as these may provide important constraints on galaxy models, particularly as kinematic measurements become possible through the detection of planetary nebulae. More than a dozen stellar streams are already known and this is probably indicative of a much larger population at very low surface brightness. Johnston et al. (2001) show that stellar streamers can survive for several gigayears and are only visible above the present optical detection limit ($\mu_V = 30\,\mathrm{mag\,arcsec^{-2}}$) for roughly 4×10^8 yr. A few galaxy groups (e.g., the Leo group) do show large-scale HI filaments that can remain visible for many Ga.

Deep CCD imaging has revealed a stellar loop around NGC 5907 (Shang et al. 1998) and a stellar feature extending from NGC 5548 (Tyson et al. 1998). The technique of photographic amplification has revealed stellar streamers in about ten sources (Malin & Hadley 1997, Calcaneo-Roldan et al. 2000, Weil et al. 1997). For these particular observations, the limiting surface brightness is $\mu_V \approx 28.5\,\mathrm{mag\,arcsec^{-2}}$. For all of these systems, we estimate that the total stream luminosities are in the range $3{-}20 \times 10^7\,\mathrm{L_\odot}$.

In a recent development, wide-field CCD cameras have revealed stellar streamers through multiband photometry of millions of individual sources. A pointillist image can then be reconstructed in narrow color intervals so as to enhance features with respect to the field. This has led to the discovery of a stellar stream in M31 (Ibata et al. 2001a) and tidal tails extending from the globular cluster Pal 5 (Odenkirchen et al. 2001). This technique has the potential to push much deeper than the direct imaging method described above.

The low surface brightness universe is notoriously difficult to observe. Modern telescope and instrument designs are simply not optimized for this part of parameter space. Many claims of diffuse light detections in the neighborhood of galaxies have been shown to arise from scattered light internal to the instrument.

In "Structures in Phase Space," below, we discuss moving groups identified within the Galaxy from proper motion and spectroscopic surveys. Their projected surface brightness is $\mu_V = 30{-}34\,\mathrm{mag\,arcsec^{-2}}$, below the limit of modern imaging techniques.

Looking farther afield, we see evidence for discrete accretion events in the making. The Galaxy is encircled by satellite galaxies that appear confined to one or two great streams across the sky (Lynden-Bell & Lynden-Bell 1995). The most renowned of these are the Magellanic Clouds and the associated HI Magellanic stream. All of these are expected to merge with the Galaxy in the distant future, largely due to the dynamical friction from the extended halo.

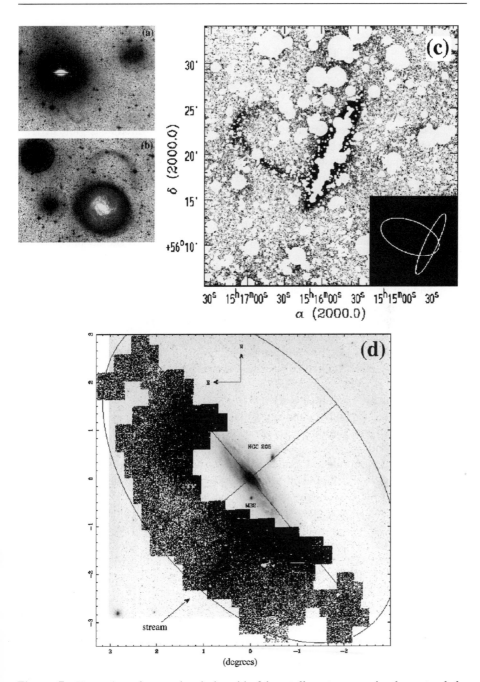

Figure 7 Examples of normal spirals with faint stellar streamers in the outer halo (see text): (*a*) M104 where the streamer is on a much larger scale than shown in Figure 3(*c*) (Malin & Hadley 1997); (*b*) M83 (Malin & Hadley 1997); (*c*) NGC 5907 (Shang et al. 1998); (*d*) M31 (Ibata et al. 2001a).

OPEN CLUSTERS   In the context of near-field cosmology, we believe that the thick disk and the old open clusters of the thin disk are among the most important diagnostics. The open clusters are the subject of an outstanding and comprehensive review by Friel (1995). Here, we summarize the properties that are most important for our purpose.

Both old and young clusters are part of the thin disk. Their key attribute is that they provide a direct time line for investigating change, which we explore in "The Gaiasphere and the Limits of Knowledge," below. The oldest open clusters exceed 10 Ga in age and constitute important fossils (Phelps & Janes 1996). In "Can disks preserve fossil information," we noted that the survival of these fossil clusters is an interesting issue in its own right. Friel (1995) finds no old open clusters within a galactocentric radius of 7 kpc; these are likely to have disrupted or migrated out of the central regions (van den Bergh & McClure 1980). It has long been recognized that open clusters walk a knife edge between survival and disruption (King 1958a,b,c).

Like field stars in the disk, Janes & Phelps (1994) find that the old cluster population (relative to Hyades) is defined by a 375-pc scale height exponential distribution, whereas young clusters have a 55-pc scale height (Figure 8a,b). Again, like the field stars, vertical abundance gradients have not been seen in open clusters (Friel & Janes 1993), although radial gradients are well established (Friel 1995, van den Bergh 2000). For old open clusters, Twarog et al. (1997) claim evidence for a stepped radial metallicity distribution where [Fe/H] $\approx$ 0 within 10 kpc, falling to [Fe/H] $\approx$ -0.3 in the outer disk. However, this effect is not seen in young objects, e.g., HII regions and B stars (Henry 1998).

In Figure 8c, both the old and young open clusters show essentially the same radial trend in metallicity. After reviewing the available observations, Friel (1995) finds no evidence for an age-metallicity relation for open clusters (Figure 8d). In agreement with Eggen & Sandage (1969), she notes that over the entire age of the disk, at any position in the disk, the oldest clusters form with compositions as enriched as those of much younger objects.

These remarkable observations appear to indicate that shortly after the main epoch of baryon dissipation, the thin disk was established at least as far out as 15 kpc. The oldest open clusters approach the age of the thick disk. Since, in "Disk Heating by Accretion," we noted that the thick disk is likely to be a snap frozen picture of the thin disk shortly after disk formation, we would expect the truncation

$\longrightarrow$

Figure 8   (a) The distribution of open clusters younger than Hyades with height from the plane as a function of Galactocentric distance R_{gc} (Friel 1995). The Sun is at 8.5 kpc. (b) The distribution of clusters with ages equal to or greater than the Hyades. (c) The open clusters exhibit a well-defined abundance gradient. (d) There is no discernible age-metallicity relation (AMR) when the cluster abundances are corrected for the radial abundance gradient.

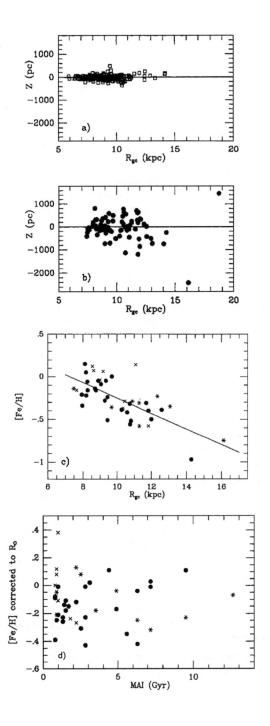

of the thick disk (see "Signatures of Global Quantities," above) to reflect the extent of the thin disk at the epoch of the event that puffed up the thick disk.

GLOBULAR CLUSTERS   We have long suspected that globular clusters are the fossil remnants of violent processes in the protogalactic era (Peebles & Dicke 1968). But there is a growing suspicion that globulars are telling us more about globulars than galactic origins (Harris 2001). The Milky Way has about 150 globular clusters with 20% lying within a few kiloparsecs of the Galactic Center. They constitute a negligible fraction of the light and mass (2%) of the stellar halo today. Their significance rests in their age. The oldest globular clusters in the outer halo have an age of 13 ± 2.5 Ga (90% confidence).

The ages of the oldest globular clusters in the inner and outer halo, the Large Magellanic Cloud and the nearby Fornax and Sgr dwarf spheroidal galaxies show a remarkable uniformity. To a precision of ± 1 Ga, the onset of globular cluster formation was well synchronized over a volume centered on our Galaxy with a radius >100 kpc (Da Costa 1999).

Globular cluster stars are older than the oldest disk stars, e.g., white dwarfs and the oldest red giants. These clusters are also more metal poor than the underlying halo light in all galaxies and at all radii (Harris 1991), but again there are exceptions to the rule. Since Morgan's (1950) and Kinman's (1959) classic work, we have known that there are two distinct populations of globular clusters in the Galaxy. The properties that we associate with these two populations today were derived by Zinn (1985) who showed that they have very different structure, kinematics and metallicities. The halo population is metal poor ($[Fe/H] < -0.8$) and slowly rotating with a roughly spherical distribution; the disk population is metal rich ($[Fe/H] > -0.8$) and in rapid rotation.

A major development has been the discovery of young globular clusters in disturbed or interacting galaxies, e.g., NGC 1275 (Holtzman et al. 1992), NGC 7252 (Whitmore et al. 1993) and the Antennae (Whitmore & Schweizer 1995). Schweizer (1987) first suspected that globular clusters were formed in mergers. Later, Ashman & Zepf (1992) predicted that the HST would reveal young globular clusters through their compact sizes, high luminosities and blue colors. The very high internal densities of globular clusters today must partly reflect the conditions when they were formed. Harris & Pudritz (1994) present a model for globular clusters produced in fragmenting giant molecular clouds, which are of the right mass and density range to resemble accretion fragments in the Searle-Zinn model.

Globular clusters have been elegantly referred to as "canaries in a coal mine" (Arras & Wasserman 1999). They are subject to a range of disruptive effects, including two-body relaxation and erosion by the tidal field of their host galaxy, and the tidal shocking that they experience as their orbits take them through the galactic disk and substructure in the dark halo. In addition to self-destruction through stellar mass loss, tidal shocking may have been very important in the early universe (Gnedin et al. 1999). If globular clusters originally formed in great numbers, the disrupted clusters may now contribute to the stellar halo (Norris & Ryan 1989, Oort 1965). Halo field stars and globular clusters in the Milky Way have similar mean metallicities (Carney 1993); however, the metallicity distribution of the halo field stars extends to much lower metallicity ([Fe/H] $\simeq -5$) than that of the globular clusters ([Fe/H] $\simeq -2.2$). We note again the remarkable similarity in the metallicity range of the globular clusters and the thick disk ($-2.2 \lesssim$ [Fe/H] $\lesssim -0.5$).

In the nucleated dwarf elliptical galaxies (Binggeli et al. 1985), the nucleus typically provides about 1% of the total luminosity; globular clusters could be considered as the stripped nuclei of these satellite objects without exceeding the visible halo mass (Zinnecker & Cannon 1986, Freeman 1993). It is an intriguing prospect that the existing globular clusters could be the stripped relics of an ancient swarm of protogalactic stellar fragments, i.e., the original building blocks of the Universe.

In the Searle-Zinn picture, globular clusters are intimately linked to *gas-rich*, protogalactic infalling fragments. Multiple stellar populations have recently been detected in ω Cen, the most massive cluster in the Galaxy (Lee et al. 1999). How did ω Cen retain its gas for a later burst? It now appears that it was associated with a gas-rich dwarf, either as an in situ cluster or as the stellar nucleus. The present-day cluster density is sufficiently high that it would have survived tidal disruption by the Galaxy, unlike the more diffuse envelope of this dwarf galaxy. The very bound retrograde orbit supports the view that ω Cen entered the Galaxy as part of a more massive system whose orbit decayed through dynamical friction.

If globular clusters are so ancient, why are the abundances of the most metal-poor population as high as they are? Because it does not take much star formation to increase the metal abundance up to [Fe/H] $= -1.5$ (Frayer & Brown 1997), the cluster abundances may reflect low levels of star formation even before the first (dark + baryon) systems came together.

Old age is not necessarily associated with low metallicity (compare "Timescales and Fossils" above). We recall that CO has been detected at $z \sim 5$ (Yun et al. 2000). Hamann & Ferland (1999) demonstrate that stellar populations at the highest red-shift currently observed appear to have solar or super-solar metallicity. We believe that there is no mystery about high abundances at high redshift. The dynamical times in the cores of these systems are short, so there has been time for multiple generations of star formation and chemical enrichment. In this sense, the cores of high redshift galaxies need not be relevant to the chemical properties of the globular clusters, although both kinds of objects were probably formed at about the same time.

The first generation of globular clusters may have been produced in merger-driven starbursts when the primordial fragments came together for the first time. If at least some fragments retained some of their identity while the halo was formed, a small number of enrichment events per fragment would ensure a Poissonian scatter in properties between globular clusters, and multiple populations within individual clusters (Searle & Zinn 1978).

STRUCTURES IN PHASE SPACE   One class of systems that exhibit coherence in velocity space are the open clusters associated with the disk. Here the common space motion of the stars with respect to the Sun is perceived as a convergence of the proper motions to a single point (strictly speaking, minimum volume) on the sky (Boss 1908; see de Zeeuw et al. 1999 for a recent application). More than a dozen such systems have been identified this way. However, these are all young open clusters largely associated with the Gould belt. With sufficiently precise kinematics, it may be possible to identify open clusters that have recently dispersed, particularly if the group is confined to a specific radial zone by resonances in the outer disk. For example, Feltzing & Holmberg (2000) show that the metal-rich ([Fe/H] ≈ 0.2) moving group HR 1614, thought to be 2 Ga old, can be identified in the Hipparcos data set.

Recently, attention has turned to a diverse set of moving groups that are thought to be associated with the stellar halo and in some instances are clearly fossils associated with accretion events in the distant past. The evidence for these groups dates back to the discovery of the halo itself. Shortly before the publication of the landmark ELS paper, Eggen & Sandage (1959) discovered that the nearby high-velocity star, Groombridge 1830, belongs to a moving group now passing through the Galactic disk.

In a long series of papers, Eggen went on to identify a number of moving groups, some of which appear to encompass the solar neighborhood, and others that may be associated with the halo. The relevant references are given by Taylor (2000). Various authors have noted that many of the groups are difficult to confirm (Griffin 1998, Taylor 2000). More systematic surveys over the past few decades have identified a number of moving populations associated with the halo (Freeman 1987, Majewksi 1993), although the reality of some of these groups is still debated. The reality of these groups is of paramount importance in the context of halo formation. Majewski et al. (1996) suspect that much or all of the halo could exhibit phase-space clumping with data of sufficient quality.

In recent years, the existence of kinematic sub-structure in the galactic halo has become clear. Helmi et al. (1999) identified 88 metal-poor stars within 1 kiloparsec of the Sun from the Hipparcos astrometric catalogue. After deducing accurate 3-D space motions, they found a highly significant group of 8 stars that appear clumped in phase space and confined to a highly inclined orbit.

The most dramatic evidence is surely the highly disrupted Sgr dwarf galaxy identified by Ibata et al. (1994, 1995). These authors used multi-object spectroscopy to uncover an elongated stellar stream moving through the plane on the far side of the Galaxy. The Sgr dwarf is a low mass dwarf spheroidal galaxy about 25 kpc from the Sun that is presently being disrupted by the Galactic tidal field. The long axis of the prolate body (axis ratios $\sim$ 3:1:1) is about 10 kpc, oriented perpendicular to the Galactic plane along $\ell = 6°$ and centered at $b = -15°$. Sgr contains a mix of stellar populations, an extended dark halo (mass $\geq 10^9$ M$_\odot$) and at least four globular clusters (Ibata et al. 1997). The Sgr stream has since been recovered by several photometric surveys (Vivas et al. 2001, Newberg et al. 2002, Ibata et al. 2001c).

N-body simulations have shown that stellar streams are formed when low mass systems are accreted by a large galaxy (e.g., Harding et al. 2001). Streamers remain dynamically cold and identifiable as a kinematic substructure long after they have ceased to be recognizable in star counts against the vast stellar background of the galaxy (Tremaine 1993, Ibata & Lewis 1998, Johnston 1998, Helmi & White 1999).

Within the Galaxy, moving groups can be identified with even limited phase-space information (de Bruijne 1999, de Zeeuw et al. 1999). This also holds for satellites orbiting within the spherical halo, since the debris remains in the plane of motion for at least a few orbits (Lynden-Bell & Lynden-Bell 1995, Johnston et al. 1996). But a satellite experiencing the disk potential no longer conserves its angular momentum and its orbit plane undergoes strong precession (Helmi & White 1999). In Figure 9, we show the sky projection of a satellite 8 Ga after disruption.

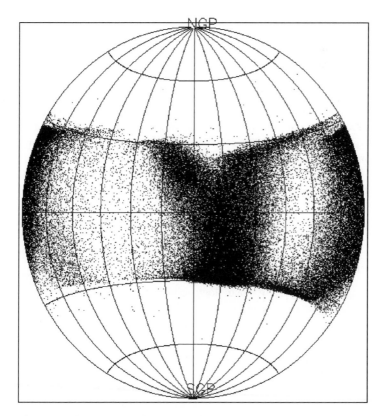

Figure 9 A satellite in orbit about the Milky Way as it would appear after 8 Ga. While stars from the disrupted satellite appear to be dispersed over a very wide region of sky, it will be possible to deduce the parameters of the original event using special techniques (see text). (We acknowledge A. Helmi and S. White for this image.)

These more complex structures are usually highly localized and therefore easy to recognize in the space of conserved quantities like energy and angular momentum for individual stars.

The evolution in phase space of a disrupting satellite is well behaved as its stars become phase mixed. Its phase space flow obeys Liouville's theorem, i.e., the flow is incompressible. Highly intuitive accounts are given elsewhere (Carlberg 1986, Tremaine 1999, Hernquist & Quinn 1988). It should be possible to recognize partially phase-mixed structures that cover the observed space, although special techniques are needed to find them.

Four astrometric space missions are planned for the next decade. These are the proposed German DIVA mission ($\sim$2003); the FAME mission ($\sim$2005) and the pointed SIM mission ($\sim$2005); and the ESA Gaia mission ($\sim$2009) which will observe a billion stars to V $\sim$ 20, with accuracy 10μas at a V $\sim$15. The web sites for these missions are at: http://www.ari.uni-heidelberg.de/diva/, http://aa.usno.navy.mil/FAME/, http://sim.jpl.nasa.gov/, http://astro.estec.esa.nl/GAIA/.

The astrometric missions will derive 6-dimensional phase space coordinates and spectrophotometric properties for millions of stars within a 20 kiloparsec sphere—the Gaiasphere. The ambitious Gaia mission will obtain distances for up to 90 million stars with better than 5% accuracy, and measure proper motions with an accuracy approaching microarcsec per year. If hierarchical CDM is correct, there should be thousands of coherent streamers that make up the outer halo, and hundreds of partially phase-mixed structures within the inner halo. A satellite experiencing the disk potential no longer conserves its angular momentum and its orbit plane undergoes strong precession (see Figure 10c,d, color insert). In Figure 10a,b, Helmi et al. (1999) demonstrate the relative ease with which Gaia will identify substructure within the stellar halo.

THE GAIASPHERE AND THE LIMITS OF KNOWLEDGE

Introduction

The ultimate goal of cosmology, both near and far, must be to explain how the Universe has arrived at its present state. It is plausible—although difficult to accept—that nature provides fundamental limits of knowledge, in particular, epochs where the sequence of events are scrambled. Our intuition is that any phase dominated by relaxation or dissipation probably removes more information than it retains.

But could some of the residual inhomogeneities from prehistory have escaped the dissipative process at an early stage? We may not know the answer to this question with absolute certainty for many years. In the absence of certainty, we consider what might be the likely traces of a bygone era prior to the main epoch of baryon dissipation.

Chemical Signatures

A major goal of near-field cosmology is to tag or to associate individual stars with elements of the protocloud. For many halo stars, and some outer bulge stars, this may be possible with phase space information provided by Gaia. But for much of the bulge and the disk, secular processes cause the populations to become relaxed (i.e., the integrals of motion are partially randomized). In order to have any chance of unravelling disk formation, we must explore chemical signatures in the stellar spectrum. Ideally, we would like to tag a large sample of representative stars with a precise time and a precise site of formation.

Over the past four decades, evidence has gradually accumulated (Figure 11) for a large dispersion in metal abundances $[X_i/Fe]$ (particularly n-capture elements) in low metallicity stars relative to solar abundances (Wallerstein et al. 1963, Pagel 1965, Spite & Spite 1978, Truran 1981, Luck & Bond 1985, Clayton 1988, Gilroy et al. 1988, McWilliam et al. 1995, Norris et al. 1996, Burris et al. 2000). Elements like Sr, Ba and Eu show a 300-fold dispersion, although $[\alpha/Fe]$ dispersions are typically an order of magnitude smaller.

In their celebrated paper, Burbidge et al. (1957—B^{2}FH) demonstrated the likely sites for the synthesis of slow (s) and rapid (r) n-capture elements. The s-process elements (e.g., Sr, Zr, Ba, Ce, La, Pb) are thought to arise from the He-burning phase of intermediate to low mass (AGB) stars (M < 10 M$_\odot$), although at the lowest metallicities, trace amounts are likely to arise from high mass stars (Burris et al. 2000, Rauscher et al. 2001).

In contrast to the s-process elements, the r-process elements (e.g., Sm, Eu, Gd, Tb, Dy, Ho) cannot be formed during quiescent stellar evolution. While some doubts remain, the most likely site for the r-process appears to be SN II, as originally suggested by B^{2}FH (see also Wallerstein et al. 1997). Therefore, r-process elements measured from stellar atmospheres reflect conditions in the progenitor cloud. In support of Gilroy et al. (1988), McWilliam et al. (1995) state that the very large scatter means that n-capture element abundances in ultra-metal-poor stars are products of one or very few prior nucleosynthesis events that occurred in the very early, poorly mixed galactic halo, a theme that has been developed by many authors (e.g., Audouze & Silk 1995, Shigeyama & Tsujimoto 1998, Argast et al. 2000, Tsujimoto et al. 2000).

Supernova models produce different yields as a function of progenitor mass, progenitor metallicity, mass cut (what gets ejected compared to what falls back towards the compact central object), and detonation details. The α elements are mainly produced in the hydrostatic burning phase within the pre-supernova star. Thus α yields are not dependent on the mass cut or details of the fallback/explosion mechanism which leads to a much smaller dispersion at low metallicity.

There is no known age-metallicity relation that operates over a useful dynamic range in age and/or metallicity. (This effect is only seen in a small subset of hot metal-rich stars—see "Is There an Age-Metallicity Relation?" above). Such a relation would require the metals to be well mixed over large volumes of the ISM.

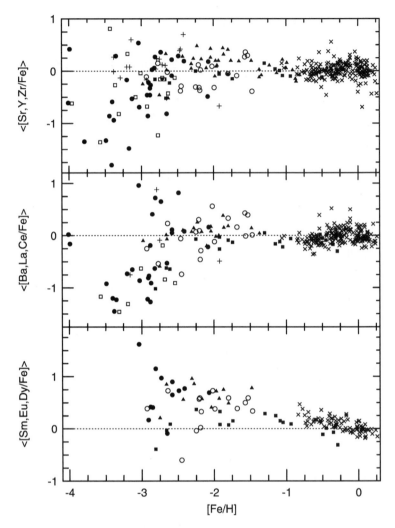

Figure 11   Mean relative abundance ratios of light s-process elements (top panel), heavy s-process elements (middle panel), and r-process elements (bottom panel) as functions of [Fe/H]. In each panel, the dotted horizontal lines represent the solar abundance ratios of these elements. The references for the data points are given in Wallerstein et al. (1997). (We acknowledge C. Sneden for this figure.)

For the forseeable future, it seems that only a small fraction of stars can be dated directly (see "Stellar Age Dating" above).

Reconstructing Ancient Star Groups

We now conjecture that the heavy element metallicity dispersion may provide a way forward for tagging groups of stars to common sites of formation. With sufficiently detailed spectral line information, it is feasible that the chemical tagging will allow temporal sequencing of a large fraction of stars in a manner analogous to building a family tree through DNA sequencing.

Most stars are born within rich clusters of many hundreds to many thousands of stars (Clarke et al. 2000, Carpenter 2000). McKee & Tan (2002) propose that high-mass stars form in the cores of strongly self-gravitating and turbulent gas clouds. The low mass stars form within the cloud outside the core, presumably at about the same time or shortly after the high-mass stars have formed. The precise sequence of events which give rise to a high-mass star is a topic of great interest and heated debate in contemporary astrophysics (e.g., Stahler et al. 2000).

A necessary condition for chemical tagging is that the progenitor cloud is uniformly mixed in key chemical elements before the first stars are formed. Another possibility is that a few high-mass stars form shortly after the cloud assembles, and enrich the cloud fairly uniformly. Both scenarios would help to ensure that long-lived stars have identical abundances in certain key elements before the onset of low-mass star formation.

For either statement to be true, an important requirement is that open clusters of any age have essentially zero dispersion in some key metals with respect to Fe. There has been very little work on heavy element abundances in open clusters. The target clusters must have reliable astrometry so as to minimize pollution from stars not associated with the cluster (Quillen 2002).

If our requirement is found not to be true, then either the progenitor clouds are not well mixed or high-mass stars are formed after most low-mass stars. A more fundamental consequence is that a direct unravelling of the disk into its constituent star groups would be impossible; in other words, the epoch of dissipation cannot be unravelled after all.

Consider the (extraordinary) possibility that we *could* put many coeval star groups back together over the entire age of the Galaxy. This would provide an accurate age for the star groups either through the color-magnitude diagram, or through association with those stars within each group that have [n-capture/Fe] $\gg 0$ and can therefore be radioactively dated. This would provide key information on the chemical evolution history for each of the main components of the Galaxy.

But what about the formation site? The kinematic signatures will identify which component of the Galaxy the reconstructed star group belongs to, but not specifically where in the Galactic component (e.g., radius) the star group came into existence. For stars in the thin disk and bulge, the stellar kinematics will have been much affected by the bar and spiral waves; it will no longer be possible to estimate their birthplace from their kinematics. Our expectation is that the derived family tree will severely restrict the possible scenarios involved in the dissipation process. In this respect, a sufficiently detailed model may be able to locate each star group within the simulated time sequence.

In addition to open clusters, we have already argued that the thick disk is an extremely important fossil of the processes behind disk formation. The thick disk is thought to be a snap-frozen relic of the early disk, heated vertically by the infall of an intermediate mass satellite. Chemical tagging of stars that make up the thick disk would provide clues on the formation of the first star clusters in the early disk.

Chemical Abundance Space

An intriguing prospect is that reconstructed star clusters can be placed into an evolutionary sequence, i.e., a family tree, based on their chemical signatures. Let us suppose that a star cluster has accurate chemical abundances determined for a large number n of elements (including isotopes). This gives it a unique location in an n-dimensional space compared to m other star clusters within that space. We write the chemical abundance space as $\mathcal{C}(\text{Fe/H}, X_1/\text{Fe}, X_2/\text{Fe}, \ldots)$ where X_1, $X_2 \ldots$ are the independent chemical elements that define the space (i.e., elements whose abundances are not rigidly coupled to other elements).

The size of n is unlikely to exceed about 50 for the foreseeable future. Hill et al. (2002) present exquisite data for the metal-poor star CS 31082-001, where abundance estimates are obtained for a total of 44 elements, almost half the entire periodic table (see also Cayrel et al. 2001). In Figure 12, we show what is now possible for another metal-poor star, CS 22892-052 (Sneden et al. 2001a). The α elements and r-process elements, and maybe a few canonical s-process elements at low [Fe/H], provide information on the cloud abundances prior to star formation, although combinations of these are likely to be coupled (Heger & Woosley 2001, Sneden et al. 2001a). There are 24 r-process elements that have been clearly identified in stellar spectra (Wallerstein et al. 1997).

The size of m is likely to be exceedingly large for the thin disk where most of the baryons reside. For a rough estimate, we take the age of the disk to be 10 Ga. If there is a unique SN II enrichment event every 100 years, we expect of order 10^8 formation sites. Typically, a SN II event sweeps up a constant mass of $5 \times 10^4\,\text{M}_\odot$ (Ryan et al. 1996, Shigeyama & Tsujimoto 1998). Simple chemical evolution models indicate that this must be of the right order to explain the metallicity dispersion at low [Fe/H] (Argast et al. 2000). Roughly speaking, there have been 10^3 generations of clouds since the disk formed, with about 10^5 clouds in each star-forming generation, such that cloud formation and dispersal cycle on a 10^7 year timescale (Elmegreen et al. 2000).

Whereas the total number of star clusters over the lifetime of the thin disk is very large, the size of m for the stellar halo (Harding et al. 2001), and maybe the thick disk (Kroupa 2002), is likely to be significantly smaller. Our primary interest is the oldest star clusters. Reconstructing star clusters within the thick disk is a particularly interesting prospect since the disk is likely to have formed within 1–1.5 Ga of the main epoch of baryon dissipation (Prochaska et al. 2000).

The task of establishing up to 10^8 unique chemical signatures may appear to be a hopeless proposition with current technology. But it is worth noting that more than 60 of the chemical elements ($Z > 30$) arise from n-capture processes. Let us suppose that half of these are detectable for a given star. We would only need to be able to measure two distinct abundances for each of these elements in order to

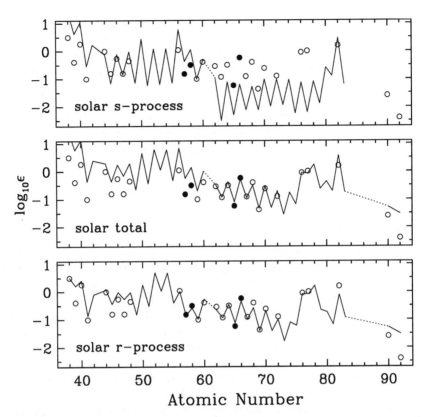

Figure 12 CS 22892-052 n-capture abundances (points) taken from Sneden et al. (2000) and scaled solar system abundances (solid and dashed lines) taken from Burris et al. (2000). Many of the heavy elements conform to the solar system r-process abundance pattern, although some elements show the hallmark of the s-process. This figure was originally presented in Sneden et al. (2001a).

achieve 10^9 independent cells in C-space. If many of the element abundances are found to be *rigidly* coupled, of course the parameter space would be much smaller.

It may not be necessary to measure as many as 30 elements if some can be found which are highly decoupled and exhibit large relative dispersions from star to star. Burris et al. (2000) demonstrate one such element pair, i.e., [Ba/Fe] and [Sr/Fe]. It is difficult at this stage to suggest which elements are most suited to chemical tagging. In part, this depends on the precise details and mechanism of formation of the n-capture elements at low [Fe/H].

The element abundances $[X_i/Fe]$ show three main peaks at $Z \approx 26$, $Z \approx 52$, and $Z \approx 78$; the last two peaks are evident in Figure 12. There have been suggestions that the r-process gives rise to random abundance patterns (e.g., Goriely & Arnould 1996), although this is not supported by new observations of a few metal-poor stars.

Heavy r-process elements around the second peak compared to the Sun appear to show a universal abundance pattern (Sneden et al. 2000, Cayrel et al. 2001, Hill et al. 2002). However, Hill et al. find that the third peak and actinide elements ($Z \geq 90$) are decoupled from elements in the second peak. We suspect that there may be a substantial number of suitable elements (~ 10) which could define a sufficiently large parameter space.

Our ability to detect structure in C-space depends on how precisely we can measure abundance differences between stars. It may be possible to construct a large database of differential abundances from echelle spectra, with a precision of 0.05 dex or better; differential abundances are preferred here to reduce the effects of systematic error.

Chemical Trajectories

Our simple picture assumes that a cloud forms with a unique chemical signature, or that shortly after the cloud collapses, one or two massive SN IIs enrich the cloud with unique yields which add to the existing chemical signature. The low-mass population forms with this unique chemical signature. If the star-formation efficiency is high ($\gtrsim 30\%$), the star group stays bound although the remaining gas is blown away. If the star-formation efficiency is low ($\lesssim 10\%$), the star cluster disperses along with the gas. In a closed box model, the dispersed gas reforms a cloud at a later stage.

In the closed box model, each successive generation of supernovae produces stellar populations with progressive enrichments. These will lie along a trajectory in C-space which can be identified in principle using minimum spanning tree methods (Sedgewick 1992). The overall distribution of the trajectories will be affected by fundamental processes like the star formation efficiency, the star formation timescale, the mixing efficiency, the mixing timescale, and the satellite galaxy infall rate.

As we approach solar levels of metallicity in [Fe/H], the vast number of trajectories will converge. By [Fe/H] ≈ -2.5, AGB stars will have substantially raised the s-process element abundances; by [Fe/H] ≈ -1, Type Ia supernovae will have raised the Fe-group abundances. Star clusters that appear to originate at the same location in this C-space may simply reflect a common formation site, i.e., the resolution limit we can expect to achieve in configuration space. The ability to identify common formation sites rests on accurate differential abundance analyses (Edvardsson et al. 1993, Prochaska et al. 2000).

Even with a well-established family tree based on chemical trajectories in the chemical C-space, this information may not give a clear indication of the original location within the protocloud or Galactic component. This will come in the future from realistic baryon dissipation models. Forward evolution of any proposed model must be able to produce the chemical tree.

However, the C-space will provide a vast amount of information on chemical evolution history. It should be possible to detect the evolution of the cluster mass function with cosmic time (Kroupa 2002), the epoch of a starburst phase and/or associated mass ejection of metals to the halo (Renzini 2001), and/or satellite infall (Noguchi 1998).

As we go back in time to the formation of the disk, we approach the chemical state laid down by population III stars. The lack of stars below [Fe/H] ≈ -5 suggests that the protocloud was initially enriched by the first generation of stars (Argast et al. 2000). However, the apparent absence of any remnants of population III remains a puzzle: Its stars may have had a top-heavy initial mass function, or have dispersed into the intra-group medium of the Local Group. If one could unravel the abundances of heavy elements at the time of disk formation, this would greatly improve the precision of nucleo-cosmochronology (see "Stellar Age Dating" above).

Candidates for Chemical Tagging

Chemical tagging will not be possible for all stars. In hot stars, our ability to measure abundances is reduced by the stellar rotation and lack of transitions for many ions in the optical. The ideal candidates are the evolved FGK stars that are numerous and intrinsically bright. These can be observed at echelle resolutions ($R > 30,000$) over the full Gaiasphere. Moreover, giants have deep, low-density atmospheres that produce strong low-ionization absorption lines compared to higher gravity atmospheres. Even in the presence of significant line blending, with sufficient signal, it should be possible to derive abundance information by comparing the fine structure information with accurate stellar synthesis models. Detailed abundances of large numbers of F and G subgiants would be particularly useful, if it becomes possible to make such studies, because direct relative ages can be derived for these stars from their observed luminosities.

It is not clear at what [Fe/H] the r-process elements become swamped by the ubiquitous Fe-group and s-process elements. At a resolution of $R \sim 10^5$, many r-process elements can be seen in the solar spectrum, although the signal-to-noise ratio of about 1000 is needed, and even then the spectral lines are often badly blended (Kurucz 1991, 1995). Travaglio et al. (1999) suggest that the s-process does not become significant until [Fe/H] ≈ -1 because of the need for pre-existing seed nuclei (Spite & Spite 1978, Truran 1981), although Pagel & Tautvaisiene (1997) argue for some s-process production at [Fe/H] ~ -2.5. Prochaska et al. (2000) detected Ba, Y and Eu in a snapshot survey of thick disk G dwarfs in the solar neighborhood with $-1.1 \lesssim$ [Fe/H] $\lesssim -0.5$. This survey only managed to detect a few transitions in each element, although their spectral coverage was redward of 440 nm with SNR ≈ 100 per pixel at $R \simeq 50,000$. Longer exposures with $R \sim 10^5$ and spectral coverage down to 300 nm would have detected more heavy elements.

Summary

In our view, observations of nucleosynthetic signatures of metal-poor stars provide a cornerstone of near-field cosmology. Success in this arena requires major progress across a wide front, including better atomic parameters (Truran et al. 2001), improved supernova models, better stellar synthesis codes and more realistic galaxy formation models. There are no stellar evolutionary models that lead to a self-consistent detonation and deflagration in a core-collapse supernova event or, for that matter, detonation in a thermonuclear explosive event. Realistic chemical production at the onset of the supernova stage requires a proper accounting of a large number of isotope networks (400–2500) that cannot be adequately simulated yet. Modern computers have only recently conquered relatively simple α networks involving 13 isotopes. The inexorable march of computer power will greatly assist here.

There is also a key experimental front both in terms of laboratory simulations of nucleosynthesis, and the need for major developments in astronomical instrumentation (see "Epilogue: Challenges for the Future"). Many authors (e.g., Sneden et al. 2001b) have stressed the importance of greatly improving the accuracy of transition probabilities and reaction rates for both heavy and light ion interactions. This will be possible with the new generation of high-intensity accelerators and radioactive-beam instruments (Käppeler et al. 1998, see Manuel 2000).

Progress on all fronts will require iteration between the different strands. Already, relative r-process and α element abundances for metal-poor stars have begun to constrain the yields for different stellar masses and associated mass cuts of progenitor supernovae (Mathews et al. 1992, Travaglio et al. 1998, Ishimaru & Wanajo 2000).

It is an intriguing thought that one day we may be able to identify hundreds or thousands of stars throughout the Gaiasphere that were born within the same cloud as the Sun.

EPILOGUE: CHALLENGES FOR THE FUTURE

Throughout this review, we have identified fossil signatures of galaxy formation and evolution which are accessible within the Galaxy. These signatures allow us to probe back to early epochs. We believe that the near-field universe has the same level of importance as the far-field universe for a comprehensive understanding of galaxy formation and evolution.

We have argued that understanding galaxy formation is primarily about understanding baryon dissipation within the CDM hierarchy; to a large extent, this means understanding the formation of disks. The question we seek to address is whether this can ever be unravelled in the near or far field. Dynamical information was certainly lost at several stages of this process, but we should look for preserved signatures of the different phases of galaxy formation.

Far-field cosmology can show how the light-weighted, integrated properties of disks change with cosmic time. While light-weighted properties provide some constraint on simulations of the future, they obscure some of the key processes during dissipation. The great advantage of near field studies is the ability to derive ages and detailed abundances for individual stars within galaxies of the Local Group.

We have addressed the issue of information content within the Gaiasphere. The detailed information that is possible on ages, kinematics, and chemical properties for a billion stars—which we see as the limit of observational knowledge over the next two decades—may reveal vast complexity throughout the disk. It may not be possible to perceive the sequence of events directly. However, we are optimistic that future dissipational models may provide unique connections with the observed complexity.

It is clear that detailed high resolution abundance studies of large samples of galactic stars will be crucial for the future of fossil astronomy. Christlieb et al. (2000) find that strong r-process enhanced stars can be identified with $R = 20,000$ and $SNR = 30$ pix^{-1} from the Eu lines. Both UVES and HDS can reach this sensitivity for a $B = 15$ star in just 20 min. But the detailed abundance work requires a substantial increase in the resolving power. Cayrel et al. (2001) and Hill et al. (2002) demonstrate the exquisite quality and capability of high resolution spectroscopy for CS 31082-001 where they achieve a $SNR \simeq 300$ in just four hours with UVES at $R \simeq 60,000$. (See Figure 12 for another excellent example.) But these are bright stars with some of the most extreme overabundances of r-process elements observed to date.

Gaia will provide accurate distances, ages and space motions for a vast number of stars, separate with great precision the various Galactic components, and identify most of the substructure in the outer bulge and halo. High resolution spectrographs like UVES on the VLT, HDS on Subaru, and HIRES on Keck are starting to reveal the rich seam of information in stellar abundances.

We must stress that in order to access a representative sample of the Gaiasphere, this will require a new generation of ground-based instruments, in particular, a multi-object echelle spectrograph with good blue response on a large aperture telescope. We close with a brief discussion of what is required.

As an example, the FGK sub-giants and giants are a characteristic population which could be studied over the full extent of the Gaiasphere, as discussed in the previous section. Typical stars will have magnitudes around 17–18, which is at the limit of the state-of-the-art spectrometer UVES at $R \simeq 60,000$.

We now consider what it would take to achieve high resolution spectroscopy for a representative sample of stars within the Gaiasphere. Our baseline instrument UVES achieves cross-dispersed echelle spectroscopy in two wavelength ranges (300–500 nm, 420–1100 nm). For a limiting resolution of $R \simeq 60,000$ for a single night exposure, the sensitivity limit is $U \approx 18.0$ and $V \approx 19.5$ in the blue and red arms. UVES now allows multi-object echelle spectroscopy (red arm) from fiber inputs provided by the Fiber Large Array Multi-Element Spectrograph (FLAMES). This will enable the simultaneous observation of eight objects over a 25′ field of view.

Existing multi-object spectrographs are mostly used redward of 450 nm because of the fundamental limits of conventional optical fibers. Normal fibers transmit light through total internal reflection but blue light is Rayleigh scattered below 450 nm. Recently, photonic crystal (microstructured) fibers threaded with air channels

(Cregan et al. 1999) have been shown to be highly transmissive down to the atmospheric cut-off. This is a technical breakthrough for blue multi-object spectroscopy.

We believe there is a real need for a high-resolution spectrograph that can reach hundreds or even thousands of stars in a square degree or more. The Gemini Wide Field proposal currently under discussion provides an opportunity for this kind of instrument (S. Barden, personal communication). Such an instrument will be expensive and technically challenging, but we believe this must be tackled if we are to ever unravel the formation of the Galaxy.

ACKNOWLEDGMENTS

The philosophy behind this review has emerged from discussions dating back to the spring of 1988 when KCF and JBH were visiting the Institute of Advanced Study at Princeton. At that time, there was a quorum of galaxy dynamicists at the IAS whose work continues to inspire and excite us. Our thanks go to John Bahcall for this opportunity. We thank Michael Perryman and the Gaia team for the inspiration of the Gaia science mission. We have greatly benefitted from excellent reviews by E. Friel, J. Sellwood, and G. Wallerstein and collaborators. Most recently, we acknowledge the inspiration of colleagues at the 2001 Dunk Island conference, in particular, Tim de Zeeuw, Mike Fall, Ivan King, John Kormendy, John Norris, Jerry Sellwood, Pieter van der Kruit, and Ewine van Dishoeck. We have benefited from discussions with Vladimir Avila-Reese, Rainer Beck, Bob Kurucz, Ruth Peterson, Tomek Plewa, and Jason Prochaska. We are indebted to Allan Sandage for many constructive comments. Finally, we thank the editor for suggesting the main title New Galaxy for this review.

The *Annual Review of Astronomy and Astrophysics* is online at
http://astro.annualreviews.org

LITERATURE CITED

Abe F, Bond IA, Carter BS, Dodd RJ, Fujimoto M, et al. 1999. *Astron. J.* 118:261–72

Argast D, Samland M, Gerhard OE, Thielemann F-K. 2000. *Astron. Astrophys.* 356:873–87

Arimoto N, Yoshii Y. 1987. *Astron. Astrophys.* 173:23–38

Arnaboldi M, Freeman KC, Hui X, Capaccioli M, Ford H. 1994. *ESO Messenger* 76:40–44

Arnould M, Goriely S. 2001. See Von Hippel et al. 2001, pp. 252–61

Arp H. 1962. *Ap. J.* 136:66–74

Arras P, Wasserman I. 1999. *MNRAS* 306:257–78

Ashman KM, Zepf SE. 1992. *Ap. J.* 384:50–61

Audouze J, Silk J. 1995. *Ap. J.* 451:L49–52

Babul A, Rees M. 1992. *MNRAS* 255:346–50

Baugh C, Cole S, Frenk C. 1996. *MNRAS* 283:1361–78

Bell E, de Jong R. 2000. *MNRAS* 312:497–520

Bertelli G, Nasi E. 2001. *Astron. J.* 121:1013–23

Bica E, Alloin D, Schmidt A. 1990. *MNRAS* 242:241–49

Binggeli B, Sandage A, Tammann GA. 1985. *Astron. J.* 90:1681–771

Binney J, Lacey C. 1988. *MNRAS* 230:597–627

Bland-Hawthorn J. 2002. In *The Dynamics, Structure and History of Galaxies*, ed. G Da Costa, H Jerjen, 273:155–66. San Francisco: Publ. Astron. Soc. Pac.

Bland-Hawthorn J, Freeman K. 2000. *Science* 287:79–84

Bland-Hawthorn J, Maloney PR. 2001. In *Extragalactic Gas at Low Redshift, ASP Conf. Ser.*, ed. J Mulchaey, J Stocke, 254:267–82. San Francisco: Publ. Astron. Soc. Pac.

Bland-Hawthorn J, Veilleux S, Cecil GN, Putman ME, Gibson BK, Maloney PR. 1998. *MNRAS* 299:611–24

Blitz L, Spergel DN, Teuben PJ, Hartmann D, Burton WB. 1999. *Ap. J.* 514:818–43

Blumenthal G, Faber S, Primack J, Rees M. 1984. *Nature* 311:517–25

Boss L. 1908. *Astron. J.* 26:31–36

Bower R, Kodama T, Terlevich A. 1998. *MNRAS* 299:1193–208

Braun R, Burton WB. 1999. *Astron. Astrophys.* 341:437–50

Bullock J, Kravtsov A, Weinberg D. 2000. *Ap. J.* 539:517–21

Burbidge EM, Burbidge GR, Fowler WA, Hoyle F. 1957. *Rev. Mod. Phys.* 29:547–650

Bureau M, Freeman KC, Pfitzner DW, Meurer GR. 1999. *Astron. J.* 118:2158–71

Burris DL, Pilachowski CA, Armandroff TE, Sneden C, Cowan JJ, Roe H. 2000. *Ap. J.* 544:302–19

Burstein D, Faber SM, Gaskell CM, Krumm N. 1984. *Ap. J.* 287:586–609

Butcher H. 1987. *Nature* 328:127–31

Calcaneo-Roldan C, Moore B, Bland-Hawthorn J, Sadler EM. 2000. *MNRAS* 314:324–33

Caldwell N, Rose J. 1998. *Astron. J.* 115:1423–32

Carlberg RG. 1986. *Ap. J.* 310:593–96

Carlberg RG, Sellwood JA. 1985. *Ap. J.* 292:79–89

Carney BW. 1993. In *The Globular Cluster–Galaxy Connection, ASP Conf. Ser.*, ed. GH Smith, JP Brodie, 48:234–45. San Francisco: Publ. Astron. Soc. Pac.

Carney BW, Aguilar L, Latham DW, Laird JB. 1990. *Astron. J.* 99:201–20

Carney BW, Laird JB, Latham DW, Aguilar LA. 1996. *Astron. J.* 112:668–92

Carpenter JM. 2000. *Astron. J.* 120:3139–61

Carraro G, Ng Y, Portinari L. 1998. *MNRAS* 296:1045–56

Castro S, Rich RM, Grenon M, Barbuy B, McCarthy J. 1997. *Astron. J.* 114:376–87

Cayrel R, Hill V, Beers TC, Barbuy B, Spite M, et al. 2001. *Nature* 409:691–92

Chaboyer B. 1998. *Phys. Rep.* 307:23–30

Chaboyer B, Demarque P, Kernan P, Krauss L. 1998. *Ap. J.* 494:96–110

Chaboyer B, Demarque P, Sarajedini A. 1996. *Ap. J.* 459:558–69

Charlot S, Silk J. 1994. *Ap. J.* 432:453–63

Chiappini C, Matteucci F, Romano D. 2001. *Ap. J.* 554:1044–58

Chiba M. 2002. *Ap. J.* 565:17–23

Chiba M, Beers T. 2000. *Astron. J.* 119:2843–65

Christensen-Dalsgaard J. 1986. *Proc. IAU Symp.* 123:295

Christlieb N, Beers TC, Hill V, Primas F, Rhee J, et al. 2001. See von Hippel et al. 2001, pp. 298–300

Cianci S. 2002. PhD thesis. Univ. Sydney

Clarke CJ, Bonnell IA, Hillenbrand LA. 2000. See Mannings et al. 2000, pp. 151–77

Clayton D. 1987. *Ap. J.* 315:451–59

Clayton D. 1988. *MNRAS* 234:1–36

Cohen JG, Christlieb N, Beers TC, Gratton R, Carretta E. 2002. astro-ph/0204082

Cole S, Aragon-Salamanca A, Frenk CS, Navarro JF, Zepf SE. 1994. *MNRAS* 271:781–806

Colin P, Avila-Reese V, Valenzuela O. 2000. *Ap. J.* 542:622–30

Concannon KD, Rose JA, Caldwell N. 2000. *Ap. J.* 536:L19–22

Courteau S, de Jong RS, Broeils AH. 1996. *Ap. J.* 457:L73–76

Cowan JJ, McWilliam A, Sneden C, Burris DL. 1997. *Ap. J.* 480:246–54

Cregan RF, Mangan BJ, Knight JC, Birks TA, Russell PS, et al. 1999. *Science* 285:1537–39

Creze M, Chereul E, Bienayme O, Pichon C. 1998. *Astron. Astrophys.* 329:920–36

Da Costa G. 1999. In *The Third Stromlo Symposium: The Galactic Halo, ASP Conf. Ser.*, ed. B Gibson, T Axelrod, M Putman, 165:153–66. San Francisco: Publ. Astron. Soc. Pac.

Dalcanton JJ, Spergel DN, Summers FJ. 1997. *Ap. J.* 482:659–76

Davies RL, Kuntschner H, Emsellem E, Bacon R, Bureau M, et al. 2001. *Ap. J.* 548:L33–36

de Bruijne J. 1999. *MNRAS* 306:381–93

de Grijs R, Kregel M, Wesson KH. 2001. *MNRAS* 324:1074–86

Dehnen WA. 2000. *Astron. J.* 119:800–12

de Jong R, Lacey C. 2000. *Ap. J.* 545:781–97

Dekel A, Silk J. 1986. *Ap. J.* 303:39–55

de Zeeuw PT, Hoogerwerf R, de Bruijne JHJ, Brown AGA, Blaauw A. 1999. *Astron. J.* 117:354–99

Dressler A. 1980. *Ap. J.* 236:351–65

Durrell PR, Harris WE, Pritchet CJ. 2001. *Astron. J.* 121:2557–71

Edmunds MG. 1990. *Nature* 348:395–96

Edvardsson B, Andersen J, Gustafsson B, Lambert DL, Nissen PE, Tomkin J. 1993. *Astron. Astrophys.* 275:101–52

Efstathiou G, Moody S, Peacock JA, Percival WJ, Baugh C, Bland-Hawthorn J, et al. 2002. *MNRAS* 330:L29–35

Eggen OJ. 1977. *Ap. J.* 215:812–26

Eggen OJ, Lynden-Bell D, Sandage AR. 1962. *Ap. J.* 136:748–66

Eggen OJ, Sandage AR. 1959. *MNRAS* 119:255–77

Eggen OJ, Sandage AR. 1969. *Ap. J.* 158:669–84

Ellis RS, Abraham RG, Brinchmann J, Menanteau F. 2000. *Astron. Geophys.* 41/2:10–16

Elmegreen BG, Efremov Y, Pudritz RE, Zinnecker H. 2000. See Mannings et al. 2000, pp. 179–215

Exter K, Barlow MJ, Walton NA, Clegg RES. 2001. *Astrophys. Space Sci.* 277:199–99

Faber S. 1973. *Ap. J.* 179:731–54

Fall SM. 1983. In *Internal Kinematics and Dynamics of Galaxies*, ed. E Athanassoula, pp. 391–98. Dordrecht: Reidel

Fall SM, Efstathiou G. 1980. *MNRAS* 193:189–206

Feltzing S, Holmberg J. 2000. *Astron. Astrophys.* 357:153–63

Feltzing S, Holmberg J, Hurley JR. 2001. *Astron. Astrophys.* 377:911–24

Font A, Navarro J, Stadel J, Quinn T. 2001. *Ap. J.* 563:L1–4

Fowler WA, Hoyle F. 1960. *Astron. J.* 65:345–45

Frayer DT, Brown RL. 1997. *Ap. J. Suppl.* 113:221–43

Freeman KC. 1970. *Ap. J.* 160:811–30

Freeman KC. 1987. *Annu. Rev. Astron. Astrophys.* 25:603–32

Freeman KC. 1991. In *Dynamics of Disk Galaxies*, ed. B Sundelius, p. 15. Göteborg: Univ. Göteborg

Freeman KC. 1993. In *The Globular Cluster–Galaxy Connection, ASP Conf. Ser.*, ed. GH Smith, JP Brodie, 48:608–14. San Francisco: Publ. Astron. Soc. Pac.

Freeman KC, Illingworth G, Oemler A Jr. 1983. *Ap. J.* 272:488–508

Friel E. 1995. *Annu. Rev. Astron. Astrophys.* 33:381–414

Friel E, Janes KA. 1993. *Astron. Astrophys.* 267:75–91

Fry AM, Morrison HL, Harding P, Boroson TA. 1999. *Astron. J.* 118:1209–19

Gilmore G, Reid N. 1983. *MNRAS* 202:1025–47

Gilmore G, Wyse RFG, Jones JB. 1995. *Astron. J.* 109:1095–111

Gilmore G, Wyse RFG, Kuijken K. 1989. *Annu. Rev. Astron. Astrophys.* 27:555–627

Gilroy K, Sneden C, Pilachowski CA, Cowan JJ. 1988. *Ap. J.* 327:298–320

Gimenez A, Favata F. 2001. See von Hippel et al. 2001, pp. 304–6

Gnedin OY, Lee HM, Ostriker JP. 1999. *Ap. J.* 522:935–49

Goetz M, Koeppen J. 1992. *Astron. Astrophys.* 262:455–67

Gomez A, Grenier S, Udry S, Haywood M, Meillon K, et al. 1997. In *Hipparcos–Venice '97*, pp. 621–24. ESA

Goriely S, Arnould M. 1996. *Astron. Astrophys.* 312:327–37

Goriely S, Arnould M. 2001. *Astron. Astrophys.* 379:1113–22

Gough DO. 1987. *Nature* 326:257–59

Gough DO. 2001. See von Hippel et al. 2001, pp. 304–6

Gratton RG, Fusi Pecci F, Carretta E, Clementini G, Corsi C, Lattanzi M. 1997. *Ap. J.* 491:49–71

Grebel EK. 2001. *Astrophys. Space Sci.* 277: 231–39

Gregg MD, West MJ. 1998. *Nature* 396:549–52

Grenon M. 1999. *Astrophys. Space Sci.* 265: 331–36

Griffin R. 1998. *Observatory* 118:223–25

Guenther DB. 1989. *Ap. J.* 339:1156–59

Guenther DB, Demarque P. 1997. *Ap. J.* 484:937–59

Hamann F, Ferland G. 1999. *Annu. Rev. Astron. Astrophys.* 37:487–531

Harding P, Morrison HL, Olszewski EW, Arabadjis J, Mateo M, et al. 2001. *Astron. J.* 122:1397–419

Harris WE. 2001. astro-ph/0108355

Harris WE. 1991. *Annu. Rev. Astron. Astrophys.* 29:543–79

Harris WE, Pudritz RE. 1994. *Ap. J.* 429:177–91

Hartkopf WI, Yoss KM. 1982. *Astron. J.* 87:1679–709

Heger A, Woosley SE. 2001. *Ap. J.* 567:532–43

Helmi A, Springel V, White SDM. 2001. astro-ph/0110546

Helmi A, White SDM. 1999. *MNRAS* 307:495–517

Helmi A, White SDM, de Zeeuw PT, Zhao H-S. 1999. *Nature* 402:53–55

Helmi A, Zhao H-S, de Zeeuw PT. 1999. In *The Third Stromlo Symposium: The Galactic Halo, ASP Conf. Ser.*, ed. BK Gibson, TS Axelrod, ME Putman, 165:125–29. San Francisco: Publ. Astron. Soc. Pac.

Henry RBC. 1998. In *Abundance Profiles: Diagnostic Tools for Galaxy History,* ed. D Friedli, M Edmunds, C Robert, L Drissen, 47:59. San Francisco: Publ. Astron. Soc. Pac.

Hermit S, Santiago BX, Lahav O, Strauss MA, Davis M, et al. 2001. *MNRAS* 283:709–20

Hernquist L, Quinn PJ. 1988. *Ap. J.* 331:682–98

Hill V, Plez B, Beers TC, Nordström B, Andersen J, et al. 2002. astro-ph/0203462

Hirshfeld A, McClure R, Twarog BA. 1978. In *The HR Diagram: The 100th Anniversary of Henry Norris Russell,* ed. AG Davis Philip, DS Hayes, p. 163. Dordrecht: Reidel

Hogan CJ, Dalcanton JJ. 2001. *Ap. J.* 561:35–45

Holtzman JA, Faber SM, Shaya EJ, Lauer TR, Grothe J, et al. 1992. *Ap. J.* 103:691–702

Hopkins AM, Irwin MJ, Connolly AJ. 2001. *Ap. J.* 558:L31–33

Hubble E, Humason M. 1931. *Ap. J.* 74:43–80

Ibata R, Gilmore G, Irwin MJ. 1994. *Nature* 370:194–96

Ibata R, Gilmore G, Irwin MJ. 1995. *MNRAS* 277:781–800

Ibata R, Irwin M, Lewis GF, Ferguson AMN, Tanvir N. 2001a. *Nature* 412:49–52

Ibata R, Irwin MJ, Lewis GF, Stolte A. 2001c. *Ap. J.* 547:L133–36

Ibata R, Lewis GF. 1998. *Ap. J.* 500:575–90

Ibata R, Lewis GF, Irwin MJ, Totten E, Quinn T. 2001b. *Ap. J.* 551:294–311

Ibata R, Wyse RFG, Gilmore G, Irwin MJ, Suntzeff NB. 1997. *Astron. J.* 113:634–55

Ikeuchi S. 1986. *Astrophys. Space Sci.* 118: 509–14

Ishimaru Y, Wanajo S. 2000 In *The First Stars,* ed. A Weiss, TG Abel, V Hill, pp. 189–93. Berlin: Springer-Verlag

Jablonka P, Martin P, Arimoto N. 1996. *Astron. J.* 112:1415–22

Janes KA, Phelps RL. 1994. *Astron. J.* 108:1773–85

Jenkins A, Binney J. 1990. *MNRAS* 245:305–17

Jenkins A, Frenk CS, White SDM, Colberg JM, Cole S, et al. 2001. *MNRAS* 321:372–84

Johnson JA, Bolte M. 2001. *Ap. J.* 554:888–902

Johnston KV, Sackett PD, Bullock JS. 2001. *Ap. J.* 557:137–49

Johnston KV. 1998. *Ap. J.* 495:297–308

Johnston KV, Hernquist L, Bolte M. 1996. *Ap. J.* 465:278–87

Johnston KV, Spergel DN, Haydn C. 2002. *Ap. J.* 570:656–64

Jones L, Smail I, Couch W. 2000. *Ap. J.* 528:118–22

Jones L, Worthey G. 1995, *Ap. J.* 446:L31–35

Just A. 2001. In *Disks of Galaxies: Kinematics, Dynamics and Perturbations*, ed. E Athanassoula, A Bosma, I Puerari. San Francisco: Publ. Astron. Soc. Pac. In press

Kamionkowski M, Liddle AR. 2000. *Phys. Rev. Lett.* 84:4525–28

Käppeler F, Thielemann FK, Wiescher M. 1998. *Annu. Rev. Nucl. Part. Sci.* 48:175–251

Karlsson T, Gustafsson B. 2001. *Astron. Astrophys.* 379:461–81

Kauffmann G, White SDM, Guiderdoni B. 1993. *MNRAS* 264:201–18

Kauffmann G. 1996. *MNRAS* 281:487–92

Kauffmann G, Charlot S. 1998. *MNRAS* 294: 705–17

King IR. 1958a. *Astron. J.* 63:109–13

King IR. 1958b. *Astron. J.* 63:114–17

King IR. 1958c. *Astron. J.* 63:465–73

Kinman TD. 1959. *MNRAS* 119:538–58

Klypin A, Kravtsov AV, Valenzuela O, Prada F. 1999. *Ap. J.* 522:82–92

Kochanek C. 1996. *Ap. J.* 457:228–43

Kroupa P. 2002. *MNRAS* 330:707–18

Kuntschner H, Davies RL. 1998. *MNRAS* 295:L29–33

Kuntschner H. 2000. *MNRAS* 315:184–208

Kurucz RL. 1991. In *The Solar Interior and Atmosphere*, ed. AN Cox, WC Livingston, M Matthews, p. 663. Tucson: Univ. Ariz. Press

Kurucz RL. 1995. In *Laboratory and Astronomical High Resolution Spectra*, ed. AJ Sauval, R Blomme, N Grevesse, p. 17. San Francisco: Publ. Astron. Soc. Pac.

Lacey C, Fall M. 1983. *MNRAS* 204:791–810

Lacey C, Fall M. 1985. *Ap. J.* 290:154–70

Larson R. 1974. *MNRAS* 169:229–46

Lee Y-W, Joo J-M, Sohn Y-J, Rey S-C, Rey S-C, Lee H-C, Walker AR. 1999. *Nature* 402:55–57

Lewis GF, Bland-Hawthorn J, Gibson BK, Putman ME. 2000. *PASP* 112:1300–4

Lin D, Pringle J. 1987. *Ap. J.* 320:L87–91

Lineweaver CH. 1999. *Science* 284:1503–7

Liu W, Chaboyer B. 2000. *Ap. J.* 544:818–29

Luck RE, Bond HE. 1985. *Ap. J.* 292:559–77

Lynden-Bell D, Kalnajs A. 1972. *MNRAS* 157: 1–30

Lynden-Bell D, Lynden-Bell RM. 1995. *MNRAS* 275:429–42

Mac Low M, Ferrara A. 1999. *Ap. J.* 513:142–55

Majewski SR. 1993. *Annu. Rev. Astron. Astrophys.* 31:575–638

Majewski SR, Hawley SL, Munn JA. 1996. In *Formation of the Halo . . . Inside & Out*, *ASP. Conf. Ser.*, ed. H Morrison, A Sarajedini, 92:119–29. San Francisco: Publ. Astron. Soc. Pac.

Majewski SR, Ostheimer JC, Kunkel WE, Patterson RJ. 2000. *Astron. J.* 120:2550–68

Malin DF, Carter D. 1980. *Nature* 285:643–45

Malin DF, Hadley B. 1997. *Proc. Astron. Soc. Austr.* 14:52–58

Maloney P. 1993. *Ap. J.* 414:41–56

Mannings V, Boss AP, Russell SS, eds. 2000. *Protostars and Planets IV*. Tucson: Univ. Ariz. Press

Manuel O. 2000. *Origins of Elements in the Solar System*. New York: Kluwer

Marquez A, Schuster WJ. 1994. *Astron. Astrophys. Suppl.* 108:341–58

Mateo M. 1998. *Annu. Rev. Astron. Astrophys.* 36:435–506

Mathews GJ, Bazan G, Cowan JJ. 1992. *Ap. J.* 391:719–35

Matteucci F, Francois P. 1989. *MNRAS* 239: 885–904

McKee CF, Tan JC. 2002. *Nature* 416:59–61

McGaugh SS, Schombert JM, Bothun GD, de Blok WJG. 2000. *Ap. J.* 533:L99–102

McWilliam A, Preston GW, Sneden C, Searle L. 1995. *Astron. J.* 109:2736–56

McWilliam A, Rich RM. 1994. *Ap. J. Suppl.* 91:749–91

Metcalf B. 2001. astro-ph/0109347

Meusinger H, Stecklum B, Reimann H-G. 1991. *Astron. Astrophys.* 245:57–74

Molla M, Ferrini F, Diaz A. 1996. *Ap. J.* 466: 668–85

Moore B, Calcaneo-Roldan C, Stadel J, Quinn T, Lake G, et al. 2001. *Phys. Rev. D* 64: 063508–19

Moore B, Ghigna S, Governato F, Lake G, Quinn T, et al. 1999. *Ap. J.* 524:L19–22

Moore B, Katz N, Lake G. 1996. *Ap. J.* 457:455–59

Morgan WW. 1959. *Astron. J.* 64:432

Morrison HL. 1993. *Astron. J.* 106:578–90

Mushotzky R. 1999. In *The Hy-Redshift Universe, ASP Conf. Ser.*, ed. AJ Bunker, WJM van Breugel, 193:323–35. San Francisco: Publ. Astron. Soc. Pac.

Newberg HJ, Yanny B, Rockosi CM, Grebel EK, Rix H-W, et al. 2002. *Ap. J.* 569:245–74

Noguchi M. 1998. *Nature* 392:253–55

Norberg P, Baugh CM, Hawkins E, Maddox S, Peacock JA, et al. 2001. *MNRAS* 328:64–70

Norris JE, Ryan SG. 1989. *Ap. J.* 336:L17–19

Norris JE, Ryan SG, Beers TC. 1996. *Ap. J. Suppl.* 107:391–421

Norris JE, Ryan SG, Beers TC. 1997. *Ap. J.* 488:350–63

Odenkirchen M, Grebel EK, Rockosi CM, Dehnen W, Ibata R, et al. 2001. *Ap. J.* 548:L165–69

Oort J. 1965. In *Galactic Structure*, ed. A Blaauw, M Schmidt, pp. 455–511. Chicago: Univ. Chicago Press

Oort J. 1966. *Bull. Astron. Inst. Neth.* 18:421

Ortolani S, Renzini A, Gilmozzi R, Marconi G, Barbuy B, et al. 1995. *Nature* 377:701–3

Oswalt TD, Smith JA, Wood MA, Hintzen P. 1996. *Nature* 382:692

Pagel BEJ. 1965. *R. Obs. Bull.* 104:127–51

Pagel BEJ. 1989. In *Evolutionary Phenomena in Galaxies*, pp. 201–23. Cambridge: Cambridge Univ. Press

Pagel BEJ, Patchett BE. 1975. *MNRAS* 172:13–40

Pagel BEJ, Tautvaisiene G. 1997. *MNRAS* 288:108–16

Peebles PJE. 1970. *Astron. J.* 75:13–20

Peebles PJE. 1971. *Physical Cosmology.* Princeton: Princeton Univ. Press

Peebles PJE. 1974. *Ap. J.* 189:L51–53

Peebles PJE, Dicke RH. 1968. *Ap. J.* 154:891–908

Peebles PJE, Seager S, Hu W. 2000. *Ap. J.* 539:L1–4

Peletier R, de Grijs R. 1998. *MNRAS* 300:L3–6

Perryman MAC, de Boer KS, Gilmore G, Høg E, Lattanzi MG, Lindegren L, et al. 2001. *Astron. Astrophys.* 369:339–63

Pfenniger D, Combes F, Martinet L. 1994. *Astron. Astrophys.* 285:79–93

Phelps R, Janes K. 1996. *Astron. J.* 111:1604–8

Pitts E, Tayler RJ. 1989. *MNRAS* 240:373–95

Pohlen M, Dettmar R-J, Lütticke R. 2000. *Astron. Astrophys.* 357:L1–4

Press WH, Schechter P. 1974. *Ap. J.* 187:425–38

Preston GW, Beers TC, Shectman SA. 1994. *Astron. J.* 108:538–54

Pritchet C, van den Bergh S. 1994. *Astron. J.* 107:1730–36

Prochaska JX, Naumov SO, Carney BW, McWilliam A, Wolfe AM. 2000. *Astron. J.* 120:2513–49

Quillen AC. 2002. astro-ph/0202253

Quillen AC, Garnett D. 2001. In *Galaxy Disks and Disk Galaxies, ASP Conf. Ser.*, ed. G Jose, SJ Funes, EM Corsini, 230:87–88. San Francisco: Publ. Astron. Soc. Pac.

Quinn PJ. 1984. *Ap. J.* 279:596–609

Quinn PJ, Goodman J. 1986. *Ap. J.* 309:472–95

Quinn PJ, Hernquist L, Fullagar D. 1993. *Ap. J.* 403:74–93

Quinn PJ, Zurek W. 1988. *Ap. J.* 331:1–18

Quinn T, Katz N, Efstathiou G. 1996. *MNRAS* 278:49–54

Rees MJ. 1986. *MNRAS* 218:25–30

Reid IN. 1998. In *Highlights of Astronomy*, ed. J Andersen, 11A:562. Dordrecht: Kluwer

Renzini A. 2000. In *From Extrasolar Planets to Cosmology*, ed. J Bergeron, A Renzini, p. 168. Berlin: Springer-Verlag

Renzini A. 2001. In *Chemical Enrichment of Intracluster and Intergalactic Medium, ASP. Conf. Ser.*, ed. F Matteucci, R Fusco-Femiano. San Francisco: Publ. Astron. Soc. Pac.

Rich RM. 2001. See von Hippel et al. 2001, pp. 216–25

Rix H-W, Zaritsky D. 1995. *Ap. J.* 447:82–102

Rocha-Pinto HJ, Scalo J, Maciel WJ, Flynn C. 2000. *Astron. Astrophys.* 358:869–85

Rodgers AW, Harding P, Sadler EM. 1981. *Ap. J.* 244:912–18

Rose J. 1994. *Astron. J.* 107:206–29

Ryan SG, Norris JE, Beers TC. 1996. *Ap. J.* 471:254–78

Rutherford E. 1904. Radiation and emanation of radium. In *The Collected Works of Lord Rutherford Vol 1.* London: Allen & Unwin 1962–65

Sandage A. 1990. *J. R. Astron. Soc. Can.* 84:70–88

Sandage A, Cacciari C. 1990. *Ap. J.* 350:645–61

Sandage A, Visvanathan N. 1978. *Ap. J.* 225:742–50

Schmidt M. 1963. *Ap. J.* 137:758–69

Schoenmakers RHM, Franx M, de Zeeuw PT. 1997. *MNRAS* 292:349–64

Schommer RA, Suntzeff NB, Olszewski EW, Harris HC. 1992. *Astron. J.* 103:447–59

Schweizer F. 1987. In *Nearly Normal Galaxies: From the Planck Time to the Present*, ed. S Faber, pp. 18–25. New York: Springer-Verlag

Searle L. 1977. In *The Evolution of Galaxies & Stellar Populations*, ed. BM Tinsley, RB Larson, p. 219. New Haven: Yale

Searle L, Zinn R. 1978. *Ap. J.* 225:357–79

Sedgewick R. 1992. *Algorithms in C^{++}.* Menlo Park, CA: Addison-Wesley

Sellwood JA. 1999. In *Astrophysical Disks*, ed. J Sellwood, J Goodman, p. 327. San Francisco: Publ. Astron. Soc. Pac.

Sellwood JA, Kosowsky A. 2002. In *The Dynamics, Structure and History of Galaxies*, ed. G Da Costa, H Jerjen, 273:243–53 San Francisco: Publ. Astron. Soc. Pac.

Sellwood JA. 2001. In *Disks of Galaxies: Kinematics, Dynamics and Perturbations*, ed. E Athanassoula, A Bosma, I Puerari. San Francisco: Publ. Astron. Soc. Pac. In press

Shang Z, Brinks E, Zheng Z, Chen J, Burstein D, et al. 1998. *Ap. J.* 504:L23–26

Sheth RK, Tormen G. 1999. *MNRAS* 308:119–26

Shigeyama T, Tsujimoto T. 1998. *Ap. J.* 507: L135–39

Sneden C, Cowan JJ, Beers TC, Truran JW, Lawler JE, Fuller G. 2001a. See von Hippel et al. 2001, pp. 235–43

Sneden C, Cowan JJ, Ivans II, Fuller GM, Burles S, et al. 2000. *Ap. J.* 533:L139–42

Sneden C, Lawler JE, Cowan JJ. 2001b. astro-ph/0109194

Spergel D, Steinhardt P. 2000. *Phys. Rev. Lett.* 84:3760–63

Spite M, Spite F. 1978. *Astron. Astrophys.* 67: 23–31

Spitzer L, Schwarzschild M. 1953. *Ap. J.* 118: 106–12

Stahler SW, Palla F, Ho PTP. 2000. See Mannings et al. 2000, pp. 327–51

Strömgren B. 1987. In *The Galaxy*, ed. G Gilmore, R Carswell, pp. 229–46. Dordrecht: Reidel

Talbot RJ, Arnett WD. 1971. *Ap. J.* 170:409–22

Taylor BJ. 2000. *Astron. Astrophys.* 362:563–79

Thoul A, Weinberg D. 1996. *Ap. J.* 465:608–16

Tinsley B. 1980. *Fund. Cosmic Phys.* 5:287–388

Trager S, Faber SM, Worthey G, Gonzalez J. 2000. *Astron. J.* 120:1645–76

Trager S, Worthey G, Faber SM, Bustein D, Gonzales J. 1998. *Ap. J. Suppl.* 116:1–28

Travaglio C, Galli D, Gallino R, Busso M, Ferrini F, et al. 1999. *Ap. J.* 521:691–702

Travaglio C, Gallino R, Zinner E, Amari S, Woosley S. 1998. *Meteorit. Planet. Sci.* 33: A155–56

Tremaine SD. 1993. In *Back to the Galaxy*, ed. SS Holt, F Verter, pp. 599–609. New York: AIP

Tremaine SD. 1999. *MNRAS* 307:877–83

Tripicco MJ, Bell RA. 1995. *Astron. J.* 110: 3035–49

Truran JW. 1981. *Astron. Astrophys.* 97:391–93

Truran JW, Burles S, Cowan J, Sneden C. 2001. See von Hippel et al. 2001, pp. 226–34

Tsikoudi V. 1980. *Ap. J. Suppl.* 43:365–77

Tsujimoto T, Shigeyama T, Yoshii Y. 2000. *Ap. J.* 531:L33–36

Tyson JA, Fischer P, Guhathakurta P, McIlroy P, Wenk R, et al. 1998. *Astron. J.* 116:102–10

Tully RB, Fisher JR. 1977. *Astron. Astrophys.* 54:661–73

Tully RB, Verheijen MA, Pierce MJ, Huang J-S, Wainscoat RJ. 1996. *Astron. J.* 112:2471–99

Twarog BA. 1980. *Ap. J.* 242:242–59

Twarog BA, Ashman KM, Anthony-Twarog BJ. 1997. *Astron. J.* 114:2556–85

Ulrich RK. 1986. *Ap. J.* 306:L37–40

Unavane M, Wyse RFG, Gilmore G. 1996. *MNRAS* 278:727–36

van den Bergh S. 1962. *Astron. J.* 67:486–90

van den Bergh S. 2000. *The Galaxies of the Local Group.* Cambridge: Cambridge Univ. Press

van den Bergh S, McClure RD. 1980. *Astron. Astrophys.* 88:360–62

van der Kruit P. 2002. In *The Dynamics, Structure and History of Galaxies*, ed. G Da Costa, H Jerjen, 273:7–18 San Francisco: Publ. Astron. Soc. Pac.

Vazdekis A. 1999. *Ap. J.* 513:224–41

Vivas AK, Zinn R, Andrews P, Bailyn C, Baltay C, et al. 2001. *Ap. J.* 554:L33–36

von Hippel T, Simpson C, Manset N, eds. 2001. *Astrophysical Ages and Time Scales, ASP Conf. Ser.*, San Francisco: Publ. Astron. Soc. Pac.

Walker IR, Mihos JC, Hernquist L. 1996. *Ap. J.* 460:121–35

Wallerstein G, Greenstein JL, Parker R, Helfer HL, Aller LH. 1963. *Ap. J.* 137:280–300

Wallerstein G, Iben I, Parker P, Boesgaard AM, Hale GM, et al. 1997. *Rev. Mod. Phys.* 69:995–1084

Weil M, Bland-Hawthorn J, Malin DF. 1997. *Ap. J.* 490:664–81

Weinberg S. 1977. *The First Three Minutes. A Modern View of the Origin of the Universe.* London: Andre Deutsch

Weiner BJ, Vogel SN, Williams TB. 2001. In *Extragalactic Gas at Low Redshift, ASP Conf. Ser.*, ed. J Mulchaey, J Stocke, 254:256–66. San Francisco: Publ. Astron. Soc. Pac.

Westin J, Sneden C, Gustafsson B, Cowan J. 2000. *Ap. J.* 530:783–99

White M, Croft RAC. 2000. *Ap. J.* 539:497–504

White SDM. 1976. *MNRAS* 177:717–33

White SDM. 1996. In *Cosmology and Large Scale Structure*, ed. R Schaeffer, J Silk, M Spiro, J Zinn-Justin, p. 349. Amsterdam: Elsevier

White SDM, Rees MJ. 1978. *MNRAS* 183:341–58

Whitmore BC, Schweizer F. 1995. *Astron. J.* 109:960–80

Whitmore BC, Schweizer F, Leitherer C, Borne K, Robert C. 1993. *Astron. J.* 106:1354–70

Widrow L, Dubinski J. 1998. *Ap. J.* 504:12–26

Worthey G. 1994. *Ap. J. Suppl.* 95:107–49

Worthey G, Dorman B, Jones L. 1996. *Astron. J.* 112:948–53

Worthey G, Faber SM, Gonzales J, Burstein D. 1994. *Ap. J. Suppl.* 94:687–722

Wyse RFG, Silk J. 1989. *Ap. J.* 339:700–11

Yokoi K, Takahashi K, Arnould M. 1983. *Astron. Astrophys.* 117:65–82

Yun MS, Carilli CL, Kawabe R, Tutui Y, Kohno K, Ohta K. 2000. *Ap. J.* 528:171–78

Zheng Z, Shang Z, Su H, Burstein D, Chen J, et al. 1999. *Astron. J.* 117:2757–80

Zinn R. 1985. *Ap. J.* 293:424–44

Zinnecker H, Cannon RD. 1986. In *Star Forming Galaxies and Related Objects*, ed. D Kunth, TX Thuan, J Thanh Van, p. 155. Paris: Ed. Front.

Zurek W, Quinn P, Salmon J. 1988. *Ap. J.* 330:519–34

Annu. Rev. Astron. Astrophys. 2002. 40:539–77
doi: 10.1146/annurev.astro.40.120401.150547
Copyright © 2002 by Annual Reviews. All rights reserved

THE EVOLUTION OF X-RAY CLUSTERS OF GALAXIES

Piero Rosati,[1,3] Stefano Borgani,[2] and Colin Norman[3,4]

[1]*European Southern Observatory, D-85748 Garching bei München, Germany;
email: prosati@eso.org*
[2]*Dipartimento di Astronomia dell'Università di Trieste, via Tiepolo 11, I-34131 Trieste,
Italy; email: borgani@ts.astro.it*
[3]*Department of Physics and Astronomy, The Johns Hopkins University, Baltimore,
Maryland 21218; email: norman@stsci.edu*
[4]*Space Telescope Science Institute, Baltimore, Maryland 21218*

Key Words   cosmology, intracluster medium, temperature, masses, dark matter

■ **Abstract**   Considerable progress has been made over the past decade in the study of the evolutionary trends of the population of galaxy clusters in the Universe. In this review we focus on observations in the X-ray band. X-ray surveys with the *ROSAT* satellite, supplemented by follow-up studies with *ASCA* and *Beppo-SAX*, have allowed an assessment of the evolution of the space density of clusters out to $z \approx 1$ and the evolution of the physical properties of the intracluster medium out to $z \approx 0.5$. With the advent of *Chandra* and *Newton-XMM* and their unprecedented sensitivity and angular resolution, these studies have been extended beyond redshift unity and have revealed the complexity of the thermodynamical structure of clusters. The properties of the intracluster gas are significantly affected by nongravitational processes including star formation and Active Galactic Nuclei (AGN) activity. Convincing evidence has emerged for modest evolution of both the bulk of the X-ray cluster population and their thermodynamical properties since redshift unity. Such an observational scenario is consistent with hierarchical models of structure formation in a flat low-density universe with $\Omega_m \simeq 0.3$ and $\sigma_8 \simeq 0.7$–0.8 for the normalization of the power spectrum. Basic methodologies for construction of X-ray–selected cluster samples are reviewed, and implications of cluster evolution for cosmological models are discussed.

INTRODUCTION

Galaxy clusters arise from the gravitational collapse of rare high peaks of primordial density perturbations in the hierarchical clustering scenario for the formation of cosmic structures (e.g., Peebles 1993, Coles & Lucchin 1995, Peacock 1999). They probe the high-density tail of the cosmic density field, and their number density is highly sensitive to specific cosmological scenarios (e.g., Press & Schechter 1974, Kofman et al. 1993, Bahcall & Cen 1993, White et al. 1993a). The space

0066-4146/02/0922-0539$14.00

density of clusters in the local universe has been used to measure the amplitude of density perturbations on ~10-Mpc scales. Its evolution, which is driven by the growth rate of density fluctuations, essentially depends on the value of the matter density parameter Ω_m[1] (e.g., Oukbir & Blanchard 1992, Eke et al. 1998, Bahcall et al. 1999). Figure 1 (see color insert) shows how structure formation proceeds and the cluster population evolves in two cosmological models characterized by different values of Ω_m. High- and low-density universes show largely different evolutionary patterns, which demonstrate how the space density of distant clusters can be used as a powerful cosmological diagnostic. What cosmological models actually predict is the number density of clusters of a given mass at varying redshifts. The cluster mass, however, is never a directly observable quantity, although several methods exist to estimate it from observations.

Determining the evolution of the space density of clusters requires counting the number of clusters of a given mass per unit volume at different redshifts. Therefore, three essential tools are required for its application as a cosmological test: (*a*) an efficient method to find clusters over a wide redshift range, (*b*) an observable estimator of the cluster mass, and (*c*) a method to compute the selection function or equivalently the survey volume within which clusters are found.

Clusters form via the collapse of cosmic matter over a region of several megaparsecs. Cosmic baryons, which represent approximately 15% of the mass content of the Universe, follow the dynamically dominant dark matter during the collapse. As a result of adiabatic compression and shocks generated by supersonic motions during shell crossing and virialization, a thin hot gas permeating the cluster gravitational potential well is formed. For a typical cluster mass of $10^{14}-10^{15} M_\odot$ this gas reaches temperatures of several 10^7 K, becomes fully ionized, and therefore emits via thermal bremsstrahlung in the X-ray band.

Observations of clusters in the X-ray band provide an efficient and physically motivated method of identification, which fulfills the three requirements above. The X-ray luminosity, which uniquely specifies the cluster selection, is also a good probe of the depth of the cluster gravitational potential. For these reasons most of the cosmological studies based on clusters have used X-ray–selected samples. X-ray studies of galaxy clusters provide: (*a*) an efficient way of mapping the overall structure and evolution of the Universe and (*b*) an invaluable means of understanding their internal structure and the overall history of cosmic baryons.

X-ray cluster studies made substantial progress at the beginning of the 1990s with the advent of new X-ray missions. First, the all-sky survey and the deep pointed observations conducted by the *ROSAT* satellite have been a goldmine for the discovery of hundreds of new clusters in the nearby and distant Universe.

---

[1]The matter-density parameter is defined as $\Omega_m = \bar{\rho}/\rho_c$, where $\bar{\rho}$ is the cosmic mean matter density; $\rho_c = 1.88\,10^{-29}h^2\,\mathrm{g\,cm^{-3}}$ is the critical density; h and h_{50} denote the Hubble constant H_0 respectively in units of 100 and 50 km s^{-1} Mpc^{-1}. Ω_Λ is referred to as the contribution to the total mass-energy density of the Universe associated with the cosmological constant Λ.

Follow-up studies with the *ASCA* and *Beppo-SAX* satellites revealed hints of the complex physics governing the intracluster gas. In addition to gas heating associated with gravitational processes, star formation processes, energy feedback from supernovae, and galactic nuclear activity are now understood to play an important role in determining the thermal history of the intracluster medium (ICM), its X-ray properties, and its chemical composition. Studies utilizing the current new generation of X-ray satellites, *Chandra* and *Newton-XMM*, are radically changing our X-ray view of clusters. The large collecting area of *Newton-XMM*, combined with the superb angular resolution of *Chandra*, have started to unveil the interplay between the complex physics of the hot ICM and detailed processes of star formation associated with cool baryons.

The aim of this article is to provide an up-to-date review of the methodology used to construct X-ray–selected cluster samples and to investigate their evolutionary properties. We emphasize the evolution of the space density of clusters and their physical parameters. Additional reviews of galaxy clusters include Forman & Jones (1982) and Sarazin (1988) for historical reviews of X-ray properties of galaxy clusters; Bahcall (1988) and Borgani & Guzzo (2001) for large-scale structure studies of galaxy clusters; Fabian (1994) for the physics of cooling flows in clusters; Mulchaey (2000) for the X-ray properties of galaxy groups; Birkinshaw (1999) and Carlstrom et al. (2001) for cluster studies with the Sunyaev-Zel'dovich effect; Mellier (1999) for studies of the mass distribution of clusters via gravitational lensing; and van Dokkum & Franx (2001) for the study of galaxy populations in clusters.

PHYSICAL PROPERTIES OF GALAXY CLUSTERS

Clusters of galaxies were first identified as large concentrations in the projected galaxy distribution (Abell 1958, Zwicky et al. 1966, Abell et al. 1989), containing hundreds to thousands of galaxies over a region of the order of ~ 1 Mpc. The first observations showed that such structures are associated with deep gravitational potential wells, containing galaxies with a typical velocity dispersion along the line of sight of $\sigma_v \sim 10^3$ km s^{-1}. The crossing time for a cluster of size R can be defined as

$$t_{\mathrm{cr}} = \frac{R}{\sigma_v} \simeq 1 \left(\frac{R}{1\,\mathrm{Mpc}} \right) \left(\frac{\sigma_v}{10^3\,\mathrm{km\,s^{-1}}} \right)^{-1} \mathrm{Gyr}. \quad (1)$$

Therefore, in a Hubble time, $t_H \simeq 10\,h^{-1}$ Gyr, such a system has enough time in its internal region, $\lesssim 1\,h^{-1}$ Mpc, to dynamically relax—a condition that cannot be attained in the surrounding, ~ 10 Mpc, environment. Assuming virial equilibrium, the typical cluster mass is

$$M \simeq \frac{R\sigma_v^2}{G} \simeq \left(\frac{R}{1\,h^{-1}\,\mathrm{Mpc}} \right) \left(\frac{\sigma_v}{10^3\,\mathrm{km\,s^{-1}}} \right)^2 10^{15}\,h^{-1} M_\odot. \quad (2)$$

Smith (1936) first noticed in his study of the Virgo cluster that the mass implied by cluster galaxy motions was largely exceeding that associated with the optical light component. This was confirmed by Zwicky (1937), and was the first evidence of the presence of dark matter.

X-Ray Properties of Clusters

Observations of galaxy clusters in the X-ray band have revealed a substantial fraction ($\sim$15%) of the cluster mass to be in the form of hot diffuse gas, permeating its potential well. If this gas shares the same dynamics as member galaxies, then it is expected to have a typical temperature

$$k_B T \simeq \mu m_p \sigma_v^2 \simeq 6 \left(\frac{\sigma_v}{10^3 \, \text{km s}^{-1}} \right)^2 \text{keV},　\tag{3}$$

where m_p is the proton mass and μ is the mean molecular weight ($\mu = 0.6$ for a primordial composition with a 76% fraction contributed by hydrogen). Observational data for nearby clusters (e.g., Wu et al. 1999) and for distant clusters (see Figure 2) actually follow this relation, although with some scatter and with a few outliers. This correlation indicates that the idealized picture of clusters as relaxed structures in which both gas and galaxies feel the same dynamics is a reasonable

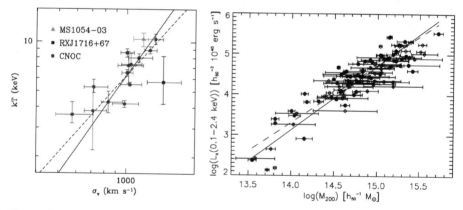

Figure 2 (*Left*) The relation between galaxy velocity dispersion, σ_v, and intracluster medium temperature, T, for distant ($z > 0.15$) galaxy clusters. Velocity dispersions are taken from Carlberg et al. (1997a) for CNOC clusters and from Girardi & Mezzetti (2001) for MS1054-03 and RXJ1716 + 67. Temperatures are taken from Lewis et al. (1999) for CNOC clusters, from Donahue et al. (1998) for MS1054-03, and from Gioia et al. (1999) for RXJ1716 + 67. The solid line shows the relation $k_B T = \mu m_p \sigma_v^2$, and the dashed line is the best-fit to the low-z T-σ_v relation from Wu et al. (1999). (*Right*) The low-z relation between X-ray luminosity and the mass contained within the radius encompassing an average density $200\rho_c$ (from Reiprich & Böhringer 2002). The two lines are the best log-log linear fit to the two data sets indicated with filled and open circles.

representation. There are some exceptions that reveal the presence of a more complex dynamics.

At the high energies implied by Equation 3, the ICM behaves as a fully ionized plasma, whose emissivity is dominated by thermal bremsstrahlung. The emissivity for this process at frequency ν scales as $\epsilon_\nu \propto n_e n_i g(\nu, T) T^{-1/2} \exp(-h\nu/k_B T)$, where n_e and n_i are the number density of electrons and ions, respectively, and $g(\nu, T) \propto \ln(k_B T / h\nu)$ is the Gaunt factor. Whereas the pure bremsstrahlung emissivity is a good approximation for $T \gtrsim 3$ keV clusters, a further contribution from metal emission lines should be taken into account when considering cooler systems (e.g., Raymond & Smith 1977). By integrating the above equation over the energy range of the X-ray emission and over the gas distribution, one obtains X-ray luminosities $L_X \sim 10^{43}$–10^{45} erg s^{-1}. These powerful luminosities allow clusters to be identified as extended sources out to large cosmological distances.

Assuming spherical symmetry, the condition of hydrostatic equilibrium connects the local gas pressure p to its density ρ_{gas} according to

$$\frac{dp}{dR} = -\frac{GM(<R)\rho_{\text{gas}}(R)}{R^2}. \tag{4}$$

By inserting the equation of state for a perfect gas, $p = \rho_{\text{gas}} k_B T / \mu m_p$, into Equation (4), one can express $M(<R)$, the total gravitating mass within R, as

$$M(<R) = -\frac{k_B T R}{G \mu m_p} \left(\frac{d \log \rho_{\text{gas}}}{d \log R} + \frac{d \log T}{d \log R} \right). \tag{5}$$

If R is the virial radius, then at redshift z we have $M \propto R^3 \bar{\rho}_0 (1+z)^3 \Delta_{vir}(z)$, where $\bar{\rho}_0$ is the mean cosmic density at present time and $\Delta_{vir}(z)$ is the mean overdensity within a virialized region (see also Equation 13, below). For an Einstein-de-Sitter cosmology, Δ_{vir} is constant and therefore, for an isothermal gas distribution, Equation 5 implies $T \propto M^{2/3}(1+z)$. Such relations show how quantities such as ρ_{gas} and T, which can be measured from X-ray observations, are directly related to the cluster mass. Thus, in addition to providing an efficient method to detect clusters, X-ray studies of the ICM allow one to measure the total gravitating cluster mass, which is the quantity predicted by theoretical models for cosmic structure formation.

A popular description of the gas density profile is the β-model,

$$\rho_g(r) = \rho_{g,0} \left[1 + \left(\frac{r}{r_c} \right)^2 \right]^{-3\beta/2}, \tag{6}$$

which was introduced by Cavaliere & Fusco-Femiano (1976; see also Sarazin 1988 and references therein) to describe an isothermal gas in hydrostatic equilibrium within the potential well associated with a King dark-matter density profile. The parameter β is the ratio between kinetic dark-matter energy and thermal gas energy (see Equation 3). This model is a useful guideline for interpreting cluster emissivity, although over limited dynamical ranges. Now, with the *Chandra* and *Newton-XMM* satellites, the X-ray emissivity can be mapped with high angular resolution and over larger scales. These new data have shown that Equation 6 with a unique β value cannot always describe the surface brightness profile of clusters (e.g., Allen et al. 2001).

Kaiser (1986) described the thermodynamics of the ICM by assuming it to be entirely determined by gravitational processes, such as adiabatic compression during the collapse and shocks due to supersonic accretion of the surrounding gas. As long as there are no preferred scales both in the cosmological framework (i.e., $\Omega_m = 1$ and power-law shape for the power spectrum at the cluster scales) and in the physics (i.e., only gravity acting on the gas and pure bremsstrahlung emission), then clusters of different masses are just a scaled version of each other, because bremsstrahlung emissivity predicts $L_X \propto M\rho_{gas}T^{1/2}$, $L_X \propto T_X^2(1+z)^{3/2}$ or, equivalently $L_X \propto M^{4/3}(1+z)^{7/2}$. Furthermore, if we define the gas entropy as $S = T/n^{2/3}$, where n is the gas density assumed fully ionized, we obtain $S \propto T(1+z)^{-2}$.

It was soon recognized that X-ray clusters do not follow these scaling relations. As we discuss in "Cosmology with X-Ray Clusters," below, the observed luminosity-temperature relation for clusters is $L_X \propto T^3$ for $T \gtrsim 2$ keV and possibly even steeper for $T \lesssim 1$ keV groups. This result is consistent with the finding that $L_X \propto M^\alpha$ with $\alpha \simeq 1.8 \pm 0.1$ for the observed mass-luminosity relation (e.g., Reiprich & Böhringer 2002) (see right panel of Figure 2). Furthermore, the low-temperature systems are observed to have shallower central gas-density profiles than the hotter systems, which turns into an excess of entropy in low-T systems with respect to the $S \propto T$ predicted scaling (e.g., Ponman et al. 1999, Lloyd-Davies et al. 2000).

A possible interpretation for the breaking of the scaling relations assumes that the gas has been heated at some earlier epoch by feedback from a nongravitational astrophysical source (Evrard & Henry 1991). This heating would increase the entropy of the ICM, place it on a higher adiabat, prevent it from reaching a high central density during the cluster gravitational collapse and, therefore, decrease the X-ray luminosity (e.g., Balogh et al. 1999, Tozzi & Norman 2001, and references therein). For a fixed amount of extra energy per gas particle, this effect is more prominent for poorer clusters, i.e., for those objects whose virial temperature is comparable with the extra-heating temperature. As a result, the self-similar behavior of the ICM is expected to be preserved in hot systems, whereas it is broken for colder systems. Both semi-analytical works (e.g., Cavaliere et al. 1998, Balogh et al. 1999, Wu et al. 2000, Tozzi et al. 2001) and numerical simulations (e.g., Navarro et al. 1995, Brighenti & Mathews 2001, Bialek et al. 2001, Borgani et al. 2001a) converge to indicate that ~ 1 keV per gas particle of extra energy is required. A visual illustration of the effect of preheating is reported in Figure 3 (see color insert), which shows the entropy map for a high-resolution simulation of a system with mass comparable to that of the Virgo cluster for different heating schemes (Borgani et al. 2001b). The effect of extra energy injection is to decrease the gas density in central cluster regions and to erase the small gas clumps associated with accreting groups.

The gas-temperature distributions in the outer regions of clusters are not affected by gas cooling. These temperature distributions have been studied with the *ASCA* and *Beppo-SAX* satellites. General agreement about the shape of the temperature profiles has still to be reached (e.g., Markevitch et al. 1998, White 2000, Irwin & Bregman 2000). De Grandi & Molendi (2002) analyzed a set of 21 clusters with *Beppo-SAX* data and found the gas to be isothermal out to $\sim 0.2 R_{\rm vir}$, with a significant temperature decline at larger radii. Such results are not consistent with the temperature profiles obtained from cluster hydrodynamical simulations (e.g., Evrard et al. 1996), thus indicating that some physical process is still lacking in current numerical descriptions of the ICM. Deep observations with *Newton-XMM* and *Chandra* will allow the determination of temperature profiles over the whole cluster virialized region.

X-ray spectroscopy is a powerful means for analyzing the metal content of the ICM. Measurements of over 100 nearby clusters have yielded a mean metallicity $Z \sim 1/3\, Z_\odot$, largely independent of the cluster temperature (e.g., Renzini 1997 and references therein). The spatial distribution of metals has recently been studied in detail with *ASCA* and *Beppo-SAX* data (e.g., White 2000, De Grandi & Molendi 2001). This field will receive a major boost over the next few years, particularly with *Newton-XMM*, which with a tenfold improvement in collecting area and much better angular resolution, will be able to map the distribution of different metals in the ICM such as Fe, S, Si, and O.

Cooling in the Intracluster Medium

In order to characterize the role of cooling in the ICM, it is useful to define the cooling timescale, which for an emission process characterized by a cooling function $\Lambda_c(T)$, is defined as $t_{cool} = k_B T /(n \Lambda(T))$, n being the number density of gas particles. For a pure bremsstrahlung emission, $t_{cool} \simeq 8.5 \times 10^{10}$ yr $(n/10^{-3}\,{\rm cm}^{-3})^{-1}$ $(T/10^8 K)^{1/2}$ (e.g., Sarazin 1988). Therefore, the cooling time in central cluster regions can be shorter than the age of the Universe. A substantial fraction of gas undergoes cooling in these regions and consequently drops out of the hot diffuse, X-ray emitting phase. Studies with the *ROSAT* and *ASCA* satellites indicate that the decrease of the ICM temperature in central regions has been recognized as a widespread feature among fairly relaxed clusters (see Fabian 1994 and references therein). The canonical picture of cooling flows predicted that as the high-density gas in the cluster core cools down, the lack of pressure support causes external gas to flow in, thus creating a superposition of many gas phases, each one characterized by a different temperature. Our understanding of the ICM cooling structure is now undergoing a revolution thanks to the much-improved spatial and spectral resolution provided by *Newton-XMM*. Recent observations have shown the absence of metal lines associated with gas at temperature $\lesssim 3$ keV (e.g., Peterson et al. 2001, Tamura et al. 2001), in stark contrast with the standard cooling flow prediction for the presence of low-temperature gas (e.g., Böhringer et al. 2001, Fabian et al. 2001a).

Radiative cooling has also been suggested as an alternative to extra heating to explain the lack of ICM self-similarity (e.g., Bryan 2000, Voit & Bryan 2002). When the recently shocked gas residing in external cluster regions leaves the hot phase and flows in, it increases the central entropy level of the remaining gas. The decreased amount of hot gas in the central regions causes a suppression of the X-ray emission (Pearce et al. 2000, Muanwong et al. 2001). This solution has a number of problems. Cooling in itself is a runaway process, leading to a quite large fraction of gas leaving the hot diffuse phase inside clusters. Analytical arguments and numerical simulations have shown that this fraction can be as large as $\sim$50%, whereas observational data indicates that only $\lesssim$10% of the cluster baryons are locked into stars (e.g., Bower et al. 2001, Balogh et al. 2001). This calls for the presence of a feedback mechanism, such as supernova explosions (e.g., Menci & Cavaliere 2000, Finoguenov et al. 2000, Pipino et al. 2001, Kravtsov & Yepes 2000) or Active Galactic Nuclei (AGN) (e.g., Valageas & Silk 1999, Wu et al. 2000, Yamada & Fujita 2001), which given reasonable efficiencies of coupling to the hot ICM, may be able to provide an adequate amount of extra energy to balance overcooling.

OBSERVATIONAL FRAMEWORK

Optically Based Cluster Surveys

Abell (1958) provided the first extensive, statistically complete sample of galaxy clusters. Based on pure visual inspection, clusters were identified as enhancements in the galaxy surface density on Palomar Observatory Sky Survey (POSS) plates, by requiring that at least 50 galaxies were contained within a metric radius $R_A = 3h_{50}^{-1}$ Mpc and a predefined magnitude range. Clusters were characterized by their richness and estimated distance. The Abell catalog has been for decades the prime source for detailed studies of individual clusters and for characterizing the large-scale distribution of matter in the nearby Universe. The sample was later extended to the Southern hemisphere by Corwin and Olowin (Abell et al. 1989) by using UK Schmidt survey plates. Another comprehensive cluster catalog was compiled by Zwicky and collaborators (Zwicky et al. 1966), who extended the analysis to poorer clusters using criteria less strict than Abell's in defining galaxy overdensities.

Several variations of the Abell criteria for defining clusters were used in an automated and objective fashion when digitized optical plates became available. The Edinburgh-Durham Southern Galaxy Catalog, constructed from the COSMOS scans of UK Schmidt plates around the Southern Galactic Pole, was used to compile the first machine-based cluster catalog (Lumsden et al. 1992). In a similar effort, the Automatic Plate Measuring machine galaxy catalog was used to build a sample of $\sim$1000 clusters (Maddox et al. 1990, Dalton et al. 1997).

Projection effects in the selection of cluster candidates have been much debated. Filamentary structures and small groups along the line of sight can mimic a moderately rich cluster when projected onto the plane of the sky. In addition, the background galaxy distribution against which two-dimensional overdensities are selected is far from uniform. As a result, the background subtraction process can produce spurious low-richness clusters during searches for clusters in galaxy catalogs. N-body simulations have been extensively used to build mock galaxy catalogs from which the completeness and spurious fraction of Abell-like samples of clusters can be assessed (e.g., van Haarlem et al. 1997). All-sky, X-ray selected surveys have significantly alleviated these problems and fueled significant progress in this field, as discussed below.

Optical plate material deeper than the POSS was successfully employed to search for more distant clusters with purely visual techniques (Kristien et al. 1978, Couch et al. 1991, Gunn et al. 1986). By using red-sensitive plates, Gunn and collaborators were able to find clusters out to $z \simeq 0.9$. These searches became much more effective with the advent of charge coupled device (CCD) imaging. Postman et al. (1996) were the first to carry out a V&I-band survey over 5 $\deg^2$ (the Palomar Distant Cluster Survey, PDCS) and to compile a sample of 79 cluster candidates using a matched-filter algorithm. This technique enhances the contrast of galaxy overdensities at a given position, utilizing prior knowledge of the luminosity profile typical of galaxy clusters. Olsen et al. (1999) used a similar algorithm to select a sample of 35 distant cluster candidates from the European Southern Observatory Imaging Survey (EIS) I-band data. A simple and equally effective counts-in-cell method was used by Lidman & Peterson (1996) to select a sample of 104 distant cluster candidates over 13 $\deg^2$. All these surveys, by using relatively deep I-band data, are sensitive to rich clusters out to $z \sim 1$. A detailed spectroscopic study of one of the most distant clusters at $z = 0.89$ discovered in this way is reported in Lubin et al. (2000).

Dalcanton (1996) proposed another method of optical selection of clusters, in which drift scan imaging data from relatively small telescopes is used to detect clusters as positive surface brightness fluctuations in the background sky. Gonzales et al. (2001) used this technique to build a sample of $\sim$1000 cluster candidates over 130 $\deg^2$. Spectroscopic follow-up observations will assess the efficiency of this technique.

The advantage of carrying out automated searches based on well-defined selection criteria (e.g., Postman et al. 1996) is that the survey selection function can be computed, thus enabling meaningful statistical studies of the cluster population. For example, one can quantify the probability of detecting a galaxy cluster as a function of redshift for a given set of other parameters, such as galaxy luminosity function, luminosity profile, luminosity and color evolution of cluster galaxies, and field galaxy number counts. A comprehensive report on the performance of different cluster detection algorithms applied to two-dimensional projected distributions can be found in Kim et al. (2002).

The success rate of finding real bound systems in optical surveys is generally relatively high at low redshift ($z < 0.3$) (Holden et al. 1999), but it degrades rapidly at higher redshifts, particularly if only one passband is used, as the field galaxy population overwhelms galaxy overdensities associated with clusters. The simplest way to counteract this effect is to observe in the near-infrared bands ($\gtrsim 1\ \mu m$). The cores of galaxy clusters are dominated by red, early-type galaxies at least out to $z \simeq 1.3$, for which the dimming effect of the K-correction is particularly severe. In addition, the number counts of the field galaxy population are flatter in the near-IR bands than in the optical. Thus, by moving to z, J, H, and K bands, one can progressively compensate the strong K-correction and enhance the contrast of (red) cluster galaxies against the background (blue) galaxy distribution. An even more effective way to enhance the contrast of distant clusters is to use some color information, so that only overdensities of galaxies with peculiar red colors can be selected from the field. With a set of two or three broad band filters, which sample the rest frame UV and optical light at different redshifts, one can separate out early-type galaxies that dominate cluster cores from the late-type galaxy population in the field. The position of the cluster red sequence in color-magnitude diagrams and red clumps in color-color diagrams can also be used to provide an accurate estimate of the cluster redshift by modeling the relatively simple evolutionary history of early-type galaxies.

The effectiveness of this method was clearly demonstrated by Stanford et al. (1997), who found a significant overdensity of red galaxies with $J - K$ and $I - K$ colors typical of $z > 1$ ellipticals and were able to spectroscopically confirm this system as a cluster at $z = 1.27$ (see also Dickinson 1997). With a similar color enhancement technique and follow-up spectroscopy, Rosati et al. (1999) confirmed the existence of an X-ray–selected cluster at $z = 1.26$. Gladders & Yee (2000) applied the same technique in a systematic fashion to carry out a large area survey in R and z bands (the Red Sequence Survey), which is currently under way and promises to unveil rare, very massive clusters out to $z \sim 1$.

By increasing the number of observed passbands, one can further increase the efficiency of cluster selection and the accuracy of their estimated redshifts. In this respect, a significant step forward in mapping clusters in the local Universe will be made with the five-band photometry provided by the Sloan Digital Sky Survey (SDSS, York et al. 2000). The data will allow clusters to be efficiently selected with photometric redshift techniques and will ultimately allow hundreds of clusters to be searched directly in redshift space. The next generation of wide field (> 100 deg^2) deep multicolor surveys in the optical and especially the near-infrared will powerfully enhance the search for distant clusters.

X-Ray Cluster Surveys

The *Uhuru* X-ray satellite, which carried out the first X-ray sky survey (Giacconi et al. 1972), revealed a clear association between rich clusters and bright X-ray sources (Gursky et al. 1971, Kellogg et al. 1971). *Uhuru* observations also established that X-ray sources identified as clusters were among the most luminous in the sky (10^{43-45} erg s^{-1}), were extended, and showed no variability. Felten et al. (1966) first suggested that the X-ray originated as thermal emission from diffuse hot intracluster gas (Cavaliere et al. 1971). This was later confirmed when the

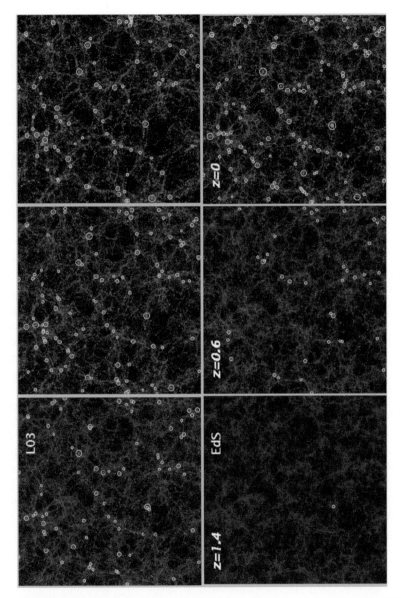

Figure 1 The evolution of the cluster population from N-body simulations in two different cosmologies (from Borgani & Guzzo 2001). Left panels describe a flat, low-density model with $\Omega_m = 0.3$ and $\Omega_\Lambda = 0.7$ (L03); right panels are for an Einstein–de-Sitter model (EdS) with $\Omega_m = 1$. Superimposed on the dark matter distribution, the yellow circles mark the positions of galaxy clusters with virial temperature $T > 3$ keV, the size of the circles is proportional to temperature. Model parameters have been chosen to yield a comparable space density of nearby clusters. Each snapshot is $250h^{-1}$ Mpc across and $75h^{-1}$ Mpc thick (comoving with the cosmic expansion).

Figure 3 Map of gas entropy from hydrodynamical simulations of a galaxy cluster (from Borgani et al. 2001a). (*Left*) gravitational heating only. (*Right*) entropy floor of 50 keV/cm² imposed at $z = 3$, corresponding to about 1 keV/part. Light colors correspond to low entropy particles, and dark blue corresponds to high-entropy gas.

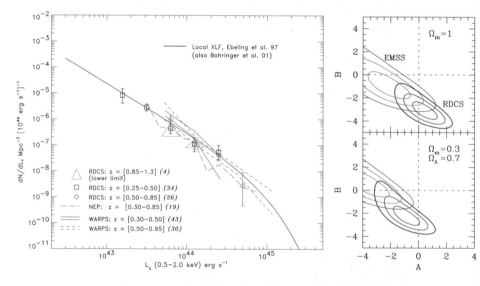

Figure 9 (*Left*) The latest compilation of distant X-ray luminosity functions (XLFs), [*ROSAT* Deep Cluster Survey (RDCS), Rosati et al. 2000; North Ecliptic Pole (NEP), Gioia et al. 2001; Wide Angle (*ROSAT*) Pointed X-Ray Survey (WARPS), Jones et al. 2000; an Einstein–de-Sitter universe with $H_0 = 50$ $250h^{-1}$ Mpc is adopted]. (*Right*) Maximum-likelihood contours (1, 2, and 3 σ confidence level) for the parameters A and B defining the XLF evolution for the RDCS and Extended Medium Sensitivity Survey (EMSS) samples (for two different cosmologies): $\phi^* = \phi_0 (1 + z)^A$, $L^* = L_0^*(1 + z)^B$ (see Equation \ref{eq:xlf}).

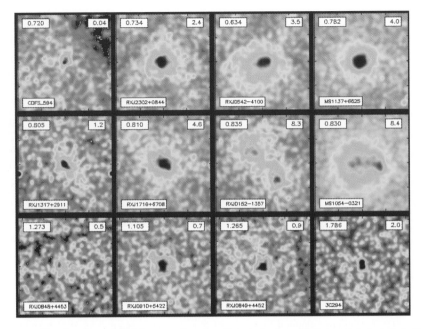

Figure 10 *Chandra* archival images of twelve distant clusters at $0.7 < z < 1.3$. Labels indicate redshifts (*upper left*) and X-ray luminosities (*upper right*) in the rest frame (0.5-2) keV band. All fields are 2 Mpc across; the X-ray emission has been smoothed at the same physical scale of 70 kpc ($h = 0.7$, $\Omega_m = 0.3$, $\Omega_\Lambda = 0.7$).

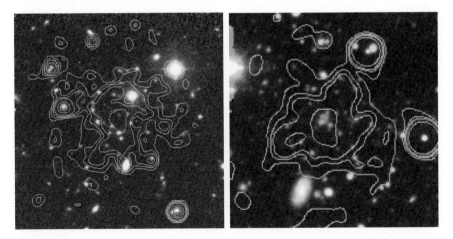

Figure 11 Color composite images combining optical and near-IR imaging of two X-ray–selected clusters at $z > 1$. Overlaid contours map the X-ray mission detected by *Chandra/ACIS-I*. (*Left*) RXJ0910+5422 at $z = 1.11$ (Stanford et al. 2002); (*right*) RXJ0849+4452 at $z = 1.26$ (Rosati et al. 1999, Stanford et al. 2001a). The two fields are 1.5 arcmin across ($\simeq 1h_{50}^{-1} Mpc$ at these redshifts).

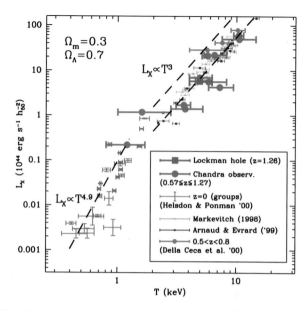

Figure 14 The (bolometric) luminosity-temperature relation for nearby and distant clusters and groups compiled from several sources (see Borgani et al. 2001b, Holden et al. 2002). The two dashed lines at $T > 2$ keV indicate the slope $\alpha = 3$, with normalization corresponding to the local L_x–T relation (*lower line*) and to the relation of Equation \ref{eq:lt} computed at $z = 1$ for $A = 1$. The dashed line at $T < 1$ keV shows the best-fitting relation found for groups by Helsdon & Ponman (2000).

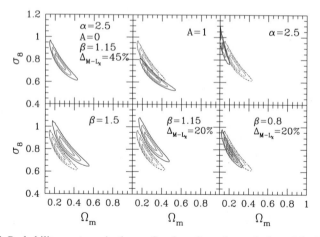

Figure 15 Probability contours in the σ_8–Ω_m plane from the evolution of the X-ray luminosity distribution of RDCS clusters. The shape of the power spectrum is fixed to $\Gamma = 0.2$. Different panels refer to different ways of changing the relation between cluster virial mass, M, and X-ray luminosity, L_x, within theoretical and observational uncertainties (see also Borgani et al. 2001b). The upper left panel shows the analysis corresponding to the choice of a reference parameter set. In each panel, we indicate the parameters that are varied, with the dotted contours always showing the reference analysis.

first high-quality X-ray spectra of clusters were obtained with the HEAO-1 A2 experiment (e.g., Henriksen & Mushotzsky 1986). These spectra were best fit by a thermal bremsstrahlung model, with temperatures in the range of $2 \times 10^7 - 10^8$ keV, and revealed the 6.8 keV iron K α line, thus showing that the ICM was a highly ionized plasma pre-enriched by stellar processes.

The HEAO-1 X-ray Observatory (Rothschild et al. 1979) performed an all-sky survey with much improved sensitivity compared with *Uhuru*, and provided the first flux-limited sample of extragalactic X-ray sources in the 2–10 keV band, with a limiting flux of 3×10^{-11} erg cm^{-2} s^{-1} (Piccinotti et al. 1982). Among the 61 extragalactic sources discovered outside the galactic plane ($|b| > 20°$), 30 were identified as galaxy clusters, mostly in the Abell catalog. This first X-ray flux-limited sample allowed an estimate of the cluster X-ray luminosity function (XLF) in the range of $L_X = 10^{43} - 3 \cdot 10^{45}$ erg s^{-1}. The derived space density of clusters (all at $z < 0.1$) is fairly close to current values. An earlier determination of the XLF based on optically selected Abell clusters (McKee et al. 1980) and the same HEAO-1 A2 data gave similar results.

The Piccinotti et al. sample was later augmented by Edge et al. (1990), who extended the sample using the *Ariel V* catalog (McHardy et al. 1981) and revised the identifications of several clusters using follow-up observations by the *Einstein* and *EXOSAT* satellites. With much improved angular resolution, these new X-ray missions allowed confused sources to be resolved and fluxes to be improved. The resulting sample included 55 clusters with a flux limit a factor of two fainter than in the original Piccinotti catalog.

Confusion effects in the large beam ($\gtrsim 1°$) early surveys, such as *HEAO-1* and *Ariel V*, had been the main limiting factor in cluster identification. With the advent of X-ray imaging with focusing optics in the 1980s, particularly with the *Einstein Observatory* (Giacconi et al. 1979), it was soon recognized that X-ray surveys offer an efficient means of constructing samples of galaxy clusters out to cosmologically interesting redshifts.

First, the X-ray selection has the advantage of revealing physically bound systems, because diffuse emission from a hot ICM is the direct manifestation of the existence of a potential well within which the gas is in dynamical equilibrium with the cool baryonic matter (galaxies) and the dark matter. Second, the X-ray luminosity is well correlated with the cluster mass (see right panel of Figure 2). Third, the X-ray emissivity is proportional to the square of the gas density ("Physical Properties of Galaxy Clusters," above); hence, cluster emission is more concentrated than the optical bidimensional galaxy distribution. In combination with the relatively low surface density of X-ray sources, this property makes clusters high-contrast objects in the X-ray sky and alleviates problems due to projection effects that affect optical selection. Finally, an inherent fundamental advantage of X-ray selection is the ability to define flux-limited samples with well-understood selection functions. This leads to a simple evaluation of the survey volume and therefore to a straightforward computation of space densities. Nonetheless, there are some important caveats described below.

Pioneering work in this field was carried out by Gioia et al. (1990a) and Henry et al. (1992) based on the *Einstein Observatory* Extended Medium Sensitivity Survey (EMSS) (Gioia et al. 1990b). The EMSS survey covered over 700 square degrees using 1435 imaging proportional counter (IPC) fields. A highly complete spectroscopic identification of 835 serendipitous sources led to the construction of a flux-limited sample of 93 clusters out to $z = 0.58$. By significantly extending the redshift range probed by previous samples (e.g., Edge et al. 1990), the EMSS allowed the cosmological evolution of clusters to be investigated. Several follow-up studies have been undertaken such as the CNOC survey (e.g., Yee et al. 1996) and gravitational lensing (Gioia & Luppino 1994).

The *ROSAT* satellite, launched in 1990, allowed a significant step forward in X-ray surveys of clusters. The *ROSAT-PSPC* ("Position Sensitive Proportional Counter") detector, in particular, with its unprecedented sensitivity and spatial resolution, as well as low instrumental background, made clusters high-contrast extended objects in the X-ray sky. The *ROSAT* All-Sky Survey (RASS) (Trümper 1993) was the first X-ray imaging mission to cover the entire sky, thus paving the way to large contiguous-area surveys of X-ray–selected nearby clusters (e.g., Ebeling et al. 1997, 1998, 2000, 2001; Burns et al. 1996; Crawford et al. 1995; De Grandi et al. 1999; Böhringer et al. 2000, 2001). In the northern hemisphere the largest compilations with virtually complete optical identification include the Bright Cluster Sample (BCS, Ebeling et al. 1998), its extension (Ebeling et al. 2000b), and the Northern *ROSAT* All-Sky Survey (NORAS, Böhringer et al. 2000). In the southern hemisphere the *ROSAT*-ESO flux limited X-ray (REFLEX) cluster survey (Böhringer et al. 2001) has completed the identification of 452 clusters, the largest, homogeneous compilation to date. Another ongoing study, the Massive Cluster Survey (MACS, Ebeling et al. 2001) is aimed at targeting the most luminous systems at $z > 0.3$ that can be identified in the RASS at the faintest flux levels. The deepest area in the RASS, the North Ecliptic Pole (NEP, Henry et al. 2001), which *ROSAT* scanned repeatedly during its All-Sky Survey, was used to carry out a complete optical identification of X-ray sources over an 81 deg^2 region. This study yielded 64 clusters out to redshift $z = 0.81$.

In total, surveys covering more than 10^4 deg^2 have yielded over 1000 clusters, out to redshift $z \simeq 0.5$. A large fraction of these are new discoveries, whereas approximately one third are identified as clusters in the Abell or Zwicky catalogs. For the homogeneity of their selection and the high degree of completeness of their spectroscopic identifications, these samples are now becoming the basis for a large number of follow-up investigations and cosmological studies.

After the completion of the all-sky survey, *ROSAT* conducted thousands of pointed observations, many of which (typically those outside the galactic plane not targeting very bright or extended X-ray sources) can be used for a serendipitous search for distant clusters. It was soon realized that the good angular resolution of the *ROSAT-PSPC* allowed screening of thousands of serendipitous sources and the selection of cluster candidates solely on the basis of their flux and spatial extent. In the central 0.2 deg^2 of the *PSPC* field of view, the point spread function (PSF) is well approximated by a Gaussian with FWHM = 30–45″. Therefore a cluster with a canonical core radius of 250 h^{-1} kpc (Forman & Jones 1992) should be resolved out to $z \sim 1$, as the corresponding angular distance always exceeds 45″ for current values of cosmological parameters (important surface brightness biases are discussed below).

ROSAT-PSPC archival pointed observations were intensively used for serendipitous searches of distant clusters. These projects, which are now completed or nearing completion, include the RIXOS survey (Castander et al. 1995), the *ROSAT* Deep Cluster Survey (RDCS, Rosati et al. 1995, 1998), the Serendipitous High-Redshift Archival *ROSAT* Cluster survey (SHARC, Collins et al. 1997, Burke et al. 1997), the Wide Angle *ROSAT* Pointed X-Ray Survey of clusters (WARPS, Scharf et al. 1997, Jones et al. 1998, Perlman et al. 2002), the 160 deg^2 large area survey (Vikhlinin et al. 1998b), and the *ROSAT* Optical X-Ray Survey (ROXS, Donahue et al. 2001). *ROSAT*-HRI (High Reolution Imager) pointed observations, which are characterized by a better angular resolution, although with higher instrumental background, have also been used to search for distant clusters in the Brera Multiscale Wavelet (BMW) catalog (Campana et al. 1999).

A principal objective of all these surveys has been the study of the cosmological evolution of the space density of clusters. Results are discussed in "The Space Density of X-Ray Clusters" and "Cosmology with X-Ray Clusters," below. In Figure 4 we give an overview of the flux limits and surveyed areas of all major cluster surveys carried out over the past two decades. RASS-based surveys have the advantage of covering contiguous regions of the sky so that the clustering properties of clusters (e.g., Collins et al. 2000, Mullis et al. 2001) and the power spectrum of their distribution (Schücker et al. 2001a) can be investigated. They also have the ability to unveil rare, massive systems, albeit over a limited redshift and X-ray luminosity range. Serendipitous surveys, or general surveys, which are at least a factor of ten deeper but cover only a few hundreds square degrees, provide complementary information on lower luminosities and more common systems and are well suited for studying cluster evolution on a larger redshift baseline. The deepest pencil-beam surveys, such as the Lockman Hole with *XMM* (Hasinger et al. 2000) and the Chandra Deep Fields (Giacconi et al. 2002, Bauer et al. 2002), allow the investigation of the faintest end of the XLF (poor clusters and groups) out to $z \sim 1$.

Strategies and Selection Functions for X-Ray Surveys

Ideally, one would like to use selection criteria based on X-ray properties alone to construct a flux-limited sample with a simple selection function. The task of separating clusters from the rest of the X-ray source population is central to this work. At the *ROSAT* flux limit ($\sim 1 \times 10^{-14}$ erg cm^{-2} s^{-1} for clusters) $\sim$10% of extragalactic X-ray sources are galaxy clusters. A program of complete optical identification is very time-consuming, as only spectroscopy can establish in many cases whether the X-ray source is associated with a real cluster. The EMSS and NEP samples, for example, were constructed in this way. In some cases the hardness

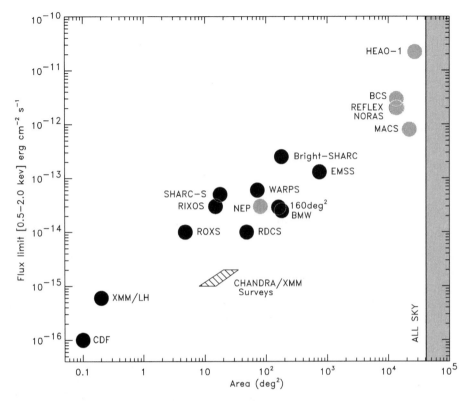

Figure 4   Solid angles and flux limits of X-ray cluster surveys carried out over the past two decades. References are given in the text. Dark filled circles represent serendipitous surveys constructed from a collection of pointed observations. Light shaded circles represent surveys covering contiguous areas. The hatched region is a predicted locus of future serendipitous surveys with *Chandra* and *Newton-XMM*.

ratio (a crude estimate of the source's X-ray spectral energy distribution) is used to screen out sources that are incompatible with thermal spectra or to resolve source blends. With the angular resolution provided by *ROSAT*, however, it became possible to select clusters on the basis of their spatial extent. This is particularly feasible with pointed observations, as opposed to all-sky survey data, which are characterized by a broader PSF and shallower exposures, so that faint and/or high redshift clusters are not always detected as extended (e.g., Ebeling et al. 1997, Böhringer et al. 2001).

In constructing RASS-based samples (shaded circles in Figure 4) most of the authors had to undertake a complete optical identification program of $\sim10^4$ sources using POSS plates or CCD follow-up imaging to build a sample of cluster candidates. Whereas a sizable fraction of these systems can be readily identified in previous cluster catalogs (primarily Abell's), spectroscopy is needed to measure redshifts of newly discovered systems or to resolve ambiguous identifications. We recall that optically selected, X-ray confirmed samples, such as the X-ray Brightest Abell-like Clusters (XBACS, Ebeling et al. 1996), while useful for studying optical–X-ray correlations, lead to incomplete flux-limited samples. Many of the low X-ray luminosity systems (poor clusters or groups) are missed in the optical selection even though they lie above the X-ray flux limit of the RASS.

Most of the *ROSAT* serendipitous surveys (dark circles in Figure 4) have adopted a very similar methodology but somewhat different identification strategies. Cluster candidates are selected from a serendipitous search for extended X-ray sources above a given flux limit in deep *ROSAT*-PSPC pointed observations. Moderately deep CCD imaging in red passbands (or in near-IR for the most distant candidates) is used to reveal galaxy overdensities near the centroid of X-ray emission. Extensive spectroscopic follow-up programs associated with these surveys have led to the identification of roughly 200 new clusters or groups and have increased the number of clusters known at $z > 0.5$ by approximately a factor of ten.

An essential ingredient for the evaluation of the selection function of X-ray surveys is the computation of the sky coverage: the effective area covered by the survey as a function of flux. In general, the exposure time, as well as the background and the PSF, are not uniform across the field of view of X-ray telescopes (owing to their inherent optical design), which introduces vignetting and a degradation of the PSF at increasing off-axis angles. As a result, the sensitivity to source detection varies significantly across the survey area so that only bright sources can be detected over the entire solid angle of the survey, whereas at faint fluxes the effective area decreases. An example of survey sky coverage is given in Figure 5

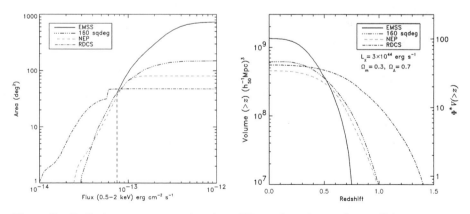

Figure 5 (*Left*) sky coverage as a function of X-ray flux of several serendipitous surveys. (*Right*) Corresponding search volumes, $V(>z)$, for a cluster of given X-ray luminosity ($L_X = 3 \times 10^{44}$ erg s^{-1} [0.5–2 keV] $\simeq L_X^*$). On the right axis the volume is normalized to the local space density of clusters, ϕ^*.

(*left*). By integrating the volume element of the Friedmann-Robertson-Walker metric, $dV/d\Omega\, dz(z, \Omega_m, \Omega_\Lambda)$ (e.g., Carroll et al. 1992), over these curves, one can compute the volume that each survey probes above a given redshift z, for a given X-ray luminosity ($L_X = 3 \times 10^{44}$ erg s^{-1} $\simeq L_X^*$, the characteristic luminosity, in the figure). The resulting survey volumes are shown in Figure 5 (*right*). By normalizing this volume to the local space density of clusters (ϕ^*, see below) one obtains the number of L^* volumes accessible in the survey above a given redshift. Assuming no evolution, this yields an estimate of the number of typical bright clusters one expects to discover.

By covering different solid angles at varying fluxes, these surveys probe different volumes at increasing redshift and therefore different ranges in X-ray luminosities at varying redshifts. The EMSS has the greatest sensitivity to the most luminous, yet rare, systems, but only a few clusters at high redshift lie above its bright flux limit. Deep *ROSAT* surveys probe instead the intermediate-to-faint end of the XLF. As a result, they have led to the discovery of many new clusters at $z > 0.4$. The RDCS has pushed this search to the faintest fluxes yet, providing sensitivity to the highest redshift systems with $L_X \lesssim L_X^*$ even beyond $z = 1$. The WARPS and particularly the 160 deg^2 survey have covered larger areas at high fluxes, thus better studying the bright end of the XLF out to $z \simeq 1$.

Particular emphasis is given in these searches to detection algorithms that are designed to examine a broad range of cluster parameters (X-ray flux, surface brightness, morphology) and to deal with source confusion at faint flux levels. The traditional detection algorithm used in X-ray astronomy for many years, the sliding cell method, is not adequate for this purpose. A box of fixed size is slid across the image, and sources are detected as positive fluctuations that deviate significantly from Poissonian expectations based on a global background map (the latter being constructed from a first scan of the image). Although this method works well for point-like sources, it is less suited to extended, low-surface brightness sources, which can consequently be missed, leading to a significant incompleteness in flux-limited cluster samples.

The need for more general detection algorithms, not only geared to the detection of point sources, became important with *ROSAT* observations, which probe a much larger range in surface brightness than previous missions (e.g., *Einstein*). A popular alternative approach to source detection and characterization developed specifically for cluster surveys is based on wavelet techniques (e.g., Rosati et al. 1995, Vikhlinin et al. 1998b, Lazzati et al. 1999, Romer et al. 2000). Wavelet analysis is essentially a multiscale analysis of the image based on an quasi-orthonormal decomposition of a signal via the wavelet transform, which enables significant enhancement of the contrast of sources of different sizes against nonuniform backgrounds. This method, besides being equally efficient at detecting sources of different shapes and surface brightnesses, is well suited to dealing with confusion effects and allows source parameters to be measured without knowledge of the background. Another method that has proved to be well suited for the detection of extended– and low–surface brightness emission is based on Voronoi Tessellation and Percolation (Scharf et al. 1997 and references therein).

Besides detection algorithms, which play a central role in avoiding selection effects, there are additional caveats to be considered when computing the selection function of X-ray cluster surveys. For example, the sky coverage function (Figure 5) depends not only on the source flux but in general on the extent or surface brightness of cluster sources (Rosati et al. 1995, Scharf et al. 1997, Vihklinin et al. 1998). This effect can be tested with extensive simulations by placing artificial clusters (typically using β-profiles) in the field and measuring the detection probability for different cluster parameters or instrumental parameters.

More generally, as in all flux-limited samples of extended sources (e.g., optical galaxy surveys), one has to make sure the sample does not become surface-brightness (SB) limited at very faint fluxes. As the source flux decreases, clusters with smaller mean SB have a higher chance of being missed because their signal-to-noise ratio is likely to drop below the detection threshold. SB dimming at high redshifts (SB $\propto (1 + z)^{-4}$) can thus create a serious source of incompleteness at the faintest flux levels. This depends critically on the steepness of the SB profile of distant X-ray clusters and its evolution. Besides simulations of the detection process, the most meaningful way to test these selection effects is to verify that derived cluster surface or space densities do not show any trend across the survey area (e.g., a decrease in regions with higher background, low exposures, degraded PSF). The task of the observer is to understand what is the fiducial flux limit above which the sample is truly flux-limited and free of SB effects. This fiducial flux limit is typically a factor of 2–3 higher than the minimum detectable flux in a given survey.

An additional source of sample contamination or misidentification may be caused by clusters hosting X-ray bright AGN or by unrelated point sources projected along the line of sight of diffuse cluster emission. The former case does not seem to be a matter of great concern because bright AGN have been found near the center of clusters in large compilations (Böhringer et al. 2001) in less than 5% of the cases. However, the latter effect can be significant in distant and faint *ROSAT* selected clusters, for which high resolution *Chandra* observations (Stanford et al. 2001, 2002) have revealed up to 50% flux contamination in some cases.

Concerning selection biases, a separate issue is whether using X-ray selection, one might miss systems that, although virialized, have an unusually low X-ray luminosity. These systems would be outliers in the $L_X - M$ or $L_X - T$ relation ("Deriving Ω_m from Cluster Evolution," below). Such hypothetical systems are at odds with our physical understanding of structure formation and would require unusual mechanisms that would (*a*) lead galaxies to virialize but the gaseous component not to thermalize in the dark matter potential well, (*b*) allow energy sources to dissipate or remove the gas after collapse, or (*c*) involve formation scenarios in which only a small fraction of the gas collapses. Similarly, systems claimed to have an unusually high mass-to-optical luminosity ratio, M/L, such as MG2016 + 112 from *ASCA* observations (Hattori et al. 1998), have not held up. MG2016 + 112 was later confirmed to be an ordinary low-mass cluster at $z = 1$ by means of near-infrared imaging (Benítez et al. 1999) and spectroscopic (Soucail et al. 2001) follow-up studies. Chartas et al. (2001) have completely revised the nature of the X-ray emission with *Chandra* observations. Comparing optical and X-ray techniques for cluster detection, Donahue et al. (2001) carried out an optical/X-ray joint survey in the same sky area (ROXS). They found no need to invoke an X-ray faint population of massive clusters.

Other Methods

X-ray and optical surveys have been by far the most exploited techniques for studying the distribution and evolution of galaxy clusters. It is beyond the scope of this paper to review other cluster-finding methods, which we only summarize here for completeness:

- *Search for galaxy overdensities around high-z radio galaxies or AGN*   Searches are conducted in near-IR or narrow-band filters or by means of follow-up X-ray observations. Although not suited for assessing cluster abundances, this method has provided the only examples of possibly virialized systems at $z > 1.5$ (e.g., Pascarelle et al. 1996, Dickinson 1997, Crawford & Fabian 1996, Hall & Green 1998, Pentericci et al. 2000, Fabian et al. 2001b, Venemans et al. 2002).

- *Sunyaev-Zeldovich (SZ) effect*   Clusters are revealed by measuring the distortion of the cosmic microwave background (CMB) spectrum owing to the hot ICM. This method does not depend on redshift and provides reliable estimates of cluster masses. It will possibly be one of the most powerful methods to find distant clusters in the years to come. At present, serendipitous surveys with interferometric techniques (e.g., Carlstrom et al. 2001) cannot cover large areas (i.e., more than ~ 1 deg^2), and their sensitivity is limited to the most X-ray–luminous clusters.

- *Gravitational lensing*   In principle this is a powerful method to discover mass concentrations in the universe through the statistical distortion of background galaxy images (see Mellier 1999 for a review).

- *Search for clusters around bent-double radio sources*   Radio galaxies with bent lobes are often associated with dense ICM and are therefore good tracers of rich cluster environments (e.g., Blanton et al. 2001).

- *Clustering of absorption line systems*   This method has led to a few detections of "proto-clusters" at $z \gtrsim 2$ (e.g., Francis et al. 1996). The most serious limitation of this technique is the small sample volume.

THE SPACE DENSITY OF X-RAY CLUSTERS

Local Cluster Number Density

The determination of the local ($z \lesssim 0.3$) cluster abundance plays a crucial role in assessing the evolution of the cluster abundance at higher redshifts. The cluster

XLF is commonly modeled with a Schechter function:

$$\phi(L_X)dL_X = \phi^* \left(\frac{L_X}{L_X^*}\right)^{-\alpha} \exp\left(-L_X/L_X^*\right) \frac{dL_X}{L_X^*}, \tag{7}$$

where α is the faint-end slope, L_X^* is the characteristic luminosity, and ϕ^* is directly related to the space-density of clusters brighter than L_{min}: $n_0 = \int_{L_{min}}^{\infty} \phi(L)dL$. The cluster XLF in the literature is often written as: $\phi(L_{44}) = K \exp(-L_X/L_X^*)L_{44}^{-\alpha}$, with $L_{44} = L_X/10^{44}$ erg s^{-1}. The normalization K, expressed in units of 10^{-7} Mpc^{-3} (10^{44} erg s^{-1})$^{\alpha-1}$, is related to ϕ^* by $\phi^* = K (L_X^*/10^{44})^{1-\alpha}$.

Using a flux-limited cluster sample with measured redshifts and luminosities, a binned representation of the XLF can be obtained by adding the contribution to the space density of each cluster in a given luminosity bin ΔL_X:

$$\phi(L_X) = \left(\frac{1}{\Delta L_X}\right) \sum_{i=1}^{n} \frac{1}{V_{max}(L_i, f_{lim})}, \tag{8}$$

where V_{max} is the total search volume defined as

$$V_{max} = \int_0^{z_{max}} S[f(L, z)] \left(\frac{d_L(z)}{1+z}\right)^2 \frac{cdz}{H(z)}. \tag{9}$$

$S(f)$ is the survey sky coverage, which depends on the flux $f = L/(4\pi d_L^2)$, $d_L(z)$ is the luminosity distance, and $H(z)$ is the Hubble constant at z (e.g., Peebles 1993, p. 312). We define z_{max} as the maximum redshift out to which the object is included in the survey. Equations 8 and 9 can be easily generalized to compute the XLF in different redshift bins.

In Figure 6 we summarize the progress made in computing $\phi(L_X)$ using primarily low-redshift *ROSAT*-based surveys. This work improved the first determination of the cluster XLF (Piccinotti et al. 1982) (see "X-Ray Cluster Surveys," above). The BCS and REFLEX cover a large L_X range and have good statistics at the bright end, $L_X \gtrsim L_X^*$ and near the knee of the XLF. Poor clusters and groups ($L_X \lesssim 10^{43}$ erg s^{-1}) are better studied using deeper surveys such as the RDCS. The very faint end of the XLF has been investigated using an optically selected, volume-complete sample of galaxy groups detected a posteriori in the RASS (Burns et al. 1996).

From Figure 6 we note the very good agreement among all these independent determinations. Best-fit parameters are consistent with each other with typical values $\alpha \simeq 1.8$ (with 15% variation), $\phi^* \simeq 1 \times 10^{-7} h_{50}^3$ Mpc^{-3} (with 50% variation), and $L_X^* \simeq 4 \times 10^{44}$ erg s^{-1} (0.5–2 keV). Residual differences at the faint end are probably the result of cosmic variance effects, because the lowest luminosity systems are detected at very low redshifts where the search volume becomes small (see Böhringer et al. 2002b). Such an overall agreement is quite remarkable considering that all these surveys used completely different selection techniques and independent datasets. Evidently, systematic effects associated with different selection functions are relatively small in current large cluster surveys. This situation

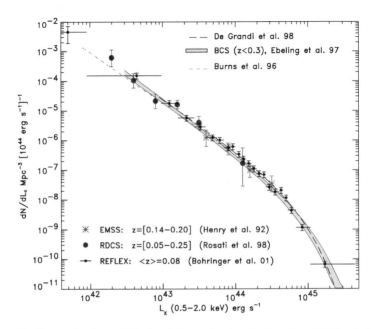

Figure 6   Determinations of the local X-ray luminosity function (XLF) of clusters from different samples (an Einstein–de-Sitter universe with $H_0 = 50 \, \text{km s}^{-1} \, \text{Mpc}^{-1}$ is adopted). For some of these surveys, only best fit curves to XLFs are shown.

is in contrast with that for the galaxy luminosity function in the nearby Universe, which is far from well established (Blanton et al. 2001). The observational study of cluster evolution has indeed several advantages with respect to galaxy evolution, despite its smaller number statistics. First, a robust determination of the local XLF eases the task of measuring cluster evolution. Second, X-ray spectra constitute a single parameter family based on temperature, and K-corrections are much easier to compute than in the case of different galaxy types in the optical bands.

The Cluster Abundance at Higher Redshifts and its Evolution

A first analysis of the EMSS cluster sample (Gioia et al. 1990a) revealed negative evolution of the XLF—a steepening of the high-end of XLF indicating a dearth of high luminosity clusters at $z > 0.3$. This result was confirmed by Henry et al. (1992) using the complete EMSS sample with an appropriate sky coverage function. Edge et al. (1990) found evidence of a strong negative evolution already at redshifts <0.2 using an HEAO-1–based cluster sample (see "X-Ray Cluster Surveys," above). The very limited redshift baseline made this result somewhat controversial until it was later ruled out by the analysis of the first RASS samples (Ebeling et al. 1997). The *ROSAT* deep surveys extended the EMSS study on cluster evolution. Early results (Castander et al. 1995) seemed to confirm and even to reinforce the

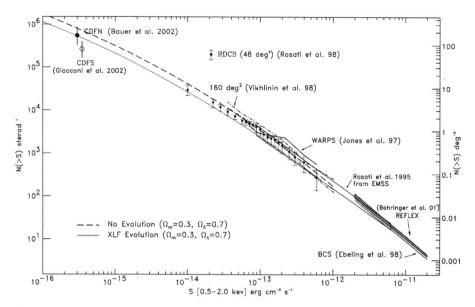

Figure 7   The cluster cumulative number counts as a function of X-ray flux (log N − log S) as measured from different surveys.

evidence of negative evolution. This claim, based on a sample of 12 clusters, was later recognized to be the result of sample incompleteness and an overestimate of the solid angle covered at low fluxes and its corresponding search volume (Burke et al. 1997, Rosati et al. 1998, Jones et al. 1998).

If cluster redshifts are not available, X-ray flux-limited samples can be used to trace the surface density of clusters at varying fluxes. In Figure 7 we show several determinations of the cumulative cluster number counts stretching over five decades in flux. This comparison shows a good agreement (at the 2σ level) among independent determinations (see also Gioia et al. 2001). The slope at bright fluxes is very close to the Euclidean value of 1.5 (as expected for a homogeneous distribution of objects over large scales), whereas it flattens to $\simeq 1$ at faint fluxes. The slope of the Log N − Log S is mainly determined by the faint-to-moderate part of the XLF, but it is rather insensitive to the abundance of the most luminous, rare systems. The fact that the observed counts are consistent with no-evolution predictions, obtained by integrating the local XLF, can be interpreted as an indication that a significant fraction of the cluster population does not evolve with redshift (Rosati et al. 1995, 1998; Jones et al. 1998, Vikhlinin et al. 1998a). We have included the recent data from the Chandra Deep Fields North (Bauer et al. 2002) and South (Giacconi et al. 2002), which have extended the number counts by two decades. Note that cosmic variance may be significant because these are only two, albeit deep, pencil beam fields ($\lesssim 0.1$ deg^2). Serendipitous surveys with *Chandra* and *XMM* (see Figure 4) will fill the gap between these measurements and the *ROSAT* surveys. The no-evolution curves in Figure 7 are computed by integrating the BCS local XLF (Ebeling et al. 1997) according to the evolutionary model in Figure 9 (see color insert).

A much improved picture of the evolution of the cluster abundance emerged when, with the completion of spectroscopic follow-up studies, several cluster samples were used to compute the XLF out to $z \simeq 0.8$. These first measurements are summarized in Figure 8. Although binned representations of the XLF are not straightforward to compare, it is evident that within the error bars there is little, if any, evolution of the cluster space density at $L_X([0.5–2] \text{ keV}) \lesssim 3 \times 10^{44} \text{ erg s}^{-1} \simeq L_X^*$ out to redshift $z \simeq 0.8$. These results (Burke et al. 1997, Rosati et al. 1998, Jones et al. 1998, Vikhlinin et al. 1998a, Nichols et al. 1999) extended the original study of EMSS to fainter luminosities and larger redshifts and essentially confirmed the EMSS findings in the overlapping X-ray luminosity range. The ability of all these surveys to adequately study the bright end of the XLF is rather limited because there is not enough volume to detect rare systems with $L_X > L_X^*$. The 160-deg^2 survey by Vikhlinin et al. (1998), with its large area, did however, confirm the negative evolution at $L_X \gtrsim 4 \times 10^{44} \text{ erg s}^{-1}$. Further analyses of these datasets have confirmed this trend, i.e., an apparent drop of super-L_X^* clusters at $z \gtrsim 0.5$ (Nichol et al. 1999 from the Bright–SHARC survey; Rosati et al. 2000 from RDCS; Gioia et al. 2001 from NEP). These findings, however, were not confirmed by Ebeling et al. (2000a) in an analysis of the WARPS sample.

The evolution of the bright end of the XLF has remained a hotly debated subject for several years. The crucial issue in this debate is to properly quantify the statistical significance of any claimed evolutionary effect. The binned representation of the XLF in Figure 8 can be misleading and can even lead to biases (Page & Carrera 2000). The full information contained in any flux-limited cluster sample can be more readily recovered by analyzing the unbinned (L_X, z) distribution with a maximum-likelihood approach, which compares the observed cluster distribution on the (L_X, z) plane with that expected from a given XLF model. Rosati et al. (2000) used this method by modeling the cluster XLF as an evolving Schechter function: $\phi(L) = \phi_0(1+z)^A L^{-\alpha} \exp(-L/L^*)$, with $L^* = L_0^*(1+z)^B$; where A and B are two evolutionary parameters for density and luminosity; ϕ_0 and L_0^* are the local XLF values (Equation 7). Figure 9 (see color insert) shows an application of this method to the RDCS and EMSS sample and indicates that the no-evolution case ($A = B = 0$) is excluded at more than 3σ levels in both samples when the most luminous systems are included in the analysis. However, the same analysis confined to clusters with $L_X < 3 \times 10^{44} \text{ erg s}^{-1}$ yields an XLF consistent with no evolution. In Figure 9 we also report the latest determinations of the XLF out to $z \sim 1$.

In summary, by combining all the results from *ROSAT* surveys, one obtains a consistent picture in which the comoving space density of the bulk of the cluster population is approximately constant out to $z \simeq 1$, but the most luminous ($L_X \gtrsim L_X^*$), presumably most massive clusters were likely rarer at high redshifts ($z \gtrsim 0.5$). Significant progress in the study of the evolution of the bright end of the

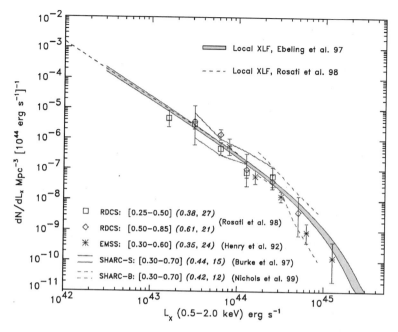

Figure 8 The X-ray luminosity function (XLF) of distant clusters out to $z \simeq 0.8$ compiled from various sources and compared with local XLFs (an Einstein–de-Sitter universe with $H_0 = 50 \, \text{km s}^{-1} \, \text{Mpc}^{-1}$ is adopted). Numbers in parenthesis give the median redshift and number of clusters in each redshift bin. RDCS, *ROSAT* Deep Cluster Survey; EMSS, Extended Medium Sensitivity Survey; SHARC, Serendipitous High-Redshift Archival *ROSAT* Cluster.

XLF would require a large solid angle and a relatively deep survey with an effective solid angle of $\gg 100 \, \text{deg}^2$ at a limiting flux of $10^{-14} \, \text{erg cm}^{-2} \, \text{s}^{-1}$.

The convergence of the results from several independent studies illustrates remarkable observational progress in determining the abundance of galaxy clusters out to $z \sim 1$. At the beginning of the *ROSAT* era, until the mid 1990s, controversy surrounded the usefulness of X-ray surveys of distant galaxy clusters, and many believed that clusters were absent at $z \sim 1$. This prejudice arose from an over-interpretation of the early results of the EMSS survey. Gioia et al. (1990a) did point out that the evolution of the XLF was limited only to the very luminous systems, but this important caveat was often overlooked. The original controversy concerning cluster evolution inferred from optical and X-ray data finds an explanation in light of the *ROSAT* results. Optical surveys (Couch et al. 1991, Postman et al. 1996) have shown no dramatic decline in the comoving volume density of rich clusters out to $z \simeq 0.5$. This was considered to be in contrast with the EMSS findings. However, these optical searches covered limited solid angles (much smaller than the EMSS) and therefore did not adequately probe the seemingly evolving high end of the cluster mass function.

Distant X-ray Clusters: The Latest View from *Chandra*

With its unprecedented angular resolution, the *Chandra* satellite has revolutionized X-ray astronomy, allowing studies with the same level of spatial details as in optical astronomy. *Chandra* imaging of low redshift clusters has revealed a complex thermodynamical structure of the ICM down to kiloparsec scales (e.g., Markevitch et al. 2000, Fabian et al. 2000). At high redshifts, deep *Chandra* images still have the ability to resolve cluster cores and to map ICM morphologies at scales below 100 kpc. Moreover, temperatures of major subclumps can be measured for the first time at $z > 0.6$.

Figure 10 (see color insert) is an illustrative example of the unprecedented view that *Chandra* can offer of distant clusters. We show 12 archival images of clusters at $0.7 < z < 1.7$, all covering 2 Mpc (projected at the cluster redshift) and smoothed at the same physical scale (a Gaussian FWHM of 70 kpc). Point-like sources in each field were removed. The intensity (in false colors) is proportional to the square root of the X-ray emission, so that they roughly map the gas density distribution in each cluster. The images are arranged in three redshift bins ($\sim$0.7, 0.8, >1) in each row, with X-ray luminosities increasing from left to right. The upper left image shows one of the highest redshift groups known to date, a system discovered in the megasecond exposure of the Chandra Deep Field South (Giacconi et al. 2002) with a core of a few arcseconds. A close inspection of these images reveals a deviation from spherical symmetry in all systems. Some of them are elongated or have cores clearly displaced with respect to the external diffuse envelope (e.g., Holden et al. 2002).

Three of the most luminous clusters at $z \simeq 0.8$ (RXJ1716: Gioia et al. 1999; RXJ0152: Della Ceca et al. 2000, Ebeling et al. 2000a; MS1054: Jeltema et al. 2001) show a double core structure both in the distribution of the gas and in their member galaxies. It is tempting to interpret these morphologies as the result of ongoing mergers, although no dynamical information has been gathered to date to support this scenario. In a hierarchical cold dark matter formation scenario, one does expect the most massive clusters at high redshift to be accreting subclumps of comparable masses and the level of substructure to increase at high redshifts. With current statistical samples, however, it is difficult to draw any robust conclusion about the evolution of ICM substructure, which is also found to be a large fraction of the low-z cluster population (e.g., Schücker et al. 2001b).

The third row in Figure 10 shows the most distant clusters observed with *Chandra* to date. The first three systems are also among the most distant X-ray–selected clusters discovered in the *ROSAT* Deep Cluster Survey (Stanford et al. 2001, 2002), at the very limit of the *ROSAT* sensitivity. RXJ0848 and RXJ0849 are only 5 arcmin apart on the sky (the Lynx field) and are possibly part of a superstructure at $z = 1.26$, consisting of two collapsed, likely virialized clusters (Rosati et al. 1999). Follow-up *Chandra* observations of the Lynx field (Stanford et al. 2001) have yielded for the first time information on ICM morphologies in $z > 1$ clusters and allowed a measurement of their temperatures (see Figure 14, color insert), implying masses of $(0.5-1) \times 10^{15} \, h_{50}^{-1} \, M_{\odot}$. The discovery and the study of these remote systems have the strongest leverage on testing cosmological models.

In Figure 11 (see color insert) we show color composite optical/near-IR images of two clusters at $z > 1$, with overlaid *Chandra* contours. Already at these large lookback times, the temperature and surface brightness profiles of these systems are similar to those of low redshift clusters. Moreover, the morphology of the gas, as traced by the X-ray emission, is well correlated with the spatial distribution of member galaxies, similar to studies at lower redshifts. This suggests that there are already at $z > 1$ galaxy clusters in an advanced dynamical stage of their formation, in which all the baryons (gas and galaxies) have had enough time to thermalize in the cluster potential well. Another example of a $z > 1$ cluster was reported by Hashimoto et al. (2002), using XMM observations of the Lockman Hole.

At $z > 1.3$, X-ray selection has not yielded any cluster based on *ROSAT* data. Follow-up X-ray observations of distant radio galaxies have been used to search for diffuse hot ICM (e.g., Crawford & Fabian 1996). A relatively short *Chandra* observation of the radio galaxy 3C294 at $z = 1.789$ (*bottom right*) in Figure 10 (Fabian et al. 2001b) has revealed an extended envelope around the central point source, which is the most distant ICM detected so far. Deeper observations are needed to accurately measure the temperature of this system.

COSMOLOGY WITH *X*-RAY CLUSTERS

The Cosmological Mass Function

The mass distribution of dark matter halos undergoing spherical collapse in the framework of hierarchical clustering is described by the Press-Schechter distribution (PS) (Press & Schechter 1974). The number of such halos in the mass range $[M, M + dM]$ can be written as

$$n(M, z) \, dM = \frac{\bar{\rho}}{M} f(\nu) \frac{d\nu}{dM} \, dM, \qquad (10)$$

where $\bar{\rho}$ is the cosmic mean density. The function f depends only on the variable $\nu = \delta_c(z)/\sigma_M$ and is normalized so that $\int f(\nu) \, d\nu = 1$. $\delta_c(z)$ is the linear-theory overdensity extrapolated to the present time for a uniform spherical fluctuation collapsing at redshift z. This quantity conveys information about the dynamics of fluctuation evolution in a generic Friedmann background. It is convenient to express it as $\delta_c(z) = \delta_0(z) \, [D(0)/D(z)]$, where $D(z)$ is the linear fluctuation growth factor, which depends on the density parameters contributed by matter, Ω_m, and by cosmological constant, Ω_Λ (e.g., Peebles 1993). The quantity $\delta_0(z)$ has a weak dependence on Ω_m and Ω_Λ (e.g., Kitayama & Suto 1997). For a critical-density Universe it is $\delta_0 = 1.686$, independent of z.

The r.m.s. density fluctuation at the mass scale M, σ_M, is connected to the fluctuation power spectrum, $P(k)$, by the relation

$$\sigma_M^2 = \frac{1}{2\pi^2} \int_0^\infty dk\, k^2\, P(k) W^2(kR).$$ (11)

The dependence of the power spectrum on the wavenumber k is usually written as $P(k) \propto k^{n_{pr}} T^2(k)$, where $T(k)$ is the transfer function, which depends both on the cosmological parameters of the Friedmann background and on the cosmic matter constituents (e.g., fraction of cold, hot, and baryonic matter; number of relativistic species) (see Kolb & Turner 1989). For a pure cold dark matter model $T(k)$ depends to a good approximation only on the shape parameter $\Gamma = \Omega_m h$ (e.g., Bardeen et al. 1986), whereas a correction to this dependence needs to be introduced to account for the presence of the baryonic component (e.g., Eisenstein & Hu 1999). The Harrison-Zel'dovich spectrum is generally assumed with the primordial index, $n_{pr} = 1$, consistent with the most recent analyses of the CMB anisotropies (de Bernardis et al. 2001 and references therein). The amplitude of $P(k)$ is usually expressed in terms of σ_8, the r.m.s. density fluctuation within a top-hat sphere of $8\,h^{-1}$ Mpc radius. Finally, in Equation 11 $W(x)$ is the Fourier representation of the window function, which describes the shape of the volume from which the collapsing object is accreting matter. The comoving fluctuation size R is connected to the mass scale M, as $R = (3M/4\pi \bar{\rho})^{1/3}$ for the top-hat window, i.e., $W(x) = 3(\sin x - x \cos x)/x^3$.

In their original derivation of the cosmological mass function, Press & Schechter (1974) obtained the expression $f(\nu) = (2\pi)^{-1/2} \exp(-\nu^2/2)$ for Gaussian density fluctuations. Despite its subtle simplicity (e.g., Monaco 1998), the PS mass function has served for more than a decade as a guide to constrain cosmological parameters from the mass function of galaxy clusters. Only with the advent of the last generation of N-body simulations, which are able to span a very large dynamical range, have significant deviations of the PS expression from the exact numerical description of gravitational clustering been noticed (e.g., Gross et al. 1998, Governato et al. 1999, Jenkins et al. 2001, Evrard et al. 2001). Such deviations are interpreted in terms of corrections to the PS approach. For example, incorporating the effects of nonspherical collapse (Sheth et al. 2001) generalizes the above PS expression for $f(\nu)$ to

$$f(\nu) = \sqrt{\frac{2a}{\pi}}\, C \left(1 + \frac{1}{(a\nu^2)^q} \right) \exp\left(-\frac{a\nu^2}{2} \right),$$ (12)

where $a = 0.707$, $C = 0.3222$, and $q = 0.3$ (Sheth & Tormen 1999). The above equation reduces to the PS expression for $a = 1$, $C = 1/2$, and $q = 0$. Fitting formulae for $f(\nu)$, which reproduce N-body results to an accuracy of about 10% (e.g., Evrard et al. 2001), are now currently used to derive cosmological constraints from the evolution of the cluster population.

In practical applications the observational mass function of clusters is usually determined over about one decade in mass. Therefore, it probes the power spectrum

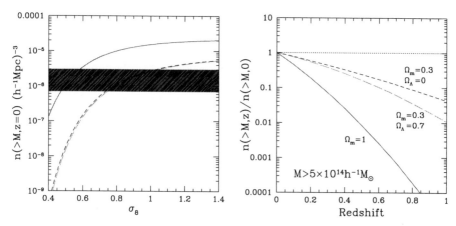

Figure 12 The sensitivity of the cluster mass function to cosmological models. (*Left*) The cumulative mass function at $z = 0$ for $M > 5 \times 10^{14} h^{-1} M_\odot$ for three cosmologies, as a function of σ_8, with shape parameter $\Gamma = 0.2$; solid line, $\Omega_m = 1$; short-dashed line, $\Omega_m = 0.3$, $\Omega_\Lambda = 0.7$; long-dashed line, $\Omega_m = 0.3$, $\Omega_\Lambda = 0$. The shaded area indicates the observational uncertainty in the determination of the local cluster space density. (*Right*) Evolution of $n(>M, z)$ for the same cosmologies and the same mass-limit, with $\sigma_8 = 0.5$ for the $\Omega_m = 1$ case and $\sigma_8 = 0.8$ for the low-density models.

over a relatively narrow dynamical range and does not provide strong constraints on the shape Γ of the power spectrum. Using only the number density of nearby clusters of a given mass M, one can constrain the amplitude of the density perturbation at the physical scale $R \propto (M / \Omega_m \rho_{crit})^{1/3}$ containing this mass. Because such a scale depends both on M and on Ω_m, the mass function of nearby ($z \lesssim 0.1$) clusters is only able to constrain a relation between σ_8 and Ω_m. In the left panel of Figure 12 we show that, for a fixed value of the observed cluster mass function, the implied value of σ_8 from Equation 12 increases as the density parameter decreases.

Determinations of the cluster mass function in the local Universe using a variety of samples and methods indicate that $\sigma_8 \Omega_m^\alpha = 0.4$–$0.6$ where $\alpha \simeq 0.4$–0.6, almost independent of the presence of a cosmological constant term providing spatial flatness (e.g., Bahcall & Cen 1993, Eke et al. 1996, Girardi et al. 1998, Viana & Liddle 1999, Blanchard et al. 2000, Pierpaoli et al. 2001, Reiprich & Böhringer 2002, Seljak 2002, Viana et al. 2002). It is worth pointing out that formal statistical uncertainties in the determination of σ_8 from the different analyses are always far smaller ($\lesssim 5\%$) than the above range of values. This suggests that current discrepancies on σ_8 are likely to be ascribed to systematic effects, such as sample selection and different methods used to infer cluster masses. We comment more on such differences in the following section. Completely independent constraints on a similar combination of σ_8 and Ω_m can be obtained with measurements of the cosmic gravitational lensing shear (e.g., Mellier 1999). The most recent results give $\sigma_8 \Omega_m^{0.6} = 0.45 \pm 0.05$ (van Waerbecke et al. 2001 and references therein).

The growth rate of the density perturbations depends primarily on Ω_m and, to a lesser extent, on Ω_Λ, at least out to $z \sim 1$, where the evolution of the cluster population is currently studied. Therefore, following the evolution of the cluster space density over a large redshift baseline, one can break the degeneracy between σ_8 and Ω_m. This is shown in a pictorial way in Figure 1 and quantified in the right panel of Figure 12: Models with different values of Ω_m, which are normalized to yield the same number density of nearby clusters, predict cumulative mass functions that progressively differ by up to orders of magnitude at increasing redshifts.

Deriving Ω_m from Cluster Evolution

An estimate of the cluster mass function is reduced to the measurement of masses for a sample of clusters, stretching over a large redshift range, for which the survey volume is well known.

Velocity dispersions for statistical samples of galaxy clusters have been provided by the ESO Nearby Abell Cluster Survey (ENACS; Mazure et al. 2001) and, more recently, by the 2dF survey (de Propris et al. 2002). Application of this method to a statistically complete sample of distant X-ray–selected clusters has been pursued by the CNOC (Canadian Network for Observational Cosmology) collaboration (e.g., Yee et al. 1996). The CNOC sample includes 16 clusters from the EMSS in the redshift range $0.17 \leq z \leq 0.55$. Approximately 100 redshifts of member galaxies were measured for each cluster, thus allowing an accurate analysis of the internal cluster dynamics (Carlberg et al. 1997b). The CNOC sample has been used to constrain Ω_m through the M/L_{opt} method (e.g., Carlberg et al. 1997b), yielding $\Omega_m \simeq 0.2 \pm 0.05$. Attempts to estimate the cluster mass function $n(>M)$ using the cumulative velocity dispersion distribution, $n(>\sigma_v)$ were made (Carlberg et al. 1997b). This method, however, provided only weak constraints on Ω_m owing to the narrow redshift range and the limited number of clusters in the CNOC sample (Borgani et al. 1999, Bahcall et al. 1997). The extension of such methodology to a larger and more distant cluster sample would be extremely demanding from the observational point of view, which explains why it has not been pursued thus far.

A conceptually similar, but observationally quite different, method to estimate cluster masses is based on the measurement of the temperature of the intracluster gas (see "Physical Properties of Galaxy Clusters," above). Based on the assumption that gas and dark matter particles share the same dynamics within the cluster potential well, the temperature T and the velocity dispersion σ_v are connected by the relation $k_B T = \beta \mu m_p \sigma_v^2$, where $\beta = 1$ would correspond to the case of a perfectly thermalized gas. If we assume spherical symmetry, hydrostatic equilibrium, and isothermality of the gas, the solution of Equation 5 provides the link between the total cluster virial mass, M_{vir}, and the ICM temperature:

$$k_B T = \frac{1.38}{\beta} \left(\frac{M_{vir}}{10^{15} \, h^{-1} \, M_\odot} \right)^{2/3} [\Omega_m \Delta_{vir}(z)]^{1/3} (1+z) \, \text{keV}. \qquad (13)$$

$\Delta_{vir}(z)$ is the ratio between the average density within the virial radius and the mean cosmic density at redshift z ($\Delta_{vir} = 18\pi^2 \simeq 178$ for $\Omega_m = 1$) (see Eke et al. 1996 for more general cosmologies). Equation 13 is fairly consistent with hydrodynamical cluster simulations with $0.9 \lesssim \beta \lesssim 1.3$ (e.g., Bryan & Norman 1998, Frenk et al. 2000; see however Voit 2000). Such simulations have also demonstrated that cluster masses can be recovered from gas temperature with a $\sim$20% precision (e.g., Evrard et al. 1996).

Observational data on the $M_{vir} - T$ relation show consistency with the $T \propto M_{vir}^{2/3}$ scaling law, at least for $T \gtrsim 3$ keV clusters (e.g., Allen et al. 2001), but with a $\sim$40% lower normalization. As for lower-temperature systems, Finoguenov et al. (2001) found some evidence for a steeper slope. Such differences might be due to a lack of physical processes in simulations. For example, energy feedback from supernovae or AGNs and radiative cooling (see "Physical Properties of Galaxy Clusters," above) can modify the thermodynamical state of the ICM and the resulting scaling relations.

Measurements of cluster temperatures for flux-limited samples of clusters were made using modified versions of the Piccinotti et al. sample (e.g., Henry & Arnaud 1991). These results have been subsequently refined and extended to larger samples with the advent of *ROSAT*, *Beppo–SAX*, and especially, *ASCA*. With these data, one can derive the X-ray temperature function (XTF), which is defined analogously to Equation 7. XTFs have been computed for both nearby (e.g., Markevitch 1998; see Pierpaoli et al. 2001 for a recent review) and distant (e.g., Eke et al. 1998, Donahue & Voit 1999, Henry 2000) clusters and used to constrain cosmological models. The mild evolution of the XTF has been interpreted as a case for a low-density Universe, with $0.2 \lesssim \Omega_m \lesssim 0.6$ (see Figure 13). The starting point in the computation of the XTF is inevitably a flux-limited sample for which $\phi(L_X)$ can

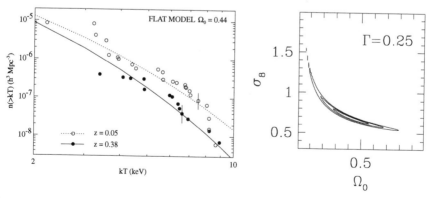

Figure 13   (*Left*) The cumulative X-ray temperature function for the nearby cluster sample by Henry & Arnaud (1991) and for a sample of moderately distant clusters (from Henry 2000). (*Right*) Probability contours in the $\sigma_8 - \Omega_m$ plane from the evolution of the X-ray temperature function (adapted from Eke et al. 1998).

be computed. Then the $L_X - T_X$ relation is used to derive a temperature limit from the sample flux limit (e.g., Eke et al. 1998). A limitation of the XTFs presented so far is the limited sample size (with only a few $z \gtrsim 0.5$ measurements), as well as the lack of a homogeneous sample selection for local and distant clusters. By combining samples with different selection criteria one runs the risk of altering the inferred evolutionary pattern of the cluster population. This can give results consistent even with a critical-density Universe (Colafrancesco et al. 1997, Viana & Liddle 1999, Blanchard et al. 2000).

Another method to trace the evolution of the cluster number density is based on the XLF. The advantage of using X-ray luminosity as a tracer of the mass is that L_X is measured for a much larger number of clusters within samples with well-defined selection properties. As discussed in "Observational Framework," above, the most recent flux-limited cluster samples now contain a large ($\sim$100) number of objects, which are homogeneously identified over a broad redshift baseline, out to $z \simeq 1.3$. This allows nearby and distant clusters to be compared within the same sample, i.e., with a single selection function. The potential disadvantage of this method is that it relies on the relation between L_X and M_{vir}, which is based on additional physical assumptions and hence is more uncertain than the $M_{vir} - \sigma_v$ or the $M_{vir} - T$ relations.

A useful parameterization for the relation between temperature and bolometric luminosity is

$$L_{bol} = L_6 \left(\frac{T_X}{6\,\text{keV}} \right)^\alpha (1+z)^A \left(\frac{d_L(z)}{d_{L,EdS}(z)} \right)^2 10^{44} h^{-2} \,\text{erg s}^{-1}, \qquad (14)$$

with $d_L(z)$ as the luminosity-distance at redshift z for a given cosmology. Several independent analyses of nearby clusters with $T_X \gtrsim 2$ keV consistently show that $L_6 \simeq 3$ is a stable result and $\alpha \simeq 2.5$–3 (e.g., White et al. 1997, Wu et al. 1999 and references therein). For cooler groups, $\lesssim 1$ keV, the $L_{bol} - T_X$ relation steepens, with a slope of $\alpha \sim 5$ (e.g., Helsdon & Ponman 2000).

The redshift evolution of the $L_X - T_X$ relation was first studied by Mushotzky & Scharf (1997), who found that data out to $z \simeq 0.4$ are consistent with no evolution for an Einstein–de-Sitter model (i.e., $A \simeq 0$). This result was extended to higher redshifts using cluster temperatures out to $z \simeq 0.8$ as measured with *ASCA* and *Beppo-SAX* data (Donahue et al. 1999, Della Ceca et al. 2000, Henry 2000). The lack of a significant evolution seems to hold beyond $z = 1$ according to recent *Chandra* observations of very distant clusters (Borgani et al. 2001b, Stanford et al. 2001, Holden et al. 2002), as well as *Newton-XMM* observations in the Lockman Hole (Hashimoto et al. 2002). Figure 14 (see color insert) shows a summary of the observational results on the $L_X - T$ relation. The high redshift points generally lie around the local relation, thus demonstrating that it is reasonable to assume $A \lesssim 1$, implying at most a mild positive evolution of the $L_{bol} - T_X$ relation. Besides the relevance for the evolution of the mass-luminosity relation, these results also have profound implications for the physics of the ICM (see "Physical Properties of Galaxy Clusters," above).

Kitayama & Suto (1997) and Mathiesen & Evrard (1998) analyzed the number counts from different X-ray flux-limited cluster surveys (Figure 7) and found that resulting constraints on Ω_m are rather sensitive to the evolution of the mass-luminosity relation. Sadat et al. (1998) and Reichart et al. (1999) analyzed the EMSS and found results to be consistent with $\Omega_m = 1$. Borgani et al. (2001b) analyzed the RDCS sample to quantify the systematics in the determination of cosmological parameters induced by the uncertainty in the mass-luminosity relation (Borgani et al. 1998). They found $0.1 \lesssim \Omega_m \lesssim 0.6$ at the 3σ confidence level, by allowing the $M - L_X$ relation to change within both the observational and the theoretical uncertainties. In Figure 15 (see color insert), we show the effect of changing in different ways the parameters defining the $M - L_X$ relation, such as the slope α and the evolution A of the $L_X - T$ relation (see Equation 14), the normalization β of the $M - T$ relation (see Equation 13), and the overall scatter Δ_{M-L_X}. We assume flat geometry here, i.e., $\Omega_m + \Omega_\Lambda = 1$. In general, constraints of cosmological models based on cluster abundance are not very sensitive to Ω_Λ (see Figure 12). To a first approximation, the best fit Ω_m has a slight dependence on Ω_Λ for open geometry: $\Omega_m \simeq \Omega_{m,fl} + 0.1(1 - \Omega_{m,fl} - \Omega_\Lambda)$, where $\Omega_{m,fl}$ is the best fit value for flat geometry.

The constraints on Ω_m shown in Figure 15 are in line with the completely independent constraints derived from the baryon fraction in clusters, f_{bar}, which can be measured with X-ray observations. If the baryon density parameter, Ω_{bar}, is known from independent considerations (e.g., by combining the observed deuterium abundance in high-redshift absorption systems with predictions from primordial nucleosynthesis), then the cosmic density parameter can be estimated as $\Omega_m = \Omega_{bar}/f_{bar}$ (e.g., White et al. 1993b). For a value of the Hubble parameter $h \simeq 0.7$, this method yields $f_{bar} \simeq 0.15$ (e.g., Evrard 1997, Ettori 2001). Values of f_{bar} in this range are consistent with $\Omega_m = 0.3$ for the currently most favored values of the baryon density parameter, $\Omega_{bar} \simeq 0.02 \, h^{-2}$, as implied by primordial nucleosynthesis (e.g., Burles & Tytler 1998) and by the spectrum of CMB anisotropies (e.g., de Bernardis et al. 2001, Stompor et al. 2001, Pryke et al. 2002).

Figure 15 demonstrates that firm conclusions about cosmological parameters can be drawn from available samples of X-ray clusters. In keeping with most of the analyses in the literature, based on independent methods, a critical density model cannot be reconciled with data. Specifically, $\Omega_m < 0.6$ at 3σ level even within the full range of current uncertainties in the relation between mass and X-ray luminosity.

A more delicate issue is whether one can use the evolution of galaxy clusters for high-precision cosmology, e.g., $\lesssim 10\%$ accuracy. Serendipitous searches of distant clusters from XMM and Chandra data will eventually lead to a significant increase of the number of high $-z$ clusters with measured temperatures. Thus, the main limitation will lie in systematics involved in comparing the mass inferred from observations with that given by theoretical models. A point of concern, for example, is that constraints on σ_8 from different analyses of the cluster abundance differ by up to 30% from each other. While a number of previous studies found $\sigma_8 \simeq 0.9-1$

for $\Omega_m = 0.3$ (e.g., Pierpaoli et al. 2001 and references therein), the most recent analyses point toward a low power spectrum normalization, $\sigma_8 \simeq 0.7$ for $\Omega_m = 0.3$ (Borgani et al. 2001b, Reiprich & Böhringer 2002, Seljak 2002, Viana et al. 2002).

A thorough discussion of the reasons for such differences would require an extensive and fairly technical review of the analysis methods applied so far. For instance, a delicate point concerns the different recipes adopted for the mass-temperature and mass-luminosity conversions. The $M - T$ relation, usually measured at some fixed overdensity from observational data, seems to have a lower normalization than that calibrated from hydrodynamical simulations (e.g., Finoguenov et al. 2001, Allen et al. 2001, Ettori et al. 2002). In turn, this provides a lower amplitude for the mass function implied by an observed XTF and, therefore, a smaller σ_8. Several uncertainties also affect the $L_X - T$ relation. The derived slope depends on the temperature range over which the fit is performed. We are also far from understanding the nature of its scatter, i.e., how much it is due to systematics and how much it is intrinsic, inherent to complex physical conditions in the gas. For example, the contribution of cooling flows is known to increase the scatter in the $L_X - T$ relation (e.g., Markevitch 1998, Allen & Fabian 1998, Arnaud & Evrard 1999). Adding such a scatter in the mass-luminosity conversion increases the amplitude of the mass-function, especially in the high-mass tail, thus decreasing the required σ_8. As an illustrative example, we show in Figure 15 how constraints in the $\sigma_8 - \Omega_m$ plane move as we change the scatter and the amplitude of the $M - L_X$ relation in the analysis of the RDCS. The upper left panel shows the result for the same choice of parameters as in the original analysis by Borgani et al. (2001b), which gives $\sigma_8 \simeq 0.7$ for $\Omega_m = 0.3$. The central lower panel shows the effect of decreasing the scatter of the $M - L_X$ relation by 20%, in keeping with the analysis by Reiprich & Böhringer (2002, see also Ettori et al. 2002). Such a reduced scatter causes σ_8 to increase by about 20%. Finally, if the normalization of the $M - T$ relation is decreased by $\sim$30% with respect to the value suggested by hydrodynamical cluster simulations (lower right panel), σ_8 is again decreased by $\sim$20%. In light of this discussion, a 10% precision in the determination of fundamental cosmological parameters, such as Ω_m and σ_8 lies in the future. With forthcoming datasets the challenge will be in comparing observed clusters with the theoretical clusters predicted by Press-Schechter–like analytical approaches or generated by numerical simulations of cosmic structure formation.

OUTLOOK AND FUTURE WORK

Considerable observational progress has been made in tracing the evolution of global physical properties of galaxy clusters as revealed by X-ray observations. The *ROSAT* satellite has significantly contributed to providing the statistical samples necessary to compute the space density of clusters in the local Universe and its evolution. A great deal of optical spectroscopic studies of these samples has consolidated the evidence that the bulk of the cluster population has not evolved significantly since $z \sim 1$. However, the most X-ray luminous, massive systems do evolve. Similarly, the thermodynamical properties of clusters as indicated by statistical correlations, such as the $L_X - T_X$ relation, do not show any strong evolution. Moreover, the *Chandra* satellite has given us the first view of the gas distribution in clusters at $z > 1$; their X-ray morphologies and temperatures show that they are already in an advanced stage of formation at these large lookback times.

These observations can be understood in the framework of hierarchical formation of cosmic structures, with a low density parameter, $\Omega_m \sim 1/3$, dominated by cold dark matter: Structure formation started at early cosmic epochs, and a sizable population of massive clusters was in place already at redshifts of unity. In addition, detailed X-ray observation of the intracluster gas shows that the physics of the ICM needs to be regulated by additional nongravitational processes.

With *Chandra* and *Newton-XMM*, we now realize that physical processes in the ICM are rather complex. Our physical models and numerical simulations are challenged to explain the new level of spatial details in the density and temperature distribution of the gas and in the interplay between heating and cooling mechanisms. Such complexities need to be well understood physically before we can use clusters as high-precision cosmological tools, particularly at the beginning of an era in which cosmological parameters can be derived rather accurately by combining methods that measure the global geometry of the Universe [the CMB spectrum, type Ia Supernovae (e.g., Leibundgut 2001)] and the large-scale distribution of galaxies (e.g., Peacock et al. 2001). It remains remarkable that the evolution of the cluster abundance, the CMB fluctuations, the type Ia Supernovae, and large scale structure—all completely independent methods—converge toward $\Omega_m \simeq 0.3$ in a spatially flat Universe ($\Omega_m + \Omega_\Lambda = 1$). Further studies with the current new X-ray facilities will help considerably in addressing the issue of systematics discussed above, although some details of the ICM in $z \gtrsim 1$ clusters, such as temperature profiles or metallicity, will remain out of reach until the next generation of X-ray telescopes. Direct measurements of cluster masses at $z \gtrsim 1$ via gravitational lensing techniques will soon be possible with the *Advanced Camera for Surveys* (Ford et al. 1998) on-board the *Hubble Space Telescope*, which offers an unprecedented combination of sensitivity, angular resolution and field of view.

The fundamental question remains as to the mode and epoch of formation of the ICM. When and how was the gas preheated and polluted with metals? What is the epoch when the first X-ray clusters formed, i.e., the epoch when the accreted gas thermalizes to the point at which they would lie on the $L_X - T$ relation (Figure 14, color insert)? Are the prominent concentrations of star-forming galaxies discovered at redshift $z \sim 3$ (Steidel et al. 1998) the progenitors of the X-ray clusters we observed at $z \lesssim 1$? If so, cluster formation should have occurred in the redshift range of 1.5–2.5. Although the redshift boundary for X-ray clusters has receded from $z = 0.8$ to $z = 1.3$ recently, a census of clusters at $z \simeq 1$ has just begun and the search for clusters at $z > 1.3$ remains a serious observational challenge. Using high-z radio galaxies as signposts for proto-clusters has been the only viable method so far to break this redshift barrier. These searches have also led to the discovery of extended $Ly\alpha$ nebulae around distant radio galaxies (e.g., Venemans et al. 2002), very similar to those discovered by Steidel et al. (2000) in correspondence with large-scale structures at $z \simeq 3$. The nature of such nebulae is still not completely understood; however, they could represent the early phase of collapse of cool gas through mergers and cooling flows.

In this review we have not treated the formation and evolution of the galaxies in clusters. This must be linked to the evolution of the ICM and the fact that we are still treating the two aspects as separate points to the difficulty in drawing a comprehensive unified picture of the history of cosmic baryons in their cold and hot phase. Multiwavelength studies are undoubtedly essential to reach such a unified picture. When surveys exploiting the Sunyaev-Zeldovich effect (e.g., Carlstrom et al. 2001) over large solid angles become available, one will be able to observe very large volumes at $z > 1$. In combination with a deep, large area X-ray survey [e.g., Wide Field X-ray Telescope (WFXT) (Burrows et al. 1992)] and an equivalent deep near-IR survey [e.g., Primordial Explorer (PRIME) (Zheng et al. 2002)], this could reveal the evolutionary trends in a number of independent physical parameters including the cluster mass, the gas density and temperature, the underlying galactic mass, and star formation rates. Advances in instrumentation and observational technique will make this approach possible and will provide vital input for models of structure formation and tight constraints on the underlying cosmological parameters.

ACKNOWLEDGMENTS

We acknowledge useful discussions with Hans Böhringer, Alfonso Cavaliere, Guido Chincarini, Roberto Della Ceca, Stefano Ettori, Gus Evrard, Isabella Gioia, Luigi Guzzo, Brad Holden, Silvano Molendi, Chris Mullis, and Adam Stanford. We thank Paolo Tozzi for his help in producing Figure 10. P.R. thanks Riccardo Giacconi for continuous encouragement of this work. P.R. is grateful for the hospitality of the Astronomical Observatory of Trieste. S.B. and C.N. acknowledge the hospitality and support of ESO in Garching.

The *Annual Review of Astronomy and Astrophysics* is online at
http://astro.annualreviews.org

LITERATURE CITED

Abell GO. 1958. *Ap. J. Suppl.* 3:211–78

Abell GO, Corwin HG Jr, Olowin RP. 1989. *Ap. J. Suppl.* 70:1–138

Allen SW, Fabian AC. 1998. *MNRAS* 297:L57–62

Allen SW, Schmidt RW, Fabian AC. 2001. *MNRAS* 328:L37–41

Allen SW, Schmidt RW, Fabian AC. 2002. *MNRAS*. In press astro-ph/0111368

Arnaud M, Evrard CE. 1999. *MNRAS* 305:631–40

Bahcall NA. 1988. *Annu. Rev. Astron. Astrophys.* 26:631–86

Bahcall NA, Cen R. 1993. *Ap. J. Lett.* 407:L49–52

Bahcall NA, Fan X, Cen R. 1997. *Ap. J. Lett.* 485:L53–56

Bahcall NA, Ostriker JP, Perlmutter S, Steinhardt PJ. 1999. *Science* 284:1481–88

Balogh ML, Babul A, Patton DR. 1999. *MNRAS* 307:463–79

Balogh ML, Pearce FR, Bower RG, Kay ST. 2001. *MNRAS* 326:1228–34

Bardeen JM, Bond JR, Kaiser N, Szalay AS. 1986. *Ap. J.* 304:15–61

Bauer FE, Alexander DM, Brandt WN, Hornschemeier AE, Miyaji T, Garmire DP. 2002. *Astron. J.* In press

Benítez N, Broadhurst T, Rosati P, Courbin F, Squires G, et al. 1999. *Ap. J.* 527:31–41

Bialek JJ, Evrard AE, Mohr JJ. 2001. *Ap. J.* 555:597–612

Birkinshaw M. 1999. *Phys. Rep.* 310:97–195

Blanchard A, Sadat R, Bartlett JG, Le Dour M. 2000. *Astron. Astrophys.* 362:809–24

Blanton MR, Dalcanton J, Eisenstein D, Loveday J, Strauss MA, et al. 2001. *Astron. J.* 121:2358–80

Böhringer H, Matsushita K, Churazov E, Ikebe Y, Chen Y. 2002a. *Astron. Astrophys.* 382:804–20

Böhringer H, Schuecker P, Guzzo L, Collins CA, Voges W, et al. 2001. *Astron. Astrophys.* 369:826–50

Böhringer H, Voges W, Huchra JP, McLean B, Giacconi R, et al. 2000. *Ap. J. Suppl.* 129:435–74

Böhringer H, Collins CA, Guzzo L, Schuecker P, Voges W, et al. 2002b. *Ap. J.* 566:93–102

Borgani S, Girardi M, Carlberg RG, Yee HKC, Ellingson E. 1999. *Ap. J.* 527:561–72

Borgani S, Guzzo L. 2001. *Nature* 409:39–45

Borgani S, Governato F, Wadsley J, Menci N, Tozzi P, et al. 2001a. *Ap. J. Lett.* 559:L71–74

Borgani S, Rosati P, Tozzi P, Norman C. 1998. *Ap. J.* 517:40–53

Borgani S, Rosati P, Tozzi P, Stanford, Eisenhardt PE, et al. 2001b. *Ap. J.* 561:13–21

Bower RG, Benson AJ, Bough CL, Cole S, Frenk CS, Lacey CG. 2001. *MNRAS* 325:497–508

Brighenti F, Mathews WG. 2001. *Ap. J.* 553:103–20

Bryan GL. 2000. *Ap. J. Lett.* 544:L1–4

Bryan GK, Norman ML. 1998. *Ap. J.* 495:80–99

Burke DJ, Collins CA, Sharples RM, Romer AK, Holden BP, et al. 1997. *Ap. J. Lett.* 488:L83–86

Burles S, Tytler D. 1998. *Space Sci. Rev.* 84(1/2):65–75

Burns JO, Ledlow MJ, Loken C, Klypin A, Voges W, et al. 1996. *Ap. J. Lett.* 467:L49–52

Burrows CJ, Burg R, Giacconi R. 1992. *Ap. J.* 392:760–65

Campana S, Lazzati D, Panzera MR, Tagliaferri G. 1999. *Ap. J.* 524:423–33

Carlberg RG, Morris SL, Yee HKC, Ellingson E, 1997b. *Ap. J. Lett.* 479:L19–23

Carlberg RG, Yee HKC, Ellingson E. 1997a. *Ap. J.* 478:462–75

Carlstrom JE, Joy M, Grego L, Holder G, Holzapfel WL, et al. 2001. In *Constructing the Universe with Clusters of Galaxies*, ed. F Durret, G Gerbal. astro-ph/0103480

Carroll SM, Press WH, Turner EL. 1992. *Annu. Rev. Astron. Astrophys.* 30:499–542

Castander FJ, Bower RG, Ellis RS, Aragon-Salamanca A, Mason KO, et al. 1995. *Nature* 377:39–41

Cavaliere A, Fusco-Femiano R. 1976. *Astron. Astrophys.* 49:137–44

Cavaliere A, Gursky H, Tucker WH. 1971. *Nature* 231:437–38

Cavaliere A, Menci N, Tozzi P. 1998. *Ap. J.* 501:493–508

Chartas G, Bautz M, Garmire G, Jones C, Schneider DP. 2001. *Ap. J. Lett.* 550:L163–66

Colafrancesco S, Mazzotta P, Vittorio N. 1997. *Ap. J.* 488:566–71

Coles P, Lucchin F. 1995. *Cosmology: The Origin and Evolution of Cosmic Structure*, Chichester, UK: Wiley

Collins CA, Burke DJ, Romer AK, Sharples RM, Nichol RC. 1997. *Ap. J. Lett.* 479:L117–20

Collins CA, Guzzo L, Böhringer H, Schücker P, Chincarini G, et al. 2000. *MNRAS* 319:939–48

Couch WJ, Ellis RS, MacLaren I, Malin DF. 1991. *MNRAS* 249:606–28

Crawford CS, Edge AC, Fabian AC, Allen SW, Böhringer H, et al. 1995. *MNRAS* 274:75–84

Crawford CS, Fabian AC. 1996. *MNRAS* 282:1483–88

Dalcanton JJ. 1996. *Ap. J.* 466:92–103

Dalton GB, Maddox SJ, Sutherland WJ, Efstathiou G. 1997. *MNRAS* 289:263–84

de Bernardis P, Ade PAR, Bock JJ, Bond JR, Borrill J. 2001. *Ap. J.* 561:13–21

De Grandi S, Böhringer H, Guzzo L, Molendi S, Chincarini G, et al. 1999. *Ap. J.* 514:148–63

De Grandi S, Molendi S. 2001. *Ap. J.* 551:153–59

De Grandi S, Molendi S. 2002. *Ap. J.* 567:163–77

Della Ceca R, Scaramella R, Gioia IM, Rosati P, Fiore F, Squires G. 2000. *Astron. Astrophys.* 353:498–506

de Propris R, The 2dFGRS Team. 2002. *MNRAS.* 329:87–101

Dickinson M, 1997. In *The Early Universe with the VLT*, ed. J Bergeron, p. 274. Berlin: Springer

Donahue M, Mack J, Scharf C, Lee P, Postman M, et al. 2001. *Ap. J. Lett.* 552:L93–96

Donahue M, Voit GM. 1999. *Ap. J. Lett.* 523:L137–40

Donahue M, Voit GM, Scharf CA, Gioia IM, Mullis CR, et al. 1999. *Ap. J.* 527:525–34

Ebeling H, Edge AC, Allen SW, Crawford CS, Fabian AC, Huchra JP. 2000b. *MNRAS* 318:333–40

Ebeling H, Edge AC, Bohringer H, Allen SW, Crawford CS, et al. 1998. *MNRAS* 301:881–914

Ebeling H, Edge AC, Fabian AC, Allen SW, Crawford CS, Böhringer H. 1997. *Ap. J. Lett.* 479:L101–4

Ebeling A, Edge AC, Henry JP. 2001. *Ap. J.* 553:668–76

Ebeling H, Jones LR, Perlman E, Scharf C, Horner D, et al. 2000a. *Ap. J.* 534:133–45

Ebeling H, Voges W, Böhringer H, Edge AC, Huchra JP, Briel UG. 1996. *MNRAS* 281:799–829

Edge AC, Stewart GC, Fabian AC, Arnaud KA. 1990. *MNRAS* 245:559–69

Eke VR, Cole S, Frenk CS. 1996. *MNRAS* 282:263–80

Eke VR, Cole S, Frenk CS, Henry JP. 1998. *MNRAS* 298:1145–58

Eisenstein DJ, Hu W. 1999. *Ap. J.* 511:5–15

Ettori S. 2001. *MNRAS.* 323:L1–5

Ettori S, De Grandi S, Molendi S. 2002. *Astron. Astrophys.* Submitted

Evrard AE. 1997. *MNRAS* 292:289–97

Evrard AE, Henry JP. 1991. *Ap. J.* 383:95–103

Evrard AE, MacFarland TJ, Couchmam HMP, Colberg JM, Yoshida N, et al. 2001. *Ap. J.* In press astro-ph/0110246

Evrard AE, Metzler CR, Navarro JF. 1996. *Ap. J.* 469:494–507

Fabian AC. 1994. *Annu. Rev. Astron. Astrophys.* 32:277–318

Fabian AC, Crawford CS, Ettori S, Sanders JS. 2001b. *MNRAS* 332:L11–15

Fabian AC, Mushotzky RF, Nulsen PEJ, Peterson JR. 2001a. *MNRAS* 321:L20–24

Fabian AC, Sanders JS, Ettori S, Taylor GB, Allen SW, et al. 2000. *MNRAS* 318:L65–68

Felten JE, Gould RJ, Stein WA, Woolf NJ. 1966. *Ap. J.* 146:955–58

Finoguenov A, David LP, Ponman TJ. 2000. *Ap. J.* 544:188–203

Finoguenov A, Reiprich TH, Böhringer H. 2001. *Astron. Astrophys.* 368:749–59

Ford HC and the ACS Science Team. 1998. In *Space Telescopes and Instruments V*, ed. PY Bely, JB Breckinridge, Proc. SPIE, 3356:234

Forman W, Jones C. 1982. *Annu. Rev. Astron. Astrophys.* 20:547–85

Francis PJ, Woodgate BE, Warren SJ, Moller P, Mazzolini M, et al. 1996. *Ap. J.* 457:490–99

Frenk PCS, White PSDM, Bode P, Bond JR, Bryan GL, et al. 2000. *Ap. J.* 525:554–82

Giacconi R, Branduardi G, Briel U, Epstein A, Fabricant D, et al. 1979. *Ap. J.* 230:540–50

Giacconi R, Murray S, Gursky H, Kellogg E, Schreier E, Tananbaum H. 1972. *Ap. J.* 178:281–308

Giacconi R, Zirm A, JunXian W, Rosati P, Nonino M, et al. 2002. *Ap. J. Suppl.* 139:369–410

Gioia IM, Henry JP, Maccacaro T, Morris SL, Stocke JT, Wolter A. 1990a. *Ap. J. Lett.* 356:L35–38

Gioia IM, Henry JP, Mullis CR, Ebeling H, Wolter A. 1999. *Astron. J.* 117:2608–16

Gioia IM, Henry JP, Mullis CR, Voges W, Briel UG. 2001. *Ap. J. Lett.* 553:L109–12

Gioia IM, Luppino GA. 1994. *Ap. J. Suppl.* 94:583–614

Gioia IM, Maccacaro T, Schild RE, Wolter A, Stocke JT, et al. 1990b. *Ap. J. Suppl.* 72:567–619

Girardi M, Borgani S, Giuricin G, Mardirossian F, Mezzetti M. 1998. *Ap. J.* 506:45–52

Girardi M, Mezzetti M. 2001. *Ap. J.* 540:79–96

Gladders MD, Yee HKC. 2000. *Astron. J.* 120:2148–62

Gonzales AH, Zaritsky D, Dalcanton JJ, Nelson A. 2001. *Ap. J. Suppl.* 137:117–38

Governato F, Babul A, Quinn T, Tozzi P, Baugh CM, et al. 1999. *MNRAS* 307:949–66

Gursky H, Kellogg E, Murray S, Leong C, Tananbaum H, Giacconi R. 1971. *Ap. J.* 167:L81–84

Gross MAK, Somerville RS, Primack JR, Holtzman J, Klypin A. 1998. *MNRAS* 301:81–94

Gunn JE, Hoessel JG, Oke JB. 1986. *Ap. J.* 306:30–37

Hashimoto Y, Hasinger G, Aranud M, Rosati P, Miyaji T. 2002. *Astron. Astrophys.* 381:841–47

Hattori M, Matuzawa H, Morikawa K, Kneib J-P, Yamashita K, et al. 1998. *Ap. J.* 503:593–98

Hall PB, Green RF. 1998. *Ap. J.* 558:558–84

Hasinger G, Altieri B, Arnaud M, Barcons X, Bergeron J, et al. 2001. *Astron. Astrophys.* 365:L45–50

Helsdon SF, Ponman TJ. 2000. *MNRAS* 315:356–70

Henriksen MJ, Mushotzky RF. 1986. *Ap. J.* 302:287–95

Henry JP. 2000. *Ap. J.* 534:565–80

Henry JP, Arnaud KA. 1991. *Ap. J.* 372:410–18

Henry JP, Gioia IM, Maccacaro T, Morris SL, Stocke JT, Wolter A. 1992. *Ap. J.* 386:408–19

Henry JP, Gioia IM, Mullis CR, Voges W, Briel UG, et al. 2001. *Ap. J. Lett.* 553:L109–12

Holden BP, Nichol RC, Romer AK, Metevier A, Postman M, et al. 1999. *Astron. J.* 118:2002–13

Holden B, Stanford SA, Squires GK, Rosati P, Tozzi P, et al. 2002. *Astron. J.* In press

Irwin JA, Bregman JN. 2000. *Ap. J.* 538:543–54

Kaiser N. 1986. *MNRAS* 222:323–45

Kellogg E, Gursky H, Leong C, Schreier E, Tananbaum H, Giacconi R. 1971. *Ap. J.* 165:L49–54

Kim RSJ, Kepner JV, Postman M, Strauss MA, Bahcall NA, et al. 2002. *Astron. J.* 123:20–36

Kitayama T, Suto Y. 1997. *Ap. J.* 490:557–63

Kofman LA, Gnedin NJ, Bahcall NA. 1993. *Ap. J.* 413:1–19

Kolb KT, Turner MS. 1989. *The Early Universe.* Reading, MA: Addison-Wesley

Kravtsov AV, Yepes G. 2000. *MNRAS* 318:227–38

Kristian J, Sandage A, Westphal JA. 1978. *Ap. J.* 221:383–94

Jeltema TE, Canizares CR, Bautz MW, Malm MR, Donahue M, Garmire GP. 2001. *Ap. J.* 562:124–32

Jenkins A, Frenk CS, White SDM, Colberg J, Cole S, et al. 2001. *MNRAS* 321:372–84

Jones LR, Ebeling H, Scharf C, Perlman E, Horner D, et al. 2000. In *Constructing the Universe with Clusters of Galaxies*, ed. F Durret, D Gerbal; CD-Rom, website

Jones LR, Scharf C, Ebeling H, Perlman E, Wegner G, et al. 1998. *Ap. J.* 495:100–14

Lazzati D, Campana S, Rosati P, Panzera MR, Tagliaferri G. 1999. *Ap. J.* 524:414–22

Leibundgut B. 2001. *Annu. Rev. Astron. Astrophys.* 39:67–98

Lewis AD, Ellingson E, Morris SL, Carlberg RG. 1999. *Ap. J.* 517:587–608

Lidman EL, Peterson BA. 1996. *Ap. J.* 112:2454–70

Lloyd-Davies EJ, Ponman TJ, Cannon DB. 2000. *MNRAS* 315:689–702

Lubin LM, Brunner R, Metzger MR, Postman M, Oke JB. 2000. *Ap. J.* 531:L5–8

Lumsden SL, Nichol RC, Collins CA, Guzzo L. 1992. *MNRAS* 258:1–22

Maddox SJ, Efstathiou G, Sutherland WJ, Loveday J. 1990. *MNRAS* 242:43p–47p

Markevitch M. 1998. *Ap. J.* 504:27–34

Markevith M, Forman WR, Sarazin CL, Vikhlinin A. 1998. *Ap. J.* 503:77–96

Markevitch M, Ponman TJ, Nulsen PEJ, Bautz

MW, Burke DJ, et al. 2000. *Ap. J.* 541:542–49

Mathiesen B, Evrard AE. 1998. *MNRAS* 295:769–80

Mazure A, Katgert P, den Hartog R, Biviano A, Dubath P, et al. 2001. *Astron. Astrophys.* 310:31–48

McHardy IM, Lawrence A, Pye JP, Pounds KA. 1981. *Ap. J.* 197:893–919

McKee JD, Mushotzky RF, Boldt EA, Holt SS, Marshall FE, et al. 1980. *Ap. J.* 242:843–56

Mellier Y. 1999. *Annu. Rev. Astron. Astrophys.* 37:127–89

Menci N, Cavaliere A. 2000. *MNRAS* 311:50–62

Monaco P. 1998. *Fund. Cosm. Phys.* 19:157–317

Muanwong O, Thomas PA, Kay ST, Pearce FR, Couchman HMP. 2001. *Ap. J.* 552:L27–30

Mulchaey JS. 2000. *Annu. Rev. Astron. Astrophys.* 38:289–337

Mullis CR, Henry JP, Gioia IM, Böhringer H, Briel UG, et al. 2001. *Ap. J. Lett.* 553:L115–18

Mushotzky RF, Scharf CA. 1997. *Ap. J. Lett.* 482:L13–16

Navarro JF, Frenk CS, White SDM. 1995. *MNRAS* 275:720–40

Nichol RC, Romer AK, Holden BP, Ulmer MP, Pildis RA, et al. 1999. *Ap. J.* 521:L21–24

Olsen LF, Scodeggio M, da Costa LN, Slijkhuis R, Benoist C, et al. 1999. *Astron. Astrophys.* 345:363–68

Oukbir J, Blanchard A. 1992. *Astron. Astrophys.* 262:L21–24

Page MJ, Carrera FJ. 2000. *MNRAS* 311:433–40

Pascarelle SM, Windhorst RA, Drivers SP, Ostrander EJ, Keel WC. 1996. *Ap. J. Lett.* 456:L21–24

Pearce FR, Thomas PA, Couchman HMP, Edge AC. 2000. *MNRAS* 317:1029–40

Peacock JA. 1999. *Cosmological Physics.* Cambridge, UK: Cambridge Univ. Press

Peacock JA, Cole S, Norberg P, Baugh CM, Bland-Hawthorn J, et al. 2001. *Nature* 4120:169–73

Peebles PJE. 1993. *Physical Cosmology.* Princeton, NJ: Princeton Univ. Press

Pentericci L, Kurk JD, Röttgering HJA, Miley GK, van Breugel W, et al. 2000. *Astron. Astrophys.* 361:L25–28

Perlman ES, Horner DJ, Jones LR, Scharf CA, Ebeling H, et al. 2002. *Ap. J. Suppl.* In press

Peterson JR, Paerels FBS, Kaastra JS, Arnaud M, Reiprich TH, et al. 2001. *Astron. Astrophys.* 365:L104–9

Piccinotti G, Mushotzky RF, Boldt EA, Holt SS, Marshall FE, et al. 1982. *Ap. J.* 253:485–503

Pierpaoli E, Scott D, White M. 2001. *MNRAS* 325:77–88

Pipino A, Matteucci F, Borgani S, Biviano A. 2002. *New Astron.* In press

Ponman TJ, Cannon DB, Navarro JF. 1999. *Nature* 397:135–37

Postman M, Lubin LM, Gunn JE, Oke JB, Hoessel JG, et al. 1996. *Astron. J.* 111:615–41

Press WH, Schechter P. 1974. *Ap. J.* 187:425–38

Pryke C, Halverson NW, Leitch EM, Kovac J, Carlstrom JE, et al. 2002. *Ap. J.* 368:46–51

Raymond JC, Smith BW. 1977. *Ap. J. Suppl.* 35:419–39

Reichart DE, Nichol RC, Castander FJ, Burke DJ, Romer AK, et al. 1999. *Ap. J.* 518:521–32

Reiprich TH, Böhringer H. 2002 *Ap. J.* 567:716–40

Renzini A. 1997. *Ap. J.* 488:35–43

Romer AK, Nichol RC, Holden BP, Ulmer MP, Pildis RA, et al. 2000. *Ap. J. Suppl.* 126:209–69

Rosati P, Borgani S, Della Ceca R, Stanford SA, Eisenhardt PR, Lidman C. 2000. In *Large Scale Structure in the X-Ray Universe*, ed. M Plionis, I Georgantopoulos, p. 13. Paris: Atlantisciences

Rosati P, Della Ceca R, Burg R, Norman C, Giacconi R. 1995. *Ap. J. Lett.* 445:L11–14

Rosati P, Della Ceca R, Burg R, Norman C, Giacconi R. 1998. *Ap. J. Lett.* 492:L21–24

Rosati P, Stanford SA, Eisenhardt PR, Elston R, Spinrad H, et al. 1999. *Astron. J.* 118:76–85

Rothschild R, et al. 1979. *Space Sci. Instr.* 4:265

Sadat R, Blanchard A, Oukbir J. 1998. *Astron. Astrophys.* 329:21–29

Sarazin C. 1988. *X-Ray Emission from Clusters of Galaxies.* Cambridge: Cambridge Univ. Press

Scharf CA, Jones LR, Ebeling H, Perlman E, Malkan M, et al. 1997. *Ap. J.* 477:79–92

Schuecker P, Böhringer H, Guzzo L, Collins CA, Neumann DM, et al. 2001a. *Astron. Astrophys.* 368:86–106

Schuecker P, Böhringer H, Reiprich TH, Feretti L. 2001b. *Astron. Astrophys.* 378:408–27

Seljak U. 2002. *MNRAS.* Submitted. astroph/0111362

Sheth RK, Mo HJ, Tormen G. 2001. *MNRAS* 323:1–12

Sheth RK, Tormen G. 1999. *MNRAS* 308:119–26

Smith S. 1936. *Ap. J.* 83:23–30

Stanford SA, Elston R, Eisenhardt PR, Spinrad H, Stern D, Dey A. 1997. *Astron. J.* 114:2232–39

Stanford SA, Holden BP, Rosati P, Eisenhardt PR, Stern D, et al. 2002. *Astron. J.* 123:619–26

Stanford SA, Holden B, Rosati P, Tozzi P, Borgani S, et al. 2001. *Ap. J.* 552:504–7

Steidel CC, Adelberger KL, Dickinson M, Giavalisco M, Pettini M, Kellogg M. 1998. *Ap. J.* 492:428–38

Steidel CC, Adelberger KL, Shapley AE, Pettini M, Dickinson M, Giavalisco M. 2000. *Ap. J.* 532:170–82

Stompor S, Abroe M, Ade P, Balbi A, Barbosa D, et al. 2001. *Ap. J. Lett.* 561:L7–10

Soucail G, Kneib J-P, Jaunsen AO, Hjorth J, Hattori M, Yamada T. 2001. *Astron. Astrophys.* 367:741–47

Tamura T, Kaastra JS, Peterson JR, Paerels FBS, Mittaz JPD, et al. 2001. *Astron. Astrophys.* 365:L87–92

Tozzi P, Norman C. 2001. *Ap. J.* 546:63–84

Tozzi P, Scharf C, Norman C. 2001. *Ap. J.* 542:106–19

Trümper J. 1993. *Science* 260:1769–71

Valageas P, Silk J. 1999. *Astron. Astrophys.* 350:725–42

van Dokkum PG, Franx M. 2001 *Ap. J.* 553:90–102

van Haarlem MP, Frenk CS, White SDM. 1997. *MNRAS* 287:817–32

van Waerbeke L, Mellier Y, Radovich M, Bertin E, Dantel-Fort M, et al. 2001. *Astron. Astrophys.* 374:757–69

Venemans BP, Kurk JD, Miley GK, Röttgering HJA, van Breugel W, et al. 2002 *Ap. J.* 569:L11–14

Viana PTP, Liddle AR. 1999. *MNRAS* 303:535–45

Viana PTP, Nichol RC, Liddle AR. 2002. *Ap. J.* 569:L75–78

Vikhlinin A, McNamara BR, Forman W, Jones C, Quintana H, Hornstrup A. 1998b. *Ap. J.* 502:558–81

Vikhlinin A, McNamara BR, Forman W, Jones C, Hornstrup A, Quintana H, et al. 1998a. *Ap. J. Lett.* 498:21–24

Voit GM. 2000. *Ap. J.* 543:113–23

Voit GM, Bryan GL. 2002. *Nature* 414:425–27

White DA. 2000. *MNRAS* 312:663–88

White DA, Jones C, Forman W. 1997. *MNRAS* 292:419–67

White SDM, Efstathiou G, Frenk CS. 1993a. *MNRAS,* 262:1023–28

White SDM, Navarro JF, Evrard AE, Frenk CS. 1993b. *Nature* 366:429–33

Wu KKS, Fabian AC, Nulsen PEJ. 2000. *MNRAS* 318:889–912

Wu X-P, Xue Y-J, Fang L-Z. 1999. *Ap. J.* 524:22–30

Yamada M, Fujita Y. 2001. *Ap. J.* 553:145–48

Yee HKC, Ellingson E, Carlberg RG. 1996. *Ap. J. Suppl.* 102:269–87

York DG, Adelman J, Anderson JE Jr, Anderson SF, Annis J, et al. 2000. *Astron. J.* 120:1579–87

Zheng W, Ford H, Tsvetanov Z, Davidsen A, Szalay A, et al. 2002. In *Lighthouses of the Universe.* In press

Zwicky F. 1937. *Ap. J.* 86:217–46

Zwicky F, Herzog E, Wild P. 1966. *Catalogue of Galaxies and of Clusters of Galaxies.* Pasadena: Calif. Inst. Technol.

Annu. Rev. Astron. Astrophys. 2002. 40:579–641
doi: 10.1146/annurev.astro.40.121301.111837
Copyright © 2002 by Annual Reviews. All rights reserved

LYMAN-BREAK GALAXIES

Mauro Giavalisco

Space Telescope Science Institute, 3700 San Martin Drive, Baltimore, Maryland 21218;
email: mauro@stsci.edu

Key Words   cosmology, galaxy observations, galaxy formation, galaxy evolution, distances, redshifts

■ **Abstract**   In this paper we review the properties of Lyman-break galaxies, namely starburst galaxies at high redshifts, approximately in the range $2.5 < z < 5$, identified by the colors of their far ultraviolet spectral energy distribution around the 912 Å Lyman continuum discontinuity. The properties of forming galaxies in the young universe are very important to constrain the history of galaxy evolution and the formation of the Hubble sequence, and until recently, they have remained largely unexplored. The Lyman-break technique has broken an impasse in the exploration of galaxies at high redshift that lasted for about two decades, and within a few years has yielded large and well-controlled samples of star-forming, but otherwise normal, galaxies at $z > 2.5$, including ~ 1000 spectroscopic redshifts and another few thousands of robust candidates. This dataset has allowed us an unprecedented look at fundamental properties of galaxies at 20% of the Hubble time or less. In this paper, we discuss the nature of the Lyman-break galaxies and their properties, including star-formation rate, stellar and total mass, chemical abundance, morphology, and interstellar medium (ISM) kinematics, and outline their contribution to the stellar content of the universe and their connection to the galaxies observed in the present-day universe. We also discuss what the properties of these galaxies, in particular their spatial clustering, imply about the mechanisms of galaxy formation and about the relationship between the underlying distribution of dark matter and the activity of star formation.

INTRODUCTION

The formation and evolution of galaxies, namely the sequence of events that led from the first stars that formed some time ago at very high redshifts to the diversity of the galaxy population observed in the present-day universe and to the assembly of the Hubble sequence, are among astronomy's grand questions that still await complete explanation. In part this is due to the fact that we do not know much about the nature of the dominant forms of mass and energy in the universe, other than that mass interacts with the radiation primarily through gravity and that the energy that currently would accelerate the cosmic expansion has an unknown equation of state and an unknown origin. These two primary cosmological components affect the formation of the large-scale structure of the universe and the timescale of its

0066-4146/02/0922-0579$14.00

evolution, including galaxies and their properties. In addition, galaxies also evolve as astrophysical systems through the conversion of gas into stars and the subsequent release of chemically enriched material by supernovae, and because we observe the cosmic structures primarily through the light of the stars that have formed in them, and because we do not understand star formation very well, we are often incapable of interpreting evidence of evolution as due to cosmology or astrophysics, with the result that we cannot constrain either. To complicate this state of affairs even further, we have long been missing empirical information on galaxies' properties at crucial epochs, primarily at high redshifts during the formation of the first structures, but also at later times when the Hubble sequence formed and galaxies diversified into their present variety.

On the bright side is the fact that we have identified gravitational instability as the most likely candidate for the primary physical process responsible for the formation of structure in the universe, at least on the scale of galaxies and clusters (Eggen et al. 1962, Sandage, et al. 1970, Peebles 1971, Press & Schechter 1974, White & Rees 1978). As a result, we have a theoretical framework that guides our investigation: Galaxies form when density perturbations in the dark matter distribution collapse under gravitational instability and when gas condenses and cools at the bottom of the potential wells of these "halos" and forms stars (White & Rees 1978). Although we do not know the nature of the mass that dominates the universe, the physics of gravity is sufficiently general that we are able to predict in simple terms for all plausible dark matter candidates how cosmic structures form, their mass spectrum, dynamics, spatial abundance, and clustering properties.

Where the major conceptual problem with galaxy formation resides is in the physics of star formation, namely how the baryonic matter formed the stars within the dark matter halos and how the properties of the latter, primarily mass, central density, and angular momentum, have affected the evolution of the former. The hydrodynamics of the gas, the role of magnetic fields, the action of feedback by the supernovae and stellar winds, the role of merging and interactions with other galaxies are the physical processes that deeply affect the activity of star formation and that are very difficult to model. Even simple models of galaxy formation and evolution generally require too large a number of free parameters and uncontrolled assumptions to result in useful predictions that can be compared with the observations.

Given that theoretical guidance is so uncertain, direct empirical information is essential in guiding the investigation, and this was recognized early (e.g., Eggen et al. 1962; Tinsley 1973a,b; Quirk & Tinsley 1973). In particular, the identification of the first galaxies that appeared in the universe was quickly identified as a fundamental step to achieve before progress could be made (Partridge & Peebles 1967a,b; Tinsley 1972a,b, 1973b; Davis & Wilkinson 1974; Meier 1976a,b). These early works promoted a searching campaign for the most distant and youngest galaxies, or primeval galaxies as they came to be generically called by the mid-1970s, which came to occupy a central role in observational extragalactic research during the following three decades, and that is still driving much of the effort at the present.

Toward the end of 1995, the frontier of galaxy redshift surveys was abruptly moved forward from $z \sim 1$ to $z \sim 4$, opening to the empirical investigation about 90% of cosmic time, and more importantly, the early phases of galaxy evolution. The observational breakthrough that made this possible was the relatively new technique (actually, a rejuvenated old idea, as we shall see later) of selecting target galaxies by means of their observed colors rather than by the more conventional flux selection. This simple method not only turned out to be very effective in targeting sources in preassigned redshift ranges, but it also proved to be surprisingly efficient, yielding large and well-controlled samples with a relatively modest investment of observing time. As a result, a wealth of high quality information on the properties of galaxies at redshifts $z > 2$, including their star-formation activity, morphology, luminosity function, spatial clustering, stellar population, and chemical enrichment was rapidly piled up, breaking an observational empasse that lasted for two decades. This progress took place over a relatively short period of time— we went from knowing of no "normal" galaxies with $z > 1$ by the end of 1995 to having samples approaching 10^3 galaxies with redshifts that reach up to $z \sim 6$ (and a plausible candidate to $z \sim 11$) by the end of the millennium. As a consequence, the full impact of the new discoveries will require additional time and work before being understood in the context of an evolutionary history of galaxies, particularly because a comparatively large gap in the observations still exists in the crucial redshift range $1 \lesssim z \lesssim 2$, where only sparse samples are available, and where the Hubble sequence likely assembled. Nevertheless, the redshift range and the region of the parameter space that the high-redshift surveys have opened are somewhat disjointed from those of previous works that it seems appropriate, now that the initial excitement of the discovery has settled down a little, to review what has been accomplished in the attempt to identify general trends and provide guidelines for future works.

In this paper we review the properties of Lyman-break galaxies (LBGs), namely of starburst, but otherwise normal, galaxies identified at high redshifts ($z > 2$) by means of the colors of their ultraviolet spectral energy distribution in proximity of the 912 Å Lyman continuum discontinuity. Thanks to the high efficiency of this Lyman-break technique (LBT), these galaxies provide the richest source of information on galaxies' properties in the high redshift universe. This review expands those by Stern & Spinrad (1999) who discussed high-redshift galaxy searches in general, and by Ferguson et al. (2000) who discussed the Hubble Deep Field survey. It also complements the reviews by McCarthy (1993) on high-redshift radio galaxies, and by Ellis (1997) on the general "faint blue galaxy" populations. This last work in particular summarizes the status of the many faint redshift surveys that studied the evolution of galaxy populations over the redshift range $z \lesssim 1$, covering approximately the past two thirds of the age of the universe, and largely complements this paper.

Two important results emerged from the surveys around $z \sim 1$. On the one hand, they have shown that bright galaxies (e.g., $L \gtrsim 0.3L^*$) of the early and middle types of the Hubble sequence were largely in place by those epochs, with the same structural components that characterize them today, and have moderately evolved from then to the present time (Brinchmann & Ellis 2000, McCarthy et al. 2001, Cimatti et al. 2002). The luminosity function of these galaxies was consistently found to have evolved by only a modest amount during this time, dimming by about $\approx$0.7–1 mag from $z \sim 1$ to the present (Steidel et al. 1994; Lilly et al. 1995; Ellis et al. 1996; Brinchmann et al. 1998; Schade et al. 1999; Cohen et al. 1999, 2000). At the same time, imaging taken by the Hubble Space Telescope (HST) confirmed that the counts of galaxies morphologically selected to be of early through mid types are characterized by modest evolution relative to the extrapolation of the present luminosity function (Schade et al. 1995, 1999; Abraham et al. 1996, 1999; Driver et al. 1995a,b; Lilly et al. 1998; Brinchmann et al. 1998). On the other hand, deep imaging with HST and spectroscopic surveys of field galaxies have provided convincing evidence that the substantial amount of evolution that is observed to have taken place in the overall galaxy population over the same cosmic time is primarily contributed by faint blue galaxies that have been identified as irregular galaxies, very similar to the later types of the Hubble sequence (Tyson 1988; Broadhurst et al. 1988; Lilly et al. 1991; Cowie et al. 1994, 1995a,b; Colless et al. 1994; Lilly et al. 1995, 1996; Glazebrook et al. 1995b; Abraham et al. 1996; Driver et al. 1995a,b, 1998; Heyl et al. 1997; Glazebrook et al. 1998; Ellis 1998; Hogg et al. 1998; Brinchmann et al. 1998; Cohen et al. 1999; Brinchmann & Ellis 2000).

Initially, it was thought that the faint blue galaxies included a population of high-redshift galaxies (e.g., $z > 2$) made very luminous, and hence easily detectable, by enhanced activity of star formation (e.g., Tyson 1988, Colless et al. 1993, Ellis 1997 and references therein). However, when the first spectroscopic surveys were carried out, it became clear that the bulk of these sources is placed at relatively modest redshifts, e.g., $\langle z \rangle \sim 0.5$ for $B \leq 24$ (Broadhurst et al. 1988, Colles et al. 1991, Lilly et al. 1991). Spectroscopic incompleteness was initially put forward as an explanation for the lack of detection of a high-redshift tail in the redshift distribution, but deeper surveys with a high degree of completeness essentially ruled out this possibility (Cowie et al. 1994, 1995a; Glazebrook et al. 1995a). Whereas the nature and evolutionary fate of the rapidly evolving faint blue population still remains unclear (Ellis 1997), the important conclusion that emerged from the observations at $z \approx 1$ is that they do not seem to be primarily responsible for the formation of the brighter and earlier members of the Hubble sequence, namely the E/S0 galaxies and the early-type spirals, because these were already formed during the same epoch when the blue galaxies were evolving very rapidly (McCarthy et al. 2001, Brinchmann & Ellis 2000, Cimatti et al. 2002). The structure and the bulk of the stellar populations of the bright galaxies, and with them the Hubble sequence, must have formed prior to $z \sim 1$, and by the mid-1990s it became clear that in order to understand how and when these systems formed, it was necessary to push the investigation to earlier epochs.

This review is organized as follows. The section titled "The Search for Primeval Galaxies" contains an historical review of the searches of star-forming galaxies at very high redshift and the ideas behind the techniques used to carry out the observations. In the section titled "The Lyman-Break Galaxy Surveys," we illustrate the first two major successful surveys for such sources, which have identified the largest samples to date by exploiting the information on the colors of the rest-frame UV spectral energy distribution of the candidates. In the section titled "Other UV-Selected Star-Forming Galaxies at High Redshift," we compare the observations of these Lyman-break galaxies to those from other surveys at even higher redshift and to galaxies identified by means of their LYα emission line. In the section titled "The Nature of Lyman-Break Galaxies," we review the primary astrophysical properties of these galaxies as deduced from the current data, while in the sections "Star Formation Rates," and "Other Starburst Galaxies at High Redshift," we discuss the important issue of the relative contribution of Lyman-break galaxies and other early star-forming galaxies to the cosmic stellar budget. The first, still rather crude attempts to estimate elemental abundances, stellar mass and age of Lyman-break galaxies are discussed in the sections "Chemical Abundances," and "Stellar Mass and Age," while in the section titled "Morphology," we review the results of the observations at high angular resolution of these sources made with the Hubble Space Telescope. In the two sections "Clustering and Large Scale Structure," and "The Mass Spectrum," we discuss the implications that the observed spatial clustering of Lyman-break galaxies has regarding the relationship between the visible structure and the underlying mass distribution, and the evolution of the galaxy mass spectrum. In the last section of the paper, titled "Looking at the Big Picture," we attempt to put our current understanding of the properties of Lyman-break galaxies and other galaxies at similarly high redshift in the more general context of galaxy formation and evolution. When we need to specify a cosmological model to transform observed quantities into physical quantities we will use the cosmological model specified by $\Omega_m = 0.3$, $\Omega_\Lambda = 0.7$, $H_0 = 65$ km^{-1} Mpc^{-1}, which seems in good agreement with the recent observations of the cosmic microwave background fluctuations and distant supernovae (Tegmark & Zaldarriaga 2000, Turner 2001). We will point out when this assumption is critical to some conclusion. At other times, to facilitate the comparison with other works, the dependence of some derived quantities on the Hubble constant is expressed as a function of $h = H_0/(100$ km s^{-1} Mpc$^{-1})$, and it will be obvious when this occurs. Throughout the paper magnitudes are expressed in the AB scale of Oke & Gunn (1983).

THE SEARCH FOR PRIMEVAL GALAXIES

Targeted observational efforts to search for primeval galaxies started in the mid-1970s. Most of the initial work consisted of either imaging or spectroscopy at optical wavelengths that attempted to identify forming galaxies at high redshifts from their UV spectroscopic features. From the observational standpoint, two considerations provided the ground for the development of these observing strategies, namely (a) that the interstellar medium of these galaxies did not contain significant amounts of dust (as expected from very young systems), implying that their UV spectrum was characterized by little or no reddening and by an intense Lyα-emission line; and (b) that the same spectrum also had a very pronounced 912 Å Lyman continuum discontinuity, which is produced both in the stellar atmospheres of massive stars and also by photoelectric absorption by neutral hydrogen in the interstellar medium.

The conceptual idea that motivated much of these works was that if the total mass of what are today bright galaxies had already assembled (although not necessarily collapsed) by the time they started to form their stars at high redshifts, then there should be a range of cosmic times when a whole population of young galaxies could be observed because of its pronounced activity of star formation. The most obvious candidates for these nascent galaxies were identified in the progenitors of the present-day spheroids, i.e., elliptical galaxies and bulges of spirals. The red spectral energy distribution and narrow dispersion of colors of these systems have been interpreted as evidence that their stellar populations are very old and, therefore, must have formed at high redshifts during a relatively short period of time and, hence, almost coevally (Baade 1958, Tinsley & Gunn 1976, Ellis et al. 1997; see also Wyse et al. 1997). One possibility is that this happened during the monolithic gravitational collapse of the whole structure, when a large fraction of the gas was converted into stars within a few free-fall timescales (Eggen et al. 1962). For systems with total mass comparable to local bright galaxies ($M \sim 10^{12}$ $M_\odot$) passbands this time is of the order of $\sim 10^8$ yr, and if their whole stellar mass, $M \sim 10^{11}$ $M_\odot$, formed during such a short burst, then the star-formation rates must have been very high, of the order of $\sim 10^3$ $M_\odot$ yr^{-1}. Even if observed at redshifts $z \gtrsim 3$, the nascent spheroids would appear rather luminous to optical and near-infrared (near-IR) observations, where the UV radiation is redshifted. For example, an unreddened galaxy with a star-formation rate of 10^3 $M_\odot$ yr^{-1} would have magnitude $\mathcal{R} = 5 \log(h) + 20.7$ in the adopted cosmology.

Partridge & Peebles (1967a,b) placed the formation of the bulk of the spheroids at very high redshifts ($10 < z < 30$), but because they assumed that the star formation occurred before they collapsed in a dissipationless "violent relaxation," they concluded that the objects would probably appear at too low a surface brightness to be detectable with the (then) available instrumentation. The authors, however, predicted a strong Lyα emission to be present in the spectra of the galaxies, and also suggested the existence of the Lyman peak in the spectrum of these primeval galaxies as an additional possible feature to identify them. In an unpublished 1966 paper, Weymann (R. J. Weymann, unpublished manuscript) suggested that the bulk of the star formation could also take place during or after the gravitational collapse, in which case the objects would appear much more compact—even stellar—than what was initially proposed by Partridge & Peebles and therefore would have correspondingly higher surface brightness.

Meier (1976a,b) used population synthesis models to compute the expected spectra of forming galaxies, and predicted their positions in color-color diagrams (e.g., $U - B$ vs. $B - V$) at various redshifts. His models differed from the earlier ones in that they included the effects of dissipation and postulated that star formation occurs at all stages in the early assembly of a galaxy. He pointed out that by looking at their colors, one can select primeval galaxy candidates among those objects whose redshifted Lyman limit is bracketed by the two shorter wavelengths, which appear red in that color but remain blue in the other one. He also proposed that, because primeval galaxies may occupy an extended range of redshifts, three or more filters and more than one color-color diagram are necessary if one is interested in covering the corresponding period of cosmic epochs. Although Meier's models did not include the effects of the cosmic opacity in the observed colors of distant galaxies [the paper appeared before the works by Sargent & Young on quasistellar object (QSO) absorption systems of the next decade, e.g., Sargent et al. (1980) and Young et al. (1982)], they were remarkably foreseeing, predicting redshifts, sizes, and apparent magnitudes that agree reasonably well with what was discovered 20 years later using essentially the same technique.

The first observational searches for primeval galaxies (and the first time that such a terminology was used) were conducted by Partridge (1974) and Davis & Wilkinson (1974) who, following the initial theoretical predictions by Partridge & Peebles (1967a,b), used photoelectric and photographic BVR photometry to look for red, extended sources with low surface brightness as young galaxy candidates, reporting no identifications. Both papers explicitly mention the possibility of a Lyman break in the spectrum of the galaxies caused by stellar atmosphere as a possible, additional identifying feature. Following these pioneering works, Koo & Kron (1980) used a charge-coupled device (CCD) and photographic slitless spectroscopy to look for Lyman-break and Lyα candidates at $z \sim 5$, also reporting no detections. It is interesting that despite the fact the idea behind the Lyman-break technique was essentially developed in its broad lines by the mid-1970s, the strategy behind the observations that began in the mid-1980s, well after these pioneering works, and lasted for a decade focused mainly on the search for redshifted Lyα emission.

The first galaxies discovered at significant redshifts were radio galaxies at $1.5 \lesssim z \lesssim 1.8$ (see the review by McCarthy 1993). These were actually found because of their strong radio luminosity, although redshift identifications were made with optical telescopes thanks to the intense UV emission lines, including the Lyα one, redshifted into the optical window (Spinrad & Djorgovski 1984a,b; Spinrad et al. 1985). After these initial discoveries, the number of radio sources identified as high-redshift galaxies steadily increased, as improved instrumentation and larger telescopes became available, and radio galaxies were identified up to redshifts as high as $z = 5.19$ (Lilly 1988; Spinrad 1989; Chambers et al. 1988, 1990; McCarthy 1991; McCarthy et al. 1990, 1991, 1992; Rawlings et al. 1990; Windhorst et al. 1991, 1992; Miley et al. 1992; Eales et al. 1993a,b; Eales & Rawlings 1993; Lacy et al. 1994; Spinrad et al. 1995; Van Breugel et al. 1999). Whereas high-redshift radio galaxies may very well be special cases of primeval galaxies, their UV-spectral energy distribution is dominated by features that are commonly observed in other types of active nuclei, such as high ionization emission lines and Lyα emission with very large equivalent width, but not in normal (i.e., without a dominant or detectable AGN activity) star-forming galaxies (Meier & Terlevich 1981, Eales & Rawlings 1990, McCarthy 1993, Dey et al. 1997, Ivison et al. 1998). Furthermore, they are very rare, with surface densities of $\sim 2 \times 10^{-3}$ radio sources per square arcmin at $S_{1.4~GHz} = 10$ mJy. Because the relationship between their properties and those expected from normal star-forming galaxies has not yet been clarified, in particular including the role played by the central nucleus in the star-formation activity and how to separate its contribution from the observed spectral energy distribution (McCarthy 1993), high-redshift radio galaxies came to be regarded by most as not representative of the general case of primeval galaxies.

Searches for high-redshift galaxies with narrowband imaging tuned to redshifted Lyα emission were first carried out by Djorgovski et al. (1985). These authors targeted fields around known distant QSOs, looking for companion galaxies at the same redshift of the quasar. The idea behind this technique is that, because galaxies are clustered in space, the probability of detection is enhanced in the volumes nearby a known high-redshift galaxy, in this case, the host to the QSO. These observations indeed yielded a successful detection of a Lyα-emitting source at $z = 3.215$ around the quasar PKS 1614 + 051. Another detection of a Lyα emitter at $z = 3.27$ around the gravitational lens MG 2016 + 112 was reported a year later by Schneider et al. (1986). In the following years, a few more similar detections were reported by other groups (Steidel et al. 1991, Hu & Cowie 1987). Other detections of galaxy-like objects include low surface brightness diffuse nebulosity around QSOs, which are thought to be part of the host galaxies (Heckman et al. 1991a,b). However, additional searches around 26 other QSOs by Djorgovski et al. (1987) and Hu & Cowie (1987) reported negative results, casting doubts on the effectiveness of this technique. The same strategy has also been adopted to look for companions of QSO-damped Lyα-absorption systems at high redshifts, which are thought to be caused by protogalactic disks (Wolfe et al. 1993, Wolfe & Prochaska 2000a,b). Except for a handful of detections (Macchetto et al. 1993, Giavalisco et al. 1994, Moller & Warren 1993, Francis et al. 1996, Djorgovski et al. 1996, Warren & Moller 1996, Fynbo et al. 1999, Francis et al. 2001), the results have been generally negative (Foltz et al. 1986, Smith et al. 1989, Wolfe 1989, Deharveng et al. 1990) or contradictory (Hunstead et al. 1990, Wolfe et al. 1992), or they have yielded cases where the emission has been attributed to the QSO itself more than to independent star-formation activity (Moller & Warren 1993).

Until very recently, the search for a population of Lyα-emitting star-forming galaxies at high redshifts in the field, namely conducted in inconspicuous "blank" regions of the sky as opposed to the targeted search described above, has been generally unsuccessful. A number of groups have carried out searches in the redshift interval $2 \lesssim z \lesssim 6$ by means of narrowband and long-slit spectroscopy (Pritchet & Hartwick 1990, Lowenthal et al. 1990, Djorgovski & Thompson 1992, De Propris et al. 1993, Thompson & Djorgovski 1995, Thompson et al. 1995), reporting no detections. Selective dust obscuration has been proposed as a possible explanation for the dearth of detections, since Lyα is a resonant line and hence the optical path of the photons in IGM is largely increased by resonant scattering (Neufeld 1991). Even a modest amount of dust is sufficient, in this case, to significantly extinguish the Lyα emission (Charlot & Fall 1991, 1993). This problem does not affect searches conducted using the Balmer lines or other optical nebular lines such as [OII] and [OIII], which at the redshifts of interest are shifted to near-IR wavelengths (Thompson et al. 1994). Interestingly, however, direct near-IR searches did not result in a higher rate of detections than their optical counterparts (Pahre

& Djorgovski 1995, Thompson et al. 1996, Teplitz et al. 1998), and only a small number of candidates have been selected, a few of which are confirmed at high redshifts (Teplitz et al. 1999, Beckwith et al. 1998).

Only recently, thanks to the sensitivity afforded by the 8-m class telescopes and the large area of modern CCD arrays, deep optical narrowband surveys routinely return Lyα emitters at high redshifts, both around QSOs and in the field (Hu & McMahon 1996; Hu et al. 1996, 1998; Cowie & Hu 1998; Steidel et al. 2000; Kudritzki et al. 2000; Rhoads et al. 2000; Rhoads & Malhotra 2001; Malhotra & Rhoads 2002; Stiavelli et al. 2002). Because of their double sensitivity thresholds of flux level and equivalent width (only galaxies with a relatively large equivalent width can be detected in these surveys, no matter how bright the continuum is) and the very narrow redshift range that each filter can probe at a given time, these surveys are generally less efficient than surveys based on continuum emission. However, they nicely complement the continuum surveys in that they can reach sources with much fainter continuum luminosity, which is particularly useful to constrain the faint end of the luminosity function, which would not be otherwise accessible, and also to study large concentrations of galaxies at the same redshifts, e.g., high-redshift clusters (Steidel et al. 2000).

The Lyman-Break Technique

Another important feature in the UV spectrum of star-forming galaxies is the Lyman continuum discontinuity at 912 Å or Lyman break. The feature forms in the stellar atmosphere of massive stars as a result of the hydrogen ionization edge and is quite pronounced, with a discontinuity of an order of magnitude in the luminosity density. The break in the stellar continuum is made more pronounced by the photoelectric absorption of the interstellar HI gas, which is abundant in young galaxies (Heckman 2000), and also by intervening HI gas (Steidel et al. 1995, 1999; Madau 1995). Spectroscopy of the rest-frame far-UV continuum found direct evidence of a break in local (Leitherer et al. 1995; see also Hurwitz et al. 1997) as well as distant star-forming galaxies (Steidel et al. 2001). Finally, the UV spectrum of sources at high redshifts is also subject to additional opacity owing to line blanketing by the intervening Lyα forest that dims the continuum between 912 Å and 1216 Å by an amount that depends on redshifts, and it is a factor of ≈ 2 at $z = 3$ (e.g., Madau 1995). This effectively creates another spectroscopic feature that helps in the identification of high-redshift galaxies from color photometry.

An additional effect that contributes to shape the UV spectrum emerging from star-forming galaxies is the presence of dust, since this often accompanies intense star formation. Dust attenuates and reddens the UV-spectral energy distribution, and to make quantitative predictions of the expected colors of galaxies at high redshifts it is necessary to calculate its effects. Different from the opacity created by neutral hydrogen, reddening by dust randomly varies by large amounts from galaxy to galaxy owing to varying amounts of dust and, in general, a varying reddening curve. Interestingly, however, the integrated UV and optical light of local starburst galaxies seems characterized by its own specific reddening curve (Calzetti 1997), which has a relatively weak dependence on wavelength, implying that only moderate reddening occurs even in relatively highly obscured galaxies. In practice, dust does not seem to introduce significant confusion between the colors of high-redshift star-forming galaxies and those of other foreground sources (e.g., elliptical galaxies). Also, the number of distant galaxies missed by commonly adopted color selection criteria because too reddened by dust is apparently small (Steidel et al. 1999, Adelberger & Steidel 2000).

The far-UV spectrum of star-forming galaxies at redshift $z > 2$ is observed at optical wavelengths, allowing surveys with ground-based telescopes and sensitive CCD detectors. Figure 1 (see color insert) graphically illustrates the idea behind the Lyman-break technique using a set of two colors based on three optical passbands applied to the case of galaxies at $z \sim 3$ and $z \sim 4$. The bluest filter, U and G in the two cases respectively, samples blueward of the Lyman break, the intermediate filter, G and $\mathcal{R}$ respectively, the Lyα-forest region, and the reddest filter, $\mathcal{R}$ and I respectively, the opacity-free spectrum.

It is clear that the choice of the passbands determines the redshift selection function of the survey, namely the redshift range of sensitivity. In practice, this is primarily determined by the bluest passband, which for ground-based surveys is the U one. This filter, coupled to filters such as the G and the $\mathcal{R}$ bands, yields a redshift selection function peaked at $z \sim 3$ and approximately covering the range $2.5 \lesssim z \lesssim 3.5$. This also is the range where the technique is most sensitive, as a result of the fortuitous combination of the color distribution of foreground galaxies, (which represent the noise), the luminosity function of the targets, and the sensitivity afforded by current technology at optical wavelengths. Galaxy candidates in this redshift range, dubbed "U–band dropouts" because they obviously appear very faint in this passband,[1] are culled from the faints counts with a yield (i.e., the rate of successfully spectroscopically confirmed high-redshift galaxies) that approaches 90% at $\mathcal{R} \sim 25.5$, and essentially no contamination by interlopers. Redder passbands return galaxies at correspondingly higher redshifts, for example B– or G–band dropouts at $z \sim 4$ and V–band dropouts at $z \sim 5$, although the efficiency decreases with redshifts and the contamination increases, e.g., $\approx 50\%$ and $\sim 20\%$, respectively for G–band dropouts at $z \sim 4$ (Steidel et al. 1999).

The first modern search for galaxies at high redshift using the Lyman-break technique was carried out by Guhathakurta et al. (1990), who took deep CCD imaging in UB_jR to identify U–band dropouts in the color-color diagram ($U - B_j$) vs. ($B_j - R$). They identified candidates down to $B_j \sim 27$, none of which received follow-up spectroscopic observations for redshift confirmation. Based on the luminosity distribution of the candidates, they estimated that the upper limits to the fraction of the deep counts with $B_j < 26$ that is at $z > 2.5$ is <7%.

---

[1] It has become customary to refer to Lyman-break galaxy candidates as dropouts, and specify the passband that they effectively drop out from when compared to redder bands, because of the Lyman discontinuity. Thus, $z \sim 3$ candidates are frequently cited as U–band dropouts, $z \sim 4$ candidates as B–band (or G–band) dropouts, and so on.

Steidel & Hamilton (1992, 1993) used the same technique to identify the galaxies responsible for QSO absorption systems at high redshifts. The idea behind this strategy was that if optically thick QSO absorption systems (i.e., damped systems and Lyman-limit systems) at high redshifts are caused by relatively massive galaxies (Wolfe et al. 1986, Turnshek et al. 1989, Lanzetta et al. 1991), as is the case at $z \sim 1$ (Steidel et al. 1994), once detected these can be used as templates to fine-tune a more general search. Furthermore, galaxy clustering could enhance the probability of finding galaxies in proximity to known ones. For their search they used a filter system optimized to construct colors with maximized signal-to-noise ratio, the $U_n G \mathcal{R}$ set shown in Figure 1, and defined a set of color selection criteria based on predictions from spectral population synthesis, including the effects of the intervening photoelectric and line blanketing absorption by neutral hydrogen (estimated from the spectra of high-redshifts QSOs). For example, they calculated that a relatively unreddened star-forming galaxy at $z \sim 3$ would be observed with colors $(U_n - G) \geq 1.6$ and $(G - \mathcal{R}) \leq 1.2$. They applied these criteria to deep CCD imaging around two QSOs with damped Lyα absorption systems, at $z \sim 3.4$ and 3.0, and despite the very poor seeing conditions, identified a sample of 23 objects with $\mathcal{R} \leq 25$, including the candidate damped systems, identified after subtracting the QSO PSF from the G and $\mathcal{R}$ images.

One of the candidates was spectroscopically confirmed at $z = 3.428$ by Giavalisco, Macchetto & Sparks et al. (1994) during the spectroscopic follow-up of a sample of Lyα-emitting candidates at $z = 3.4$ from narrowband imaging around the same QSO (Q0000-263), confirming the effectiveness of color selection in identifying galaxies at high redshift. Unfortunately, spectroscopic observations from 4-m telescopes of the other candidates were inconclusive. Giavalisco et al. (1994, unpublished manuscript) obtained significantly deeper $U_n G \mathcal{R}$ imaging in the same QSO field in good seeing conditions using the NTT telescope. Using the same selection criteria of Steidel & Hamilton, they obtained a new, higher quality sample of candidates (including again the confirmed galaxy and the candidate damped Lyα absorber of the QSO). Steidel et al. (1996b) expanded this sample with additional candidates obtained with the 5-m Hale telescope on Mount Palomar and followed them up with spectroscopy taken at the 10-m Keck telescope with the LRIS spectrograph (Oke et al. 1995), confirming that U–band dropouts include star-forming but otherwise normal galaxies (i.e., with no detectable AGN activity) with redshifts in the range $2.2 \lesssim z \lesssim 3.6$.

LYMAN-BREAK GALAXY SURVEYS

The Color Selection

Two major and largely complementary surveys for Lyman-break galaxies (LBGs) were started after the successful spectroscopic identification of the initial candidates by Steidel et al. (1996a,b), one conducted from the ground and the other from HST. The ground-based survey is the continuation of the discovery works described above, which has also been expanded to galaxies at $z \sim 4$, and today it includes ~ 1000 spectroscopic redshifts and several thousands of other candidates (Steidel et al. 1999). In the following we shall refer to it as the ground-based survey. The HST survey is the Hubble Deep Field (HDF), which has been conducted during two separate observing campaigns, targeting one field in the Northern sky and the other in the Southern (Williams et al. 1996, 2000; Casertano et al. 2000), achieving the deepest multicolor images of the sky at optical wavelengths with the highest angular resolution. Designed to be sensitive to LBGs, the survey returned a sample with flux limit $V \sim 27.5$ (Madau et al. 1996, Dickinson 1998, Cristiani et al. 2000), extending the reach of the ground-based one about 2.5 mag fainter, while at the same time providing information on the morphology of the galaxies. Ferguson et al. (2000) discuss in greater detail the HDF survey and the samples of LBGs derived from it. Spectroscopic confirmation of the brightest LBG candidates of the HDF sample have been obtained by Steidel et al. (1996a) and Lowenthal et al. (1997) with the Keck telescope soon after the HDF North observations. Cristiani et al. (2000) obtained spectra of candidates from the HDF South sample with the European Southern Observatory (ESO) 8-m Very Large Telescope (VLT).

The selection of galaxies in a preassigned redshift interval by means of their broadband colors is essentially equivalent to very low resolution spectroscopy, with a resolving power $\lambda/\Delta\lambda \approx 4$. The technique, therefore, is not useful to measure the redshifts of the galaxies, which is the goal of the conceptually similar but significantly more sophisticated technique of photometric redshifts (e.g., Connolly et al. 1997, Fernandez-Soto et al. 1999), but rather to identify large numbers of candidates, distributed in the targeted redshift range according to a characteristic probability distribution function. The technique is very efficient when strong spectral features are present in the spectra of the candidates, such as in the case of the LBGs.

The candidates are selected according to their position in a color-color plane, as illustrated in Figure 1 (see color insert) for the cases of $z \sim 3$ and $z \sim 4$. The expected colors in a given passband set can be calculated from templates or population synthesis models of star-forming galaxies (Bruzual & Charlot 1993), under assumptions of star-formation history and dust reddening. To first approximation, young burst with no dust has a relatively steep UV spectrum, with spectral index approximately $\alpha \sim -0.5$, if the spectrum is modeled as $f_\nu \propto \lambda^{-\alpha}$, which corresponds to colors that are bluer than the flat spectrum ($\alpha = 0$). Continuous star formation has $\alpha \sim 0$, whereas a decaying burst can be very red, depending on its age. The effect of dust in the interstellar medium (ISM) of the galaxies is commonly modeled as

$$F_a(\lambda) = F_i(\lambda) \times 10^{-0.4 \times A_\lambda \times E(B-V)}, \qquad (1)$$

(Calzetti 2001) where F_a and F_i are the attenuated and intrinsic spectra, respectively, the color excess $E(B-V)$ parametrizes the amount of dust, and the attenuation function A_λ describes the reddening as a wavelength-dependent attenuation. Because this function generally increases with decreasing wavelength, shorter wavelengths are selectively more attenuated than longer wavelengths. To calculate the reddened spectrum for an assigned value of $E(B - V)$, the function A_λ is required. This is generally not known for galaxies at high redshifts and, unfortunately, rather different results are obtained using local reddening curves, such as the Galactic, the Small Magellan Cloud (SMC), the Large Magellan Cloud (LMC) ones, or the starburst attenuation function that describes the integrated extinction properties of local starburst galaxies (Calzetti 1997). Luckily, there is some evidence that the Calzetti law might also apply to starbursts at higher redshifts, including LBGs (Meurer et al. 1997, 1999; Adelberger & Steidel 2000, Pettini et al. 2000, Calzetti & Giavalisco 2001), and thus this is generally adopted for them as well, although a final test of the robustness of this assumption is still lacking. Finally, the calculation of the expected colors also includes the cosmic opacity, namely line blanketing and the photoelectric absorption of ionizing radiation by local and intervening HI gas, since the column density of the intervening gas (cosmic opacity) depends on the redshift, and it has a nonnegligible effect on the observed spectra $z \gtrsim 2$ (Madau 1995).

The dispersion of intrinsic spectral shape and dust reddening of LBGs is not known, and sets a limit to one's ability to accurately define a region of the color-color plane that selects all cases of star-forming galaxies. Furthermore, photometric errors scatter the observed colors around their intrinsic values by an amount that depends on the quality of the data and mixes them with the colors of interlopers at lower redshifts. Thus, the definition of a region to use in practice for candidate selection is somewhat arbitrary, and reflects the compromise between one's desire to include as many diverse galaxies as possible while keeping the contamination by interlopers to a minimum (Steidel et al. 1995, 1999; Madau et al. 1996; Dickinson 1998). With different passbands, the optimal color selection criteria will be different, and no definitive method has yet been established. The criteria adopted for the ground-based survey have been fine-tuned for the $U_n G \mathcal{R}$ and $G \mathcal{R} I$ photometric systems, and have also been optimized by taking advantage of the availability of large samples of spectroscopic redshifts. For candidates at $z \sim 3$ and ~ 4 (U– and G–band dropouts) of the ground-based survey, these color criteria are

$$(U_n - G) \geq 1 + (G - \mathcal{R}); \quad (U_n - G) \geq 1.6; \quad (G - \mathcal{R}) \leq 1.2, \qquad (2)$$

and

$$(G - \mathcal{R}) \geq 2.0; \quad (G - \mathcal{R}) \geq 2 \times (\mathcal{R} - I) + 1.5; \quad (\mathcal{R} - I) \leq 0.6, \qquad (3)$$

respectively. Figure 1 shows the corresponding regions in the color-color diagrams as shaded areas together with the tracks of star-forming galaxies (the UV spectrum of continuous star formation with age $T = 0.1$ Gyr has been assumed) for selected amounts of dust reddening as they are placed at progressively higher redshifts. The triangles correspond to no reddening, i.e., $E(B - V) = 0$, the squares to $E(B - V) = 0.15$, the pentagons to $E(B - V) = 0.3$, and the hexagons to $E(B - V) = 0.45$. The figure also shows the colors of galactic stars, from early to late types, and for the case of G–band dropouts only, of elliptical galaxies in the redshift range $0.5 < z < 1.6$ (no evolution assumed). Stars of the G and K types contaminate the U–band dropouts (by about 3.4% at $\mathcal{R} = 25.5$), whereas for the G–band dropouts the predominant contamination comes from early-type galaxies at intermediate redshifts ($\approx 20\%$ down to $I = 24$).

Dickinson (1998) has discussed the case of the HDF survey, for which the WFPC2 four-band photometric system U_{300}, B_{450}, V_{606}, and I_{814} has been adopted. The number of LBG candidates that have been selected by various groups varies from sample to sample depending on the adopted criteria. For example, Madau et al. (1996) defined conservative criteria in order to select $z > 2$ galaxies while avoiding significant risk of contamination from objects at lower redshifts. These criteria, however, miss some of the HDF galaxies that have been spectroscopically confirmed to have $z \gtrsim 2$ by other groups (e.g., Lowenthal et al. 1997). Using the redshift information currently available in the HDF (Steidel et al. 1996a, Lowenthal et al. 1997, Cohen et al. 1996, Cowie et al. 1996) to fine-tune the selection, Dickinson (1998) built a sample of LBGs from the HDF-N by applying criteria similar to the ground-based ones discussed above, which for the HDF filters are written as

$$(U_{300} - B_{450}) \geq 1.0 + (B_{450} - V_{606}); \quad (B_{450} - V_{606}) \leq 1.2; \quad V_{606} \leq 27 \quad (4)$$

and

$$(B_{450} - V_{606}) > 1.5; \quad (B_{450} - V_{606}) > 1.7 \times (V_{606} - I_{814}) + 0.7$$

$$(B_{450} - V_{606}) < 3.5 \times (V_{606} - I_{814}) + 1.5; \quad (V_{606} - I_{814}) < 1.5; \quad V_{606} \leq 27.7$$

$$(5)$$

for the U_{300}– and B_{450}–dropouts, respectively. The first selection window successfully recovers all 25 spectroscopically confirmed LBGs in the HDF at $2 < z < 3.5$, whereas the second includes the only spectroscopically confirmed B_{450}–dropout (Dickinson 1998; see also Ferguson et al. 2000). As is the case for the ground-based samples, the dominant source of contaminants is a very small number of galactic stars, but these are easily recognized in the WFPC2 images and excluded (all obvious stars with the above colors have also already been observed spectroscopically, as it turns out).

The extensive program of spectroscopy of the LBG candidates carried out during the ground-based survey has demonstrated that the Lyman-break technique is highly efficient. Multiband imaging from 4-m class telescopes returns photometry of sufficient accuracy for color selection of candidates down to flux limits where follow-up spectroscopy with 8-m class telescopes is still practical, namely $\mathcal{R} \sim 25.5$ for U–band dropouts and $I \sim 25$ for G–band dropouts. Down to these limits the ground-based survey currently includes about 2000 U–band dropout candidates from nine different pointings in the sky with approximately 1000 secured redshifts, and about 200 G–band dropout candidates with ~ 50 secured redshifts. The surface density of candidates that satisfy Equation 2 and Equation 3 down to these flux limits is

$$\Sigma_{U_n}(25.5) = 1.21 \pm 0.06 \text{ galaxies/arcsec}^{-2} \qquad (6)$$

for the U–band dropouts and

$$\Sigma_G(25) = 0.47 \pm 0.02 \text{ galaxies/arcsec}^{-2}, \qquad (7)$$

where, in both cases, the error bars reflect both the Poisson counting and the observed field-to-field variations (Steidel et al. 1999, Giavalisco & Dickinson 2001). At the flux limits reached by the HDF survey, namely $V \sim 27$ for the U–band dropouts and $I \sim$ for the B–band ones, the surface densities are

$$\Sigma_{U_{300}}(27) = 29.4 \pm 3.5 \text{ galaxies/arcsec}^{-2} \qquad (8)$$

and

$$\Sigma_{B_{450}}(25) = 8.7 \pm 5 \text{ galaxies/arcsec}^{-2}, \qquad (9)$$

respectively (Dickinson 1998, Giavalisco & Dickinson 2001). Of course, most of the HDF candidates are too faint for spectroscopic observations, even with the largest telescopes available.

The surface density of the ground-based samples is relatively high and well matched to the field of view and multiplexing capabilities of modern multi-object spectrographs, which makes the compilation of large samples of LBGs relatively economical in terms of telescope time. For example, by covering 1 square degree of sky, which is a relatively easy task with today's large CCD arrays, one secures $\sim$4400 U–band dropouts and $\sim$1700 G–band dropouts. These can be followed-up with spectroscopy in less than two dozen nights. The spectroscopic yield, namely the fraction of spectroscopic targets that are confirmed in the expected redshift range, is also high. For U–band dropouts the observed yield is $\approx$85%, with a $\sim$3.4% contamination by galactic stars (Steidel et al. 1999) and the remainder of the targets unidentified. However, the true yield is certainly higher, because the missed identifications are invariably due to lack of S/N, very likely as the result of observational accidents such as slit misalignements or insufficient exposure time (in no case has a spectroscopically confirmed U–band dropout turned out to be an interloper). The yield of G–band dropouts is about 45%, and this is the combined result of the fact that the targets are fainter and securing the redshifts is more difficult because the relevant spectral features have now moved from the range 4500–6000 Å to 6500–7500 Å, where the sky is 1.5 to 2 times brighter. Furthermore, early-type galaxies at intermediate redshifts provide a contamination of $\approx$20%. Taken together with the lower surface density of candidates, this highlights the difficulty of extending the Lyman-break surveys to higher redshift intervals.

The Redshift Distribution

Figure 2 shows the redshift distribution of the spectroscopically confirmed galaxies of the ground-based survey together with the redshift distribution function

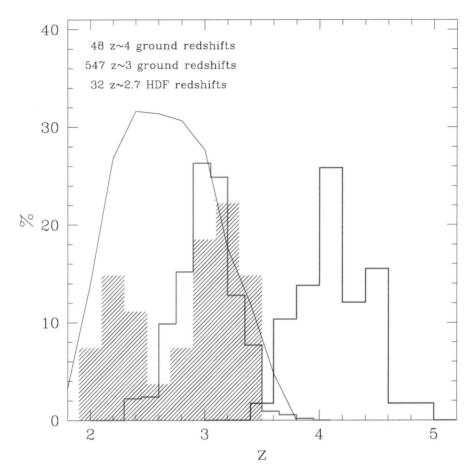

Figure 2   Histograms of the spectroscopic redshifts of U–band ($z \sim 3$) and G–band ($z \sim 4$) dropouts. The thick histograms represent galaxies from the ground-based survey, the shaded histogram galaxies from the HDF (U–band dropouts only). The thin continuous line is the expected redshift distribution function of the HDF. Only 32 redshifts of HDF U–band dropouts have been measured so far, and the histogram is clearly affected by small number statistics; nonetheless there is good agreement with the predictions. Note the wider- and lower-redshift distribution of the HDF U–band dropouts compared to their ground-based counterparts, due to the bluer and wider F300W passband compared to the U_n one.

calculated from Monte Carlo simulations of the color selection. The cutoff at low redshift of the bell-shaped curve occurs because most galaxies in that redshift region are not red enough in the bluest color [either $(U_n - G)$ or $(G - \mathcal{R})$ at $z \sim 3$ or ~ 4, respectively] to satisfy the color selection, because the Lyman break is still within in the bluest passband. The cutoff at higher redshifts occurs because there galaxies appear redder because of the stronger line blanketing of the Lyman continuum and eventually fail to satisfy the color criteria. Figure 2 also shows the predicted redshift selection function of the HDF U–band dropouts (thin continuous curve) together with the histogram of the observed 32 spectroscopic redshifts (shaded area). As expected, because the effective wavelength of the U_{300} filter is ≈600 Å shorter than that of the U_n filter, the HDF redshift distribution is broader and extends to lower redshifts than the ground-based counterpart. For the B–band dropouts of the HDF sample, there is only tentative spectroscopic confirmation of only one candidate (Dickinson 1998), an object with a single line emission, which is asymmetric and has absorption on the blue side, similar to what was observed in LBGs at $z \sim 3$ and 4. This strongly suggests that the line is indeed redshifted Lyα emission, placing the galaxy at $z = 4.02$.

Throughout the sampled redshift windows, galaxies with excessively red UV colors remain undetected by the Lyman-break color selection criteria. This occurs preferentially at the high-redshift end because the stronger line blanketing and the fact that the galaxies appear fainter both reduce the useful color dynamic range. In other words, the distribution of the intrinsic colors of the detected galaxies is bluer at the higher end of the redshift distribution. Red UV colors are the result of either reddening by dust or of an aging burst or both, and it is of great interest to estimate how many of such red galaxies are being missed by the surveys. Steidel et al. (1999) and Adelberger & Steidel (2000) used Monte Carlo simulations to estimate the magnitude of the selection effect against red LBGs as a function of redshift in the ground-based survey, and they concluded that such galaxies are actually intrinsically rare. For example, star-forming galaxies with dust reddening as large as $0.3 < E(B - V) < 0.45$ could be detected by the survey if they were present, but they have never been observed. Galaxies with very red colors have been detected by near-IR surveys (McCarthy et al. 2001), but their redshift distribution is not yet constrained, although at bright magnitudes it does not seem to overlap with that of the LBGs. For example, Cimatti et al. 2002 showed that about half of spectroscopically identified galaxies with $R - K < 5$ down to $K \sim 19.5$ are very dusty star-forming galaxies at $0.6 < z < 1.6$, with the other half being elliptical galaxies at the same redshifts. However, because no rest-frame UV spectra of these dusty starbursts have been recorded, it is not clear whether they would be observed as LBGs if they were placed at $z \sim 3$.

With large spectroscopic samples available, Steidel et al. (1999) estimated the luminosity functions of LBGs at $z \sim 3$ and ~ 4 from the ground-based survey, and found that they agree very well, both in shape and in normalization, over the common range of absolute luminosity. This is shown in Figure 3 (see color insert), where for simplicity the luminosity functions are computed in the Einstein de Sitter cosmology(with $H_0 = 50$ km^{-1} Mpc^{-1}), although the agreement is insensitive to the choice of the world model. This shows that, at least at the relatively bright luminosity probed from the ground, there is no evidence of evolution in the LBG population over this redshift interval contrary to claims based on the fainter and much smaller HDF sample alone (Madau et al. 1996, 1998; Pozzetti et al. 1998).

Unfortunately, because of the limited number of spectroscopic redshifts available for the HDF sample, it is not possible to estimate its luminosity function in the same way as it is done from the ground. Note that, to first approximation, the distribution of apparent magnitude of LBGs can be used as a proxy for the luminosity function because the redshift distribution function is rather peaked around the average redshift of the survey (both at $z \sim 3$ and ~ 4), and thus the dispersion of luminosity distance of galaxies with given apparent magnitude is relatively small. Thus, in a given bin of apparent magnitude, the difference in absolute luminosity between a galaxy placed at either end of the redshift distribution function and one placed at the average redshift is $\sim 50\%$, and the apparent magnitude can be used to trace the absolute magnitude.

A luminosity function constructed in this way (for simplicity, plotted as a function of apparent magnitude) and including both the ground-based and HDF data, is plotted in Figure 3, *right* (see color insert). The top panel shows the U–band dropouts;[2] there is very good consistency between the ground-based and the HDF data in the range of overlapping luminosity, and a Schechter function (Schechter 1976) provides a good model of the data (the best fit and its parameters are also shown in the figure). The faint end slope is significantly steeper than that of the optical luminosity function of the local mix of galaxies, and similar to that of the luminosity function of local and moderate redshift late-type and irregular galaxies (Folkes et al. 1999), UV-selected galaxies (Sullivan et al. 2000), and Hα-selected galaxies (Sullivan et al. 2000, Tresse & Maddox 1998, Gallego et al. 1995). This is consistent with the general trend that star-forming galaxies and galaxies of later types (at any redshift) have a steeper slope. The characteristic luminosity m^*, $\mathcal{R} = 24.5$, is a relatively faint apparent magnitude, but at $z = 3$ it corresponds to an absolute luminosity at $\lambda = 1700$ of $M = -21.2$. The characteristic volume density $\phi^* = 1.6 \times 10^{-2} \, h^3 \, \mathrm{Mpc}^{-3}$ directly compares with that of the K–band local luminosity function, $\phi_K^* = 1.08 \times 10^{-2}$ (Cole et al. 2001), showing that the volume density of LBGs is similar to that of present-day galaxies. Note that the cosmic volume probed by the $z = 3$ and $z = 4$ surveys is very large. For example, at $z = 3$ the width of the redshift distribution function corresponds to $2600 \, h^{-3} \, \mathrm{Mpc}^3$ per square arcmin, and a survey that covers 1 square degree of sky probes a volume of $9.4 \times 10^6 \, h^{-3} \, \mathrm{Mpc}^3$, namely $1.5 \times 10^5 \, (\phi^*)^{-1}$.

The luminosity function of the G– and B–band dropouts at $z \sim 4$ is shown in the bottom panel. In this case, it is difficult to assess the consistency between the ground-based data and the HDF because there is very little overlap between the ranges of luminosities spanned by the two samples (only a single HDF B–band dropout is bright enough to be included in the ground-based sample); however, the data suggest at least a broad consistency. The $z = 4$ data are not good enough for a meaningful fit to the Schechter function; however, it is possible to compare them

---

[2]The U_{300} sample from the HDF has been put onto the same magnitude scale of the U_n sample using the approximation $\mathcal{R} \sim (V_{606} + I_{814})/2$ (Steidel et al. 1996a), and all the HDF magnitudes have been made fainter by the amount $\Delta m = 0.25$ to account for its smaller mean redshift ($\bar{z} \sim 2.6$ vs. $\bar{z} \sim 3$).

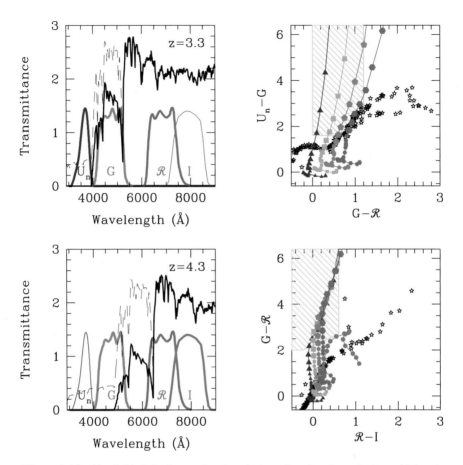

Figure 1 The idea behind the Lyman-break technique. Star-forming galaxies at high red-shifts are identified by the colors of the spectral region around the 912 Å continuum discontinuity obtained through a suitable set of filters (shown here is the U_nGRI set used for the ground-based survey). Redshifted synthetic model spectra of continuous star formation (Bruzual & Charlot 1993), which include the effects of cosmic opacity (Madau 1995), are plotted together with the filters' transmittance (*left panels*). The two upper panels show the *U*–band dropouts at $z \sim 3$, defined with the U_nGR filters; the lower ones the *G*–band dropouts at $z \sim 4$, defined through the *GRI* set. The candidates are selected from their position in the color-color space, as shown in the right panels. The curves represent galaxies placed at progressively higher redshifts, starting at $z = 0.5$ with step $\Delta z = 0.1$. The case of continuous star formation is used, and the four tracks correspond to different amounts of dust obscuration, namely $E(B-V) = 0$, i.e., no dust (*triangles*); $E(B-V) = 0.15$ (*squares*); $E(B-V) = 0.3$ (*pentagons*); $E(B-V) = 0.45$ (*hexagons*). In all cases, the starburst attenuation law by Calzetti et al. (1997) is used. Larger symbols denote the two redshift intervals $2.6 < z < 3.5$ and $3.8 < z < 5$. The locus of stellar colors is also shown. The shaded areas correspond to Equation 3.2 and 3.3, respectively.

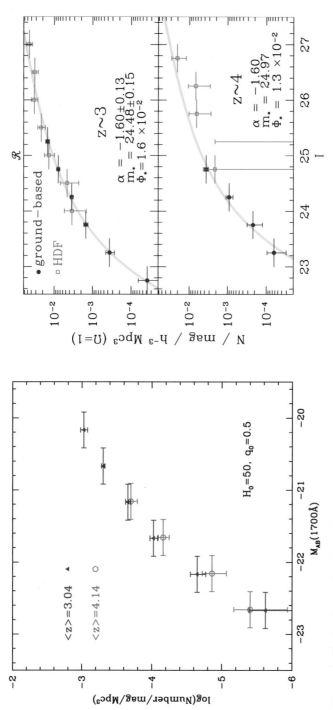

Figure 3 The luminosity function of U-band (*upper panel*) and G-band dropouts (*lower panel*). Because of the relatively narrow range of luminosity distance spanned by the samples of LBGs, apparent magnitudes can be used as proxy of absolute ones. The data points, which range over 4.5 magnitudes, are derived from the ground-based and HDF surveys, as indicated in the figure. For the U-band dropouts the figure also shows the best Schechter-function fit to the data and its parameters. At z ~ 4 the data are not good enough for a fit, and in this case the z ~ 3 best-fit luminosity function has been "redshifted" and overimposed to the data (see text for the procedure). The apparent shallower low-luminosity end is very likely the effect of cosmic variance fluctuations in the small HDF sample. In fact, the comparison shows no significative evidence of evolution in the luminosity function of Lyman-break galaxies between z ~ 3 and z ~ 4 (reproduced from Steidel et al. 1999).

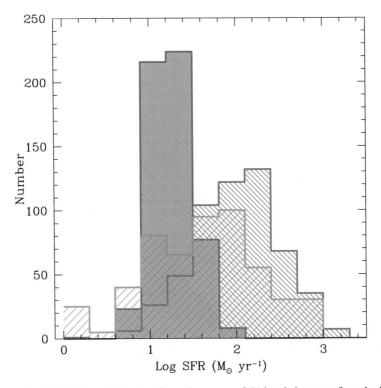

Figure 6 Distributions of the star formation rates of *U*–band dropouts from both the ground-based and HDF surveys normalized on arbitrary scales. The dark-shaded histogram shows the rates calculated from the UV luminosity without any correction for dust obscuration. The medium-shaded histogram shows the rates corrected for dust obscuration assuming a continuous star formation with age $T = 0.1$ Gyr as the underlying spectral energy distribution and the starburst obscuration law with the value of $E(B-V)$ determined by comparing the observed colors with the predicted ones. The light-shaded histogram shows the rates derived from the fitting procedures by Papovich, Dickinson & Ferguson (2001) and Shapley et al. (2001), as described in the text. In all cases the Madau's (1995) cosmic opacity was used in the calculation. There is broad agreement between the two techniques. Note that the corrected rates are distributed over three orders of magnitude, including galaxies that form stars as slowly as the Milky Way or as rapidly as required to put together the stellar mass of an L^* galaxy in 1 Gyr or less.

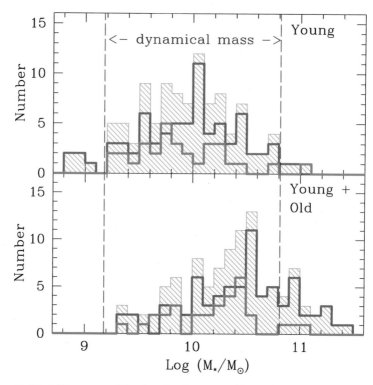

Figure 9 (*Top*) Histogram of the fits of stellar mass of LBGs. The dark gray curve is the ground-based sample, the light gray curve the HDF, while the shaded histogram is the sum of the two. (*Bottom*) Histogram of the sum of the young stellar mass plus the old one (see text) with the same coding as above. The 2-population fit has only been done for the HDF sample; for the ground-based sample the mass of the old population has been taken equal to 4 times that of the young population, the average value found in the HDF. Also shown is the range of values of dynamical mass derived from the kinematics of the optical nebular emission lines (see text).

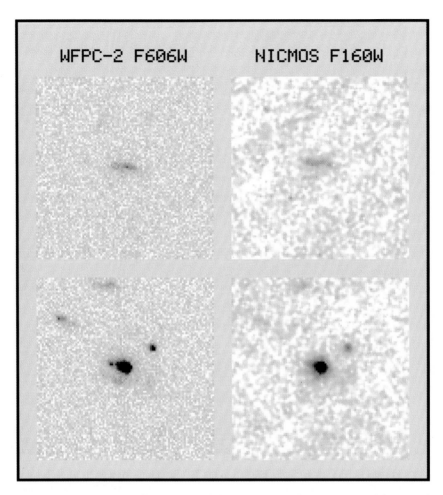

Figure 10 Two examples of regular morphology encountered among LBGs of the ground-based sample observed with *HST*. In each case the left panel shows optical images taken with WFPC2 in the F702W filter, corresponding roughly to the rest-frame UV at $\lambda = 1700$ Å; the right panel shows near-IR images in the *H* band (F160W) taken with NICMOS, and corresponding to rest-frame $\lambda = 4000$ Å. (*Top*) A disk-like galaxy at $z = 3.23$ with $R = 25.2$. (*Bottom*) A spheroid-like galaxy, with an $r^{1/4}$ profile and relatively compact and regular morphology, at $z = 2.96$ with $R = 22.8$. Note that the morphology is essentially independent on the wavelength.

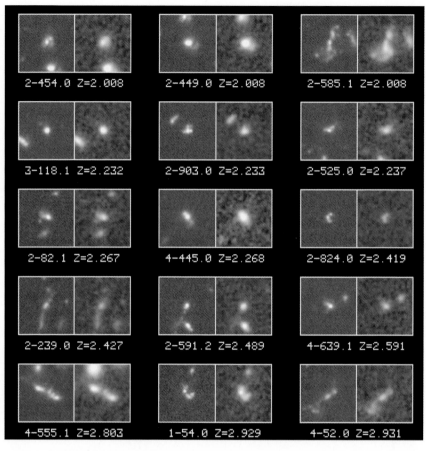

Figure 11 *HST* images of LBGs from the HDF sample. On the left of each panel is an optical *BVI* (F450W, F606W and F814W) color image observed with WFPC2, while on the right there is the corresponding near-IR *IJH* (F814W, F125W and F160W) taken with NICMOS. Note that in almost all cases the morphology is independent on wavelength. Also note that no Hubble types can be recognized in these images, which, on the other hand, show a remarkable variety, from small compact and relatively regular objects to highly fragmented, diffuse and irregular ones (reproduced from Dickinson 1998).

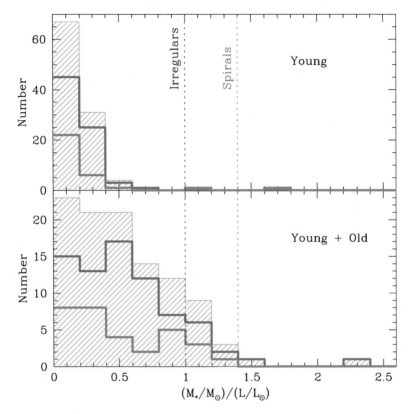

Figure 12 Histogram of the stellar mass-to-light ratio (rest-frame *B* band) of LBGs derived from the fitting procedure described in the text. The green histogram represents the ground-based sample, the red histogram the HDF one, and the shaded histogram the sum of the two samples. Also plot is the typical value for Irregular and Spiral galaxies from Fukugita, Hogan & Peebles (1998). Note the LBGs have smaller stellar mass-to-light ratio than local late type galaxies.

to the $z = 3$ fit by redshifting m^* by the relative distance modulus and by adjusting ϕ^* to take into account the slight difference in the cosmic volume probed by the $z = 3$ and $z = 4$ selection criteria. This "redshifted" luminosity function matches the $z = 4$ data points at the bright end fairly well, although it seems to overpredict the number of fainter HDF galaxies by about a factor of two, apparently implying a luminosity function with a shallower slope than at $z \sim 3$. Steidel et al. (1999) suggest that this likely is an effect of the small size of the HDF sample, and consequently, of the large sample variance due to strong spatial clustering (more on this later). In conclusion, there is no evidence of evolution between $z \sim 3$ and $z \sim 4$, and in fact, the null hypothesis that there is no evolution cannot be rejected with confidence with the current data.

OTHER UV-SELECTED STAR-FORMING GALAXIES AT HIGH REDSHIFT

Lyman-Break Galaxies at $z > 4$

A number of groups have identified LBGs at redshifts $z > 4$, mostly selecting them from the HDF, but the numbers are too small to study them with the same detail as was done at $z = 3$ and 4. For some candidates there are tentative spectroscopic confirmations, although almost all spectra have either low S/N or show only one emission line to securely confirm the redshifts.

Spinrad et al. (1998) obtained Keck spectra for two faint V–band dropouts selected from the HDF with $V_{606} = 28.1$ and $I_{814} = 25.6$. The galaxies had been previously identified by Fernandez-Soto et al. (1999), who measured photometric redshifts $z_{ph} = 5.28$ for both of them. The Keck spectra show a relatively sharp and pronounced discontinuity in otherwise blue, featureless continua, similar to what was observed in other LBGs at $z \sim 3$ and 4, consistent with both galaxies being at $z = 5.34 \pm 0.01$.

Weymann et al. (1998) obtained a Keck spectrum of another V–band dropout from the HDF. As in the other case, the object had been previously selected by Lanzetta et al. (1998) as a candidate, with an estimated photometric redshift of $z_{ph} = 6.8$. The Keck spectrum shows a single emission line that, if interpreted as Lyα, places the galaxy at $z = 5.60$. As is often the case, the line profile is asymmetric, with clear evidence of absorption on the blue side, supporting the redshift interpretation.

Combining their deep HST Near Infrared Camera and Multi-Object Spectrometer (NICMOS) photometry of the HDF in J and H with the existing WFPC2 optical data, Dickinson et al. (2000) reported the detection of a J–band dropout. The optical through near-IR colors of this object (in the observed frame) can be matched by those of a dusty galaxy at $z \gtrsim 2$, by those of an old elliptical galaxy at $z \gtrsim 3$, or by an object at $z \gtrsim 10$ where the cosmic opacity has suppressed the light observed shortward of 1.1 μm.

A number of other spectroscopic observations of dropout candidates at $z > 4$, selected either from the HDF or other ground-based surveys, have been reported in the past few years, although most remain unpublished at the time of this writing (see Stern & Spinrad 1999).

Other galaxies or galaxy candidates at $z > 4$ have been identified from the putative Lyα emission by means of narrowband imaging (Hu et al. 1998, 1999; Thommes et al. 1998), serendipitous long-slit spectroscopy (Dey et al. 1998, Hu et al. 1998) or slitless spectroscopy from HST (Chen et al. 1999). If confirmed, the latter discovery would hold the record of the most distant galaxy, with a probable redshift of $z = 6.68$ (based on a spectrum with a single emission line), although recent observations cast doubt on this identification (Chen et al. 2000).

Lyα Galaxies

High-redshift galaxies identified by the strength of their Lyα emission are currently routinely detected thanks to the combination of telescopes with larger aperture and wide-field detector arrays, including cases with redshift as high as $z \sim 5.7$. The recent surveys mostly consist of narrow-band imaging (Cowie & Hu 1998; Hu et al. 1998, 1999; Thommes et al. 1998; Kudritzki et al. 2000; Steidel et al. 2000; Rhoads et al. 2000; Rhoads & Malhotra 2001; Malhotra & Rhoads 2002; Stiavelli et al. 2002), but serendipitous detections with long-slit spectroscopy (Pascarelle et al. 1996a, 1998; Dey et al. 1998, Hu et al. 1998) have also been reported. These galaxies are similar to the LBGs in that they also are characterized by a high UV luminosity that is relatively unobscured by dust, although the proportions of Lyα and continuum emission might differ. The selection function of narrow-band surveys, which cover a much narrower redshift range than continuum-selected surveys, results from a relatively complex combination of equivalent width sensitivity and flux sensitivity, and only the fraction of galaxies with a relatively large equivalent width, e.g., rest-frame $W_\alpha \gtrsim 20$ Å, can be detected and included in the samples. The distribution of Lyα equivalent width of LBGs (Steidel et al. 2000, Shapley et al. 2001), which are selected independently of their Lyα emission, shows that down to the flux limit of the ground-based survey this fraction is small. Therefore, narrow-band surveys miss a large fraction of the galaxies that are identified by the continuum emission. However, they are sensitive to galaxies with fainter continua than LBGs, provided that they have large enough Lyα emission, and in this sense, they provide a nice complement to continuum-selected surveys. Furthermore, they become competitive with continuum-selected ones at very high redshifts (e.g., $z > 5$), where the galaxies are faint and the continuum selection is inefficient. Even at lower redshifts, they are also very useful to map the spatial distribution of galaxies in narrow redshift intervals, e.g., where large concentrations of galaxies have already been identified. In such cases, the Lyα selection allows one to build large samples that would be much more expensive (in terms of telescope time) or even impossible to collect with the Lyman-break technique.

Pascarelle et al. (1996a, 1998), Francis et al. (1997), and Campos et al. (1999) discovered large concentrations of Lyα emitters at $z \sim 2.4$, possibly protogroups or clusters, with narrowband imaging. Steidel et al. (2000) carried out deep Lyα narrow-band imaging of a large concentration of LBGs at $z \sim 3.1$ discovered along the line of sight of one of the ground-based survey fields (Steidel et al. 1998), possibly a protocluster or supercluster. The Lyα selection returned both galaxies that were previously identified as LBGs as well as galaxies that were not, because their continuum is too faint. Based on the luminosity distribution of the Lyα sample, they suggested LBGs and Lyα galaxies belong to a similar population, although it is possible that at fainter continuum magnitudes the Lyα emission is intrinsically larger than that observed in LBGs. Kudritzki et al. (2000) probed similar redshifts, and Cowie & Hu (1998), and Hu et al. (1998) also extended the search to $z \sim 4.5$. Both these teams observed that the Lyα emitters seem to contribute comparable amounts to the cosmic star formation at their redshifts as LBGs, and concluded that Lyα searches are better suited to study faint, young, and relatively dust-free star-forming galaxies at $z \gtrsim 4$, where color selection becomes progressively more difficult.

Hu et al. (1998) report the detection of galaxies with very large equivalent width ($W_\alpha \gg 100$ Å rest-frame) and continua too faint to be detected by current continuum searches (except the HDF, which, however, covers too little an area to be effective). Rhoads et al. (2000) and Rhoads & Malhotra (2001) also find large numbers of similar objects. These Lyα galaxies with large equivalent width are very rare in LBG samples at $z \sim 3$ (e.g., Steidel et al. 2000, Shapley et al. 2001) and perhaps even at $z \sim 4$ (Steidel et al. 1999), possibly because their continua are too faint. The Lyα samples, however, seem to include too few galaxies with small W_α, i.e., in the 50–100 Å range (rest-frame), which represent a considerable fraction of the equivalent width distribution of LBGs (Steidel et al. 2000, Shapley et al. 2001), even if current narrowband surveys are sensitive to these objects. Malhotra & Rhoads (2002) suggest that these very large Lyα equivalent widths occur during the early stages of evolution of starburst galaxies, when the first generations of massive stars have, essentially, null metal content and, possibly, also a top-heavy Initial Mess Function (IMF). Unfortunately, there is currently too little spectroscopy of narrow-band-selected galaxies to constrain their nature with confidence and in particular, to estimate the contribution of AGN. This situation should improve very rapidly, however.

THE NATURE OF LYMAN-BREAK GALAXIES

Starburst Galaxies at ∼10% of the Hubble Time

Since the first spectroscopic identifications, it became immediately clear that the sources identified by the color selection criteria discussed above are starburst galaxies (Steidel et al. 1996a,b; Lowenthal et al. 1997; Trager et al. 1997). At the redshifts of the LBGs, $z \geq 2$, the far-UV spectrum is observed at optical wavelengths, and sensitive CCD ground-based spectroscopy from large-aperture telescopes reveals a great deal about the nature of these galaxies, even at the relatively low spectral resolution used for redshift identification. This is illustrated in Figure 4, where examples of spectra of U–band dropouts obtained with the Keck telescope and the Low Resolution Imaging Spectrometer (LRIS) instrument (Oke et al. 1995)

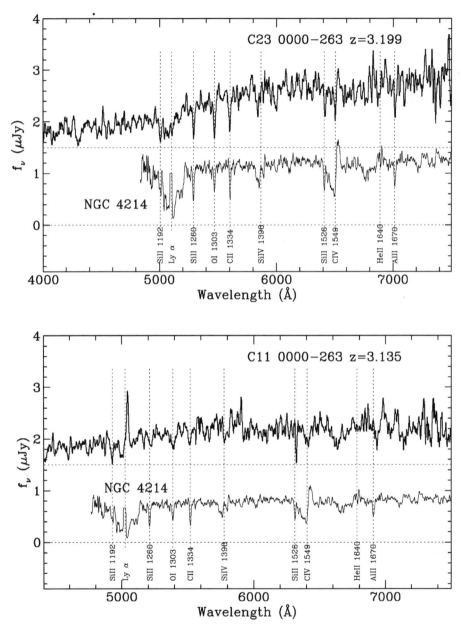

Figure 4 Examples of spectra of Lyman-break galaxies compared to the starburst galaxy NGC 4214. The high-redshift spectra have been obtained with the Keck telescope and the LRIS spectrograph; the spectrum of NGC 4214 has been recorded by Leitherer et al. (1996) with HST and the GHRS. The spectra are shown in the observer's frame, at the indicated redshift. The figure illustrates the similarity of the spectra of Lyman-break galaxies with that of local starburst galaxies, but also the diversity found among the LBG spectra (reproduced from Steidel et al. 1996b).

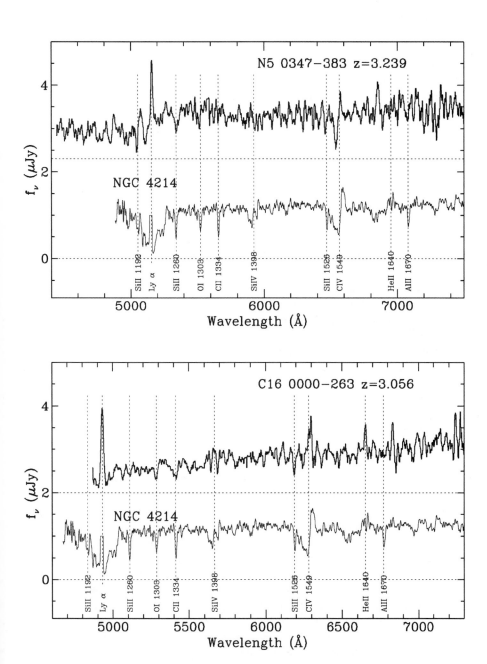

Figure 4   (*Continued*)

are compared to the local starburst galaxy NGC 4214 observed with HST and the GHRS spectrograph (Leitherer et al. 1996).

The similarity between the spectra of the LBGs and that of local galaxies is striking. Both have strong blue continua, denoting production of massive young stars of the O and B types. Similar to their local counterparts, the LBG spectra generally are redder than the predictions for dust-free spectral energy distributions of star formation from stellar population synthesis (Bruzual & Charlot 1993, Leitherer & Heckman 1995), with no obvious indications that the high-redshift galaxies are characterized by bluer spectra. Despite the copious production of ionizing photons, the Lyα properties of LBGs (Steidel et al. 2000) are also rather similar to those of nearby star-forming galaxies (e.g., Giavalisco et al. 1996a), and the distribution of rest-frame equivalent width is symmetrical around the median value $W_{\alpha,median} = 0$ (Steidel et al. 2000, Shapley et al. 2001). When the line is observed in emission, the equivalent width is typically much weaker (usually by factors of more than 10) than the ionization-bound, dust-free expectations, $W_\alpha \sim 150$ Å, for continuous star formation (Charlot & Fall 1991, 1993). Only a rather small fraction of the galaxies have large equivalent widths, e.g., larger than 20 Å, even if the color selection introduces no bias against objects with large equivalent width, which generally also have unobscured UV continua. If such objects exist in large numbers, they must have continua significantly fainter than the flux limit of the LBG samples, $\mathcal{R} = 25.5$.

Another characteristic of the LBGs' UV spectra in common with local starbursts is the presence of strong interstellar absorption lines due to low-ionization stages of C, O, Si, and Al, and of prominent high-ionization stellar lines of He II, C IV, Si IV, and N V. These stellar and interstellar lines are actually the most distinctive spectral features, which make confirmation of LBG candidates with optical spectroscopy relatively easy and economical in terms of telescope time, given the sensitivity and multiplexing capability of current instrumentation, even in the absence of strong emission lines such as Lyα. The interstellar lines, which can have rest-frame equivalent widths as large as $W_0 \simeq 2$–3.5 Å, are evidence that the interstellar medium of these galaxies has obviously undergone some degree of chemical enrichment. However, they cannot be used, at least in these low-resolution spectra, to measure the metallicity, because they are undoubtedly heavily saturated, and hence are much more sensitive to the kinematics of the gas than to its column density. The velocity widths are generally large, with FWHM in the range ≥ 400–700 km s^{-1}, implying velocity fields of the order of $\sigma_v = 200$–400 km s^{-1}, very likely large-scale outflows driven by the release of mechanical energy of supernovae and winds from massive stars (Pettini et al. 2001). Evidence of such large outflows, which are common in starburst galaxies (Heckman 1996), also comes from the relative kinematics of interstellar absorption lines, Lyα emission lines, and optical nebular emission lines of the HII gas (observed with near-IR spectroscopy), which suggests the presence of large-scale gas shells expanding with velocity as high as ~ 600 km s^{-1} (Pettini et al. 2001, Frye et al. 2002).

Despite the morphological similarity of the spectra, the UV emission of the LBGs is much more intense than that of the local star-forming regions. For example, an L^* LBG at $z \sim 3$, observed with apparent magnitude $\mathcal{R} = 24.5$, has $M_{1700} \sim$ -21.2 (Steidel et al. 1999). This corresponds to a continuum specific luminosity at 1700 Å of $L_{1700} = 5.5 \times 10^{40} \, h^{-2} \, \text{erg s}^{-1} \, \text{Å}^{-1}$, approximately 1500 times higher than the knot of star formation in NGC 4214 and ≈ 100 times higher than the brightest such knots seen in nearby starburst galaxies. Assuming an initial mass function (IMF) with a Salpeter index $x = 1.33$ and extending from $0.1_\odot$ to 100 $M_\odot$, the far-UV continuum of the typical LBG is produced by the equivalent of $\approx 6 \times 10^5$ O5 stars. Despite this difference in scale, however, the dominant characteristics of the far-UV spectra in common with both low-redshift and high-redshift starbursts seem to point to a similarity of physical conditions in the regions of star formation.

Figure 4 is also illustrative of the diversity of properties encountered in the LBG spectra (Steidel et al. 1999; see also Pettini et al. 2000), most noticeably the strength and morphology of the Lyα line, of the interstellar absorption lines of Si II, O I, and C II, and of the P-Cygny C IV features. For example, the spectrum on the top panel does not have Lyα in emission and shows a broad absorption feature around the wavelength of the line. It also has relatively narrow interstellar absorption lines of Si II, O I, and C II, and a relatively well detected C IV feature. The spectrum in the bottom of the figure has Lyα in emission together with much broader interstellar lines, and a different morphology of the absorption component of the C IV feature. In general, while there is a large variety in the strengths of the lines that are predominantly of stellar origin (C IV $\lambda 1549$, Si IV $\lambda\lambda 1393$, 1402, and He II $\lambda 1640$), these features seem to be weaker than in present-day starbursts. Steidel et al. (1996b) suggest that this is probably an abundance effect; these lines are formed predominantly in the winds of massive stars, where both mass-loss rates and wind terminal velocities are known to depend sensitively on metallicity (e.g., Walborn et al. 1995).

The optical spectra of LBGs have recently been studied with near-IR spectroscopy (Pettini et al. 1998, 2000; Kobulnicky & Koo; Teplitz et al. 2000a,b). The observations targeted the optical nebular emission lines of [OII]$\lambda 3727$, Hβ, and [OIII]$\lambda\lambda 4959$, 5007, because the continuum emission generally is too faint to be detected with useful signal-to-noise ratio. As expected from starburst galaxies, these features are generally strong and although narrow, they have been resolved in every case, raising the possibility of deriving dynamical information on the galaxies. Despite the still relatively modest signal-to-noise ratios, these features provide a wealth of information on the activity of star formation, extinction, metallicity, and kinematics of the gas in LBGs, as we discuss in the next sections.

STAR-FORMATION RATES

An obvious important issue in understanding both the evolution of the individual galaxies and their contribution to the cosmic stellar mass budget is the star-formation activity of the LBGs. In the absence of dust obscuration, the far-UV luminosity of star-forming galaxies, which is dominated by the integrated light of the short-lived massive stars (O and early B types) formed in the burst, is simply proportional to the instantaneous rate of star formation (Kennicutt 1998). The conversion between UV luminosity density and star-formation rate can be computed from population synthesis models (e.g., Bruzual & Charlot 1993), and the optimal wavelength range to derive the constant of proportionality is $1250 \lesssim \lambda \lesssim 2500$, most of which is covered by the LBG surveys. Calibrations have been derived by several authors (Buat 1989, Deharveng et al. 1994, Leitherer & Heckman 1995, Meurer et al. 1995, Cowie et al. 1997, Madau et al. 1998), and the results agree with each other within a factor of ≈ 2, when converted to a common reference wavelength and initial mass function. The difference between the various derivations reflects differences in the used stellar libraries and/or different assumptions about the timescale of the star formation, namely if longer or comparable to the lifetime of O and B stars. Assuming Salpeter's (1995) IMF with mass limits between 0.1 and 100 $M_\odot$, and that the duration of the star-formation activity is significantly longer than the lifetime of the stars, i.e., $T > 5 \times 10^7$ yr (continuous star-formation approximation), Madau et al. (1998) derived the relationship:

$$\text{SFR } M_\odot \text{ yr}^{-1} = 1.4 \times 10^{-28} L_\nu(\lambda_{1500}), \tag{10}$$

where L_ν is the luminosity density in units of erg s^{-1} Hz^{-1}. The luminosity of an instantaneous burst is larger, and use of this relationship in such cases would overestimate the star-formation rate.

Unfortunately, the direct UV emission from the forming stars is seldom observed, because they are often embedded in an ISM rich in dust, which reprocesses the UV radiation and re-radiates it over a huge wavelength range, from the UV itself to radio wavelengths (Calzetti 2001, Adelberger & Steidel 2000). For example, in virtually all starburst galaxies in the local universe, most of the energy powered by star formation emerges at far-IR wavelengths as thermal emission by dust. Thus, to measure the star-formation rate one is left with the options of either carrying out the observations in the far-IR or correcting the UV data for dust obscuration. The former is a very hard task, because far-IR observations are comparatively much more difficult and less sensitive than optical ones. Adelberger & Steidel (2000) suggest that as far as the detection of the galaxies is concerned, targeting the UV emission from star-forming galaxies is still the most economical and efficient way to detect them, despite the large fraction of energy destroyed by dust and the uncertainty involved in estimating it.

In general, it is very difficult to infer the amount of dust obscuration from the observed UV-spectral energy distribution, because different geometries of the dust spatial distribution result in different observed colors for the same amount of dust and the same intrinsic spectrum. Luckily, the case of UV-bright starburst galaxies is simpler because these systems show a tight correlation between the slope of the UV continuum and the total dust obscuration as traced by the ratio of the far-IR to bolometric luminosity (Calzetti 1997, Meurer et al. 1999). This property allows one to calibrate the observed UV colors in terms of dust obscuration and hence estimate the intrinsic UV luminosity and the star-formation rate. However, Ultra Luminous Infrared Galaxies (ULIRGs) markedly deviate from such a relationship (Goldader et al. 2002), whereas galaxies with infrared properties intermediate between the ULRIGs, and the UV-bright ones (Alonso-Herrero et al. 2001, Forster et al. 2001) exhibit a less pronounced deviation. The behavior of LBGs is not currently known.

The simple behavior of the dust opacity in starburst galaxies is very likely due to the combined effects of the specific geometry of the dust that forms in the starburst environment and of the duration of the burst. The diffuse nature of the starbursts in UV-bright galaxies favors the onset of large-scale galactic winds that are conducive to the formation of a foreground screen of dust, which apparently is the main factor behind the proportionality between UV-spectral index and dust obscuration (Calzetti 1997, 2001) observed in these systems. Diffuse starbursts also seem to last significantly longer than the lifetime of O and B stars (Calzetti 1997, Tremonti et al. 2001), implying a relatively narrow dispersion of intrinsic UV-spectral energy distribution, namely that of continuous star formation. These conditions are not verified in the case of the ULIRGs, where the high concentration of gas and dust required to fuel large nuclear starbursts favors a combination of a clumpy mixed and foreground distribution of dust and stars as well as intense bursts of short duration (Genzel et al. 1998).

The UV and optical spectral energy distribution and far-IR emission (Barger et al. 1999b, Chapman et al. 2000, Van der Werf et al. 2001) of LBGs suggest that their dust obscuration properties are similar to those of UV-bright starburst galaxies (Adelberger & Steidel 2000, Calzetti 2001). In this case, the starburst attenuation law can be used to infer the reddening and obscuration correction. Figure 5 shows the $G - \mathcal{R}$ colors of LBGs as a function of redshift together with the predicted colors for continuous star formation ($\tau = 0.1$ Gyr) for selected values of $E(B - V)$ and for an unreddened burst of star formation ($\tau = 0.001$ Gyr). For a given redshift, almost all LBGs are redder than the unobscured burst, which represents the blue envelope of the distribution, and most of them are redder than the unobscured continuous star formation, with the median value in the range $0.15 \lesssim E(B - V) \lesssim 0.2$. For reference, using the starburst attenuation law, this corresponds to an attenuation at 1700 Å (the rest-frame wavelength that corresponds to the $\mathcal{R}$ band at $z = 3$) in the range $4 < C(1700) < 7$, where $C(\lambda) = 10^{0.4 \times A_\lambda \times E(B-V)}$. Note that a number of galaxies have colors that appear artificially bluer because of the presence of Lyα emission with large equivalent width. Also, selection effects bias the color distribution toward bluer values, implying that the fraction of redder galaxies, either because intrinsically redder or because at higher redshift, is higher than observed, although this effect is probably small (Steidel et al. 1999).

Figure 6 (see color insert) compares the star-formation rates derived from the observed UV fluxes (heavily shaded histogram) with those corrected from dust obscuration in two cases. In the first case (medium-shaded histogram), the correction is calculated from the observed UV colors under the assumption of the starburst attenuation law and that the unattenuated spectrum is that of continuous

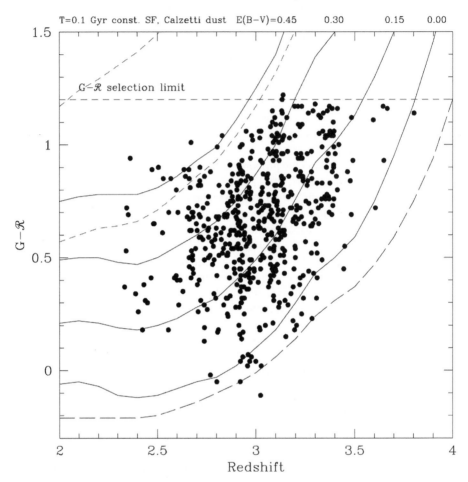

T=0.1 Gyr const. SF, Calzetti dust E(B−V)=0.45   0.30   0.15   0.00

Figure 5 Scatter plot of the $G - \mathcal{R}$ color versus redshift of U–band dropouts. The curves show the predicted colors for various choices of star formation history and dust obscuration. The continuous and dashed curves show the case of continuous star formation with Salpeter IMF and age $T = 0.1$ Gyr with the starburst attenuation law (*continuous curves*) and SMC extinction law (*dashed curves*). The amount of obscuration is parameterized by the $E(B - V)$ color excess, as labeled. For the SMC curves only the first three cases from $E(B - V) = 0$ through $E(B - V) = 0.3$ (*the curve to the upper left*) can be seen. The dotted curve shows a burst with $T = 0.001$ Gyr with no dust obscuration, which defines the blue envelope of the color distribution. Some galaxies have bluer colors because of the strong Lyα emission in the G band. The median color excess is $E(B - V) \approx 0.17$ and most galaxies are between $E(B - V) \sim 0.1$ and $E(B - V) \sim 0.35$. The cosmic opacity is calculated as prescribed by Madau (1995).

star formation. In the second case (lightly shaded histogram), the unobscured UV luminosity (and hence the star-formation rate) is calculated from fitting broadband photometry covering from the rest UV to the optical to population-synthesis models (Papovich et al. 2001, Shapley et al. 2001) to derive the best fitting intrinsic spectral energy distribution, extinction, age, and stellar mass, under the assumption of initial mass function, metallicity, star-formation history, and extinction law (we describe this procedure in more detail later). It is interesting that there is a broad agreement between the results from the two techniques, given the large uncertainties involved. After the correction, most rates exceed 10^2 M$_\odot$ yr^{-1}, with some reaching as high as $\sim 10^3$ M$_\odot$ yr^{-1}. At these rates, $\sim 10^{11}$ M$_\odot$ worth of stars, the value observed today in $L > L^*$ galaxies, can be assembled in ~ 1 Gyr. Shapley et al. (2001) suggest that there is a statistically significant correlation between the star-formation rate and age of the star-formation activity, where younger bursts have higher rates and larger dust obscuration.

Independent information on the star-formation rate and extinction can be obtained from the optical nebular emission lines, in particular Hβ and Hα, where the effect of extinction is less pronounced than in the UV. Teplitz et al. (2000b) obtained high S/N near-IR spectra of the lensed LBG MS1512–cB58, and measured the flux of optical nebular lines, from [O II]λ3727 to Hα and [N II]λ6583. They report that the star-formation rate derived from Hα is ≈ 4 times smaller than that derived from the UV continuum, despite an extinction of $E(B-V) \sim 0.27$ measured from the Balmer decrement. They suggest that this discrepancy is probably due to the different calibrations of star-formation rates used for the Hα and the UV continuum, (e.g., Kennicutt 1998), implying that the magnitude of this uncertainty is comparable to the relative effects of reddening at optical and UV wavelengths. For comparison, with the same value of $E(B-V)$, the starburst law predicts that the extinction at 1700 Å is ≈ 4 times larger than that at 6563 Å, whereas the SMC extinction law predicts a factor ≈ 7. Kobulnicky & Koo (2000) observed two galaxies, one Lyα-emitting galaxy at $z = 2.3$ and one LBG at 2.9, and reported the star-formation rates derived from the Balmer lines to be 2 to 3 times larger than those derived from the UV continuum, although no correction for extinction was made because it was not possible to measure the Balmer decrement. Moorwood et al. (2000) obtained spectra for six galaxies, selected from Hα-emitting galaxies (not LBGs), at $z \sim 2.2$ and found that in two cases the star-formation rates derived from the line flux are ≈ 4 times larger than those obtained from the UV continuum, whereas in the other four cases the rates are approximately the same.

Pettini et al. (2000) observed a comparatively larger sample, including 14 galaxies selected from the ground-based survey. They find that the star-formation rates derived from the Hβ flux are, on average, no larger than those derived from the UV continuum. Whereas a few galaxies can be found that have larger Hβ rates than the UV ones, looking at the whole sample reveals that this is just an effect of the scatter of the data. Thus, both indicators uncorrected for dust extinction are equally affected by the systematics introduced by dust, and statistically yield the same star-formation rates. Also, the ratio of the two estimators does not show any obvious tendency to correlate with the UV-spectral index derived from the observed UV color $G - \mathcal{R}$, which in local starburst galaxies is a good tracer of the dust extinction (Calzetti et al. 1994). This is consistent with the results from the other observations of the nebular lines of LBGs, and adds strong evidence that the attenuation law of LBGs is relatively gray, adding support to the notion that the starburst attenuation law provides a good description of the reddening properties of these galaxies, also.

Cosmic Star Formation

Madau et al. (1996) have summarized the available empirical information on the evolution of the cosmic activity of star formation with time in terms of the UV luminosity density as a function of redshifts (Madau diagram). Recalling that UV luminosity is a good tracer of the star-formation rate (modulo the obscuration by dust), this can also be expressed as star-formation density per unit of comoving volume. For LBGs, if the luminosity function remains approximately constant within the redshift interval targeted by a given set of color selection criteria, the star-formation density can be estimated as an average of the integral of the luminosity function over the probed cosmic volumes

$$\rho^*(z) = \int f^{-1}(z)\,dz \int S_\lambda L_\lambda \phi(L_\lambda, z)\,dL_\lambda, \tag{11}$$

where $f(z)$ is the redshift selection function and S_λ converts the luminosity L_λ into the star-formation rate. It is important to note that the observed UV luminosity function of the LBGs at $z \sim 3$ implies that $\gtrsim 50\%$ of the total contribution to the star-formation density by galaxies with $\mathcal{R} \leq 27$ is produced by galaxies with $\mathcal{R} \lesssim 25$.

The original compilation of data by Madau et al. (1996, 1998) covered the redshift range $0 < z < 5$ using the LBGs from the Hubble Deep Field in combination with galaxies from other spectroscopic and photometric surveys at lower redshifts. For the LBGs they assumed that the adopted color selection criteria for the $U-$ and $B-$band dropouts uniformly probe the redshift intervals $2.0 < z < 3.5$ and $3.5 < z < 4.5$, respectively, estimated by convolving the WFPC2 photometric system with population synthesis spectra (i.e., $f(z) = 1$ in these intervals). They estimated the LBG luminosity functions from the magnitude distribution of the HDF sample (Dickinson 1998) and did not attempt any correction for dust extinction. The corresponding Madau plot shows a rapid increase of the star-formation activity from $z = 0$ to $z \sim 1$ (Lilly et al. 1995, 1996), where a broad peak seems to be present, followed by a decrease at higher redshifts (see Figure 7, triangles).

This peak, however, does not seem confirmed by the data of the ground-based survey (Steidel et al. 1999). The data points at $z \sim 3$ and $z \sim 4$ from the ground-based survey, which covers ~ 200 times more area than the HDF and benefits from

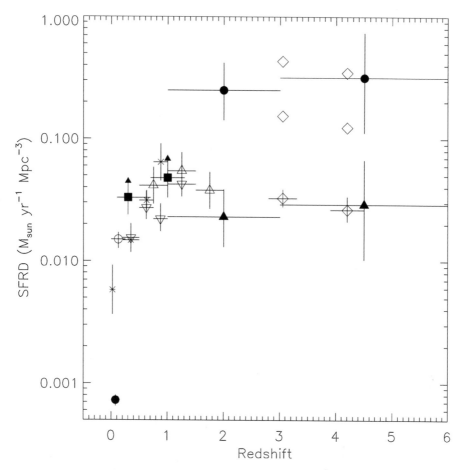

Figure 7 Star-formation density traced by a variety of star-forming galaxies at different redshifts (Madau diagram). The *open diamonds* with error bars are the data points at $z = 3$ and 4 of LBGs from the ground-based survey with no dust correction included (Steidel et al. 1999). The *filled triangles* are SCUBA galaxies with flux >6 mJy, whereas the *filled circles* are SCUBA galaxies integrated down to 0.2 mJy. The two sets of *open diamonds* with no error bars represent two different estimates of the dust obscuration correction for the LBGs, as explained in the text (reproduced from Barger et al. 2000).

almost 1000 redshifts, are consistent with a constant star-formation density at least up to $z \sim 5$ (Steidel et al. 1999). The difference from the previous work includes (*a*) a proper measure of the volume density, which has been derived using the observed redshift distributions; (*b*) an estimate of the effects of incompleteness with Monte Carlo simulations performed on the data themselves; and (*c*) an attempt to include the effect of dust obscuration. For consistency, data points at all redshifts have been corrected with the same technique, assuming the same mean $E(B-V) \sim 0.15$, the Calzetti (1997) starburst extinction law, and the implied correction at $z \sim 1$ is consistent with that derived by Glazebrook et al. (1998) from Hα spectroscopy and with those derived by Tresse & Maddox (1998) from direct UV observations at $z \sim 0.3$.

It is important to observe that the ground-based data points at $z \sim 3$ and $z \sim 4$ do not support the peak at $z \sim 1$ even without corrections for dust obscuration. This result is particularly interesting, because it actually does not depend on the assumptions about the luminosity function. Even if the integral in Equation 10 is extended over the observed luminosity, as opposed to the extrapolated luminosity, the two data points would be the same within the errors. Steidel et al. (1999) calculated the star-formation density at each redshift by integrating the luminosity functions down to $0.1L^*$, which corresponds to $\mathcal{R} \sim 27$ at $z \sim 3$. Because of the small size of the HDF sample, the luminosity function at $z \sim 3$ and (especially) $z \sim 4$ have not been convincingly measured down to such faint luminosity. This is one of the major uncertainties that affect this type of comparison, although if the slope is not significantly steeper than $\alpha \sim -1.6$ (as the data at $z \sim 3$ seem to show), the contribution of galaxies with $m > m^*$ to the luminosity density is relatively small. Finally, note that Lanzetta et al. (2002) suggest that current estimates of the the star-formation density at high redshift have been underestimated because of the strong bias introduced by the $(1 + z)^4$ surface brightness dimming, and that the cosmic star-formation density increases monotonically with redshift up to the highest observed redshifts.

OTHER STARBURST GALAXIES AT HIGH REDSHIFTS

In the past few years, faint surveys at wavelengths of 450 and 850 μm with the Submillimeter Common-User Bolometer Array (SCUBA) at the James Clerk Maxwell telescope (JCMT) have identified starburst galaxies at high redshifts by targeting their far-IR emission, which at redshift of a few is redshifted to the submillimeter spectral region (Smail et al. 1997; Barger et al. 1998, 1999a, 2000; Hughes et al. 1998; Blain et al. 1999a,b; Eales et al. 1999; Lilly et al. 1999). In the local universe, most of the UV luminosity produced by young, massive stars in starburst galaxies, which dominate the total bolometric luminosity, is absorbed by dust and re-radiated as far-IR wavelengths (Heckman et al. 1998; see also Calzetti 2001, Adelberger & Steidel 2000). The spectral energy distribution of the reprocessed radiation reaches a maximum in the range $60 \lesssim \lambda \lesssim 100$ μm (Sellgren 1984, Helou 1986) due to the thermal emission of hot dust heated by stellar radiation at UV and optical wavelengths. If starbursts at high redshifts are similar to the local ones, observations at submillimeter wavelengths will detect them up to $z \lesssim 10$ thanks to the negative k-correction that results when the spectral region redward of the peak is redshifted to ~ 850 μm, which more than compensates for the dimming due to the luminosity distance. For example, a dusty star-forming galaxy observed at $\sim$850 μm would require approximately constant sensitivity to be observed within the range $2 \lesssim z \lesssim 10$ (Blain et al. 1998). Such observations would be ideal to derive an accurate census of the star-formation activity of the universe at those epochs were it not for the fact that they reach limited sensitivity due to the generally high opacity of the atmosphere in most of those bands and the very low angular resolution of the instrumentation (at 850 μm, SCUBA has a beam size of $\sim$14 arcsec), which also hampers the secure identification of the sources with their optical and near-IR counterparts (e.g., Adelberger & Steidel 2000). In comparison, despite most of the luminosity being absorbed, UV observations are much more efficient at detecting and identifying the galaxies, except that reconstructing the bolometric luminosity is highly uncertain.

One of the most important questions is what is the nature of the SCUBA sources, namely if mostly starburst or mostly AGN, and what is their contribution to the cosmic star-formation activity at high redshift compared to that of the LBGs? Because the direct identification of SCUBA sources with optical and near-IR counterparts is generally very uncertain, the redshift distribution of the SCUBA sources is poorly constrained.[3] In some cases, likely optical and near-IR candidate counterparts to SCUBA sources have been identified (Hughes et al. 1998, Lilly et al. 1999, Smail et al. 1999, Scott et al. 2000), and in a few cases, the additional detection of these counterparts in high-resolution radio images makes the identification more likely (Barger et al. 1999b, 2000), including bona fide LBGs at redshifts $z \sim 2.5$. Other likely identifications rely on the optical images alone, and the redshifts of most of these galaxies are significantly lower than those of the LBGs (Lilly et al. 1999, Scott et al. 2000).

Because the far-IR radiation does not suffer significant reprocessing, and it is an indicator of the star-formation rate (Kennicutt 1998, Adelberger & Steidel 2000, Calzetti 2001), a very useful upper limit to the total amount of star formation at high redshifts is set by the value of the diffuse background at 850 μm measured

---

[3]Note that if starburst galaxies at $z \gtrsim 6$ significantly contribute to the SCUBA population, they are unlikely to have detectable counterparts in current imaging or spectroscopic surveys, even if relatively unobscured. If such galaxies have a luminosity function similar to that of the LBGs at $z \sim 3$ and 4, they would appear extremely faint in even the deepest images and certainly beyond current spectroscopic capabilities. Regardless of the likelihood of the proposed identifications, another caveat to keep in mind is that there is neither evidence nor compelling reason to believe that in every case, the same energy output that powers a SCUBA source also powers its optical counterpart (rest-frame UV), even in the same galaxy. One could have, for example, galaxies hosting nuclear starburst regions heavily obscured by dust with no detectable UV radiation that are very bright in the far-IR, which also have circumnuclear UV-bright starburst regions with low dust obscuration (Kennicutt 1998). Such galaxies would be detected in both submillimetric and optical surveys, but in this case the observed UV and far-IR luminosity are not quantitatively related. In this regard, note that Blain et al. (1999a) suggested that violent, dust-obscured star formation might represent a transient phase (triggered, for example, by a close encounter or a merging event) in the life of an otherwise relatively more quiescent starburst galaxy.

by the Cosmic Background Explorer (COBE) (Puget et al. 1996, Hauser et al. 1998, Fixsen et al. 1998, Schlegel et al. 1998). At bright flux levels, e.g., >6 mJy, a fraction of the order of 75% of the SCUBA sources are powerful starbursts in the redshift range $2 < z < 5$, the high-redshift equivalent of ULIRGs, although they contribute only ~5% of the background (Barger et al. 2000). Down to the confusion limit of 2 mJy, discrete SCUBA sources contribute 20–30% of the background (Barger et al. 1999a), although it is difficult to understand their nature, because of the poor resolution. Of the handful of optical identifications that are based on extensive multiwavelength data sets, approximately one half are AGN, and a few have been associated with star-forming galaxies (Frayer et al. 1999, Dey et al. 1999, Chapman et al. 2000, Frayer et al. 2000, Ivison et al. 2000). Barger et al. (2000) have shown that bright sources with flux >6 mJy contribute as much star formation as LBGs down to $\mathcal{R} \sim 27$ uncorrected for dust obscuration (Steidel et al. 1999), whereas integrating along the SCUBA luminosity distribution down to ~0.5 mJy, the contribution is about one order of magnitude larger (Smail et al. 1998, Barger et al. 2000), even if LBGs are about one order of magnitude more abundant than SCUBA galaxies (e.g., Lilly et al. 1998). At these flux levels, the contribution to the background is estimated to be ~90% (Blain et al. 1999b), implying that the faint SCUBA sources host most of the star formation at high redshift. Correcting the UV luminosity of LBGs for dust obscuration yields a similar star-formation density to that of the SCUBA galaxies (see Figure 7).

To compare the contribution of the cosmic star-formation density of LBGs and SCUBA galaxies one needs to know the amount of overlap between the two classes of sources, i.e., the contribution of the LBGs to the SCUBA faint counts and to the submillimeter background. Empirically, the situation is very uncertain. Adelberger & Steidel (2000) suggest that LBGs should have fainter far-IR emission than ULIRGs, which are powered by huge nuclear starbursts (Sanders & Mirabel 1996) and have their UV emission largely obscured by vast amounts of dust. Indeed, only a small fraction of LBGs observed with SCUBA have been detected at flux levels of a few mJy (Barger et al. 1999b, Chapman et al. 2000, Sawicki 2001, Van der Werf et al. 2001). At these flux levels, the submillimeter counts are probably dominated by AGN and cosmologically distant ULIRGs with redshift comparable to that of LBGs (Hughes et al. 1998, Barger et al. 1998, Lilly et al. 1999, Smail et al. 1998, Ivison et al. 2000), supporting the notion that LBGs have properties closer to those of UV-bright starbursts (Aldeberger & Steidel 2000, Calzetti 2001) and that they significantly contribute to the SCUBA counts mostly at the faintest levels currently explored.

Modeling the contribution of LBGs to the SCUBA counts is also uncertain, because it requires essentially uncontrolled assumptions about the properties of starburst galaxies at high redshifts. Calzetti et al. (2000) estimate that the detectability with SCUBA of LBGs strongly depends on the far-IR Spectral Energy Distribution (FIR SED), in particular the submillimeter/total-FIR ratio, and suggest that metal-poor LBGs (e.g., <1/7 solar) might not be detectable at all. Adelberger & Steidel (2000) noted that the contribution of starbursts to the submillimeter background is produced in a large redshift range, approximately $1 < z < 5$, and calculated that if these galaxies have (*a*) far-IR SED in the same range as observed in the local universe; (*b*) obey to the same relationship between UV spectral index and bolometric dust luminosity (Meurer et al. 1999), then they are responsible for $\approx75\%$ of the background and provide most of the faint SCUBA counts.

Barger et al. (2000) showed that SCUBA submillimeter sources brighter than 6 mJy at $1 < z < 3$ contribute as much as LBGs uncorrected for dust obscuration (see Figure 7), and that their contribution to the background is only $\sim5\%$, indicating that these sources, which are all ULIRGS, do not host the bulk of star formation at high redshifts. By the same token, however, this shows that the contribution to the background by LBGs at $z > 2$ is significantly larger, because dust does obscure the UV luminosity of these sources. SCUBA sources brighter than ~2 mJy contribute to $\sim40\%$ of the background, and they certainly include ULIRGS at $z > 1$ (Arp 220 at $z = 3$ would be observed with a 850 μm flux of 2 mJy). However, ULIRGS are not the only starbursts contributing to the submillimeter counts at these flux levels, because at least a few LBGs have also been detected (Barger et al. 1999b, 2000; Chapman et al. 2000; van der Werf et al. 2001), although a robust estimate of the fraction of LBGs that are also SCUBA sources at these flux level is not known. SCUBA sources down to the faintest 850 μm flux levels observed account for a major fraction, and possibly for all, of the submillimeter background (Blain et al. 1999, Barger et al. 2000), indicating that the galaxies that produced the bulk of the stars at $z > 2$ are to be searched among them.

The key question is, therefore, how many LBGs contribute to these faint SCUBA counts, and for this no direct empirical constraints exist. Peacock et al. (2000) show that the large fluctuations of diffuse light in deep SCUBA images are comparable to the strength of the spatial clustering of LBGs, arguing that faint SCUBA galaxies and LBGs largely coincide. Adelberger & Steidel (2000) and Papovich et al. (2001) predict that if LBGs obey the same relationship of proportionality between the UV-spectral index β and the ratio of UV bolometric luminosity to dust bolometric luminosity and also have a similar far-IR spectral energy distribution as local starburst galaxies, then they dominate the faint SCUBA counts. The first assumption is a direct consequence of the spatial distribution of dust in the ISM of starburst galaxies (Calzetti 1997, 2001; Calzetti et al. 2000), and it seems reasonable to expect that it also is verified by LBGs (Adelberger & Steidel 2000); the validity of the second assumption, however, is much more difficult to assess, because it depends on the specific physical conditions of the ISM of the galaxies, primarily the temperature and composition of the dust (Kennicutt 1998, Calzetti 2001). These could be different in the high-redshift galaxies, which have a much larger star-formation rate per unit volume than their local counterparts (Meurer et al. 1997). The star-formation density of LBGs at $2 < z < 4$ corrected for dust obscuration is roughly comparable to that of the faint SCUBA galaxies. This is illustrated in Figure 7, which shows the star-formation density of LBGs at $z = 3$ and 4 after correction for dust obscuration. Two dust-corrected points are plotted in the figure at each redshift to illustrate the uncertainty that affects the correction. The lower value corresponds to an average multiplicative correction of a factor 5 estimated from the observed distribution of UV colors, under the assumption of continuous star formation (Steidel et al. 1999, Adelberger & Steidel 2000); the upper value corresponds to a correction of a factor 13, and is derived from fitting multiwavelength photometry to stellar population synthesis models (Papovich et al. 2001, Shapley et al. 2001, see also Meurer et al. 1999).

CHEMICAL ABUNDANCES

Measuring the chemical abundances of LBGs is very important in understanding the evolutionary link to present-day galaxies. We have discussed earlier that although the presence of strong interstellar absorption lines is evidence that these galaxies have undergone chemical enrichment, the features cannot be used to measure it because they are heavily saturated. Under these circumstances, values of metallicity anywhere between 1/1000 of solar and solar are compatible with the line strengths, and observations at higher resolution—observations of intrinsically weaker lines are required to measure elemental abundances (Pettini & Lipman 1995). These have been attempted by Pettini et al. (2000) for the gravitationally lensed galaxy cB58, whose bright apparent magnitude allowed them to reach comparatively high S/N ratio. The resolution of the recorded spectrum of cB58, although ten times higher than that of the spectra obtained for redshift identification, is still an order of magnitude lower than that normally required to measure elemental abundances from interstellar lines. In this case, one can consider the weakest ($W_0 \leq 0.5$ Å) gaseous lines in the spectrum (Ni II, Si II, and S II), and derive metallicity by assuming that the saturation effects are unimportant to estimate the column density of the observed species. Similar to what is frequently observed in LBGs, the Lyα line in cB58 contains both an emission and an absorption component, the later characterized by a damped profile. Thus, one can directly estimate the column density of HI and use it to compute the metal abundances in the interstellar gas. This yields values between one third and one fifth of solar, in qualitative agreement with the metallicity of the stars from the morphology of the C IV P-Cygni profile.

More recently, Pettini et al. (2002) obtained a spectrum of cB58 with a resolution of 58 km s^{-1}, and found that the ISM of the galaxy is already highly enriched in elements from Type II supernovae, such as O, Mg, Si, P, and S, which have abundances of $\sim$2/5 their solar value. However, N and the Fe-peak elements Mn, Fe, and Ni are under-abundant by a factor of 3. This implies that the burst that has enriched the ISM of cB58 must be young, of the order of the timescale for release of N from intermediate mass stars, namely $\leq$300 Myr and comparable with the dynamical timescale, suggesting that the galaxy is perhaps a young spheroid.

Direct measures of the metal enrichment of the nebular gas from optical line ratios have been presented by Pettini et al. (2001), Kobulnicky & Koo (2000), and Teplitz et al. (2000b), who obtained near-IR spectra of a sample of LBGs. They used the ratio of [OII] + [OIII] relative to Hβ, known as the R_{23} index (Pagel 1986, McGaugh 1991, Kobulnicky et al. 1999) to measure the oxygen abundance relative to hydrogen of the HII gas. The method is relatively easy and generally accurate within $\sim$0.2 dex. Unfortunately, in the region of values of R_{23} occupied by the LBGs, the calibration of the index onto a scale of oxygen abundance is a double valued function, and for a given value of R_{23}, two values of the oxygen abundance are derived that bracket the true one, effectively defining an interval of uncertainty that is larger than the magnitude of the random error. This degeneracy can be broken with measures of Hα and N [II] flux, which unfortunately are generally unavailable from ground-based observations at the redshifts of the LBGs. This is illustrated in Figure 8, which plots the calibration of the 4 LBGs from the paper by Pettini et al. The figure shows that for three galaxies the metallicity is constrained to within an order of magnitude, between 1/10 solar and near solar. In one lucky case, the R_{23} index is located near the apex of the double valued function, yielding the measure $[O/H] = 0.3$ solar. A similar case has been reported by Kobulnicky & Koo (2000). Of course, whereas these measures are in general agreement with the measures of cB58 discussed above, given the very small sample size, they should not be used to conclude that all LBGs have a metallicity of about one third solar.

Also, note that no correction for dust obscuration has been made to the line fluxes discussed here, because no measure of the Balmer decrement could be made without the Hα line. Dust obscuration, which in starbursts affects the line fluxes about two times more strongly than the continuum for a given amount of extinction (Calzetti et al. 1994), causes the measures of R_{23} to be overestimated. This means that the data point in Figure 8 should be moved to the left of the plot (by an unknown amount), toward a region of larger uncertainty for the final value of [O/H]. Thus, in conclusion, while it is probably fair to say that LBGs are more metal rich than damped Lyα absorbers at the same redshifts and more metal poor than luminous QSOs (Pettini et al. 2000, 2001), it is not possible to say much more about their metallicity other than constraining it between $\sim$1/10 solar and solar value.

Finally, it is worth mentioning that Pettini et al. (2001) and Kobulnicky & Koo (2000) compared the position of LBGs relative to the well-known correlation between metallicity and B–band absolute luminosity observed for star-forming and late-type galaxies. This comparison shows that LBGs are either too luminous for their metallicity or too metal poor for their luminosity. Keeping the metallicity constant, a dimming of about 5 magnitudes is required to bring the LBGs in line with the other galaxies. This likely reflects the fact that LBGs have a rather low M/L ratio, due to their very high activity of star formation.

STELLAR MASS AND AGE

Measuring the stellar mass assembled in LBGs and the age of the stars at $z \sim 3$ and higher is very important in constraining the link to galaxies observed at later cosmic times. These measures require observations at rest-frame optical and preferably

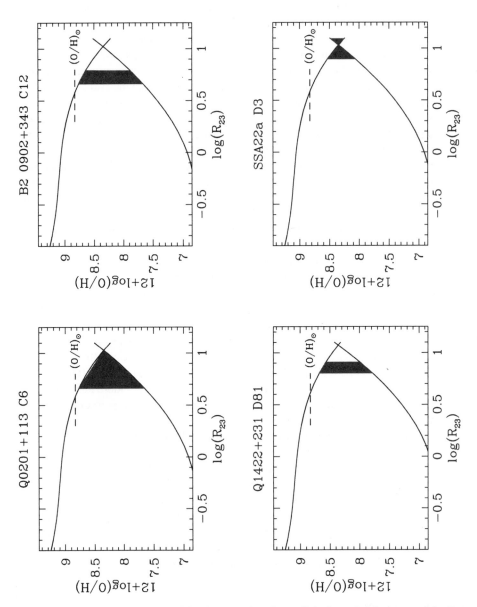

Figure 8   Calibration diagram of the Oxygen abundance (in solar units) in terms of the R_{23} index. The uncertainty in the measure of R_{23} is the width of the shaded band (see text). The value of the solar metallicity is marked by a *horizontal dashed line* [reproduced from Pettini et al. (2001)].

longer wavelengths, because the rest-frame UV spectrum alone provides information only on the young, massive stars of the newly formed stellar population. The current estimates come from fitting UV through optical broadband photometry to stellar population synthesis models to derive the intrinsic parameters of the forming and formed stellar populations, including the stellar mass. The details of the fitting procedure vary, but in general, the stellar populations are modeled with a set of parameters. Some of these parameters, such as the total length of the burst, the age of the burst, and the extinction are adjusted during the fit; others, such as the IMF, the metallicity, and the extinction law, remain fixed and their effect on the results is explored by varying them within preassigned intervals and repeating the fitting procedures.

Sawicki & Yee (1998) first applied this technique to a sample of 17 LBGs galaxies from the HDF–N (Williams et al. 1996) using the optical WFCP2 photometry augmented by ground-based near-IR JHK photometry of medium depth obtained by Dickinson et al. (compare Dickinson 1998). They concluded that on average these galaxies are young starbursts observed at an average age of $\sim$25 Myr after the onset of star formation and highly reddened by dust, with mean $E(B - V) \sim 0.3$ and $A_{1700} \sim$3.5 mag. This seems to indicate that the typical LBG of their sample could only produce a very small fraction, approximately 5%, of the total stellar mass of a present-day L^* galaxy. However, the LBGs from the HDF are rather faint, and the ground-based IR data, which were obtained from a 4-m class telescope, could only reach relatively modest flux levels. As a consequence, the fits by Sawicki & Yee are characterized by large uncertainty. The reason is that because the UV light of these galaxies is dominated by the short-lived massive stars of the forming populations, to constrain the mass in smaller stars one important requirement is to have high S/N rest-frame optical photometry. With most of the information in the UV alone, it is simply not possible to break the degeneracy between age, metallicity, and dust extinction.

Photometry of LBGs in J and H obtained with HST and NICMOS in the HDF (Papovich et al. 2001) and with Keck in J and K (Shapley et al. 2001) reach very deep flux levels and are better suited to estimate the contribution to the optical luminosity by formed—as opposed to forming—stellar populations. The fitting procedures and other details of these two works differ a little, but they both reach the same general conclusions. Papovich et al. derived both the total burst duration and its age (effectively, the star-formation history) from the fit, while Shapley et al. adopted the continuous star-formation approximation and fitted the age. Also, Papovich et al. restricted their analysis to LBGs with spectroscopic redshifts in the range $2 < z < 3.5$ to ensure adequate samples of the rest-frame optical wavelengths, and considered models with varying metallicity and IMF, while Shapley et al. adopted solar metallicity and the Salpeter function throughout their analysis.

The two teams consistently found that the stellar mass is generally well constrained by the fitting procedure, and the results are also relatively insensitive to the assumptions of IMF and metallicity, within a factor of $\approx$2 (recall that from direct analysis of the spectrum of cB58 Pettini et al. (2000) found no evidence of IMF substantially different from the Salpeter one). Fitted values of the instantaneous rate of star formation have comparatively larger uncertainties, but they still provide useful information on the star-formation activity of the population. However, the technique does not seem to be useful to constrain the star-formation history of the individual galaxies, in particular the age, since the errors on this parameter are so large as to make the fitted value useless.

The top panel of Figure 9 (see color insert) shows the histogram of the fitted stellar mass from both the ground-based and HDF samples (shaded area). The dark gray histogram represents the ground-based sample alone, whereas the light gray one represents the HDF sample. It is important to keep in mind that the fitted stellar mass is that of the forming stellar population (i.e., of the current starburst), and hence, only a lower limit to that of the formed populations assembled in previous bursts. Using a two-population fit (the forming one plus a maximally old one), Papovich et al. (2001) estimate that one can miss, on average, four times the mass of the young population, and derive the mass of the old population for each galaxy of the HDF sample. The histogram of the sum of the mass of the young plus old populations of the HDF sample is shown with the light gray curve in the bottom panel of Figure 9. Unfortunately, no similar information is available for the ground-based sample; we hence have crudely estimated the old population mass for this sample by multiplying the young one by a factor of 4, the average value found in the HDF. The corresponding histogram is plotted with the dark gray curve, whereas the sum of the HDF plus ground-based samples is represented by the shaded histogram. Also plotted is the range of dynamical mass derived from the kinematics of the optical nebula emission lines. Notice that even the young masses alone span the range from the typical mass of an L^* galaxy today, $M^* = 10^{11} M_\odot$ (Cole et al. 2001) to that of dwarf galaxies.

MORPHOLOGY

At redshift $z \sim 3$, cameras on board HST can resolve spatial scales of $\sim 0.5\, h^{-1}$ kpc, sufficient to resolve the size of individual starburst regions (Kennicutt 1989, 1998) and provide good dynamic range to characterize the overall morphology of LBGs, e.g., ≈ 500 resolution elements for an L^* LBG in HDF-like images. Initial HST works consisted of optical imaging sampling the far-UV morphology of the galaxies. After the installation of the NICMOS instrument on HST the optical morphology has also been studied with near-IR imaging. The samples included galaxies from the ground-based survey (Giavalisco et al. 1995, 1996a, 1996c) and the HDF itself (Steidel et al. 1996a; Abraham et al. 1996; Lowenthal et al. 1997; Dickinson 1998, 2000), approximately covering the redshift range $2 \lesssim z \lesssim 4$. Other studies included HST imaging of Lyα galaxies (Pascarelle et al. 1996a,b) at similar redshifts.

These works found that whereas LBGs are characterized by a relatively wide dispersion of morphological properties, there are also a number of common features that seem to be typical of the population. These galaxies generally are smaller, more compact and more irregular than local galaxies of comparable *B*-band luminosity, and they have comparatively bluer integrated colors than the local mix, with UV to optical SED similar to that of the Irregular Hubble types (Papovich et al. 2001). Although color gradients are clearly observed within the galaxies, their morphology depends mildly on the wavelength, remaining essentially unchanged from the far-UV to the optical window. In general, their morphology cannot be classified in terms of the traditional Hubble types (other than pointing to a generic similarity with Ellipticals or Irregular types for some of them), because the structural components of bulge, disk, spiral arms, and bars are not identified in these galaxies.

The mix of morphologies ranges from relatively regular systems to irregular to fragmented and multiple systems, and both the UV and optical images show typical half-light radii in the range $r_{1/2} \sim 3$–$5 \ h^{-1}$ kpc. Such sizes are significantly larger than those of dwarf galaxies, and similar to those of large bulges and elliptical galaxies of intermediate luminosity in the local universe. (See, for example, the compilation of data on the structural parameters of spheroids by Bender et al. 1992.) In a number of cases, the isophotal analysis reveals a relatively smooth morphology with either an $r^{1/4}$ or exponential profile and an overall appearance reminiscent of an elliptical galaxy or a spiral one. Figure 10 (see color insert) shows two such examples selected from the ground-based survey at both UV and optical rest-frame wavelengths. In a few cases there are tentative detections of rotation curves from spectroscopy of the optical nebular lines (Pettini et al. 2001, Moorwood et al. 2000), suggesting that rotationally supported disks might be present among LBGs. Unfortunately, no spectra have been taken yet of those galaxies for which existing HST imaging suggests a disk-like morphology. Thus, it is quite possible that the unknown inclination of the disk and the lack of alignment with the major axis have resulted in underestimating both the amount of rotation and the spatial extent of the rotation curve. Also, note that in other cases, deep spectroscopy of LBGs with an apparent edge-on disk-like morphology failed to reveal any evidence of rotation (H. Spinrad, private communication).

More frequently, however, the galaxies have disturbed or fragmented morphologies, with one bright core or multiple knots embedded in diffuse nebulosity, reminiscent of merging events or interactions. Diffuse, low-surface brightness galaxies seem to be comparatively rare, even though the HST images have enough sensitivity to detect them if they were present. Also rare are highly elongated galaxies, e.g., like the chain galaxies found by Cowie et al. (1995b), which seem to suggest that overall LBGs are characterized by a relatively high degree of spherical symmetry (Giavalisco et al. 1996c). Figure 11 (see color insert) shows color images of LBGs from the HDF at both UV and optical rest-frame wavelengths and illustrates both their morphological diversity and the weak dependence of morphology on wavelength. Regardless of the underlying (unknown) kinematics of the galaxies, it is interesting to observe that very often the bulk of the light of LBGs, including both the single galaxies and multiple ones (Giavalisco et al. 1996c, Steidel et al.

1996a), come from compact structures, suggesting that the sites of the most intense star formation in these galaxies are dense, virialized structures.

Based on their morphology and size, the closest counterparts to most LBGs in the local universe are either late-type spiral and irregular galaxies or merging systems (Lowenthal et al. 1997, Trager et al. 1997, Sawicki & Yee 1998). However, LBGs have a much higher star-formation activity than local galaxies, and this fact could be responsible, at least in part, for their apparent morphology, since even at rest-frame optical wavelengths the brightest regions of star formation dominate the integrated light. In local galaxies, the morphology of these regions is often different from that of the underlying galaxy, and tend to be fragmented and irregular (Giavalisco et al. 1996b; Hibbard & Vacca 1997; Kuchinski et al. 2001, 2000; Marcum et al. 2001) and in general is not representative of that of the whole structure. Thus, the observed morphology of LBGs could primarily reflect that of their most active regions of star formation, while that of the underlying galaxy remains unconstrained. The weak dependence of the morphology of LBGs on wavelength is generally consistent with this interpretation. We note, however, that the constraints on the relative proportions of old and young stars in LBGs are rather loose (Papovich et al. 2001), approximately $0 \lesssim m_{old}/m_{new} \lesssim 7$; on average, as much as four times the stellar mass of the forming population could still be present in LBGs and still its morphology be outshined by that of the forming population, thus remaining substantially unobserved in the images.

Finally, we conclude by stressing the importance that morphological studies will have in understanding the mechanisms of star formation in these galaxies when complemented with kinematic information. A key example is testing the occurrence of merging in these systems, which is predicted to be the primary trigger of star formation in systems according to the hierarchical cosmologies (e.g., Somerville et al. 2000, 2001). This requires the study of the kinematics of the apparent fragments that compound the galaxies on scales of a few kiloparsec, and it will be made possible by integral field spectrographs attached to large-aperture space telescopes.

CLUSTERING AND LARGE-SCALE STRUCTURES AT $z \sim 3$

A very important avenue of research opened by the high efficiency of the Lyman-break technique is the possibility of studying galaxy clustering and large-scale structures in the young universe. The large samples yielded by the technique have well-controlled selection effects, and their redshift distribution function is measured with high accuracy, which makes them well suited for applications such as the correlation function and count statistics.

The strength of the spatial clustering of LBGs has been measured at $z \sim 3$ and 4 in a number of works, generally finding consistent results. Early reports of strong spatial clustering came from Giavalisco et al. (1994b) who observed a concentration of LBG candidates around the $z = 3.39$ damped Lyα absorber of quasar Q000–2619 and estimated a spatial correlation length of $\sim 2.5\ h^{-1}$ Mpc. Steidel et al. (1998) reported strong clustering in redshift space of spectroscopically confirmed LBGs and noticed a large spike of LBGs at $z \sim 3.01$ in one of their survey areas (the SSA22a field), which they suggested is likely a protocluster or possibly even a supercluster. Giavalisco et al. (1998) and Giavalisco & Dickinson (2001) measured the angular correlation function from a sample of $\sim 1000\ U$–band dropouts from the ground-based survey and inverted it using the Limber transform (Peebles 1980, Efstathiou et al. 1991) to derive the spatial one, finding a correlation length $r_0 \sim 4\ h^{-1}$ Mpc. This method is particularly efficient and accurate when, as in the case of the U–band dropouts, the redshift distribution function of the sample is known and the contamination by interlopers is essentially negligible. In this case, one can use the photometric samples, i.e., galaxies with and without spectroscopic redshifts, and take advantage of the larger numbers. Connolly et al. (1998) and Arnouts et al. (1999) used the same technique with samples from the HDF selected by photometric redshifts and reported strong clustering at $z > 2$. Adelberger et al. (1998) derived the spatial correlation length from the fluctuations of the number counts in spatial cells and found $r_0 \sim 6\ h^{-1}$ Mpc, higher than the value derived from the angular clustering, but consistent within the errors [a new application of the technique to a larger sample yielded $r \sim 4\ h^{-1}$ Mpc, (K.L. Adelberger, private communication)]. This method has the advantage of including the information on the clustering along the radial (redshift) direction, which is lost in the angular projection, but it requires the knowledge of the redshift of the individual galaxies and hence it relies on smaller samples. More recently, Ouchi et al. (2001) extended these studies to redshift $z \sim 4$ using a sample of $\approx 1200\ B$–band dropouts from deep imaging with the Subaru telescope, and found a correlation length $r_0 \sim 3\ h^{-1}$ Mpc.

Although the size of current LBGs samples is such that clustering measures could actually be made, these samples are still small in absolute terms. For example, the sample used by Giavalisco et al. (1998) only covers ~ 0.3 square degrees; in comparison, local samples are two orders of magnitude larger (e.g., 2dF). As a consequence, large fluctuations of volume density are still observed from sample to sample (e.g., Adelberger et al. 1998, Porciani & Giavalisco 2002). This means that the average volume density of LBGs is only broadly known, and the measure of the correlation is affected by bias, because its strength is normalized to the observed volume density and not to the true one, which is unknown. The direction of this integral constraint bias, which is notoriously difficult to estimate, is that the observed correlation is lower than the true one (Hamilton 1988). Porciani & Giavalisco (2002) carried out numerical simulations to test for the presence of the bias in the current samples and concluded that the measures are affected at least at the 15% level. Because the bias also affects the shape of the correlation (its slope, in the case of the power law model), detailed comparisons with the theory require larger samples. Another consequence of the relatively small samples is that the traditional statistics used in these works could yield crude measures of the shape of the correlation function, i.e., simultaneously give r_0 and the slope γ in the traditional power law approximation $\xi(r) = (r/r_0)^{-\gamma}$ (Peebles 1980).

Although the measure of r_0 was found to be robust, with a relative error $\sim 30\%$ (Giavalisco et al. 1998), the slope was much more loosely constrained, $\gamma = 2.0 \pm 0.7$ (Giavalisco et al. 1998, Giavalisco & Dickinson 2001) to compare with the local value $\gamma = 1.77 \pm 0.3$ (from 2dF). Porciani & Giavalisco (2002) use statistics based on the counts of galaxies in angular cells to minimize the effects of the shot noise. They again find $r_0 \sim 4\,h^{-1}$ Mpc but, more importantly, also find evidence that a power law correlation function is a good approximation for the LBGs and estimate the slope to be $\gamma = 1.5 \pm 0.3$, apparently shallower than the local one.

Regardless of these problems, however, because the effect of the integral constraint is to bias the measures toward lower values, the conclusion that LBGs have strong spatial clustering seems a robust one. A correlation length $r_0 \sim 4\,h^{-1}$ Mpc or higher at $z \sim 3$ directly compares with that found in local surveys, $r_0 = 5.7\,h^{-1}$ Mpc (e.g., the SDSS measure). This cannot be explained in terms of the gravitational evolution of clustering alone in any reasonable cosmology that is normalized to reproduce the clustering of galaxies and clusters at $z = 0$ (e.g., Eke et al. 1996). This conclusion does not strongly depend on the shape of the power spectrum used to calculate the evolution or on the adopted cosmology; it is simply a direct consequence of the assumption that gravity drove the formation of structure. To reconcile the weak clustering of the mass predicted at $z \sim 3$ by the gravitational instability and the cluster normalization with the observed correlation function of LBGs, one has to conclude that the spatial distribution of LBGs is biased with respect to that of the mass, namely they preferentially trace regions of the mass-density distribution that are significantly more clustered than the average.

This conclusion has profound physical implications, because the theory predicts that such biased regions are collapsed perturbations of the mass-density field, or halos (Kaiser 1984, Bardeen et al. 1986). Thus, the strong clustering would imply a direct association between the most active regions of star formations and the halos, and this conclusion is solely based on the assumption of gravity as the primary physical interaction responsible for the formation of cosmic structures and on the empirical evidence that LBGs are strongly clustered in space.

The key paradigm behind our understanding of galaxy formation is that they form within dark matter halos when gas condenses and cools in the bottom of their potential well and is converted into stars (White & Rees 1978). The strong clustering observed among LBGs suggests that this paradigm can be tested, at least in broad lines, on empirical grounds. In the local universe the test is difficult, because the effects of bias are not very pronounced and the galaxies are evolved. For example, values of the bias estimated for the local mix of galaxies are in the range $b \sim 1$–1.5 (Peacock et al. 2001, Hamilton & Tegmark 2000), although galaxies of different type and luminosity cluster differently, with earlier type and more luminous galaxies being in general more clustered than later type and fainter ones. At high redshifts, the bias is much stronger and the galaxies are observed during a relatively early phase of formation, and measuring the clustering properties of LBGs offers the opportunity of testing fundamental ideas of galaxy formation.

The clustering properties of the dark matter are well understood from the theoretical point of view and the clustering and abundance of halos are relatively easily calculated, given a cosmological model, namely a power spectrum and the cosmological parameters (Mo & Fukugita 1996, Mo & White 1996, Kauffmann et al. 1997, Jing & Suto 1998, Governato et al. 1998, Baugh et al. 1999). The problem is that predicting the properties of the observable halos, namely the galaxies, is still highly uncertain, because it requires assumptions on the poorly understood mechanisms of star formation.

Some general features of the theory can be tested without modeling star formation, or with only very general assumptions. For example, Adelberger et al. (1998) showed that volume density and clustering strength of LBGs are very similar to those predicted for massive Cold Dark Matter (CDM) halos, roughly with $M > 10^{11} M_\odot$. Giavalisco & Dickinson (2001) found that fainter samples of LBGs (i.e., with fainter UV luminosity) cluster less strongly than brighter ones, and explored the scaling relationship between volume density and clustering strength, finding it to be in good agreement with that of the halos. This scaling law depends very weakly on the details of the relationship between star-formation activity and the properties of the halos, if the scatter between UV luminosity and total mass is small. Thus, this finding seems to imply that (*a*) LBGs trace the halos very closely, i.e., star formation occurs with high efficiency in collapsed structures; and (*b*) the star-formation rate is, on average, regulated by the mass of the galaxy, because less massive halos correlate less strongly than more massive ones. A further implication is that the mass spectrum of the observed samples of LBGs is approximately 10^{10}–$10^{12} M_\odot$. These conclusions, however, are all based on the evidence that the faint HDF sample of LBGs is less strongly clustered than the brighter ground-based one, with a correlation length ~3 times smaller. The HDF sample covers a volume of space nearly two orders of magnitude smaller than that of the ground-based survey, and although numerical simulations suggest that the detection of such weak clustering is significant at the ~96% confidence level, it needs confirmation with new observations.

More detailed comparisons require more sophisticated modeling of the star formation in the forming galaxies. In general, both N-body simulations that include the gas hydrodynamics (Katz et al. 1999) and semianalytical models (Governato et al. 1998, Baugh et al. 1998, Somerville et al. 2001, Wechsler et al. 2001) can successfully reproduce both the strong clustering and volume density of LBGs, suggesting that the paradigm of galaxy formation is correct in its broad lines. Differences mostly arise in the predictions of the mass spectrum of the galaxies, and these are clearly due to the different assumptions of the mechanisms of star formation. Semianalytical models that require merging as the primary trigger of star formation (Somerville et al. 2001, Wechsler et al. 2001, Mustakas & Somerville 2002) generally predict that the visible galaxies are systems of small mass that are substructures of more massive halos. N-body simulations where star formation is described in terms of the hydrodynamics of the baryonic gas tend to predict a larger mass spectrum, but the robustness of this prediction depends on the mass resolution, which is still relatively poor. Some of these properties can, in principle, be tested by the observations, but the current samples are still too small for this. For example, the merging models predict a relatively high abundance of galaxies with close separations (Wechsler et al. 2001, Bullock et al. 2002) and a very mild dependence of the clustering strength with luminosity, whereas the observations are still too uncertain. Porciani & Giavalisco (2002) find a deficiency of close pairs (at the 98% level), at $z \sim 3$ that they interpret as evidence of the physical size of halos, whereas Ouchi et al. (2001) find an excess of pairs at $z \sim 4$; Giavalisco & Dickinson (2001) find weak clustering of faint LBGs, whereas Arnouts et al. (1999) claim the detection of strong clustering at $z > 2$ in the HDF, although based on a very small sample of 39 photometric redshifts.

It is also of interest to compare the high bias inferred for LBGs to that of the "faint blue galaxies" at $0.5 \lesssim z \lesssim 1$ (Efstathiou et al. 1991, Efstathiou 1995; see also Ellis 1997, Le Fevre et al. 1996, Carlberg et al. 1997). These galaxies, which as the LBGs host the bulk of the cosmic star-formation activity (as traced by the UV light) at their epoch (Lilly et al. 1995, Madau et al. 1996), have correlation lengths comparable to that of LBGs and about one half that of the local galaxies. However, because the clustering of the mass at $z \sim 1$ is larger than at $z \sim 3$, their bias is smaller than that of the LBGs by approximately a factor of 2. In the local universe, the average bias is estimated to be in the range 1 to 1.5 (Peacock et al. 2001, Peacock 1997, Katz et al. 1999, Hamilton & Tegmark 2000). Thus, the bias of galaxies that host most of the star formation at their epoch appears to be decreasing in going from high to low redshift, in qualitative agreement with the predictions of biased galaxy formation that galaxies form first and more efficiently in the denser and more massive peaks of the mass-density field. Star formation then propagates to the less dense and less massive halos during the course of evolution.

Clusters at High Redshifts?

A number of groups have reported detection of concentrations of galaxy-size objects at $z > 2$ that could be young clusters of galaxies observed at an early stage of evolution. Most of these detections come from narrow-band imaging targeting Lyα emission from companions of known high-redshift objects such as QSO, QSO absorbers, and radio galaxies (see, among others, Pascarelle et al. 1996a,b, 1998; Campos et al. 1999; Carilli et al. 1998a,b; Pentericci et al. 2000; Francis et al. 1997; see also Dickinson 1997, 2000). Lyman-break galaxy surveys have also returned candidate protoclusters, observed as marked overdensities, or spikes, in the redshift distribution of the galaxies in a given field. Virtually every field has yielded at least one such detection (Steidel et al. 1998, Adelberger et al. 1998).

One of the most conspicuous of these candidate protoclusters or superclusters of LBGs, located at $z = 3.09$, has been studied in greater detail. The structure extends spatially over several h^{-1} Mpc (comoving) and, down to a flux limit of $\mathcal{R} = 25.5$, includes 67 LBGs. Follow-up narrow-band imaging of the cluster that targets the Lyα emission of its galaxies revealed many more members, and showed that it possibly contains two substructures. These are physically coincident with two sources of very large ($\sim$100 h^{-1} kpc), diffuse Lyα emission. One of these hosts a source that is very red and very faint in the rest-frame UV and optical wavelengths (Steidel et al. 2000) but rather bright at submillimeter wavelength (Chapman et al. 2001) consistent with a massive proto elliptical galaxy.

THE MASS SPECTRUM

A key measure in galaxy evolution is the evolution of the mass spectrum. As is the case at any redshift, measuring the mass of galaxies is affected by large uncertainties, and at $z \sim 3$ the situation is made worse by the generally increased difficulty of carrying out the relevant observations. Currently, there are no robust constraints to the mass spectrum or even individual masses of LBGs from direct measures. Crude estimates of mass for a few LBGs are available from the kinematic of optical nebular emission lines, and from the measure of the stellar mass under the assumption of the stellar-to-total mass ratio. Other crude estimates of the mass spectrum come from the observed spatial clustering.

Line Emission Kinematics and the Total Mass

A number of groups (Pettini et al. 1998, 2001; Kobulnicky & Koo 2000; Teplitz et al. 2000a,b; Moorwood et al. 2000) observed the optical emission lines of [OII], Hβ, and [OIII] with near-IR spectroscopy for about 20 LBGs with the goal of deriving dynamical information on these galaxies from their kinematics. The lines have been resolved in every case, and all groups report values of the FWHM of the lines in the range 200–400 km s^{-1}, which if interpreted as due to rotation, corresponds to rotation velocity in the range of 60–120 km s^{-1} (similar values are obtained if the linewidth is interpreted as due to velocity dispersion). Pettini et al. (2001) and Moorwood et al. (2000) also find possible evidence of rotation curves in the spatially resolved spectra of two of their galaxies.

Images at rest-frame optical wavelengths of LBGs taken with HST/NICMOS show that this velocity comes from radii of a few kpc, implying that it is smaller than the rotation velocity of local bright galaxies at similar radii (e.g., Sofue & Rubin 2001). Assuming that the line broadening is due to ordered gravitational motions, either due to rotation or velocity dispersion, and using the observed size of LBGs from HST imaging, all groups derived virial mass $M_{vir} = v^2 \times r_{1/2}/G$ in the range 10^{10} to 10^{11} M$_\odot$. This is between $1/10$ and $1/100$ of the total mass within the optical diameter of a typical L^* galaxy in the local universe and compares to the mass of the bulge of a bright spiral (e.g., Dwek et al. 1995, Sofue & Rubin 2001) and low- to intermediate-luminosity ellipticals (Bender et al. 1992).

It is important to realize that these measures are very likely affected by large and uncontrolled systematics, and must be used with caution. For example, if LBGs are rotation-supported disks, the limited dynamic range in surface brightness of these spectra due to the $(1 + z)^4$ dimming would result in only the central part of the rotation curve being reasonably well observed. Because the spectra are spatially only partially resolved, the total flux is dominated by the central regions of the galaxy. Thus, the velocity width is effectively luminosity-weighted, and dominated by the brightest regions. The derived rotation velocity, therefore, will be systematically lower than its true value. Furthermore, to measure the rotation velocity one needs to align the spectrograph's slit with the semi-major axis of the galaxy and also to estimate the inclination of the disk on the plane of the sky, otherwise the observed velocity is underestimated. This requires morphological information from suitable high-resolution imaging, e.g., HST that was not available for any but one of the galaxies observed by Pettini et al. (2001) and Kobulnicky & Koo (2000), and in both cases, no clear evidence of a disk-like morphology was found.

Another serious problem is that it is unknown if nebular emission lines can be used as dynamical indicators in starburst galaxies. One source of uncertainty is the extent to which the brightest emitting HII regions trace the dynamics of the whole structure. If they segregate in the central regions, this leads to underestimating the rotation velocity and hence the mass. Another problem is the kinematic structure of the HII gas, namely the relative importance of integrated ordered motions of gravitational origin vs. nongravitational motions and turbulence generated by the activity of star formation, such as super-winds and SNe ejecta. If nongravitational motions significantly contribute to the observed linewidth, they lead to overestimation of the rotation velocity. Weedman (1983) noted that the width of emission lines in starburst nuclei were small compared to expectations for gas in virial equilibrium in the bulge of spiral galaxies. Lehnert & Heckman (1996) also found no correlation between the FWHM of the linewidths measured from the nuclear spectra and the galaxy rotations in a sample of IR-selected edge-on starburst galaxies and conclude that the narrow nuclear lines arise because the gas in nuclei where most of the starburst activity takes place does not sample the full range of the rotation curves. On the one hand, Kabulnicky & Gebhardt (2000) find that gaseous and stellar kinematic tracers in a sample of local late-type galaxies with ongoing star formation have comparable line broadening. In particular, they find the [OII]λ3727 and the HI 21-cm emission line to have the same width, with an observed dispersion of only $\sim$20 km s^{-1}. Pisano et al. (2001), on the other hand, find that in nearby compact blue galaxies the linewidth of the [OII] line is systematically narrower than that of the HI line, with the ratio of the two widths being on average $W_{[OII]}/W_{HI} \sim 0.6$. By comparing HI and H$\alpha$ kinematics, Barton & van Zee (2001) showed that the optical line grossly underestimates the rotation velocity of galaxies with centrally concentrated starbursts. They also showed that even the spatial size is usually underestimated in these systems, which together with the measure of the velocity, leads to grossly underestimating the mass.

Overall, all these works suggest that, at least in local galaxies, nebular line kinematics tends to underestimate the virial mass, and that there is no evidence that strong nongravitational motions contribute to the kinematics of the lines. Whether this condition applies to LBGs as well is not known. Pettini et al. (2001) compared the redshifts of interstellar gas, HII gas, and Lyα of LBGs and concluded that large-scale outflows of several hundreds km s^{-1} exist in these systems. However, it is not known what this implies for the kinematics of the nebular gas.

For pressure-supported systems, projection and slit alignment are not factors. Unfortunately though, virtually nothing is known about the kinematics of the HII gas in these systems. Because the gas is highly collisional, for example, it is not inconceivable that cloud-cloud collisions in the dense central regions shock the HII gas and bring it to high temperature ($T \sim 10^6$ K), resulting in depressed line emission in the core of the galaxies. In this case, the observed linewidth would only carry information on the dynamics of the less dense and more external regions, where the stellar velocities are smaller, biasing the measure of the velocity dispersion toward low values. Both the observed rest-frame UV (Giavalisco et al. 1996b,c Steidel et al. 1996a, Lowenthal et al. 1997, Dickinson 1998) and optical (Calzetti & Giavalisco 2001) images taken with HST show light profiles and morphology that are consistent with spheroids in some cases or disks in others. However, often the morphology is simply irregular and fragmented, essentially offering no clues as to the dynamical status of the galaxies. In one case, spectroscopy of a bright galaxy with an apparent edge-on, disk-like morphology (HDF 4-555, Spinrad et al. 1999; H. Spinrad, private communication) failed to return conclusive results.

Stellar Mass and Total Mass

Crude estimates of the total mass of individual LBGs can be obtained from the measure of stellar mass discussed above (Papovich et al. 2001, Shapley et al. 2001), with assumptions of the stellar-to-total mass ratio (shown here as m_{s2t}). An L^* galaxy in the local universe contains $M_{star} \sim 10^{11}$ M$_\odot$ worth of stars (Cole et al. 2001) and has a total mass around $M_{tot} \sim 10^{12}$ M$_\odot$, although the latter is known only approximately. It actually depends on the radius within which the mass is being measured, because most rotation curves are nearly flat up to the largest observed radii. For example, the Milky Way has a total mass of $\sim 10^{11}$ M$_\odot$ interior to the orbit of the Sun (Sofue & Rubin (2001) and $\approx 6 \times 10^{11}$ M$_\odot$ interior to the distance of the LMC (Wilkinson & Evans 1999). Rotation curves of galaxies of similar luminosity give comparable mass values (Sofue & Rubin 2001). Measures based on weak lensing (Fischer et al. 2000) probe the mass within a few hundred kpc radii and yield larger values, $\approx 3 \times 10^{12}$ M$_\odot$ for an L^* galaxy (Smith et al. 2001, Wilson et al. 2001a,b), which is consistent with estimates of the Galaxy mass based on its satellite galaxies (Zaritsky & White 1994).

Thus, local galaxies have stellar-to-total mass ratios in the range $m_{s2t} \approx 10–30$, depending on the radial distance that the specific technique targets.[4] Assuming

---

[4] An upper limit to this number can be set from the ratio of the cosmic mass density $\Omega_m \sim 0.3$ to the baryon density $\Omega_b \sim 0.03$ (e.g., Balbi et al. 2000, 2001) multiplied by the fraction of baryons locked in star $f_b \sim 17\%$ (Fukugita et al. 1998), which yields $m_{s2t} \approx f_s \times \Omega_m/\Omega_b = 60$.

$m_{s2t} = 10$, namely the value found within the optical diameter of local galaxies, the stellar mass fits by Papovich et al. (2001) and by Shapley et al. (2001) translate into a total mass of LBGs in the range 10^{10} to 10^{12} $M_\odot$.

Measures of the total mass-to-light ratio at large radial distances ($r \lesssim 260$ kpc) for galaxies at moderate redshifts as a function of luminosity and galaxy type have become recently available from weak lensing observations (McKay et al. 2001). In the (rest-frame) B and V bands, this ratio depends on the galaxy type, and it increases toward earlier types. However, at wavelengths larger than about 6000 Å, it appears to reach a minimum at ≈ 100 (in solar units) and to be independent of the galaxy type, indicating that variations of stellar population composition have become negligible. LBGs, however, are forming stars at higher rates than starburst galaxies at moderate redshift (Madau et al. 1996, Steidel et al. 1999) and have a lower stellar mass-to-light ratio than local spirals and irregulars, approximately by a factor of 2 (see Figure 12, color insert). In other words, they are on average twice as luminous for a given amount of stellar mass. Thus, an estimate of the total mass-to-light ratio for these galaxies is derived by dividing by 2 the value 100 at 6000 Å discussed above, namely ~ 50 (in solar units). The total mass is then estimated by multiplying this number by the observed rest-frame B–band luminosity reported by Papovich et al. (2001) and Shapley et al. (2001). The mass found in this way is in the range $10^{11} < M_T < 10^{13}$ $M_\odot$ (see Figure 13), significantly larger than the estimates of dynamical mass discussed above. In the local universe, this mass range is typical of relatively massive galaxies.

Spatial Clustering and the Mass Spectrum

Another crude estimate of the mass spectrum of LBGs comes from their spatial clustering. This estimate relies on the interpretation that the observed large correlation length of LBGs at $z \sim 3$ and $z \sim 4$ (Steidel et al. 1998, Giavalisco et al. 1998, Adelberger et al. 1998, Arnouts et al. 1999, Giavalisco & Dickinson 2001, Porciani & Giavalisco 2002, Ouchi et al. 2001) is due to the clustering bias of collapsed structures (halos) relative to the average mass density distribution (Mo & White 1996, Jing & Suto 1998, Governato et al. 1998, Somerville et al. 2001). As we discussed in "Clustering and Large-Scale Structures at $z \sim 3$," above, for a given power spectrum, the volume density, and the spatial correlation length of the halos, to first order, are only a function of the mass of the halos (Press & Schechter 1974, Mo & White 1996). The estimate is done by solving for the mass spectrum of halos that simultaneously reproduces the observed volume density (corrected for incompleteness) and correlation length of LBGs (Giavalisco et al. 1998, Adelberger et al. 1998, Bullock et al. 2002, Giavalisco & Dickinson 2001, Porciani & Giavalisco 2002). For the U–band dropouts of the ground-based and HDF surveys, it is found approximately $5 \times 10^{10} < M < 5 \times 10^{12}$ $M_\odot$. This is in broad agreement with the mass derived from the stellar mass distribution, and therefore, in disagreement with that derived from the dynamical measures.

Key elements in this derivation are robust measures of the spatial correlation length and volume density, and also knowledge of how many LBGs are observed, on average, in each halo, i.e., the amount of substructure (one LBG per halo was assumed above). This last parameter is important, because significant substructure means that many LBGs are bound satellites of massive halos. In such a case, whereas such halos still have to be rather massive to satisfy the observed strong clustering ($M \gtrsim 10^{11} M_\odot$), the implied mass of the individual galaxies is generally small, e.g., $\gtrsim 10^9 M_\odot$ (Kolatt et al. 1999, Somerville et al. 2001, Wechsler et al. 2001, Bullock et al. 2002, Moustakas & Somerville 2002). Currently, the amount of substructure is poorly constrained. Porciani & Giavalisco (2002) report the possible deficiency of close pairs of LBGs of the ground-based sample, which would imply that bright LBGs have no multiplicity. However, Ouchi et al. (2001) report an excess of close pairs at $z \sim 4$, which would imply the opposite conclusion. The multiplicity will be quantified as a function of luminosity in the upcoming large-area surveys of LBGs from HST. This will considerably help reconstruct the mechanisms of formation of massive galaxies, i.e., whether a significant fraction of the stellar and total mass were already in place at $z \sim 3$ or whether the total mass was but the stellar mass was not and it accreted later by hierarchical merging of small satellites.

LOOKING AT THE BIG PICTURE

The large samples afforded by the Lyman-break technique have given us the first systematic and quantitative picture of normal galaxies in the high-redshift universe, selected because they are forming stars and not by properties related to their nuclear activity. Sources selected by their prominent $Ly\alpha$ or submillimeter emission also include normal galaxies in this sense. However, except for a handful of cases, their continuum emission is too faint or their optical counterparts too uncertain to allow us to study their properties at the same level of detail that has been possible to do with LBGs.

We have seen that normal galaxies at high redshifts look very different from the normal galaxies of the present-day universe, and the question is which evolutionary path, if any, links these two distant populations together. In this regard, one thing to keep in mind is that stars that have formed at $z > 2$ are, by necessity, very old today. If we see them in today's normal galaxies, they must be part of the oldest stellar populations, and hence be the components of the spheroids (elliptical and S0 galaxies, and bulges) as well as of the disks of early spiral galaxies (Dressler & Gunn 1992; Wyse et al. 1998; Renzini 1998, 1999). Today, between 50% and 70% of all the stars are in spheroids and the remainder in the disks of spirals and in later-type galaxies (Schechter & Dressler 1987, Persic & Salucci 1992, Fukugita et al. 1998). Thus, if the association between LBGs and today's old stellar populations is correct, LBGs must have hosted a large fraction of the star formation that took place at their epoch. The flatness of the Madau diagram (Madau et al. 1998, Steidel et al. 1999) or even its possible increase toward the highest observed redshifts (Lanzetta et al. 2001), implies that LBGs at $z > 2$ are responsible for a large fraction, between ≈ 30 and 60% depending on the uncertain corrections of dust obscuration, of all the cosmic star formation that can be observed through its UV emission. What needs to be established to test the consistency of this picture is how much star formation at high redshift remains undetected with UV observations because it is hidden by large amounts of dust.

Powerful starburst galaxies such as Arp 220 and other ULRIGs, for example, would not be observed as LBGs if placed at $z > 2$, because their UV-spectral energy distribution is too reddened and obscured by dust to satisfy the color selection criteria. At submillimeter wavelengths, they can be observed up to redshift $z \sim 10$ or so. In the local universe, galaxies such as Arp 220 contribute only a small fraction of the total stellar production, which is dominated by starburst galaxies, spirals, and other later types. These galaxies were evolving very rapidly in the past, however, and their volume density, and correspondingly, their contribution to the cosmic star-formation density increased by almost a factor 10^3 from $z = 0$ to $z \sim 2$, as illustrated in Figure 7, reproduced from Barger et al. (2000). Whereas UV-bright starburst galaxies also show considerable evolution over the same epoch (Madau et al. 1996, Steidel et al. 1999), this is not as strong as that of the ULRIGs, because the star-formation density traced by the UV light increases about 10 times from $z = 0$ to $z \sim 1$. Thus, it is not completely clear which of the two species of star-forming galaxies has given the largest contribution to the cosmic star-formation activity.

As we discussed in "Other Starburst Galaxies at High Redshifts," if the measure of the submillimeter background and of the faint SCUBA number counts, and if the correction for dust obscuration of LBGs are not all grossly wrong, faint SCUBA galaxies and LBGs seem to trace the same population of star-forming galaxies. At the highest submillimeter flux (>6 mJy), a large fraction of the sources are ULRIGs, hence very faint (and red) in the UV and not detected as LBGs because of the extreme dust obscuration. At fainter flux (approximately >2 mJy), the fraction of ULRIGs decreases and some UV-bright starbursts appear among the SCUBA counts, while at faint flux (<2 mJy), the counts are dominated by UV-bright starbursts observed as LBGs. The data, however, are hardly accurate enough to accept or reject such an interpretation with high confidence. Others have proposed that the SCUBA sources include a population of heavily obscured powerful starburst galaxies at high redshift that are not significantly present in the UV-selected surveys. These high-redshift ULRIGs would be responsible for the submillimeter background and hosted the bulk of star formation at their epoch. They would be the progenitors of the massive spheroids (Barger et al. 1998; Lilly et al. 1999; Smail et al. 1999, 2000) that formed either during the monolithic collapse of a massive structure or during a major merging event. In comparison, LBGs would be the sites of more quiescent and prolonged star formation that would lead to spheroids of smaller mass and to the old stellar populations presently observed in disks. Progress in this area will have to wait for new submillimiter facilities of increased sensitivity and angular resolution.

The morphology of LBGs only offers limited clues to discriminate between these two scenarios, but it is not inconsistent with either. The sizes of the galaxies are consistent with the interpretation that they have formed the central regions of today's bright galaxies, particularly considering that they are very likely lower limits of the true ones owing to the large $(1 + z)^4$ surface brightness dimming. Light profiles that, in some cases, are reminiscent of elliptical galaxies, and in other cases, of exponential disks (see Figure 10, color insert), with possible evidence of rotation curves are also broadly consistent with this scenario. More generally, however, these galaxies show a fragmented and irregular morphology (Giavalisco et al. 1996c, Steidel et al. 1996a) that cannot be classified in terms of the Hubble sequence. This is in line with the general result that, at any redshift, the morphology of starburst galaxies (or at least of the starbursting stellar population that dominated the UV/optical luminosity) is considerably more fragmented and less regular than the underlying optical morphology (Giavalisco et al. 1996b; Hibbard & Vacca 1997; Kuchinski et al. 2001, 2000; Marcum et al. 2001) and is not very useful to constrain their subsequent evolution. It does tell, however, that the morphological differentiation of galaxies into the Hubble sequence occurred later than the epochs at which LBGs are being observed, which are still poorly studied from an empirical point of view.

The estimate of the spectrum of stellar mass (Shapley et al. 2001, Papovich et al. 2001) implies that a significant fraction of LBGs, already at $z \sim 3$, have stellar masses not much smaller than those found in today's bright galaxies, such as large bulges or elliptical galaxies of intermediate luminosity (see Figure 10). Combining the stellar mass distribution with the estimate of the rest-frame B–band luminosity function (Shapley et al. 2001) shows that a substantial number of stars have assembled in LBGs at $z > 2$, which supports the notion of an evolutionary link between LBGs and today's spheroids. Dickinson et al. (2002, submitted) estimate that between $\approx 1/10$ and $\approx 1/3$ of the current cosmic stellar mass density has been assembled in LBGs at $z > 2$. The suggestion is that the most massive LBGs could have already assembled the bulk of the stellar mass and the body of the final structure at $z \sim 3$, and have evolved almost passively from then. This is consistent with the discovery of passively evolving old galaxies at redshift as high as $z \sim 1.6$ (Cimatti et al. 2002) and with the observation that some LBGs have the star-formation history and metal enrichment pattern of massive bursts of star formation (Pettini et al. 2002). The less massive LBGs can have either given origin to todays' smaller galaxies or merged into larger systems. Recall, however, that the frequent occurrence of fragmented morphology among LBGs does not necessarily imply that they are starbursts triggered by merging and interactions, although this can be true in some cases. The fragments that are often observed in the images of LBGs, which are compact star-forming regions that segregate a considerable fraction of the total star-formation rate of the galaxy (Giavalisco et al. 1996c, Steidel et al. 1996a), are virtually always observed on spatial scales of a few kiloparsec, corresponding to less than an arcsec in the images. Models of merging-triggered starbursts for LBGs predict an excess of galaxy pairs over tens to hundreds kiloparsec (Bullock et al. 2002), i.e., over tens of arcsec, but there is no clear evidence that this is observed in the data. Porciani & Giavalisco (2002) claim a deficiency of pairs of U–band dropouts over these scales relative to the small-scale extrapolation of the angular correlation function, whereas Ouchi et al. (2001) claim an excess of pairs of B–band dropouts.

The estimate of the total mass (baryonic plus nonbaryonic) remains highly uncertain. Taken at face value, measures of the dynamical mass from the kinematics of optical emission lines yield small values, between 10^9 to 10^{10} $M_\odot$, or about $1/1000$ to $1/100$ that of present-day L^* galaxies, and comparable to the stellar mass of bulges. However, it is unknown if the optical nebular lines can be used as reliable dynamical indicators in these galaxies, and the few works that have explored this issue in local star-forming spiral galaxies have shown that these spectral features generally underestimate the magnitude of the rotation traced by the HI gas. Furthermore, because of the lack of morphological information on the galaxies for which dynamical measures have been attempted, the estimates of the mass are only lower limits.

The clustering properties imply a larger mass range, in the range $10^{11} < M < 10^{13}$ $M_\odot$, about a factor of 100 larger than the dynamical estimates (Giavalisco & Dickinson 2001, Porciani & Giavalisco 2002, Wechsler et al. 2002, Bullock et al., Moustakas & Somerville 2002). This conclusion, however, depends on the assumption that there is essentially a one-to-one correspondence between LBGs and dark matter halos, at least at the observed luminosity. If more substructure is present, i.e., if more LBGs are associated with the same halo, the above mass range is only valid for the halos, whereas the visible galaxies (which are bound satellites of the massive halos) would be considerably less massive, e.g., $M \gtrsim 10^9$ $M_\odot$ (Somerville et al. 2001, Wechsler et al. 2001, Bullock et al. 2002, Somerville & Moustakas 2002). These conclusions seem robust, regardless of the specific assumptions on star-formation activity and the properties of the halos. Whereas current direct constraints to the amount of substructure are not conclusive (Porciani & Giavalisco 2002, Ouchi et al. 2001, Bullock et al. 2002), this situation should rapidly improve with upcoming new surveys. Regardless however, the conclusion that the clustering properties of LBGs imply that the bulk of star formation at $z \sim 3$ has occurred in association with relatively massive systems seems to hold. Finally, despite the large uncertainty involved, the spectrum of total mass inferred from the stellar mass (under assumptions of M/L ratio) is comparable to that derived from the clustering properties (see Figure 13).

Although estimates of the age of the stellar populations from fitting star-formation history are highly uncertain (Casey et al. 2001, Shapley et al. 2001), the range of values that has been found is generally consistent (given the observed star-formation rates) with both the presence of massive galaxies, and with the presence of smaller galaxies or even subgalactic fragments that will merge into large systems. The claim by Shapley et al. (2001) of a correlation between age, star-formation rate, and obscuration is in qualitative agreement with this picture. According to this interpretation, galaxies undergo an initial phase of very intense but short-lived star formation where their UV radiation is relatively highly obscured by large amounts of dust, and then they enter a phase with lower—but still high in absolute terms—star-formation rates that lasts for $\sim 10^9$ yr and is characterized by lower dust obscuration. This is consistent with the assembly of the spheroidal

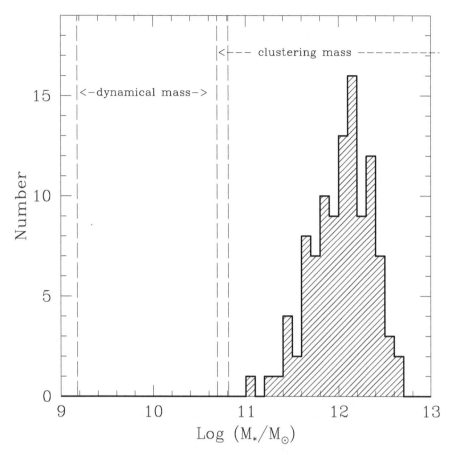

Figure 13 Histogram of the total mass of LBGs derived from the total mass-to-light ratio of field galaxies observed from weak gravitational lensing experiments (McKey et al. 2002) corrected by the ratio of the stellar mass-to-light ratio of LBGs to that of irregular galaxies (see text). Also shown are the range of dynamical mass derived from the kinematics of the optical nebular emission lines and the range of mass derived from the clustering properties.

component of spiral galaxies and the subsequent formation of the disk (Eggen et al. 1962). The crude estimate of metallicity currently available for LBGs are also qualitatively consistent with a large fraction of the stellar mass of the oldest populations of present-day galaxies being assembled at $z > 2$.

A rather important point is that the current data indicate that the UV-luminous galaxies at $z \sim 4$ are similar to those at $z \sim 3$. Whereas there is not a large wavelength basis for measuring the UV continua of the $z \gtrsim 4$ galaxies, the assumption that they have the same distribution of intrinsic SED observed at $z \sim 3$ results in a predicted redshift and luminosity distribution that are consistent with the spectroscopic redshifts and the observed luminosity function. Thus, there is no empirical ground to support a decline of the star-formation activity for redshifts $z > 1$, a fact that pushes the detection of the onset of the dark age of the universe at $z > 5$, in qualitative agreement with the recent suggestion based on the Gunn-Peterson (1965) troughs of distant QSO that the reionization might have happened at $z > 6$ (Becker et al. 2001, Djorgovsky et al. 2001).

Another very important result is the strong spatial clustering of LBGs. This has provided empirical evidence of the general robustness of one of the fundamental ideas about galaxy formation, namely the association between massive virialized structures (the halos) and the activity of star formation, by showing evidence of the reality of galaxy biasing. This is much more difficult to test in the local universe, where the light on average traces the distribution of mass (Peacock et al. 2001). A key observation in this regard is the report that the clustering strength of LBGs scales with their UV luminosity (Giavalisco & Dickinson 2001), which seems to show that the scaling law of the clustering strength and volume density of the galaxies is similar to that predicted for the halos by the gravitational instability theory. It also implies that, on average, the star-formation rate of the galaxies depends more strongly on their total mass, namely on the local gravity, than it does on external accidents such as interactions and merging. Currently, this evidence is based on the measure of the spatial correlation function of the HDF sample, which is small and has an associated large uncertainty. Given its importance, it will be the subject of a careful test during upcoming, much larger surveys from the ground and HST.

Surveys of LBGs have finally opened a wide window on the universe of galaxies, covering most (if not all) of the cosmic time where galaxies are plausibly expected to exist, and have demonstrated that at least our basic ideas about how galaxies form are holding up reasonably well. There is enormous potential for progress in the area of understanding galaxy and structure formation in extending these surveys with larger samples, thus reducing the effects of cosmic variance that still makes many of the results that we have been discussing uncertain, and covering a larger redshift range to study their evolution. Detector technology is maturing to the point where a systematic exploration of the redshift range $1 < z < 10$ is becoming realistic, and telescopes and instrumentation, either space-borne or ground-based, seem to promise to let us do just this with less effort than it took us to confirm the first color-selected galaxy at redshift ~3.5 not even ten years ago.

The *Annual Review of Astronomy and Astrophysics* is online at
http://astro.annualreviews.org

LITERATURE CITED

Abraham RG, Merrifield MR, Ellis RS, Tanvir NR, Brinchmann J. 1999. *MNRAS* 308:569

Abraham RG, Van Der Bergh S, Glazebrook K, Ellis RS, Santiago X, et al. 1996. *Ap. J. Suppl.* 107:1

Adelberger KL, Steidel CC. 2000. *Ap. J.* 544:218

Adelberger KL, Steidel CC, Giavalisco M, Dickinson M, Pettini M, Kellogg M. 1998. *Ap. J.* 505:18

Alonso-Herrero A, Engelbracht CW, Rieke MJ, Rieke GH, Quillen AC. 2001. *Ap. J.* 546:952

Arnouts S, Cristiani S, Moscardini L, Matarrese S, Lucchin F, et al. 1999. *MNRAS* 310:540

Baade W. 1958. In *Stellar Populations*, ed. DJK O'Connell, p. 3. Amsterdam: North Holland

Balbi A, Ade P, Bock J, Borrill J, Boscaleri A, et al. 2000. *Ap. J.* 545:1

Balbi A, Ade P, Bock J, Borrill J, De Bernardis P, et al. 2001. *Ap. J.* 558:145

Bardeen JM, Bond JR, Kaiser N, Szalay AS. 1986. *Ap. J.* 304:15

Barger AJ, Aragon-Salamanca A, Smail I, Ellis RS, Couch WJ, et al. 1998. *Ap. J.* 501:522

Barger AJ, Cowie LL, Richards EA. 2000. *Astron. J.* 119:2092

Barger AJ, Cowie LL, Sanders DB. 1999a. *Ap. J.* 518:5

Barger AJ, Cowie LL, Smail I, Ivison RJ, Blain AW, Kneib JP. 1999b. *Astron. J.* 117:2656

Barton EJ, van Zee L. 2001. *Ap. J.* 550:35

Baugh CM, Benson AJ, Cole S, Frenk CS, Lacey CG. 1999. *MNRAS* 305:21

Baugh CM, Cole S, Frenk CS, Lacey CG. 1998. *Ap. J.* 498:504

Becker RH, Fan XH, White RL, Strauss MA, Narayanan VK, et al. 2001. *Astron. J.* 122:2850

Beckwith SV, Thompson D, Mannucci F, Djorgovski SG. 1998. *Astron. J.* 116:1591

Bender R, Burnstein D, Faber SM. 1992. *Ap. J.* 399:462

Blain AW, Ivison RJ, Smail I. 1998. *MNRAS* 296:29

Blain AW, Smail I, Ivison RJ, Kneib JP. 1999a. *MNRAS* 302:632

Blain AW, Smail I, Ivison RJ, Kneib JP. 1999b. *Ap. J.* 512:87

Brinchmann J, Abraham R, Schade D, Tresse L, Ellis RS, et al. 1998. *Ap. J.* 499:112

Brinchmann J, Ellis RS. 2000. *Ap. J.* 536:L77

Broadhurst TJ, Ellis RS, Shanks T. 1988. *MNRAS* 235:827

Bruzual G, Charlot S. 1993. *Ap. J.* 405:538

Buat V. 1989. *Astron. Astrophys.* 220:49

Bullock JS, Wechsler RH, Somerville RS. 2002. *MNRAS* 329:246

Calzetti D. 1997. *Astron. J.* 113:162

Calzetti D. 2001. *Publ. Astron. Soc. Pac.* 113:1449

Calzetti D, Armus L, Bohlin RC, Kinney AL, Korneef J, Storchi-Bergmann T. 2000. *Ap. J.* 533:682

Calzetti D, Giavalisco M. 2001. *Ap. Space Sci.* 277:609

Calzetti D, Kinney AL, Storchi-Bergmann T. 1994. *Ap. J.* 429:582

Campos A, Yahil A, Windhorst RA, Richards EA, Pascarelle S, et al. 1999. *Ap. J.* 511:1

Carilli CL, Harris DE, Pentericci L, Rottgering HJA, Miley GK, Bremer MN. 1998a. *Ap. J.* 496:57

Carilli CL, Harris DE, Pentericci L, Rottgering HJA, Miley GK, Bremer MN. 1998b. *Ap. J.* 494:143

Carlberg RG, Cowie LL, Songaila A, Hu EM. 1997. *Ap. J.* 484:538

Casertano S, de Mello D, Dickinson M, Ferguson HC, Fruchter AS, et al. 2000. *Astron. J.* 120:2747

Chambers KC, Miley GK, van Breugel W. 1990. *Ap. J.* 363:21

Chambers KC, Miley GK, van Breugel W. 1988. *Ap. J.* 327:47

Chapman SC, Lewis GF, Scott D, Richards E, Borys C, et al. 2001. *Ap. J.* 548:17

Chapman SC, Scott D, Steidel CC, Borys C, Halpern M, et al. 2000. *MNRAS* 319:318

Charlot S, Fall SM. 1991. *Ap. J.* 378:471

Charlot S, Fall SM. 1993. *Ap. J.* 415:580

Chen H-W, Lanzetta KM, Pascarelle S. 1999. *Nature* 398:586

Chen H-W, Lanzetta KM, Pascarelle S, Yahota N. 2000. *Nature* 408:6812

Cimatti A, Daddi E, Mignoli M, Pozzetti L, Renzini A, et al. 2002. *Astron. Astrophys.* 381:L68

Cohen JG, Blandford R, Hogg DW, Pahre MA, Shopbell PL. 1999. *Ap. J.* 512:30

Cohen JG, Cowie LL, Hogg DW, Songaila A, Blandford R, et al. 1996. *Ap. J.* 471:5

Cohen JG, Hogg DW, Blandford R, Cowie LL, Hu E, et al. 2000. *Ap. J.* 538:29

Cole S, Norberg P, Baugh C, Frenk CS, Bland-Hawthorn J, et al. 2001. *MNRAS* 326:55

Colless M, Ellis RS, Broadhurst TJ, Taylor K, Peterson BA. 1993. *MNRAS* 261:19

Colless M, Ellis RS, Shaw G, Taylor K. 1991. *MNRAS* 253:686

Colless M, Schade D, Ellis RS, Broadhurst TJ. 1994. *MNRAS* 267:1108

Connolly AJ, Szalay AS, Brunner RJ. 1998. *Ap. J.* 499:125

Connolly AJ, Szalay AS, Dickinson M, Subbarao MU, Brunner RJ. 1997. *Ap. J.* 486:11

Cowie LL, Gardner JP, Hu HM, Songaila A, Hodapp KW, Wainscoat RJ. 1994. *Ap. J.* 434:11

Cowie LL, Hu EM. 1998. *Astron. J.* 115:1319

Cowie LL, Hu EM, Songaila A. 1995a. *Nature* 377:603

Cowie LL, Hu EM, Songaila A. 1995b. *Astron. J.* 110:1576

Cowie LL, Hu EM, Songaila A, Egami E. 1997. *Ap. J.* 481:9

Cowie LL, Songaila A, Hu EM, Cohen JG. 1996. *Astron. J.* 112:839

Cristiani S, Appenzeller I, Arnouts S, Nonino M, Aragon-Salamanca A, et al. 2000. *Astron. Astrophys.* 359:489

Davis M, Wilkinson DT. 1974. *Ap. J.* 192:251

Deharveng JM, Bowyer S, Buat V. 1990. *Astron. Astrophys.* 236:351

Deharveng JM, Sassen TP, Buat V, Bowyer S, Lampton M, Wu X. 1994. *Astron. Astrophys.* 289:715

De Propris R, Pritchet CJ, Hartwick FDA, Hickson P. 1993. *Astron. J.* 105:1243

Dey A, Graham JR, Ivison RJ, Smail I, Wright GS, Liu MC. 1999. *Ap. J.* 519:610

Dey A, Spinrad H, Stern D, Graham JR, Chaffee FH. 1998. *Ap. J.* 498:93

Dey A, Van Breugel W, Vacca WD, Antonucci R. 1997. *Ap. J.* 490:698

Dickinson M. 1997. In *The Early Universe with the VLT*, ed. J Bergeron, p. 274. Berlin: Springer-Verlag

Dickinson M. 1998. STScI May Symp. Ser. 11, p. 219. New York: Cambridge Univ. Press

Dickinson M. 2000. *Philos. Trans. R. Astron. Soc. London Ser. A* 358(1772):2001

Dickinson M. 2000. In *Constructing the Universe with Clusters of Galaxies*, ed. F Durret, D Gerbal. Paris: *IAP*

Dickinson M, Hanley C, Elston R, Eisenhardt PR, Stanford SA, et al. 2000. *Ap. J.* 531:624

Djorgovski SG, Castro S, Stern D, Mahabal AA. 2001. *Ap. J.* 560:5

Djorgovski SG, Pahre MA, Bechtold J, Elston R. 1996. *Nature* 382:234

Djorgovski SG, Spinrad H, McCarthy P. 1985. *Ap. J.* 299:1

Djorgovski SG, Strauss MA, Spinrad H, McCarthy PJ, Perley RA. 1987. *Astron. J.* 93:1318

Djorgovski SG, Thompson DJ. 1992. *IAU Symp. No. 149*, p. 337. Dordrecht: Kluwer

Dressler A, Gunn JE. 1992. *Ap. J. Suppl.* 78:1

Driver SP, Fernandez-Soto A, Couch WJ, Odewahn SC, Windhorst RA, et al. 1998. *Ap. J.* 496:93

Driver SP, Windhorst RA, Griffiths RE. 1995a. *Ap. J.* 543:48

Driver SP, Windhorst RA, Ostrander EJ, Keel WC, Griffiths RE, Ratnatunga KU. 1995b. *Ap. J.* 449:23

Dwek E, Arendt RG, Hauser MG, Kelsall T, Lisse CM, et al. 1995. *Ap. J.* 445:716

Eales SA, Lilly S, Gear W, Dunne L, Bond JR, et al. 1999. *Ap. J.* 515:518

Eales SA, Rawlings S. 1990. *MNRAS* 243:1

Eales SA, Rawlings S. 1993. *Ap. J.* 411:67

Eales SA, Rawlings S, Dickinson M, Spinrad H, Hill G, Lacy M. 1993a. *Ap. J.* 409:578

Eales SA, Rawlings S, Puxley P, Rocca-Volmerange B, Kunth K. 1993b. *Nature* 363:140

Efstathiou G. 1995. *MNRAS* 272:L25

Efstathiou G, Bernstein G, Katz N, Tyson AJ, Guhathakurta P. 1991. *Ap. J.* 380:L47

Eggen OJ, Lynden-Bell D, Sandage AR. 1962. *Ap. J.* 136:748

Eke VR, Cole S, Frenk CS. 1996. *MNRAS* 282:263

Ellis RS. 1997. *Annu. Rev. Astron. Astrophys.* 35:389

Ellis RS. 1998. *Nature* 395:3

Ellis RS, Colles M, Broadhurst T, Heyl J, Glazebrook K. 1996. *MNRAS* 280:235

Ellis RS, Smail I, Dressler A, Couch WJ, Oemler A Jr, et al. 1997. *Ap. J.* 483:582

Ferguson HC, Dickinson M, Williams R. 2000. *Annu. Rev. Astron. Astrophys.* 38:667

Fernandez-Soto A, Lanzetta KM, Yahil A. 1999. *Ap. J.* 513:34

Fischer P, McKay TA, Sheldon E, Connolly A, Stebbins A, et al. 2000. *Astron. J.* 120:1198

Fixsen DJ, Dwek E, Mather JC, Bennett CL, Shafer RA. 1998. *Ap. J.* 508:123

Folkes S, Ronen S, Price I, Lahav O, Colless M, et al. 1999. *MNRAS* 308:459

Foltz CB, Chafee FH, Weymann RJ. 1986. *Astron. J.* 92:247

Forster Schreiber NM, Genzel R, Lutz D, Kunze D, Stemberg A. 2001. *Ap. J.* 552:544

Francis PJ, Woodgate BE, Warren S, Moller P, Mazzolini M, et al. 1996. *Ap. J.* 457:490

Francis PJ, Williger GM, Collins NR, Palunas P, Malumuth EM, et al. 2001. *Ap. J.* 554:1001

Francis PJ, Woodgate BE, Danks AC. 1997. *Ap. J.* 482:25

Frayer DT, Ivison RJ, Smail I, Yun MS, Armus L. 1999. *Astron. J.* 118:139

Frayer DT, Smail I, Ivison RJ, Scoville NZ. 2000. *Astron. J.* 120:1668

Frye B, Broadhurst TJ, Benitez N. 2002. *Ap. J.* 568:558

Fukugita M, Hogan CJ, Peebles PJE. 1998. *Ap. J.* 503:518

Fynbo JU, Moller P, Warren SJ. 1999. *MNRAS* 305:849

Gallego J, Zamorano J, Aragon-Salamanca A, Rego M. 1995. *Ap. J.* 455:L1

Genzel R, Lutz D, Sturm E, Egami E, Kunze D, et al. 1998. *Ap. J.* 498:579

Giavalisco M, Dickinson M. 2001. *Ap. J.* 550:177

Giavalisco M, Koratkar A, Calzetti D. 1996a. *Ap. J.* 466:831

Giavalisco M, Livio M, Bohlin R, Macchetto F, Stecher T. 1996b. *Astron. J.* 112:369

Giavalisco M, Macchetto FD, Madau P, Sparks WB. 1995. *Ap. J.* 441:L13

Giavalisco M, Macchetto FD, Sparks WB. 1994a. *Astron. Astrophys.* 288:103

Giavalisco M, Steidel CC, Adelberger KL, Dickinson ME, Pettini M, Kellogg M. 1998. *Ap. J.* 503:543

Giavalisco M, Steidel CC, Macchetto FD. 1996c. *Ap. J.* 470:189

Giavalisco M, Steidel CC, Szalay AS. 1994b. *Ap. J.* 425:5

Glazebrook K, Abraham R, Santiago B, Ellis R, Griffiths R. 1998. *MNRAS* 297:885

Glazebrook K, Ellis R, Colles M, Broadhurst T, Allington-Smith J, Tanvir N. 1995a. *MNRAS* 273:157

Glazebrook K, Ellis R, Santiago B, Griffiths R. 1995b. *MNRAS* 275:19

Goldader JD, Meurer G, Heckman TM, Seibert M, Sanders DB, et al. 2002. *Ap. J.* 568:651

Governato F, Baugh CM, Frenk CS, Cole S, Lacey CG, et al. 1998. *Nature* 392:359

Guhathakurta P, Tyson JA, Majewski SR. 1990. *Ap. J.* 357:9

Gunn JE, Peterson BA. 1965. *Ap. J.* 142:1633

Hamilton AJ. 1988. *Ap. J.* 331:L59

Hamilton AJ, Tegmark M. 2000. *MNRAS* 312:28

Hauser M, Arendt R, Kelsall T, Dwek E, Odergard N, et al. 1998. *Ap. J.* 508:25

Heckman TM. 1996. In *The Interplay Between Star Formation the ISM and Galaxy Evolution*, ed. D Kunth, B Guiderdoni, M Heydari-Malayeri, T Xu Thuan, p. 159. Paris: Editions Frontieres

Heckman TM. 2000. *Philos. Trans. R. Astron. Soc. London Ser. A* 358(1772):2077

Heckman TM, Lehnert MD, Miley GK, van Breugel W. 1991a. *Ap. J.* 381:373

Heckman TM, Miley GK, Lehnert MD, van Breugel W. 1991b. *Ap. J.* 370:78

Heckman TM, Robert C, Leitherer C, Garnett DR, Van der Rydt F. 1998. *Ap. J.* 503:646

Helou G. 1986. *Ap. J.* 311:L33

Heyl J, Colless M, Ellis R, Broadhurst T. 1997. *MNRAS* 285:613

Hibbard JE, Vacca WD. 1997. *Astron. J.* 114:1741

Hogg DW, Cohen JG, Blandford R, Pahre MA. 1998. *Ap. J.* 504:622

Hu EM, Cowie LL. 1987. *Ap. J.* 317:L7

Hu EM, Cowie LL, McMahon RG. 1998. *Ap. J.* 502:99

Hu EM, McMahon RG. 1996. *Nature* 382:281

Hu EM, McMahon RG, Cowie LL. 1999. *Ap. J.* 522:9

Hu EM, McMahon RG, Egami E. 1996. *Ap. J.* 459:53

Hughes DH, Serjeant S, Dunlop J, Rowan-Robinson M, Blain A, et al. 1998. *Nature* 394:241

Hunstead RW, Fletcher AB, Pettini M. 1990. *Ap. J.* 356:23

Hurwitz M, Jelinsky P, Dixon WVD. 1997. *Ap. J.* 481:31

Ivison RJ, Dunlop JS, Hughes DH, Archibald EN, Stevens JA, et al. 1998. *Ap. J.* 494:211

Ivison RJ, Smail I, Barger AJ, Kneib JP, Blain AW, et al. 2000. *MNRAS* 315:209

Jing YP, Suto Y. 1998. *Ap. J.* 494:5

Kaiser N. 1984. *Ap. J.* 284:9

Katz N, Hernquist L, Weinberg DH. 1999. *Ap. J.* 523:463

Kauffmann G, Nusser A, Steinmetz M. 1997. *MNRAS* 286:795

Kennicutt RC. 1989. *Ap. J.* 344:685

Kennicutt RC. 1998. *Annu. Rev. Astron. Astrophys.* 36:189

Kobulnicky HA, Gebhardt K. 2000. *Astron. J.* 119:1608

Kobulnicky HA, Kennicutt RC, Pizagno JL. 1999. *Ap. J.* 514:544

Kobulnicky HA, Koo DC. 2000. *Ap. J.* 545:712

Kolatt TS, Bullock JS, Somerville RS, Sigad Y, Jonsson P, et al. 1999. *Ap. J.* 523:109

Koo DC, Kron RG. 1980. *Publ. Astron. Soc. Pac.* 545:537

Kuchinski LE, Freedman WL, Madou BF, Trewhella M, Bohlin RC, et al. 2000. *Ap. J. Suppl.* 131:441

Kuchinski LE, Madore BF, Freedman WL, Trewehella M. 2001. *Astron. J.* 122:729

Kudritzki RP, Mendez RH, Feldmeier JJ, Ciardulo R, Jacoby GH, et al. 2000. *Ap. J.* 536:19

Lacy M, Miley G, Rawlings S, Sauders R, Dickinson M, et al. 1994. *MNRAS* 271:504

Lanzetta KM, Wolfe AM, Turnshek DA, Lu L, McMahon RG, Hazard C. 1991. *Ap. J. Suppl.* 77:1

Lanzetta KM, Yahata N, Pascarelle S, Chen H-W, Fernandez-Soto A. 2002. *Ap. J.* 570:492

Lanzetta KM, Yahil A, Fernandez-Soto A. 1998. *Astron. J.* 116:1066

Le Fèvre O, Hudon D, Lilly SJ, Crampton D, Hammer F, Tresse L. 1996. *Ap. J.* 461:534

Lehnert MD, Heckman TM. 1996. *Ap. J.* 472:546

Leitherer C, Ferguson HC, Heckman TM, Lowenthal JD. 1995. *Ap. J.* 454:19

Leitherer C, Heckman TM. 1995. *Ap. J. Suppl.* 96:9

Leitherer C, Vacca W, William D, Conti PS, Filippenko AV, et al. 1996. *Ap. J.* 465:717

Lilly SJ. 1988. *Ap. J.* 333:161

Lilly SJ, Cowie LL, Gardner JP. 1991. *Ap. J.* 369:79

Lilly SJ, Eales SA, Gear WKP, Hammers F, Le Fevre O, et al. 1999. *Ap. J.* 518:641

Lilly SJ, Le Fevre O, Hammer F, Crampton D. 1996. *Ap. J.* 460:1

Lilly SJ, Schade D, Ellis R, Le Fevre O, Brinchmann J, et al. 1998. *Ap. J.* 500:75

Lilly SJ, Tresse L, Hammer F, Crampton D, Le Fèvre O. 1995. *Ap. J.* 455:108

Lowenthal JD, Hogan CJ, Leach RW, Schmidt GD, Foltz CB. 1990. *Ap. J.* 357:3

Lowenthal JD, Koo DC, Guzman R, Gallego J, Phillips AC, et al. 1997. *Ap. J.* 481:673

Macchetto F, Lipari S, Giavalisco M, Turnshek D, Sparks WB. 1993. *Ap. J.* 404:51

Madau P. 1995. *Ap. J.* 441:18

Madau P, Ferguson HC, Dickinson ME, Giavalisco M, Steidel CC, Fruchter A. 1996. *MNRAS* 283:1388

Madau P, Pozzetti L, Dickinson M. 1998. *Ap. J.* 498:106

Malhotra S, Rhoads J. 2002. *Ap. J.* 565:71

Marcum PM, O'Connell RW, Fanelli MN, Cornett RH, Waller WH, et al. 2001. *Ap. J. Suppl.* 132:129

Matarrese S, Coles P, Lucchin F, Moscardini L. 1997. *MNRAS* 286:115

McCarthy PJ. 1991. *Astron. J.* 102:518

McCarthy PJ. 1993. *Annu. Rev. Astron. Astrophys.* 31:639

McCarthy PJ, Carlberg RG, Chen HW, Marzke RO, Firth AE, et al. 2001. *Ap. J.* 560:131

McCarthy PJ, Kapahi VK, van Breugel W, Subrahmanya CR. 1990. *Astron. J.* 100:1014

McCarthy PJ, Persson SE, West SC. 1992. 386:52

McCarthy PJ, van Breugel W, Kapahi VK, Subrahmanya CR. 1991. *Astron. J.* 102:522

McGaugh SS. 1991. *Ap. J.* 380:140

McKay TA, Sheldon ES, Racusin J, Fischer P, Seljak V, et al. 2002. *Ap. J.* In press. astro-ph/0108013

Meier DL. 1976a. *Ap. J.* 207:343

Meier DL. 1976b. *Ap. J.* 203:103

Meier DL, Terlevich R. 1981. *Ap. J.* 246:L109

Meurer GR, Heckman TM, Calzetti D. 1999. *Ap. J.* 521:64

Meurer GR, Heckman TM, Lehnert MD, Leitherer C, Lowenthal J. 1997. *Astron. J.* 114:54

Meurer GR, Heckman TM, Leitherer C, Kinney AL, Robert C, Garnett DR. 1995. *Astron. J.* 110:2665

Miley G, Chambers KC, van Breugel W, Macchetto F. 1992. *Ap. J.* 401:69

Mo HJ, Fukugita M. 1996. *Ap. J.* 467:L9

Mo HJ, White SDM. 1996. *MNRAS* 282:347

Moller P, Warren S. 1993. *Astron. Astrophys.* 270:43

Moorwood AFM, van der Werf PP, Cuby JG, Oliva E. 2000. *Astron. Astrophys.* 362:9

Moustakas LA, Somerville RS. 2002. *Ap. J.* In press. astro-ph/0110584

Neufeld DA. 1991. *Ap. J.* 370:85

Oke JB, Cohen JG, Carr M, Cromez J, Dingizian A, et al. 1995. *Publ. Astron. Soc. Pac.* 107:3750

Oke JB, Gunn JE. 1983. *Ap. J.* 266:713

Ouchi M, Shimasaku K, Akamura S, Doi M, Fuzusawa H, et al. 2001. *Ap. J.* 558:83

Pagel BEJ. 1986. *Publ. Astron. Soc. Pac.* 98:1009

Pahre MA, Djorgovski GS. 1995. *Ap. J.* 449:1

Papovich C, Dickinson M, Ferguson HC. 2001. *Ap. J.* 559:620

Partridge RB. 1974. *Ap. J.* 192:241

Partridge RB, Peebles PJE. 1967a. *Ap. J.* 148:377

Partridge RB, Peebles PJE. 1967b. *Ap. J.* 147:868

Pascarelle SM, Windhorst RA, Driver SP, Ostrander EJ, Keel WC. 1996a. *Ap. J.* 456:21

Pascarelle SM, Windhorst RA, Keel WC. 1998. *Astron. J.* 116:2659

Pascarelle SM, Windhorst RA, Keel WC, Odewan SC. 1996b. *Nature* 383:45

Peacock JA. 1997. *MNRAS* 284:885

Peacock JA, Cole S, Norberg P, Baugh CM, Bland-Hawthorn J, et al. 2001. *Nature* 410:169

Peacock JA, Rowan-Robinson M, Blain AW, Dunlop JS, Efstathiou A, et al. 2000. *MNRAS* 318:535

Peebles PJE. 1971. *Physical Cosmology.* Princeton: Princeton Univ. Press

Peebles PJE. 1980. *The Large-Scale Structure of the Universe.* Princeton: Princeton Univ. Press

Pentericci L, Kurk JD, Röttgering HJ, Miley GK, van Breugel W, et al. 2000. *Astron. Astrophys.* 361:25

Persic M, Salucci P. 1992. *MNRAS* 258:14

Pettini M, Kellogg M, Steidel CC, Dickinson M, Adelberger KL, Giavalisco M. 1998. *Ap. J.* 508:539

Pettini M, Lipman K. 1995. *Astron. Astrophys.* 297:63

Pettini M, Rix SA, Steidel CC, Adelberger KL, Hunt MP, Shapley A. 2002. *Ap. J.* 569:742

Pettini M, Shapley AE, Steidel CC, Cuby J-G, Dickinson M, et al. 2001. *Ap. J.* 554:981

Pettini M, Steidel CC, Adelberger KL, Dickinson M, Giavalisco M. 2000. *Ap. J.* 528:96

Pisano DJ, Kobulnicky HA, Guzman R, Gallego J, Bershady MA. 2001. *Astron. J.* 122:1194

Porciani C, Giavalisco M. 2002. *Ap. J.* 565:24

Pozzetti L, Madau P, Zamorani G, Ferguson HC, Bruzual GA. 1998. *MNRAS* 298:1133

Press WH, Schechter P. 1974. *Ap. J.* 187:425

Pritchet CJ, Hartwick FDA. 1990. *Ap. J.* 355:11

Puget J-L, Abergel A, Bernard J-P, Boulanger F, Burton WB, et al. 1996. *Astron. Astrophys.* 308:5

Quirk WJ, Tinsley BM. 1973. *Ap. J.* 179:69

Rawlings S, Eales S, Warren S. 1990. *MNRAS* 243:14

Renzini A. 1998. *Astron. J.* 115:2459

Renzini A. 1999. *Ap. Space Sci.* 267:357

Rhoads JE, Malhotra S. 2001. *Ap. J.* 563:L5

Rhoads JE, Malhotra S, Dey A, Stern D, Spinrad H, Jannuzi BT. 2000. *Ap. J.* 545:L85

Salpeter EE. 1995. *The Physics of the Interstellar Medium and Intergalactic Medium, ASP Conf. Ser.*, ed. A Ferrara, CF McKee, PR Shapiro, 80:264. San Francisco: Publ. Astron. Soc. Pac.

Sandage A, Freeman KC, Stokes NR. 1970. *Ap. J.* 160:831

Sanders DB, Mirabel IF. 1996. *Annu. Rev. Astron. Astrophys.* 34:749

Sargent WLW, Young PJ, Boksemberg A, Tytler D. 1980. *Ap. J. Suppl.* 42:41

Sawicki M. 2001. *Astron. J.* 121:240

Sawicki M, Yee HKC. 1998. *Astron. J.* 115: 1329

Schade D, Lilly SJ, Crampton D, Ellis R, Le Fèvre O, et al. 1999. *Ap. J.* 525:31

Schade D, Lilly SJ, Crampton D, Hammer F, Le Fèvre O, Tresse L. 1995. *Ap. J.* 451:L1

Schechter P. 1976. *Ap. J.* 203:297

Schechter PL, Dressler A. 1987. *Astron. J.* 94: 563

Schlegel D, Finkbeiner D, Davis M. 1998. *Ap. J.* 500:525

Schneider D, Gunn J, Turner E, Lawrence C, Hewitt J, et al. 1986. *Astron. J.* 91:991

Scott D, Lagache G, Borys C, Chapman SC, Halpern M, et al. 2000. *Astron. Astrophys.* 357:5

Sellgren K. 1984. *Ap. J.* 277:623

Shapley AE, Steidel CC, Adelberger KL, Dickinson M, Giavalisco M, Pettini M. 2001. *Ap. J.* 562:95

Smail I, Ivison RJ, Blain AW. 1997. *Ap. J.* 490:5

Smail I, Ivison RJ, Blain AW, Kneib JP. 1998. *Ap. J.* 507:21

Smail I, Ivison RJ, Kneib JP, Cowie LL, Blain AW, et al. 1999. *MNRAS* 308:1061

Smail I, Ivison RJ, Owen FN, Blain AW, Kneib JP. 2000. *Ap. J.* 528:612

Smith DR, Bernstein GM, Fischer P, Jarvis M. 2001. *Ap. J.* 551:643

Smith HE, Cohen RD, Burns JE, Moore DJ, Uchida BA. 1989. *Ap. J.* 347:87

Sofue Y, Rubin V. 2001. *Annu. Rev. Astron. Astrophys.* 39:137

Somerville RS, Lemson G, Kolatt TS, Dekel A. 2000. *MNRAS* 316:479

Somerville RS, Primack JR, Faber SM. 2001. *MNRAS* 320:504

Spinrad H. 1989. In *The Epoch of Galaxy Formation, ASI Ser. C*, 264:39. Dordrecht: Kluwer

Spinrad H, Dey A, Graham JR. 1995. *Ap. J.* 438:51

Spinrad H, Dey A, Stern D, Bunker A. 1999. In *The Most Distant Galaxies*, ed. HJA Rottgering, PN Best, MD Lehnert, p. 257. Amsterdam: Neth. R. Acad. Arts Sci.

Spinrad H, Djorgovski S. 1984a. *Ap. J.* 285:49

Spinrad H, Djorgovski S. 1984b. *Publ. Astron. Soc. Pac.* 96:795

Spinrad H, Filippenko AV, Wyckoff S, Stocke JT, Wagner MR, Lawrie DG. 1985. *Ap. J.* 299:L7

Spinrad H, Stern D, Bunker A, Dey A, Lanzetta K, et al. 1998. *Astron. J.* 116:2617

Steidel CC, Adelberger KL, Dickinson ME, Giavalisco M, Pettini M, Kellogg M. 1998. *Ap. J.* 492:428

Steidel CC, Adelberger KL, Giavalisco M, Dickinson ME, Pettini M. 1999. *Ap. J.* 519:1

Steidel CC, Adelberger KL, Shapley AE, Pettini M, Dickinson M, Giavalisco M. 2000. *Ap. J.* 532:170

Steidel CC, Dickinson M, Persson SE. 1994. *Ap. J.* 437:75

Steidel CC, Dickinson M, Sargent WLW. 1991. *Astron. J.* 101:1187

Steidel CC, Giavalisco M, Dickinson M, Adelberger KL. 1996a. *Astron. J.* 112:352

Steidel CC, Giavalisco M, Pettini M, Dickinson M, Adelberger K. 1996b. *Ap. J.* 462:17

Steidel CC, Hamilton D. 1992. *Astron. J.* 104: 941

Steidel CC, Hamilton D. 1993. *Astron. J.* 105: 2017

Steidel CC, Pettini M, Adelberger KL. 2001. *Ap. J.* 546:665

Steidel CC, Pettini M, Hamilton D. 1995. *Astron. J.* 110:2519

Stern D, Spinrad H. 1999. *Publ. Astron. Soc. Pac.* 111:1475

Stiavelli M, Scarlatta C, Panagia N, Treu T,

Bertin G, Bertola F. 2001. *Ap. J.* 561: 37

Sullivan M, Treyer M, Ellis RS, Bridges TJ, Milliard B, Donas J. 2000. *MNRAS* 312:442

Tegmark M, Zaldarriaga M. 2000. *Ap. J.* 544:30

Teplitz HI, Malkan MA, Steidel CC, McLean IS, Becklin EE, et al. 2000a. *Ap. J.* 542:18

Teplitz HI, McLean IS, Becklin EE, Figer DF, Gilbert AM, et al. 2000b. *Ap. J.* 533:65

Teplitz HI, Malkan MA, McLean IS. 1999. *Ap. J.* 514:33

Teplitz HI, Malkan MA, McLean IS. 1998. *Ap. J.* 506:519

Thommes E, Meisenheimer K, Fockenbrock R, Hippelein H, Roeser H-J, Beckwith S. 1998. *MNRAS* 293:L6

Thompson D, Djorgovski SG. 1995. *Astron. J.* 110:982

Thompson D, Djorgovski S, Beckwith SV. 1994. *Astron. J.* 107:1

Thompson D, Djorgovski SG, Trauger J. 1995. *Astron. J.* 110:963

Thompson D, Mannucci F, Beckwith SV. 1996. *Astron. J.* 112:1794

Tinsley BM. 1972a. *Ap. J.* 178:319

Tinsley BM. 1972b. *Astron. Astrophys.* 20:383

Tinsley BM. 1973a. *Ap. J.* 186:35

Tinsley BM. 1973b. *Astron. Astrophys.* 24:89

Tinsley BM, Gunn JE. 1976. *Ap. J.* 302:52

Trager SC, Faber SM, Dressler A, Oemler A. 1997. *Ap. J.* 485:92

Tremonti CA, Calzetti D, Leitherer C, Heckman TM. 2001. *Ap. J.* 555:322

Tresse L, Maddox SJ. 1998. *Ap. J.* 495:691

Turner M. 2001. *Publ. Astron. Soc. Pac.* 113: 653·

Turnshek DA, Wolfe AM, Lanzetta KM, Briggs FH, Cohen RD, et al. 1989. *Ap. J.* 344:567

Tyson JA. 1988. *Ap. J.* 96:1

Van Breugel W, De Breuck C, Stanford SA, Stern D, Rottgering H, Miley G. 1999. *Ap. J.* 518:61

van der Werf PP, Kraiberg Knudsen K, Labbe I, Franx M. 2002. In *The Far-Infrared and Submillimeter Spectral Energy Distribution of Active and Starburst Galaxies*, ed. P Barthel, B Wilkes, I Van Bemmel. Amsterdam: Elsevier. In press

Walborn NR, Lennon DJ, Haser SM, Kudritzki RP, Voles SA. 1995. *Publ. Astron. Soc. Pac.* 107:104

Warren SJ, Moller P. 1996. *Astron. Astrophys.* 311:25

Wechsler RH, Somerville RS, Bullock JS, Kolatt TS, Primack JR, et al. 2001. *Ap. J.* 554:85

Weedman DW. 1983. *Ap. J.* 266:479

Weymann RJ, Stern D, Bunker A, Spinrad H, Chaffee FH, et al. 1998. *Ap. J.* 505:95

White SM, Rees MJ. 1978. *MNRAS* 183:341

Wilkinson MI, Evans NW. 1999. *MNRAS* 310:645

Williams RE, Blacker B, Dickinson M, Dixon W, Ferguson HC, et al. 1996. *Astron. J.* 112: 1335

Williams RE, Baum S, Bergeron L, Bernstein N, Blacker B, et al. 2000. *Astron. J.* 120:2735

Wilson G, Kaiser N, Luppino G. 2001a. *Ap. J.* 556:601

Wilson G, Kaiser N, Luppino G, Cowie LL. 2001b. *Ap. J.* 555:572

Windhorst R, Burstein D, Mathis DF, Neuschaefer LW, Bertola F, et al. 1991. *Ap. J.* 380:362

Windhorst R, Mathis DF, Keel WC. 1992. *Ap. J.* 400:1

Wolfe AM. 1989. *NATO Adv. Res. Workshop, ASI Ser. C* 264:101. Dordrecht: Kluwer

Wolfe AM, Lanzetta KM, Turnshek DA, Oke JB. 1992. *Ap. J.* 385:151

Wolfe AM, Prochaska JX. 2000a. *Ap. J.* 545: 603

Wolfe AM, Prochaska JX. 2000b. *Ap. J.* 545: 591

Wolfe AM, Turnshek DA, Lanzetta KM, Lu L. 1993. *Ap. J.* 404:480

Wolfe AM, Turnshek DA, Smith HE, Cohen RD. 1986. *Ap. J. Suppl.* 61:249

Wyse RFG, Gilmore G, Franz M. 1997. *Annu. Rev. Astron. Astrophys.* 35:637

Young PJ, Sargent WLW, Boksemberg A. 1982. *Ap. J. Suppl.* 48:455

Zaritsky D, White SDM. 1994. *Ap. J.* 435:599

Annu. Rev. Astron. Astrophys. 2002. 40:643–80
doi: 10.1146/annurev.astro.40.060401.093803
Copyright © 2002 by Annual Reviews. All rights reserved

COSMOLOGY WITH THE SUNYAEV-ZEL'DOVICH EFFECT

John E. Carlstrom[1], Gilbert P. Holder[2], and Erik D. Reese[3]

[1]Center for Cosmological Physics, Department of Astronomy and Astrophysics,
Department of Physics, Enrico Fermi Institute, University of Chicago,
5640 S. Ellis Avenue, Chicago, Illinois 60637; email: jc@hyde.uchicago.edu
[2]Institute for Advanced Study, Princeton, New Jersey 08540; email: holder@ias.edu
[3]Chandra Fellow, Department of Physics, University of California, Berkeley,
California 94720; email: reese@cfpa.berkeley.edu

Key Words   galaxy clusters, cosmic microwave background, structure formation,
surveys

■ Abstract   The Sunyaev-Zel'dovich effect (SZE) provides a unique way to map
the large-scale structure of the universe as traced by massive clusters of galaxies.
As a spectral distortion of the cosmic microwave background, the SZE is insensitive to the redshift of the galaxy cluster, making it well-suited for studies of clusters at all redshifts, and especially at reasonably high redshifts ($z > 1$) where the
abundance of clusters is critically dependent on the underlying cosmology. Recent
high signal-to-noise detections of the SZE have enabled interesting constraints on the
Hubble constant and on the matter density of the universe using small samples of
galaxy clusters. Upcoming SZE surveys are expected to find hundreds to thousands of
new galaxy clusters, with a mass selection function that is remarkably uniform with
redshift. In this review we provide an overview of the SZE and its use for cosmological studies, with emphasis on the cosmology that can, in principle, be extracted
from SZE survey yields. We discuss the observational and theoretical challenges that
must be met before precise cosmological constraints can be extracted from the survey
yields.

INTRODUCTION

The Sunyaev-Zel'dovich effect (SZE) offers a unique and powerful observational
tool for cosmology. Recently, there has been considerable progress in detecting and
imaging the SZE. Efforts over the first two decades after the SZE was first proposed
in 1970 (Sunyaev & Zel'dovich 1970, 1972) yielded few reliable detections. Over
the past decade new detectors and observing techniques have allowed high quality
detections and images of the effect for more than 50 clusters with redshifts as high
as one. The next generation of SZE instruments that are being built or planned
will be orders of magnitude more efficient. Entering the fourth decade of SZE

observations, we are now in position to exploit fully the power of the SZE by obtaining detailed images of a set of clusters to understand the intracluster medium (ICM), by obtaining large SZE samples of clusters to determine statistically robust estimates of the cosmological parameters and, most importantly, by conducting large untargeted SZE surveys to probe the high redshift universe. These surveys will provide a direct view of the growth of large-scale structure and will provide large catalogs of clusters that extend past $z \sim 2$ with remarkably uniform selection functions.

The physics of the SZE has been covered well in previous reviews (Birkinshaw 1999, Rephaeli 1995, Sunyaev & Zel'dovich 1980a), with Birkinshaw (1999) and Carlstrom et al. (2000) providing recent reviews of the observations. In this review we look to the near future, using recent observations as a guide to what we can expect.

The SZE is best known for allowing the determination of cosmological parameters when combined with other observational diagnostics of clusters of galaxies such as X-ray emission from the intracluster gas, weak and strong lensing by the cluster potential, and optical galaxy-velocity dispersion measurements. For example, cluster distances have been determined from the analysis of SZE and X-ray data, providing independent estimates of the Hubble constant. A large homogeneous sample of galaxy clusters extending to high redshift should allow a precise measure of this number, as well as a measure of the angular diameter distance relation to high redshift, where it is highly sensitive to cosmological parameters. Similarly, the SZE and X-ray measurements will allow tight constraints on cluster gas-mass fractions that can be used to estimate Ω_M, assuming the composition of clusters represents a fair sample of the universal composition. The observed redshift dependence of the gas-mass fraction can also be used to constrain cosmological parameters as well as test speculative theories of dark-matter decay.

The most unique and powerful cosmological tool provided by the exploitation of the SZE will likely be the direct measurement of the evolution of the number density of galaxy clusters by deep, large-scale SZE surveys. The redshift evolution of the cluster density is critically dependent on the underlying cosmology, and in principle can be used to determine the equation of state of the "dark energy." SZE observations are particularly well suited for deep surveys because the important parameter that sets the detection limit for such a survey is the mass of the cluster; SZE surveys will be able to detect all clusters above a mass limit independent of the redshift of the clusters. This remarkable property of SZE surveys is due to the fact that the SZE is a distortion of the cosmic microwave background (CMB) spectrum. Whereas the CMB suffers cosmological dimming with redshift, the ratio of the magnitude of the SZE to the CMB does not; it is a direct, redshift-independent measurement of the ICM column density weighted by temperature, i.e., the pressure integrated along the line of sight. The total SZE flux detected will be proportional to the total temperature-weighted mass (total integrated pressure) and, of course, inversely proportional to the square of the angular diameter distance.

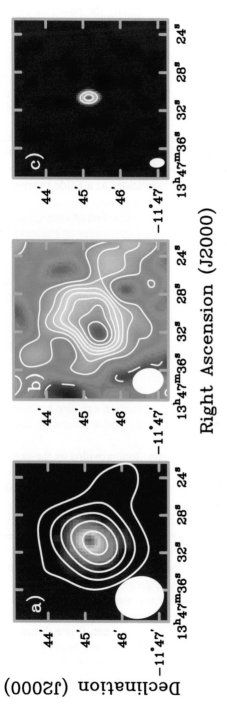

Figure 5 Interferometric images of the SZE of galaxy cluster RXJ 1347-1145, emphasizing different spatial scales. The Full Width Half Maximum (FWHM) ellipse of the synthesized beam is shown in the lower left corner of each panel. (*a*) Point source—subtracted SZE image (contours) overlaid on ROSAT X-ray image (false color). The contours are multiples of $185\mu K$ ($\sim 2\sigma$), and negative contours are shown as solid lines. A $1500\,\lambda$ half-power radius Gaussian taper was applied to the u,v data, resulting in a $63'' \times 80''$ synthesized beam. The X-ray image is High Resolution Imager (HRI) raw counts smoothed with a Gaussian with $\sigma = 6''$ and contains roughly 4000 cluster counts. (*b*) Higher-resolution point source—subtracted SZE image (both contours and false color). A $3000\,\lambda$ half-power radius Gaussian taper was applied, resulting in a $40'' \times 50''$ synthesized beam. The contours are multiples of $175\mu K$ ($\sim 1\sigma$). (*c*) Image of the point source made using projected baselines greater than $3000\,\lambda$. This map has a synthesized beam of $15'' \times 24''$, corresponding to a $\sim 1200\mu K$ brightness sensitivity. The contours are multiples of 15σ. The data was taken with the BIMA mm-array operating at 28.5 GHz. Single-dish maps of the SZE toward RXJ 1347-1145 have also been made with the IRAM 30-m and the Nobeyama 45-m telescopes (Pointecouteau et al. 2001; Komatsu et al. 2001).

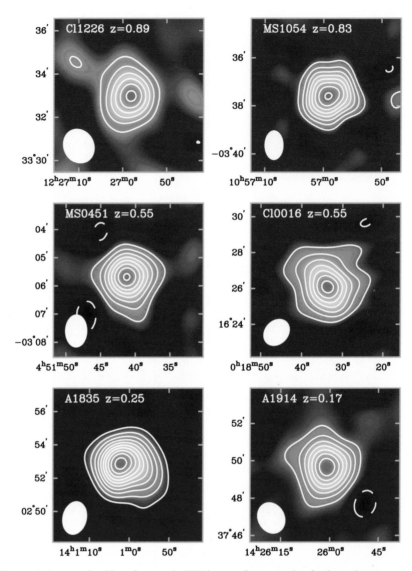

Figure 6 Deconvolved interferometric SZE images for a sample of galaxy clusters over a large redshift range ($0.17 \leq z \leq 0.89$). The contours are multiples of 2σ, and negative contours are shown as solid lines. The Full Width Half Maximum (FWHM) ellipse of the synthesized beam is shown in the lower left corner of each panel. The noise level σ ranges from $25\mu K$ to $70\mu K$ for the clusters shown. Radio point sources were removed from a large fraction of the images shown. The interferometer was able to separate the point-source emission from the SZE by using the high-resolution data obtained with long baselines. All of the clusters shown have similar high X-ray luminosities and, as can be seen, the strength of the SZE signals is similar despite the factor of five in redshift, illustrating the independence of the SZE on redshift.

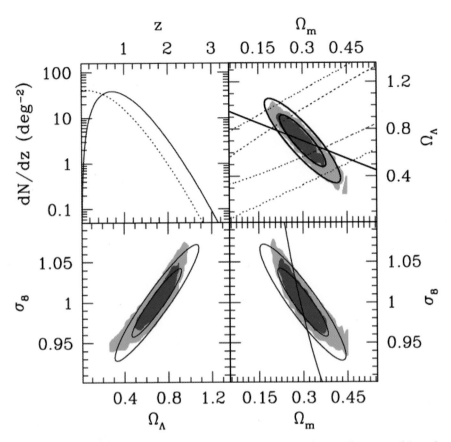

Figure 11 Expected constraints in the Ω_M–Ω_Λ (*top right*), Ω_M–σ_8 (*bottom right*), and Ω_Λ–σ_8 (*bottom left*) planes from the analysis of an SZE survey covering 12 square degrees in which all clusters above $10^{14}h^{-1}$ M$_\odot$ are detected and the redshifts are known. The top left panel shows the expected redshift distribution of clusters (*solid line*) and the cumulative $N(> z)$ (*dotted line*). In the other panels, the 68% confidence regions are shown by the darkest shaded regions, and 95% confidence by the lighter regions. The solid contours correspond to the same confidence regions derived from an approximate method. In each panel the dimension not shown has been marginalized over rather than kept fixed. In the upper right panel the broken line diagonal ellipses are the constraints based on the analyses of type Ia Supernova at 68% and 95% confidence (Riess et al. 1998, Perlmutter et al. 1999). The diagonal line at $\Omega_M + \Omega_\Lambda = 1$ is for a flat universe as suggested by recent cosmic microwave background anisotropy measurements (Miller et al. 1999, de Bernardis et al. 2000, Hanany et al. 2000, Pryke et al. 2002). The solid line in the lower right panel shows the approximate direction of current constraints (Viana & Liddle 1999).

Adopting a reasonable cosmology and accounting for the increase in the universal matter density with redshift, the mass limit for a given SZE survey flux sensitivity is not expected to change more than a factor of $\sim$2–3 for any clusters with $z > 0.05$.

SZE surveys therefore offer an ideal tool for determining the cluster-density evolution. Analyses of even a modest survey covering $\sim$10 square degrees will provide interesting constraints on the matter density of the universe. The precision with which cosmological constraints can be extracted from much larger surveys, however, will be limited by systematics due to our insufficient understanding of the structure of clusters, their gas properties, and their evolution.

Insights into the structure of clusters will be provided by high-resolution SZE observations, especially when combined with other measurements of the clusters. Fortunately, many of the cluster properties derived directly from observational data can be determined in several different ways. For example, the gas-mass fraction can be determined by various combinations of SZE, X-ray, and lensing observations. The electron temperature, a direct measure of a cluster's mass, can be measured directly through X-ray spectroscopy or determined through the analysis of various combinations of X-ray, SZE, and lensing observations. Several of the desired properties of clusters are therefore over-constrained by observation, providing critical insights to our understanding of clusters, and critical tests of current models for the formation and evolution of galaxy clusters. With improved sensitivity, better angular resolution, and sources out to $z \sim 2$, the next generation of SZE observations will provide a good view of galaxy cluster structure and evolution. This will allow, in principle, the dependence of the cluster yields from large SZE surveys on the underlying cosmology to be separated from the dependence of the yields on cluster structure and evolution.

We outline the properties of the SZE in the next section and provide an overview of the current state of the observations in "Status of Observations." This is followed in "Sky Surveys with the Sunyaev-Zel'dovich Effect" by predictions for the expected yields of upcoming SZE surveys. In "Cosmology from Sunyaev-Zel'dovich Effect Survey Samples" we provide an overview of the cosmological tests that will be possible with catalogs of SZE-selected clusters. This is followed by a discussion of backgrounds, foregrounds, contaminants, and theoretical uncertainties that could adversely affect cosmological studies with the SZE and a discussion of observations that could reduce or eliminate these concerns. Throughout the paper, h is used to parametrize the Hubble constant by $H_0 = 100\,h$ km s^{-1} Mpc^{-1}.

THE SUNYAEV-ZEL'DOVICH EFFECT

Thermal Sunyaev-Zel'dovich Effect

The SZE is a small spectral distortion of the CMB spectrum caused by the scattering of the CMB photons off a distribution of high-energy electrons. We focus only on the SZE caused by the hot thermal distribution of electrons provided by the ICM of galaxy clusters. CMB photons passing through the center of a massive cluster have only a $\approx 1\%$ probability of interacting with an energetic ICM electron. The resulting inverse Compton scattering preferentially boosts the energy of the CMB photon by roughly $k_B T_e / m_e c^2$, causing a small ($\lesssim 1$ mK) distortion in the CMB spectrum. Figure 1 shows the SZE spectral distortion for a fictional cluster that is over 1000 times more massive than a typical cluster to illustrate the small effect. The SZE appears as a decrease in the intensity of the CMB at frequencies below $\lesssim 218$ GHz and as an increase at higher frequencies.

The derivation of the SZE can be found in the original papers of Sunyaev & Zel'dovich (Sunyaev & Zel'dovich 1970, 1972), in several reviews (Sunyaev & Zel'dovich 1980a, Rephaeli 1995, Birkinshaw 1999), and in a number of more recent contributions that include relativistic corrections (see below for references). This review discusses the basic features of the SZE that make it a useful cosmological tool.

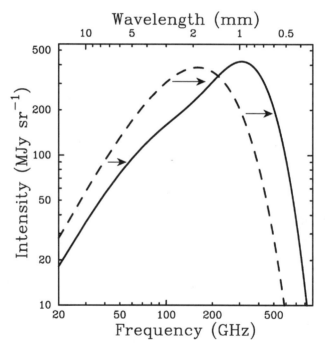

Figure 1   The cosmic microwave background (CMB) spectrum, undistorted (*dashed line*) and distorted by the Sunyaev-Zel'dovich effect (SZE) (*solid line*). Following Sunyaev & Zel'dovich (1980a) to illustrate the effect, the SZE distortion shown is for a fictional cluster 1000 times more massive than a typical massive galaxy cluster. The SZE causes a decrease in the CMB intensity at frequencies $\lesssim 218$ GHz and an increase at higher frequencies.

The SZE spectral distortion of the CMB expressed as a temperature change ΔT_{SZE} at dimensionless frequency $x \equiv \frac{h\nu}{k_B T_{CMB}}$ is given by

$$\frac{\Delta T_{SZE}}{T_{CMB}} = f(x)\, y = f(x) \int n_e \frac{k_B T_e}{m_e c^2} \sigma_T \, d\ell, \tag{1}$$

where y is the Compton y-parameter, which for an isothermal cluster equals the optical depth, τ_e, times the fractional energy gain per scattering, σ_T is the Thomson cross-section, n_e is the electron number density, T_e is the electron temperature, k_B is the Boltzmann constant, $m_e c^2$ is the electron rest mass energy, and the integration is along the line of sight. The frequency dependence of the SZE is

$$f(x) = \left(x \frac{e^x + 1}{e^x - 1} - 4 \right)(1 + \delta_{SZE}(x, T_e)), \tag{2}$$

where $\delta_{SZE}(x, T_e)$ is the relativistic correction to the frequency dependence. Note that $f(x) \to -2$ in the nonrelativistic and Rayleigh-Jeans (RJ) limits.

It is worth noting that $\Delta T_{SZE}/T_{CMB}$ is independent of redshift, as shown in Equation 1. This unique feature of the SZE makes it a potentially powerful tool for investigating the high-redshift universe.

Expressed in units of specific intensity, common in millimeter SZE observations, the thermal SZE is

$$\Delta I_{SZE} = g(x) I_0 y, \tag{3}$$

where $I_0 = 2\,(k_B T_{CMB})^3/(hc)^2$ and the frequency dependence is given by

$$g(x) = \frac{x^4 e^x}{(e^x - 1)^2} \left(x \frac{e^x + 1}{e^x - 1} - 4 \right)(1 + \delta_{SZE}(x, T_e)). \tag{4}$$

ΔT_{SZE} and ΔI_{SZE} are simply related by the derivative of the blackbody with respect to temperature, $|dB_\nu/dT|$.

The spectral distortion of the CMB spectrum by the thermal SZE is shown in Figure 2 (*solid line*) for a realistic massive cluster ($y = 10^{-4}$) in units of intensity (*left panel*) and RJ brightness temperature (*right panel*). The RJ brightness is shown because the sensitivity of a radio telescope is calibrated in these units. It is defined simply by $I_\nu = (2 k_B \nu^2/c^2) T_{RJ}$, where I_ν is the intensity at frequency ν, k_B is Boltzmann's constant, and c is the speed of light. The CMB blackbody spectrum, $B_\nu(T_{CMB})$, multiplied by 0.0005 (*dotted line*) is also shown for comparison. Note that the spectral signature of the thermal effect is distinguished readily from a simple temperature fluctuation of the CMB. The kinetic SZE distortion is shown by the dashed curve (see "Kinetic Sunyaev-Zel'dovich Effect," below). In the nonrelativistic regime it is indistinguishable from a CMB temperature fluctuation.

The gas temperatures measured in massive galaxy clusters are around $k_B T_e \sim$ 10 keV (Mushotzky & Scharf 1997, Allen & Fabian 1998) and are as high as $\sim$17 keV in the galaxy cluster 1E 0657-56 (Tucker et al. 1998). The mass is expected to scale with temperature roughly as $T_e \propto M^{2/3}$. At these temperatures

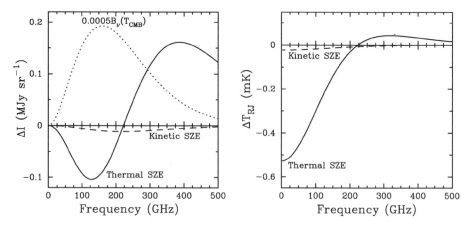

Figure 2 Spectral distortion of the cosmic microwave background (CMB) radiation due to the Sunyaev-Zel'dovich effect (SZE). The left panel shows the intensity and the right panel shows the Rayleigh Jeans brightness temperature. The thick solid line is the thermal SZE and the dashed line is the kinetic SZE. For reference the 2.7 K thermal spectrum for the CMB intensity scaled by 0.0005 is shown by the dotted line in the left panel. The cluster properties used to calculate the spectra are an electron temperature of 10 keV, a Compton y parameter of 10^{-4}, and a peculiar velocity of 500 km s^{-1}.

electron velocities are becoming relativistic, and small corrections are required for accurate interpretation of the SZE. There has been considerable theoretical work that includes relativistic corrections to the SZE (Wright 1979; Fabbri 1981; Rephaeli 1995; Rephaeli & Yankovitch 1997; Stebbins 1997; Itoh et al. 1998; Challinor & Lasenby 1998, 1999; Sazonov & Sunyaev 1998a,b; Nozawa et al. 1998b; Molnar & Birkinshaw 1999; Dolgov et al. 2001). All of these derivations agree for $k_B T_e \lesssim 15$ keV, appropriate for galaxy clusters. For a massive cluster with $k_B T_e \sim 10$ keV ($k_B T_e/m_e c^2 \sim 0.02$), the relativistic corrections to the SZE are on the order of a few percent in the RJ portion of the spectrum but can be substantial near the null of the thermal effect. Convenient analytical approximations to fifth order in $k_B T_e/m_e c^2$ are presented in Itoh et al. (1998).

Particularly relevant for finding clusters with an SZE survey is the integrated SZE signal. Because the SZE signal is the integrated pressure, integrating over the solid angle of the cluster provides a sum of all of the electrons in the cluster weighted by temperature. This provides a relatively clean measure of the total thermal energy of the cluster. Integrating the SZE over the solid angle of the cluster, $d\Omega = dA/D_A^2$, gives

$$\int \Delta T_{SZE}\, d\Omega \propto \frac{N_e \langle T_e \rangle}{D_A^2} \propto \frac{M \langle T_e \rangle}{D_A^2}, \tag{5}$$

where N_e is the total number of electrons in the clusters, $\langle T_e \rangle$ is the mean electron temperature, D_A is the angular diameter distance, and M is the mass of the cluster (either gas or total mass as $M_{gas} = M_{total} f_g$, where f_g is the gas-mass fraction). The integrated SZE flux is simply the temperature-weighted mass of the cluster divided by D_A^2. The angular diameter distance $D_A(z)$ is fairly flat at high redshift. Also, a cluster of a given mass will be denser and therefore hotter at high redshift because the universal matter density increases as $(1 + z)^3$. Therefore, one expects an SZE survey to detect all clusters above some mass threshold with little dependence on redshift (see "Mass Limits of Observability," below).

The most important features of the thermal SZE are that (a) it is a small spectral distortion of the CMB of order ~ 1 mK, which is proportional to the cluster pressure integrated along the line of sight [Equation 1]; (b) it is independent of redshift; (c) it has a unique spectral signature with a decrease in the CMB intensity at frequencies $\lesssim 218$ GHz and an increase at higher frequencies; and (d) the integrated SZE flux is proportional to the temperature-weighted mass (total thermal energy) of the cluster, implying that SZE surveys will have a mass threshold nearly independent of redshift.

Kinetic Sunyaev-Zel'dovich Effect

If the cluster is moving with respect to the CMB rest frame, there will be an additional spectral distortion due to the Doppler effect of the cluster bulk velocity on the scattered CMB photons. If a component of the cluster velocity, v_{pec}, is projected along the line of sight to the cluster, then the Doppler effect will lead to an observed distortion of the CMB spectrum referred to as the kinetic SZE. In the nonrelativistic limit the spectral signature of the kinetic SZE is a pure thermal distortion of magnitude

$$\frac{\Delta T_{SZE}}{T_{CMB}} = -\tau_e \left(\frac{v_{pec}}{c} \right), \qquad (6)$$

where v_{pec} is along the line of sight; i.e., the emergent spectrum is still described completely by a Planck spectrum, but at a slightly different temperature, lower (higher) for positive (negative) peculiar velocities (Sunyaev & Zel'dovich 1972, Phillips 1995, Birkinshaw 1999) (see Figure 2).

Relativistic perturbations to the kinetic SZE are due to the Lorentz boost to the electrons provided by the bulk velocity (Nozawa et al. 1998a, Sazonov & Sunyaev 1998a). The leading term is of order $(k_B T_e/m_e c^2)(v_{pec}/c)$ and for a 10 keV cluster moving at 1000 km s^{-1} the effect is about an 8% correction to the nonrelativistic term. The $(k_B T_e/m_e c^2)^2 (v_{pec}/c)$ term is only about 1% of the nonrelativistic kinetic SZE, and the $(v_{pec}/c)^2$ term is only 0.2%.

Polarization of the Sunyaev-Zel'dovich Effect

The scattering of the CMB photons by the hot intracluster medium (ICM) electrons can result in polarization at levels proportional to powers of (v_{pec}/c) and τ_e. The largest polarization is expected from the anisotropic optical depth to a given location in the cluster. For example, toward the outskirts of a cluster one expects to see a concentric (radial) pattern of the linear polarization at frequencies at which the thermal SZE is positive (negative). Sazonov & Sunyaev (1999) presented plots of the polarization pattern. Nonspherical morphology for the electron distributions will lead to considerably complicated polarization patterns. The peak polarization of this signal will be of order τ_e times the SZE signal, i.e., of order $0.025(k_B T_e/m_e c^2)\tau_e^2$ times the CMB intensity. For a massive cluster with $\tau_e = 0.01$, the effect would be at the 0.1 μK level toward the edge of the cluster. In principle, this effect could be used to measure the optical depth of the cluster and therefore separate T_e and τ_e from a measurement of the thermal SZE (see Equation 1).

It can be shown that polarization of the SZE comes entirely from the quadrupole component of the local radiation field experienced by the scattering electron. In the case above, the quadrupole component at the outskirts of the cluster is caused by the anisotropy in the radiation field in the direction of the cluster center due to the SZE. Sunyaev and Zel'dovich discussed polarization due to the motion of the cluster with respect to the CMB and transverse to our line of sight (Sunyaev & Zel'dovich 1980b; see also Sazonov & Sunyaev 1999). In this case, the quadrupole comes from the Doppler shift. They found the largest terms to be of order $0.1\tau_e (v_{pec}/c)^2$ and $0.025\tau_e^2(v_{pec}/c)$ of the CMB intensity. The latter term, second order in τ_e, can be thought of as imperfect cancellation of the dipole term due to the anisotropic optical depth. Using $\tau_e = 0.01$ and a bulk motion of 500 km s^{-1} results in polarization levels of order 10 nK, far beyond the sensitivity of current instrumentation.

The CMB as seen by the cluster electrons will have a quadrupole component and therefore the electron scattering will lead to linear polarization. This mechanism could possibly be used to trace the evolution of the CMB quadrupole if polarization measurements could be obtained for a large number of clusters binned in direction and redshift (Kamionkowski & Loeb 1997; Sazonov & Sunyaev 1999). Sazonov and Sunyaev calculated the expected polarization level and found that the maximum CMB quadrupole–induced polarization is 50 ($\tau_e/0.01$) nK, somewhat higher than the expected velocity-induced terms discussed above. The effect is again too small to expect detection in the near future. However, by averaging over many clusters, detecting this polarization might be possible with future satellite missions.

STATUS OF OBSERVATIONS

In the 20 years following the first papers by Sunyaev and Zel'dovich (Sunyaev & Zel'dovich 1970, 1972) there were few firm detections of the SZE despite a considerable amount of effort (Birkinshaw 1991). Over the past several years, however, observations of the effect have progressed from low S/N detections and upper limits to high confidence detections and detailed images. In this section we briefly review the current state of SZE observations.

The dramatic increase in the quality of the observations is due to improvements both in low-noise detection systems and in observing techniques, usually using specialized instrumentation to carefully control the systematics that often prevent one from obtaining the required sensitivity. The sensitivity of a low-noise radio receiver available 20 years ago should have easily allowed the detection of the SZE toward a massive cluster. Most attempts, however, failed due to uncontrolled systematics. Now that the sensitivities of detector systems have improved by factors of 3 to 10, it is clear that the goal of all modern SZE instruments is the control of systematics. Such systematics include, for example, the spatial and temporal variations in the emission from the atmosphere and the surrounding ground, as well as gain instabilities inherent to the detector system used.

The observations must be conducted on the appropriate angular scales. Galaxy clusters have a characteristic scale size of order a megaparsec. For a reasonable cosmology, a megaparsec subtends an arcminute or more at any redshift; low redshift clusters will subtend a much larger angle—for example the angular extent of the Coma cluster ($z = 0.024$) is of order a degree (core radius $\sim 10'$) (Herbig et al. 1995). The detection of extended low–surface brightness objects requires precise differential measurements made toward widely separated directions on the sky. The large angular scale presents challenges to control offsets due to differential ground pick-up and atmospheric variations.

Sources of Astronomical Contamination and Confusion

In designing an instrument for SZE observation, one also needs to take into account several sources of possible contamination and confusion from astronomical sources. One such source is anisotropy of the CMB itself (see Figure 3). For distant clusters with angular extents of a few arcminutes or less it is not a serious problem, as the CMB anisotropy is expected (Hu & White 1997) and indeed found to be damped considerably on these scales (Church et al. 1997, Subrahmanyan

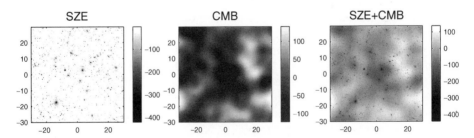

Figure 3   Illustration of the characteristic angular scales of primary CMB anisotropy and of the SZE. The images each cover one square degree and the gray scales are in μK. (*Left*) An image of the SZE from many galaxy clusters at 150 GHz (2 mm) from a state-of-the-art hydrodynamic simulation (Springel et al. 2001). The clusters appear point-like at this angular scale. (*Center*) A realization of CMB anisotropy for a ΛCDM cosmology. (*Right*) The combination of the CMB and SZE signals. Note, the SZE can be distinguished readily from primary CMB anisotropy, provided the observations have sufficient angular resolution.

et al. 2000, Dawson et al. 2001, see also Holzapfel et al. 1997, LaRoque et al. 2002 for CMB limits to SZE contamination). For nearby clusters, or for searches for distant clusters using beams larger than a few arcminutes, the intrinsic CMB anisotropy must be considered. The unique spectral behavior of the thermal SZE can be used to separate it from the intrinsic CMB in these cases. Note, however, that for such cases it will not be possible to separate the kinetic SZE effects from the intrinsic CMB anisotropy without relying on the very small spectral distortions of the kinetic SZE due to relativistic effects.

Historically, the major source of contamination in the measurement of the SZE has been radio point sources. It is obvious that emission from point sources located along the line of sight to the cluster could fill in the SZE decrement, leading to an underestimate. The radio point sources are variable and therefore must be monitored. Radio emission from the cluster member galaxies, from the central cD galaxy in particular, is often the largest source of radio point source contamination, at least at high radio frequencies (Cooray et al. 1998, LaRoque et al. 2002). The typical spectral index of the radio point sources is $\alpha \sim 0.7$ for $S_\nu \propto \nu^{-\alpha}$, where S_ν is the point-source flux. In the RJ limit the SZE flux is proportional to ν^2 and, therefore, point sources are much less of an issue at higher radio frequencies.

Although it is most likely that insufficient attention to radio point sources would lead to an underestimate of the SZE effect, it could also lead to an overestimate. The most obvious example is if unaccounted point sources are in the reference fields surrounding the cluster. An effect caused by gravitational lensing has also been pointed out for low-frequency observations in which the flux from many point sources must be taken into account before a reliable measure of the SZE can be made. Essentially, gravitational lensing increases the efficiency of detecting point sources toward the center of the cluster, which could lead to an overestimate of the SZE decrement (Loeb & Refregier 1997). This effect should be negligible at frequencies greater than roughly 30 GHz.

At frequencies near the null of the thermal SZE and higher, dust emission from extragalactic sources as well as dust emission from our own galaxy must be considered. Dust emission from our Galaxy rises steeply as $\nu^{2+\beta}$, with the observed dust opacity index β found to be $0 < \beta < 2$ over the frequencies of interest.

At the angular scales and frequencies of interest for most SZE observations, contamination from diffuse Galactic dust emission is easily compensated for and will not usually be significant. Consider instead the dusty extragalactic sources such as those that have been found toward massive galaxy clusters with the SCUBA bolometer array (Smail et al. 1997). Spectral indices for these sources are estimated to be ~ 1.5–2.5 (Blain 1998, Fischer & Lange 1993). Sources with 350 GHz (850 μm) fluxes greater than 8 mJy are common, and all clusters surveyed had multiple sources with fluxes greater than 5 mJy. A 10 mJy source at 350 GHz corresponds to $\Delta T_{CMB} = 345$ μK for a 1′ beam, or a Compton y-parameter of 6×10^{-5}. The same source scaled to 270 GHz, assuming a ν^2 spectrum corresponds to $\Delta T_{CMB} = 140$ μK at 270 GHz for a 1′ beam and a y-parameter of 6×10^{-5}. Scaling to the SZE thermal null at 218 GHz gives 3.9 mJy, which corresponds to a $\Delta T_{CMB} = 85\ \mu K$ for a $1'$ beam. This in turn translates directly to an uncertainty in a measurement of the cluster-peculiar velocity (Equation 6); for a massive cluster with an optical depth of 0.01 and an electron temperature of 10 keV, 85 μK corresponds to a peculiar velocity of 930 km s^{-1}. The contamination is more severe for less massive clusters, with the dependence scaling as $\Delta v_{pec} \propto \tau_e^{-1} \propto R^2/M \propto M^{-1/3} \propto T_e^{-1/2}$. The contamination scales inversely with the beam area.

As with SZE observations at radio frequencies, the analyses of high-frequency observations must also consider the effects of point sources and require either high dynamic angular range, large spectral coverage, or both to separate the point source emission from the SZE.

Single-Dish Observations

All observations sensitive enough to observe the SZE are differential. The primary issue for single-dish observations is how to switch the beam on the sky without introducing systematics comparable to the SZE. This beam switching can be accomplished in several ways, including but not limited to Dicke switching between feeds and chopping mirrors, which switch or sweep the beam on the sky. With a single-dish telescope, modulation of the beam sidelobes can lead to an offset. This offset can be removed if it remains stable enough to be measured on some portion of the sky without a cluster. However, temperature variations of the optics and features on the ground will cause the offset to change as the source is tracked. Therefore, it has become common practice to observe leading and trailing fields when they have the same position with respect to the ground as the source. In this way, any constant or linear drift in offset can be removed at the price of observing efficiency and sensitivity.

The first measurements of the SZE were made with single-dish radio telescopes at centimeter wavelengths. Advances in detector technology made the measurements possible, although early observations appear to have been plagued by systematic errors that led to irreproducible and inconsistent results. Eventually, successful detections using beam switching techniques were obtained. During this period the pioneering work of Birkinshaw and collaborators with the Owens Valley Radio Observatory (OVRO) 40-m telescope stands out for its production of results that built confidence in the technique (Birkinshaw et al. 1978a,b; Birkinshaw 1991). More recently, leading and trailing beam switching techniques have been used successfully with the OVRO 5-m telescope at 32 GHz to produce reliable detections of the SZE in several intermediate redshift clusters (Herbig et al. 1995, Myers et al. 1997, Mason et al. 2001). The SEST 15-m and IRAM 30-m telescopes have been used with bolometric detectors at 140 GHz and chopping mirrors to make significant detections of the SZE effect in several clusters (Andreani et al. 1996, 1999; Desert et al. 1998; Pointecouteau et al. 1999, 2001). The Nobeyama 45-m telescope has also been been used at 21 GHz, 43 GHz, and 150 GHz to detect and map the SZE (Komatsu et al. 2001, 1999).

In the Sunyaev-Zel'dovich Infrared Experiment (SuZIE) pixels in a six-element 140-GHz bolometer array are electronically differenced by reading them out in a differential bridge circuit (Holzapfel et al. 1997). Differencing in this way makes the experiment insensitive to temperature and amplifier gain fluctuations that produce 1/f noise. This increased low-frequency stability allows SuZIE to observe in a drift-scanning mode in which the telescope is fixed and the rotation of the earth moves the beams across the sky. Using this drift-scanning technique, the SuZIE experiment has produced high signal-to-noise strip maps of the SZE emission in several clusters (Holzapfel et al. 1997, Mauskopf et al. 2000a).

Because of the high sensitivity of bolometric detectors at millimeter wavelengths, single-dish experiments are ideally suited for the measurement of the SZE spectrum. By observing at several millimeter frequencies, these instruments should be able to separate the thermal and kinetic SZEs from atmospheric fluctuations and sources of astrophysical confusion. One of the first steps to realizing this goal is the measurement of SZE as an increment. So far, there have been only a few low signal-to-noise detections at a frequency of approximately 270 GHz. The main reason for the lack of detection is the increased opacity of the atmosphere at higher frequencies. Holzapfel et al. (1997) reported a detection of Abell 2163 with the SuZIE instrument at 270 GHz. Andreani et al. (1996) claimed detections of the SZE increment in two clusters, although observations of a third cluster appear to be contaminated by foreground sources or systematic errors (Andreani et al. 1999).

Figure 4 shows the measured SZE spectrum of Abell 2163, spanning the decrement and increment with data obtained from different telescopes and techniques (Holzapfel et al. 1997, Desert et al. 1998, LaRoque et al. 2002). The SZE spectrum is a good fit to the data demonstrating the consistency and robustness of modern SZE measurements.

Single-dish observations of the SZE are just beginning to reach their potential, and the future is very promising. The development of large-format millimeter-wavelength bolometer arrays will increase the mapping speed of current SZE experiments by orders of magnitude. The first of this new generation of instruments is the BOLOCAM 151-element bolometer array (Mauskopf et al. 2000b, Glenn et al. 1998), which will soon begin routine observations at the Caltech Submillimeter Observatory. BOLOCAM will observe in drift-scanning mode and produce differences between bolometer signals in software. To the extent that atmospheric fluctuations are common across a bolometric array, it will be possible to realize the intrinsic sensitivity of the detectors. Operating from high astronomical sites with stable atmospheres and exceptionally low precipitable water vapor, future large-format bolometer arrays will have the potential to produce high signal-to-noise SZE images and search for distant SZE clusters with unprecedented speed.

Interferometric Observations

The stability and spatial filtering inherent to interferometry have been exploited to make high-quality images of the SZE. The stability of an interferometer is due to its

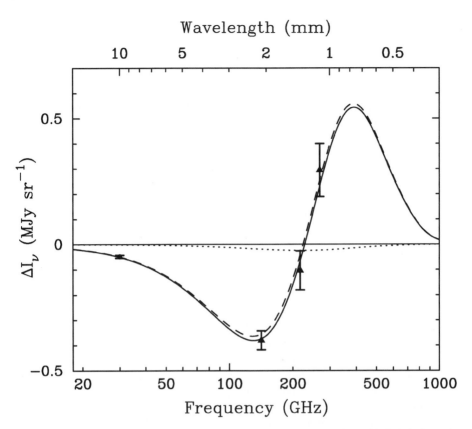

Figure 4 The measured SZE spectrum of Abell 2163. The data point at 30 GHz is from the Berkeley-Illinois-Maryland-Association (BIMA) array (LaRoque et al. 2002), at 140 GHz it is the weighted average of Diabolo and SuZIE measurements (Desert et al. 1998, Holzapfel et al. 1997) (*filled square*), and at 218 GHz and 270 GHz from SuZIE (Holzapfel et al. 1997) (*filled triangles*). The best fit thermal and kinetic SZE spectra are shown by the dashed and dotted lines, respectively, with the spectra of the combined effect shown by the solid line. The limits on the Compton y-parameter and the peculiar velocity are $y_0 = 3.56^{+0.41+0.27}_{-0.41-0.19} \times 10^{-4}$ and $v_p = 410^{+1030+460}_{-850-440}$ km s^{-1}, respectively, with statistical followed by systematic uncertainties at 68% confidence (Holzapfel et al. 1997, LaRoque et al. 2002).

ability to perform simultaneous differential sky measurements over well-defined spatial frequencies.

An interferometer measures the time-averaged correlation of the signals received by a pair of telescopes; all interferometric arrays can be thought of as a collection of $n(n-1)/2$ two-element interferometers. For each pair of telescopes, the interferometer effectively multiplies the sky brightness at the observing frequency by a cosine, integrates the product, and outputs the time-average amplitude of the product (see Thompson et al. 2001). In practice the signals are split and two correlations are performed, one with a 90 degree relative phase shift, so that the output of the interferometer, referred to as the visibility, is the complex Fourier transform (amplitude and phase) of the sky brightness. The interferometer is therefore only sensitive to angular scales (spatial frequencies) near B/λ, where the baseline B is the projected separation of the telescopes as seen by the source and λ is the observation wavelength. The interferometer response is essentially insensitive to gradients in the atmospheric emission or other large-scale emission features.

There are several other features that allow an interferometer to achieve extremely low systematics. For example, only signals that correlate between array elements will lead to detected signal. For most interferometers this means the bulk of the sky noise for each element will not lead to signal. Amplifier gain instabilities for an interferometer will not lead to large offsets or false detections, although if severe they may lead to somewhat noisy signal amplitude. To remove the effects of offsets or drifts in the electronics as well as the correlation of spurious (noncelestial) sources of noise, the phase of the signal received at each telescope is modulated, and the proper demodulation is applied to the output of the correlator.

The spatial filtering of an interferometer also allows the emission from radio point sources to be separated from the SZE emission. This is possible because at high angular resolution ($\lesssim 10''$) the SZE contributes very little flux. This allows one to use long baselines—which give high angular resolution—to detect and monitor the flux of radio point sources while using short baselines to measure the SZE. Nearly simultaneous monitoring of the point sources is important, as they are often time variable. The signal from the point sources is then easily removed, to the limit of the dynamic range of the instrument from the short-baseline data, which are sensitive also to the SZE.

Figure 5 (see color insert) illustrates the spatial filtering of an interferometer with data on the galaxy cluster RXJ 1347–1145 from the BIMA interferometer outfitted with 30-GHz receivers (e.g., Carlstrom et al. 1996, Carlstrom et al. 2000). Figure 5a shows the point source–subtracted SZE image (contours) overlaid on a ROSAT X-ray image (color scale). Spatial scales typical of galaxy clusters were stressed before deconvolution. Figure 5b depicts a higher-resolution SZE image, showing the range of spatial scales measured by the interferometer. Figure 5c shows the long-baseline (high angular resolution) data only. The bright on-center point source is readily apparent.

For the reasons given above, interferometers offer an ideal way to achieve high brightness sensitivity for extended low–surface brightness emission, at least at radio wavelengths. Most interferometers, however, were not designed for imaging low–surface brightness sources. Interferometers have been built traditionally to obtain high angular resolution and thus have employed large individual elements for maximum sensitivity to small-scale emission. As a result, special purpose interferometric systems have been built for imaging the SZE (Jones et al. 1993, Carlstrom et al. 1996, Padin et al. 2000). All of them have taken advantage of low-noise High Electron Mobility Transistor (HEMT) amplifiers (Pospieszalski et al. 1995) to achieve high sensitivity.

The first interferometric detection (Jones et al. 1993) of the SZE was obtained with the Ryle Telescope. The Ryle Telescope was built from the 5-Kilometer Array, consisting of eight 13-m telescopes located in Cambridge, England, operating at 15 GHz with east-west configurations. Five of the telescopes can be used in a compact east-west configuration for imaging of the SZE (Jones et al. 1993; Grainge et al. 1993, 1996, 2000; Saunders et al. 2000; Grainger et al. 2001; Grainge et al. 2001; Jones et al. 2001).

The OVRO and BIMA SZE imaging project uses 30-GHz (1 cm) low-noise receivers mounted on the OVRO[1] and BIMA[2] mm-wave arrays in California. They have produced SZE images toward 60 clusters to date (Carlstrom et al. 1996, 2000; Patel et al. 2000; Grego et al. 2000; Reese et al. 2000; Grego et al. 2001; Joy et al. 2001; LaRoque et al. 2002; Reese et al. 2002). Figure 6 (see color insert) shows a sample of their SZE images. Figure 6 also clearly demonstrates the independence of the SZE on redshift. All of the clusters shown have similar high X-ray luminosities and, as can be seen, the strength of the SZE signals are similar despite the factor of five in redshift. The OVRO and BIMA arrays support two-dimensional configuration of the telescopes, including extremely short baselines, allowing good synthesized beams for imaging the SZE of clusters at declinations greater than ~ -15 degrees.

The Ryle Telescope, OVRO, and BIMA SZE observations are insensitive to the angular scales required to image low-redshift clusters, $z \ll 0.1$. Recently, however, the Cosmic Background Imager (CBI) (Padin et al. 2000) has been used to image the SZE in a few nearby clusters (Udomprasert et al. 2000). The CBI is composed of 13 0.9-m telescopes mounted on a common platform with baselines spanning 1–6 m. Operating in 10 1-GHz channels spanning 26–36 GHz, it is sensitive to angular scales spanning $3'-20'$. The large field of view of the CBI, 0.75 degrees FWHM, makes it susceptible to correlated contamination from terrestrial sources, i.e., ground emission. To compensate, they have adopted the same observing strategy as for single-dish observations (see "Single-Dish Observations," above), and subtract data from leading and trailing fields offset by ± 12.5 minutes in right ascension from the cluster.

Interferometric observations of the SZE, as for single-dish observations, are just beginning to demonstrate their potential. Upcoming instruments will be over an order of magnitude more sensitive. The OVRO-BIMA SZE imaging team is now building the Sunyaev-Zel'dovich Array (SZA), consisting of eight 3.5-m telescopes outfitted with 26–36-GHz and 85–115-GHz low-noise receivers and employing an 8-GHz wideband correlator. The SZA is expected to be operational by the end of 2003. It will be deployed with the existing six 10.4-m OVRO

---

[1]An array of six 10.4-m telescopes located in Owens Valley, CA, and operated by Caltech.
[2]An array of ten 6.1-m mm-wave telescopes located at Hat Creek, CA, and operated by the Berkeley-Illinois-Maryland-Association.

telescopes and nine 6.1-m BIMA telescopes at a new high site if the site is ready in time. If not, the array will be deployed in the Owens Valley site with the existing OVRO telescopes. The SZA will operate both in a dedicated survey mode and as a fully heterogeneous array with the larger telescopes. The heterogeneous array will provide unprecedented imaging of the SZE at high resolution.

The Ryle Telescope SZE team is also building the ArcMinute Imager, consisting of ten 3.7-m telescopes operating at 15 GHz near the Ryle Telescope in Cambridge. The team does not plan to operate the ArcMinute Imager as a heterogeneous array with the Ryle Telescope, but the Ryle Telescope would be used for concurrent point-source monitoring.

Additionally, plans have been discussed to reconfigure the CBI to 90 GHz. With its 13 0.9-m telescopes and 10-GHz bandwidth, the CBI would be a formidable SZE survey machine. A similar fixed platform interferometer, the Array for Microwave Background Anisotropy (AMiBA), is also being built with 19 1.2-m telescopes to operate at 90 GHz. AMiBA, like the reconfigured CBI, would also be ideally suited for performing SZE surveys at moderate resolution.

This next generation of interferometric SZE instruments will conduct deep SZE surveys covering tens and possibly hundreds of square degrees. Although not as fast as planned large-format bolometric arrays, the interferometers will be able to provide more detailed imaging. In particular, the high resolution and deep imaging provided by the SZA/OVRO/BIMA heterogeneous array (referred to as CARMA, the Combined ARray for Millimeter Astronomy) operating at 90 GHz will provide a valuable tool for investigating cluster structure and its evolution. As discussed in "Challenges for Interpreting Sunyaev-Zel'dovich Effect Surveys," below, such studies are necessary before the full potential of large SZE surveys for cosmology can be realized.

SKY SURVEYS WITH THE SUNYAEV-ZEL'DOVICH EFFECT

With recent developments in instrumentation and observing strategies, it will soon be possible to image large areas with high sensitivity, enabling efficient and systematic SZE searches for galaxy clusters.

The primary motivation for large surveys for galaxy clusters using the SZE is to obtain a cluster catalog with a well-understood selection function that is a very mild function of cosmology and redshift. There are two primary uses for such a catalog. The first is to use clusters as tracers of structure formation, allowing a detailed study of the growth of structure from $z \sim 2$ or 3 to the present day. The second use is for providing a well-understood sample for studies of individual galaxy clusters, either as probes of cosmology or for studies of the physics of galaxy clusters.

Numerous authors have presented estimates of the expected yields from SZE surveys (Korolev et al. 1986, Bond & Myers 1991, Bartlett & Silk 1994, Markevitch et al. 1994, De Luca et al. 1995, Bond & Myers 1996, Barbosa et al.

1996, Colafrancesco et al. 1997, Aghanim et al. 1997, Kitayama et al. 1998, Holder et al. 2000b, Bartlett 2000, Kneissl et al. 2001). Results from the diverse approaches to calculating the cluster yields are in broad agreement.

Cluster Abundance

The number of clusters expected to be found in SZE surveys depends sensitively on the assumed cosmology and detector specifications. Estimates of the order of magnitude, however, should be robust and able to give a good indication of the expected scientific yields of surveys for galaxy clusters using the SZE.

In calculating the number of clusters expected in a given survey, three things are needed:

(1) the volume per unit solid angle as a function of redshift,

(2) the number density of clusters as a function of mass and redshift, and

(3) an understanding of the expected mass range that should be observable with the particular SZE instrument and survey strategy.

The physical volume-per-unit redshift per unit solid angle is given by (Peebles 1994)

$$\frac{dV}{d\Omega dz} = D_A^2 \, c \frac{dt}{dz},\tag{7}$$

where $dt/dz = 1/[H(z)(1+z)]$ and $H(z)$ is the expansion rate of the universe. The comoving volume is simply the physical volume multiplied by $(1+z)^3$.

The number density of clusters as a function of mass and redshift can be either derived by applying the statistics of peaks in a Gaussian random field (Press & Schechter 1974, Bond et al. 1991, Sheth et al. 2001) to the initial density perturbations or taken from large cosmological N-body simulations (Jenkins et al. 2001). The mass function is still not understood perfectly, with small but important differences between competing estimates, especially at the high-mass end of the spectrum. Precise cosmological studies will require an improved understanding, but reasonably accurate results can be obtained with the standard Press-Schechter (Press & Schechter 1974) mass function, with the comoving number density between masses M and $M + dM$ given by

$$\frac{dn(M,z)}{dM} = -\sqrt{\frac{2}{\pi}} \frac{\bar{\rho}}{M^2} \frac{d\ln\sigma(M,z)}{d\ln M} \frac{\delta_c}{\sigma(M,z)} \exp\left[\frac{-\delta_c^2}{2\sigma^2(M,z)}\right],\tag{8}$$

Where $\bar{\rho}$ is the mean background density of the universe today, $\sigma^2(M,z)$ is the variance of the density field when smoothed on a mass scale M, and δ_c (typically ~ 1.69) is the critical overdensity for collapse in the spherical collapse model (Peebles, 1980).

Smoothing the density field on a mass scale corresponds to finding the comoving volume that encloses a given mass for a region at the mean density of the universe and smoothing the density field over this volume. The variance $\sigma^2(M,z)$ is separable as $\sigma(M, z) \equiv \sigma(M) D(z)$, where $\sigma(M)$ is the variance in the initial density field and $D(z)$ is the linear growth function that indicates how the amplitude of the density field has grown with time. For a universe with $\Omega_M = 1$ this growth function is simply proportional to the scale factor.

For a universe composed only of matter and vacuum energy (either with or without spatial curvature), accurate fitting functions for the growth function can be found in Carroll et al. (1992) or can be found as a straightforward one-dimensional numerical integral, using the solution found by Heath (1977). For a more exotic universe, for example one with dark energy not in the form of a cosmological constant, the growth function requires solution of a two-dimensional ordinary differential equation.

The Press-Schechter formulation has the advantage of making it clear that the abundance of very massive objects is exponentially suppressed, showing that massive clusters are expected to be rare. The amount of suppression as a function of redshift is sensitive to the linear growth function $D(z)$, which is itself sensitive to cosmological parameters. Structure grows most efficiently when the universe has $\Omega_M \sim 1$, so the growth function as a function of z should give a good indication of the epoch when either curvature, vacuum energy, or dark energy started to become dynamically important.

The exponential dependence of the cluster abundance makes SZE surveys a potentially powerful probe of cosmology. This is shown in Figure 7, which shows the relative importance of volume and number density. A difference in cosmology can cause a difference in volume of a few tens of percent, whereas the corresponding

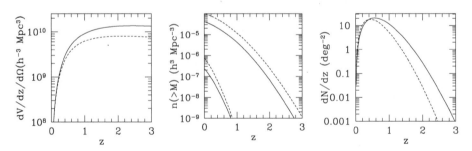

Figure 7 Comoving volume element (*left*) and comoving number density (*center*) for two cosmologies, $(\Omega_M, \Omega_\Lambda) = (0.3, 0.7)$ (*solid*) and $(0.5, 0.5)$ (*dashed*). (*Middle*) The normalization of the matter power spectrum was taken to be $\sigma_8 = 0.9$ and the Press-Schechter mass function was assumed. The lower set of lines in the middle panel correspond to clusters with mass greater than $10^{15} h^{-1} M_\odot$ while the upper lines correspond to clusters with mass greater than $10^{14} h^{-1} M_\odot$. The right panel corresponds to the cluster redshift distribution per square degree for clusters with masses greater than $10^{14} h^{-1} M_\odot$, with the normalization of the power spectrum adjusted to produce the same local cluster abundance for both cosmologies ($\sigma_8 = 0.75$ for $\Omega_M = 0.5$). Note that in this case fewer clusters are predicted at high redshift for the higher density cosmology.

change in comoving number density is typically a factor of a few. This plot also shows the rapid decline in the number density of the cluster abundance with redshift and its steep dependence on mass, both of which are due to the exponential suppression of high peaks.

The cosmology with the higher mass density can be seen in Figure 7 to have a higher abundance at $z = 0$ for a fixed normalization of the power spectrum, i.e., σ_8. This is primarily because a given cluster mass will correspond to a slightly smaller size for a universe with higher matter density. The matter power spectrum rises toward smaller scales, so a fixed amplitude of the power spectrum on a specific scale will lead to a higher density cosmology having more power on a given mass scale. Choosing a slightly lower value of σ_8 for the cosmology with the higher density removes this offset in the cluster abundance at $z = 0$ and leads to a lower cluster abundance at higher redshifts, i.e., if the cluster abundance is normalized at $z = 0$ for all cosmological models, the higher-density models will have relatively fewer clusters at high redshift. This can be seen in the right panel of Figure 7, where the redshift distribution per square degree has been normalized to give the same number of clusters above a mass of $10^{14}\ h^{-1}\ M_\odot$ at $z = 0$ by lowering the normalization of the power spectrum from $\sigma_8 = 0.90$ to 0.75.

Mass Limits of Observability

The range of masses to which a survey is sensitive is set by the effective beam size and sensitivity of the instrument as well as the cluster profile on the sky. In the case of a beam that is larger than the cluster, a survey is limited by SZE flux. From Equation 5, a flux limit corresponds to constant $N_e T/D_A^2$. The dependence on angular diameter distance rather than luminosity distance leads to a relative factor of $(1 + z)^4$ when compared with a usual flux limit for emission from a distant source. Past $z \sim 1$ the angular diameter distance is slowly varying, and the gas temperature for a fixed mass should be higher than at $z = 0$, because the clusters are more dense and therefore more tightly bound, i.e., smaller. As a result, at $z > 1$ the limiting mass for an SZE survey is likely to be gently declining with redshift (Holder et al. 2000b, Bartlett 2000, Kneissl et al. 2001). Nearby clusters ($z < 0.2$) are likely to be at least partially resolved by most SZE surveys, making the mass selection function slightly more difficult to estimate robustly. It is not expected that the mass threshold of detectability should change more than a factor of ~ 2–3 for clusters with $z > 0.05$, making an SZE-selected catalog remarkably uniform in redshift in terms of its mass selection function.

The expected cluster profiles are not well known because there are very few known clusters at $z > 0.5$, and these are much more massive than the typical clusters expected to be found in deep SZE surveys. The total SZE flux from a cluster should be fairly robust against changes in cluster profiles owing to substructure or merging. Broadly speaking, the SZE is providing an inventory of hot electrons. The characteristic temperature in a cluster is set by virial considerations, because the electrons are mainly heated by shocks due to infall. Kinetic energy of infalling

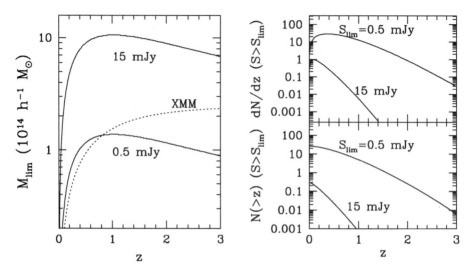

Figure 8 (*Left*) Mass limits as a function of redshift for a typical wide-field type of survey (equivalent to ~15 mJy at 30 GHz) and a typical deep survey (~0.5 mJy). The approximate XMM-Newton serendipitous survey limit is also shown. (*Right*) Differential (*top*) and cumulative (*bottom*) counts as a function of redshift for the two SZE surveys shown at left, assuming a ΛCDM cosmology (Holder et al. 2000a).

gas should be converted into thermal energy in a time shorter than the Hubble time, suggesting that the thermal energy per particle must necessarily be on the order of GM/R. Therefore, it is very difficult to substantially alter the expected SZE flux for a cluster of a given mass.

The mass limit of detection corresponding to an SZE flux–limited survey is shown in the left panel of Figure 8. The two types of surveys shown correspond to deep ground-based imaging of a few tens of square degrees down to μK sensitivities with arcminute resolution or wide-field surveys (similar to the *Planck Surveyor* satellite) with ~5′–10′ resolution.

In contrast to the total integrated SZE flux, the concentration of SZE flux is very model dependent, with very compact clusters having high central decrements (or increments) but subtending a relatively small solid angle. The integrated SZE flux is thus a potentially very powerful criterion for controlling selection effects in samples for cluster studies. The clusters can first be found using integrated SZE flux and then investigated with high-resolution SZE imaging.

Estimates of SZE Source Counts

The expected source counts are shown in the right panels of Figure 8. From the considerations discussed above, it should be clear that the exact numbers will depend on cosmology and observing strategy. A robust conclusion is that upcoming deep surveys should find tens of clusters per square degree. Less deep surveys, such as the all-sky *Planck Surveyor* satellite survey should detect a cluster in every few square degrees. The resulting catalogs should be nearly uniformly selected in mass, with the deep catalogs extending past $z \sim 2$.

COSMOLOGY FROM SUNYAEV-ZEL'DOVICH EFFECT SURVEY SAMPLES

In this section we review the use of the SZE for cosmological studies and provide an update on current constraints. Emphasis is given to the cosmology that can, in principle, be extracted from SZE survey yields, as well as the observational and theoretical challenges that must be met before the full potential of SZE surveys for cosmology can be realized.

Future SZE surveys, with selection functions that are essentially a uniform mass limit with redshift (see "Sky Surveys with the Sunyaev-Zel'dovich Effect," above), will provide ideal cluster samples for pursuing cosmology. A large catalog of distant clusters will enable studies of large-scale structure using the same methods as are applied to large catalogs of galaxies. The SZE surveys will also provide a direct view of the high-redshift universe. Should clusters exist at redshifts much higher than currently predicted, i.e., if the initial mass fluctuations were non-Gaussian, they will be found by SZE surveys but missed in even the deepest X-ray observations planned.

The SZE survey samples can be used to increase the precision of the more traditional applications of the SZE to extract cosmological parameters, such as cluster distance measurements and the Hubble constant, the ratio Ω_B/Ω_M, and cluster-peculiar velocities. These are discussed in "Distance Determinations, Hubble Constant," "Cluster Gas-Mass Fractions, Ω_M," and "Cluster Peculiar Velocities," respectively. The ability to derive these parameters depends primarily on the ability to use the SZE and other cluster observables to constrain or even over-constrain cluster properties. Precise SZE measurements will allow tests of the underlying assumptions in these derivations. For example, high resolution imaging of SZE, X-ray, and lensing will allow detailed tests of the assumption of hydrostatic equilibrium (Miralda-Escude & Babul 1995b, Loeb & Mao 1994, Wu & Fang 1997, Squires & Kaiser 1996, Allen & Fabian 1998). The sample yields will also allow the determination of global properties of clusters and their relationship to observables. For example, it is already possible to estimate cluster gas temperatures without X-ray data using current SZE data (Joy et al. 2001).

The new frontier for SZE cosmology will be in exploiting the ability of future SZE surveys to measure cleanly the number density of clusters and in its evolution in time. The redshift distribution of galaxy clusters is critically sensitive to Ω_M and the properties of the dark energy. For sufficiently large and deep SZE surveys, it is possible, in principle, to extract the equation of state of the dark energy. This is discussed in "Energy Densities in the Universe and Growth of Structure," and the theoretical and observational challenges are outlined in "Challenges for Interpreting Sunyaev-Zel'dovich Effect Surveys."

Distance Determinations, Hubble Constant

Several years after the SZE was first proposed (Sunyaev & Zel'dovich 1970, 1972) it was recognized that the distance to a cluster could be determined with a measure of its SZE and X-ray emission (Cavaliere et al. 1977, Gunn et al. 1978, Silk & White 1978, Cavaliere & Fusco-Femiano 1978, Birkinshaw 1979). The distance is determined by exploiting the different density dependencies of the SZE and X-ray emission. The SZE is proportional to the first power of the density; $\Delta T_{SZE} \sim \int d\ell n_e T_e$, where n_e is the electron density, T_e is the electron temperature, and $d\ell$ is along the line of sight. The distance dependence is made explicit with the substitution $d\ell = D_A d\zeta$, where D_A is the angular diameter distance of the cluster.

The X-ray emission is proportional to the second power of the density; $S_x \sim \int d\ell n_e^2 \Lambda_{eH}$, where Λ_{eH} is the X-ray cooling function. The angular diameter distance is solved for by eliminating the electron density,[3] yielding

$$D_A \propto \frac{(\Delta T_0)^2 \Lambda_{eH0}}{S_{x0} T_{e0}^2} \frac{1}{\theta_c},$$ (9)

where these quantities have been evaluated along the line of sight through the center of the cluster (subscript 0), and θ_c refers to a characteristic scale of the cluster along the line of sight, whose exact meaning depends on the density model adopted. Only the characteristic scale of the cluster in the plane of the sky is measured, so one must relate the line-of-sight and plane-of-sky characteristic scales. For detailed treatments of this calculation see Birkinshaw & Hughes (1994) and Reese et al. (2000). Combined with the redshift of the cluster and the geometry of the universe, one may determine the Hubble parameter, with the inverse dependencies on the observables as that of D_A. With a sample of galaxy clusters, one fits the cluster distances versus redshift to the theoretical angular diameter distance relation, with the Hubble constant as the normalization (e.g., see Figure 9).

There are two explicit assumptions made in SZE and X-ray distance determinations. The first one, mentioned above, is that the characteristic scale of the cluster along the line of sight must be related (usually assumed equal) to the scale in the plane of the sky. Typically, spherical symmetry is assumed for the cluster geometry because for a large sample of clusters one would expect $\langle \theta_c / \theta_c^{sky} \rangle = 1$, at least in the absence of selection effects. This assumption is supported by simulations as well (Sulkanen 1999). The second assumption is that $\langle n_e^2 \rangle^{1/2}$ equals $\langle n_e \rangle$ along the line of sight, i.e., that the clumping factor

$$C \equiv \frac{\langle n_e^2 \rangle^{1/2}}{\langle n_e \rangle}$$ (10)

equals unity. If significant substructure exists in galaxy clusters, the derived Hubble constant will be overestimated by a factor of C^2.

---

[3]Similarly, one could eliminate D_A in favor of the central density, n_{e0}.

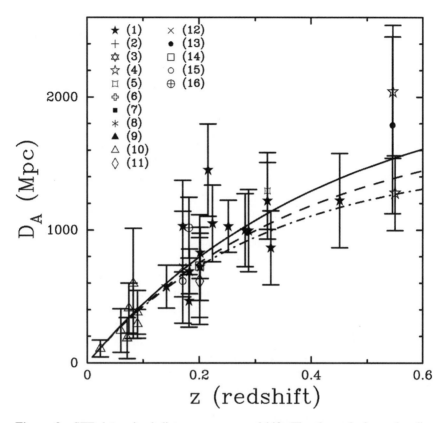

Figure 9 SZE-determined distances versus redshift. The theoretical angular diameter distance relation is plotted for three different cosmologies, assuming $H_0 = 60$ km s^{-1} Mpc^{-1}. $\Omega_M = 0.3$, $\Omega_\Lambda = 0.7$ (*solid line*), $\Omega_M = 0.3$, $\Omega_\Lambda = 0$ (*dashed line*), and $\Omega_M = 1.0$, $\Omega_\Lambda = 0$ (*dot-dashed line*). The clusters are beginning to trace out the angular diameter distance relation. References: (1) Reese et al. 2002; (2) Pointecouteau et al. 2001; (3) Mauskopf et al. 2000a; (4) Reese et al. 2000; (5) Patel et al. 2000; (6) Grainge et al. 2000; (7) Saunders et al. 2000; (8) Andreani et al. 1999; (9) Komatsu et al. 1999; (10) Mason et al. 2001, Mason 1999, Myers et al. 1997; (11) Lamarre et al. 1998; (12) Tsuboi et al. 1998; (13) Hughes & Birkinshaw 1998; (14) Holzapfel et al. 1997; (15) Birkinshaw & Hughes 1994; (16) Birkinshaw et al. 1991.

To date, there are 38 distance determinations to 26 different galaxy clusters from analyses of SZE and X-ray observations. In Figure 9 we show all SZE-determined distances from high signal-to-noise SZE experiments. The uncertainties shown are statistical at 68% confidence. There are currently three samples of clusters with SZE distances: (*a*) a sample of 7 nearby ($z < 0.1$) galaxy clusters observed with the OVRO 5-m telescope (Myers et al. 1997, Mason et al. 2001); (*b*) a sample of 5 intermediate redshift ($0.14 < z < 0.3$) clusters from the Ryle telescope interferometer (Jones et al. 2001); and (*c*) a sample of 18 clusters with $0.14 < z < 0.83$ from interferometric observations by the OVRO and BIMA SZE imaging project (Reese et al. 2002). A fit to the ensemble of 38 SZE-determined distances yields $H_0 = 60 \pm 3$ km s^{-1} Mpc^{-1} for an $\Omega_M = 0.3$, $\Omega_\Lambda = 0.7$ cosmology, where only the statistical uncertainty is included (at 68% confidence). The systematic uncertainty, discussed below, is of order 30% and clearly dominates. Because many of the clusters are at high redshift, the best-fit Hubble constant will depend on the cosmology adopted; the best-fit Hubble constant shifts to 56 km s^{-1} Mpc^{-1} for an open $\Omega_M = 0.3$ universe and to 54 km s^{-1} Mpc^{-1} for a flat $\Omega_M = 1$ geometry.

The prospects for improving both the statistical and systematic uncertainties in the SZE distances in the near future are promising. Note, from Equation 9 that the error budget in the distance determination is sensitive to the absolute calibration of the X-ray and SZE observations. Currently, the best absolute calibration of SZE observations is $\sim 2.5\%$ at 68% confidence based on observations of the brightness of the planets Mars and Jupiter. Efforts are now underway to reduce this uncertainty to the 1% level (2% in H_0). Uncertainty in the X-ray intensity scale also adds another shared systematic. The accuracy of the ROSAT X-ray intensity scale is debated, but a reasonable estimate is believed to be $\sim 10\%$. It is hoped that the calibration of the Chandra and XMM-Newton X-ray telescopes will greatly reduce this uncertainty.

The largest systematic uncertainties are due to departures from isothermality, the possibility of clumping, and possible point-source contamination of the SZE observations (for detailed discussion of systematics see, e.g., Birkinshaw, 1999; Reese et al. 2000, 2002). Chandra and XMM-Newton are already providing temperature profiles of galaxy clusters (e.g., Nevalainen et al. 2000, Markevitch et al. 2000, Tamura et al. 2001). The unprecedented angular resolution of Chandra will provide insight into possible small-scale structures in clusters. In addition, multiwavelength studies by existing radio observatories, e.g., the Very Large Array, can shed light on the residual point source contamination of the radio wavelength SZE measurements. Therefore, though currently at the 30% level, many of the systematics can and will be addressed through both existing X-ray and radio observatories and larger samples of galaxy clusters provided from SZE surveys.

The beauty of the SZE and X-ray technique for measuring distances is that it is completely independent of other techniques and that it can be used to measure distances at high redshifts directly. Because the method relies on the well-understood physics of fully ionized plasmas, it should be largely independent of cluster evolution. Inspection of Figure 9 already provides confidence that a large survey of SZE distances consisting of perhaps a few hundred clusters with redshifts extending to one and beyond would allow the technique to be used to trace the expansion history of the universe, providing a valuable independent check of the recent determinations of the geometry of the universe from type Ia supernova (Riess et al. 1998, Perlmutter et al. 1999) and CMB primary anisotropy experiments (Pryke et al. 2002, Netterfield et al. 2002, Stompor et al. 2001).

Cluster Gas-Mass Fractions, Ω_M

The ICM contains most of the baryons confined to the cluster potential with roughly an order of magnitude more baryonic mass than that observed in the galaxies themselves (White et al. 1993, Forman & Jones 1982). The gas mass fraction, f_g, is therefore a reasonable estimate of the baryonic mass fraction of the cluster. It should also be a reasonable approximation of the universal baryon mass fraction, $f_B \equiv \Omega_B/\Omega_M$, because it is not believed that mass segregation occurs on the large scales from which massive clusters condense: ~ 1000 Mpc^3. The cluster gas fraction is actually a lower limit, $f_g \leq f_B$, because a small fraction of baryons ($\sim 10\%$) are likely lost during the cluster formation process (White et al. 1993, Evrard 1997) and we cannot rule out the possibility of additional reservoirs of baryons in galaxy clusters that have yet to be detected.

A measurement of f_B leads directly to an estimate of Ω_M, given a determination of Ω_B. Recent reanalysis of big bang nucleosynthesis (BBN) predictions with careful uncertainty propagation (Nollett & Burles 2000; Burles et al. 1999) along with recent deuterium-to-hydrogen ratio measurements in Lyman α clouds (Burles & Tytler 1998a,b) constrain the baryon density to be $\Omega_B h^2 = 0.019 \pm 0.0012$ at 68% confidence. Recent CMB primary anisotropy experiments provide an additional independent determination of $\Omega_B h^2$ consistent with the Lyman α cloud result (Pryke et al. 2002, Netterfield et al. 2002, Stompor et al. 2001).

The gas-mass is measured directly by observations of the SZE, provided the electron temperature is known. The total gravitating mass can be determined by assuming hydrostatic equilibrium and using the distribution of the gas and, again, the electron temperature. The SZE-derived gas fraction will therefore be proportional to $\Delta T_{SZE}/T_e^2$. Alternatively, the total gravitating mass can be determined by strong lensing (on small scales) or weak lensing (on large scales). Recently there has been considerable work on SZE gas fractions using total mass determinations derived under the assumption of hydrostatic equilibrium.

SZE derived cluster gas-mass fractions have been determined for two samples of clusters, and the results were used to place constraints on Ω_M: a sample of 4 nearby clusters (Myers et al. 1997) and a sample of 18 distant clusters (Grego et al. 2001). Both analyses used a spherical isothermal β-model for the ICM. The nearby sample was observed with the Owens Valley 5.5-m telescope at 32 GHz as part of an SZE study of an X-ray flux–limited sample (Myers et al. 1997). In this study the integrated SZE was used to normalize a model for the gas density from published X-ray analyses and then compared to the published total masses to determine the gas-mass fraction. For three nearby clusters, A2142, A2256 and the Coma cluster, a gas-mass fraction of $f_g h = 0.061 \pm 0.011$ at radii of 1–$1.5\,h^{-1}$ Mpc is found; for the cluster Abell 478 a gas-mass fraction of $f_g h = 0.16 \pm 0.014$ is reported.

The high redshift sample of 18 clusters ($0.14 < z < 0.83$) was observed interferometrically at 30 GHz using the OVRO and BIMA SZE imaging system (Grego et al. 2001). In this study, the model for the gas density was determined directly by the SZE data. X-ray emission-weighted electron temperatures were used, but no X-ray imaging data was used. The gas fractions were computed from the data at a 1′ radius, where they are best constrained by the observations. Numerical simulations suggest, however, that the gas-mass fraction at r_{500} (the radius inside of which the mean density of the cluster is 500 times the critical density) should reflect the universal baryon fraction (Evrard 1997, Evrard et al. 1996, David et al. 1995). The derived gas fractions were therefore extrapolated to r_{500} using scaling relations from cluster simulations (Evrard 1997). The resulting mean gas mass fractions are $f_g h = 0.081^{+0.009}_{-0.011}$ for $\Omega_M = 0.3$, $\Omega_\Lambda = 0.7$, $f_g h = 0.074^{+0.008}_{-0.009}$ for $\Omega_M = 0.3$, $\Omega_\Lambda = 0.0$ and $f_g h = 0.068^{+0.009}_{-0.008}$ for $\Omega_M = 1.0$, $\Omega_\Lambda = 0.0$. The uncertainties in the electron temperatures contribute the largest component to the error budget.

The angular diameter distance relation $D_A(z)$ enters the gas fraction calculation and introduces the cosmology dependence of the results of the high z sample. In addition, the simulation scaling relations used to extrapolate the gas fractions to r_{500} have a mild dependence on cosmology. Figure 10 shows the constraints on Ω_M implied by the measured gas-mass fractions assuming a flat universe ($\Omega_\Lambda \equiv 1 - \Omega_M$) and $h = 0.7$ to calculate D_A and the r_{500} scaling factor. The upper limit to Ω_M and its associated 68% confidence interval is shown as a function of Ω_M. The measured gas-mass fractions are consistent with a flat universe and $h = 0.7$ when Ω_M is less than 0.40, at 68% confidence. For the measurements to be consistent with $\Omega_M = 1.0$ in a flat universe, the Hubble constant must be very low, h less than ∼0.30.

To estimate Ω_M, we need to account for the baryons contained in the galaxies and those lost during cluster formation. The galaxy contribution is assumed to be a fixed fraction of the cluster gas, with the fraction fixed at the value observed in the Coma cluster, $M_g^{true} = M_g^{obs}(1 + 0.20h^{3/2})$ (White et al. 1993). Simulations suggest that the baryon fraction at r_{500} will be a modest underestimate of the true baryon fraction $f_g(r_{500}) = 0.9 \times f_B$(universal) (Evrard 1997). These assumptions lead to $f_B = (f_g(1 + 0.2h^{3/2})/0.9)$. Using this to scale the gas fractions derived from the high z SZE cluster sample and assuming $h = 0.7$ and a flat cosmology leads to the constraints illustrated in Figure 10 with a best estimate $\Omega_M \sim 0.25$ (Grego et al. 2001).

Cluster gas-mass fractions can also be determined from cluster X-ray emission in a similar manner as from SZE measurements. However, there are important differences between X-ray- and SZE-determined gas fractions. For example, the X-ray emission is more susceptible to clumping of the gas, C, because it is proportional to the ICM density squared. On the other hand, the X-ray–derived gas mass is essentially independent of temperature for the ROSAT 0.1–2.4-keV band used in the analyses (Mohr et al. 1999), whereas the SZE derived gas mass is proportional to T_e^{-2}.

Currently X-ray data for low-redshift clusters is of exceptional quality, far surpassing SZE data. X-ray–based gas-mass fractions have been measured to cluster radii of 1 Mpc or more (e.g., White & Fabian 1995, David et al. 1995, Neumann &

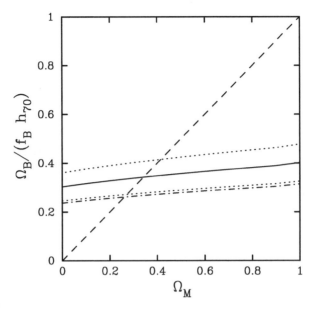

Figure 10 Limits on Ω_M from SZE-measured cluster gas fractions (Grego et al. 2001). Upper limit on the total matter density, $\Omega_M \leq \Omega_B/(f_B h_{70})$ (*solid line*) and its associated 68% confidence region (*dotted lines*) as a function of cosmology with $\Omega_\Lambda \equiv 1 - \Omega_M$. The intercept between the upper dotted line and the dashed line $\Omega_M = \Omega_B/(f_B h_{70})$ gives the upper limit to Ω_M at 68% confidence. The dot-dashed line shows the total matter density when the baryon fraction includes an estimate of the contribution from baryons in galaxies and those lost during cluster formation. The intercept of the dot-dashed line and the dashed line gives the best estimate of $\Omega_M \sim 0.25$, assuming a flat universe with $h = 0.7$.

Bohringer 1997, Squires et al. 1997, Mohr et al. 1999). A mean gas-mass fraction within r_{500} of $f_g h^{3/2} = 0.0749 \pm 0.0021$ at 90% confidence was derived from X-ray data from a large, homogeneous, nearby sample of clusters with $T_e > 5$ keV (Mohr et al. 1999). The gas-mass fractions derived from SZE measurements depend differently on the cosmology assumed than those derived from X-ray images, and this should be noted when comparing the results. Qualitatively, the comparison does not suggest any large systematic offsets. In fact, for a cold dark matter (ΛCDM) cosmology, solving for h from the combination of the Grego et al. (2001) and Mohr et al. (1999) results gives $h = 0.85^{+0.30}_{-0.20}$ at 68% confidence. This is significant because a large clumping factor, $C \gg 1$ (see Equation 10), has been suggested as an explanation for the high gas-mass fractions in clusters (White et al. 1993, Evrard 1997).

Cluster gas-mass fractions can also be measured by comparing SZE-derived gas masses and weak lensing–derived total masses. The comparison is particularly interesting, as both are measures of projected mass distributions. In addition, gas-mass fractions can be derived without assuming a model for the cluster structure and without assuming hydrostatic equilibrium. Comparisons of SZE and lensing data have only been done for a few clusters to date (Holder et al. 2000a). However, as for the SZE, the quality and quantity of weak lensing observations toward galaxy clusters are rapidly increasing, and several weak lensing surveys are underway. Holder et al. (2000a) demonstrated that gas-mass fractions can be determined from the analysis of SZE and weak lensing measurements without need to parameterize the ICM distribution. Furthermore, by comparing this mass fraction with one derived by assuming hydrostatic equilibrium, it is possible to solve for the ICM electron temperature and the angular diameter distance.

SZE surveys will provide a large catalog of galaxy clusters at redshifts $z > 1$. The increased sensitivity and larger angular dynamic range of the next generation of SZE instruments will allow measurements of cluster gas fractions to r_{500} directly, greatly increasing the precision of the gas-mass fractions. Moreover, extending the gas fraction analyses to high redshift will enable studies of the evolution of cluster structure. It should, for example, be straightforward to test speculative theories of dark matter decay (Cen 2001).

Cluster Peculiar Velocities

The kinetic SZE is a unique and potentially powerful cosmological tool, as it provides the only known way to measure large-scale velocity fields at high redshift (see "Kinetic Sunyaev-Zel'dovich Effect," above). To obtain an accurate measure of the peculiar velocity of a galaxy cluster, sensitive multifrequency SZE observations are required to separate the thermal and kinetic effects. From inspection of Figure 2, it is clear that measurements of the kinetic SZE are best made at frequencies near the null of the thermal effect at ∼218 GHz. However, as discussed in "Sources of Astronomical Contamination and Confusion" above, contamination by CMB temperature fluctuations as well as other sources makes it difficult to determine accurately the peculiar velocity for a given cluster. There have been only a few recent attempts to measure the kinetic SZE.

The first interesting limits on the peculiar velocity of a galaxy cluster were reported in Holzapfel et al. (1997). They used the Sunyaev-Zel'dovich Infrared Experiment (SuZIE) to observe Abell 2163 ($z = 0.202$) and Abell 1689 ($z = 0.183$) at 140 GHz (2.1 mm), 218 GHz (1.4 mm), and 270 GHz (1.1 mm). These observations include and bracket the null in the thermal SZE spectrum. Using a β model, with the shape parameters (θ_c, β) from X-ray data, they found $v_{pec} = +490^{+1370}_{-880}$ km s^{-1} for Abell 2163 and $v_{pec} = +170^{+815}_{-630}$ km s^{-1} for Abell 1689, where the uncertainties are at 68% confidence and include both statistical and systematic uncertainties. These results are limited by the sensitivity of the SZE observations, which were limited by differential atmospheric emission. The SuZIE data for Abell 2163 were reanalyzed with the addition of higher-frequency measurements that are sensitive to emission from Galactic dust in the direction of the cluster (Lamarre et al. 1998). More recently LaRoque et al. (2002) also reanalyzed all of the available data for

Abell 2163, including a new measurement obtained with the OVRO and BIMA SZE imaging system at 30 GHz (1 cm). As shown in Figure 4, the data is well fitted by parameters similar to the original values from Holzapfel et al. (1997). The agreement between the measurements using different instruments and techniques is striking.

The intrinsic weakness of the kinetic SZE and the degeneracy of its spectrum with that of primary CMB fluctuations make it exceedingly difficult to use it to measure the peculiar velocity of a single cluster. It may be possible, however, to determine mean peculiar velocities on extremely large scales by averaging over many clusters.

Energy Densities in the Universe and Growth of Structure

The evolution of the abundance of galaxy clusters is a sensitive probe of cosmology (see "Sky Surveys with the Sunyaev-Zel'dovich Effect," above). Measurements of the cluster masses and number density as a function of redshift can be used to constrain the matter density, Ω_M, and, for sufficiently large samples, the equation of state of the dark energy. X-ray surveys have already been used to constrain Ω_M (e.g., Borgani et al. 2001, Viana & Liddle 1999, Bahcall & Fan 1998, Oukbir et al. 1997), but they have been limited by sample size and their reduced sensitivity to high-redshift clusters. SZE surveys offer the attractive feature of probing the cluster abundance at high redshift as easily as the local universe; as discussed in "Mass Limits of Observability," above, the sensitivity of an SZE survey is essentially a mass limit (Bartlett & Silk 1994, Barbosa et al. 1996, Holder et al. 2000b, Bartlett 2000, Kneissl et al. 2001).

The simple mass selection function of SZE surveys could allow the source count redshift distribution to be used as a powerful measure of cosmological parameters (Barbosa et al. 1996, Haiman et al. 2001, Holder et al. 2001, Weller et al. 2001, Benson et al. 2002) and the structure formation paradigm in general. As an example, we show in Figure 11 (see color insert) the expected constraints in Ω_M–Ω_Λ–σ_8 parameter space from the analysis of a deep SZE survey (Carlstrom et al. 2000; Holder et al. 2000b, 2001). The shaded regions show the result of a Monte Carlo method for estimating confidence regions. Many realizations of a fiducial model (Ω_M=0.3, Ω_Λ=0.7, σ_8=1) were generated and fit in the three-dimensional parameter space, and the shaded regions indicate regions which contain 68% and 95% of the resulting best fits. The contours show confidence regions from a Fisher matrix analysis, where the confidence regions are assumed to be Gaussian ellipsoids in the parameter space.

The next generation of dedicated telescopes equipped with large-format bolometer detector arrays offers the possibility of conducting SZE surveys over thousands of square degrees with $\lesssim 10 \, \mu K$ sensitivity. As shown in Figure 12, the yields from such a survey should, in principle (see next section), enable highly accurate estimation of cosmological parameters, notably of the matter density and the properties of the dark energy. Most importantly, the degeneracies in the constraints of the

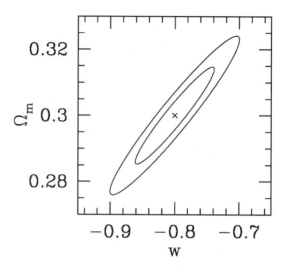

Figure 12   Expected constraints on the matter density Ω_M and the dark energy equation of state w from the analysis of an SZE survey covering several thousand square degrees in which all clusters above $2.5 \times 10^{14}\ h^{-1}\ M_\odot$ are detected and the redshifts are known. The normalization of the power spectrum has been marginalized over, and contours show 68% and 95% confidence regions for two parameters. Note that no systematic errors have been assumed in deriving the cosmological constraints. As discussed in the text, considerable observational and theoretical work needs to be done before such tight constraints could be extracted from large-scale SZE surveys.

cosmological parameters derived from SZE survey yields are very different from those expected from distance measures or CMB measurements. This simply arises because clusters are probing a fundamentally different physical effect, the growth rate of structure, rather than distance. Both growth and distance are related to the expansion history of the universe, but the two measures are effectively sensitive to different moments of the expansion rate.

A generic prediction of inflation is that the primordial fluctuations should be Gaussian. With cluster surveys probing the highest peaks of the density field, non-Gaussianity in the form of an excess of high peaks should be readily apparent, especially at high redshift (Benson et al. 2002). Cluster surveys are therefore probing both the structure formation history of the universe and the nature of the primordial fluctuations. In this way, cluster surveys are emerging as the next serious test of the cold dark matter paradigm.

Challenges for Interpreting Sunyaev-Zel'dovich Effect Surveys

In order to realize the full potential of the evolution of the cluster number density as a cosmological probe, a strong understanding of the physics of galaxy clusters will be required. As fully collapsed objects, the complete physics of galaxy clusters is highly nonlinear and complex. The size of such massive objects, however, makes them insensitive to disruption from most physical mechanisms. Nevertheless, there are several important aspects of gas dynamics that could affect interpretation of SZE galaxy cluster surveys.

The SZE is only sensitive to free electrons; any process that removes electrons from the optically thin ICM can affect the magnitude of the SZE for a given total mass. For example, cooling of the ICM, star formation, or heating of the ICM from supernovae can affect the observed SZE (Holder & Carlstrom 1999, Springel et al. 2001, Holder & Carlstrom 2001, Benson et al. 2002). If the cooling or star formation is not dependent on cluster mass or redshift, this can be simply calibrated and accounted for in deriving the survey selection function. The most promising theoretical path for understanding such processes is through high resolution cosmological simulations that include the relevant gas dynamics; such simulations are only now becoming feasible. Figure 13 shows the possible effects of such gas dynamics (Holder & Carlstrom 2001). A simple model has been adopted for this figure, where some combination of heating or cooling has reduced the number of hot electrons in the central regions of galaxy clusters by an amount modeled through the effects of an "entropy floor" (Ponman et al. 1999). The curves in this figure show the extreme cases of either no heating or cooling (no entropy floor) or extreme gas evolution, with an assumed value for the minimum entropy that is roughly a factor of two larger than is required for consistency with the observations (Ponman et al. 1999). Changing the evolution of the ICM could mimic changes in cosmological parameters at levels much larger than the expected statistical errors. With detailed imaging of the ICM using SZE and X-ray telescopes, the effects of heating or cooling should be apparent, so the possible systematic errors owing to heating or cooling should in practice be much smaller than 10% in Ω_M.

Although the SZE-mass relation is easy to understand theoretically in general terms, the details of the normalization and redshift evolution will require additional studies of an at least moderately sized SZE cluster catalog. For example, simple scaling arguments can be used to estimate the expected relation between cluster mass and temperature, but the exact cosmological dependence could be sensitive to merger rates as a function of cosmology or other nonlinear effects.

Furthermore, the mapping between the initial density field and the number density of clusters of a given mass as a function of redshift, i.e., the mass function, is not known perfectly at this time. The current generation of large cosmological simulations offers hope for a much better understanding in coming years. An important use for a large scale SZE cluster survey will be to characterize this mass function and test the reliability of various analytical (Press & Schechter 1974, Bond et al. 1991, Sheth et al. 2001) and numerical (Jenkins et al. 2001) estimates.

In order to exploit fully the potential of SZE surveys, the limiting mass as a function of redshift for galaxy clusters will have to be understood to an accuracy of better than 5%, and uncertainties in the mass function must be reduced to better than 10% (Holder et al. 2001). The former will require a concerted observational

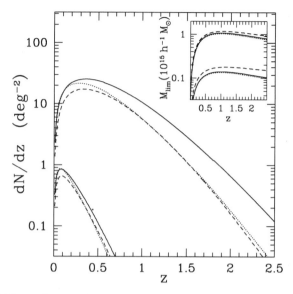

Figure 13   Effects of gas evolution on cluster survey yields. In the inset the top group of lines correspond to mass limits for an SZE survey similar to the *Planck Surveyor* satellite survey, with the uppermost line indicating the expected mass limit for a model with significant gas heating while the lower line in the top group shows the expected mass limit of detection for the case of no cluster gas heating. The cosmology chosen is $\Omega_M = 0.3$, $\Omega_\Lambda = 0.7$, $\sigma_8 = 1$. The lower set of lines show the same effects for a deep SZE survey. The main panel shows the expected redshift distributions for the mass limits in the inset. In the main panel the top group of lines correspond to the deep surveys and the lower lines correspond to *Planck Surveyor*. The solid line within each set shows the expected counts for the case of no heating, and the dashed line shows the effect of gas heating. The dotted curve, i.e., the middle curve in each set, shows a model with no heating but with $\Omega_M = 0.33$ and σ_8 modified to keep the same number of clusters at $z = 0$ (Holder & Carlstrom 2001). The assumptions of no heating and the very high value of heating for this plot are extreme and should bracket the true gas evolution.

effort and is likely to be the most difficult to achieve; the latter requirement is not far from current uncertainties (Jenkins et al. 2001). Note that it is not required that we know the mass of each cluster in the catalog to this accuracy, but only that we can characterize the cluster detection efficiency as a function of mass to this level.

Although some of the theory of the physics of galaxy clusters is not known, there are plenty of observational diagnostics that can be used in the interpretation of SZE surveys. High-resolution SZE imaging of high-redshift clusters will provide information on the relative importance of gas dynamics (Holder & Carlstrom 2001) to the observed properties of galaxy clusters. The main effect of most gas processes should be to cause the gas to be less centrally concentrated, either because the low-entropy gas has been removed (Bryan 2000) or because the gas has gained entropy from nongravitational heating.

It may also be possible with the next generation of large-format multifrequency bolometer arrays to exploit the small relativistic corrections to the SZE spectrum to determine the gas temperature. Such measurements will allow an understanding of temperature structure in the gas, even without follow-up X-ray observations. A direct SZE-weighted temperature would be more directly relevant to the SZE observations in determining the gas mass, and comparisons with X-ray spectral temperatures, when possible, could provide valuable information on the temperature structure along the line of sight. Temperature information, combined with high-resolution imaging, allows a reconstruction of the cluster potential and therefore can provide important constraints on the gas-mass fraction as a function of radius as well as a diagnostic for the effects of gas cooling.

In terms of the properties of the cluster catalog produced by an SZE survey, it is difficult to avoid the conclusion that the survey limit will be mainly dependent only on cluster mass and that this mass limit will be relatively flat with redshift beyond $z \sim 0.1$. This will make the resulting catalog especially useful for studies of the physics of galaxy clusters. The requirements for understanding the cluster mass selection function for such studies are much less strict than those required for cosmological studies. The results from detailed studies of individual clusters in the catalog will naturally feed back and improve our understanding of the survey-selection function.

SUMMARY

The SZE is emerging as a powerful tool for cosmology. Over the past several years, detection of the SZE toward massive galaxy clusters has become routine, as has high-quality imaging at moderate angular resolution of order an arcminute. Measurements of the effect already have been used to place interesting constraints on the Hubble constant and, through measurements of cluster gas-mass fractions, the matter density of the universe, Ω_M.

The next step is to exploit the redshift independence of the SZE signal to conduct blind surveys for galaxy clusters. The limit for such a survey is essentially a mass limit that is remarkably uniform with redshift. The cluster catalog from such an unbiased survey could be used to greatly increase the precision and redshift range of present SZE constraints on the Hubble constant and Ω_M and could, for example, allow $D_A(z)$ to be determined to high redshift ($z \sim 2$).

The most powerful use of the SZE for cosmology will be the measurement of the evolution of the abundance of galaxy clusters. SZE surveys are ideally suited for this because they are able to probe the abundance at high redshift as easily as the local universe. The evolution of the abundance of galaxy clusters is a sensitive probe of cosmology. For example, the yields from a deep SZE survey covering only 10 square degrees would be able to place interesting constraints on Ω_M, Ω_Λ, and σ_8.

A generic prediction of inflation is that the primordial density fluctuations should be Gaussian. Non-Gaussianity in the form of an excess of high-mass clusters should be readily apparent, especially at high redshift, from SZE survey yields. SZE cluster surveys will therefore probe both the structure formation history of the universe and the nature of the primordial fluctuations. In this way, cluster surveys are emerging as the next serious test of the cold dark matter paradigm.

Current SZE observations, while routine, require substantial integration time to secure a detection; a prohibitively long time would be required to conduct blind surveys over a large region of sky with the instruments now available. However, the next generation of instruments now being built or planned will be substantially faster. Dedicated interferometric arrays being built will be able to conduct deep SZE surveys over tens of square degrees. Heterogeneous arrays, such as the SZA combined with the OVRO array, will also allow detailed high-resolution follow-up SZE observations of the resulting cluster catalog.

A dedicated, low–noise, single-dish telescope with $\sim 1'$ resolution, equipped with a next generation, large format bolometric array receiver (~ 1000 elements) and operating from a superb site would be able to conduct a deep SZE survey over thousands of square degrees. The statistics provided by the yields from such a large survey ($\sim 10^4$ clusters) in the absence of systematic effects and assuming redshifts are known would be sufficient to determine precise constraints on Ω_M, Ω_Λ, σ_8 and even set meaningful constraints on the equation of state of the dark energy.

The possible systematics that could affect the yields of SZE surveys are presently too large to realize the full potential of a deep SZE survey covering thousands of square degrees. The systematics include, for example, the uncertainties on the survey mass detection limit owing to unknown cluster structure and cluster gas evolution, as well as the uncertainties in the theoretical mapping between the initial density field and the number density of clusters of a given mass as a function of redshift, i.e., the mass function.

These systematics can begin to be addressed through detailed follow-up observations of a moderate area SZE survey (tens of square degrees). High-resolution SZE, X-ray, and weak lensing observations will provide insights into evolution and structure of the cluster gas. Numerical simulations directly compared and normalized to the SZE yields should provide the necessary improvement in our understanding of the mass function.

It is not unreasonable to consider the possibility of a space-based telescope operating at centimeter through submillimeter wavelengths with high angular resolution (<1 arcminute) and good spectral coverage. For studies of the SZE this would allow simultaneous determinations of electron column densities, temperatures, and peculiar velocities of galaxy clusters. Such a satellite would make detailed images of the cosmic microwave background, while also providing important information on the high-frequency behavior of radio point sources and the low-frequency behavior of dusty extragalactic submillimeter sources. The upcoming *Planck Surveyor* satellite is a first step in this direction; it should provide an

SZE all-sky survey, although at moderate ($\sim$5 arcminute) resolution. Such a survey should find on the order of 10^4–10^5 clusters, most of them at redshift $z < 1$.

We can look forward to the SZE emerging further as a unique and powerful tool in cosmology over the next several years as the next generation of SZE instruments come online and SZE surveys become a reality.

ACKNOWLEDGMENTS

We thank M. Joy and W. Holzapfel for their considerable input to this review and W. Hu, S. LaRoque, A. Miller, J. Mohr, and D. Nagai for their comments on the manuscript. We also thank M. White and C. Pryke for assistance with Figure 3. This work was supported in part by NASA LTSA account NAG5-7986 and NSF account AST-0096913. JEC also acknowledges support from the David and Lucile Packard Foundation and the McDonnell Foundation. EDR acknowledges support from a NASA GSRP fellowship (NGT5-50173) and a Chandra Fellowship (PF1-20020).

The *Annual Review of Astronomy and Astrophysics* is online at
http://astro.annualreviews.org

LITERATURE CITED

Aghanim N, De Luca A, Bouchet FR, Gispert R, Puget JL. 1997. *Astron. Astrophys.* 325:9–18

Allen SW, Fabian AC. 1998. *MNRAS* 297:L57–62

Andreani P, Böhringer H, Dall'Oglio G, Martinis L, Shaver P, et al. 1999. *Ap. J.* 513:23–33

Andreani P, Pizzo L, dall'Oglio G, Whyborn N, Boehringer H, et al. 1996. *Ap. J.* 459:L49–52

Bahcall N, Fan X. 1998. *Ap. J.* 504:1–6

Barbosa D, Bartlett J, Blanchard A, Oukbir J. 1996. *Astron. Astrophys.* 314:13–17

Bartlett JG. 2000. *Astron. Astrophys.* Submitted. astro-ph/0001267

Bartlett JG, Silk J. 1994. *Ap. J.* 423:12–18

Benson AJ, Reichardt C, Kamionkowski M. 2001. *MNRAS.* 331:71–84

Birkinshaw M. 1979. *MNRAS* 187:847–62

Birkinshaw M. 1991. In *Physical Cosmology*, ed. A Blanchard, L Celnikier, M Lachi'eze-Rey, J Tran Thanh Van, p. 177. Gif-sur-Yvette, France: Editions Frontieres

Birkinshaw M. 1999. *Phys. Rep.* 310:97–195

Birkinshaw M, Gull SF, Northover KJE. 1978a. *Nature* 275:40–41

Birkinshaw M, Gull SF, Northover KJE. 1978b. *MNRAS* 185:245–62

Birkinshaw M, Hughes JP. 1994. *Ap. J.* 420:33–43

Birkinshaw M, Hughes JP, Arnaud KA. 1991. *Ap. J.* 379:466–81

Blain AW. 1998. *MNRAS* 297:502–10

Bond J, Kaiser N, Cole S, Efstathiou G. 1991. *Ap. J.* 379:440–60

Bond JR, Myers ST. 1991. In *Primordial Nucleosynthesis and Evolution of Early Universe*, ASSL Vol. 169, ed. K Sato, p. 305–23

Bond JR, Myers ST. 1996. *Ap. J. Suppl.* 103:63–79

Borgani S, Rosati P, Tozzi P, Stanford SA, Eisenhardt PR, et al. 2001. *Ap. J.* 561:13–21

Bryan GL. 2000. *Ap. J.* 544:L1–5

Burles S, Nollett K, Truran J, Turner M. 1999. *Phys. Rev. Lett.* 82:4176–79

Burles S, Tytler D. 1998a. *Ap. J.* 499:699–712

Burles S, Tytler D. 1998b. *Ap. J.* 507:732–44

Carlstrom JE, Joy M, Grego L, Holder G, Holzapfel WL, et al. 2000. In *Constructing the Universe with Clusters of Galaxies*, ed. F Durret, G Gerbal, E43:1–28. IAP

Carlstrom JE, Joy M, Grego LE. 1996. *Ap. J.* 456:L75–78

Carlstrom JE, Joy MK, Grego L, Holder GP, Holzapfel WL, et al. 2000. *Physica Scripta Volume T* 85:148–55

Carroll S, Press W, Turner E. 1992. *Annu. Rev. Astron. Astrophys.* 30:499–542

Cavaliere A, Danese L, de Zotti G. 1977. *Ap. J.* 217:6–15

Cavaliere A, Fusco-Femiano R. 1978. *Astron. Astrophys.* 70:677–84

Cen R. 2001. *Ap. J.* 546:L77–80

Challinor A, Lasenby A. 1998. *Ap. J.* 499:1–6

Challinor A, Lasenby A. 1999. *Ap. J.* 510:930–33

Church SE, Ganga KM, Ade PAR, Holzapfel WH, Mauskopf PD, et al. 1997. *Ap. J.* 484: 523–37

Colafrancesco S, Mazzotta P, Rephaeli Y, Vittorio N. 1997. *Ap. J.* 479:1–16

Cooray AR, Grego L, Holzapfel WL, Joy M, Carlstrom JE. 1998. *Aston. J.* 115:1388–99

David L, Jones C, Forman W. 1995. *Ap. J.* 445: 578–90

Dawson KS, Holzapfel WL, Carlstrom JE, Joy M, LaRoque SJ, et al. 2001. *Ap. J.* 533:L1–4

de Bernardis P, Ade PAR, Bock JJ, Bond JR, Borrill J, et al. 2000. *Nature* 404:955–59

De Luca A, Desert FX, Puget JL. 1995. *Astron. Astrophys.* 300:335–45

Desert F, Benoit A, Gaertner S, Bernard J, Coron N, et al. 1998. *New Astron.* 3:655–69

Dolgov AD, Hansen SH, Pastor S, Semikoz DV. 2001. *Ap. J.* 554:74–84

Evrard A. 1997. *MNRAS* 292:289–97

Evrard AE, Metzler CA, Navarro JF. 1996. *Ap. J.* 469:494–507

Fabbri R. 1981. *Astrophys. Space Sci.* 77:529–37

Fischer ML, Lange AE. 1993. *Ap. J.* 419:433–39

Forman W, Jones C. 1982. *Annu. Rev. Astron. Astrophys.* 20:547–85

Glenn J, Bock JJ, Chattopadhyay G, Edgington SF, Lange AE, et al. 1998. *Proc. SPIE* 3357:326–34

Grainge K, Grainger WF, Jones ME, Kneissl R, Pooley G, et al. 2001. *MNRAS* 329:890–96

Grainge K, Jones M, Pooley G, Saunders R, Baker J, et al. 1996. *MNRAS* 278:L17–22

Grainge K, Jones M, Pooley G, Saunders R, Edge A. 1993. *MNRAS* 265:L57–58

Grainge K, Jones ME, Pooley G, Saunders R, Edge A, et al. 2000. *MNRAS.* Submitted. astro-ph/9904165

Grainger WF, Das R, Grainge K, Jones ME, Kneissl R, et al. 2001. *MNRAS.* Submitted. astro-ph/0102489

Grego L, Carlstrom JE, Joy MK, Reese ED, Holder GP, et al. 2000. *Ap. J.* 539:39–51

Grego L, Carlstrom JE, Reese ED, Holder GP, Holzapfel WL, et al. 2001. *Ap. J.* 552:2–14

Haiman Z, Mohr JJ, Holder GP. 2001. *Ap. J.* 553:545–61

Hanany S, Ade P, Balbi A, Bock J, Borrill J, et al. 2000. *Ap. J.* 545:L5–9

Heath DJ. 1977. *MNRAS* 179:351–58

Herbig T, Lawrence CR, Readhead ACS, Gulkis S. 1995. *Ap. J.* 449:L5–8

Holder G, Carlstrom J. 1999. In *Microwave Foregrounds*, ed. A de Oliveira-Costa, M Tegmark, pp. 199–216. San Francisco: Astron. Soc. Pac., astro-ph/9904220

Holder G, Haiman Z, Mohr JJ. 2001. *Ap. J.* 560: L111–14

Holder GP, Carlstrom JE. 2001. *Ap. J.* 558:515–19

Holder GP, Carlstrom JE, Evrard AE. 2000a. In *Constructing the Universe with Clusters of Galaxies*, ed. F Durret, G Gerbal, E45:1–7. IAP

Holder GP, Mohr JJ, Carlstrom JE, Evrard EA, Leitch EM. 2000b. *Ap. J.* 544:629–35

Holzapfel WL, Ade PAR, Church SE, Mauskopf PD, Rephaeli Y, et al. 1997a. *Ap. J.* 481: 35–48

Holzapfel WL, Wilbanks TM, Ade P, Church SE, Fischer M, et al. 1997b. *Ap. J.* 479:17–30

Hu W, White M. 1997. *Ap. J.* 479:568–79

Hughes JP, Birkinshaw M. 1998. *Ap. J.* 501:1–14

Itoh N, Kohyama Y, Nozawa S. 1998. *Ap. J.* 502:7–15

Jenkins A, Frenk CS, White SDM, Colberg JM, Cole S, et al. 2001. *MNRAS* 321:372–84

Jones M, Saunders R, Alexander P, Birkinshaw M, Dilon N, et al. 1993. *Nature* 365:320–23

Jones ME, Edge AC, Grainge K, Grainger WF, Kneissl R, et al. 2001. *MNRAS.* Submitted. astro-ph/0103046

Joy M, Laroque S, Grego L, Carlstrom JE, Dawson K, et al. 2001. *Ap. J.* 551:L1–4

Kamionkowski M, Loeb A. 1997. *Phys. Rev. D* 56:4511–13

Kitayama T, Sasaki S, Suto Y. 1998. *Publ. Astron. Soc. Jpn.* 50:1–11

Kneissl R, Jones ME, Saunders R, Eke VR, Lasenby AN, et al. 2001. *MNRAS* 328:783–94

Komatsu E, Kitayama T, Suto Y, Hattori M, Kawabe R, et al. 1999. *Ap. J.* 516:L1–4

Komatsu E, Matsuo H, Kitayama T, Hattori M, Kawabe R, et al. 2001. *Publ. Astron. Soc. Jpn.* 53:57–62

Korolev VA, Syunyaev RA, Yakubtsev LA. 1986. *Sov. Astron. Lett.* 12:141–47

Lamarre JM, Giard M, Pointecouteau E, Bernard JP, Serra G, et al. 1998. *Ap. J.* 507:L5–8

LaRoque SJ, Carlstrom JE, Reese ED, Holder GP, Holzapfel WL, et al. 2002. *Ap. J.* Submitted, astroph/02041341

Loeb A, Mao S. 1994. *Ap. J.* 435:L109–12

Loeb A, Refregier A. 1997. *Ap. J.* 476:L59–62

Markevitch M, Blumenthal GR, Forman W, Jones C, Sunyaev RA. 1994. *Ap. J.* 426:1–13

Markevitch M, Ponman TJ, Nulsen PEJ, Bautz MW, Burke DJ, et al. 2000. *Ap. J.* 541:542–49

Mason BS. 1999. *An improved measurment of the Hubble constant using the Sunyaev-Zel'dovich effect.* Ph.D. thesis. Philadelphia: Univ. Pennsylvania

Mason BS, Myers ST, Readhead ACS. 2001. *Ap. J.* 555:L11–15

Mauskopf PD, Ade PAR, Allen SW, Church SE, Edge AC, et al. 2000a. *Ap. J.* 538:505–16

Mauskopf PD, Rownd BK, Edgington SF, Hristov VV, Mainzer AK, et al. 2000b. In *Imaging at Radio through Submillimeter Wavelengths, Astron. Soc. Pac., Conf. Ser.,* ed. J Mangum, p. E46. San Francisco: Publ. Astron. Soc. Pac.

Miller AD, Caldwell R, Devlin MJ, Dorwart WB, Herbig T, et al. 1999. *Ap. J.* 524:L1–4

Miralda-Escude J, Babul A. 1995. *Ap. J.* 449: 18–27

Mohr J, Mathiesen B, Evrard A. 1999. *Ap. J.* 517:627–49

Molnar SM, Birkinshaw M. 1999. *Ap. J.* 523: 78–86

Mushotzky RF, Scharf CA. 1997. *Ap. J.* 482: L13–16

Myers ST, Baker JE, Readhead ACS, Leitch EM, Herbig T. 1997. *Ap. J.* 485:1–21

Netterfield CB, Ade PAR, Bock JJ, Bond JR, Borrill J, et al. 2002. *Ap. J.* Submitted. astro-ph/0104460

Neumann DM, Bohringer H. 1997. *MNRAS* 289:123–35

Nevalainen J, Markevitch M, Forman W. 2000. *Ap. J.* 536:73–78

Nollett KM, Burles S. 2000. *Phys. Rev. D* 61: 123505

Nozawa S, Itoh N, Kohyama Y. 1998a. *Ap. J.* 508:17–24

Nozawa S, Itoh N, Kohyama Y. 1998b. *Ap. J.* 507:530–57

Oukbir J, Bartlett JG, Blanchard A. 1997. *Astron. Astrophys.* 320:365–77

Padin S, Cartwright JK, Mason BS, Pearson TJ, Readhead ACS, et al. 2000. *Ap. J.* 549:L1–5

Patel SK, Joy M, Carlstrom JE, Holder GP, Reese ED, et al. 2000. *Ap. J.* 541:37–48

Peebles P. 1980. *The Large Scale Structure of the Universe.* Princeton, NJ: Princeton Univ. Press

Peebles P. 1994. *Physical Cosmology.* Princeton, NJ: Princeton Univ. Press

Perlmutter S, Aldering G, Goldhaber G, Knop R, Nugent P, et al. 1999. *Ap. J.* 517:565–86

Phillips PR. 1995. *Ap. J.* 455:419–20

Pointecouteau E, Giard M, Benoit A, Désert FX, Aghanim N, et al. 1999. *Ap. J.* 519: L115–18

Pointecouteau E, Giard M, Benoit A, Désert FX, Bernard JP, Coron N, Lamarre JM. 2001. *Ap. J.* 552:42–48

Ponman TJ, Cannon DB, Navarro, JF. 1999. *Nature* 397:135–37

Pospieszalski MW, Lakatosh WJ, Nguyen LD, Lui M, Liu T, et al. 1995. *IEEE MTT-S Int. Microw. Symp.* 1121–24

Press W, Schechter P. 1974. *Ap. J.* 187:425–38

Pryke C, Halverson NW, Leitch EM, Kovac J, Carlstrom JE, et al. 2002. *Ap. J.* 568:46–51

Reese ED, Carlstrom JE, Joy M, Mohr JJ, Grego L, et al. 2002. *Ap. J.* Submitted. astro-ph/0205350

Reese ED, Mohr JJ, Carlstrom JE, Joy M, Grego L, et al. 2000. *Ap. J.* 533:38–49

Rephaeli Y. 1995. *Annu. Rev. Astron. Astrophys.* 33:541–80

Rephaeli Y, Yankovitch D. 1997. *Ap. J.* 481: L55–58

Riess AG, Filippenko AV, Challis P, Clocchiattia A, Diercks A, et al. 1998. *Astron. J.* 116: 1009–38

Saunders R, Kneissl R, Grainge K, Jones ME, Maggi A, et al. 2000. *MNRAS.* Submitted. astro–ph/9904168

Sazonov SY, Sunyaev RA. 1998a. *Ap. J.* 508:1–5

Sazonov SY, Sunyaev RA. 1998b. *Astron. Lett.* 24:553–67

Sazonov SY, Sunyaev RA. 1999. *MNRAS* 310: 765–72

Sheth RK, Mo HJ, Tormen G. 2001. *MNRAS* 323:1–12

Silk J, White SDM. 1978. *Ap. J.* 226:L103–6

Smail I, Ivison RJ, Blain AW. 1997. *Ap. J.* 490: L5–8

Springel V, White M, Hernquist L. 2001. *Ap. J.* 549:681–87

Squires G, Kaiser N. 1996. *Ap. J.* 473:65–80

Squires G, Neumann DM, Kaiser N, Arnaud M, Babul A, et al. 1997. *Ap. J.* 482:648–58

Stebbins A. 1997. Preprint. astro-ph/9709065

Stompor R, Abrob M, Ade P, Balbi A, Barbosa D, et al. 2001. *Ap. J.* 561:L7–10

Subrahmanyan R, Kesteven MJ, Ekers RD, Sinclair M, Silk J. 2000. *MNRAS* 315:808–22

Sulkanen ME. 1999. *Ap. J.* 522:59–65

Sunyaev RA, Zel'dovich YB. 1970. *Comments Astrophys. Space Phys.* 2:66–74

Sunyaev RA, Zel'dovich YB. 1972. *Comments Astrophys. Space Phys.* 4:173–78

Sunyaev RA, Zel'dovich YB. 1980a. *Annu. Rev. Astron. Astrophys.* 18:537–60

Sunyaev RA, Zel'dovich YB. 1980b. *MNRAS* 190:413–20

Tamura T, Kaastra JS, Peterson JR, Paerels FBS, Mittaz JPD, et al. 2001. *Astron. Astrophys.* 365:L87–92

Thompson AR, Moran JM, Swenson GW. 2001. *Interferometry and Synthesis in Radio Astronomy.* New York: Wiley-Intersci.

Tsuboi M, Miyazaki A, Kasuga T, Matsuo H, Kuno N. 1998. *Proc. Astron. Soc. Jpn.* 50: 169–73

Tucker W, Blanco P, Rappoport S, David L, Fabricant D, et al. 1998. *Ap. J.* 496:L5–8

Udomprasert PS, Mason BS, Readhead ACS. 2000. In *Constructing the Universe with Clusters of Galaxies,* ed. F Durret, G Gerbal, E48:1–5. IAP

Viana P, Liddle A. 1999. *MNRAS* 303:535–45

Weller J, Battye R, Kneissl R. 2001. *Phys. Rev. Lett.* Submitted. astro-ph/0110353

White DA, Fabian AC. 1995. *MNRAS* 273:72–84

White S, Navarro J, Evrard A, Frenk C. 1993. *Nature* 366:429–33

Wright EL. 1979. *Ap. J.* 232:348–51

Wu X, Fang L. 1997. *Ap. J.* 483:62–67

SUBJECT INDEX

A

Abell catalog, 546
Abell 119, 328
Abell 2163, 654–55
ABMD
  See Anti-ballistic missile defense
Absorption
  by a statistically large number of clumps or filaments, 398–99
  by a statistically small number of clumps or filaments, 399
  by a uniform external medium, 398
Absorption line systems clustering of, 556
Acceleration scale, 265–68
  dark halos with, 272–74
  global Newtonian mass-to-K'-band-luminosity ratio, 266
  near-infrared Tully-Fisher relation, 268
Accretion disk formation, 475–77
  in binary PNe progenitor, 476
Accretion disk of the Sun Kuiper Belt objects as relics from, 63–101
Accretion disks, 361, 476
  self-protective, 34
  stability of, 374–77
Accretion time, 151, 356, 368
Acoustic oscillations idealized, 180
Acoustic peaks, 176–93
  angular diameter distance, 181

baryon loading, 184–85
basics, 178–81
damping, 187–88
gravitational forcing, 183–84
idealized acoustic oscillations, 180
initial conditions, 181–83
integral approach, 190–92
parameter sensitivity, 192–93
polarization, 188–90
radiation driving, 185–86
Active Galactic Nuclei (AGN), 323, 342, 539, 546, 585, 611–12
Activity cycles
  in stellar rotation, binarity, and magnetic cycles, 240
Adiabatic expansion, 323
Adiabatic wind-blown bubbles, 460
Advanced Research Projects Agency (ARPA), 12–13
Afterglows, 137–39, 154
  of GRBs, 415–16
AGBs
  evolution, 439, 449
  luminous, 441
Age-metallicity relation (AMR), 508
AGN
  See Active Galactic Nuclei
Albedos, 77–78
  colors, 78–81
  of KBOs, 77–78
  of KBOs beyond 50 AU, 76
  plot of albedos vs. diameter, 78
  spectral properties, 81–83

Alfvén speeds, 29, 41
Algol triple system, 252
Algol-type stars, 234
  close binaries, 219, 229, 231
ALMA
  See Atacama Large Millimeter Array
Alternative models of GRBs, 150
Alternative to dark matter basics of modified Newtonian dynamics, 265–74
  cosmology and the formation of structure, 306–11
  modified Newtonian dynamics as, 263–317
  pressure-supported systems, 285–97
  rotation curves of spiral galaxies, 274–85
Ambipolar diffusion, 349
American Physical Society, 14
AMiBA
  See Array for Microwave Background Anisotropy
AMR
  See Age-metallicity relation
Anase
  and FLIERS, 472–73
Angular diameter distance, 181
Anisotropy
  jet, 146
  pitch-angle distribution, 335
Anti-ballistic missile defense

681

(ABMD), 14
Apparent red magnitude
of Kuiper Belt objects, 72
Aquadric Lagrangian
(AQUAL) theory, 301–2
AQUAL
See Aquadric Lagrangian
theory
Arecibo Ionospheric
Observatory, 12
305m telescope, 54
Arecibo Radio Telescope, 18
Ariel V catalog, 549
ARPA
See Advanced Research
Projects Agency
Array for Microwave
Background Anisotropy
(AMiBA), 658
ASCA satellite, 539, 545,
567
Aspherical nebulae
the GISW model, 461
Asteroseismology, 494
Astronomical contamination
and confusion
sources of, 651–53
Astrophysics
nuclear, 7–8
Atacama Large Millimeter
Array (ALMA), 58
Atmospheres
overturning convectively,
122
Atmospheres and their
relation to the interior of
giant planets, 117–25
condensates, 119–22
EGP atmospheres, 124–25
expected compositions,
118
Galileo probe results,
122–24
T_{eff}-T_{10} relation, 118–19
Automatic Plate Measuring
machine galaxy catalog,
546

Axisymmetric state
direct breakup of binary
stars from, 372–73

B
B-I vs. V-J for KBOs, 80
B-V vs. B-I for KBOs, 79
B-Z (Blandford & Znajek)
mechanism, 151, 153
Baade, Walter, 10–11, 22
Balmer line emission, 511,
586, 607
BALs
See Broad Absorption Line
outflows
Bandpowers
estimation, 208–10
extracting, 208
Baryon crisis
magnetic support of cluster
gas and, 338–39
Baryon dissipation, 524, 528
Baryon loading, 184–85
BATSE
See Burst and Transient
Experiment
BBMN, 494
BBN
See Big band
nucleosynthesis
predictions
Bekenstein-Milgrom theory,
297–300, 312
scaled growth rate of the
$m = 2$ instability in
Newtonian disks, 299
Beppo-SAX satellite, 138,
155, 332, 415–16, 541,
545, 567–68
Narrow Field Instruments,
409
Berkeley-Illinois-Maryland-
Association (BIMA)
inferometer, 655–57,
666
Bessel function, 197
Betelgeuse, 241–42

Bethe, Hans, 5–8, 16
Bethe-Salpeter equation, 7
BHs
See Black holes
Biermann battery effect,
341–42
Big band nucleosynthesis
(BBN) predictions, 667
Big Bang, 8, 495
Bilobed PNe, 443
BIMA
See Berkeley-Illinois-
Maryland-Association
inferometer
Binarity, 239–40
activity cycles, 240
radio emission and stellar
rotation, 239–40
Binary and rapid rotation in
the nuclei of PNe,
455–56
Binary KBOs, 86–87
Binary stars
effects of, 474
possibility of resulting,
366–67
Binary systems, 403
Bipolar AGNS, 440
Bipolar HII regions, 42
Bipolar molecular outflows,
38–41
CO outflow characteristics,
40
Bipolar nebulae, 443, 448
Bipolar rotating emission line
jets (BRETS), 445, 472
Birkhoff theorem, 300
Bispectrum, 204
Black hole accretion, 17–18
Black holes (BHs), 150–51
Blastwave-CSM interaction
region, 399, 405
Blastwave expansion
from SN1993J, 408
Blastwave model of GRB
afterglows, 142–45
observed lightcurves of the

afterglow of GRB
970228, 144
Blue supergiant (BSG) phase,
405
BOK Globule CB230, 356
BOLOCAM instrument, 654
Bolometric detectors, 654,
675
Bolometric luminosity, 604
Boltzmann constant, 220
Boltzmann equation, 187
Bondi-Hoyle-Littleton theory
of accretion for motion
through a uniform gas, 17
Borrelly comet, 88
Bow shock hypothesis, 43, 45
Brehmstrahlung, 220–21
radiation from, 336
BRETS
See Bipolar rotating
emission line jets
Broad Absorption Line
outflows (BALs), 342
Brown dwarf-like evolution,
126
Brown dwarfs, 104–5, 115
silicate clouds in, 121
trajectories of, 120
Brute force likelihood
analyses, 205
BSG
See Blue supergiant phase
Bulge
of the galaxy, 495–96
Burst and Transient
Experiment (BATSE),
137–38, 423
Burst rate scales, 154
Butterfly PNe, 443, 445, 447,
451–52

C

Calculated average dust
temperature, 33
Calibration diagram of
oxygen abundance, 616
Caltech Submillimeter

Observatory
BOLOCAM instrument,
654
Calzetti law, 591
Canada-France-Hawaii
Telescope (CFHT), 66
Canadian Network for
Observational
Cosmology (CNOC),
566
Candidates for chemical
tagging, 527
CANGAROO-III, 162
Capture
of binary stars, 361–62
Carbon stars, 455
CARE
See Compressible analog
of the Riemann Ellipsoids
CARMA
See Combined Array for
Research in
Millimeter-Wave
Astronomy
Cat's Eye Nebula, 439,
444–45
CBI
See Cosmic Background
Imager experiment
CCD
See Charge-coupled device
detection
CDM
See Cold dark matter
paradigm
CE
See Common envelope
evolution
Centaurs, 77, 80, 88
relationship of KBOs to,
87–89
Center for Radiophysics and
Space Research, 14
Cepheid-based distance scale,
284–85
CFHT
See Canada-France-Hawaii

Telescope
CGRO
See Compton Gamma-Ray
Observatory
Champagne-type flows, 45
Chandra X-Ray Observatory
(CHANDRA), 58, 157,
448, 539, 541, 543,
555–56, 562–63, 569,
571, 666
Charge-coupled device
(CCD) detection, 63,
547, 587–88, 593, 599
cameras, 66
Charon, 78, 87
Chemical abundance space,
524–26
CS 22892-052 n-capture
abundances, 525
Chemical abundances of
LBGs, 614–15
Chemical signatures, 521–22
mean relative abundance
ratios of process
elements, 522
Chemical trajectories, 526–27
Chemistry
of UC HII regions, 48–49
Cherenkov detector, 162
Cheshire Cat galaxies, 20
Chevalier model, 400
Chiron, 88
CIA
See Collision-induced
absorption
Circumstellar medium (CSM)
structure, 388, 390–94,
398–401, 405–7
stellar wind-formed, 434
Classic KBOs, 65–67
eccentricity of, 66
inclination of, 67
Classic T Tau (CTT) phase,
245–46, 248
Classical PNe, 443–48
extended S-rays in PNe,
448

halos, 445–46
knots and jets, 444–45
large-scale kinematics and proper motions, 446–47
nebular momenta, 447–48
organizing PNe by morphological type, 443–44
Classical theories of rotating bodies, 113
Classical view of fission, 370–71
CLF
See Cumulative luminosity function
Cloud formation, 121
Cluster Abell 85 central region at different wavelengths, 334
Cluster abundance, 659–61
comoving volume element and number density, 660
Cluster abundance at higher redshifts and its evolution, 558–62, 624–25
X-ray luminosity function of distant clusters, 561
Cluster center sources, 323–29
maximum absolute RM as a function of estimated cooling flow rate, 326
observed position angles of the linearly polarized radio emission, 327
Cluster cumulative number counts, 559
Cluster gas-mass fractions, Ω_M, 667–70
limits on Ω_M from SZE-measured cluster gas fractions, 669
Cluster magnetic fields, 319–48
cold fronts, 337–38
Faraday rotation, 323–31

field origin, 340–44
GZK limit, 339
inverse Compton X-ray emission, 331–37
magnetic support of cluster gas and the baryon crisis, 338–39
synchrotron radiation, 320–23
synthesis, 339–40
Cluster mergers, 343
Cluster modulation, 201
Cluster-peculiar velocities, 670–71
Clustering and large-scale structures at $z \sim 3$, 620–25
clusters at high redshifts, 624–25
Clustering of absorption line systems, 556
CMB
See Cosmic microwave background
CMBFAST code, 178, 191
release of, 211
CMR
See Color-magnitude relation
CNOC
See Canadian Network for Observational Cosmology
CO outflow characteristics, 40
COBE
See Cosmic Background Explorer
Coherent emission, 223
Coherent radio bursts, 225
Cold dark matter (CDM) paradigm, 263, 294, 489, 562, 623, 669
Cold fronts, 337–38
Collapsing cloud core
nearly homologous, 363–64
nonhomologous, 364–65
typical conditions in, 358

Collimation
of planetary nebulae, 467
Collision-induced absorption (CIA), 104
Color-magnitude relation (CMR), 509
Color selection
in Lyman-break galaxy surveys, 589–93
Colors
predicted, for various choices of star formation history, 606
Colors of Kuiper Belt objects, 78–81
B-I vs. V-J for KBOs, 80
B-V vs. B-I for KBOs, 79
Coma cluster, 320–22, 329, 335, 651, 667
Combined Array for Research in Millimeter-Wave Astronomy (CARMA), 58, 658
Comets
relationship of KBOs to, 87–89
Common envelope (CE) evolution, 474–75
Comoving volume element and number density, 660
Companion stars, 455
Comparative distributions of UC HII regions, 53
Compressible analog of the Riemann Ellipsoids (CARE), 373
Compton Catastrophe, 410
Compton drag epoch, 194–95
Compton Gamma-Ray Observatory (CGRO), 137, 431
Compton photosphere, 141
Compton scattering, 141
See also Inverse-Compton (IC) scattering
Compton Telescope

(Comptel), 137
Compton y-parameter, 647, 652, 655
Comptonization, 148
Condensates, 119–22
Continuity equation, 183
Continuum
  discontinuities, 582
  luminosity, 587
Cooling in the intracluster medium, 545–46
Cooling theory
  application of to Jupiter and Saturn, 126–28
  luminosity vs. age for EGPs, 127
  temperatures vs. age for EGPs, 128–30
Correlation between quiescent radio and X-ray emissions, 231–33
Cosmic Background Explorer (COBE), 174, 612
  maps, 204, 496
  normalization, 194–95
Cosmic Background Imager (CBI) experiment, 177
  reconfiguring, 658
Cosmic microwave background (CMB), 645–52
  polarization field, 175–76
  spectral distortion of, 648
  spectrum, 645–47
  spectrum fluctuations, 569, 571
  temperature field, 173–75
Cosmic microwave background (CMB) anisotropies, 171–216, 331
  acoustic peaks, 176–93
  beyond the peaks, 193–204
  data analysis, 205–11
  observables, 172–76
Cosmic rays
  of GRBs, 159–63

Cosmic star formation, 608–10
  star formation density, 609
Cosmochronology, 493–95
Cosmological effects of GRBs, 430
Cosmological mass function, 563–66
  sensitivity of the cluster mass function to cosmological models, 565
Cosmological paradigm standard, 172–73
Cosmological parameter estimation, 210–11
Cosmological setting, 154–59
  schematic GRB afterglow, 158
  schematic GRB from internal shocks, 157
Cosmology and the formation of structure, 306–11
  spherically symmetric over-densities in a low-density baryonic universe, 310
Cosmology from Sunyaev-Zel'dovich effect survey samples, 663–75
  challenges for interpreting, 672–75
  cluster gas-mass fractions, Ω_M, 667–70
  cluster peculiar velocities, 670–71
  distance determinations, Hubble constant, 664–66
  energy densities in the universe and growth of structure, 671–72
Cosmology with the Sunyaev-Zel'dovich effect, 643–80
  sky surveys with the Sunyaev-Zel'dovich effect, 658–63

status of observations, 650–58
  the Sunyaev-Zel'dovich effect, 645–50
Cosmology with X-ray clusters, 563–70
  the cosmological mass function, 563–66
  deriving Ω_m from cluster evolution, 566–70
Coulomb effects, 126, 224
Coulomb mean free path (mfp), 337
Coulomb scattering, 176
Courant-Freidrichs-Lewey condition, 362
Cramer-Rao inequality, 207
Critical mass infall rate, 36
CS 22892-052 n-capture abundances, 525
CSM
  See Circumstellar medium
CTT
  See Classic T Tau phase
Cumulative luminosity function (CLF), 71–72
Cumulative X-ray temperature function for nearby cluster sample, 567
Cygnus A observations, 324, 328–29

D
DAC
  See Diamond-anvil cell measurements
Damping, 177, 187–88
  diffusion, 186
Dark energy, 173, 193
Dark halos
  with an acceleration scale, 272–74
  of the galaxy, 498
DASI
  See Degree Angular Scale Interferometer

Data analysis, 205–11
  bandpower estimation,
    208–10
  cosmological parameter
    estimation, 210–11
  data pipeline and radical
    compression, 205
  mapmaking, 206–8
Data pipeline and radical
  compression, 205
Deceleration of blastwave
  expansion, 407
  from SN1987A, 408
  from SN1993J, 408
Degree Angular Scale
  Interferometer (DASI),
    211
Delayed breakup of binary
  stars, 367–77
  direct breakup from an
    axisymmetric state,
    372–73
  nonaxisymmetric
    instabilities in rapidly
    rotating, equilibrium
    cores, 368–71
  slow contraction of a
    rapidly rotating ellipsoid,
    373–74
  stability of accretion disks,
    374–77
Density distribution of Sa to
  Sm galaxies, 504
Density modulation
  general, 202
Deriving Ω_m from cluster
  evolution, 566–70
  cumulative X-ray
    temperature function for
    nearby cluster sample,
    567
De Sitter space, 304
Diabolo measurements, 655
Diamond-anvil cell (DAC)
  measurements, 107, 109
Dicke switching, 653
Differential time lags for the

arrival of GRB pulses,
    150
Diffusion damping, 186
Dinematics, line emission,
  and the total mass,
    625–27
Disk heating, 505
Disk of the galaxy, 496–97
Disk-plus-bulge
  generalizations, 298
Dissociation
  of hydrogen, 109
Distance determinations
  with the Hubble constant,
    664–66
Distant X-ray clusters,
    562–63
Distribution of open clusters,
    515
Distribution of UC HII
  regions, 51–54
  comparative distributions,
    53
  distribution in Galactic
    coordinates of all
    candidate regions, 52
DIVA mission, 520
Doppler boost, 150
Doppler effect, 190, 201–2,
    649
  large angle, 200
  suppressed, 201
  See also Modified Doppler
    effects
Doppler shifts
  supersonic, 445
  temperature, 175
Doppler spectroscopic
  detections, 132
Doppler-velocity mapping,
    446–47
Dropouts, 588n, 592–93
  J-band, 597
  spectroscopic redshifts of,
    594
Dust destruction
  by UV protons, 156

Dust ring
  around HR4796A, 94
Dwarf spheroidal systems,
    290–91
Dynamos
  and the challenge to the
    magneto-centrifugal
    models, of planetary
    nebulae, 469–70
Dyson, Freeman, 6

E

Eccentricity
  of classic KBOs, 66
Echelle spectrograph, 446
Eddington, Arthur, 12
Edgeworth, Kenneth, 64
Edgeworth-Kuiper Belt
  See Kuiper Belt
Edinburgh-Durham Southern
  Galaxy catalog, 546
Effective adiabatic exponent
  γ, 359–61
EGPs
  See Extrasolar giant planets
EGRET, 162
  See Energetic Gamma Ray
    Experiment
Einstein, Albert, 217
Einstein-de-Sitter model, 568
Einstein Observatory
  Extended Medium
    Sensitivity Survey, 550
Einstein satellite, 549
EIS
  See European Southern
    Observatory Imaging
    Survey
Electron cyclotron maser
  emission, 223, 233
Electron-degeneracy effects,
    126
Electron density-size relation,
    46
Electron lifetimes, 333–35
  cluster Abell 85 central
    region at different

wavelengths, 334
Electron pitch-angle
distribution
relativistic, 335
Elementary formulae for
radiation and particles,
219–20
Elliptical nebulae, 443
Embedded sources
background and, 329–30
EMSS
See Extended Medium
Sensitivity Survey
ENACS
See ESO Nearby Abell
Cluster Survey
Energetic Gamma Ray
Experiment (EGRET),
137, 148
Energy
dark, 173, 193
Energy densities in the
universe and growth of
structure, 671–72
expected constraints on the
matter density Ω_m, 672
Energy scale of inflation, 198
Enrichment of the interstellar
gas, 10–12
"Entropy floor," 673
Environment, 154–59
schematic GRB afterglow,
158
schematic GRB from
internal shocks, 157
Equations
of planetary nebulae,
457–59
Equatorial models, 238
Eridani, 93
ESA Gaia mission, 520
ESO
See European Southern
Observatory
ESO Nearby Abell Cluster
Survey (ENACS), 566
Estimates

of mass-loss rates for
RSNe, 400
of SZE source counts,
662–63
Euclidean analysis, 180
Euler equation, 178–79, 183,
187
European Southern
Observatory (ESO)
Imaging Survey (EIS), 547
Very Large Array, 218,
281, 329, 337, 388, 419,
421
EUV observations, 231
Evolution
to the main sequence of
star formation, 247–48
Evolution of Jupiter and
Saturn
final stages of, 131
Evolution of the Kuiper belt,
90–93
model cumulative
luminosity functions, 92
size and velocity
distributions, 91
Evolution of X-ray clusters of
galaxies, 539–77
cosmology with X-ray
clusters, 563–70
observational framework,
546–56
physical properties of
galaxy clusters, 541–46
research needed, 570–72
space density of X-ray
clusters, 556–63
Evolution to the ultracompact
state, 28–41
hot cores as precursors to
UC HII regions, 32–41
the natal material, 28–30
prestellar cores, 30–31
Evolutionary trajectory of
central region of
protostellar gas cloud,
354

EXOCET satellite, 549
Expansion velocity
of radio supernovae, 410
Expected compositions, 118
Expected constraints on the
matter density Ω_m, 672
Extended Medium Sensitivity
Survey (EMSS), 550,
554, 558, 560–61, 566,
569
Extended S-rays in PNe, 448
Extrasolar giant planets
(EGPs), 103–5, 115
atmospheres, 124–25
synthetic spectra of isolated
Jupiter-mass planets, 125

F
Faber-Jackson relation, 273,
311
Fabry-Perot interferometer,
446
Falsification of modified
Newtonian dynamics
with rotation curves,
282–85
inferred mass-to-light
ratios for the UMa spirals,
283
FAME mission, 520
Faraday rotation, 323–31
background and embedded
sources, 329–30
cluster center sources,
323–29
high redshift sources,
330–31
measures of, 323–24,
329–30, 332
Far-field cosmology, 488–89,
528
Far-IR Spectral Energy
Distribution (FIR SED)
colors, 44, 48
detectability, 612–13
flux densities, 44, 48
Fast Low-Ionization

Emission Regions (FLIERs), 444–45
Feynman, Richard, 6, 17
FGK stars, 527, 529
Fiber Large Array Multi-Element Spectrograph (FLAMES), 529
Field origin, 340–44
FIR SED
   See Far-IR Spectral Energy Distribution
Fireball shock
   afterglow model, 144
   scenario, 159
First order signatures of galaxy formation, 503–10
   age-metallicity relation, 508
   effects of environment and internal evolution, 508–10
   possibility of disks preserving fossil information, 503–8
   structure of the disk, 503
Fish law, 271
Fisher matrix, 207, 210, 671
Fitting parameters for RSNe, 398
FK Com class radio sources, 219, 224
FLAMES
   See Fiber Large Array Multi-Element Spectrograph
FLIERs
   See Fast Low-Ionization Emission Regions
Formation mechanisms
   of the Kuiper belt, 89–90
   possible, of binary stars, 361
Fourier amplitude, 190
Fourier modes, 177, 179, 189

Fourier space, 178
Fourier wavenumber, 174
Fowler, Willy, 8–9
Free oscillations
   question of, 116–17
Freestreaming power, 175
Fresnel diffraction spot, 96
Friedmann background, 563–64
Friedmann equation, 308–9, 311
Friedmann-Lemaitre models, 309
Full-width at half-maximum (FWHM), 51, 322, 626

G

GAIA, 96–97
Gaiasphere, 527
   candidates for chemical tagging, 527
   chemical abundance space, 524–26
   chemical signatures, 521–22
   chemical trajectories, 526–27
   and the limits of knowledge, 520–28
   reconstructing ancient star groups, 522–24
Galactic acceleration field, 290
Galactic hosts, 154–59
   schematic GRB afterglow, 158
   schematic GRB from internal shocks, 157
Galactic population and distribution, 50–57
   distribution and number of UC HII regions, 51–54
   galactic temperature and abundance gradients, 54–57
   spiral structure using UC HII regions, 54

Galactic system
   accretions to objects crossing, 17
   chemical evolution history, 494
   distribution of dark matter in, 264
   energetics of massive stars in, 54
   estimating mass, 627
   mapping region around, 208
   overdensities, 556
Galactic temperature and abundance gradients, 54–57
   the galactocentric electron temperature gradient, 56
   spiral arms suggested, 55
Galaxies
   and dark matter, 18–19
   See also New galaxies
Galaxy-galaxy lensing, 306
Galaxy turbulent wakes, 343
Galaxy velocity dispersions, 542
Galilei invariant, 303
Galileo probe, 113, 121
   Helium Abundance Detector, 122
   Mass Spectrometer, 122–23
   results, 122–24
Gamma-ray bursts (GRBs), 132, 409, 415–34
   with afterglows, 142–45, 415–16, 429, 433
   alternative models, 150
   cosmic rays, neutrinos, GeV-TeV photons, and gravity waves, 159–63
   cosmological effects, 430
   cosmological setting, galactic hosts and environment, 154–59
   the fireball shock scenario, 139–42

interstellar scintillation, 429–30
multiple origins, 433–34
parameterization, 428–29
progenitors, 150–54
radio detections, 416–28
radio emission from, 415–34
relativistic effects, 430–31
relativistic fireball, 432–33
relativistic jets and hypernovae, 433
slow-soft class of, 388
standard model developments and issues, 145–50
theoretical models, 137–69, 431–34
typical GRB lightcurve observed with BATSE, 138
See also Radio gamma-ray bursts
Gamow, George, 10
Gaussian fluctuations, 174, 208
Gaussian statistics, 71
GCs
See Globular clusters
Gell-Mann, Murray, 7
Gemini Wide Field proposal, 530
General predictions of modified Newtonian dynamics, 268–72
line rotation curves of a low surface brightness galaxy, 270
General properties of pressure-supported systems, 285–87
line-of-sight velocity dispersion vs. characteristic radius, 286
General Relativity, 265
Generalist looking backward, 1–25

black hold accretion, 17–18
the enrichment of the interstellar gas, 10–12
galaxies and dark matter, 18–19
helium-burning, 8–10
high pressure statistical mechanics, 12–13
hydrodynamics of the interstellar medium, 18–19
the initial mass function, 10–12
the JASON group, 13–15
the neuromuscular junction, 15–17
NSF astronomy grants, 19–22
nuclear astrophysics, 7–8
plasma physics, 12–13
quantum electrodynamics, 4–6
relativistic bound states, 7–8
upbringing and schooling, 2–4
Generalized Interacting Stellar Wind (GISW) models, 442–43, 445, 450, 454, 457–58, 477, 479
GeV region, 142, 148, 162
GeV-TeV photons of GRBs, 159–63
GHRS spectrograph, 600–2
Giant Meter Wave Radio Telescope, 337
Giant molecular clouds (GMCs), 28–29, 352–53
Giant planets
atmospheres and their relation to the interior, 117–25
gravity field as a window to the interior, 112–16
high-pressure phase diagrams, 105–12

historical background, 103–5
orbital evolution of giant planets, 131–32
question of free oscillations, 116–17
theories of, 103–36
thermal evolution of giant planets, 125–31
trajectories of, 120
Gibbs free energy, 119
Gibbs phase rule, 110
GISW
See Generalized Interacting Stellar Wind models; Generalized interacting stellar wind models
GLAST, 162
Global Newtonian mass-to-K'-band-luminosity ratio, 266
Globular clusters, 516–17
Globular clusters (GCs), 352–53
and molecular clouds, 291–93
MOND mass-to-light ratio for dwarf spheroidal satellites, 292
GMCs
See Giant molecular clouds
Gould belt, 518
Gravitational assembly, 156
Gravitational collapse, 263
Gravitational forcing, 183–84
Gravitational instability, 580
Gravitational lensing, 199–200, 556
and no-go theorems, 304–6
Gravitational secondaries, 196–200
integrated Sachs-Wolfe effect, 196–97
Rees-Sciama and moving halo effects, 197
Gravitational waves, 198–99
and the energy scale of

inflation, 198
Gravity field
    connection to classical
        theories of rotating
        bodies, 113
    tests of expected interior
        rotation and convection
        state, 113–16
    as a window to the interior,
        112–16
Gravity harmonics
    Jupiter zonal, 114, 116
Gravity waves, 159–63
GRB lightcurve
    observed with BATSE,
        typical, 138
GRBs
    See Gamma-ray bursts
GRB000301C, 425, 427–28
GRB970508, 415–17,
        428–30, 433
GRB980329, 419–21, 423,
        428, 430
GRB980425, 387–88, 405,
        409–15, 428–29, 431
    of radio supernovae, 409
GRB980519, 421–22, 433
GRB991208, 423–24
GRB991216, 423, 425–26
Gunn-Peterson trough, 200
Gyromagnetic emission,
        221–23
    gyroresonance emission,
        221–22
    synchrotron emission from
        a power-law electron
        distribution, 222–23
Gyromagnetic emission
    mechanisms, 226–28
    radio spectra, 227
Gyroresonance emission,
        221–22
    coronal, 233
Gyroresonance mechanism,
        227
Gyrosynchrotron emission
    from a power-law electron

distribution, 222
    from a thermal plasma, 222
Gyrosynchrotron radiation,
        228, 239
    quiescent, 229
GZK limit, 339

H

Hale telescope, 589
    LRIS spectrograph, 589
Half Width at Half Max
        (HWHM), 66
Halley comet, 88, 96
Halos, 445–46
Hawking radiation, 304
HB 283447, 235, 248
HDF
    See Hubble Deep Field
    survey
HD209458b, 129–31
HEAO-1 X-ray Observatory,
        549
Heisenberg, Werner, 5
Helium Abundance Detector,
        122
Helium-burning, 8–10
Helix Nebula, 445
Helmholtz free energy, 108
HEMT
    See High Electron Mobility
    Transistor amplifiers
Herbig stars, 219, 249
    Ae/Be, 249
Hertzsprung-Russell Diagram
        (HRD), 219
HESS, 162
HETE-2 spacecraft, 155
Heterogeneous condensation,
        119
He^2-104, 440, 447
He^3-1475, 453
Hierarchical fragmentation,
        365
High Electron Mobility
        Transistor (HEMT)
        amplifiers, 657
High-pressure phase

diagrams, 105–12
    hydrogen and helium,
        theory, 111–12
    hydrogen, theory compared
        with experiment, 106–11
High pressure statistical
        mechanics, 12–13
High redshifts
    clusters at, 624–25
    sources of, 330–31
High surface brightness
        (HSB) galaxies, 272,
        275–76, 279, 281
HII regions, 403, 627
    bipolar, 42
Historical background of
        giant planets, 103–5
    discovery of EGPs, 105
    thermal properties based on
        heat flow measurements,
        104–5
HK project, 240
Homogeneous cooling theory,
        127
Hot Big Bang, 306
Hot cores as precursors to UC
        HII regions, 32–41
    bipolar molecular outflows,
        38–41
    hyper-compact HII regions,
        41
    outflows and accretion
        disks, 37–38
    radio free-free emission,
        34–37
    spectral energy
        distributions, 34
    temperature structure,
        32–34
Hoyle, Fred, 9, 20
HR 1099, 240
HRD
    See Hertzsprung-Russell
    Diagram
HR4796A, 93–94
HSB
    See High surface

brightness galaxies
HST
  See Hubble Space
    Telescope
Hubble constant, 180,
    192–93, 644, 664–66,
    668
Hubble deceleration, 309–10
Hubble Deep Field (HDF)
    survey, 491, 581, 590,
    592, 595–97, 608–9,
    617–19, 621, 623–24
Hubble flows, 272, 310, 462
Hubble sequence, 582
  assembly of, 579
Hubble Space Telescope
    (HST), 71, 415, 439–40,
    582, 597–98, 600–2,
    617–18, 625
Hubble 5, 440, 447
Hugoniot, 111
HVCs, 502
HWHM
  See Half Width at Half
    Max
Hydra A, 328
Hydrodynamic (Biermann)
    battery effect, 341
Hydrodynamic jet collimation
    mechanisms, 470–71
Hydrodynamic modeling, 446
  of the interstellar medium,
    18–19
Hydrodynamic models of
    inertial confinement,
    459–64
  aspherical nebulae, the
    GISW model, 461
  morphologies and
    kinematics of GISW
    model, 463
  numerical models, 461–62
  spherical interacting winds
    models, 459–60
  theoretical models and
    Hubble flows, 462–64
Hydrogen

isochores for, 109
  phase diagram of, 106
  theory compared with
    experiment, 106–11
Hydrogen and helium
  immiscibility of a solar
    mixture of, 112
  theory of, 111–12
Hydrogen dissociation and
    metallization, 109
Hyper-compact HII regions,
    41
Hypersonic plasma beams,
    470

I

IAU working group, 126
IC
  See Inverse-Compton limit
IC 3568, 442–43
IC 5249, 496
ICECUBE, 160–61
ICM
  See Intercluster medium;
    Intracluster medium
Imaging proportional counter
    (IPC) fields, 550
IMF
  See Initial mass function
Immiscibility
  of a solar mixture of
    hydrogen and helium, 112
Inclination
  of classic KBOs, 67
Incoherence
  temporal, 183
Incoherent flares, 224
Inertial confinement models
    of planetary nebulae, 457
Inferred mass-to-light ratios
    for the UMa spirals, 283
Information preserved
  since dark matter
    virialized, 499–503
  since the main epoch of
    baryon dissipation,
    503–10

Infrared Array Camera
    (IRAC), 54
Infrared Astronomical
    Satellite (IRAS), 423,
    425
  Point Source Catalog, 51
Initial conditions
  of planetary nebulae,
    457–59, 467
Initial mass function (IMF),
    10–12, 29
  stellar, 29, 599, 603
Instability
  brink of, 350
Integrated RM
  plotted as a function of
    source impact parameter,
    330
Integrated Sachs-Wolfe
    (ISW) effect, 196–97,
    200–1
Interacting stellar wind (ISW)
    models, 441–42, 448
Intercluster medium (ICM),
    320, 324
Interferometric observations,
    654–58
  SZE spectrum of Abell
    2163, 655
Interior rotation and
    convection state
  Jupiter zonal gravity
    harmonics, 114, 116
  Jupiter zonal wind speeds,
    115
  tests of expected, 113–16
Internal structure
  of the Kuiper Belt, 85–86
Interstellar gas
  enrichment of, 10–12
Interstellar medium
  and hydrodynamics, 18–19
Interstellar medium (ISM),
    29, 393, 432
  local, 29
Interstellar scintillation (ISS),
    419, 421

of GRBs, 429–30
Intracluster medium (ICM),
541, 545, 644
evolution of, 572
substructure of, 562–63
temperature of, 566–67
Inverse-Compton (IC) limit,
431
Inverse-Compton (IC)
scattering, 142, 149,
223, 323
Inverse-Compton (IC) X-ray
emission, 331–37
electron lifetimes, 333–35
reconciling IC- and
RM-derived fields,
335–37
RXTE spectrum of the
Coma cluster, 331–37
Ionization modulation, 202
Ionized gas of UC HII
regions, 42–46
low-density extended
halos, 43–46
morphologies and
lifetimes, 42–43
Ionized hydrogen (HII) along
the line of sight, 403–4
Ionized winds from hot stars,
243–45
VLA map of the WR 147
system, 244
IPC
See Imaging proportional
counter fields
IRAC
See Infrared Array Camera
IRAM telescope, 653
IRAS
See Infrared Astronomical
Satellite
IRAS 17106-3046, 451
Irregular nebulae, 443
ISM
See Interstellar medium
Isochores
for hydrogen, 109

Isolated giant planets
T_{eff}-T_{10}-g surface for
calculating evolution of,
119
ISS
See Interstellar scintillation
ISW
See Integrated Sachs-Wolfe
effect; Interacting stellar
wind models

J

J-band dropouts, 597
Jacobi-Dedekind (JD)
sequence, 369–71
Jacobi ellipsoid, 85
James Clerk Maxwell
Telescope (JCMT), 49,
421, 610
Submillimeter
Common-User Bolometer
Array galaxies, 609–13
JASON group, 13–15
JCMT
See James Clerk Maxwell
Telescope
JD
See Jacobi-Dedekind
sequence
Jeans instability, 351, 358–59
typical conditions in
collapsing cloud core,
358
Jeans length, 309
Jeans-unstable gas clouds,
363
Jet anisotropy, 146
Jet development in a 14 $M_\odot$
collapsar, 154
Jets and limb-brightening
effects, 146–47
Jupiter
metallic-hydrogen layers,
104
Jupiter-family comets, 87–88
Jupiter zonal gravity
harmonics, 114, 116

K

KBOs
See Kuiper Belt objects
Keck spectra, 597
Keck telescope, 529, 589,
599–600
Kelvin-Helmholtz timescale,
36
Kenyon and Luu model
of the formation of the
Kuiper belt, 89–90
Keplerian orbital velocity, 93
Key timescales, 354–56
scales of interest, 355
Kinetic Sunyaev-Zel'dovich
effect, 649
Kirchhoff's law, 220
Klein-Gordon equation, 198
Knots and jets, 444–45
Konus satellites, 137, 423
Kuiper, Gerard, 64
Kuiper Belt objects (KBOs),
63–101
binary KBOs, 86–87
evolution of the Kuiper
belt, 90–93
formation of the Kuiper
belt, 89–90
internal structure, 85–86
physical properties, 77–85
population characteristics,
71–75
radial extent of the classical
Kuiper Belt, 75–77
relationship to other solar
system bodies, 87–89
relationship with
circumstellar disks, 93–95
research needed, 95–97
structure of the Kuiper
Belt, 65–70
Kuiper Belt V-R colors, 82
Kuzmin disks, 298

L

Lagrangian-based theory of
MOND, 297

Large-angle polarization, 201
Large Magellanic Cloud
    (LMC), 516, 591
Large Millimeter Telescope
    (LMT), 58
Large-scale kinematics
    and proper motions,
    446–47
Large Synoptic Survey
    Telescope, 96
Las Campanas Redshift
    Survey, 293
Laser Interferometric
    Gravitational Wave
    Observatory (LIGO),
    163
LBGs
    See Lyman-break galaxies
LBT
    See Lyman-break
    technique
Lebovitz's revised version of
    the fission theory, 371
Light-curves, 158, 229
    break, 147
LIGO
    See Laser Interferometric
    Gravitational Wave
    Observatory
Limber equation, 197
Limits on Ω_M
    from SZE-measured cluster
    gas fractions, 669
Linear modulation, 202
Line emission dinematics
    and the total mass,
    625–27
Line-of-sight velocity
    dispersion
    vs. characteristic radius,
    286
Line rotation curves
    of a low surface brightness
    galaxy, 270
LIS
    See Low-ionization
    structures

LMC
    See Large Magellanic
    Cloud
LMT
    See Large Millimeter
    Telescope
Local cluster number density,
    556–58
    local X-ray luminosity
    function of clusters, 558
Local Group, 21, 249, 497,
    499, 510, 528
Local X-ray luminosity
    function (XLF) of
    clusters, 558
Lockman Hole, 563, 568
Look-back time
    as a function of redshift for
    different world models,
    490
Lorentz factors, 139, 141,
    143, 145, 153, 159, 336,
    432–33, 649
Lorentz invariance, 302
Lorentz transforms, 331
Loss times, 224
Low-density extended halos,
    43–46
    apparent at lower spatial
    resolutions, 44
    the electron density-size
    relation, 46
    proposed model for, 45
Low-ionization structures
    (LIS), 444–45
Low-mass stars
    on the main sequence of
    star formation, 248–49
Low Resolution Imaging
    Spectrometer (LRIS)
    instrument, 589,
    599–600
Low surface brightness (LSB)
    galaxies, 273–76, 281,
    299
    line rotation curves of, 270
Low surface brightness (LSB)

structures in galaxies,
    512–13
    normal spirals with faint
    stellar streamers in the
    outer halo, 513
Lower atmospheres of cool
    stars, 240–41
Lower self-adjoint (LSA)
    series
    of Riemann ellipsoids, 369,
    372–74, 379
Lower spatial resolutions
    low-density extended halos
    apparent at, 44
LRIS
    See Low Resolution
    Imaging Spectrometer
    instrument
LSA
    See Lower self-adjoint
    series
LSB
    See Low surface brightness
    galaxies
Luminosity function, 73
    cumulative, 71–72
    isotropic-equivalent, 150,
    154
Luminosity vs. age for EGPs,
    127
Luminous elliptical galaxies,
    287–90
    mass-velocity dispersion
    relation, 289
Lyα forest clouds
    super-clusters and,
    295–97
Lyα galaxies, 571, 598–99
    absorption lines in, 584,
    586–88, 595, 598–99,
    602, 605, 614, 625
Lyman α absorption, 155
Lyman-break galaxies
    (LBGs), 579–641
    chemical abundances,
    614–15
    clustering and large-scale

structures at $z \sim 3$, 620–25
the mass spectrum, 625–29
morphology, 618–20
nature of Lyman-break
    galaxies, 599–603
normal galaxies in the high
    redshift universe, 629–34
search for primeval
    galaxies, 583–89
star formation rates,
    603–10
starburst galaxies at high
    redshifts, 610–14
starburst galaxies at $\sim 10\%$
    of the Hubble time,
    599–603
stellar mass and age,
    615–18
UV-selected star forming
    galaxies at high redshift,
    597–99
at $z > 4$, 597–98
Lyman-break galaxy surveys,
    589–97
the color selection, 589–93
the redshift distribution,
    593–97
Lyman-break technique
    (LBT), 581–82, 587–89,
    629
Lyr system, 251

M
$M_\odot$ values, 357
Maclaurin sequence, 369
Madau diagram, 608, 629
Magellanic Clouds, 512
MAGIC, 162
Magnetars, 158
Magnetic cycles, 239–40
    activity cycles, 240
    radio emission and stellar
        rotation, 239–40
Magnetic fields in PNe,
    455–56
Magnetic support of cluster
    gas and the baryon

crisis, 338–39
Magnetized Wind Blown
    Bubble (MWBB), 458,
    465
    model for the ant nebula,
        Menzel 3, 480
    models of planetary
        nebulae, challenges to,
        467
Magneto-centrifugal models
    of planetary nebulae,
    467–69
Magneto hydrodynamic
    (MHD)
    effects, 457, 459
    entrainment, 41
    models of jet formation,
        471–72
    models of PN formation,
        464–67
    numerical MWBB models,
        465–67
    waves, 29, 151
Magnetospheric models,
    237–39
    equatorial model for, 238
Major processes involved in
    subsequent evolution of
    galaxies, 510–20
    globular clusters, 516–17
    low surface brightness
        structures in galaxies,
        512–13
    open clusters, 514–16
    star formation history,
        510–11
    structures in phase space,
        518–20
Malmquist bias, 38
MAP
    See Microwave Anisotropy
        Probe satellite
Mapmaking, 206–8
Masers
    electron cyclotron
        emission, 223, 233
    menthol, 49

of UC HII regions, 49–50
Mass limits of observability,
    661–62
    redshift for two SZE
        surveys, 662
    redshift for typical
        wide-field survey type,
        662
Mass-loss rate, 401–3
    of presumed red supergiant
        progenitor to SN1993J,
        402
    of Type IIb SN1993J, 402
Mass-loss rate from radio
    absorption, 394–400
    absorption by a statistically
        large number of clumps or
        filaments, 398–99
    absorption by a statistically
        small number of clumps
        or filaments, 399
    absorption by a uniform
        external medium, 398
    estimated mass-loss rates
        for RSNe, 400
    fitting parameters for
        RSNe, 398
    radio "light curves" for
        SN1988Z, 397
    of Type Ib SN1983N, 395
    of Type Ic SN1990B, 395
    of Type II SN1979C, 396
    of Type II SN1980K, 396
Mass-loss rate from radio
    emission, 400–1
Mass outflow rate of bipolar
    molecular outflows, 38
Mass Spectrometer, 122–23
Mass spectrum, 625–29
    of LBGs, 625–29
    line emission dinematics
        and the total mass,
        625–27
    spatial clustering and the
        mass spectrum, 628–29
    stellar mass and total mass,
        627–28

Mass-to-light ratios
inferred, for the UMa
spirals, 283
Mass-velocity dispersion
relation, 289
Massive star formation
evolution to the
ultracompact state, 28–41
galactic population and
distribution, 50–57
research needed, 57–58
UC HII regions, 41–50
ultra-compact HII regions,
27–62
Matter power spectrum,
194–96
cosmological implications,
195–96
physical description,
194–95
Maximum absolute RM as a
function of estimated
cooling flow rate, 326
Maximum size of KBOs
decrease in, 76–77
Mean free path (mfp), 337
Mean orbital elements
of resonant KBOs, 68
Mean relative abundance
ratios of process
elements, 522
Menthol masers, 49
Menzel 3, 440, 445, 480
MERLIN network, 235, 247
Metal-poor stars, 524
See also Super metal-rich
disk stars
Metallization
hydrogen, 109
MeV radiation, 142
mfp
See Mean free path
MHD
See Magneto
hydrodynamic
Micro-lensing, 312
Microflaring at radio

wavelengths, 229–30
light curves, 229
Microwave Anisotropy Probe
(MAP) satellite, 174
Mid-Course Space
Experiment (MSX), 54
Milky Way, 51, 516
mapping, 97, 496
satellite dwarf companions
of, 271–72
satellite in orbit about,
519
Minkowski space, 303
Model-cumulative luminosity
functions, 92
Models of planetary nebulae,
457–59
basic physical elements,
459
inertial confinement
models, 457
possible pPNe and PNe
formation mechanisms,
458
self confinement models,
457
Models of radio supernovae,
390–93
general properties, 390–91
parameterized radio light
curves, 391–93
Modified Doppler effects,
201–2
cluster modulation, 201
general density
modulation, 202
ionization modulation, 202
linear modulation, 202
Modified Newtonian
dynamics (MOND)
an acceleration scale,
265–68
as an alternative to dark
matter, 263–317
basics of, 265–74
controversy, 19–20
cosmology and the

formation of structure,
306–11
dark halos with an
acceleration scale, 272–74
dynamical mass of clusters
of galaxies, 294
fits to the rotation curves of
spiral galaxies, 282
fits to the rotation curves of
the Ursa Major galaxies,
279–80
general predictions,
268–72
mass-to-light ratio for
dwarf spheroidal
satellites, 292
as a modification of
general relativity, 300–3
as a modification of
Newtonian inertia, 303–4
pressure-supported
systems, 285–97
rotation curves of spiral
galaxies, 274–85
Molecular environment of
UC HII regions, 47–50
chemistry, 48–49
masers, 49–50
physical properties, 47–48
Momentum conservation, 178
MOND
See Modified Newtonian
dynamics
Monte Carlo calculations, 16,
109, 671
simulating, 74, 594–95,
609
Morphologies
of LBGs, 618–20
optical and IR continuum
observations, 449–50
Morphologies and kinematics
of GISW model, 463
Morphologies and lifetimes of
UC HII regions, 42–43
examples of UC HII
morphological types, 42

Morphologies of pPNe
with AGB nuclei, 450
with post-AGB nuclei,
450–51
Mt. Palomar Observatory
Hale telescope, 589
See Palomar
Observatory Sky
Survey
Moving halo effects
Rees-Sciama and, 197
MSX
See Mid-Course Space
Experiment
Mühsam, H., 217
Multiple long-slit
high-dispersion
spectrograph, 446
Multiple origins
of GRBs, 433–34
MWBB
See Magnetized Wind
Blown Bubble
MyCn18, 440, 445
Myr accretion period, 86

N

Narrow Field Instruments
(NFI), 409
Natal material, 28–30
National Science Board, 14
National Science Foundation
(NSF), 20–21
astronomy grants, 19–22
NEAR, 423, 425
Near-field cosmology,
488–89
goals of, 492–93
Near Infrared Camera and
Multi-Object
Spectrometer
(NICMOS) photometry,
597, 617–18, 625
Near-infrared Tully-Fisher
relation, 268
Nearly homologous
collapses, 363–64

Nebulae
aspherical, the GISW
model, 461
Nebular features
anase and FLIERS, 472–73
bipolar rotating emission
line jets, 472
hydrodynamic jet
collimation mechanisms,
470–71
MHD models of jet
formation, 471–72
theories of planetary
nebulae, 470–73
Nebular momenta, 447–48
Nebular types, 443
Neptune
radial migration of, 69
See also Trans-Neptunian
objects
Neupert effect, 230–31
Neuromuscular junction,
15–17
Neutrinos
of GRBs, 159–63
massive, 196
TeV neutrino production
by photomeson
interactions, 162
Neutron stars (NSs)
binaries, 150
black holes, 150
New galaxies
gaiasphere, and the limits
of knowledge, 520–28
goals of near-field
cosmology, 492–93
near-field and far-field
cosmology, 488–89
research needed, 528–30
signatures of their
formation, 487–537
stellar age dating, 493–95
structure of galaxies,
495–98
timescales and fossils,
491–92

working model of galaxy
formation, 489–91
Newton-Raphson method, 21
Newton-XMM observations,
539, 541, 543, 545, 563,
568–69, 666
Newtonian curvature, 183
Newtonian dynamical mass,
263
of clusters of galaxies, 294
Newtonian-MOND objects
mixed, 288
Newtonian potential, 183,
194–95
Newtonian theory, 263
NFI
See Narrow Field
Instruments
NGC 3132, 442–43
NGC 4214, 600–2
NGC 4762, 496
NGC 6543, 440, 446–47
NICMOS
See Near Infrared Camera
and Multi-Object
Spectrometer photometry
Nobeyama telescope, 653
Non-Friedmannian universe,
308
Non-Gaussianity, 204
Nonaxisymmetric instabilities
in rapidly
equilibrium cores, 368–71
Nonaxisymmetric instabilities
in rapidly-rotating
equilibrium cores
classical view of fission,
370–71
Lebovitz's revised version
of the fission theory, 371
phase-space diagram for
ellipsoidal configurations,
369
Nonbaryonic CDM, 307
Nonhomologous collapses,
364–65
Nonlocal thermodynamic

equilibrium (NonLTE),
55–56
Normal galaxies in the high
redshift universe,
629–34
total mass of LBGs, 633
Normal spirals with faint
stellar streamers in the
outer halo, 513
Normalization
COBE, 194–95
Northern VLA Sky Survey
(NVSS), 321–22
NSF
See National Science
Foundation
NSs
See Neutron stars
NTT telescope, 589
Nuclear astrophysics, 7–8
Nucleo-cosmochronology,
493–95
Number density
comoving volume element
and, 660
Numerical models, 461–62
MWBB, 465–67
NVSS
See Northern VLA Sky
Survey

O

OB stars, 243
Observables, 172–76
CMB polarization field,
175–76
CMB temperature field,
173–75
standard cosmological
paradigm, 172–73
Observation of the T Tau
system with the
MERLIN interferometer,
247
Observational framework,
546–56
clustering of absorption

line systems, 556
gravitational lensing, 556
optically-based cluster
surveys, 546–48
search for clusters around
bent-double radio
sources, 556
search for galaxy
overdensities around
high-z radio galaxies, 556
strategies and selection
functions for X-ray
surveys, 551–56
Sunyaev-Zeldovich effect,
556
X-ray cluster surveys,
548–51
Observational support for
collimation mechanisms,
455–56
binary and rapid rotation in
the nuclei of PNe, 455–56
magnetic fields in PNe,
455–56
Observed cumulative
luminosity function, 74
Observed lightcurves of the
afterglow of GRB
970228, 144
Observed position angles of
the linearly polarized
radio emission, 327
Observed supernovae, 389
Occam's razor, 313
Occultations, 96
ODE
See Ordinary differential
equation
OH231.8 + 4.2, 452–53
OIR
See Optical/infrared data
Oort Cloud, 64, 71, 87
Oort-Spitzer rocket effect,
472
Ooty Synthesis Radio
Telescope (OSRT), 334
Open clusters, 514–16

distribution of, 515
Öpik, Ernst, 9, 22
Oppenheimer, Robert, 6
Optical/infrared (OIR) data,
416, 419, 423, 425, 428
Optically-based cluster
surveys, 546–48
Optimal statistics, 205
Orbital evolution of giant
planets, 131–32
Oriented Scintillation
Spectrometer
Experiment (OSSE), 137
Origin of binary stars,
349–85
background, 349–51
basic physical principles,
351–61
capture, 361–62
delayed breakup, 367–77
possible formation
mechanisms, 361
prompt fragmentation,
362–67
Orion cluster
high stellar density of, 32
Orion IRc2 outflow, 39
Oscillations
quasi-periodic, 225
OSRT
See Ooty Synthesis Radio
Telescope
OSSE
See Oriented Scintillation
Spectrometer Experiment
Outflow collimation, 467
Outflow momenta of pPNe,
451–52
Outflows and accretion disks,
37–38
mass outflow rate of
bipolar molecular
outflows, 38
Owens Valley Radio
Observatory (OVRO),
653, 657–58, 665–66,
671, 676

P

Palomar Observatory Sky Survey (POSS), 546–47, 552

Parameter sensitivity, 192–93

Parameterization of GRBs, 428–29

Parameterized radio light curves, 391–93
the supernova and its shocks, 393

Parameters for GRB radio afterglows, 418

Partitioning processes, 118

Peak radio luminosities and distances, 407–8

Peak suppression, 200

Peculiar Emission Line Algols (PELAs), 251

PELAs
See Peculiar Emission Line Algols

Perseus-Pisces filament, 295–96

Perturbations
density, 580
gravitational potential, 183
linear, 196
photon-baryon density, 186
relativistic, 649
temperature, 183
theory of, 111–12
See also Quadrupole gravitational perturbation

Pfaffenzeller phase-excluded region, 111–12

Phase diagram of hydrogen, 106

Phase-space diagram for ellipsoidal configurations, 369

Phased lightcurve of KBO (20000) Varuna, 84

Phenomenology
of quiescent emission from coronae of cool stars, 226

Pholus, 88

Photon-baryon density perturbations, 186

Photon-baryon fluid, 187

Photosphere
Compton, 141

Physical basis of MOND Bekenstein-Milgrom theory, 297–300
gravitational lensing and no-go theorems, 304–6
modified Newtonian dynamics as a modification of general relativity, 300–3
modified Newtonian dynamics as a modification of Newtonian inertia, 303–4

Physical elements
of planetary nebulae, 459

Physical parameter estimates of radio supernovae, 413–14

Physical parameters, 351–54
evolutionary trajectory of central region of protostellar gas cloud, 354
scales covering 40 orders of magnitude in length, 353

Physical parameters of binary stars, 351–54

Physical principles of binary stars, 351–61
implications of the viral theorem, 356–58
importance of the effective adiabatic exponent γ, 359–61
the Jeans instability, 358–59
key timescales, 354–56
physical parameters, 351–54

Physical properties
of UC HII regions, 47–48

Physical properties of galaxy clusters, 541–46
cooling in the intracluster medium, 545–46
X-ray properties of clusters, 542–45

Physical properties of Kuiper Belt objects, 77–85
albedos, 77–78
rotation, 84–85

Physics
of planetary nebulae, 457–59

Pian source, 409

Pink loop, 406

Planck function, 331

Planck spectrum, 649

Planck Surveyor satellite, 663, 674

Planetary nebulae (PNe), 439–86
basic models, 457–59
classical, 443–48
collimation, initial conditions, and challenges to MWBB models, 467
dynamos and the challenge to the magneto-centrifugal models, 469–70
historical setting, 440–41
interacting stellar wind concept, 441–42
magneto-centrifugal models, 467–69
MHD models of PN formation, 464–67
observational results and questions, 453–56
observational support for collimation mechanisms, 455–56
organizing by morphological type, 443–44
physics, initial conditions, and equations, 457–59
PN morphologies, 439–40
possible formation

mechanisms of, 458
protoPNe, 448–53
research needed, 477–80
self-confinement, 464–67
shapes and shaping of,
    439–86
stellar origins of disks and
    tori, 473–77
theoretical models, 456–77
theories of nebular
    features, 470–73
Planetesimals, 70–71, 123
Plasma physics, 12–13
    hypersonic beams, 470
Pleiades Moving Group, 249
Plot of albedos vs. diameter,
    78
Plutinos, 69
Pluto, 78, 87
Pluto-Charon and Triton
    relationship to KBOs, 87
PMS
    See Pre-main sequence
        stellar population
PNe
    See Planetary nebulae
Pointing matrix, 206
Poisson equation, 194–95,
    297–98
Poisson statistics, 71, 554
Polarization, 188–90
Polarization of the
    Sunyaev-Zel'dovich
    effect, 649–50
Polarization spectra, 176
Population characteristics of
    Kuiper Belt objects,
    71–75
    size distribution, 73–74
    surface density, 71–73
    total mass, 74–75
Population synthesis studies,
    266
    See also Stellar population
POSS
    See Palomar Observatory
        Sky Survey

Possibility of disks preserving
    fossil information,
    503–8
Power-law electrons
    accelerated, 144
    population of, 231
Power-law index, 173
    approximation, 621
Power-law spectra, 140, 142
Poynting-Robertson, 94
pPNe
    See ProtoPNe
PPT, 108–10
Pre-main sequence (PMS)
    stellar population,
    349–50, 362
Precursors of UC HIIs
    (PUCHs), 32–34, 37
Predicted colors
    for various choices of star
        formation history, 606
Predicted orbital distributions
    of resonant KBOs, 68
Press-Schecter formulation,
    570, 660
Pressure-supported systems,
    285–97
    dwarf spheroidal systems,
        290–91
    general properties, 285–87
    globular clusters and
        molecular clouds, 291–93
    luminous elliptical
        galaxies, 287–90
    rich clusters of galaxies,
        293–95
    small groups of galaxies,
        293
    super-clusters and Lyα
        forest clouds, 295–97
Prestellar cores (PSCs),
    30–31
    spectral energy distribution
        in, 31
Primeval galaxies
    search for, 583–89
Primordial Explorer

(PRIME), 572
Progenitors of GRBs, 150–54
    jet development in a
        14 $M^{\odot}$ collapsar, 154
Prompt flashes
    and reverse shocks,
        147–48
Prompt fragmentation of
    binary stars, 362–67
    nearly homologous
        collapses, 363–64
    nonhomologous collapses,
        364–65
    possibility of prompt
        fragmentation, 365–66
    possibility of resulting
        binaries, 366–67
ProtoPNe (pPNe), 448–53
    He^{3}-1475, 453
    morphologies with AGB
        nuclei, 450
    morphologies with
        post-AGB nuclei, 450–51
    OH231.8 + 4.2, 452–53
    optical and IR continuum
        observations, 449–50
    outflow momenta, 451–52
    possible formation
        mechanisms of planetary
        nebulae, 458
Protostars
    low-mass, 37
Protostars and Planets IV, 28
Protostellar accretion phase,
    246
PSCs
    See Prestellar cores
PUCHs
    See Precursors of UC HIIs

Q
QCD
    See Quark-hadron phase
        transition
QSOs
    See Quasistellar objects
Quadrupole temperature

anisotropy, 188
Quantum electrodynamics,
  4–6
Quark-hadron (QCD) phase
  transition, 341–42
Quasistellar objects (QSOs)
  absorption systems,
  585–86, 588–89, 615
  radio, 341
QUEST survey, 73
Quiescent emission
  from coronae of cool stars,
  226–28
  and flares, 228
  gyromagnetic emission
  mechanisms, 226–28
  phenomenology of, 226

R

Radial extent of the classical
  Kuiper Belt, 75–77
  decrease in the maximum
  size of KBOs, 76–77
  steep density drop-off in
  the Kuiper Belt near 50
  AU, 77
  systematically lower
  albedos of KBOs beyond
  50 AU, 76
Radial velocity technique,
  105
Radiation
  residual, 191, 196
  synchrotron, 320–23
Radiation driving, 185–86
  and diffusion damping, 186
Radiation processes
  efficiencies and pairs,
  148–49
Radiative wind-blown
  bubbles, 460
Radio coronal structure,
  234–39
Radio detections of GRBs,
  416–28
  GRB970508, 417
  GRB980329, 420

GRB980519, 422
GRB991216, 426
parameters for GRB radio
  afterglows, 418
000301C, 427
Radio dynamic spectra, 225
Radio eclipses and rotational
  modulation, 234–35
  magnetospheric models,
  237–39
  very long baseline
  interferometry, 235–37
Radio emission, 387–438
  from brown dwarfs, 251
  from gamma-ray bursters,
  415–34
  from radio supernovae,
  388–408, 410
  from
  SN1998bw/GRB980425,
  414–15
  in stellar rotation, binarity,
  and magnetic cycles,
  239–40
Radio emission from
  chromospheres and
  winds, 240–45
  ionized winds and
  synchrotron emission
  from hot stars, 243–45
  lower atmospheres of cool
  stars, 240–41
  winds from cool stars,
  241–43
Radio emission from
  low-mass young stellar
  objects, 245–47
  observation of the T Tau
  system with the MERLIN
  interferometer, 247
Radio emission from stellar
  atmospheres
  bremsstrahlung, 220–21
  coherent emission, 223
  elementary formulae for
  radiation and particles,
  219–20

gyromagnetic emission,
  221–23
  loss times, 224
  theories of, 219–23
  wind emission, 223–24
Radio flares and coronal
  heating, 228–33
  correlation between
  quiescent radio and X-ray
  emissions, 231–33
  microflaring at radio
  wavelengths, 229–30
  quiescent emission and
  flares, 228
Radio flares and the Neupert
  effect, 230–31
Radio flares from cool stars,
  224–25
  coherent radio bursts, 225
  incoherent flares, 224
  radio dynamic spectra, 225
Radio free-free emission,
  34–37
  critical mass infall rate, 36
  SEDs of hot cores, 35
Radio gamma-ray bursts
  (RGRBs), 428–29
Radio halos, 320–22
  Westerbork Synthesis
  Radio Telescope radio
  image of the Coma cluster
  region, 321
Radio light curves, 411–13
  of GRB980425, 413
  of SN1998bw, 411, 413
Radio "light curves" for
  SN1988Z, 397
Radio photosphere, 245
Radio relics, 322–23
Radio spectra
  of gyromagnetic emission
  mechanisms, 227
Radio supernovae (RSNe),
  388–408
  background, 409–10
  expansion velocity, 410
  mass-loss rates for, 400

models, 390–93
observed supernovae, 389
physical parameter
estimates, 413–14
radio light curves, 411–13
SN1998bw/GRB980425,
409, 414–15
Radio supernovae (RSNe)
emissions, 388–408, 410
binary systems, 403
change in mass-loss rate,
401–3
circumstellar medium
structure, 405–7
deceleration of blastwave
expansion, 407
ionized hydrogen (HII)
along the line of sight,
403–4
mass-loss rate from radio
absorption, 394–400
mass-loss rate from radio
emission, 400–1
peak radio luminosities and
distances, 407–8
rapid presupernova stellar
evolution, 404–5
Radio surveys
and the radio
Hertzsprung-Russell
Diagram (HRD), 219
Radio synchotron radiation,
335
Radiogenic heat, 85
Rapid presupernova stellar
evolution, 404–5
Rapidly rotating ellipsoid
slow contraction of,
373–74
Rarefaction wave, 364
RASS
See ROSAT satellite
All-Sky Survey
Rayleigh-Jeans
approximation, 220
Rayleigh-Jeans brightness
temperature, 648

Rayleigh-Jeans limit, 202–3,
652
Rayleigh scattering, 529
Rayleigh-Taylor modes, 464
RDCS sample, 569–70
Reconciliation
of IC- and RM-derived
fields, 335–37
Reconstruction
of ancient star groups,
522–24
Red giant stars
energy production in, 8–9
Red magnitude
apparent, of Kuiper Belt
objects, 72
Redshift
for two SZE sui veys, 662
for typical wide-field
survey type, 662
Redshift distribution, 593–97
in Lyman-break galaxy
surveys, 593–97
spectroscopic redshifts of
dropouts, 594
Rees-Sciama
and moving halo effects,
197
Reflection spectrum of KBO
1999 DE$_9$, 83
Relation between components
of the velocity dispersion
and the stellar age, 506
Relationship of KBOs to
other solar system
bodies, 87–89
comets and centaurs, 87–89
Pluto-Charon and Triton,
87
Relationship of KBOs to
circumstellar disks,
93–95
dust ring around
HR4796A, 94
Relativistic bound states, 7–8
Relativistic effects of GRBs,
430–31

Relativistic fireball, 432–33
Relativistic jets and
hypernovae, 433
Research needed
into the evolution of X-ray
clusters of galaxies,
570–72
into Kuiper Belt objects,
95–97
into new galaxies, 528–30
into planetary nebulae,
477–80
into ultra-compact HII
regions, 57–58
Resonant capture theory, 70
Resonant KBOs, 67–70
mean orbital elements of,
68
predicted orbital
distributions of, 68
Reverse shocks
prompt flashes and, 147–48
RGRBs
See Radio gamma-ray
bursts
Rich clusters of galaxies,
293–95
MOND dynamical mass of
clusters of galaxies, 294
Newtonian dynamical mass
of clusters of galaxies,
294
Riemann ellipsoids, 368–69,
371–73, 380
compressible analog of,
373
lower self-adjoint series,
369, 372–74, 379
upper self-adjoint series,
369
RM
See Rotation measures
"Roasters," 117, 128–31
final stages of evolution of
Jupiter and Saturn, 131
Robotic Optical Transient
Search Experiment

(ROTSE) camera, 148
ROSAT satellite, 401, 540,
    545, 550–72, 656
  Deep Cluster Survey,
    562–63
  PSPC detector, 332, 334,
    550–51, 553
ROSAT satellite All-Sky
    Survey (RASS), 550,
    552, 557–58
Ross' model, 108, 110–11
Rotating gas-liquid body
    theories of, 113
Rotation, 84–85
Rotation curve fitting
  method and results of,
    274–82
  MOND fits to the rotation
    curves of spiral galaxies,
    282
  MOND fits to the rotation
    curves of the Ursa Major
    galaxies, 279–80
  rotation curve fits, 276–78
Rotation curves of spiral
    galaxies, 274–85
  falsification of modified
    Newtonian dynamics with
    rotation curves, 282–85
  method and results of
    rotation curve fitting,
    274–82
Rotation measures (RM),
    323–29
  extreme, 328
Rotation of Kuiper Belt
    objects, 84–85
  phased lightcurve of KBO
    (20000) Varuna, 84
ROTSE
  See Robotic Optical
    Transient Search
    Experiment camera
Round nebulae, 443
RR Lyrae stars, 494–95
RS CVn binaries, 219, 225,
    229, 231, 234

polarization measurements
    in, 239
  VLBI studies of, 228, 233
RSNe
  See Radio supernovae
RXTE satellites, 332
RXTE spectrum of the Coma
    cluster, 331–37
Ryle Telescope, 657–58

S

Sachs-Wolfe effects, 184, 191
Sagan, Carl, 22
Saha equations, 106, 188
  approximation, 188
Salpeter stellar mass
    spectrum, 30, 604
Salpeter value, 20, 29, 603
Satellite in orbit about the
    Milky Way, 519
SB
  See Surface-brightness
    limitation
Scaled growth rate
  of the $m = 2$ instability in
    Newtonian disks, 299
Scales covering 40 orders of
    magnitude in length, 353
Scattered KBOs (SKBOs),
    70–71
Scattering secondaries, 200–3
  Doppler effect, 201
  large-angle polarization,
    201
  modified Doppler effects,
    201–2
  peak suppression, 200
  Sunyaev-Zel'dovich effect,
    202–3
Schechter function, 596
Schematic GRB
  afterglow, 158
  from internal shocks, 157
Schmidt survey, 546
Schrödinger, Erwin, 2
Schwarzschild metrics, 110,
    305

Schwinger, Julian, 6
SCUBA
  See Submillimeter
    Common-User Bolometer
    Array galaxies
SCVH equation, 107–10
Searches
  for clusters around
    bent-double radio
    sources, 556
  for galaxy overdensities
    around high-z radio
    galaxies, 556
  Lyman-break technique,
    587–89
  for primeval galaxies,
    583–89
Second order signatures,
    510–20
SEDs
  See Spectral energy
    distributions
Self-absorbed break, 144
Self-confinement
  of planetary nebulae, 457,
    464–67
Semi-analytic modeling, 203
Sensitivity of the cluster mass
    function to cosmological
    models, 565
Serpentid stars, 251
SEST telescope, 653
Sextic potential theory, 302
SFH
  See Star formation history
SFIC
  See Shock Focused Inertial
    Confinement
Shapes and shaping of
    planetary nebulae,
    439–86
  classical PNe, 443–48
  historical setting, 440–41
  interacting stellar wind
    concept, 441–42
  observational results and
    questions, 453–56

observational support for collimation mechanisms, 455–56

PN morphologies, 439–40

protoPNe, 448–53

research needed, 477–80

theoretical models, 456–77

Shock compression experiments, 110

Shock Focused Inertial Confinement (SFIC), 470

Shock physics, 149–50

Signatures

of the CDM hierarchy, 501–3

of the environment, 499

of global quantities, 500

of the internal distribution of specific angular momentum, 500–1

Signatures of the formation of new galaxies, 487–537

gaiasphere and the limits of knowledge, 520–28

goals of near-field cosmology, 492–93

near-field and far-field cosmology, 488–89

research needed, 528–30

stellar age dating, 493–95

structure of the galaxy, 495–98

timescales and fossils, 491–92

working model of galaxy formation, 489–91

Silicate clouds in brown dwarfs, 121

SIM mission, 520

Single-dish observations, 653–54

Single-star mechanisms for producing tori, 473–74

Singular isothermic spheres, 34

Singular logatropic spheres, 34

SIRTF

See Space Infrared Telescope Facility

Size distribution of Kuiper Belt objects, 73–74

observed cumulative luminosity function, 74

selected measurements of KBOs, 73

SKA

See Square Kilometer Array

SKBOs

See Scattered KBOs

Sky coverage as a function of X-ray flux, 553

Sky surveys with the Sunyaev-Zel'dovich effect, 658–63

cluster abundance, 659–61

estimates of SZE source counts, 662–63

mass limits of observability, 661–62

SMA

See Submillimeter Array

Small groups of galaxies, 293

Small Magellanic Cloud (SMC), 591

extinction law, 421, 606

SMC

See Small Magellanic Cloud

SMOs

See Substellar-mass objects

SMR

See Super metal-rich disk stars

SN

See Supernova

SNRs

See Supernova remnants

SN1979C, 394, 396, 401, 403–5, 407, 412, 414

SN1980K, 394, 396, 401, 403–5, 412, 414

SN1983N, 395, 410

SN1987A, 403, 405–8, 410, 412, 440

blastwave expansion from, 408

SN1988Z, 394, 397, 404–5

SN1990B, 394–95

SN1993J, 401–3, 407–8

blastwave expansion from, 408

SN1998bw, 387–88, 405, 409–15, 428–29, 431

of radio supernovae, 409

SOFIA

See Stratospheric Observatory for Infrared Astronomy

Solar System mapping, 97

Solar wind drag, 94

Solid angles of X-ray cluster surveys, 552

Space density of X-ray clusters, 556–63

cluster abundance at higher redshifts and its evolution, 558–62

distant X-ray clusters, 562–63

latest view from *Chandra* satellite, 562–63

local cluster number density, 556–58

Space Infrared Telescope Facility (SIRTF), 28, 52, 58, 75, 96, 124

Infrared Array Camera, 54

Spatial clustering and the mass spectrum, 628–29

Spectra of Lyman-break galaxies, 600–1

Spectral distortion of the cosmic microwave background, 648

Spectral energy distributions

(SEDs), 30, 34, 619
colors, 44, 48
detectability, 612–13
first order signatures,
  503–10
flux densities, 44, 48
of hot cores, 35
information preserved
  since dark matter
  virialized, 499–503
information preserved
  since the main epoch of
  baryon dissipation,
  503–10
infrared, 448–49
intrinsic, 633
major processes involved
  in subsequent evolution,
  510–20
in prestellar cores, 31
second order signatures,
  510–20
zero order signatures,
  499–503
Spectral properties of Kuiper
  Belt objects, 81–83
Kuiper Belt V-R colors, 82
reflection spectrum of
  KBO 1999 DE$_9$, 83
Spectroscopic redshifts of
  dropouts, 594
Spherical harmonic analysis,
  174
Spherical interacting winds
  models, 459–60
Spherically symmetric
  over-densities in a
  low-density baryonic
  universe, 310
Spiral structure
  suggested, 55
  using UC HII regions, 54
Spiral waves, 505
Spitzer value, 337
Square Kilometer Array
  (SKA), 58
SSA

See Synchotron
  self-absorption
Stability
  of accretion disks, 374–77
Standard cosmological
  paradigm, 172–73
Standard GRB model
  developments and
  issues, 145–50
Standard model developments
  and issues, 145–50
  density, angle, and
  time-dependent injection,
  145–46
  differential time lags for
  the arrival of GRB pulses,
  150
  jets and limb-brightening
  effects, 146–47
  prompt flashes and reverse
  shocks, 147–48
  radiation processes,
  efficiencies, and pairs,
  148–49
  shock physics, 149–50
Star formation and the
  solar-stellar connection,
  245–49
  evolution to the main
  sequence, 247–48
  Herbig Ae/Be stars, 249
  low-mass stars on the main
  sequence, 248–49
  radio emission from
  low-mass young stellar
  objects, 245–47
Star formation density, 609
Star formation history (SFH),
  510–11
  predicted colors for various
  choices of, 606
Star formation rates, 154,
  603–10
  cosmic star formation,
  608–10
  of LBGs, 603–10
  predicted colors for various

choices of star formation
  history, 606
Starburst attenuation law,
  605–6
Starburst galaxies
  at high redshifts, 610–14
  spectra of Lyman-break
  galaxies, 600–1
  at ~10% of the Hubble
  time, 599–603
Starfish morphologies, 449
Stars at the interface between
  hot winds and coronae,
  250–51
Status of observations,
  650–58
  interferometric
  observations, 654–58
  single-dish observations,
  653–54
  sources of astronomical
  contamination and
  confusion, 651–53
Steady state theory, 8, 22
Steep density drop-off in the
  Kuiper Belt near 50 AU,
  77
Stellar age dating, 493–95
Stellar astrometry, 251–52
Stellar halo
  of the galaxy, 497–98
Stellar magnetic fields,
  233–34
Stellar mass and age, 615–18
  calibration diagram of
  oxygen abundance, 616
  of LBGs, 615–18
Stellar mass and total mass,
  627–28
Stellar origins of disks and
  tori, 473–77
  accretion disk formation in
  binary PNe progenitor,
  476
  common envelope
  evolution, 474–75
  effects of binary stars, 474

formation of accretion
disks, 475–77
single-star mechanisms for
producing tori, 473–74
Stellar population
pre-main sequence,
349–50, 362
Stellar radio astronomy,
217–61
quiescent emission from
coronae of cool stars,
226–38
radio coronal structure,
234–39
radio eclipses and
rotational modulation,
234–35
radio emission from brown
dwarfs, 251
radio emission from
chromospheres and
winds, 240–45
radio flares and coronal
heating, 228–33
radio flares from cool stars,
224–25
radio surveys and the radio
Hertzsprung-Russell
Diagram, 219
star formation and the
solar-stellar connection,
245–49
stars at the interface
between hot winds and
coronae, 250–51
stellar astrometry, 251–52
stellar magnetic fields,
233–34
theory of radio emission
from stellar atmospheres,
219–23
Stellar rotation, 239–40
activity cycles, 240
radio emission and stellar
rotation, 239–40
Stellar winds, 241
escaping along magnetic

fields, 237
Stokes parameters, 175–76,
189
Stratospheric Observatory for
Infrared Astronomy
(SOFIA), 28, 58
Strömgren theory, 34, 45, 52
Structure of the disk, 503
density distribution of Sa to
Sm galaxies, 504
relation between
components of the
velocity dispersion and
the stellar age, 506
Structure of the galaxy,
495–98
the bulge, 495–96
the dark halo, 498
the disk, 496–97
the stellar halo, 497–98
Structure of the Kuiper Belt,
65–70
classic KBOs, 65–67
resonant KBOs, 67–70
scattered KBOs, 70–71
Structures in phase space,
518–20
satellite in orbit about the
Milky Way, 519
Subaru Telescope Facility, 39,
96, 529
Submillimeter Array (SMA),
58
Submillimeter Common-User
Bolometer Array
(SCUBA) galaxies,
609–13, 630, 652
Substellar-mass objects
(SMOs), 126
Sunyaev-Zel'dovich effect
(SZE), 201–3, 331, 556,
572, 645–50
kinetic, 649
polarization of, 649–50
spectrum of Abell 2163,
655
thermal, 645–49

Sunyaev-Zel'dovich effect
(SZE) surveys
effects of gas evolution on
cluster survey yields, 674
interpreting, 672–75
Sunyaev-Zel'dovich Infrared
Experiment (SuZIE),
654–55, 670
Super-clusters
and Lyα forest clouds,
295–97
Super metal-rich (SMR) disk
stars, 505
Supernova remnants (SNRs),
432
Supernova (SN)
explosions, 157–58
and its shocks, 393
Type II, 406
Surface-brightness (SB)
limitation, 555
Surface density of Kuiper
Belt objects, 71–73
apparent red magnitude,
72
SuZIE
See Sunyaev-Zel'dovich
Infrared Experiment
Synchotron self-absorption
(SSA), 391–92, 412
Synchrotron emission
from hot stars, 243–45
from a power-law electron
distribution, 222–23
VLA map of the WR 147
system, 244
Synchrotron losses, 323
Synchrotron radiation,
320–23
radio halos, 320–22
radio relics, 322–23
Synchrotron self-absorbed
spectrum, 144, 148
Synthetic spectra
of isolated Jupiter-mass
planets, 125
SZE

See Sunyaev-Zel'dovich effect

T

Taiwan-America Occultation Survey (TAOS), 96
T_{eff}-T_{10} relation, 118–19
T_{eff}-T_{10}-g surface for calculating evolution of isolated giant planets, 119
trajectories of giant planets and brown dwarfs, 120
Temperature structure, 32–34
calculated average dust temperature, 33
Temperatures vs. age for EGPs, 128–30
TeV neutrino production by photomeson interactions, 162
TF
See Tully-Fisher law
Theoretical models and Hubble flows, 462–64
Theoretical models of GRBs, 431–34
multiple origins, 433–34
relativistic fireball, 432–33
relativistic jets and hypernovae, 433
Theoretical models of planetary nebulae, 456–77
basic models, 457–59
collimation, initial conditions, and challenges to MWBB models, 467
dynamos and the challenge to the magneto-centrifugal models, 469–70
magneto-centrifugal models, 467–69
MHD models of PN formation, 464–67
physics, initial conditions, and equations, 457–59
self-confinement, 464–67

stellar origins of disks and tori, 473–77
theories of nebular features, 470–73
Theories of gamma-ray bursts, 137–69
alternative models, 150
blast wave model of GRB afterglows, 142–45
cosmic rays, neutrinos, GeV-TeV photons, and gravity waves, 159–63
cosmological setting, galactic hosts and environment, 154–59
GRB phenomenology, 139–42
progenitors, 150–54
standard model developments and issues, 145–50
typical GRB lightcurve observed with BATSE, 138
Theories of giant planets, 103–36
atmospheres and their relation to the interior, 117–25
gravity field as a window to the interior, 112–16
high-pressure phase diagrams, 105–12
historical background, 103–5
orbital evolution of giant planets, 131–32
question of free oscillations, 116–17
thermal evolution of giant planets, 125–31
Theories of nebular features, 470–73
anase and FLIERS, 472–73
bipolar rotating emission line jets, 472
hydrodynamic jet

collimation mechanisms, 470–71
MHD models of jet formation, 471–72
Theories of radio emission from stellar atmospheres, 219–23
bremsstrahlung, 220–21
coherent emission, 223
elementary formulae for radiation and particles, 219–20
gyromagnetic emission, 221–23
loss times, 224
wind emission, 223–24
Thermal evolution of giant planets, 125–31
application of cooling theory to Jupiter and Saturn, 126–28
brown dwarf-like evolution, 126
"roasters," 128–31
Thermal plasma, 220
Thermal properties of EGPs based on heat flow measurements, 104–5
Thermal Sunyaev-Zel'dovich effect, 645–49
and the cosmic microwave background spectrum, 646
spectral distortion of the cosmic microwave background, 648
Thermal wind, 247
Thermodynamic equilibrium, 220
Thomson differential cross section, 188
Thomson scattering, 176, 188–89
rapid, 198
Time-dependent injection, 145–46
Timescales

and fossils, 491–92
of interest, 355
Timestream, 205–6
Tori
  single-star mechanisms for
    producing, 473–74
Total mass
  of KBOs, 74–75
  of LBGs, 633
Trajectories
  of giant planets and brown
    dwarfs, 120
Trans-Neptunian objects, 63,
    86–87
  See also Kuiper Belt
    objects
Transient spiral waves, 505
Trispectrum, 204
Tully-Fisher (TF) law, 265,
    267–68, 273, 285, 311,
    500, 503
  Cepheid-calibrated, 265,
    267–68, 273, 285
Turbulent cooling flows
  in molecular clouds, 29
2 micron all sky survey
  (2MASS), 54
Type Ib SN1983N, 395
Type Ic SN1990B, 395
Type II SN1979C, 396
Type II SN1980K, 396
Type IIb SN1993J, 402

U

UC HII regions, 41–50
  the ionized gas, 42–46
  the molecular environment,
    47–50
  warm dust cocoons, 47
Uhuru X-ray satellite, 548
ULIRGs
  See Ultra Luminous
    Infrared Galaxies
Ultra-compact (UC) HII
    regions and massive star
    formation, 27–62
  evolution to the

ultracompact state, 28–41
galactic population and
    distribution, 50–57
research needed, 57–58
Ultra Luminous Infrared
  Galaxies (ULIRGs),
    604, 612, 630
Ulysses, 423, 425
UMa spirals
  inferred mass-to-light
    ratios for, 283
Universe
  accelerated expansion of,
    308
  energy density of, 177
  expansion of, 173
Unruh radiation, 303–4
Upper self-adjoint (USA)
    series
  of Riemann ellipsoids, 369
Ursa Major cluster
  spiral galaxies in, 266, 268
USA
  See Upper self-adjoint
    series
UV protons
  dust destruction by, 156
UV-selected star forming
    galaxies at high redshift,
    597–99
  Lyα galaxies, 598–99
  Lyman-break galaxies at
    $z > 4$, 597–98
UVES instrument, 529

V

V-R color histograms, 78
Van Allen belts, 237, 250
Van der Waals bonding,
    120
Varuna, 75, 78, 84–85
Vela satellites, 137, 415
VERITAS, 162
Very Large Array (VLA),
    281, 329, 337, 388
  map of the WR 147
    system, 244

view of stellar radio
    emission, 218, 419, 421
Very Large Baseline Array
    (VLBA), 235
  image of the dMe star UV
    Cet, 236–37
Very long baseline
    interferometry (VLBI)
    studies, 49, 228, 235–37,
    388, 397, 407
  VLBA image of the dMe
    star UV Cet, 236–37
Virial equilibrium, 360
VLA
  See Very Large Array
VLBA
  See Very Large Baseline
    Array
VLBI
  See Very long baseline
    interferometry studies
VLT, 529

W

Warm dust cocoons, 47
WCD
  See Wind Compressed
    Disk mechanism
Weak-lined T Tau (WTT)
    phase, 245–46, 248
Westerbork Radio Synthesis
    Telescope (WRST), 281,
    321
  radio image of the Coma
    cluster region, 321
Weyl conformal gravity, 305
WFPC2 four-band
    photometric system,
    592, 597, 608
Wide Field X-ray Telescope
    (WFXT), 572
Wien tail, 202
Wind-blown bubbles
  classes of, 460
  See also Magnetized Wind
    Blown Bubble
Wind Compressed Disk

(WCD) mechanism, 473–74
Wind emission, 223–24
Wind-shedding stars, 219
Wind-shock zones, 243
Wind speeds
  Jupiter zonal, 115
Winds from cool stars, 241–43
  Betelgeuse, 242
Wolf-Rayet (WR) stars, 219, 243–44, 434
Working model of galaxy formation, 489–91
  look-back time as a function of redshift for different world models, 490
WR
  See Wolf-Rayet stars
WRST
  See Westerbork Radio Synthesis Telescope
WTT
  See Weak-lined T Tau phase

X
X-ray Brightest Abell-like Clusters (XBACS), 553
X-ray cluster surveys, 548–51
X-ray luminosity function (XLF), 549, 557–59, 568
  of distant clusters, 561
X-ray properties of clusters, 542–45
  galaxy velocity dispersions, 542
X-ray surveys
  strategies and selection functions for, 551–56
X-ray temperature function (XTF), 567–68, 570
X-wind theory, 39
XBACS
  See X-ray Brightest Abell-like Clusters
XLF
  See Local X-ray luminosity function of clusters; X-ray luminosity function
XMM

See Newton-XMM observations
XTF
  See X-ray temperature function

Y
YSOs, 440, 447, 470

Z
Zeeman measurements, 248
Zero-age main-sequence (ZAMS) stage, 248–49, 252, 390
Zero order signatures, 499–503
  of the CDM hierarchy, 501–3
  of the environment, 499
  of global quantities, 500
  of the internal distribution of specific angular momentum, 500–1
000301C, 427
Zonal harmonics, 113–14
Zonal winds, 114–15

CUMULATIVE INDEXES

CONTRIBUTING AUTHORS, VOLUMES 29–40

A

Alexander DR, 35:137–77
Allard F, 35:137–77
Aller LH, 33:1–17
Angel JR, 36:507–37
Antonucci R, 31:473–521
Arendt RG, 30:11–50
Arnett D, 33:115–32
Aschwanden MJ, 39:175–210

B

Bachiller R, 34:111–54
Bailyn CD, 33:133–62
Balick B, 40:439–86
Bally J, 39:403–55
Baraffe I, 38:337–77
Barcons X, 30:429–56
Barkana R, 39:19–66
Barnes JE, 30:705–42
Basri G, 38:485–519
Bastian TS, 36:131–88
Beck R, 34:153–204
Beckers JM, 31:13–62
Benz AO, 36:131–88
Bertelli G, 30:235–85
Bertschinger E,
    36:599–654
Bessell MS, 31:433–71
Bignami GF, 34:331–81
Binney J, 30:51–74
Black DC, 33:359–80
Blake GA, 36:317–68
Blanco VM, 39:xiv, 1–18
Blandford RD, 30:311–58
Bland-Hawthorn J,
    40:487–537
Bodenheimer P,
    33:199–238
Bolte M, 34:461–510

Borgani S, 40:539–77
Bothun G, 35:267–307
Bowyer S, 29:59–88;
    38:231–88
Bradt HVD, 30:391–427
Branch D, 30:359–89;
    36:17–55
Brandenburg A, 34:153–204
Bressan A, 30:235–85
Brown TM, 32:37–82
Burbidge EM, 32:1–36
Burrows A, 40:103–36
Busso M, 37:239–309
Butler RP, 36:57–97

C

Caldeira K, 32:83–114
Cameron AGW, 37:1–36
Caraveo PA, 34:331–81
Carilli CL, 40:319–48
Carlstrom JE, 40:643–80
Carr B, 32:531–90
Carroll SM, 30:499–542
Cesarsky CJ, 38:761–814
Chabrier G, 38:337–77
Chanmugam G, 30:143–84
Charnley SB, 38:427–83
Chiosi C, 30:235–85
Churchwell E, 40:27–62
Colavita MM, 30:457–98
Combes F, 29:195–237
Condon JJ, 30:575–611
Conti PS, 32:227–75
Cowan JJ, 29:447–97
Cowley AP, 30:287–310

D

Davidson K, 35:1–32
Dekel A, 32:371–418

de Vegt C, 37:97–125
de Zeeuw T, 29:239–74
Dickinson M, 38:667–715
Dodelson S, 40:171–216
Done C, 31:716–61
Draine BT, 31:373–432
Drake JJ, 38:231–88
Dultzin-Hacyan D,
    38:521–71
Duncan MJ, 31:265–95
Dwek E, 30:11–50;
    39:249–307

E

Eggen OJ, 31:1–11
Ehrenfreund P, 38:427–83
Elitzur M, 30:75–112
Ellis RS, 35:389–443
Evans NJ II, 37:311–62

F

Fabian AC, 30:429–56;
    32:277–318
Falcke H, 39:309–52
Feigelson ED,
    37:363–408
Ferguson HC, 38:667–715
Ferland G, 37:487–531
Ferrari A, 36:539–98
Fich M, 29:409–45
Filippenko AV, 35:309–55
Fishman GJ, 33:415–58
Fowler WA, 30:1–9
Frank A, 40:439–86
Franklin FA, 39:581–631
Franx M, 29:239–74;
    35:637–75
Freeman K, 40:487–537
Friel ED, 33:381–414

G

Gallino R, 37:239–309
Gary DE, 36:131–88
Gautschy A, 33:75–113;
   34:551–606
Genzel R, 38:761–814
Giavalisco M, 40:579–641
Gilliland RL, 32:37–82
Gilmore G, 35:637–75
Giovanelli R, 29:499–541
Glassgold AE, 34:241–78
Goldstein ML, 33:283–325
Gosling JT, 34:35–73
Gough D, 29:627–84
Güdel M, 40:217–61

H

Haisch B, 29:275–324
Hamann F, 37:487–531
Harris WE, 29:543–79
Hartmann L, 34:205–39
Hauschildt PH, 35:137–77
Hauser MG, 39:249–307
Haxton WC, 33:459–503
Haynes MP, 29:499–541;
   32:115–52
Henry RC, 29:89–127
Herbig GH, 33:19–73
Hernquist L, 30:705–42
Hickson P, 35:357–88
Hillebrandt W, 38:191–230
Holder GP, 40:643–80
Hollenbach DJ, 35:179–215
Holman MJ, 39:581–631
Horányi M, 34:383–418
Howard RF, 34:75–109
Hu W, 40:171–216
Hubbard WB, 40:103–36
Hudson H, 33:239–82
Humphreys RM, 35:1–32

I

Impey C, 35:267–307

J

Jewitt DC, 40:63–101
Johnston KJ, 37:97–125

K

Kahabka P, 35:69–100
Kahler SW, 30:113–41
Kellermann KI, 39:457–509
Kennicutt RC Jr, 36:189–231
Kenyon SJ, 34:205–39
Knapp GR, 36:369–433
Koo DC, 30:613–52
Kormendy J, 33:581–624
Kouveliotou C, 38:379–425
Kovalevsky J, 36:99–129
Kron RG, 30:613–52
Kudritzki RP, 38:613–66
Kulkarni SR, 32:591–639
Kulsrud RM, 37:37–64
Kwok S, 31:63–92

L

Lawrence CR, 30:653–703
Lean J, 35:33–67
Lebreton Y, 38:35–77
Lecar M, 39:581–631
Leibundgut B, 39:67–98
Lin DNC, 33:505–40;
   34:703–47
Lissauer JJ, 31:129–74
Loeb A, 39:19–66
Lunine JI, 31:217–63;
   40:103–36
Luu JX, 40:63–101

M

Maeder A, 32:227–75;
   38:143–90
Majewski SR, 31:575–638
Maraschi L, 35:445–502
Marcus PS, 31:523–73
Marcy GW, 36:57–97
Marziani P, 38:521–71
Mateo M, 34:511–50
Mateo ML, 36:435–506
Mathieu RD, 32:465–530
Matthaeus WH, 33:283–325
McCarthy PJ, 31:639–88
McCray R, 31:175–216
McGaugh SS, 40:263–317
McKee CF, 31:373–432

McWilliam A, 35:503–56
Meegan CA, 33:415–58
Melia F, 39:309–52
Mellier Y, 37:127–89
Melrose DB, 29:31–57
Mendis DA, 32:419–63
Mészáros P, 40:137–69
Meyer BS, 32:153–90
Meynet G, 38:143–90
Mikkola S, 29:9–29
Mirabel IF, 34:749–92;
   37:409–43
Monaghan JJ, 30:543–74
Montes MJ, 40:387–438
Montmerle T, 37:363–408
Moran JM, 39:457–509
Morris M, 34:645–701
Moss D, 34:153–204
Mulchaey JS, 38:289–335
Murray NW, 39:581–631
Mushotzky RF, 31:717–61

N

Narayan R, 30:311–58
Narlikar JV, 29:325–62;
   39:211–48
Niemeyer JC, 38:191–230
Norman C, 40:539–77

O

O'Connell RW, 37:603–48
O'Dell CR, 39:99–136
Ohashi T, 30:391–427
Oke JB, 38:79–111
Olszewski EW, 34:511–50
Osterbrock DE, 38:1–33
Ostriker JP, 31:689–716

P

Paczynski B, 34:419–59
Padmanabhan T, 29:325–62;
   39:211–48
Panagia N, 40:387–438
Papaloizou JCB, 33:505–40;
   34:703–47
Peale SJ, 37:533–602
Phinney ES, 32:591–639

Pinsonneault M, 35:557–605
Poland AI, 39:175–210
Pounds KA, 30:391–427;
31:717–61
Press WH, 30:499–542
Puls J, 38:613–66

Q
Quinn T, 31:265–95
Quirrenbach A, 39:353–401

R
Rabin DM, 39:175–210
Rampino MR, 32:83–114
Rana NC, 29:129–62
Rauch M, 36:267–316
Readhead ACS, 30:653–703
Reese ED, 40:643–80
Reid IN, 37:191–237
Reid MJ, 31:345–72
Reipurth B, 39:403–55
Rephaeli Y, 33:541–79
Richstone D, 33:581–624
Roberts DA, 33:283–325
Roberts MS, 32:115–52
Robinson B, 37:65–96
Rodonò M, 29:275–324
Rodriguez LF, 37:409–43
Rood RT, 32:191–226
Rosati P, 40:539–77
Rosenberg M, 32:419–63
Rossi B, 29:1–8
Rubin V, 39:137–74
Ryan J, 33:239–82

S
Saio H, 33:75–113;
34:551–606
Salpeter EE, 40:1–25
Sandage A, 37:445–86
Sanders DB, 34:749–92
Sanders RH, 40:263–317
Sandquist EL, 38:113–41

Sargent AI, 31:297–343
Savage BD, 34:279–329
Schatzman E, 34:1–34
Scott D, 32:319–70
Scoville NZ, 29:581–625
Sembach KR, 34:279–329
Serabyn E, 34:645–701
Shao M, 30:457–98
Shibazaki N, 34:607–44
Shukurov A, 34:153–204
Silk J, 32:319–70
Sofue Y, 39:137–74
Sokoloff D, 34:153–204
Sramek RA, 40:387–438
Starrfield S, 35:137–77
Stern SA, 30:185–233
Stetson PB, 34:461–510
Stevenson DJ, 29:163–93
Stringfellow GS, 31:433–71
Strong KT, 29:275–324
Sulentic JW, 38:521–71
Suntzeff NB, 34:511–50

T
Taam RE, 38:113–41
Tammann GA, 29:363–407;
30:359–89
Tanaka Y, 34:607–44
Tarter J, 39:511–48
Taylor GB, 40:319–48
Teerikorpi P, 35:101–36
Thielemann F-K, 29:447–97
Tielens AGGM, 35:179–215
Tohline JE, 40:349–85
Toomre J, 29:627–84
Townes CH, 35:xiii–xliv
Tremaine S, 29:409–45
Truran JW, 29:447–97
Turner EL, 30:499–542

U
Ulrich M-H, 35:445–502
Urry CM, 35:445–502

V
Valtonen M, 29:9–29
van de Hulst HC, 36:1–16
VandenBerg DA,
34:461–510
van den Bergh S,
29:363–407
van den Heuvel EPJ,
35:69–100
van der Klis M, 38:717–60
van Dishoeck EF,
36:317–68
van Paradijs J, 38:379–425
van Woerden H, 35:217–66
Vennes S, 38:231–88
Verbunt F, 31:93–127

W
Waelkens C, 36:233–66
Wagner SJ, 33:163–97
Wakker BP, 35:217–66
Wallerstein G, 36:369–433;
38:79–111
Wasserburg GJ, 37:239–309
Waters LBFM, 36:233–66
Weiler KW, 40:387–438
Weissman PR, 33:327–57
Welch WJ, 31:297–343
White M, 32:319–70
Wijers RAMJ, 38:379–425
Williams R, 38:667–715
Willson LA, 38:573–611
Wilson TL, 32:191–226
Witzel A, 33:163–97
Woolf N, 36:507–37
Wyse RFG, 35:637–75

Y
Young JS, 29:581–625

Z
Zensus JA, 35:607–36
Zuckerman B, 39:549–80

CHAPTER TITLES, VOLUMES 29–40

Prefatory Chapters

The Interplanetary Plasma	B Rossi	29:1–8
From Steam to Stars to the Early Universe	WA Fowler	30:1–9
Notes from a Life in the Dark	OJ Eggen	31:1–11
Watcher of the Skies	EM Burbidge	32:1–36
An Astronomical Rescue	LH Aller	33:1–17
The Desire to Understand the World	E Schatzman	34:1–34
A Physicist Courts Astronomy	CH Townes	35:xiii–xliv
Roaming Through Astrophysics	HC van de Hulst	36:1–16
Adventures in Cosmogony	AGW Cameron	37:1–36
A Fortunate Life in Astronomy	DE Osterbrock	38:1–33
Telescopes, Red Stars, and Chilean Skies	VM Blanco	39:1–18
A Generalist Looks Back	EE Salpeter	40:1–25

Solar System Astrophysics

Radioactive Dating of the Elements	JJ Cowan, F-K Thielemann, JW Truran	29:447–97
The Pluto-Charon System	SA Stern	30:185–233
Planet Formation	JJ Lissauer	31:129–74
The Atmospheres of Uranus and Neptune	JI Lunine	31:217–63
The Long-Term Dynamical Evolution of the Solar System	MJ Duncan, T Quinn	31:265–95
Jupiter's Great Red Spot and Other Vortices	PS Marcus	31:523–73
The Goldilocks Problem: Climatic Evolution and Long-Term Habitability of Terrestrial Planets	MR Rampino, K Caldeira	32:83–114
Cosmic Dusty Plasmas	DA Mendis, M Rosenberg	32:419–63
The Kuiper Belt	PR Weissman	33:327–57
Corotating and Transient Solar Wind Flows in Three Dimensions	JT Gosling	34:35–73
Charged Dust Dynamics in the Solar System	M Horányi	34:383–418
The Sun's Variable Radiation and Its Relevance for Earth	J Lean	35:33–67
Origin and Evolution of the Natural Satellites	SJ Peale	37:533–602

Organic Molecules in the Interstellar
   Medium, Comets, and Meteorites: A
   Voyage from Dark Clouds to the Early
   Earth   P Ehrenfreund,   38:427–83
           SB Charnley
Chaos in the Solar System   M Lecar, FA Franklin,   39:581–631
           MJ Holman,
           NW Murray
Kuiper Belt Objects: Relics from the
   Accretion Disk of the Sun   JX Luu, D Jewitt   40:63–101
Theory of Giant Planets   WB Hubbard,   40:103–36
           A Burrows,
           JI Lunine

Solar Physics

Flares on the Sun and Other Stars   B Haisch, KT Strong,   29:275–324
           M Rodonó
Seismic Observations of the Solar Interior   D Gough, J Toomre   29:627–84
Solar Flares and Coronal Mass Ejections   SW Kahler   30:113–41
High-Energy Particles in Solar Flares   H Hudson, J Ryan   33:239–82
Magnetohydrodynamic Turbulence in the
   Solar Wind   ML Goldstein,   33:283–325
           DA Roberts,
           WH Matthaeus
The Solar Neutrino Problem   WC Haxton   33:459–503
Solar Active Regions as Diagnostics of
   Subsurface Conditions   RF Howard   34:75–109
The Sun's Variable Radiation and Its
   Relevance for Earth   J Lean   35:33–67
Radio Emission from Solar Flares   TS Bastian, AO Benz,   36:131–88
           DE Gary
The New Solar Corona   MJ Aschwanden,   39:175–210
           AI Poland,
           DM Rabin

Stellar Physics

The Search for Brown Dwarfs   DJ Stevenson   29:163–93
Flares on the Sun and Other Stars   B Haisch, KT Strong,   29:275–324
           M Rodonó
Galactic and Extragalactic Supernova Rates   S van den Bergh,   29:363–407
           GA Tammann
Radioactive Dating of the Elements   JJ Cowan,   29:447–97
           F-K Thielemann,
           JW Truran
Globular Cluster Systems in Galaxies
   Beyond the Local Group   WE Harris   29:543–79

Magnetic Fields of Degenerate Stars	G Chanmugam	30:143–84
New Developments in Understanding the HR Diagram	C Chiosi, G Bertelli, A Bressan	30:235–85
Evidence for Black Holes in Stellar Binary Systems	AP Cowley	30:287–310
Type Ia Supernovae as Standard Candles	D Branch, GA Tammann	30:359–89
Proto-Planetary Nebulae	S Kwok	31:63–92
Origin and Evolution of X-Ray Binaries and Binary Radio Pulsars	F Verbunt	31:93–127
Supernova 1987A Revisited	R McCray	31:175–216
Millimeter and Submillimeter Interferometry of Astronomical Sources	AI Sargent, WJ Welch	31:297–343
The Faint End of the Stellar Luminosity Function	MS Bessell, GS Stringfellow	31:433–71
Asteroseismology	TM Brown, RL Gilliland	32:37–82
The r-, s-, and p-Processes in Nucleosynthesis	BS Meyer	32:153–90
Massive Star Populations in Nearby Galaxies	A Maeder, PS Conti	32:227–75
Pre-Main-Sequence Binary Stars	RD Mathieu	32:465–530
Binary and Millisecond Pulsars	ES Phinney, SR Kulkarni	32:591–639
Stellar Pulsations Across the HR Diagram: Part 1	A Gautschy, H Saio	33:75–113
Explosive Nucleosynthesis Revisited: Yields	D Arnett	33:115–32
Blue Stragglers and Other Stellar Anomalies: Implications for the Dynamics of Globular Clusters	CD Bailyn	33:133–62
Angular Momentum Evolution of Young Stars and Disks	P Bodenheimer	33:199–238
The Old Open Clusters of the Milky Way	ED Friel	33:381–414
The Solar Neutrino Problem	WC Haxton	33:459–503
Bipolar Molecular Outflows from Young Stars and Protostars	R Bachiller	34:111–54
The FU Orionis Phenomenon	L Hartmann, SJ Kenyon	34:205–39
Circumstellar Photochemistry	AE Glassgold	34:241–78
Geminga, Its Phenomenology, Its Fraternity, and Its Physics	GF Bignami, PA Caraveo	34:331–81
The Age of the Galactic Globular Cluster System	DA VandenBerg, M Bolte, PB Stetson	34:461–510

Old and Intermediate-Age Stellar
   Populations in the Magellanic Cloud   EW Olszewski,   34:511–50
                                            NB Suntzeff,
                                            M Mateo

Stellar Pulsations Across the HR Diagram:
   Part 2   A Gautschy, H Saio   34:551–606
X-Ray Novae   Y Tanaka, N Shibazaki   34:607–44
Eta Carina and Its Environment   K Davidson,   35:1–32
                                              RM Humphreys
Luminous Supersoft X-Ray Sources   P Kahabka,   35:69–100
                                             EPJ van den Heuvel

Model Atmospheres of Very Low Mass Stars
   and Brown Dwarfs   PH Hauschildt,   35:137–77
                                             S Starrfield,
                                             F Allard,
                                             DR Alexander

Optical Spectra of Supernovae   AV Filippenko   35:309–55
Abundance Ratios and Galactic Chemical
   Evolution   A McWilliam   35:503–56
Mixing in Stars   M Pinsonneault   35:557–605
Type Ia Supernovae and the Hubble Constant   D Branch   36:17–55
Detection of Extrasolar Giant Planets   GW Marcy, RP Butler   36:57–97
Herbig Ae/Be Stars   C Waelkens,   36:233–66
                                             LBFM Waters
Carbon Stars   G Wallerstein,   36:369–433
                                             GR Knapp

Astronomical Searches for Earth-Like
   Planets and Signs of Life   N Woolf, JR Angel   36:507–37
The HR Diagram and the Galactic Distance
   Scale After Hipparcos   IN Reid   37:191–237
Nucleosynthesis in Asymptomatic Giant
   Branch Stars: Relevance for Galactic
   Enrichment and Solar System Formation   M Busso, R Gallino,   37:239–309
                                             GJ Wasserburg

High-Energy Processes in Young Stellar
   Objects   ED Feigelson,   37:363–408
                                             T Montmerle

Stellar Structure and Evolution: Deductions
   from Hipparcos   Y Lebreton   38:35–77
Common Envelope Evolution of Massive
   Binary Stars   RE Taam,   38:113–41
                                             EL Sandquist
The Evolution of the Rotating Stars   A Maeder, G Meynet   38:143–90
Type Ia Supernova Explosion Models   W Hillebrandt,   38:191–230
                                             JC Niemeyer

Theory of Low-Mass Stars and Substellar
   Objects   G Chabrier, I Baraffe   38:337–77

Observations of Brown Dwarfs	G Basri	38:485–519
Mass Loss from Cool Stars: Impact on the Evolution of Stars and Stellar Populations	LA Willson	38:573–611
Winds from Hot Stars	R-P Kudritzki, J Puls	38:613–66
Millisecond Oscillations in X-Ray Binaries	M van der Klis	38:717–60
The Orion Nebula and Its Associated Population	CR O'Dell	39:99–136
Herbig-Haro Flows: Probes of Early Stellar Evolution	B Reipurth, J Bally	39:403–55
Dusty Circumstellar Disks	B Zuckerman	39:549–80
Stellar Radio Astronomy: Probing Stellar Atmospheres from Protostars to Giants	M Güdel	40:217–61
The Origin of Binary Stars	JE Tohline	40:349–85
Radio Emission from Supernovae and Gamma-Ray Bursters	KW Weiler, N Panagia, MJ Montes, RA Sramek	40:387–438

Dynamical Astronomy

The Few-Body Problem in Astrophysics	M Valtonen, S Mikkola	29:9–29
Evidence for Black Holes in Stellar Binary Systems	AP Cowley	30:287–310
Origin and Evolution of X-Ray Binaries and Binary Radio Pulsars	F Verbunt	31:93–127
Binary and Millisecond Pulsars	ES Phinney, SR Kulkarni	32:591–639
Theory of Accretion Disks I: Angular Momentum Transport Processes	JCB Papaloizou, DNC Lin	33:505–40
Geminga, Its Phenomenology, Its Fraternity, and Its Physics	GF Bignami, PA Caraveo	34:331–81
Theory of Accretion Disks II: Application to Observed Systems	DNC Lin, JCB Papaloizou	34:703–47
First Results from Hipparcos	J Kovalevsky	36:99–129
Chaos in the Solar System	M Lecar, FA Franklin, MJ Holman, NW Murray	39:581–631

Interstellar Medium

Collective Plasma Radiation Processes	DB Melrose	29:31–57
Distribution of CO in the Milky Way	F Combes	29:195–237
Molecular Gas in Galaxies	JS Young, NZ Scoville	29:581–625
Dust-Gas Interactions and the Infrared Emission from Hot Astrophysical Plasmas	E Dwek, RG Arendt	30:11–50

Astronomical Masers	M Elitzur	30:75–112
Proto-Planetary Nebulae	S Kwok	31:63–92
Millimeter and Submillimeter Interferometry of Astronomical Sources	AI Sargent, WJ Welch	31:297–343
Theory of Interstellar Shocks	BT Draine, CF McKee	31:373–432
Abundances in the Interstellar Medium	TL Wilson, RT Rood	32:191–226
The Diffuse Interstellar Bands	GH Herbig	33:19–73
Angular Momentum Evolution of Young Stars and Disks	P Bodenheimer	33:199–238
Bipolar Molecular Outflows from Young Stars and Protostars	R Bachiller	34:111–54
Circumstellar Photochemistry	AE Glassgold	34:241–78
Insterstellar Abundances from Absorption-Line Observations with the Hubble Space Telescope	BD Savage, KR Sembach	34:279–329
Eta Carina and Its Environment	K Davidson, RM Humphreys	35:1–32
Dense Photodissociation Regions (PDRs)	DJ Hollenbach, AGGM Tielens	35:179–215
High-Velocity Clouds	BP Wakker, H van Woerden	35:217–66
Chemical Evolution of Star-Forming Regions	EF van Dishoeck, GA Blake	36:317–68
Reference Frames in Astronomy	KJ Johnston, C de Vegt	37:97–125
Physical Conditions in Regions of Star Formation	NJ Evans II	37:311–62
Organic Molecules in the Interstellar Medium, Comets, and Meteorites: A Voyage from Dark Clouds to the Early Earth	P Ehrenfreund, SB Charnley	38:427–83
Herbig-Haro Flows: Probes of Early Stellar Evolution	B Reipurth, J Bally	39:403–55
Dusty Circumstellar Disks	B Zuckerman	39:549–80
Ultra-Compact HII Regions and Massive Star Formation	E Churchwell	40:27–62
Shapes and Shaping of Planetary Nebulae	B Balick, A Frank	40:439–86

The Galaxy

Chemical Evolution of the Galaxy	NC Rana	29:129–62
Distribution of CO in the Milky Way	F Combes	29:195–237
The Mass of the Galaxy	M Fich, S Tremaine	29:409–45
The Distance to the Center of the Galaxy	MJ Reid	31:345–72

The Faint End of the Stellar Luminosity
  Function | MS Bessell, | 31:433–71
  | GS Stringfellow |
Galactic Structure Surveys and the Evolution
  of the Milky Way | SR Majewski | 31:575–638
Blue Stragglers and Other Stellar
  Anomalies: Implications for the
  Dynamics of Globular Clusters | CD Bailyn | 33:133–62
The Old Open Clusters of the Milky Way | ED Friel | 33:381–414
Galactic Magnetism: Recent Developments
  and Perspectives | R Beck, | 34:153–204
  | A Brandenburg, |
  | D Moss, |
  | A Shukurov, |
  | D Sokoloff |
Gravitational Microlensing in the Local
  Group | B Paczynski | 34:419–59
The Age of the Galactic Globular Cluster
  System | DA VandenBerg, | 34:461–510
  | M Bolte, |
  | PB Stetson |
The Galactic Center Environment | M Morris, E Serabyn | 34:645–701
High-Velocity Clouds | BP Wakker, | 35:217–66
  | H van Woerden |
First Results from Hipparcos | J Kovalevsky | 36:99–129
A Critical Review of Galactic Dynamos | RM Kulsrud | 37:37–64
Sources of Relativistic Jets in the Galaxy | IF Mirabel, | 37:409–43
  | LF Rodriguez |
The Supermassive Black Hole at the
  Galactic Center | F Melia, H Falcke | 39:309–52
The New Galaxy: Signatures of its
  Formation | K Freeman, | 40:487–537
  | J Bland-Hawthorn |

Extragalactic Astronomy and Cosmology

The Cosmic Far Ultraviolet Background | S Bowyer | 29:59–88
Ultraviolet Background Radiation | RC Henry | 29:89–127
Structure and Dynamics of Elliptical
  Galaxies | T de Zeeuw, M Franx | 29:239–74
Inflation for Astronomers | JV Narlikar, | 29:325–62
  | T Padmanabhan |
Galactic and Extragalactic Supernova Rates | S van den Bergh, | 29:363–407
  | GA Tammann |
Radioactive Dating of the Elements | JJ Cowan, | 29:447–97
  | F-K Thielemann, |
  | JW Truran |

Redshift Surveys of Galaxies | R Giovanelli, MP Haynes | 29:499–541

Globular Cluster Systems in Galaxies Beyond the Local Group | WE Harris | 29:543–79
Molecular Gas in Galaxies | JS Young, NZ Scoville | 29:581–625
Warps | J Binney | 30:51–74
Astronomical Masers | M Elitzur | 30:75–112
Cosmologial Applications of Gravitational Lensing | RD Blandford, R Narayan | 30:311–58

Type Ia Supernovae as Standard Candles | D Branch, GA Tammann | 30:359–89

The Origin of the X-Ray Background | AC Fabian, X Barcons | 30:429–56
The Cosmological Constant | SM Carroll, WH Press, EL Turner | 30:499–542

Smoothed Particle Hydrodynamics | JJ Monaghan | 30:543–74
Radio Emission From Normal Galaxies | JJ Condon | 30:575–611
Evidence for Evolution in Faint Field Galaxy Samples | DC Koo, RG Kron | 30:613–52
Observations of the Isotropy of the Cosmic Microwave Background Radiation | ACS Readhead, CR Lawrence | 30:653–703

Dynamics of Interacting Galaxies | JE Barnes, L Hernquist | 30:705–42
Supernova 1987A Revisited | R McCray | 31:175–216
The Faint End of the Stellar Luminosity Function | MS Bessell, GS Stringfellow | 31:433–71

Unified Models for Active Galactic Nuclei and Quasars | R Antonucci | 31:473–521
High Redshift Radio Galaxies | PJ McCarthy | 31:639–88
Astronomical Tests of the Cold Dark Matter Scenario | JP Ostriker | 31:689–716
X-Ray Spectra and Time Variability of Active Galactic Nuclei | RF Mushotzky, C Done, KA Pounds | 31:717–61

Physical Parameters along the Hubble Sequence | MS Roberts, MP Haynes | 32:115–52

Massive Star Populations in Nearby Galaxies | A Maeder, PS Conti | 32:227–75
Cooling Flows in Clusters of Galaxies | AC Fabian | 32:277–318
Anisotropies in the Cosmic Microwave Background | M White, D Scott, J Silk | 32:319–70

Dynamics of Cosmic Flows | A Dekel | 32:371–418
Baryonic Dark Matter | B Carr | 32:531–90
Intraday Variablity in Quasars and BL Lac Objects | SJ Wagner, A Witzel | 33:163–97

Gamma-Ray Bursts	GJ Fishman, CA Meegan	33:415–58

Comptonization of the Cosmic Microwave
   Background: The Sunyaev-Zeldovich
   Effect
	Y Rephaeli	33:541–79

Inward Bound: The Search for Supermassive Black Holes in Galactic Nuclei	J Kormendy, D Richstone	33:581–624
Galactic Magnetism: Recent Developments and Perspectives	R Beck, A Brandenburg, D Moss, A Shukurov, D Sokoloff	34:153–204
Gravitational Microlensing in the Local Group	B Paczynski	34:419–59
Old and Intermediate-Age Stellar Populations in the Magellanic Cloud	EW Olszewski, NB Suntzeff, M Mateo	34:511–50
Luminous Infrared Galaxies	DB Sanders, IF Mirabel	34:749–92

Observational Selection Bias Affecting the
   Determination of the Extragalactic
   Distance Scale
	P Teerikorpi	35:101–36

Low Surface Brightness Galaxies	C Impey, G Bothun	35:267–307
Optical Spectra of Supernovae	AV Filippenko	35:309–55
Compact Groups of Galaxies	P Hickson	35:357–88
Faint Blue Galaxies	RS Ellis	35:389–443
Variability of Active Galactic Nuclei	M-H Ulrich, L Maraschi, CM Urry	35:445–502
Parsec-Scale Jets in Extragalactic Radio Sources	JA Zensus	35:607–36
Galactic Bulges	RF Wyse, G Gilmore, M Franx	35:637–75
Type Ia Supernovae and the Hubble Constant	D Branch	36:17–55
Star Formation in Galaxies Along the Hubble Sequence	RC Kennicutt Jr.	36:189–231
The Lyman Alpha Forest in the Spectra of Quasistellar Objects	M Rauch	36:267–316
Dwarf Galaxies of the Local Group	ML Mateo	36:435–506
Modeling Extragalactic Jets	A Ferrari	36:539–98
Simulations of Structure Formation in the Universe	E Bertschinger	36:599–654
Probing the Universe with Weak Lensing	Y Mellier	37:127–89

Element Abundances in Quasistellar Objects: Star Formation and Galactic Nuclear Evolution at High Redshifts	F Hamann, G Ferland	37:487–531
Far-Ultraviolet Radiation from Elliptical Galaxies	RW O'Connell	37:603–48
Extreme Ultraviolet Astronomy	S Bowyer, JJ Drake, S Vennes	38:231–88
X-ray Properties of Groups of Galaxies	JS Mulchaey	38:289–335
Gamma-Ray Burst Afterglows	J van Paradijs, C Kouveliotou, RAMJ Wijers	38:379–425
Phenomenology of Broad Emission Lines in Active Galactic Nuclei	JW Sulentic, P Marziani, D Dultzin-Hacyan	38:521–71
The Hubble Deep Fields	HC Ferguson, M Dickinson, R Williams	38:667–715
Extragalactic Results from the Infrared Space Observatory	R Genzel, CJ Cesarsky	38:761–814
The Reionization of the Universe by the First Stars and Quasars	A Loeb, R Barkana	39:19–66
Cosmological Implications from Observations of Type IA Supernovae	B Leibundgut	39:67–98
Rotation Curves of Spiral Galaxies	Y Sofue, V Rubin	39:137–74
Standard Cosmology and Alternatives: A Critical Appraisal	JV Narlikar, T Padmanabhan	39:211–48
The Cosmic Infrared Background: Measurements and Implications	MG Hauser, E Dwek	39:249–307
Theories of Gamma-Ray Bursts	P Mészáros	40:137–69
Cosmic Microwave Background Anisotropies	W Hu, S Dodelson	40:171–216
Modified Newtonian Dynamics as an Alternative to Dark Matter	R Sanders, Stacy McGaugh	40:263–317
Cluster Magnetic Fields	GB Taylor, Chris Carilli	40:319–48
Radio Emission from Supernovae and Gamma-Ray Bursters	KW Weiler, N Panagia, MJ Montes, RA Sramek	40:387–438
The Evolution of X-ray Clusters of Galaxies	P Rosati, S Borgani, C Norman	40:539–77
Lyman Break Galaxies	M Giavalisco	40:579–641

| Cosmology with the Sunyaev-Zeldovich Effect | JE Carlstrom, GP Holder, ED Reese | 40:643–80 |

High Energy Astrophysics

Collective Plasma Radiation Processes	DB Melrose	29:31–57
The Cosmic Far Ultraviolet Background	S Bowyer	29:59–88
Ultraviolet Background Radiation	RC Henry	29:89–127
The Origin of the X-Ray Background	AC Fabian, X Barcons	30:429–56
Radio Emission From Normal Galaxies	JJ Condon	30:575–611
X-Ray Spectra and Time Variability of Active Galactic Nuclei	RF Mushotzky, C Done, KA Pounds	31:717–61
Binary and Millisecond Pulsars	ES Phinney, SR Kulkarni	32:591–639
Intraday Variablity in Quasars and BL Lac Objects	SJ Wagner, A Witzel	33:163–97
High-Energy Particles in Solar Flares	H Hudson, J Ryan	33:239–82
Gamma-Ray Bursts	GJ Fishman, CA Meegan	33:415–58
Comptonization of the Cosmic Microwave Background: The Sunyaev-Zeldovich Effect	Y Rephaeli	33:541–79
X-Ray Novae	Y Tanaka, N Shibazaki	34:607–44
Luminous Supersoft X-Ray Sources	P Kahabka, EPJ van den Heuvel	35:69–100
Variability of Active Galactic Nuclei	M-H Ulrich, L Maraschi, CM Urry	35:445–502
Parsec-Scale Jets in Extragalactic Radio Sources	JA Zensus	35:607–36
Modeling Extragalactic Jets	A Ferrari	36:539–98
Sources of Relativistic Jets in the Galaxy	IF Mirabel, LF Rodriguez	37:409–43
Theories of Gamma-Ray Bursts	P Mészáros	40:137–69
Cluster Magnetic Fields	GB Taylor, Chris Carilli	40:319–48

Instrumentation and Techniques

X-Ray Astronomy Missions	HVD Bradt, T Ohashi, KA Pounds	30:391–427
Long-Baseline Optical and Infrared Stellar Interferometry	M Shao, MM Colavita	30:457–98
Adaptive Optics for Astronomy: Principles, Performance, and Applications	JM Beckers	31:13–62

Frequency Allocation: The First Forty Years   B Robinson   37:65–96
Optical Interferometry   A Quirrenbach   39:353–401
The Development of High Resolution
    Imaging in Radio Astronomy   KI Kellermann,   39:457–509
                                              JM Moran

New Areas of Research, History

X-Ray Astronomy Missions   HVD Bradt, T Ohashi,   30:391–427
                                              KA Pounds

Completing the Copernican Revolution: The
    Search for Other Planetary Systems   DC Black   33:359–80
The First 50 Years at Palomar: 1949–1999:
    The Early Years of Stellar Evolution,
    Cosmology, and High-Energy
    Astrophysics   A Sandage   37:445–86
The First 50 Years at Palomar, 1949–1999
    Another View: Instruments, Spectroscopy,
    and Spectrophotometry and the Infrared   G Wallerstein, JB Oke   38:79–111
The Development of High Resolution
    Imaging in Radio Astronomy   KI Kellermann,   39:457–509
                                              JM Moran

The Search for Extraterrestrial Intelligence
    (SETI)   J Tarter   39:511–48